GROVER E. MURRAY
STUDIES IN THE
AMERICAN SOUTHWEST

Texas Natural History

in the 21ˢᵗ Century

David J. Schmidly, Robert D. Bradley, and Lisa C. Bradley

Foreword by Fred C. Bryant
Afterword by Robert C. Dowler

TEXAS TECH UNIVERSITY PRESS

This book is typeset in Adobe Jenson Pro. The paper used in this book meets the minimum requirements of ANSI/NISO Z39.48-1992 (R1997).⊗

Designed by Hannah Gaskamp
Cover illustration by Linda M. Feltner

ISBN (cloth): 978-1-68283-070-3
Library of Congress Control Number: 2020950900

Printed in the United States of America
22 23 24 25 26 27 28 29 30 / 9 8 7 6 5 4 3 2 1

Texas Tech University Press
Box 41037
Lubbock, Texas 79409-1037 USA
800.832.4042
ttup@ttu.edu
www.ttupress.org

Dedication

This book is dedicated to the many naturalists, conservationists, and landowners who have labored to preserve Texas's wildlife heritage for future generations.

And, a special dedication is offered to Professors Robert L. Packard, J. Knox Jones Jr., Clyde Jones, and Robert J. Baker, in recognition of their efforts to establish the Natural Science Research Laboratory at Texas Tech University; to build the largest collection of mammals and associated genetic tissues in Texas; and to establish Texas Tech as a major center of research, education, and outreach in mammalogy. These deceased individuals demonstrated rare professional and institutional dedication in establishing programs and resources that have fostered a better scientific understanding of Texas mammals.

Texans are a unique group of people who reflect a multiplicity of cultures that coincide with the diverse geography of the state. But universally, you will find no group more proud of their state; given half a chance, they won't hesitate to brag or embellish on the attributes and importance of Texas, especially compared to other "lesser" states. This is evident in the words of two of Texas's favorite songwriters and musicians. Waylon Jennings stated, "when you cross that Texas line, the air starts smellin' sweeter and the water tastes like wine" (lyrics from "People Up in Texas"). Then there is Ray Wylie Hubbard's declaration, "Screw you, we're from Texas" (lyrics from "Screw You, We're From Texas"), that lets everyone know just how some Texans view the rest of the world. These words and others capture the pride and, some would say, arrogance of Texans and their relationship with their beloved state. Below are other selected quotations that highlight various views on Texas and its natural resources, as well as quotations from writers and naturalists regarding conservation of natural resources in general.

> "With an area of more than a quarter of a million square miles, with an unlimited variety of soil, climate and productions, with a capacity of growth and prosperity beyond calculation, with a steady stream of immigration converging from all parts of the world, with her vast prairies, her rivers, her streams, her enterprise, her history, her sacred memories and her teeming future, Texas is indeed the Coming Empire."
>
> —OSCAR WILDE, 1882

> "Conservation will ultimately boil down to rewarding the private landowner who conserves the public interest."
>
> —ALDO LEOPOLD, 1933

> "I have said that Texas is a state of mind, but I think it is more than that. It is a mystique closely approximating a religion."
>
> —JOHN STEINBECK, 1962

> "Each generation has its own rendezvous with the land, for despite our fee titles and claims of ownership, we are all brief tenants on this planet. By choice, or by default, we will carve out a land legacy for our heirs."
>
> —STEWART UDALL, 1963

> "We cannot win the battle to save species and environments without forging an emotional bond between ourselves and nature—for we simply will not fight to save what we do not love."
>
> —STEPHEN JAY GOULD, 1993

Contents

Illustrations

Figures

Tables

Foreword

I T IS MY GREAT HONOR TO BE ASKED TO WRITE THE FOREWORD FOR *TEXAS Natural History in the Twenty-First Century*. I grew up a hunter, which led me to a career in wildlife conservation and management—that is, hunting was the background that taught me to love the non-hunted species as much as I loved the ones I hunted. While my in-depth knowledge of the natural history of vertebrates is meager, my in-depth knowledge of the authors of this amazing work is not.

I have known Dave Schmidly for over four decades. Although my friend has a larger than life persona, to me, he is a genuine person: down to earth, caring and compassionate, and a dedicated biologist. He is a mentor and friend whom I have admired for nearly half a century. My personal crossroads with Lisa and Robert Bradley came about two decades later. When I was a faculty member at Texas Tech University, Lisa helped me with her outstanding writing and editing skills. We became close friends working together on scientific manuscripts; as a side benefit, I got to know her husband, Robert, who was then a doctoral student in Biological Sciences working under my good friend, Bob Baker. I tell you this so you understand my familiarity with the authors, even if I don't have the strongest background in natural history and taxonomy.

This book was appropriately dedicated to four of Texas Tech University's amazing scientists whom I also knew, Robert L. Packard, J. Knox Jones Jr., Clyde Jones, and Robert J. Baker. To remember these pathfinders is fitting and proper.

My ultimate goal in developing the foreword for this work on Texas natural history is to tempt the reader to explore what is contained in this timely book. What fascinates me, as much as anything, is the historical context the authors present to you. You learn about C. Hart Merriam's motivation for the original Texas biological survey conducted between 1889 and 1905. You are introduced, with personal glimpses, to all of the field agents who contributed. Some examples: diary notes from Vernon Bailey's collections of cougar and bear skulls in West Texas; William Lloyd running into a smallpox quarantine in Piedras Negras (not dissimilar to the pandemic quarantines we face with today's novel coronavirus); Louis Agassiz Fuertes's collection of a zone-tailed hawk and his later notoriety as a wildlife artist and illustrator; and the personal account of the only grizzly killed in Texas, a hunt that was accompanied by the great-grandfather of my good friend and West Texas rancher, John Means (John still has a tooth saved from the great bear, whose skull is at the Smithsonian). I find these biographies, personal glimpses, and life experiences in Texas nothing short of fascinating.

For the budding naturalist or the hardcore taxonomist, the authors offer documentation and understanding of what the Texas biota was like 120 years ago and the changes that have been seen since. From the 1905 *Biological Survey of Texas*, the only known and documented natural history of vertebrates of early Texas, to the 2002 Schmidly book *Texas Natural History: A Century of Change*, to this 2021 updated compendium, the reader is offered the most complete overview and crucial baseline to assess landscape and biotic changes.

The authors do an excellent job in providing annotations to Bailey's 1905 book, which allows those of us who are novice taxonomists the chance to "catch up" on current scientific nomenclature. And they give us the current taxonomic subspecies, if appropriate. The desert bighorn sheep is the perfect example, because the subspecies documented by Bailey has since been extirpated: the desert bighorn in Texas today is the result of transplants during the 1950s from different populations and subspecies from western North America.

The authors also address in great detail the changes to rural Texas, its wildlands and wetlands. Land conversion; development; impacts on aquatic and coastal systems, streams, and tributaries; loss of groundwater; changes in rangeland vegetation; habitat fragmentation; invasive plants and animals; and diseases and zoonoses: these are all important topics for the readers to comprehend. The authors' discussion of these changes within each ecoregion is particularly relevant.

Section III tells the story of changes to mammalian fauna—the "before and after" as the authors aptly put it. They address the significant trends in mammal populations since 1905 (Chapter 5) and the current population and conservation status of mammals and selected reptiles (Chapter 6). Chapter 5 addresses the proliferation of extinctions; declines in geographic distribution and population abundance; range expansions; increase in designations of rare, threatened, and endangered species; and commercialization of wildlife. The species accounts are extraordinary, even for ever-expanding populations of the armadillo, the icon of Texas aficionados. But the authors also point to a concern for the decline of nine-banded armadillo. Regarding non-native species, of interest and great concern to me was their mention of the greatest source of anthropogenic mortality for US birds and mammals—feral cats! Their documentation of the decline of bats in Texas borders on alarming, but has nothing to do with feral cats—wind farms and wind turbines are causative factors.

Last, Section IV addresses conservation issues, challenges, and strategies for today. Loss of biodiversity has emerged as a central issue. The authors examine projects, programs, initiatives, and policies that have been implemented or will need to be implemented. Action steps are described in Chapter 7, whereas Chapter 8 reflects new political realities and implications thereof. Near and dear to my heart is their discussion of the disgusting and negative impacts of the deer breeding industry on Texas white-tailed deer and the overarching issue of privatization of wildlife—wildlife that belongs to no one, but belongs to all of us, the citizens of Texas. In their treatise of "Concluding Thoughts," the authors are right on target by noting, "as Texas approaches the daunting challenge of conserving its biodiversity, there is an urgent need to make conservation a higher priority in state and local government actions."

The readers of this tremendous book are offered a journey of epic proportions: the journey of our understanding through a wild and untamed Texas to the region we inhabit today. The stakes could never be higher for our beloved Texas wildlife. Indeed one of the keys to saving wild species in Texas is found in the epigraph from Aldo Leopold that appropriately appears on page vii of this book: "Conservation will ultimately boil down to rewarding the private landowner who conserves the public interest."

FRED C. BRYANT, PHD
CAESAR KLEBERG WILDLIFE RESEARCH INSTITUTE
TEXAS A&M UNIVERSITY–KINGSVILLE
KINGSVILLE, TEXAS

Acknowledgments

Many people and organizations made this project possible. The Robert and Helen Kleberg Foundation, the Wray Foundation, the Helen Jones Foundation, the Plum Foundation, and The Nature Conservancy of Texas provided financial backing to access, copy, and bring to Texas Tech University all the archival materials pertaining to the biological survey of Texas. These materials are deposited either in the Southwest Collection archives or at the Natural Science Research Laboratory of Texas Tech, and they provided the basis for the publication of the first edition of the book, *Texas Natural History: A Century of Change*. The database created from that material also is the basis for much of the information included in the second edition of this book.

The President's Office and President Lawrence Schovanec of Texas Tech University provided financial support for the beautiful artwork for the book cover that was illustrated by Linda Feltner of Hereford, Arizona. We also thank the following individuals and institutions for providing historic and recent photographs for use throughout the book: American Heritage Center, University of Wyoming; Derrick Ard, Texas Tech University Center at Junction; Carlton Britton, Texas Tech University; East Texas Research Center, Steen Library, Stephen F. Austin State University; Laguna Atascosa National Wildlife Refuge, Los Fresnos, Texas; Museum of the Big Bend, Sul Ross State University; Philadelphia Academy of Natural Sciences; Southwest Collection, Texas Tech University; Michael Tewes, Texas A&M–Kingsville; Texas Parks and Wildlife Department; John and Gloria Tveten; Emily Wright, Texas Tech University; and Marianna Wright, National Butterfly Center, Mission, Texas.

The University of New Mexico, through the office of the Vice President for Research, provided DJS with the resources for travel to archival institutions to acquire the additional material needed for the second edition. We thank the graduate and undergraduate students of RDB at Texas Tech who provided assistance with literature retrieval and documentation: Laramie Lindsey, Emma Roberts, Taylor Soniat, Irene Vasquez, and Emily Wright. Staff at the Southwest Collection at Texas Tech, including J. Weston Marshall and Elia Martinez, provided assistance in scanning photographs of Texas landscapes and land uses. Personnel at the Texas Parks and Wildlife Department, including Chuck Kowaleski, Tim Siegmund, and Jonah Evans, assisted by providing information about land trends in Texas as well as natural history information about Texas species.

We thank the Smithsonian Institution, the Library of Congress, and the U.S. Biological Survey for providing access to archival materials under their care pertaining to the biological survey of Texas. Finally, we thank Texas Tech University, especially the Natural Science Research Laboratory at the Museum of Texas Tech University, the Department of Biological Sciences, and line-item support from the State of Texas, for supporting students, LCB, and RDB during various stages of this project.

Abbreviations

ACEP	Agricultural Conservation Easement Program
AFWA	Association of Fish and Wildlife Agencies
APHIS	Animal and Plant Health Inspection Service
BCI	Bat Conservation International
BSC	Biological Species Concept
CBD	Convention on Biological Diversity
CITES	Convention on International Trade in Endangered Species of Wild Fauna and Flora
CRP	Conservation Reserve Program
CSP	Conservation Stewardship Program
CWD	Chronic Wasting Disease
DFW	Dallas-Fort Worth Metroplex
EPA	Environmental Protection Agency
EQUIP	Environmental Quality Incentives Program
ESA	Endangered Species Act
GRIP	Grassland Restoration Incentive Program
GSC	Genetic Species Concept
IPBES	United Nations Intergovernmental Science-Policy Platform on Biodiversity and Ecosystem Services
IPCC	United Nations International Panel on Climate Change
IUCN	International Union for the Conservation of Nature
LIP	Landowner Incentive Program
NGO	Nongovernmental Organization
NPS	National Park Service
NRCS	Natural Resources Conservation Service

PUB   Pastures for Upland Birds Program
RCPP   Regional Conservation Partnership Program
TNC   The Nature Conservancy
TPWD   Texas Parks and Wildlife Department
TPWP   Texas Prairie Wetlands Program
TWSP   Texas Wildlife Services Program
USBS   United States Biological Survey
USDA   United States Department of Agriculture
USDOI   United States Department of Interior
USFWS United States Fish and Wildlife Service
USGS   United States Geological Survey
WHIP   Wildlife Habitat Incentive Program
WMA   Wildlife Management Associations
WMP   Wildlife Management Plans (TPWD assisted)
WMU   Wildlife Management Use Property Tax Appraisal Program
WNS   White-nose Syndrome
WRP   Wetlands Reserve Program

Preface to the Second Edition

THE FIRST EDITION OF *TEXAS NATURAL HISTORY: A CENTURY OF Change* (Schmidly, 2002) addressed the changes in the state's natural history during the twentieth century. It emphasized and included a complete reprinting of the *Biological Survey of Texas*, written by Vernon Bailey, as published in 1905 with information on the status of the state's mammal fauna at the end of the nineteenth and beginning of the twentieth century (Bailey, 1905). The 2002 edition of *Texas Natural History* was popular and quickly exceeded its original print run. In this, the second edition of the book, I am joined by colleagues Robert D. Bradley, Professor of Biological Sciences and Director of the Natural Science Research Laboratory (NSRL) at Texas Tech University (TTU), and Lisa C. Bradley, Research Associate and Production Editor for the NSRL's collections-based publication series, *Occasional Papers* and *Special Publications*, in extending the story from the end of the twentieth century and into the first two decades of the twenty-first century. To reflect the extensive revisions to this edition, which include an expanded focus on recent changes in the fauna of the state and the current and future challenges for conserving Texas's diverse wildlife resources, this edition has been renamed *Texas Natural History in the Twenty-First Century*.

I left Texas in 2003, after sixty years, to serve as President and CEO of Oklahoma State University. Then, in 2007, I moved from Oklahoma to New Mexico to serve in the same role at the University of New Mexico. In 2012, I retired from university administration and returned to my academic work in mammalogy, emphasizing Texas and the border region with Mexico. I now live near the village of Placitas in the Sandia Mountains of New Mexico, between the cities of Albuquerque and Santa Fe, in a state vastly different geographically, demographically, and philosophically from Texas. This experience has provided me with a unique perspective about what I encountered in my former home state.

Although I have not lived in Texas in the past eighteen years, I never lost my interest in the place or in being called a "Texan" (a moniker my Placitas friends often use!). Since retiring, I have continued to travel across the ecological regions of the state, observing its robust population growth, economy, and changing ecology, while also writing about its natural history, emphasizing the mammal fauna. Also, I have lectured at various universities and scientific gatherings, spoken with farmers and ranchers, business executives, conservation specialists, university scientists, and environmental writers, and, in the process, formed more thoughts, observations, opinions, and perspectives about what makes Texas so significant, and at the same time so challenging, from a conservation standpoint.

During the timeframe between the publication of the first and second editions of *Texas Natural History*, another 9 million people have been added to the population base of the state, increasing the number from 20 million in 2000 to more than 29 million today. Most of this growth, as in the latter part of the twentieth century, has taken place along and to the east of Interstate 35, which bifurcates the state beginning in Laredo and continuing to San Antonio, Austin, Dallas–Fort Worth, and on to the Oklahoma line. The growth has occurred primarily in the urban and suburban regions around the major metropolitan areas, with comparatively little change in the population in the rural sectors of the state. With this rapid growth and changing population demography, Texas continues to face increasing pressure to protect its natural resources, particularly its fauna and flora.

To provide better clarity about the context and purpose of the book, the new version has been organized into four sections with two chapters in each section. Section I explains the background and history of the biological survey that was conducted from 1889 to 1905. The reprint of Bailey's 1905 publication, *Biological Survey of Texas*, remains a vital part of the revised edition and is included in Chapter 2 of this section. Although annotations to the *Biological Survey of Texas* were placed in a separate chapter in the 2002 edition of this book, in the current edition the annotations have been incorporated into the page margins of the reprinted *Survey*. This arrangement improves the flow of the book by giving the reader immediate and easy access to current information regarding taxonomic changes, distributional changes, and other notes that update the historic accounts.

Section II contains two chapters that chronicle the changes in Texas landscapes and land uses that occurred during the twentieth century and the first two decades of the twenty-first century. Chapter 3, which is focused on the landscapes of Texas at the time of Bailey's biological survey of the state, is essentially like Chapter 4 of the first edition, with only a few changes. Chapter 4 in this edition has been updated considerably to more thoroughly describe landscape changes in the twentieth and twenty-first centuries by providing an overview of factors causing major changes to Texas landscapes.

Section III details how the mammal fauna of the state changed from the time of Bailey's work in the state to today. Chapter 6 of the first edition, "Twentieth-Century Changes in Texas Mammal Fauna," has been split into two chapters in the new edition. Chapter 5 addresses significant trends in Texas's mammal populations since the time of the biological survey, and Chapter 6 summarizes the current population and conservation status of mammals and selected reptiles in the state. These latter two chapters address the classical aspects of natural history—distribution, taxonomy, and ecology.

Advancements in the natural sciences, and particularly in wildlife science, have enhanced our capability to understand and manage our biota at a scope unimaginable during the time of Bailey and the biological survey. But even with this enhanced scientific understanding, society has continued to struggle with addressing the complexity of issues that continue to cause environmental degradation, including the loss of biodiversity. Section IV of this new edition includes two chapters that focus on recent conservation problems and issues, such as climate change, pollution, and wildlife diseases and zoonoses, that are likely to grow in importance and impact over the remainder of the twenty-first century. The two chapters also describe strategies and initiatives that, if implemented, could significantly strengthen the conservation of natural resources in the state. The conceptual basis of this discussion is that time and history, landscape, and socioeconomic community issues are tightly inter-twined and must be addressed as such to successfully sustain the state's resources.

Four publications recently have appeared that augment and provide important insights that complement this revised edition. Information from these new references, described below, has been incorporated into various aspects of *Texas Natural History in the Twenty-First Century.*

Because Texas lands are primarily under private ownership (more than 95 percent of the state is privately owned), the actions of individual landowners and citizens are import-ant in any successful effort to preserve the state's natural legacy. In 2010, David Todd and David Weisman compiled a series of oral histories, titled *The Texas Legacy Project*, about the efforts of citizens and leaders in the state to take actions that protect our wildlife, parklands, waters, and air (Todd and Weisman, 2010).

In 2015, Michelle M. Haggerty and Mary Pearl Meuth edited a large compilation of materials—collected in *Texas Master Naturalist*—written by several Texas naturalists and land-use specialists as part of a statewide curriculum for the Texas Master Naturalist Program (Haggerty and Meuth, 2015). This comprehensive volume provides an overview of the state's vast natural resources, and it graphically depicts the population and demo-graphic trends that now characterize Texas and are impacting its habitats and resources. Also, the volume addresses the uniqueness of the state and how that uniqueness impacts various conservation challenges.

In 2016, David Todd and Jonathan Ogren compiled *The Texas Landscape Project*, an extensive atlas of changes to the state's land, wildlife, water, air, energy, and built world as the state's population and economy boomed and Texans were striving to conserve the state's resources (Todd and Ogren, 2016). A fourth book, *The Natural History of Texas*, written by the late Brian R. Chapman and Eric G. Bolen and published in 2018, provides an overview of the current ecological regions and natural areas of the state, along with beautiful color photographs of the characteristic species of plants and animals and their habitats, plus the many land-use challenges that have become evident in each (Chapman and Bolen, 2018).

Finally, the literature cited section of *Texas Natural History in the Twenty-First Century* has been correspondingly expanded to include important publications, articles, and materials available from the past twenty years, and a glossary and list of abbreviations have been added to assist the reader to better understand some of the scientific terminology.

With the incorporation of this new material, it is my hope that conservationists, scientists, mammalogists, and citizens of the Lone Star State will find *Texas Natural History in the Twenty-First Century* to be a useful description and documentation of how the state's natural history has changed, continues to change, and what will be required in this new century to protect and sustain its precious faunal and floral resources.

DAVID J. SCHMIDLY

Texas Natural History

in the 21st Century

Background of the *Biological Survey of Texas*

Federal field agents employed by the US Biological Survey lived and worked under primitive conditions as they traveled throughout Texas while documenting the landscapes and wildlife of the state. Photo courtesy Biological Survey Unit, US Geological Survey, Patuxent Wildlife Research Center, National Museum of Natural History.

CONSERVATIONISTS WORKING IN TEXAS ARE FORTUNATE TO HAVE A baseline inventory of scientific information about the fauna and flora of the state: the 1905 *Biological Survey of Texas*, documented between the years 1889 and 1905, before the rapid population explosion that began in the early part of the twentieth century started the slow, gradual process of altering land cover, which subsequently led to alterations in our fauna and flora that have continued largely unabated today. Conducted by federal agents employed by the US Biological Survey, a division of the Department of Agriculture, the history of that survey is documented in the two chapters contained in this section.

Chapter 1 tells the story of the men who participated in the survey and what they observed about the fauna and flora as well as the landscapes during that period. The central characters in the story are two men who are legends in the field of US natural history and field biology: C. Hart Merriam, who organized the Biological Survey and conceived of the Texas survey, and Vernon Bailey, the chief field naturalist of the Survey, who led the expeditions to Texas and wrote the final story in the 1905 publication. There is now an excellent biography of Vernon Bailey, *Vernon Bailey Writings of a Field Naturalist on the Frontier* (Schmidly, 2018), that addresses his life and accomplishments, and provides more detail about his work in Texas. Information from that reference has been incorporated into various aspects of this new edition of our work.

A reprinting of the original *Biological Survey of Texas* is included in Chapter 2, with annotations (shaded) incorporated into the margins of the reprinted version. This archival information base can be exceptionally useful as we develop and implement future management strategies for our wildlife resources. The Survey represents one of the key information sources needed for the kind of planning, partnership, and management that will be necessary to allow future generations to enjoy the great outdoors and the natural resources that are part of our heritage as Texans.

Our conceptual understanding of the classification and taxonomy of mammals has changed dramatically since Bailey and Merriam's era. To assist the reader in clarifying the many changes in the scientific names of plants and animals since the original publication of the *Biological Survey of Texas*, four tables have been added to Chapter 2 that cross-reference the scientific names that Bailey used with the current Latin names of the organisms. In the first edition, this material was included as an appendix at the end of the book; in the new edition, it has been updated, expanded, and inserted into the relevant chapter to make it more convenient for the reader. The reader is directed to the four tables by superscripts that appear after scientific names that have changed ([P] for plants, [B] for birds, [R] for reptiles, and [M] for mammals). Chapter 2 also includes annotations for reptiles and amphibians (intentionally omitted from the first edition), primarily through the efforts of RDB, and extensively updates the information for mammals that appeared in Chapter 3 of the 2002 edition.

CHAPTER 1

Introduction

TEXAS WAS A VERY DIFFERENT PLACE AT THE TURN OF THE TWENTIETH century than it is today. In 1900, the human population was less than three million (about 11 people per square mile), compared to more than 29 million today (about 105 people per square mile) (*Texas Almanac and State Industrial Guide*, 1904; US Census Bureau, 2020). The most populous city at the time, San Antonio, had just 53,321 residents. More than 80 percent of the population lived in rural areas. Farming, ranching, and lumber production were the primary means of income for residents. Railroads and horses were the most common means of transportation. Vast areas in the western part of the state remained unsettled and relatively undisturbed. Common mammal species included gray wolves, red wolves, black bear, black-footed ferrets, pronghorn, and other species that are now extinct, endangered, or severely reduced in distribution.

This was the Texas visited by Vernon Bailey, chief field naturalist for the US Bureau of Biological Survey. From 1889 to 1905, Bailey and a crew of twelve federal field agents traversed the state, recording detailed field reports of the mammals, birds, and plants they encountered, and describing the topography, land use, and climate of each place visited. The agents also collected and preserved plant and animal specimens and took photographs of the landscapes, plants, and animals they encountered. The purpose of these biological investigations was to thoroughly survey the flora and fauna of this vast state, to determine the distribution of its diverse plant and animal species, and to assess the economic relationship of birds and mammals to farming and ranching. In 1905, the *Biological Survey of Texas* was published as a summary of the thousands of pages of field notes, reports, documentation, and photographs generated by the efforts of Vernon Bailey and the other federal agents (Bailey, 1905).

Figure 1a-b. C. Hart Merriam as a young man (a) and in later life (b). Courtesy Biological Survey Unit, US Geological Survey, Patuxent Wildlife Research Center, National Museum of Natural History.

This introductory chapter includes a brief history of the US Bureau of Biological Survey, a description of the biological survey efforts in Texas, biographies of key people involved in the Texas survey, and a profile of contemporary mammalogists who worked in the state at the same time as Bailey. In addition, there is a discussion about the history of mammalogy in Texas since Bailey's time as well as an explanation of changes in the philosophy for classifying mammals since his work was published.

C. Hart Merriam and the US Bureau of Biological Survey

The establishment in the late 1800s of the US Bureau of Biological Survey can be attributed to the dedication and passion of one man, a young naturalist by the name of Clinton Hart Merriam (Fig. 1a, b). Merriam was later to become a leading biologist in the United States and is recognized today for his outstanding contributions to the fields of ornithology, mammalogy, and natural history research. For a detailed biography of Merriam and the full history of the US Biological Survey, the reader is referred to the more comprehensive accounts of Keir Sterling (1974, 1989, 2016) and Schmidly et al. (2016b).

Born 5 December 1855, Clinton Hart Merriam (shortened by his friends and colleagues to C. Hart) grew up on his family's rural estate in Locust Grove, New York, where he became fascinated by nature. In 1872, at the age of 16, he accompanied Spencer Fullerton Baird (author of *Mammals of North America* [Baird, 1859] and the first

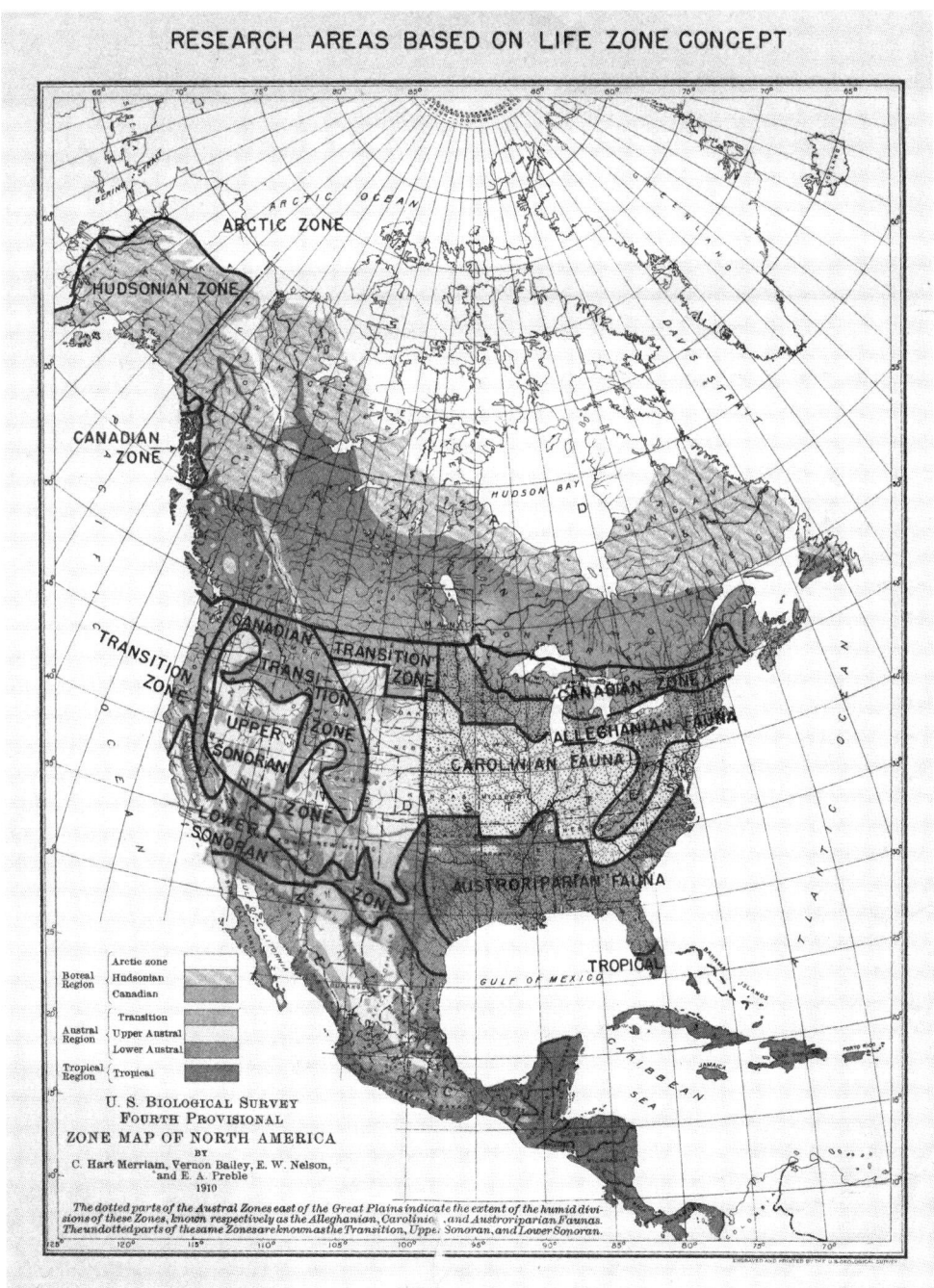

Figure 2. C. Hart Merriam's depiction of the Life Zones of North America (from Merriam et al., 1910).

Secretary of the Smithsonian Institution) on the Hayden Survey, an expedition into the Yellowstone region, as the expedition's naturalist. Merriam demonstrated a passion for collecting and a penchant for detailed record keeping even at this early date.

Merriam entered Yale's Sheffield Scientific School in the fall of 1874. In the same year, he wrote a manuscript (that was never published) that foreshadowed his later theories on life zones and geographic distribution of species (see Merriam, 1894, 1898b, Merriam et al., 1910, and Fig. 2). At the age of 22, Merriam published his first major work in natural history, "Review of the Birds of Connecticut, with remarks on their habits" (Merriam, 1877), which established his reputation as an authority in ornithology.

Merriam entered medical school at Columbia University in 1877 where he met Albert Kenrick (A. K.) Fisher, also an ornithologist. The two became lifelong friends and colleagues. After completing medical school in 1879, Merriam spent six years as a practicing physician but was never devoted to the field. Rather, he regarded his work as a doctor primarily as a source of income to support his true love of natural history study.

In 1883, Merriam and A. K. Fisher were instrumental in the formation of the American Ornithologists' Union (AOU). Merriam was appointed as chairman of the Committee on the Migration of Birds, and he organized a network of twelve hundred volunteers throughout the United States and Canada to gather bird migration data. The resulting flood of data soon overwhelmed Merriam and his small group of volunteer assistants. On the advice of long-time friend and mentor Spencer Fullerton Baird, Secretary of the Smithsonian Institution, Merriam appealed to the federal government for aid.

The funding request presented to Congress suggested that farmers could benefit from a clearer understanding of the distribution, food habits, and economic impact of birds. In truth, the economic impact of birds on agriculture was of little interest to Merriam, but he was well aware that successful funding was more likely if the proposal was justified by its benefit to the agricultural interests of the nation. The bill passed Congress, and in 1885 the Division of Entomology of the US Department of Agriculture became the Division of Entomology and Economic Ornithology (see Gardner, 2016). Merriam was appointed as Economic Ornithologist and was granted an annual research budget of $5,000.

Merriam's duties as ornithologist were described as "the study of the interrelation of birds and agriculture, an investigation of the food, habits, and migration of birds in relation to both insects and plants, and publishing reports thereon." Although Merriam's official staff in the first year consisted only of Fisher, whom Merriam had appointed as his assistant, and one clerk, Merriam soon had numerous volunteers sending in specimens for study. Among the collectors was a young farm boy named Vernon Bailey, who contributed large numbers of specimens and so impressed Merriam that he was later hired as a special field agent.

In 1886, Merriam's budget was doubled to $10,000 and his duties were expanded to include the study of the economic importance of mammals to farmers and livestock raisers. The office was separated from that of Entomology and retitled the Division of Economic Ornithology and Mammalogy. Although the purpose of the office remained the study of the economic relationships of birds and mammals to agriculture, Merriam's primary interests were with the collection and naming of species, the geographic distribution of birds and mammals, and the eventual completion of a national biological survey.

Figure 3. Vernon Bailey, C. Hart Merriam, T. S. Palmer, and A. K. Fisher during the biological survey of Death Valley, California, 1891. Courtesy National Archives, 22-WB-47Admin.-B1929.

He gradually guided the efforts of his division to these studies, and subsequently met with great opposition from the Agriculture Committee of the House and other members of Congress who felt Merriam's work was not of practical value to the government or the public. Merriam, however, contended that one could not deal with the economic effects of birds and mammals without knowing precisely which birds or mammals one was dealing with.

Despite ongoing criticism that his division was not fulfilling its true mission of economic studies, Merriam continued to direct the research efforts of his staff primarily to the study of the geographic distribution of species and biological survey work. By the late 1890s, agricultural research was a minor role of the division. To better reflect its true role, the office was renamed the Division of Biological Survey in 1896, and in 1905 it was elevated to a Bureau (Gardner, 2016).

Among the research efforts Merriam directed were several state and regional biological surveys (Fig. 3). Merriam was fiercely dedicated to the field investigation method and developed the field method used by the biological survey teams. This involved sending out parties in the area to be studied to collect mammal, bird, reptile, amphibian, and plant specimens, taking thorough notes on the life history, numbers, distribution, and economic importance of the species observed, and taking photographs of the regions studied (Sterling, 1978).

Merriam was dedicated to publishing the results of the bureau's varied lines of research. Many contributions appeared in the various series of the Department of Agriculture, such as technical bulletins, farmers' bulletins, yearbook articles, circulars, and the *Journal of Agricultural Research*. But the Bureau had three sets of publications which were its own— the *North American Fauna* series, the Biological Survey Bulletins, and the circulars of the Bureau of Biological Survey. During the first twenty years of the bureau (1885-1905), 23 *North American Fauna* were issued, as well as 22 bulletins and 48 circulars. The most valuable and enduring was the *North American Fauna* series, which consisted primarily of regional reports of extensive natural history studies and technical monographs of groups of birds and mammals.

In 1901, when Theodore Roosevelt became President of the United States, much attention became focused on conservation and preventing the destruction of natural resources. The amount of acreage in both national forests and the national parks grew tremendously as did scientific efforts in natural history. Roosevelt and Merriam were good friends and kindred zoologists, and the success and importance of the Survey was never greater than during the Roosevelt era (Cutright, 1956). Whenever his support was needed, Roosevelt would assist Merriam in protecting the mission and budget of the Survey. Nevertheless, from 1900 to 1910, the Survey was forced by demands from farmers and ranchers to devote increasing attention to the control of noxious and predatory animals. In addition, several actions by Congress during the same time transformed the mission of the agency from survey-oriented research to active management of wildlife resources and their habitats. The study of geographic distribution of species became a minor role of the bureau, and Merriam realized that his dream of a national biological survey would never come to pass. Frustrated by these events, Merriam retired as chief of the Bureau of Biological Survey in 1910, ending a government career of 25 years. Following his retirement, Merriam surprised his friends and former colleagues by dedicating himself to the study of the culture and vocabularies of the vanishing Native American tribes of California (Osgood, 1943). C. Hart Merriam died 19 March 1942, at the age of 86.

Merriam's influence in the development of the field of mammalogy deserves mention. From 1885 to 1900, revolutionary changes in mammalogy took place largely as a result of research by Merriam and his biological survey colleagues. Within 15 years, the number of known species and subspecies of American mammals nearly quadrupled. Merriam himself described no fewer than 660 new mammals; his mammal collection totaled 136,613 specimens; and his published writings included more than 600 titles. Merriam's dedication to training collectors, utilizing the most advanced methods of collection, careful preservation of specimens, taking extensive field notes, and publishing results set a standard for the field of mammalogy that is followed to this day. When the American Society of Mammalogists was formed in 1919, he was selected to serve as its first president (1919-1921).

One of Merriam's greatest contributions to mammalogy was the adoption of a technique he had learned from the study of birds—the idea of bringing together and studying in minute detail large series of specimens, all uniformly prepared, from every possible locality (Miller, 1929). This introduced the concept of local and geographic variation into the study of mammals, and linked an understanding of mammalian variation with Darwin's theory of evolution. Toward this end, Merriam was greatly assisted by the development of the "Cyclone" mousetrap that was developed in the late 1880s and which revolutionized the collecting of small mammals (see Figs. 7 and 8 later in this chapter). Using this new technology to collect mammals and applying the species concept, Merriam and the federal agents obtained volumes of specimens, and they were able to begin mapping the geographical distribution of species.

The great collection of mammals brought together by the Biological Survey was, from the beginning, cared for in the US National Museum, one of the branches of the Smithsonian. To avoid scientific duplication, the staff of the National Museum directed its efforts to the study of mammals outside of the United States (Miller, 1929), as opposed to the Biological Survey that focused primarily on US mammals. Gerrit S. Miller Jr. came to the National Museum in 1898 from the Bureau of Biological Survey, and for forty years he cared for the mammal collection (Kellogg, 1946). From 1889 until 1943, the Division of Reptiles and Batrachians at the National Museum was supervised by Leonhard Stejneger, who assisted Bailey with the section on snakes and lizards in the *Biological Survey of Texas*.

The Biological Survey persisted as a government program until 2018 when it was suddenly shuttered by the US Geological Survey (USGS), which had overseen the division since 1996 (Al Gardner, personal communication). Interior Department officials sought to close it as a way to trim that department's overall budget. At the time, it had six employees and a budget of $1.6 million that was used to maintain nearly a million bird, reptile, and mammal specimens and historic field notes housed at the Smithsonian National Museum of Natural History. Several academic and scientific societies protested the move, saying the unit's staff provided access to key specimens that shed insight on critical scientific questions, including how ecological conditions changed over more than a century in the western US and elsewhere in North America. At the time of its closure, the Biological Survey Unit was the last surviving remnant of C. Hart Merriam's Section of Biological Investigations. It was the oldest affiliated agency, having been in existence for 133 years, in the Smithsonian Institution's National Museum of Natural History. After the Biological Survey was closed, the USGS turned over management of the more than 1 million specimens in the USBS collections to the Smithsonian Institution.

The *Biological Survey of Texas*

One of C. H. Merriam's visions for the Bureau of Biological Survey was the completion of a nationwide biological survey. He focused his early efforts for this survey project, however, on the western states because they were the agricultural states, and there he could justify his survey projects to Congress by the benefit they would have to

Figure 4. A. K. Fisher, E. W. Nelson, W. H. Osgood, and Vernon Bailey, working at the National Museum in Washington, DC, early 1900s. Courtesy Biological Survey Unit, US Geological Survey, Patuxent Wildlife Research Center, National Museum of Natural History.

agricultural interests. One of the states chosen early for a survey project was Texas. This was justified by the need to study urgent economic problems such as the decimation of cotton crops by the boll weevil, the depletion of range forages by the millions of prairie dogs in the state, and the loss of sheep and goats to predation by coyotes and other predators. Merriam's conviction that an understanding of the taxonomy and distribution of mammals was basic to providing solutions to economic problems of agriculture was the impetus for a program of basic research on mammalian systematics and biogeography that continued throughout the twentieth century (Wilson and Eisenberg, 1990).

In truth, Merriam's desire to conduct a biological survey of Texas was motivated in large part by the fact that the state had such a wide variety of soil types, climates, and topography, and these factors resulted in diverse and abundant plant and wildlife resources. By nature of its geographic location in the south-central United States, Texas sits at the juncture of at least four major biomes: the deciduous forests of the southeast, the grasslands of the central plains, the deserts and mountains of the southwest, and the brushlands and tropical habitats of the south. The convergence of these ecological regions provides habitat for Texas's native terrestrial mammal and bird species, many of which reach their distributional limits in the state.

The diverse and abundant natural resources in Texas had been recognized by many early explorers and settlers who wrote excellent descriptions of the state and its natural resources up to the mid-1800s (Weniger, 1984, 1997). These accounts describe an area teeming with vast herds of bison, pronghorn, deer, and other game, filling the millions of acres of prairies, plains, and forests of the state with wildlife too plentiful for modern man to comprehend.

During the mid-1800s, however, the "taming" of Texas began in earnest. The last of the Native Americans in western Texas were killed or forced onto reservations by 1880, and the bison, once numbering an estimated 60 million, were already decimated in many areas by 1860. By the early 1880s fewer than 200 bison remained; these were saved through the efforts of legendary rancher Charles Goodnight.

In 1870, Texas was occupied by only 818,579 people (*Texas Almanac*, 1873). In 1871, a state government office titled the Bureau of Immigration of the State of Texas was created to encourage immigration from other states and foreign countries. As a result of the Bureau of Immigration's efforts and the rapid growth of railroads in the state, from 711 miles (1,144 km) in 1870 to 11,294 miles (18,176 km) in 1904, the population grew dramatically. An estimated 400,000 immigrants entered the state in 1876 alone. By the turn of the century, the population had grown to 3,048,710, an increase of 372 percent from 1870. This dramatic rise in population affected natural resources in many ways, as the state's prairies, rangelands, forests, and wildlife resources were exploited or depleted by farming, ranching, lumber production, predator control, and hunting.

The dramatic growth and change during the 1870s and 1880s may have played an additional role in Merriam's decision to conduct a biological survey of the state. In 1889, Merriam appointed Vernon Bailey to lead the investigation and to train new field agents as they joined the team. Field agents were often hired as temporary employees, but other agents were permanent employees of the Bureau and were based in Washington, DC, when not conducting field research (Fig. 4).

Much of the fieldwork was conducted in conjunction with work in other states. Field agents would often work in northern states during the summer months, and move south during the fall, winter, or early spring. However, all seasons of the year are represented in the field data for Texas. Survey efforts were conducted from 1889 to 1892, in 1894, and from 1899 to 1906 by 12 field agents who worked 2,185 man-days (Table 1). Survey work took place at more than 200 sites in all ten ecological regions of the state (Figs. 5, 6a-j).

At each site visited, the field agents prepared written reports describing the physiography of the region and annotated lists of the mammals, birds, and plants observed or collected. These reports were prepared in addition to an annual field journal for each agent, describing their day-to-day movements and activities, and a field catalog of all specimens collected (Figs. 7, 8). The agents also took more than a thousand photographs of landscapes, habitats, plants, and animals from throughout the state, many of which are provided in Chapter 3.

In addition to the "pure" science efforts, during the 1901 survey of the Trans-Pecos region, Vernon Bailey and ornithologist Harry Church Oberholser were joined in the

Table 1. Man-days[1] in Texas by each Survey contributor.

	1889	1890	1891	1892	1894	1899	1900	1901	1902	1903	1904	1905	1906	Total
Bailey	25	39	0	28	0	58	45	94	88	0	47	0	1	425
Bray	0	0	0	0	0	50	0	0	0	0	0	0	0	50
Cary	0	0	0	0	0	0	0	0	119	0	0	0	0	119
Donald	0	0	0	0	0	0	0	0	19	0	0	0	0	19
Dutcher	0	0	0	27	0	0	0	0	0	0	0	0	0	27
Fisher	0	0	0	0	6	0	0	0	0	0	0	0	0	6
Gaut	0	0	0		0	0	0	0	0	135	10	106		251
Hollister	0	0	0	0	0	0	0	0	86	0	0	0	0	86
Howell	0	0	0	0	0	0	0	0	0	51	0	114	36	201
Lloyd	0	176	233	87	0	0	0	0	0	0	0	0	0	496
Loring	0	0	0	0	108	0	0	0	0	0	0	0	0	108
Oberholser	0	0	0	0	0	0	150	123	124	0	0	0	0	397
Total	25	215	233	142	114	108	195	217	436	186	57	220	37	2185

[1] Total days from beginning to end of each trip—includes days collecting as well as days traveling between sites, writing reports, etc.

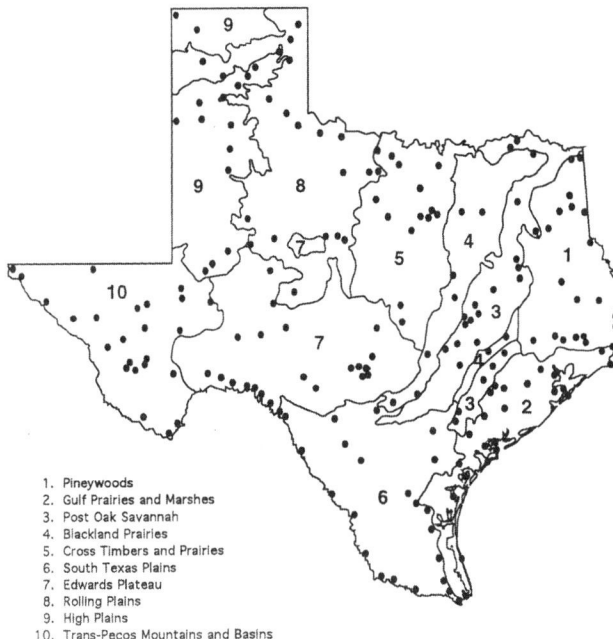

1. Pineywoods
2. Gulf Prairies and Marshes
3. Post Oak Savannah
4. Blackland Prairies
5. Cross Timbers and Prairies
6. South Texas Plains
7. Edwards Plateau
8. Rolling Plains
9. High Plains
10. Trans-Pecos Mountains and Basins

Figure 5. The ecological regions of Texas and localities visited (closed circle) by federal agents during the biological survey of Texas, 1889-1905.

a

g
h

i
j

Figure 6a-j. Series of maps depicting each ecological region of Texas, with cities and towns included as references, and sites visited by the federal agents indicated by closed circles. (a) Piney Woods, (b) Gulf Coast Prairies and Marshes, (c) Post Oak Savannah, (d) Blackland Prairies, (e) Cross Timbers and Prairies, (f) South Texas Plains, (g) Edwards Plateau, (h) Rolling Plains, (i) Trans-Pecos Mountains and Basins, and (j) High Plains.

field by the famed wildlife artist, Louis Agassiz Fuertes. Fuertes produced some of his finest works during this trip. His portfolio, after four months in the field, included more than one hundred portraits of birds as well as a number of mammals and reptiles (Fig. 9a-d).

The working conditions for the field agents were primitive at best, and arduous or even life-threatening at worst. The agents usually traveled from Washington, DC, to the scheduled first stop by train, using vouchers provided by the Bureau to pay for their train fare. Once in Texas, however, virtually all of the agents' day-to-day living expenses were paid out of their

Figure 7. A typical night's capture of small rodents in Cyclone traps. Courtesy National Archives, 22-WB-50-B15515.

Figure 8. A sampling of rodents captured in Cyclone traps and snap traps, Seguin, Guadalupe County, 1904. Courtesy National Archives, 22-WB-62-7225.

pockets, from a salary that averaged less than $100 per month (see Schmidly, 2016b). The agents would hire a local to serve as their "camp man," who would cook for them and set up their camps in the field (Fig. 10a, b). They would also rent horses, mules, and wagons, as needed, and purchase all of their food and other supplies from a local town, then set out

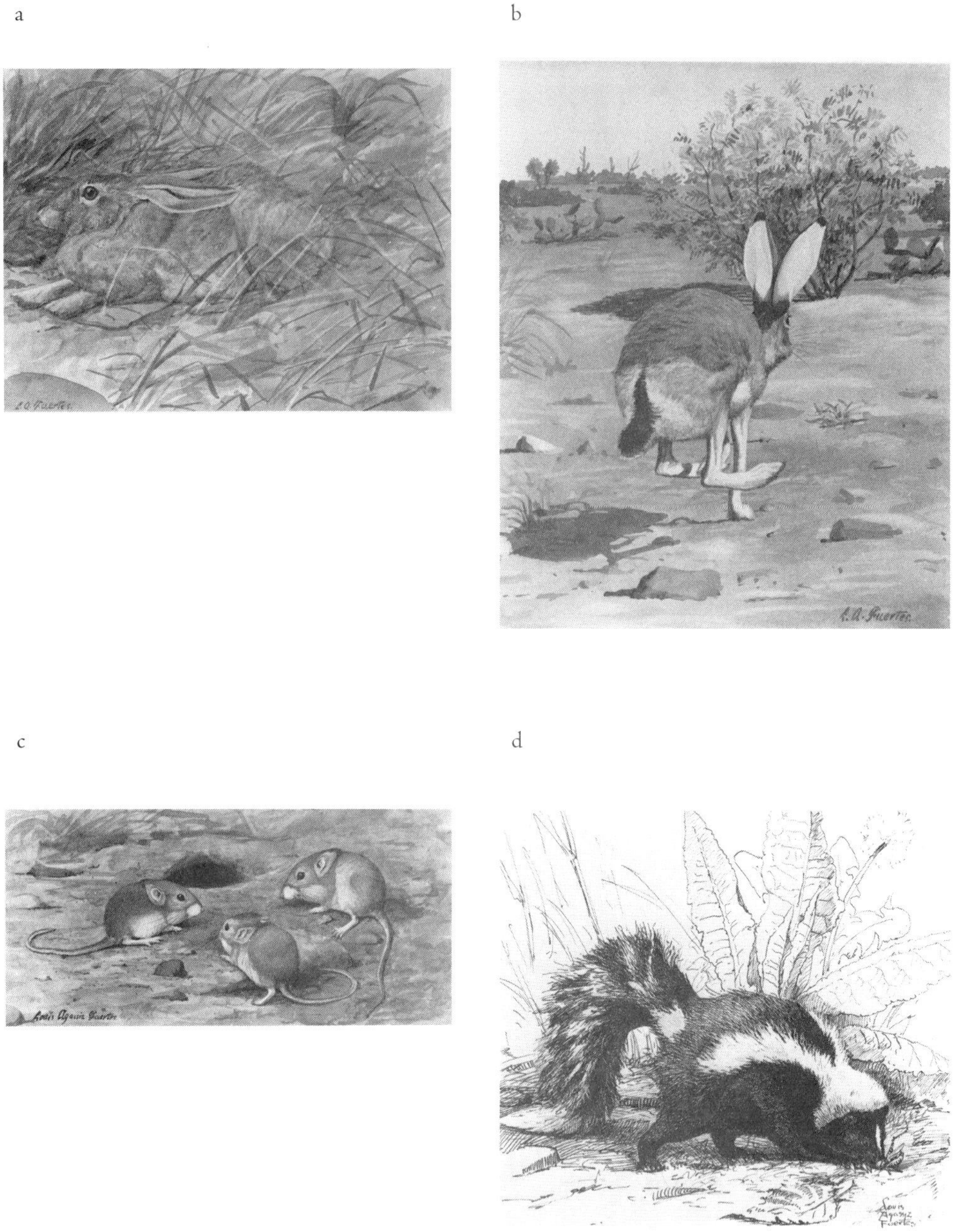

a

b

c

d

Figure 9a-d. A portfolio of mammal sketches by Louis Agassiz Fuertes: (a) cottontail rabbit, (b) jackrabbit, (c) pocket mice, and (d) striped skunk. Courtesy Philadelphia Academy of Natural Sciences.

to the field to observe and collect data and specimens. Their stay in any one locality would range from one day to two weeks or more.

At most localities, the field agents would set up camp, and travel on foot or by horseback as they collected specimens and recorded their observations (Fig. 11). The agents often befriended the local landowners and usually were welcomed to collect specimens and to camp on the property. Occasionally they were fortunate enough to be invited to supper and to stay in the landowner's home or barn for the night. More often than not, however, the field conditions for the agents were rigorous, and the field diaries describe accounts of being caught in raging floodwaters, becoming isolated in areas without food or water, losing horses and pack mules, suffering from the extreme heat, and becoming ill without a doctor or medicine available. The field agents were armed with weapons for hunting and obtaining specimens, but to our knowledge they never had to use these to defend themselves while working in Texas.

Following are a selection of accounts from field diaries of the agents. These accounts provide a vivid picture of the hardships these men endured as well as their unwavering dedication to hard work and their passion for natural history study.

Vernon Bailey, Finley's Ranch (14 miles west of Ft. Davis)

January 13, 1890. Twenty-two degrees at sunrise, clear and still. Coldest morning of the winter so far. Took a horse and started over the mountains to see the country on the north side and visit the ranches and get bear skulls. Went up the canyon about 3 miles and then climbed the right side and crossed the divide where it is 7,270 feet . . .

Followed an old government road down the canyon on the other side and reached Mr. Perkins' ranch at 1 p.m. This is about as high as Finley's – 6,000 feet. Was welcomed by four hounds and the rest of the family. After dinner went farther down the canyon and killed a *Lepus sylvaticus*[M] [note: see page 5 for explanation of superscripts]. Saw a *Spermophilus grammurus*[M]. Birds and plants are about the same as at Finley's. *Pinus flexilis*[P] and the narrow-leaved oak grow on the north side of the mountains high up. The canyons are brushy and deep. The north side of Livermore Peak is a perpendicular face of granite that looks to be 500 feet or more. In evening went fox hunting with three boys and five hounds. It was windy and dry and the dogs could not keep a trail and caught nothing. They ran two tracks for some time and made lots of music. We followed up and down the mountains over broken rocks and nearly broke our necks. It was dark and cold. Staid [*sic*] out till 1:30 a.m. It paid me well for the work just to hear the dogs among the rocks. We couldn't tell which was dogs and which echo.

January 14. Couldn't get any bear skulls at Perkins'. They had killed a number but a long way from home and did not bring in the skulls of but one and the boys smashed it. Could not induce Mr. Perkins to take a cent for my board. Said no one ever paid a cent for staying under his roof. They are tough but generous and intelligent people

a

b

Figure 10a-b. (a) US Bureau of Biological Survey field agents at a typical field camp. Courtesy Biological Survey Unit, US Geological Survey, Patuxent Wildlife Research Center, National Museum of Natural History. (b) Vernon Bailey with his field party in the Big Bend region of Texas in 1901. From left to right, Harry C. Oberholser, Louis Agassiz Fuertes, Vernon Bailey, and Murray Surber. Courtesy Biological Survey Unit, US Geological Survey, Patuxent Wildlife Research Center, National Museum of Natural History.

Figure 11. Field agents on horseback, near the entrance to one of the Painted Caves, Val Verde County, Texas, 1901. Courtesy National Archives, 22-WB-30-2790.

as are most of the ranch men. Mr. Perkins told me where I could find a panther skull at a ranch 9 miles west of there, Mr. Kerr's, so I started to follow his directions up one canyon and down another. Killed a *Spermophilus grammurus*[M] and saw another. They were in a sheltered place in the canyon where the sun shone and it was warm; were feeding under juniper trees. The one killed had its cheek pouch full of juniper berries and was very fat, though not an adult. Saw no new birds or much game of any kind. Went over a divide and down another canyon, came to Mr. Kerr's ranch about 2 p.m. Found the bones of two *Felis*[M] and got the skull of one and the upper part of skull and the tail bone and scapula of the other. They had the skin of one, but Mr. Kerr was not at home and I could not get it. It was in fine pelage, of a grayish brown and has the nose and claws and tail all on. Both the panthers were said to be female. Found a bear's skull, which is of a female that had two cubs. Went down the canyon a mile farther to Mr. Kirksey's after a bear skull that was said to be there, but couldn't find it. Came back into Mill Creek Canyon and then over the main divide home. The sun went down while I was eight miles from home with a mountain to cross and trail over the top. By good luck I did not get lost and reached Mr. Finley's at 8 o'clock. Rather enjoyed the trip, saw much country, got some good specimens, and met several pleasant people.

William Lloyd, Eagle Pass

October 21, 1890. Caught one usual *Neotoma*, owing to sharp norther (first of season) blowing up about midnight and continuing with considerable force all day making all bird life and animals quiescent. Saw three Audubon's warblers and put out 18 traps principally for *Sigmodon*.

October 22. Norther still continued but with diminished severity. Caught nothing in traps, but two coon traps set and several Cyclone carried off. Sharp frost, first of year. A few birds about but principally cactus wrens, mockingbirds and *Melanerpes*.

October 23. See a rat in afternoon and putting out a trap startled an immense frog and trying to secure it put my hands on a large rattlesnake coiled up. I was within two feet of it [for] over ten minutes [while] arranging the trap. Put my foot on it and caught it by neck and brought it back to camp. Bat flying along ground in night, medium size, first seen since 11th. Slight frost but so hot in middle of day that a rabbit was blown by blowflies in an hour after it was killed. Find large block of petrified wood, pine?

October 24. The trap put out yesterday by rat hole caught an *Onychomys*. A ranchman to whom I showed it said he had lived there six years and it was the first mouse he had seen. Saw flock of *Parus atricristatus*[B] and wounded one but could not secure it. Also saw a lone long-billed curlew, and evening a sandhill crane grazing on prairie. Caught three more rats and coyotes ran away with two Cyclone and one Climax, and two steel had fragments of lips and claws, one of a possum and the other a *Neotoma*. Into Eagle Pass and crossed river to Piedras Negras. Find a strict quarantine is established (small pox) rendering it out of question to collect on other side without delay and bother. Moved south of Eagle Pass about six miles.

Vernon Bailey, Paris, Texas

June 11, 1892. Took the 10:45 train west to Blossom, then to Reno and then to Paris, where I stopped and set traps and shot bats in evening. Shot three *Atalpha noveboracensis*[M] and three *Vesperus*[M]. Caught nothing in my traps at Clarksville and do not expect to here. The soil is hard and baked. Most of the country is grassy prairie or fields. Perhaps half timbered. Timber and prairie alternate from Clarksville to here. The large mounds extend most of the way from Clarksville to Paris. Got a set of four eggs of the Scissor-tail Flycatcher and saw the bird alive for the first time. A hot day.

June 12. Sunday, a hot day, 88 degrees in my room. Don't feel well.

June 13. Took up my traps, they had not been touched. Shot scissortails and a *Chondestes* and made up skins. Failed to get a *Sturnella*. Was sick all night and am now so weak I can hardly stand up. My eyes begin to turn yellow so I know it is only jaundice. Could hardly drag myself around to get my traps. Saved a few plants. Will take the 6 p.m. train northward. Another hot day. Left Paris at 6 p.m. and went to Arthur, 15 miles north at crossing of the Red River and stopped. Shot at bats but got none. All clay land from Paris to Arthur, some fields of cotton, oats, corn, castor plant, etc. Most of the way in uncleared scrub oak land. Arthur is composed of six or eight little houses on the bank of the Red River. A crew of men are building a saw mill.

Vernon Bailey, Chisos Mountains

June 8, 1901. Fuertes and I took our saddle horses and went around the south side of the mountain into the big gulch leading up from the south into the big canyon that cuts nearly through the mountains. Left our horses at an old sheep camp at 5,500 feet and climbed up the gulch west of us past the obelisque-like pinnacle around the base of Mt. Emory and turning south again up onto the big, forested ridge that joins the southernmost spur of the mountains. On reaching the top of this ridge the aneroid read 6,200 feet (probably 300 feet too low as it works slowly) and Mt. Emory just across the gulch to the northwest appears at least 500 feet higher. As the climb up had taken six hours and it was 7 p.m. when we reached the top we concluded not to return tonight so found a sheltering rock, made a bed of pine boughs and grass and gathered wood for a fire and camped, comfortable but hungry . . . .

June 9. The night was cool and we had to keep the fire burning to keep warm enough to sleep, but passed a comfortable night. Hunted an hour or so before starting down the mountain and then shot birds on the way down. Reached the horses at 11 a.m. and camp at 2 p.m., tired and hungry. Had only two biscuits for lunch yesterday and nothing to eat since for 26 hours, but were fortunate in finding plenty of water. Got a good lot of birds but no mammals except old skull of a bear and a deer. Found a fine old hound lying on our saddle blankets near the horses and glad to see us. Had evidently got lost so we took him to camp.

Louis Agassiz Fuertes (from a letter written to his family), Chisos Mountains

May 29, 1901. Your letter of the 21st came over with Bailey from Boquillas last night, and as we are going to send a lot of stuff over this evening, I will just send along a little note to tell you that we are all right, in good health and happiness, having replenished our larder which for the last day or two has been without coffee, baking powder, or condensed milk.

Well, I got the hawk, and had an adventure in the bargain by virtue of which I spent a delightful hour in a hole 400 feet up a 600 foot cliff till Oberholser could get to camp and back with a rope, on which to continue my journey. The bird [a zone-tailed hawk] is a Texas record, and one of the very few US records, so that when I had at last shot him, after three days straight of hunting him, it would never have done to let the splendid thing rot just because he fell over a cliff down in the canyon. I don't think I was in any danger any of the time, for when I found myself unable to go any further because of a boulder that was lodged in the fissure above me, and also at least unwilling to go down, I got O[berholser] to go for a rope. Then I sat in my comfortable hole, sang to a superb echo for a while, watched lizards and ravens and got rested for an hour, and came out all right, and the bird had by that time earned his record. I painted him fresh that afternoon, and am mighty glad of it, for all his lovely plum bloom has gone, in the skin, and he is still splendid, but nearly dead black, instead of like a rich ripe black plum.

The culmination of the survey was Bailey's 1905 publication, *Biological Survey of Texas*, which is reprinted in its entirety in Chapter 2 of this book. The report on the birds of Texas, to be authored by H. C. Oberholser, was too lengthy to include in the Survey publication and the decision was made to publish it separately. Oberholser continued to edit and revise the manuscript, however, and it was to remain unpublished until 1974, 11 years after his death (Oberholser, 1974). The information gleaned from the botanical reports and plant collections made during the Survey appeared in a number of early publications about the vegetation and plant communities of Texas (e.g., Bray, 1906). Over the years the specimens obtained during the Texas survey have been used in hundreds, if not thousands, of publications about Texas natural history.

Biographies of the Federal Field Agents

During the early years of the US Bureau of Biological Survey, the appropriations were small and the personnel limited in numbers, but the outstanding interest of the members of the staff and their devotion to duty enabled them to make noteworthy advancements in the field of science, and to place the bureau in the foremost ranks of the world's scientific organizations engaged in wildlife research at the time. The field agents who participated in the Texas survey included some individuals who would go on to distinguish themselves scientifically—Vernon Bailey, William Bray, A. K. Fisher, Ned Hollister, A. H. Howell, and Harry Oberholser—as well as some of the lesser-known agents who made only limited scientific contributions such as Merritt Cary, Gordon Donald, Basil Dutcher, James Gaut, William Lloyd, John Loring, and Clark Streator. Nonetheless, this latter group of individuals made enormous contributions to the specimen collections and physiographic descriptions. For a complete description of the lives and accomplishments of these amazing naturalists, see Schmidly et al. (2016b).

a

b

Figure 12a-b. Vernon Bailey as a young man (a) and in 1905 (b), the year of publication of the *Biological Survey of Texas*. a. Courtesy Biological Survey Unit, US Geological Survey, Patuxent Wildlife Research Center, National Museum of Natural History. b. Courtesy Library of Congress, LC-USZ62-058826.

Vernon Bailey

The following account about the life and career of Vernon Bailey has been adopted from the recent biography about him written by David Schmidly (2018). A biography also is available about his wife Florence Merriam Bailey, who was C. Hart Merriam's younger sister and a distinguished ornithologist of her time (see Kofalk, 1989).

Vernon Orlando Bailey (Fig. 12a, b) was born 21 June 1864, in Manchester, Michigan. At the age of six, he moved with his pioneer parents to a farm in Elk River, Minnesota. It was there that Bailey's interest in nature began. He began collecting birds, mammals, and reptiles, and he learned the art of taxidermy from a small book on the subject and a great deal of experimentation. In the beginning, he collected specimens primarily by trapping, and many of the trapping methods he used were of his own design. At age 11, his father gave him his first gun—a long-barreled, 16-gauge, muzzle-loading shotgun that was longer than he was tall. Bailey soon filled a room of his parents' home with preserved specimens.

Bailey developed a particular interest in shrews and devised a method for trapping them easily, but he was at a loss as to the proper identification of the various species he was collecting. Bailey was advised by an acquaintance to contact C. Hart Merriam, whose reputation had already reached Minnesota. Bailey was eager for scientific information,

and so in 1884, he wrote Merriam asking if he would identify animals if prepared specimens were sent to him. Merriam not only agreed but also offered to buy specimens from
Bailey of the different species found in Minnesota. The prices he offered seemed fabu-
lous to the young farm boy—twenty-five cents apiece for mice and shrews and a dollar
for woodchucks and skunks. He soon had a box packed for shipment, which Merriam
received eagerly, and thus began a lifelong relationship that eventually would culminate
with Bailey's marriage to Merriam's sister, Florence.

The following excerpt was written by Vernon Bailey, and describes his early correspondence with Merriam:

> I wrote him and received a prompt and courteous reply, offering to identify any
> specimens that he could and also asking if I would collect certain things for him that
> he needed. This I was more than glad to do and supplied Dr. Merriam with series of
> specimens of each of the different kinds of shrews, mice, and small mammals that I was
> finding. At first he wanted to purchase all the shrews I could get, but when the number
> ran into hundreds, he put a limit on the number of each kind that he would need. He
> also wanted to buy skulls of many of the larger mammals, including those trapped for fur,
> such as muskrat, mink, otter and foxes. Of these he wanted all he could get, but when I
> sent him the first lot, including a hundred muskrat skulls nicely cleaned and labeled, he
> said that would be enough muskrats for the present, but continued to take other species
> in moderate numbers. My greatest interest now was in hearing from Dr. Merriam and
> getting names of different animals that I was sending him, and, more important, getting
> his criticism and suggestions in regard to preparing specimens. Dr. Merriam took the
> greatest pains in correcting errors in my methods, giving careful directions for taking
> measurements, labeling, cataloguing, and keeping notes on all mammals collected. He
> wanted information on habits and abundance and many factors relating to the lives of the
> mammals, and I was able to furnish him much information that was of use to him, while
> he was giving me the training that I needed for natural history work.

In May 1887, Merriam hired Bailey as the first Field Agent of the Division of Economic
Ornithology and Mammalogy. His official salary was $40 per month, but Merriam per-
sonally supplemented this by an additional $10 per month (Sterling, 1978). Bailey's first
field experience lasted from May through November, and took him from Minnesota,
throughout the Dakota Territory, and into Montana. In November, Merriam asked
Bailey to remain in the field and move on to Texas, but Bailey elected to return home to
Elk River. Bailey came from a close and loving family, and being away from them for the
first time in his life, for a period of six months, had proven difficult for him and his family.
Bailey's continued refusal to make the trip to Texas did not please Merriam, but Merriam
did not want to officially remove Bailey from the payroll. Bailey was put on an official leave
of absence without pay, effective from 1 December to the following 1 June. In the summer
of 1888, Bailey once again set out for a season of fieldwork, a pattern that would continue
almost without change for the next 45 years.

Figure 13. Florence Merriam Bailey, wife to Vernon Bailey and sister to C. Hart Merriam, conducting field research in the Guadalupe Mountains, New Mexico. Courtesy American Heritage Center, University of Wyoming.

Bailey's first season of fieldwork in Texas began in December 1889, in El Paso. Bailey worked at several localities in the Trans-Pecos ecoregion and began moving eastward to Langtry and Del Rio. At Del Rio, however, Bailey became quite ill, and in February 1890, he abruptly returned home to Elk River to recuperate without telling Merriam. For this he was severely chastised, but Merriam "doctored up" the paperwork so that it did not appear that Bailey had abandoned his position. The matter was soon forgotten and never became a big issue between the two men.

Bailey had quickly become a key figure in the field for Merriam, and in 1890 he gained the title of chief field naturalist. Bailey was given the primary responsibility for training new field investigators and supervising the parties of field agents conducting biological survey work throughout the western US.

Bailey returned to Texas in 1892. He spent more than two months surveying the flora and fauna from Clarksville in northeastern Texas to Canadian in the Panhandle region. His third trip, from April through June of 1899, began in Galveston and continued north-westward through central Texas and into the Panhandle region. Additional trips took place in 1900, 1901, 1902, and 1904. In total, Bailey spent 425 days surveying Texas before publishing his report in 1905. Bailey's collecting trips thoroughly traversed the state and included work in every ecological region.

In 1899, Bailey married Florence Merriam, C. Hart Merriam's younger sister. Florence Merriam Bailey was a prominent ornithologist and often accompanied her brother and her husband in the field, assisting in the identification of birds during survey trips (Fig. 13). The Baileys took their first field trip together in 1900 in Texas, traveling by train to Texarkana and on to Corpus Christi where they began a 360 mile (580 km) wagon trip across the prairies to the Mexican line, camping on the King Ranch and other places en route (Kofalk, 1989). It was during this trip that Florence Bailey became aware of the slaughter of thousands of water birds for the millinery trade, a practice she was determined to stop (Doughty, 1983). She reported her findings to Witmer Stone who commented on the problem in a 1901 article published in *The Auk*.

During his career, Bailey conducted biological surveys of Texas, New Mexico, North Dakota, and Oregon, and studies of the mammals of several national parks. He contributed more than 13,000 mammal specimens to the Biological Survey collections, a large number of them new species. A bibliography of his work contains 234 titles (see Schmidly, 2016a for a complete list of his publications). He authored seven of the most significant *North American Fauna* publications—numbers 17, 25, 35, 39, 49, 53, and 55.

Bailey particularly was concerned with the humane treatment of animals and was a pioneer in the development of new live traps. He designed and perfected a trap, patented as the VerBail Trap, which was widely used in beaver restocking efforts, as well as the "foothold" trap, and he received recognition from the American Humane Association for his efforts. Bailey was a founding member of the American Society of Mammalogists and President of that organization from 1933 to 1935. He also served as President of the Biological Society of Washington and was an active member in numerous other wildlife and conservation societies.

Bailey retired from the Bureau in 1933, after 46 years of service, but remained active as a consultant and collaborator on special projects. He was planning a new expedition to Texas at the time of his death, 20 April 1942, at the age of 78 (Zahniser, 1942).

William Bray

William Bray was born 19 September 1865, in Burnside, Illinois. He received his PhD ("Vegetation of Western Texas") in botany from the University of Chicago in 1898. Bray was a professor of botany at the University of Texas, Austin, from 1897 to 1907, where he organized the Botany Department and became its first Head. He served as a collaborator with the US Forest Service from 1899 to 1909. He authored several bulletins on Texas forests, published by the US government, and various articles on plant distribution and adaptation in botanical journals.

Bray contributed 31 plant and physiography reports to the biological survey of Texas over a fifty-day period in 1899, during which time he traveled from the Gulf Coast region through the central and northwestern part of the state. In 1907, Bray left Texas and went to Syracuse University where he spent the remainder of his career in various administrative positions until retiring in 1943. Bray passed away 25 May 1953, in Syracuse, New

Figure 14. Merritt Cary in Canada, 1903. Courtesy National Archives, 22-WB-46-B6585.

York. A short account of his life was published in the *Bulletin of the Torrey Botanical Club* in 1955.

Merritt Cary

Merritt Cary (Fig. 14) was born in Nebraska on 21 December 1880. He joined the Bureau in 1902, after a few years of study at the University of Nebraska. He received his field training under the tutelage of Vernon Bailey. Cary worked in Texas from June through October 1902, for a total of 119 days in the field. During that time he prepared 31 survey reports and collected 255 mammal specimens. His travels began near

Figure 15. Basil Hicks Dutcher, date unknown. Courtesy Library of Congress, LC-USZ62-073584.

Kerrville and continued westward into the Trans-Pecos region, then northeastward through Abilene, and ended in Tarrant County.

During his career, Cary participated in fieldwork in many of the western states and in central and northern Canada. His most important works were a biological survey and list of the mammals of Colorado (*North American Fauna* No. 33, published in 1911) and the life zones of Wyoming (*North American Fauna* No. 42, published in 1917) (Henderson and Preble, 1935). He resigned in 1917 because of ill health and died not long afterward in 1918.

Gordon Donald

Very little is known about Gordon Donald other than his brief tenure at the Survey. He worked in Texas from July through September of 1902. He first met up with Bailey at Howard Lacey's ranch in July and apparently traveled with him for a few weeks until

Figure 16. Albert Kenrick Fisher at Fort Huachuca, Arizona, 1892. Courtesy Library of Congress.

Bailey sent him to Devils River to collect. Donald also surveyed the area from Paisano through Valentine and Fort Hancock, and from there he traveled with Merritt Cary eastward to Pecos, Monahans, Warfield, and Abilene. Donald contributed 131 mammals, 17 birds, and five reptiles to the collections of the Biological Survey. Nothing else is known about this gentleman. It appears that he was only temporarily employed by the Bureau, and his activities before and after the summer of 1902 are unknown.

Basil Hicks Dutcher

Basil Hicks Dutcher (Fig. 15) was born 3 December 1871, in Bergen Point, New Jersey. Dutcher became interested in natural history while a young boy, and he accompanied his father, William Dutcher, an ornithologist and long-time president of the National Audubon Society, on frequent collecting trips (Hume, 1942). In 1890, at the age of 19, Dutcher was appointed by Merriam as a temporary field agent to participate in a biological survey of Idaho. He did considerable fieldwork with Vernon Bailey, and the two became lifelong friends. In 1891, Dutcher became a member of the Bureau's Death Valley Expedition, involving a biological survey of southern California, southern Nevada, and parts of Utah and Arizona. In 1892, Dutcher was appointed to the Texas survey for his third and final summer of fieldwork for the Bureau. Dutcher spent 27 days at four localities in Texas (Stanton, Colorado City, Brazos, and Saginaw) during August and September. He collected 56 mammal specimens and prepared 16 survey reports during that time.

After graduating from the College of Physicians and Surgeons, New York, in 1895, Dutcher joined the US Army as an assistant surgeon with the rank of 1st lieutenant. He served for 25 years, and retired from active service in 1920 as a colonel. He died 16 January 1922, at the age of 50.

Albert Kenrick (A. K.) Fisher

Albert Kenrick Fisher (Fig. 16) was born 21 March 1856, in Sing Sing (now Ossining), New York. In 1879, Fisher received his MD from the College of Physicians and Surgeons in New York, where he met and befriended C. Hart Merriam. Merriam convinced him to give up further thought of medical practice and join him in the new Section of Economic Ornithology when it opened in 1885. Fisher helped out with fieldwork but especially in administration and as the liaison between the Survey and the Congress. Fisher's only contributions to the biological survey of Texas consisted of a six-day visit to Colorado City and El Paso in 1894. Fisher was promoted to assistant biologist in 1896 and assistant chief of the Bureau of Biological Survey in 1902. In 1906, Fisher was placed in charge of economic investigations, and he served in this capacity until his retirement from the Bureau in 1931. He passed away at the age of 92 on 12 June 1948, in Washington, DC. He was very widely known and liked, and to some degree he was as much of a politician as a scientist.

During his career, Fisher participated in the Death Valley Expedition (1891), the Harriman Expedition to Alaska (1899), and biological surveys of several western states from 1892 to 1898. Among the most important of Fisher's 160-plus publications were *Ornithology of the Death Valley Expedition of 1891* (Fisher, 1893a) and *Hawks and Owls of the United States* (Fisher, 1893b). Fisher was a founding member of the American Ornithologists' Union and served as its president from 1914 to 1917.

Louis Agassiz Fuertes

Louis Agassiz Fuertes (Fig. 17) was born 7 February 1874, in Ithaca, New York. Fuertes was fascinated by nature from early childhood, and collected all types of specimens, alive and dead. He was introduced at a young age to the works of John James Audubon, and soon began drawing birds himself. His first painting of a bird "from the flesh" was at the age of fourteen. His early insistence on working from live or freshly killed specimens, rather than preserved specimens, led him to actively collect birds and to become skilled at observing bird behavior as well as mimicking bird calls. In 1896, while a senior at Cornell University, Fuertes attended his first meeting of the American Ornithologists' Union. A collection of Fuertes' paintings had been displayed at the previous year's AOU Congress. At the 1896 meeting, Elliott Coues, one of the foremost ornithologists of the day and a mentor of Fuertes, introduced Fuertes to the scientific community as a young artist "on

Figure 17. Louis Agassiz Fuertes at a century plant in the Big Bend of Texas, 1901. Courtesy National Archives, 22-WB-30-3658.

whom the mantle of John James Audubon has fallen" (Peck, 1982). C. Hart Merriam was at that meeting and was among the many members of the scientific community who were impressed by the young Fuertes and his obvious talent. He immediately began commissioning works from Fuertes for the publications of the Division of Economic Ornithology and Mammalogy. In that same year, Fuertes also illustrated the book *A-Birding on a*

Figure 18. James Gaut, preparing specimens at camp, 1904. Courtesy Biological Survey Unit, US Geological Survey, Patuxent Wildlife Research Center, National Museum of Natural History.

Bronco by Florence Merriam. In 1899, Merriam recommended that Fuertes be invited to join the Harriman Expedition to Alaska. Fuertes' genial personality and commitment to hard work made him a welcome member of any expedition, and for the remainder of his life he was seldom without an invitation to travel.

In 1901, Fuertes joined Vernon Bailey and H. C. Oberholser for a four-month survey of the Trans-Pecos region of Texas. Fuertes found the experience exhilarating, for he was able to study a wider variety of birds than on any previous trip. He produced hundreds of bird portraits, as well as paintings and sketches of a number of mammals, a whip scorpion, and several lizards. Most of his works from that trip were purchased by the government for $20-30 each.

Fuertes continued to travel and paint, with major expeditions to the Bahamas, Mexico, Colombia, and Abyssinia (now Ethiopia). During his thirty-year career, Fuertes illustrated more than 35 books and approximately 50 educational leaflets, handbooks, and bulletins. He also was a regular contributor to more than a dozen popular and scholarly journals. Recognized as one of America's greatest ornithological artists, at least seven institutions now hold major collections of Fuertes' work. Tragically, Fuertes was killed at age 53 when his car was struck by a train in Unadilla, New York, on 22 August 1927.

James Gaut

James Gaut (Fig. 18), who had been acquainted with several of the more active naturalists about Washington, DC, worked as a scientist for the survey collecting specimens

Figure 19. Ned Hollister, 1922. Courtesy Library of Congress, LC-USZ62-073596.

and performing other tasks (Henderson and Preble, 1935). His mammal field notes indicate that he collected 3,860 mammals from 2 November 1896 to 25 January 1906. Gaut collected mammals in many areas of North America, including Virginia, Maryland, California, New Mexico, Texas, Oklahoma Territory, and Colorado.

Gaut's efforts in Texas began with a trip during February to June of 1903, from the Franklin Mountains (El Paso County) to Del Rio and then Alpine. He briefly visited the state in June and October of 1904, and then worked from southeastern Texas westward to Langtry and Samuels from February to June 1905. He spent a total of 251 days in Texas, preparing 35 survey reports, numerous bird specimens, and more than seven hundred mammal specimens. Gaut resigned from the bureau in 1906 and died in an automobile accident in 1914 at the age of 35.

Ned Hollister

Ned Hollister (Fig. 19) was born in Delavan, Wisconsin, on 26 November 1876. Self-taught in nature studies at an early age, Hollister took part in private studies in zoology from 1896 to 1901. His first scientific paper was published when he was just 16, and included data collected from the time he was 12 years old (Osgood, 1925). Hollister began voluntarily contributing specimens to the Bureau of Biological Survey in 1892. In 1901, Hollister visited the Smithsonian and the National Museum and met several key figures

Figure 20. Arthur Holmes Howell, 1903. Courtesy Library of Congress.

Figure 21. William Lloyd, 1929. Courtesy Library of Congress, LC-USZ62-58589.

of the bureau, including Vernon Bailey. In 1902, he was hired to work with Bailey as a temporary field assistant on the Texas survey. Hollister spent 86 days in the state during that year, and collected in the Piney Woods, Edwards Plateau, and Trans-Pecos regions. Hollister contributed 16 survey reports and 119 mammal specimens to the Survey. The following year, Hollister was again hired for a temporary position to accompany Wilfred Osgood to Alaska. He was finally appointed to a permanent position with the Bureau of Biological Survey in 1904, and participated in biological surveys of New Mexico, British Columbia, Washington, Oregon, California, Nevada, Louisiana, and Arizona. From 1910 to 1916, Hollister served as assistant curator of mammals at the US National Museum. Hollister was hired as the superintendent of the National Zoological Park in 1916 and served in that position until his death on 3 November 1924, at the age of 47.

Hollister's contributions to the Bureau's collections totaled 3,625 mammals and 1,509 birds, including the type specimens for 26 mammals. He named 162 new mammals and published more than 150 titles. Hollister was involved in the formation of the American Society of Mammalogists and served as the first editor of the *Journal of Mammalogy* from 1919 until his death.

Arthur Holmes Howell

Arthur Holmes Howell (Fig. 20) was born 3 May 1872, in Lake Grove, New York, where he grew up on a farm and developed a passion for natural history early in his life. With only a public school education, Howell was a self-taught naturalist. He was first employed by the Bureau of Biological Survey in 1895, when he served his apprenticeship in northern Montana with Vernon Bailey. During his tenure with the Bureau he conducted survey work in Alabama, Alaska, Arkansas, Florida, Georgia, Illinois, Kentucky, Louisiana, Missouri, Montana, New Mexico, North Carolina, and Texas.

Howell's work in Texas began with a two-month visit to the northern Panhandle in 1903. He returned in 1905 for a thorough four-month survey of an area extending from Falls County southward to Aransas County, and from Waller County westward to Medina County. Survey results from a third visit in 1906 were not included in the 1905 publication. Howell's three visits to the state totaled more than 200 days of fieldwork in eight ecological regions. He contributed 48 reports, 180 mammal specimens, and 76 photographs to the efforts of the Texas survey.

Howell worked for the Bureau of Biological Survey until his death 10 July 1940. During his career he published works on the birds of Arkansas and Florida and the mammals of Alabama, as well as a large number of monographic revisions of mammalian genera, including harvest mice (genus *Reithrodontomys*) and flying squirrels (genus *Glaucomys*).

William Lloyd

William Lloyd (Fig. 21) was born in 1854 of English parents in Cork, Ireland, and immigrated to the United States in 1876 (Casto, 1992). In 1881, Lloyd met John A. Loomis, owner of the Silvercliffe Ranch near Paint Rock, Concho County, and Lloyd was invited to join Loomis and two of his ranch hands on a hunt in Zavala County, Texas (Loomis, 1982). It was on this hunting trip that Lloyd's interest in ornithology began. Upon returning from that hunting trip, Lloyd was employed as a sheep herder on the Loomis ranch for two years and during that time spent many hours studying birds and educating himself about natural history from the books and journals in the ranch library. In 1885, Lloyd was elected an associate member of the American Ornithologists' Union, and in 1887 he published an annotated list of birds of the Concho Valley. Lloyd spent much of 1887 collecting birds for George B. Sennett, a businessman, conservationist, and ornithologist who devoted much of his life to collecting birds in Texas and Mexico (Maxwell, 1979a, b). Lloyd was then hired by Frederick Godman of the British Museum of Natural History to collect birds in Mexico from 1888 to 1889. In July 1889 Godman discharged Lloyd, and Lloyd returned to his ranch in Marfa.

Lloyd was an acquaintance of C. Hart Merriam's as a fellow member of the American Ornithologists' Union. In 1890, Merriam sent Bailey instructions to meet Lloyd at his ranch near Marfa for a collecting trip, with Lloyd as a protégé and prospective field agent. A letter from Bailey to his family explained that "Merriam says Lloyd can't make good skins and can't trap but wants me to show him all I can and get him started at it" (Schmidly, 2018).

Apparently, Bailey's training was successful, and Lloyd was hired as a field agent on 1 July 1890. His travels took him from his home near Marfa southeastward along the Rio Grande to Laredo and ended there on 26 December. In May of 1891, Lloyd resumed his travels beginning at Laredo and continuing southward to Brownsville and then northward along the Gulf Coast to Houston, where he ended his work for the survey in March of 1892. Lloyd's total of 496 days in the field exceeded even Bailey's efforts in Texas prior to the 1905 publication. Lloyd's work took him into four ecological regions, but most of his efforts were in the Gulf Prairies and Marshes and South Texas Plains regions. He contributed 51 survey reports and 1,181 mammal specimens to the Bureau's collection.

After 1892, little is known about William Lloyd's life, but he apparently never again worked as a naturalist. By 1898, Lloyd had moved to New Orleans where he owned a store and dealt in old books, coins, and stamps. Lloyd died in New Orleans in October 1937 (Geiser, 1956).

John Alden Loring

John Alden Loring (Fig. 22) was born 6 March 1871, in Cleveland, Ohio. He served as a field agent with the Bureau of Biological Survey from 1892 to 1897. Noted for being a most enthusiastic collector, Loring worked in many of the western states and in central Canada. His fieldwork in Texas involved 108 days, from January to May 1894. He traveled from El Paso southward to Brownsville and Hidalgo, then proceeded northward, collecting from Arcadia on the Gulf Coast through eastern Texas to Fort Worth, then westward to Tascosa in the Panhandle. He contributed 19 survey reports and more than 300 mammal specimens to the survey.

Figure 22. John Alden Loring, 1913. Courtesy Library of Congress, LC-USZ62-073590.

Figure 23. Harry Church Oberholser, 1937. Courtesy Library of Congress, LC-USZ62-117490.

From 1897 to 1901, Loring served as curator of animals at the New York Zoological Park. While conducting fieldwork in Europe for the National Museum, Loring broke all previous records by collecting and preserving the skins of 913 mammals and birds in 63 days (Palmer et al., 1954). Loring was a participant in the Smithsonian-Roosevelt Scientific Expedition to Africa, 1909-1910. Loring's publication record contains primarily popular articles and nature books for children, including *Young Folks' Nature Field Book*. Loring died 8 May 1947, in Osweego, New York, at the age of 76.

Harry Church Oberholser

Harry Church Oberholser (Fig. 23) was born 25 June 1870, in Brooklyn, New York. He entered Columbia University in 1888, but was compelled to withdraw in 1891 because of poor health. In later years he received the BA and MS (1914) and the PhD (1916) from George Washington University.

In 1895, Oberholser was appointed as an ornithological clerk in the Division of Economic Ornithology. He was promoted to assistant biologist in 1914, biologist in 1924, and senior biologist in 1928. He retired from the government in 1941 at the age of 71 but served as curator of ornithology at the Cleveland Museum of Natural History until 1947. Throughout his 46-year government career, his primary duty was the identification of birds; in later years he was often called as an expert witness to identify evidence in trials for alleged violations of federal game laws. He conducted fieldwork in many states, but most notably in the Southwest. Oberholser published nearly 900 papers during his career and was a member of 40 scientific and conservation organizations.

Oberholser's lifelong interest in the birds of Texas began with his first biological survey trip with Vernon Bailey in 1900, a five-month survey of portions of the Gulf Coast, southern Texas, and northern Texas at Henrietta. Oberholser returned for four months in 1901, when he traveled from San Angelo westward into the Trans-Pecos region and northward to the Panhandle. In 1902, Oberholser surveyed the eastern third of the state for four months. His survey efforts during those three years, a total of 397 days in the state, took him to every ecological region. Oberholser contributed 118 survey reports, 276 mammal specimens, and 710 photographs of landscapes, wildlife, and habitats to the survey.

The ornithological data gathered during Oberholser's survey trips originally were to be published as part of the *Biological Survey of Texas*, but the report soon grew too lengthy, and it was decided that a separate report on the birds would be published. However, Oberholser's work continued to grow in length, and remain unpublished through the years, as he continuously strove to update and expand the information to satisfy himself that it was suitable for release. In 1941, when Oberholser retired from government service, the manuscript had grown to more than three million words. Eventually, the University of Texas Press acquired funding from a private source to publish *The Bird Life of Texas*. They assigned an editor, Edgar B. Kincaid Jr., to assist Oberholser with the task of reducing the manuscript from three million to one million words. Oberholser continued to work on the book until shortly before his death on 25 December 1963, at the age of 93.

Figure 24. Clark Streator, 1930. Courtesy National Archives, 22-WB-46-B4414M.

Figure 25. Howard Lacey's ranch near Kerrville, Kerr County, 1906. (In this and all subsequent land-scape figures, county designations reflect current county boundaries.) Courtesy National Archives, 22-WB-30-9048.

Mr. Kincaid completed the editing of the book, and *The Bird Life of Texas* was finally published as a two-volume set in 1974 (Oberholser, 1974).

Clark P. Streator

Clark Perkins Streator (Fig. 24) was born in Ohio in 1866. In the early 1880s, Streator traveled extensively in the West Indies collecting birds for ornithologist Charles B. Cory. About 1890, he collected in British Columbia for the American Museum of Natural History. In later years, Streator collected in Mexico with E. W. Nelson and E. A. Goldman and in several western states, including Idaho, Colorado, California, Nevada, and Texas. Streator was considered an authority on the mammals of California, and he had visited the type localities for almost every mammal species in the state (Palmer et al., 1954).

Streator's contributions to the biological survey of Texas were limited to travel with William Lloyd during November and December 1890, in Maverick and Webb counties. Streator contributed 58 mammals, 11 birds, and two snakes to the Survey collection during this trip, including the type specimen for *Cratogeomys castanops angusticeps* from Eagle Pass. Streator died at Santa Cruz, California, on 28 November 1952, at the age of 86.

Local Naturalists

Bailey and the field agents relied heavily on local naturalists and landowners while conducting the survey. Bailey was particularly fond of an Englishman, Howard Lacey, who owned a ranch near Kerrville in the Hill Country of central Texas. Also, H. P. Attwater from San Antonio made numerous trips to secure important specimens when the field agents were not in the state. Both Lacey and Attwater made significant contributions to Texas natural history and brief biographies for them are included here.

Howard Lacey

Howard George Lacey was born at Wareham, Dorset, England, on 15 April 1856. At the age of 26 he immigrated to the United States and settled in the Texas Hill Country, on a ranch on Turtle Creek, about 10 miles (16 km) from Kerrville (Fig. 25). He made his livelihood by raising horses, cattle, and Angora goats, but he was very interested in natural history and soon became an authority on the fauna and flora of central Texas (Palmer et al., 1954). He maintained correspondence with naturalists in various parts of the country and welcomed many to his ranch to conduct natural history studies, including Vernon Bailey and other members of the Bureau of Biological Survey.

In his field diary of 7 May 1892, after spending several days with Lacey, Bailey wrote: "We were sorry to say good-bye to Mr. Lacey who has treated us with the greatest hospitality and helped us in many ways with our work. He is one of the best specimens of a typical, whole-hearted, generous Englishman I have met. He has a good ranch but does not give much time to farming or business. Collects butterflies and knows birds and mammals well."

Figure 26. Henry Philemon Attwater. Courtesy Library of Congress, LC-USZ62-073594.

Figure 27. Ab Carter, hog raiser and bear hunter, near a bear-gnawed tree, Tarkington, Liberty County, 1904. Courtesy National Archives, 22-WB-51-7236.

Figure 28. Ab Carter's house at Tarkington, Liberty County, 1904. Courtesy American Heritage Center, University of Wyoming.

Lacey collected many natural history specimens himself and donated a number of them to the collections of the Biological Survey and National Museum. In recognition of his contributions to natural history research, three taxa of small mammals were named for him (*Peromyscus laceianus, P. boylei laceyi*[M], and *Reithrodontomys fulvescens laceyi*). In 1919 Lacey sold his central Texas ranch and returned to England, where he died 5 March 1929.

H. P. Attwater

Henry Philemon Attwater (Fig. 26) was born in London, England, 28 April 1854. In 1873, he immigrated to Ontario, Canada, where he engaged in farming and beekeeping. Attwater soon became interested in natural history. During 1884, he made a trip to Bexar County, Texas, to collect specimens. Attwater was employed in 1884 and 1885 to prepare and exhibit natural history specimens in the Texas pavilion at the New Orleans World's Fair (Casto, 1992).

In 1889, Attwater moved to San Antonio, Texas. He soon became an authority on the natural products and resources of the state and conducted experiments in agriculture and horticulture. In 1900, he relocated to Houston when he was appointed agricultural and industrial agent of the Southern Pacific Railroad. He continued to expand his natural history collections and contributed numerous specimens to museums, including the US National Museum and the American Museum of Natural History (Palmer et al., 1954). He was always eager to assist anyone who needed information or material that he could provide. He was devoted to the protection of birds and other wildlife and he had an important influence on the development of wildlife conservation laws in the state. The Attwater's greater prairie chicken (*Tympanuchus cupido attwateri*) and several small mammals (e.g., *Peromyscus attwateri* and *Geomys attwateri*) are named in his honor, in recognition of his major contributions as a scientist and conservationist. H. P. Attwater passed away 25 September 1931, at his home in Houston.

Farmers, Ranchers, and Woodsmen

Local farmers, ranchers, and woodsmen contributed much to the wildlife information base in the Texas survey. For example, Ab Carter, a farmer and woodsman from Tarkington Prairie, in Liberty County in the southeastern part of the state, provided Bailey with information about the demise of black bear populations in that area (Figs. 27, 28). According to Carter, he and a neighbor personally had a hand in killing 182 bears within a ten-mile radius of their ranches over a two-year period.

Likewise, Mr. C. O. Finley, a rancher from Fort Davis in the Davis Mountains of west Texas (Fig. 29), provided Merriam and Bailey with the following detailed account of the killing of the only grizzly bear ever shot in Texas.

Figure 29. C. O. Finley's ranch, Davis Mountains, Jeff Davis County, 1890s. Courtesy Museum of the Big Bend, Sul Ross State University.

"The Davis Mountain Silvertip,"
by C. O. Finley, Pecos, Texas.

I have been asked to write the true story of a bear hunt on which the only grizzly bear had ever been killed in Texas.

As a prelude to this story, I will give a brief history of the beginning of what turned out to be an annual bear hunt by a number of the old pioneer ranchmen in the Davis Mountains.

In the early Nineties, some seven or eight families of us, the Means, Evans, Marleys, Jones, Mayfields, Finleys, and a number of others, all met at what is known by all of the people in the Davis Mountains as the Rock Pile, near Saw Tooth Mountain and one of the most beautiful spots for camping in the Davis Mountains. Here we met several years prior to the noted year that we killed the old grizzly. In addition to our own families we always had a number of friends from over a good part of Texas that would come from Fort Worth, Dallas, San Antonio and other parts to spend a week with us on our hunts. We also invited one or two ministers every year to our camp, which was as clean and free of bad language as was possible. Never an oath was heard nor a bottle of whiskey came to our camps.

Our party usually ran from forty to seventy-five people. Men, women, and children, and they ran from young untried hands to old tried and true bear dogs. We had from 100 to 150 saddle horses and always took three or four old time chuck wagons and plenty of good Mexican cooks, as good ones as ever cooked a meal around a campfire and they always keep plenty of warm grub to eat. We always had two day horse wranglers and two night hawks, as they were called, to herd our horses, day and night. In the morning the old cooks would be up early and have us a good hot breakfast ready before daylight, so we were always ready when our night hawks brought our horses up near camp by the time it was light enough for us to see how to rope our horses for the days ride. Such a good time we had roping and saddling our horses. All horses those days were not pets and every morning before we could get saddled and out of camp, some would be pitching with their saddles and some with their riders, and very often some "old boy" would get pitched off and his horses would have to be caught. When everyone was ready, which was never later than sunup and pretty cold in October and November in the mountains, someone would blow their horn and off we would go to the mountains, which was bear heaven.

We hardly ever failed to get a bear and sometimes three or four during the day. And every day or so someone would get lost from the bear chase and kill a deer or two. While this was all good sport, especially if you got in the chase and got to the killing of the bear (which was sometimes up a tree or a cave, or out in the open) it was awfully hard on the horses and men, who rode hard and reckless in that rough country trying to follow and keep up with the hounds.

When we got in to camp of an evening, horses, men and dogs were usually all in, but a well cooked meal, a little rest and a few cups of coffee, and we were ready for a good talk and a prayer by one of our ministers. We would then clear us off a spot of ground and dance the old square dances for an hour or so, as we always had musicians and plenty of music in the camp. Then off to bed and a good night's rest in our tents and on our old camp beds, and early the next morning we were ready for another days hunt. Some days we had fine luck and other days our luck was not so good. These hunts were made up of old men, young men, women and children old enough to ride. The women and children would stay as long as possible, but not often did they get in a chase after a bear and see him killed, but would drop out and a few of the older men would drop back to take care of them and see that they got back to camp. It was always uncertain which way a bear would run or how far. If he was fat he would probably not run over a mile or two, but would climb a tree, but if he was poor he was always hard to stop and sometimes would get away entirely. When we found an old poor bear and failed to stop him, we always lost some of our dogs and they would not get in for sometimes a day or so, all footsore and almost starved.

Well, I started this story to tell you of the jolliest old days ever spent by our party or any other parties in the Davis Mountains, which was in the fall of 1900. On the 29th day of October, we met at our old campground at the Rock Pile, about 75 of us, just as we had each year before, with every one feeling good. The horses were fat and the dogs in good shape and we all anticipated a good time, and we had it. We spent the first four days with just fairly good luck, getting a bear or two and some blacktail deer each day. On the third day of November, which was our fifth days hunt, we all left camp as usual about sunup and all went together that morning right into the mountains. We traveled some eight or ten miles crossing up near the head of what is known as Saw Mill Canyon just north of Livermore Peak, and on southeast over into the head of Limpia Canyon. This is the canyon that part of Fort Davis is on and in going from the head of Saw Mill Canyon through the mountain over into Limpia Canyon, there is a gap that was always called Bridge Spring Gap. Well, all of our party, some thirty or forty people, went on through the gap, we dropped back behind and turned south up the side of the mountain, topped out and rode along the top, parallel with the balance of our party and the dogs, who were going down Limpia Canyon. All the country is very rough. Full of canyons, bluffs, and lot of timber from shin oak thickets to pines fifty feet high. Well, when John Means and I got out on top, we were possibly a mile or more from the balance of the party. We heard a dog yelp and then another and pretty soon the whole pack was running and yelping and the dogs and Means all rode up on a four year old fat cow that had been killed up on the side of the mountain and then drug down to the hill about 100 yards into a big thicket and part of her had been eaten. Well by that time the dogs had started the old bears trail and when the bear heard the dogs, he pulled out from where he was bedded up near the cow he had killed, and ran out of the canyon south across the mountain and crossed out on top just ahead of John Means and myself, so we rode like drunk Indians to keep up in hearing of the dogs.

Just after we had crossed their trail in behind them, John looked around to the right and said "Otie, I see the old devil." He had gotten out on an open spot and stopped. Then we switched around some brush and rocks to where we could get a shot at him but when we came out to where we could get a shot, the old bear was gone and there set four dogs, the only dogs out of the whole pack that would run his trail. Well, we muched [sic] them a little and got them to go on and we got back on our horses and followed them about a mile and a half further on over some very rough canyons and down into the head of Merrill Canyon. There the old bear had stopped again. Means and I had ridden as far as we could and had to leave our horses and walk down the mountainside to where we heard the dogs barking. When we got about a quarter of a mile where they had stopped the old bear, we met the four dogs all coming back, trailing along one just behind the other to meet us. Well, we muched [sic] them up again and got them to go back to where they had stopped the bear, which was down in a deep bushy rough canyon. We located him standing with his head toward us in the brush with the dogs standing off barking. When we got down to within about 125 yards of him, we sat down, side by side, on a small bluff of rocks with our little short saddle guns, 30-30, and began pumping lead into him. We evidently hit him with our first shots as he began to pitch and bellow like a wild bull, and made a dash in the brush at all the dogs, but only succeeded in catching one old blue speckled bob-tailed hound that belonged to Bill Jones and the bear tore the dogs jaws and neck up so bad that we had to kill the old fellow. He failed to catch the other dogs as they were younger and could get out of his way. He then came back and stopped exactly in the same place he was in when we shot our first shots. We did not lose any time shooting four more shots each into him and he just melted down on his old belly. We did not know yet what we had, but knew he was an extra big bear. We took a little round and got about one hundred feet above him on the canyon and stopped to see if he was sure enough dead. While we were standing there, Means raised up his gun and said "Otie, hadn't I better put one in his old head? He looks awfully big." And I said, "No, lets don't tear his head up, as he looks awfully big and some of us might want to keep it." So about that time a little black and tan hound that belong[ed] to Joe Marley walked up and caught the bear by some hair and shook him a little, then we knew the old bear was dead, so we went down to him. When we got down we discovered the gray tips of his hair. John yelled like a Comanche and threw his hat as high in the air as he could and said, "Otie, we have got a grizzly." To say we were an excited pair is putting it very lightly. Well, we turned him over and opened him up and removed his entrails then turned him back to drain good and went back up the mountainside out on top to where we had left our horses and blew and blew our horns and finally got eight more men to come to us. The party was scattered out over the mountains trying to locate the dogs and bear. We went back to the old bear with the eight men but he was so large there was no way to take him back to camp, so we took his hide off and left his feet and head on the hide and put the hide across the biggest stoutest horse we had, and lit out for camp, with the man who had been riding the horse riding behind another man. We could not go straight through the mountains to camp and lead this horse with such a load, as we had estimated the old bear to weigh at least 800 pounds, so his hide and feet would weigh 400 or 500 pounds. We had to go around and out of the mountains on the west side and out by my ranch. The horse that was carrying the hide had the thumps and almost had lockjaw so we had to turn him loose and get a pair of little Spanish mules and an old buck board I had at the ranch to carry our hide on to camp. All this was about 20 miles from where we had killed the bear. It was about 9 o'clock that night when ten of us with our three hounds got into camp with our kill. All worn out and hungry, not having had anything to eat since before daylight that morning. You can imagine how good those old cooks and their old dutch ovens filled with good warm camp grub and that black coffee looked to us.

When we had unsaddled and turned our tired and worn out ponies loose and had unloaded and stretched the old bear hide over a big rock, we were the most excited bunch of people ever seen. Everybody was talking, some asking questions and some telling their story of what had happened and how it all happened. We had all seen a lot of this kind of thing for a good many years, but this was the biggest days hunting that had ever been pulled off in the Davis Mountains.

The next morning we cut off his feet and found that he only had three claws on one of his front feet and one on the other. George Evans, John Means, and I kept three of them and took them to El Paso and had them plainly mounted for watch fobs, and the fourth we gave to a good friend of ours by the name of Tat Hulling who ranched at that time back north of Kent on the Delaware, but who has since died, but his widow still lives in El Paso. Mr. Means and Mr. Evans always wore their bear claws for watch fobs, just as they were mounted, but they were so big and heavy that I did not like to wear mine so I had a pin put on it for Mrs. Finley to wear and she wears a pin different from any other woman.

The hide we sent to San Antonio and had it nicely dressed and gave it to Mr. L. S. Thorn, who at that time, was the General Superintendent of this western district of the T & P Railroad, who was a good friend of ours, as we all shipped a good many cattle over his road those days from Van Horn and Kent.

The head I kept Mr. Means from shooting into, I took home and put it in an old big wash pot and boiled all the meat off of it and scraped it and cleaned it up good and hung it over our front door outside and never did expect anything to be done with it other then [sic] to be seen and admired by all guests. However, that winter, a young man from the biological department of Washington came to the Davis Mountains to collect specimens of all kinds of varmints and fowls that grew in the mountains and came to our ranch and asked to stay. Mrs. Finley let him stay and during the winter he heard us talking about our past fall bear hunt and saw the bears head hanging out over the door, and when he went back to Washington the next spring, he related the story to the departments heads and they wrote me and asked me to send it to them, which I did. After keeping it several months, they sent it back with several long names attached to it that we couldn't read, but pronounced it a real grizzly bear and that there was no history of one ever having been found in Texas before. In about a year the department wrote me again and asked me to return it to them, which I did as they wanted to resurvey it. Then they wrote and asked to buy it for the Government stating that it was a very rare specimen and that they would forever preserve it and attach any record to it that we might want to keep with it. We tried to get them to send it to Dallas to be exhibited at the Fair but the Department said they kept it locked in a vault and were afraid to let it go. I have a nice letter from them which is attached hereto and will give an idea what it is and what they think of it in Washington.

In conclusion I must say a few more words about our hunts and our association with our neighboring ranchmen in the Davis Mountains. In all of our sports or meeting together, thanks to our mothers and wives, there was a spiritual side to it all that gave us the association of the most religious people and out of this camp meeting, which originated from the mind of our beloved little preacher, Mr. Bloys, and a few of our old pioneer ranch mothers in 1890, and was and is called Bloys Campmeeting, and started with only a few people of all four denominations and which has continued as a union meeting from the beginning and has grown as a great many of you know to a regular attendance each year of about 3000 people and grows a little larger every year. Most of the original starters of the meeting have gone on to their reward but a few survivors have great faith in the present generation carrying this meeting on forever.

Mammalogist Contemporaries of Merriam and Bailey

During the time that Bailey and the federal agents worked in Texas, other naturalists were at work as well. Another government exploration was underway along the Mexican boundary and the region of the Rio Grande. Dr. Edgar A. Mearns, who had been stationed at three military posts on the Texas border, was detailed in 1892 by the War Department to act as medical officer and naturalist for the International Boundary Commission. This Commission had been established in the 1880s when the United States and Mexico signed a treaty providing for an international boundary survey to relocate the existing frontier line between the two countries. The scientific work involved a biological survey of the Mexican boundary region. Three years (1892-1895) were spent in the survey of more than 700 miles (1,126 km) of the boundary. Mearns and his assistants traversed the entire border, including all of Texas, collecting plants and animals and recording detailed natural history information.

In 1907, Mearns published *Mammals of the Mexican Boundary of the United States*, recording his observations and scientific information (Mearns, 1907). He contributed more than thirty thousand plant and animal specimens to the National Museum, including seven thousand mammals (Wilson and Eisenberg, 1990), and he described fourteen taxa of Texas mammals from throughout the borderlands region. Many of the species he named have now been placed in synonymy with other taxa.

J. A. Allen was a contemporary of Merriam and Bailey who served as the curator of birds and mammals at the American Museum of Natural History in New York City. Allen purchased several private collections of mammals from south Texas (Allen, 1894), including Bexar County (Allen, 1896) and Aransas County (Allen, 1894). H. P. Attwater was well acquainted with Allen and provided him with specimens and descriptions of Texas mammals. Also, three well-known amateur ornithologists, George Sennett, F. B. Armstrong, and George Ragsdale, worked in Texas, and they also provided specimens to both Bailey and Allen. Allen was a prolific taxonomist, and he described fifteen taxa of mammals from Texas, but insofar as we are aware he never visited the state.

Life Zones and Texas Vegetation

Determining and understanding the factors that governed the geographic distribution of birds and mammals was C. Hart Merriam's scientific passion and a major justification for his extensive fieldwork and biological surveys across the United States. His first theories on this question apparently were influenced by accounts he had read as a child of the zonal pattern of life in the Andes Mountains and by his own observations around his home near the Adirondacks in New York. Even from these early years, Merriam believed that the distribution of species was determined by differences in temperature and humidity, with temperature during the breeding season being the most crucial factor. These beliefs formed the basis for Merriam's Life

Zone Theory, first articulated in the 1890 *North American Fauna* publication based on Merriam's ecological studies of the San Francisco Mountain region of Arizona (Merriam, 1890b) (see Fig. 2).

Merriam recognized four life-zone belts, designating them the Canadian, Transition, Upper Austral, and Lower Austral, the boundaries of which coincided closely with isotherms (Merriam, 1894). Each of these zones was divided, at about the 100th meridian, into eastern humid and western arid faunal areas to indicate the secondary role played by moisture and vegetation. The 100th meridian practically bisects Texas, a fact that undoubtedly had much to do with Merriam's interest in conducting an extensive biological survey of the state. Texas, with its varied geography and vegetation types, including both arid and humid conditions, was a key state for verifying Merriam's hypotheses about plant and animal distributions.

Four of the transcontinental life zones are represented in Texas as broad bands stretching across the state or as encircling rings or caps on elevated peaks and mountain ranges in the far western part of the state (Fig. 30a). The life zones, divisions, and biotic regions listed for Texas include the following (Merriam, 1898b):

Lower Austral Zone
  Lower Sonoran Division
    Gulf Strip of Texas
    Austroriparian of Eastern Texas
    Grand and Black Prairies
    Coast Prairie
    Coast Marshes
    Beaches and Marshes
    Semiarid Lower Sonoran
    Extreme Arid Lower Sonoran

Upper Austral Zone
  Upper Sonoran Division

Transition Zone (upper slopes of the Guadalupe, Davis, and Chisos Mountains)

Canadian Zone (along northeastern base of Mt. Livermore in the Davis Mountains)

Bailey often spoke of the Lower and Upper Sonoran as if they, too, were zones, but in reality they were only the arid subdivisions of the Lower and Upper Austral Zones, respectively. The humid divisions of these same zones were designated as the Austroriparian and the Carolinian, although only the former was represented in Texas. Bailey also recognized a number of unique biotic regions within the major zones and divisions. These regions were not formerly recognized in Merriam's life zone maps, but they were sufficiently unique to warrant distinction by Bailey.

Although useful in the local analysis of faunal origin, dispersal, and evolution, Merriam's efforts to explain the geographic distribution of birds and mammals based

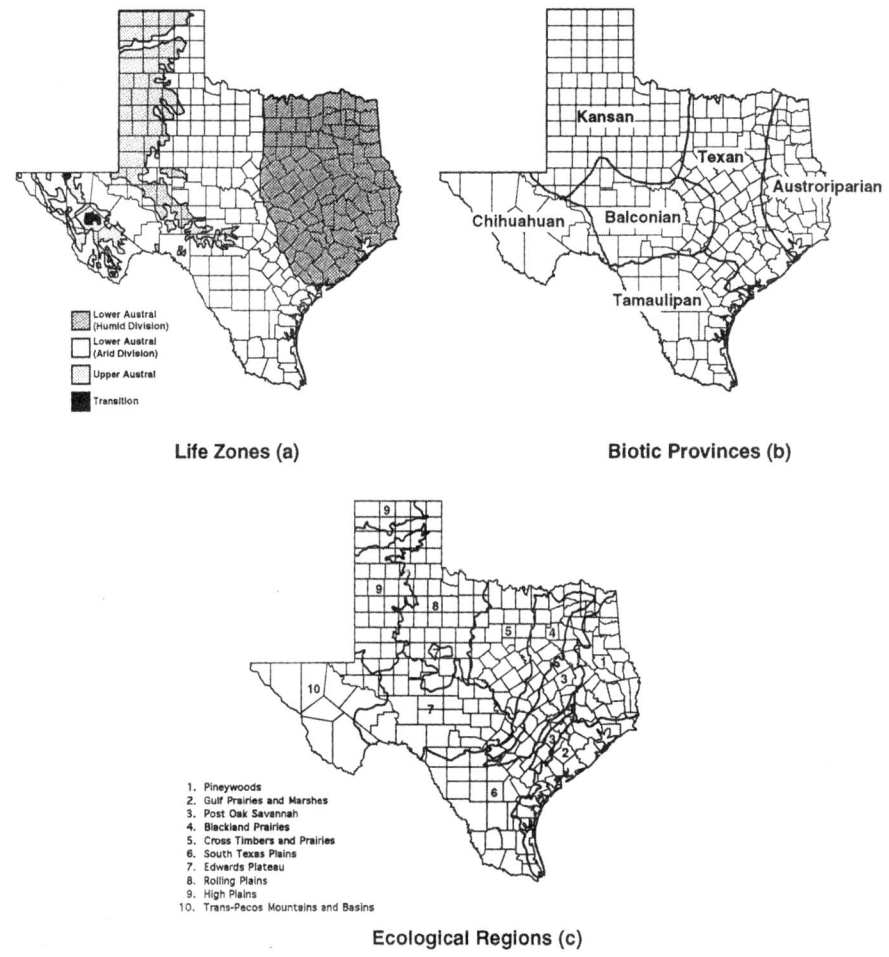

Figure 30a-c. The Life Zones (a), Biotic Provinces (b), and Ecological Regions (c) of Texas.

on temperature and humidity alone received acceptance for only a short period (Kendeigh, 1932). Investigations of biogeography by other scientists revealed flaws in Merriam's data and conclusions and demonstrated that many factors other than temperature and humidity played important roles in the geographic distribution of species (Shelford, 1932, 1945). His conclusions were considered too simplistic to explain the geographic distribution of birds and mammals in all cases. Under Merriam's system, the lower Rio Grande Valley, the eastern Panhandle, and the deserts of the Trans-Pecos are placed in the same life zone. Such diverse regions as Kerr County on the Edwards Plateau, the Hueco Mountains of the Trans-Pecos, and the High Plains of the Panhandle are placed together in another life zone. Such a system obviously is not descriptive of animal distribution and ecology in Texas and has limited utility (Blair, 1950).

Table 2. Summary of the physical and climatic characteristics of the ecological regions of Texas.[1]

	Million acres	Topography	Elevation (feet)	Annual precipitation (inches)	Frost-free days
Piney Woods	15.8	Nearly level to gently undulating	200-700	40-56	235-265
Gulf Prairies and Marshes	10.0	Nearly level	0-250	26-56	245-320
Post Oak Savannah	6.85	Nearly level to gently rolling	300-800	30-45	235-280
Blackland Prairies	12.6	Nearly level to rolling	250-700	30-45	230-280
Cross Timbers and Prairies	15.3	Gently rolling	500-1,500	25-35	230-280
South Texas Plains	20.9	Nearly level to rolling	0-1,000	18-30	260-340
Edwards Plateau	25.45	Deeply dissected hilly, stony plain	1,200-3,000	12-32	220-260
Rolling Plains	24.0	Nearly level to rolling	1,000-3,000	18-28	185-235
High Plains	19.4	Nearly level high plateau	3,000-4,500	14-21	180-220
Trans-Pecos	17.95	Mountain ranges, rough, rocky land, flat basins and plateaus	2,500-8,751	8-18	220-245

[1] Reprinted from Hatch et al. (1990).

In 1943, Lee R. Dice, an ecologist at the University of Michigan, mapped the general distribution of the biotic provinces of North America. In 1950, Frank Blair, an ecologist from the University of Texas, more accurately defined, mapped, and described the biotic provinces within Texas (Fig. 30b). A biotic province is defined as a considerable and continuous geographic area, characterized by the occurrence of one or more ecologic associations that differ, at least in proportional area covered, from the associations of adjacent provinces. In general, biotic provinces are characterized also by peculiarities of vegetative type, ecological climax, flora, fauna, climate, physiography, and soil. The utility of biotic provinces has proven far more useful in describing the fauna and flora of Texas than Merriam's life zones.

Most biologists working in Texas today use either ten or eleven ecological regions (depending on the predilection of the author), called ecoregions, in referring to the distribution of plants and animals in the state (Fig. 30c). The delineation of these regions is based on the interaction of geology, soils, physiography, climate, and the distribution of plant and animal communities (Table 2). Two recent excellent accounts of these ecoregions have been published and are used throughout this book to describe the location of work conducted by the federal agents and as a reference to the biological diversity of the state. The two references are Brian Chapman's and Eric Bolen's *The Natural History of*

Texas (Chapman and Bolen, 2018) published by the Texas A&M University Press, and a chapter written by David Riskind entitled "Ecological Regions of Texas" (Riskind, 2015) that appears in the *Texas Master Naturalist* statewide curriculum book (Haggerty and Meuth, 2015).

The Development of Mammalogy in Texas

Texas now has a long, deep history in mammalogy that has greatly strengthened its ability to monitor the status of populations and species, thereby facilitating our understanding of how to best conserve them. But it was not always this way. During the time that Bailey and the federal agents worked in Texas, there were no professionally trained mammalogists living in the state. A number of private citizens called themselves amateur naturalists, and they collaborated with the federal agents. Probably the closest person to a so-called expert would have been H. P. Attwater from San Antonio. Mr. Attwater made extensive collections in the regions about San Antonio and Aransas Pass and discovered a number of new forms (including Attwater's prairie chicken, *Tympanuchus cupido attwateri*) that were described by Dr. J. A. Allen in the *Bulletin of the American Museum of Natural History*. He also collaborated and worked with the biological survey field agents, including Vernon Bailey.

Vernon Bailey's 1905 publication was the first comprehensive study of mammals in the state and set the stage for further studies of their distribution, taxonomy, and natural history. Edgar Alexander Mearns's account of the mammals of the Mexican boundary of the US appeared in 1907, and it included considerable detail about Texas mammals along the border. An updated checklist of mammals was authored in 1926 by John K. Strecker (Strecker, 1926), curator of the museum at Baylor University, but the progress of mammalogy in Texas was slow for the first portion of the twentieth century. It was not until the decade of the Great Depression that the science started again in earnest when Texas A&M University established the Department of Wildlife Science in 1937, with William B. Davis as its head, and the Wildlife Cooperative Research Unit, helmed by Walter P. Taylor. The following year, 1938, Davis established the Texas Cooperative Wildlife Collection, the first major mammal collection in the state. Both Davis and Taylor were leading authorities on mammals, having received PhD degrees in the field under the direction of Joseph Grinnell at the University of California–Berkeley, and they jointly authored the first edition of *The Mammals of Texas* in 1947 (Taylor and Davis, 1947). In 1946, the University of Texas at Austin brought onto its faculty Frank Blair, an ecologist from the University of Michigan, where he had studied under mammalogist Lee R. Dice. Blair conducted studies on Texas mammals for more than a decade. With these moves, academic mammalogy, at both the undergraduate and graduate levels, was established at the two leading public institutions of higher education in the state.

During the latter half of the twentieth century, the science of mammalogy exploded in Texas as a major field of research and education. A detailed account of this history, including the major events and the many individuals who made significant contributions to it, is beyond the scope of this book, but highlights of the subject can be found in several sources (McCarley, 1986; Birney and Choate, 1994; Baker, 1995; Schmidly and Dixon, 1998; Bradley et al., 2005; Baker et al., 2007). Much of the second phase of growth in mammalogy was fostered at the Natural Science Research Laboratory at Texas Tech University.

The crowning achievement in the growth of mammalogy in Texas was the establishment in 1981 of a separate scientific society, the Texas Society of Mammalogists (TSM). The late Robert L. Packard of Texas Tech University was the lead organizer of the society (see Baker et al., 2007, for a complete history of TSM). Texas and California are the only states with a scientific society devoted to mammalogy, and there are more professional mammalogists living in Texas than in any other state in the country.

The mission of the Texas Society of Mammalogists, as stated in its constitution, is to "promote the study of mammals, living and fossil" in Texas and beyond. Beginning in 1983, the society has met in February of every year at the site of Texas Tech University's Junction Campus in Kimble County. Mammalogists and students from throughout the state and surrounding states convene and present papers and hold discussions about the biology and conservation of mammals, emphasizing Texas species. The annual membership of the Texas Society of Mammalogists now numbers approximately 135.

Clearly, Bailey and the federal agents planted a seed that would grow into a major scientific field of study. They probably had no idea this would happen, but it is almost certain they would be pleased with the outcome. The scope of the science of mammalogy and the scientific talent available in the state will be crucial to our ability to manage and conserve mammalian species and communities wisely in the twenty-first century. A very good synopsis of the major developments in Texas mammalogy can be found in Schmidly (2002) and updated by Schmidly and Bradley (2016).

Changes in Classification and Taxonomy of Texas Mammals

The twentieth century witnessed a dramatic shift in the philosophy of classifying mammals for the purpose of taxonomic designation. Mammalogists altered the taxonomy of mammals based on new information about the variation in populations, subspecies, and species. For this reason, there is little resemblance in the taxonomic arrangement of Texas mammals today compared to the era of Bailey and the biological survey.

Vernon Bailey and his mentor, C. Hart Merriam, were restricted to a morphological and typological concept of species and subspecies. According to their logic, if specimens representing two populations could not be distinguished morphologically, then they would not be regarded as taxonomically distinct (Merriam, 1897, 1919). If the specimens were different, especially when compared to "type" specimens (designates for a species or subspecies), depending on the degree of difference, then they would be classified as either subspecies or species. This approach came to be known as "splitting," and Merriam and his followers, including Bailey, were major proponents of the practice (Sterling, 1974).

The biological species concept (see Mayr, 1942, 1963), dominated by the recognition of geographically variable species (termed polytypic species), slowly began to take hold in the twentieth century, and many of the species recognized by Merriam and Bailey were "lumped" into single, wide-ranging species. Dr. E. Raymond Hall, a highly distinguished mammalogist at the University of Kansas from the 1940s through the 1970s, was the leading proponent of this practice in mammalogy (Schmidly and Naples, 2019).

With the advent of modern techniques of genetics and molecular biology in the last three decades of the twentieth century, new tools became available to measure genetic (and evolutionary) relatedness among populations. These techniques allowed scientists to study chromosomes (called karyology) and the sequences of genes in animals. Likewise, new sophisticated techniques of statistics allowed for more refined assessments of morphology (called morphometrics) among populations of mammals. Based on the use of these new techniques, several newly recognized species, referred to as "cryptic species," were recognized in the state. Although such species could not easily be differentiated on the basis of observed morphological characteristics, they were, in fact, genetically distinct and reproductively isolated, thus meeting the basic requirement for species distinctness. Taxonomic studies also have revealed examples of taxa that were different enough morphologically to be called separate species but proved to be almost identical genetically and thus deemed fully capable of interbreeding and producing viable offspring. Modern taxonomists typically arrange such populations as different subspecies of the same species.

The best example of cryptic species is presented by the pocket gophers of the genus *Geomys*. Bailey recognized nine taxa of *Geomys* in Texas, including five species and four subspecies, all on the basis of morphological distinctness. As the biological species concept slowly began to take hold in the twentieth century, all but two of the taxa recognized by Bailey (*G. personatus* and *G. arenarius*) were lumped into one wide-ranging species, named *G. bursarius*, which was distributed over most of the Great Plains and south-central United States, including almost all of Texas (Hall and Kelson, 1959; Hall, 1981).

Recent studies by specialists trained in cytological and molecular taxonomy have revealed the existence of six species of pocket gophers ranging over what was formerly considered the range of *G. bursarius*. These species (plains pocket gopher, *G. bursarius*; Attwater's pocket gopher, *G. attwateri*; Baird's pocket gopher, *G. breviceps*; Jones's pocket gopher, *G. knoxjonesi*; Llano pocket gopher, *G. texensis*; and Hall's pocket gopher, *G. jugossicularis*) are considered cryptic species, meaning they cannot be easily differentiated on the basis of observed morphological characteristics although they are genetically distinct and reproductively isolated. Although all of the species appear to be allopatric in range, karyotypic, electrophoretic, and mitochondrial DNA data are required to distinguish questionable specimens with confidence (Baker and Genoways, 1975; Honeycutt and Schmidly, 1979; Tucker and Schmidly, 1981; Bohlin and Zimmerman, 1982; Baker et al., 1989; Dowler, 1989; Jones et al., 1995; Elrod et al., 1996; Jolley et al., 2000; Sudman et al., 2006; Genoways et al., 2008; Chambers et al., 2009).

Similar instances of cryptic species have now been discovered in several other groups of rodents, including deer mice (genus *Peromyscus*; Schmidly, 1973b; Bradley et al., 2015), grasshopper mice (genus *Onychomys*; Hinesley, 1979), pocket mice (genus *Chaetodipus* and genus *Perognathus*; Lee and Engstrom, 1991; Lee et al., 1996) and kangaroo rats (genus *Dipodomys*; Schmidly and Hendricks, 1976), as well as shrews of the genus *Blarina* (Baumgardner et al., 1992; Reilly et al., 2005). These types of discoveries account for much of the change in the taxonomy and classification of mammals during the twentieth century.

An example of the opposite situation—that is, new taxonomic approaches have resulted in combining species formerly considered to be distinct and separate—involves the arid-land foxes of the genus *Vulpes*. For most of the twentieth century, arid-land foxes were regarded as two similar but separate species, the swift fox (*V. velox*) and the kit fox (*V. macrotis*), and that was the arrangement used by Bailey. A subsequent taxonomic study of these foxes based on advanced morphometric and protein-electrophoretic methods concluded, however, that the taxa were not sufficiently distinct to warrant separate species status (Dragoo et al., 1990). Thus, the two foxes were grouped into a single species, *V. velox*, comprising two subspecies, *V. v. velox* from the Panhandle and adjacent areas and *V. v. macrotis* from the Trans-Pecos. Ironically, using more modern molecular DNA data, these same scientists have now reversed their previous conclusion and consider these two foxes to be separate species.

Similarly, Dragoo et al. (2003) revised the taxonomic status of the hog-nosed skunks (genus *Conepatus*). Two species of hog-nosed skunks had been recognized in Texas: *C. mesoleucus* in western Texas and the Hill Country and *C. leuconotus* along the Texas coast. Using external and cranial morphology as well as mitochondrial DNA sequences, these authors concluded that hog-nosed skunks are represented by only a single species, for which the taxonomic name *C. leuconotus* has priority. The former designation of *C. mesoleucus* for the hog-nosed skunks of western and central Texas has proven to be no longer valid.

Finally, our understanding of higher taxa (above the species level) of mammals is being challenged by molecular genetics. Studies of gene sequences, subjected to phylogenetic analysis, are revealing different genera and even families of mammals. A classic example in Texas is represented by the big-eared bats of the genus *Plecotus*. In the early part of the twentieth century, these bats were arranged in the genus *Corynorhinus*, reflecting a division of the New World big-eared bats from the Old World big-eared bats of the genus *Plecotus*. Then, for reasons of morphological similarity, Charles Handley of the National Museum of Natural History lumped the two genera together and placed all species in the genus *Plecotus*, and for the next several decades, the two Texas species were arranged in that genus. In 1992, however, a phylogenetic analysis of 25 morphological and 11 karyological characters suggested that *Corynorhinus* should be afforded generic distinction from *Plecotus*; more recently, a sequence analysis of mitochondrial genes has provided additional support for that taxonomic interpretation, and now the Texas species are once again placed in the genus *Corynorhinus*. We can expect more of these kinds of classification revisions as our knowledge of mammalian gene sequences increases. Similar examples among Texas bats include *Lasiurus cinereus*, now placed in the genus *Aeorestes*, and two species of *Lasiurus* (*ega* and *xanthinus*), which now are placed in the genus *Dasypterus*.

The most significant changes in classifying mammals come from the continued use of DNA sequence data and the re-emergence of an old species concept: the genetic species concept (GSC). Sequence data from DNA offers a mechanism to examine genetic differences among populations of mammals and to determine the extent of genomic divergence between those populations. We have seen a proliferation of DNA sequence data since the latter part of the 1990s. As this technique has become less expensive and easier to perform (due to advances culminating from the Human Genome Project), it has become a staple methodology for many mammalian research laboratories.

The second significant change involves a conceptual shift away from the biological species concept (BSC), as championed by Ernst Mayr (1942, 1963), toward a phylogenetic species concept, as described by Joel Cracraft (1983). More recently, one of us (RDB) and his colleague at Texas Tech University, the late Robert J. Baker, elevated awareness of the genetic species concept proposed by Dobzhansky, Bateson, Mueller, and others during the early and mid-1900s (Bradley and Baker, 2001; Baker and Bradley, 2006). Baker and RDB defined a genetic species as a group of genetically compatible interbreeding natural populations that is genetically isolated from other such groups. Simply put, if a researcher can demonstrate consistent genetic differences between two groups, then those groups are genetically isolated and by extension are maintaining separate and distinct gene pools. In other words, the populations are not exchanging genes and are behaving as distinct species. In addition, the GSC infers that the more genetically distinct populations are, the longer the time frame required for that isolation to have taken place; thus, older pairs of species would be expected to differ by a larger amount than would pairs of young species.

The advantage of the GSC is that it focuses on genetic isolation instead of reproductive isolation, as required by the BSC. Also, DNA sequence data provides a convenient means for testing genetic isolation, whereas reproductive isolation is more difficult to measure in natural populations. In short, examination of DNA sequence data under the framework of the GSC has led to several taxonomic revisions for Texas mammals in the last twenty years. The major taxonomic changes involving Texas mammals in the twenty-first century have been summarized in the most recent edition of *The Mammals of Texas* and the interested reader is referred to that publication (see Schmidly and Bradley, 2016).

CHAPTER 2

The *Biological Survey of* Texas, 1889-1905

B AILEY'S 1905 PUBLICATION CONTAINED ONLY A SMALL PART OF THE information generated by the survey. At the conclusion of the Texas survey, and after the publication was issued, all of the accumulated materials were deposited in the Bird and Mammal Laboratory of the US Department of Agriculture, the US National Museum, the National Archives of the Smithsonian Institution in Washington, DC, or the National Archives and Records Administration in College Park, Maryland. The archival materials included scientific specimens of birds, mammals, and reptiles; museum catalogs of the scientific specimens; field trip diaries describing the travels of the field agents; detailed biological reports of plants and animals observed; reports of significant events; physiographic reports of each place visited in Texas; special correspondence with landowners and field agents; and more than 1,000 black-and-white photographs of Texas landscapes. These archival materials represent a detailed depiction of Texas natural history at the turn of the century.

By any measure possible, the biological survey was a huge success scientifically. More than 5,000 scientific specimens were obtained by the federal agents during the period from the beginning of the survey until the publication of Bailey's book in 1905. These specimens included 353 reptiles and 4,820 mammals. Forty-nine taxa of mammals were described from the specimens obtained during the survey (Table 3). Nine were described directly in Bailey's 1905 work, and the others were reported in other outlets of the time, such as the *Proceedings of the Biological Society of Washington*. C. Hart Merriam and Vernon Bailey described most of the taxa, and Bailey himself collected 20 of the type specimens. All but nine of the 49 mammals described are still recognized as valid species

Table 3. List of mammalian taxa described from specimens collected during the biological survey of Texas.

Original Taxon	Type Locality	Collector (Date Collected)	Modern Taxonomic Designation
Didelphis marsupialis texensis (s)	Brownsville, Cameron County	F.B. Armstrong (4/13/1892)	*Didelphis virginiana californica*
Tatu novemcinctum texanum*(s)	Brownsville, Cameron County	F.B. Armstrong (6/10/1892)	*Dasypus novemcinctus mexicanus*
Lepus pinetis robustus	Davis Mountains, Jeff Davis County	V. Bailey (1/6/1890)	*Sylvilagus robustus*
Eutamias cinereicollis canipes	Guadalupe Mountains, Culberson County	V. Bailey (8/24/1901)	*Tamias canipes canipes*
Tamias interpres	El Paso, El Paso County	V. Bailey (12/10/1889)	*Ammospermophilus interpres*
Spermophilus spilosoma annectens	Padre Island, Cameron County	W. Lloyd (8/24/1891)	*Xerospermophilus spilosoma annectens*
Spermophilus spilosoma arens (s)	El Paso, El Paso County	A. K. Fisher (5/10/1894)	*Xerospermophilus spilosoma canescens*
Spermophilus spilosoma marginatus	Alpine, Brewster County	V. Bailey (7/5/1901)	*Xerospermophilus spilosoma marginatus*
Spermophilus tridecemlineatus texensis	Gainesville, Cooke County	G. H. Ragsdale (4/15/1886)	*Ictidomys tridecemlineatus texensis*
Glaucomys volans texensis	Sour Lake, Hardin County	J. H. Gaut (3/15/1905)	*Glaucomys volans texensis*
Thomomys aureus lachuguilla	El Paso, El Paso County	V. Bailey (9/24/1901)	*Thomomys bottae lachuguilla*
Thomomys baileyi	Sierra Blanca, Hudspeth County	V. Bailey (12/28/1889)	*Thomomys bottae baileyi*
Thomomys baileyi spatiosus	Alpine, Brewster County	V. Bailey (5/26/1900)	*Thomomys bottae spatiosus*
Thomomys bottae guadalupensis	McKittrick Canyon, Guadalupe Mtns., Culberson County	V. Bailey (8/22/1901)	*Thomomys bottae texensis*
Thomomys bottae pervarius	Lloyd Ranch, 35 mi S Marfa, Presidio County	V. Bailey (1/20/1890)	*Thomomys bottae lachuguilla*
Thomomys fulvus texensis	Head of Limpia Creek, Davis Mtns., Jeff Davis County	V. Bailey (1/7/1890)	*Thomomys bottae texensis*
Thomomys lachuguilla confinalis	35 mi E Rock Springs, Kerr County	V. Bailey (7/11/1902)	*Thomomys bottae confinalis*
Geomys arenarius	El Paso, El Paso County	V. Bailey (12/14/1889)	*Geomys arenarius*
Geomys breviceps ammophilus (s)	Cuero, DeWitt County	V. Bailey (4/26/1899)	*Geomys attwateri*
Geomys breviceps attwateri	Rockport, Aransas County	H. H. Keays (11/18/1892)	*Geomys attwateri*
Geomys breviceps llanensis* (s)	Llano, Llano County	V. Bailey (5/15/1899)	*Geomys texensis llanensis*

Original Taxon	Type Locality	Collector (Date Collected)	Modern Taxonomic Designation
Geomys breviceps sagittallis	Clear Creek, Galveston Bay, Galveston County	W. Lloyd (3/28/1892)	*Geomys breviceps sagittalis*
Geomys personatus	Padre Island, Kleberg County	C. K. Worthen (4/11/1888)	*Geomys personatus personatus*
Geomys personatus fallax	Nueces Bay, Nueces County	W. Lloyd (11/30/1891)	*Geomys personatus fallax*
*Geomys texensis**	Mason, Mason County	I. B. Henry (12/17/1885)	*Geomys texensis*
Cratogeomys castanops angusticeps	Eagle Pass, Maverick County	C. P. Streator (11/11/1890)	*Cratogeomys castanops angusticeps*
Cratogeomys castanops perplanus	Tascosa, Oldham County	V. Bailey (6/5/1899)	*Cratogeomys castanops perplanus*
Perognathus paradoxus spilotus	Gainesville, Cooke County	G. H. Ragsdale (10/8/1886)	*Chaetodipus hispidus spilotus*
Dipodomys ambiguus	El Paso, El Paso County	V. Bailey (12/13/1889)	*Dipodomys merriami ambiguus*
Dipodomys compactus	Padre Island, Cameron County	C. K. Worthen (4/3/1888)	*Dipodomys compactus*
Dipodomys elator	Henrietta, Clay County	J. A. Loring (4/13/1894)	*Dipodomys elator*
Liomys texensis	Brownsville, Cameron County	J. A. Loring (2/19/1894)	*Liomys irroratus texensis*
*Castor canadensis texensis**	Cummings Creek, Colorado County	F. Brune (12/25/1900)	*Castor canadensis texensis*
*Reithrodontomys griseus**	San Antonio, Bexar County	H. P. Attwater (3/4/1897)	*Reithrodontomys montanus griseus*
Reithrodontomys merriami	near Alvin, Brazoria County	W. Lloyd (3/15/1892)	*Reithrodontomys humulis merriami*
*Peromyscus boylei laceyi** (s)	Turtle Creek, Kerr County	H. P. Attwater (12/4/1897)	*Peromyscus attwateri*
Peromyscus pectoralis laceianus	Turtle Creek, Kerr County	V. Bailey (5/3/1899)	*Peromyscus laceianus*
*Peromyscus taylori subater**	Columbia, Brazoria County	W. Lloyd (2/25/1892)	*Baiomys taylori subater*
Onychomys longipes	Concho County	W. Lloyd (3/11/1887)	*Onychomys leucogaster longipes*
Sigmodon ochrognathus	Chisos Mountains, Brewster County	V. Bailey (6/13/1901)	*Sigmodon ochrognathus*
Microtus mexicanus guadalupensis	Guadalupe Mountains, Culberson County	V. Bailey (8/21/1901)	*Microtus mogollonensis guadalupensis*
*Canis nebracensis texensis**	45 mi SW Corpus Christi, Nueces County	J. M. Priour (12/14/1901)	*Canis latrans texensis*
Ursus horriaeus texensis (s)	Davis Mountains, Jeff Davis County	C. O. Finley and J. Z. Means (11/2/1890)	*Ursus arctos horribilis*
Lutra canadensis texensis (s)	20 mi W Angleton, Brazoria County	B. V. Lilly (3/-/1908)	*Lontra canadensis lataxina*
Spilogale leucoparia	Mason, Mason County	Ira B. Henry (12/2/1885)	*Spilogale gracilis leucoparia*

Original Taxon	Type Locality	Collector (Date Collected)	Modern Taxonomic Designation
Conepatus leuconotus texensis	Brownsville, Cameron County	F. B. Armstrong (7/20/1892)	*Conepatus leuconotus leuconotus*
Conepatus mesoleucus mearnsi (s)	Mason, Mason County	I. B. Henry (2/20/1886)	*Conepatus leuconotus leuconotus*
*Conepatus mesoleucus telmalestes**	7 mi NE Sour Lake, Hardin County	J. H. Gaut (3/17/1905)	*Conepatus leuconotus telmalestes*
Ovis canadensis texianus (s)	Guadalupe Mountains, Culberson County	V. Bailey (9/2/1902)	*Ovis canadensis mexicanus*[1]

An asterisk (*) indicates those taxa described in Bailey (1905). Adapted from Poole and Schantz (1942), Miller and Kellogg (1955), Hall (1981), and Schmidly and Bradley (2016). (S) indicates taxa that have been placed in synonymy by modern taxonomists.

[1] *Ovis canadensis mexicanus* is probably extinct in Texas; current bighorn sheep populations probably have resulted from introductions of *O. c. canadensis* and *O. c. nelsoni*.

or subspecies in Texas today, which is a tribute to the taxonomic ability of the mammalogists working for the biological survey.

In 1992, a project was initiated to document all of the archival natural history information from the Texas biological survey (see Schmidly, 1998). The complete set of archives has now been deposited at the Southwest Collection/Special Collections Library, Texas Tech University. This collection of field reports, photographs, and other material provides a wealth of detailed information not included in Bailey's 1905 book and never before available to biologists in Texas. Those resources were utilized to augment the 2002 edition of *Texas Natural History*, and they are again included in this edition. In 2020, the Texas Tech archive was expanded to include all of the material assembled to write the biography of Vernon Bailey, including his correspondence and reports from Texas (Schmidly, 2018).

Documenting and understanding the changes in Texas's diverse and unique biota depends on reliable data about the flora and fauna of the region before it was negatively affected by humans. The availability of Bailey's 1905 publication, together with the discovery of its complete archives, gives a vital natural history picture of every region of the state as it existed more than a century ago. This information provides crucial baseline data to compare with the results of current biological surveys and to assess landscape and biotic change information useful to land managers and others seeking to improve land and ecosystem management. This chapter includes a reprinting of the *Biological Survey of Texas*, with numerical annotations that update the mammal and reptile accounts with current taxonomic and natural history information.

As noted in the Section I introductory material (p. 5), the scientific names of many of the plants and animals mentioned in Bailey's 1905 publication have changed substantially over the last 100+ years. There are various reasons for these changes, but the most common are taxonomic updates based on more recent research (e.g., elevation of subspecies to species, sinking of species to subspecies), as well as typos, misspellings, and mistaken Latinization

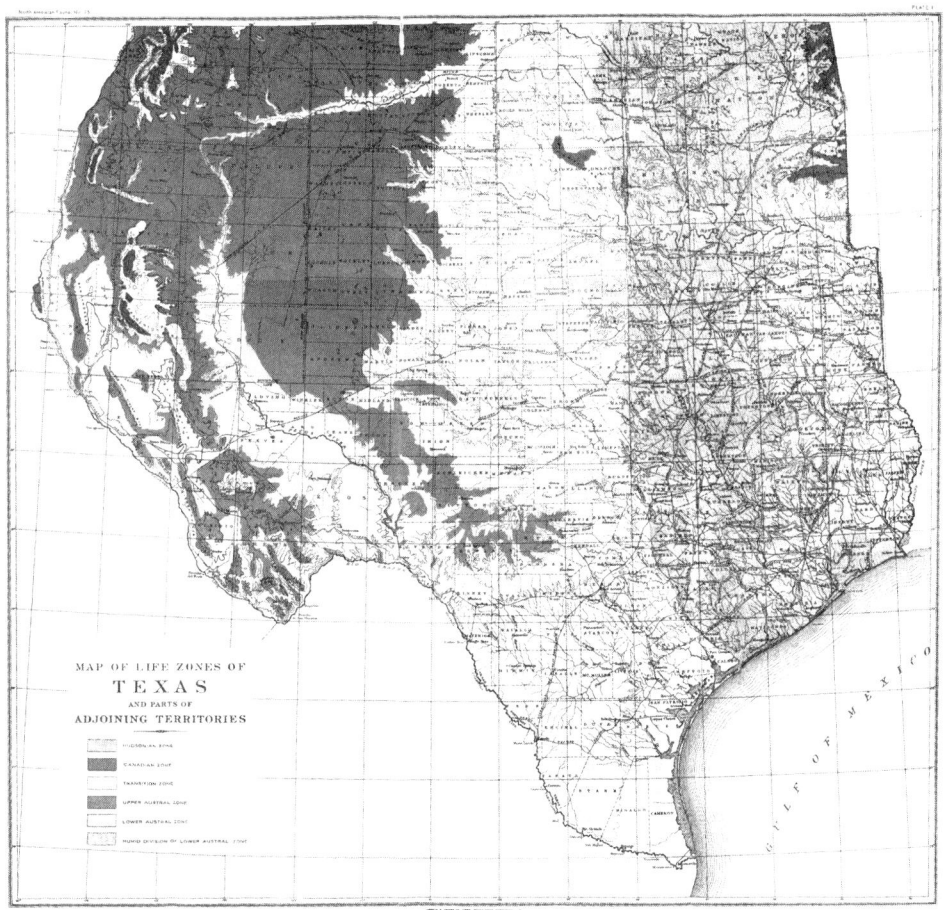

Figure 31. Map of the Life Zones of Texas and parts of adjoining territories. Fold-out from *Biological Survey of Texas*, 1905.

of words. To assist the reader, a series of tables (Tables 4-7, pp. 301–337) have been included at the end of this chapter that cross-reference the scientific names that Bailey used in the *Biological Survey of Texas* with the current (taxonomically correct and correctly spelled) Latin names of the organisms. Scientific names that have changed since the publication of the 1905 survey are indicated in the reprint, as well as in excerpts of the written accounts of the field agents that are provided in various chapters of this publication, by superscripts after the scientific names, i.e., [P] for plants, [B] for birds, [R] for reptiles, and [M] for mammals. The superscripts direct the reader to the four corresponding tables, Table 4 through Table 7, respectively. Scientific names that remain valid did not receive a superscript in the text, but they are listed in the tables, as well.

The following reprint of the *Biological Survey of Texas* is a faithful reproduction of the original book, with only a few minor exceptions:

+ As mentioned above, all scientific names that are no longer valid are followed by a superscript that directs the reader to the appropriate table of current names.

+ Annotation numbers have been inserted that correspond to the comments in the margins.

+ The Frontispiece of the original publication, designated as Plate I, was an over-sized color map of the Life Zones of Texas. That map has been recreated as a black-and-white line drawing and can be found in Chapter 1 (Figure 30a, p. 54), and a scan of the original is reproduced herein as Figure 31.

+ The remaining Plates (II–XVI) from the published *Survey* have been reproduced and can be found as a group in the mid-section of the *Biological Survey of Texas* reprint (pp. 133–147).

+ The Index of the original publication has not been reprinted here. The Index for *Texas Natural History in the Twenty-First Century* includes all the items indexed in the original publication.

The mammal survey by the federal biologists was comprehensive. Bailey documented 121 terrestrial species of Texas mammals. The only terrestrial group he failed to document in detail was bats (16 of 34 species), which is not surprising given that modern detection and capture techniques for bats (e.g., mist-nets) were not available at that time. Forty-nine new taxa were described from material collected during the survey, including nine in the 1905 publication (see Table 3).

The bird survey reported by Bailey was restricted to a simple listing of observations and specimens collected in each Life Zone. Bailey's brief treatment of birds presumably was a result of Oberholser having been assigned the lead role on the bird report, which was not published until decades later (Oberholser, 1974).

The reptile survey by Bailey was augmented by the expertise of Leonhard Stejneger, head of the Division of Reptiles and Batrachians at the United States National Museum of Natural History in Washington, DC. The treatment of reptiles was rather cursory (e.g., it included no mention of turtles or tortoises) and included only 61 of the 128 currently recognized species of snakes and lizards. Unless noted otherwise, all scientific documentation for the annotations of reptiles (annotations 6-69) have been taken from James R. Dixon's 2013 book, *Amphibians and Reptiles of Texas* (Dixon, 2013).

The comments written about each species and subspecies of mammal are based on our own experience of studying Texas mammals in conjunction with the published scientific literature about these taxa. Unless mentioned otherwise, the scientific documentation for the mammal accounts was compiled from the following publications: Taylor and Davis (1947); Davis (1960, 1966, 1974); Davis and Schmidly (1994); Dalquest and Horner (1984); Hall (1981); Manning and Jones (1998); Schmidly (1977, 1983, 1984a, 1984b, 1991, 2004); Jones and Jones (1992); Jones, J. K., Jr. et al. (1988a, b); Bradley et al. (2014); and Schmidly and Bradley (2016).

For groups other than mammals, the following scientific sources were utilized: plants—Tropicos (2019) and USDA Plants Database (2019); reptiles—Dixon (2013) and Reptile Database (2019); and birds—Avibase (2019) and Texas Bird Records (2019).

Throughout the annotations, we use the full scientific name the first time a species, subspecies, or name combination is mentioned, even if the genus name, or genus and species names, were previously mentioned in the annotation, i.e., a name is appropriately truncated only if used again in full within that annotation. For example, *Baiomys taylori subater* would represent the first usage, followed by *B. t. subater* for additional uses within the annotation. However, *Baiomys taylori taylori*, a different taxon, would be written out in full within that account. Finally, if a taxon is mentioned in subsequent annotations, the full taxonomic name is once again provided.

U. S. DEPARTMENT OF AGRICULTURE
BIOLOGICAL SURVEY

NORTH AMERICAN FAUNA

No. 25

[Actual date of publication, October 24, 1905]

BIOLOGICAL SURVEY OF TEXAS

LIFE ZONES, with Characteristic Species of Mammals, Birds, Reptiles, and Plants
REPTILES, with Notes on Distribution
MAMMALS, with Notes on Distribution, Habits, and Economic Importance

BY

VERNON BAILEY
CHIEF FIELD NATURALIST

Prepared under the direction of

Dr. C. HART MERRIAM
CHIEF OF BIOLOGICAL SURVEY

WASHINGTON
GOVERNMENT PRINTING OFFICE

1905

LETTER OF TRANSMITTAL.

U. S. DEPARTMENT OF AGRICULTURE,
BIOLOGICAL SURVEY,
Washington, D. C., July 10, 1905.

SIR: I have the honor to forward herewith, for publication as North American Fauna No. 25, a report on the results of a biological survey of Texas, by Vernon Bailey. The report consists of three sections: The first characterizes the life zones and defines the distribution areas of the State; these are mapped in detail and are accompanied by practical suggestions as to their adaptation to agricultural uses. The second comprises a brief report on the snakes and lizards, adding considerably to previous knowledge of the distribution of these groups. The third consists of a report on the mammals of the State, and contains much of a practical nature on distribution, habits, and economic relations of the several species.

The maps and illustrations are essential to the clearness and brevity of the report.

C. HART MERRIAM,
Chief, Biological Survey.

Hon. JAMES WILSON,
Secretary of Agriculture.

3

[This page intentionally left blank]

CONTENTS.

5

[This page intentionally left blank]

ILLUSTRATIONS.

PLATES.

TEXT FIGURES.

7

8   ILLUSTRATIONS.

No. 25.   NORTH AMERICAN FAUNA.   Oct., 1905.

BIOLOGICAL SURVEY OF TEXAS.

By Vernon Bailey.

INTRODUCTION.

For a number of years the Biological Survey has been collecting information and specimens bearing on the natural history of Texas. Some of the results are here brought together in a discussion of the life zones and their subdivisions and a report on the mammals and reptiles of the State. The original plan included also a report on the birds of Texas, by H. C. Oberholser, but the present paper has grown to such proportions that the bird report will be published separately.

Much of the field work has been carried on in connection with that in adjacent regions, and on several occasions it has been possible to continue parties in the field until late in the season or throughout the winter by moving them southward into Texas in the fall, or to begin work there early in the spring before the season had opened sufficiently for operations farther north. Hence, while the Texas work has the appearance of being desultory and scattered, the ground in reality has been covered with great economy of time and labor. Part of the field work has been carried on in connection with special studies of urgent economic problems, as the prairie dog, coyote, and boll weevil pests, and throughout all of it the economic status of birds and mammals has received special attention. The distribution of mammals, birds, reptiles, and plants, so far as they have an important bearing on the extent and boundaries of faunal areas, has been studied in detail in the field, and in the case of most species a sufficient number of specimens has been collected to show the variation due to climatic differences. Of many of the larger game mammals, and especially of the deer, bear, and panther, it has not been possible to secure enough material to satisfactorily establish the present geographic limits of the species and subspecies, but it is greatly to be

9

hoped that the growing interest in natural history will inspire local hunters and residents of the country to send specimens of these vanishing forms to the National Museum before it is too late. Many important problems can be solved only by aid from local naturalists or other intelligent residents of the State. The skull that is left in the woods or thrown away would often aid in solving one of these problems.

PERSONNEL OF BIOLOGICAL SURVEY WORK IN TEXAS.

In carrying on the field work in Texas the writer was assisted at different times by the following regular or temporary field naturalists of the Biological Survey: William Lloyd, Clark P. Streator, William L. Bray, Harry C. Oberholser, N. Hollister, Merritt Cary, Gordon Donald, Arthur H. Howell, and James H. Gaut.

Several local naturalists and collectors have added materially to the results of the work in Texas, and among these thanks are especially due to Mr. H. P. Attwater and Mr. Howard Lacey.

Extensive collections of mammals, birds, reptiles, batrachians, crustaceans, mollusks, and plants have been made from localities practically covering the State, and the field reports of the collectors contain a mine of important facts on habits, distribution, correlation, and economic importance of species. Much of this material has already been published by the Biological Survey in the form of bulletins and papers on economic subjects, and much still remains for use in future papers.

ACKNOWLEDGMENTS.

To Dr. C. Hart Merriam, under whose direction the work was planned and carried out, I am indebted for the use of his private collection of mammals deposited in the United States National Museum. To Mr. F. W. True, curator, and Mr. Gerrit S. Miller, jr., assistant curator of mammals in the National Museum, I am indebted for the use of the museum collection; also to Dr. J. A. Allen, curator of birds and mammals in the American Museum of Natural History; Mr. Outram Bangs, curator of mammals in the Museum of Comparative Anatomy; and Mr. Witmer Stone, curator of birds and mammals in the Philadelphia Academy of Sciences, for the loan of types and topotypes of mammals from the collections under their supervision

NEW SPECIES OF MAMMALS.

A number of new species of plants, reptiles, birds, and mammals has been found in the Texas collection. Most of these have been described and named by various specialists, but descriptions of a few previously undescribed mammals are included in the present report. They are as follows:

FAUNA AND FLORA OF TEXAS IN RELATION TO LIFE ZONES AND MINOR DISTRIBUTION AREAS.

The fauna and flora of Texas are wonderfully rich and varied, not only in abundance of individuals and species, but in the number of genera, families, and orders, some of which do not occur in any other part of the United States. This richness is due, not so much to the enormous extent of the State, as to its varied physical and climatic conditions, for it embraces areas of abundant humidity and extreme aridity, of dense forest and extensive plain, of low coast prairies and rugged mountains. Besides stretching across the aerial pathway of north and south migrating birds and bats, it lies at the threshold of the Tropics and claims a large contingent of Mexican species. On the east it includes the fauna and flora of the lower Mississippi Valley, with most of the species ranging to the Atlantic coast, and on the west reaches far into the desert region of highly specialized forms, while in the middle portion it is traversed by a wide tongue of the more northern fauna and flora of the Great Plains. In the west several mountain masses reach an altitude of 8,000 feet, with peaks rising to 8,500 and 9,500 feet. This range of altitude, together with the great extent of latitude, suffices to include within the State the full width of three of the principal life zones, Lower Austral, Upper Austral, and Transition, each with its characteristic series of plants and animals. In the Lower Rio Grande and Gulf Coast region there is, moreover, a slight overlapping of tropical species, accompanying the almost tropical climate, while high up in the Guadalupe, Davis, and Chisos mountains are mere traces of Canadian zone species.

The agricultural and commercial interests of the State are as varied as the climatic conditions on which they largely depend, and when mapped they are found in many cases closely to correspond with the areas of distribution of certain species of native plants and animals. In other words, various agricultural industries are being slowly developed by endless and costly experiment along the same lines that the native species have followed in the course of adaptation to their environment. Thus the lumbering industries of the State are prescribed by the distribution of certain species of trees. On the other hand, successful stock raising depends in part on the absence of forests and the abundance of certain grasses, and in part on the absence of certain disease-conveying parasites. Several varieties of wheat are successfully raised over a limited area near the upper edge of humid Lower Sonoran zone, but most of the State lies below the belt of small grains. Rice and sugar cane are standard crops of the semitropical coast region, and cotton is the staple for the whole Lower Sonoran zone, wherever the rainfall is sufficient to mature the crop or water is available for irrigation. Parts of the State are peculiarly adapted to the production of early fruits and winter vegetables for the northern market, but these industries are as yet more or less restricted by inadequate facilities for quick transportation.

The division of the State into wheat, cotton, and stock-raising districts is no matter of accident, nor is it a matter of choice on the part of those engaged in the various industries. While usually there is no room for doubt in the middle of each area as to the crop it is best adapted to, there is always a question along the boundaries. For instance, where does the successful production of cotton yield to that of wheat? Nature in her processes avoids sharp lines and hard-and-fast rules, but usually gives reliable averages. Even after a season of copious rainfall in a valley clothed with cactus and scrubby mesquite trees, the experienced ranchman knows better than to plow and plant with the idea that the following season will be similar; but from the character of the vegetation and of the animals present he may not only learn approximately the average amount of rainfall, but also the life zone in which he is located, with its average range of temperature and many of the crops best adapted to it. While much has been done and much more will be done to overcome arid conditions and to convert the now almost worthless desert soil into the most productive in the State, the normal conditions limiting life zones can not be materially overcome, nor can they be safely ignored. The attempt to raise cotton in Upper Sonoran zone results only in failure and loss, but enough of this zone lies within the State to produce, with the water available for irrigation, an abundance of the finest apples, as well as many other fruits and crops not adapted to lower zones.

The primary object of the present report is a careful definition of the ranges of native species of plants and animals and a correlation of these ranges into well-defined areas of distribution. In 'Life Zones and Crop Zones of the United States' Doctor Merriam has given, with as much detail as the data collected to 1898 would allow, the adaptation of various crops to the zones and their subdivisions, and has clearly set forth the practical application of the knowledge of faunal areas to agriculture. Under the heading 'Relations of the Biological Survey to Practical Agriculture,' he says:[a]

The Biological Survey aims to define and map the natural agricultural belts of the United States, to ascertain what products of the soil can and what can not be grown successfully in each, to guide the farmer in the intelligent introduction of foreign crops, and to point out his friends and his enemies among the native birds and mammals, thereby helping him to utilize the beneficial and ward off the harmful. *   *   *   *

The farmers of the United States spend vast sums of money each year in trying to find out whether a particular fruit, vegetable, or cereal will or will not thrive in localities where it has not been tested. Most of these experiments result in disappointment and pecuniary loss. It makes little difference whether the crop experimented with comes from the remotest parts of the earth or from a neighboring State, the result is essentially the same, for the main cost is the labor of cultivation and the use of the land. If the crop happens to be one that requires a period of years for the test, the loss from its failure is proportionately great.

The cause of failure in the great majority of cases is climatic unfitness. The quantity, distribution, or interrelation of heat and moisture may be at fault. Thus, while the total quantity of heat may be adequate, the moisture may be inadequate, or the moisture may be adequate and the heat inadequate, or the quantities of heat and moisture may be too great or too small with respect to one another or to the time of year, and so on. What the farmer wants to know is *how to tell in advance* whether the climatic conditions on his own farm are fit or unfit for the particular crop he has in view, and what crops he can raise with reasonable certainty. It requires no argument to show that the answers to these questions would be worth in the aggregate hundreds of thousands of dollars yearly to the American farmer. The Biological Survey aims to furnish these answers.

Agricultural colleges, experiment stations and substations, horticulturists, and countless farmers are working out the details of these problems in different parts of the country and constantly pushing their experiments into new regions. As a crop becomes an established success in one locality, a study of the zone map will show over what adjoining country it can be profitably extended. For instance, Roswell, N. Mex., where apple raising has proved a great financial success, is situated at the junction of Upper and Lower Sonoran zones, or in a mixed belt of overlapping of the two, at the western edge of the Staked Plains. By tracing this lower border of Upper Sonoran zone around the southern arm and along the eastern edge of the Staked Plains, a belt approximately 1,000 miles long of the same zonal level and climatic conditions is found, lying within the State of Texas. This is largely undeveloped agricultural land, but a considerable part of it can be irrigated, and there is every reason to believe that it will be found perfectly adapted to the varieties of apples that thrive in the Pecos Valley at Roswell. The Staked

---

[a] Life Zones and Crop Zones of the United States, by C. Hart Merriam. Bul. 10, U. S. Dept. Agr., Div. Biol. Survey, pp. 9, 12, 1898.

Plains, lying within this belt, are pure Upper Sonoran, the real home of most of the standard varieties of apples. Other northern crops, both cereals and fruits, have proved a success along this southern projection of Upper Sonoran zone, but have not been introduced as systematically as the advantages of its position seem to warrant. To quote again from Life Zones and Crop Zones, page 15, under the heading "Special value of narrow extensions of faunas," Doctor Merriam says:

> In looking at the map of the life zones it will be seen that nearly all of the belts and areas send out long arms, which penetrate far into the heart of adjoining areas. When such arms occupy suitable soils in thickly inhabited regions, so that their products may be conveniently marketed, they are of more than ordinary value, for the greater the distance from its area of principal production a crop can be made to succeed the higher price it will command. Hence, farms favorably situated in northern prolongations or islands of southern zones, or in southern prolongations or islands of northern zones, should be worth considerably more per acre than those situated within normal parts of the same zones. The obvious reason is that by growing particular crops at points remote from the usual sources of supply, and at the same time conveniently near a market, the cost of transportation is greatly reduced and the profit correspondingly increased.

Since the publication of Doctor Merriam's zone map, detailed work in Texas has enabled me to make minor corrections and to establish the zone boundaries with more precision than has been possible heretofore.

TROPICAL ELEMENT OF THE LOWER RIO GRANDE REGION.

Until recent years more thorough biological collecting had been done in the Lower Rio Grande region than in any other part of Texas, with the result of giving a somewhat exaggerated impression of the tropical element found there. Later and more systematic field work over the State, together with the extensive investigations of Nelson and Goldman in Mexico, have shown that the Texas mammals of tropical groups—as the armadillo, ocelot, jaguar, red and gray cats, and spiny pocket mouse—elsewhere range through Lower Sonoran zone, or at least its Tamaulipan subdivision, while a more critical study of these groups, based on the rapidly increasing amount of material, has resulted in every case in the specific or subspecific separation of the Texas forms. The single specimen of *Nasua,* apparently of a tropical species, from Brownsville may have been imported, and if this is so not a strictly tropical mammal reaches the border of Texas.

The close proximity to the Tropics is shown most pronouncedly by the birds of the Lower Rio Grande region. A considerable number of species, mainly tropical in distribution, reach southern Texas, where some breed regularly, while others are more or less regular visitors.

BIRDS OF MAINLY TROPICAL RANGE WHICH EXTEND INTO SOUTHERN TEXAS.

Colymbus dominicus brachypterus.[B]
Phalacrocorax vigua mexicanus.[B]
Fregata aquila.[B]
Nomonyx dominicus.
Dendrocygna autumnalis.
Guara alba.[B]
Mycteria americana.
Ajaia ajaja.[B]
Jacana spinosa.
Ortalis vetula maccalli.
Leptotila fulviventris brachyptera.[B]
Columba flavirostris.[B]
Melopelia leucoptera.[B]
Scardafella inca.[B]
Elanus leucurus.
Parabuteo unicinctus harrisi.
Buteo abbreviatus.[B]
Buteo albicaudatus sennetti.[B]
Urubitinga anthracina.[B]
Falco fusco-caerulescens.[B]

Polyborus cheriway.[B]
Glaucidium phalaenoides.[B]
Crotophaga sulcirostris.
Ceryle torquata.[B]
Ceryle americana septentrionalis.[B]
Nyctidromus albicollis merrilli.
Amizilis tzacatl.[B]
Amizilis cerviniventris chalconota.[B]
Tyrannus melancholicus couchi.[B]
Pitangus derbianus.[B]
Myiarchus mexicanus.[B]
Pyrocephalus rubineus mexicanus.[B]
Ornithion imberbe.[B]
Tangavius aeneus involucratus.[B]
Agelaius phoeniceus richmondi.
Megaquiscalus major macrourus.[B]
Arremonops rufivirgatus.[B]
Sporophila morelleti.
Vireo flavoviridis.
Geothlypis poliocephala.

A few species of reptiles supposed to be of tropical origin enter southern Texas, but the task of verifying the records and determining ranges has not been undertaken in connection with the present work.

In the case of plants, as of mammals, the tropical element of southern Texas has been overestimated. A number of species of genera that are mainly tropical extend into the Lower Rio Grande region, but very few species of known tropical range. The Texas palm (*Inodes texana* [B]Cook)[a] is found in limited numbers in the Brownsville region, but apparently nothing is known of its southern extension or zonal significance. So with other supposedly tropical forms the southern limits and zonal position have not been satisfactorily determined, but evidently no purely tropical species holds an important place in the flora of the Lower Rio Grande region. This absence or scarcity of tropical plants is fully accounted for by Professor Bray in the Botanical Gazette for August, 1901 (p. 102), as follows:

A record of sixteen years at Brownsville showed a minimum temperature of 18° (the minimum in February, 1899, was 12°) and five years without frost. At Indianola a record of fifteen years showed a minimum of 15° and four years without frost. Probably a freeze severe enough to kill tropical woody vegetation occurs in periods of ten to twelve years. The fatal temperature for tropical plants in this region is that due to northers, which bring abnormally low temperatures suddenly, and not infrequently during the growth season.

---

[a] *Sabal mexicana* Mart. of Sargent, Coulter, and Small.

[1] The prevailing taxonomy at the time Bailey published this account was to place the pygmy mice with the deer mice in the genus *Peromyscus*. However, Miller (1912) and later Packard (1960) assigned the pygmy mice to a separate genus, *Baiomys*, on the basis of a number of unique features of the skull, skeleton, and teeth. Two subspecies occur in Texas: *Baiomys taylori subater*, which was described by Bailey (1905) on the basis of specimens from East Bernard in Brazoria County; and *Baiomys taylori taylori*, which had been described on the basis of specimens collected at San Diego in Duval County. Since the early twentieth century, the subspecies *B. t. taylori* consistently has extended its range northward and eastward by invading the oak-hickory association, the blackland prairies, the cross-timbers, rolling plains, and the high plains (Fig. 32a-b) (see Chapter 6 for a discussion). The subspecies *B. t. subater* is restricted to extreme southeastern Texas.

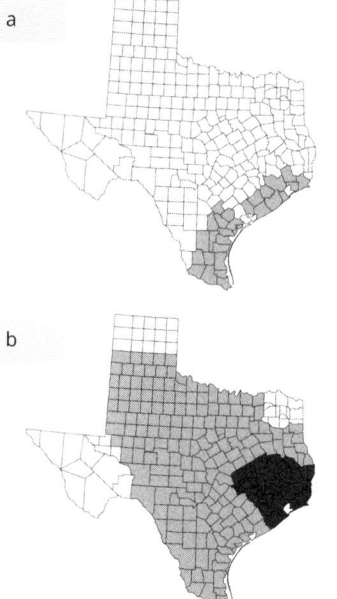

Figure 32a-b. The historic (a) and current (b) distribution of the northern pygmy mouse (*Baiomys taylori*) in Texas. Light shading depicts the current distribution of *B. t. taylori* and dark shading depicts the current distribution of *B. t. subater*.

A striking example of the fatal effects of a 'norther' was witnessed over the coast region of Texas from Galveston to Port Lavaca in the spring of 1899, following the extremely cold wave of the preceding February, when the abundant huisache trees were killed to the ground. In the Brownsville region, however, as I found in the following spring, these trees had escaped, but all of the bananas had been killed. Under such climatic conditions tropical species could hardly be expected to persist, and it is not surprising that the preponderating species of plants and mammals are those characteristic of Lower Sonoran zone. Nor is it surprising that tropical species of birds, with their greater freedom of motion, should overlap the limits of their zone slightly beyond the more stationary groups.

Bananas offer a good illustration of the partial success of a tropical fruit in this region. During a period of warm years they thrive and even bear fruit, but only to be killed by the first hard freeze. Even at Brownsville they require artificial protection to insure their living through the winter. Oranges in like manner are a partial success, but an assured success only where artificial protection can be afforded during the winter.

LOWER AUSTRAL ZONE.

By far the greater part of Texas, including all but the Staked Plains with their northern and southern extensions and the mountain elevations in the western part of the State, lies within the Lower Austral, or cotton-producing zone, the subdivisions of which within the limits of the State equal, if they do not exceed, in practical importance the more restricted intrusions of other transcontinental zones. The most important of these subdivisions of Lower Austral are the narrow Gulf strip, with semitropical climate, and the Austroriparian, or humid eastern, and Lower Sonoran, or arid western, areas, which divide the zone in Texas into approximately equal parts.

GULF STRIP OF TEXAS.

A comparatively narrow strip of country bordering the Gulf coast of Texas is characterized by a limited number of species of unquestioned tropical affinities, ranging as extensions from Mexico or Florida part or all of the way along the Gulf coast, but not extending back over the rest of Lower Austral zone. While associated with a preponderance of characteristic Lower Austral species, they mark a border of modified climatic conditions too important to be ignored. This strip has been mapped as a semitropical or Gulf strip of the Lower Austral zone, of which it is merely a subdivision.

1   In mammals the best representatives of a mainly tropical group (subgenus *Baiomys*[M] are the little *Peromyscus taylori*[M] and its subspecies *subater*,[M]which inhabit the coast prairies from Brownsville to Galveston. Among birds the caracara, a bird of wide tropical range, is common in the coast region of Texas as far east as Port Lavaca, while the jackdaws—the great-tailed and boat-tailed grackles—of the genus *Megaquiscalus*,[B]extend in one form or the other from the tropics of eastern Mexico along the Gulf coast to Florida, and breed abundantly along the whole Texas coast region.

[2] The taxonomic designation of huisache in Texas has had a complex history, being at times recognized in the genus *Acacia* and by various specific epithets. Most taxonomists today recognize the species as *Vachellia farnesiana*. The current distribution of huisache in Texas is depicted in Figure 33.

FIG. 1.—Distribution area of huisache (*Vachellia farnesiana*). **2**

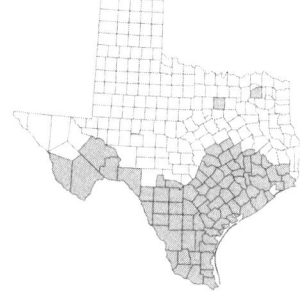

Figure 33. The current distribution of huisache (*Vachellia farnesiana*) in Texas.

In plants some of the species marking the Gulf strip extend into the tropical regions of Mexico or Florida, while others are limited to some part of this narrow strip. As stated by Professor Bray,[a] the outlines of the strip are approximately indicated in Texas by the range of *Vachellia* (=*Acacia*) *farnesiana*[P]and *Parkinsonia aculeata,* both species of partly tropical range, and to these I should add *Daubentonia longifolia* (*Sesban cavanillesii*)[P]and *Lantana camara* as equally important, while others of less extensive range in Texas are

[a] Botanical Gazette, August, 1901, 103.

Castela nicholsonii,[P] *Amyris parvifolia*,[P] *Karwinskia humboldtiana*,[P] *lbervillea lindheimeri, Castalia elegans, Yucca treculeana, Manfreda maculosa, Tillandsia baileyi, Jatropha macrorhiza* and *multifida, Malpighia glabra,* and *Solanum triquetrum.* It is worthy of note that none of these plants enter the swamp and timber country to any extent.

AUSTRORIPARIAN OF EASTERN TEXAS.

The eastern part of Texas, west to approximately the ninety-eighth meridian, agrees very closely in climate, physiography, and the bulk of its species of plants and animals with the lower Mississippi Valley. Except for the strip of coast prairie, and farther north the areas known as the Black Prairie and Grand Prairie,[a] it is largely a forested region, comprising both deciduous and coniferous trees and inhabited by forest species of birds and mammals.

While a rich though only half-developed agricultural region devoted mainly to cotton, corn, fruits, and vegetables, it still comprises extensive areas of native forest and uninhabited cypress swamps. Most of the numerous streams have wide bottom lands subject to occasional floods, from which they derive a deep rich soil especially adapted to luxuriant forest growth. These rich bottoms are largely grown up to sweet gum, sour gum, various oaks, swamp hickory, sycamore, willow, holly, and magnolia, while along the streams and in swamps and shallow lagoons the cypress, tupelo gums, and palmettoes are often the characteristic growth. Where interlaced with vines these bottom-land forests are almost impenetrable thickets. The uplands and ridges are usually more openly forested with deciduous trees, such as oaks, hickories, dogwood, and sassafras, or often densely covered with one or more of the three species of pines which furnish most of the lumber of the State. Of these *Pinus taeda* and *echinata* are distributed over the State as far west as Houston, Hockley, Trinity, and Palestine in about equal abundance. The longleaf pine (*Pinus palustris*) occupies the southeastern part of the State, and where untouched by ax or fire forms miles of dense forest of the cleanest, most uniform, and symmetrical body of pine to be found on the continent, excelling the yellow pine forests of Arizona and California in the close array of graceful trunks.

In eastern Texas many species stop short of filling the whole humid area, and when their ranges are carefully mapped are found to be absent from, or in fewer cases to be restricted to, some of the following nonforested sections: The Grand and Black prairies of the Fort Worth and Dallas region; the coast prairie; coast marshes; islands and beaches.

---

[a] Physical Geog. of the Texas Region. R. T. Hill, U. S. Geol. Survey, Topographic Atlas, p. 13, 1900.

GRAND AND BLACK PRAIRIES.

The Grand and Black prairies, lying parallel, with only the narrow strip of Lower Cross Timbers between, extend from near Austin north in a broad strip to the Red River bottoms and east to Paris, forming an extensive area over which trees and forest species are mainly restricted to narrow stream bottoms. The rich black 'waxland' soil of these prairies is almost proof against burrowing rodents, which penetrate the region only along some sandy stream bottoms, while the open country tempts jack rabbits, coyotes, and other plains species eastward slightly beyond their usual bounds. Few, if any, species are restricted to these prairies, however, and the effect on distribution is mainly negative.

Here and there island strips of rich soiled grassy prairie occur in the timbered region farther east, becoming smaller and less frequent as they recede from the Black Prairie and Grand Prairie, and in some cases these islands are inhabited by a few plains species of birds, mammals, and reptiles nearly to the eastern edge of the State. Such an example is Nevils Prairie, near Antioch, where N. Hollister found scissor-tailed flycatchers, jack rabbits, and horned toads.

COAST PRAIRIE.

Over a wide strip of level coast prairie, extending along the Gulf from western Louisiana to San Antonio Bay and irregularly beyond, the timber is restricted to relatively narrow strips in the river bottoms, while the greater part of the surface is characterized by a rich growth of grass and many flowering plants. Spreading live oaks, loaded with Spanish moss, border the prairies or grow in scattered motts over them. In addition to the strictly shore species and those of the salt marshes which occasionally range over it or follow up the rivers to the limits of the open country, a few species of birds and mammals are characteristic of these coast prairies.

The most characteristic mammals are *Didelphis v. pigra*,[M] *Peromyscus taylori*[M] and *subater*,[M] *Oryzomys palustris*,[M] *Reithrodontomys aurantius*,[M] *R. merriami*,[M] *Sigmodon h. texianus*, *Microtus ludovicianus*,[M] *Geomys sagittalis*,[M] *Lepus merriami*,[M] and *Spilogale indianola*,[M] and of these *Peromyscus taylori*[M] and *subater*,[M] *Microtus ludovicianus*,[M] and *Geomys sagittalis*[M] are, so far as known, restricted to it.

The characteristic breeding birds of the coast prairies are *Tympanuchus attwateri*,[B] *Otocoris a. giraudi*,[B] *Megaquiscalus major*[B] and *macrourus*,[B] *Ammodramus m. sennetti*, *Coturniculus s. bimaculatus*,[B] and *Geothlypis t. brachidactyla*.[B]

Among its flowering plants *Baptisia, Oenothera, Meriolix, Hartmannia, Monarda, Coreopsis, Ratibida, Grindelia, Callirhoe, Eustoma,* and *Hymenocallis* are conspicuous genera, with numerous species, while such low shrubs as *Daubentonia longifolia,* *Vachellia farnesiand, Morella cerifera, Ascyrum,* and low willows are found here and there in favorable localities.

COAST MARSHES.

Extensive marshes border the Gulf shore irregularly as far west as Port Lavaca, and recur at intervals, mainly near the mouths of the streams, to the Rio Grande. These brackish, sedgy, tide-washed marshes are inhabited by rice rats, rails, water snakes, and great numbers of crustaceans. They are favorite resorts also for numerous migrating waders and water birds.

BEACHES AND ISLANDS.

The Gulf beaches and low islands offshore have a largely maritime fauna, the most striking feature of which is the abundance of shore birds, pelicans, cormorants, gulls, and terns. Not until the long reef-like bar of Padre Island is reached do we find any restricted forms of island mammals, and here only two—*Perodipus compactus* [M] and *Geomys personatus.*

The following species and subspecies of mammals, breeding birds, reptiles, and plants occur more or less commonly in the Austroriparian or humid subdivision of Lower Austral zone in eastern Texas, but rarely, if at all, in the arid western subdivision of the zone. None of the lists are complete.

MAMMALS OF EASTERN TEXAS AUSTRORIPARIAN.

Didelphis virginiana.	*Lepus aquaticus.*[M]
Didelphis virginiana pigra.[M]	*Lepus aquaticus attwateri.*[M]
Sciuropterus volans querceti[M]	*Felis* (sp.?) (panther).[M]
Sciurus ludovicianus.[M]	*Lynx rufus texensis.*
Sciurus carolinensis.	*Canis ater.*[M]
Citellus tridecemlineatus texensis.[M]	*Vulpes fulvus*[M]
Peromyscus gossypinus.	*Urocyon cinereoargenteus floridanus.*
Peromyscus leucopus.	*Ursus luteolus*[M]
Peromyscus taylori subater.[M]	*Procyon lotor.*
Oryzomys palustris[M]	*Lutra* (canadensis?).[M]
Reithrodontomys aurantius.[M]	*Lutreola lutreocephala*[M]
Reithrodontomys merriami.[M]	*Spilogale indianola*[M]
Neotoma floridana rubida[M]	*Mephitis mesomelas*[M]
Sigmodon hispidus texensis.	*Conepatus mesoleucus telmalestes.*[M]
Microtus pinetorum auricularis.[M]	*Scalopus aquaticus.*
Microtus ludovicianus.[M]	*Blarina brevicauda carolinensis*[M]
Castor canadensis texensis.	*Blarina parva.*[M]
Geomys breviceps.	*Nycticeius humeralis.*
Geomys sagittalis[M]	*Lasiurus borealis.*
Perognathus hispidus spilotus.[M]	*Lasiurus borealis seminolus.*[M]
Lepus floridanus alacer.[M]	*Pipistrellus subflavus.*[M]

BIRDS BREEDING IN EASTERN TEXAS AUSTRORIPARIAN.

Hydranassa tricolor ruficollis.[B]
Florida caerulea.[B]
Colinus virginianus.
Tympanuchus americanus.[B]
Tympanuchus americanus attwateri.[B]
Meleagris gallopavo silvestris.
Elanoides forficatus.
Buteo lineatus.
Falco sparverius.
Syrnium v. helveolum.[B]
Bubo virginianus.
Megascops asio.
Campephilus principalis.
Dryobates pubescens.[B]
Dryobates villosus auduboni.[B]
Dryobates borealis.[B]
Ceophloeus pileatus.[B]
Melanerpes erythrocephalus.
Centurus carolinus.[B]
Colaptes auratus.
Antrostomus carolinensis.[B]
Chordeiles (virginianus?).[B]
Chordeiles virginianus chapmani.[B]
Trochilus colubris.[B]
Coccyzus americanus.
Tyrannus tyrannus.
Myiarchus crinitus.
Contopus virens.
Empidonax virescens.
Cyanocitta cristata.
Agelaius phoeniceus.
Agelaius phoeniceus floridanus.
Icterus galbula.

Quiscalus quiscula aeneus.
Megaquiscalus major.[B]
Spizella socialis.[B]
Spizella pusilla.
Peucaea aestivalis bachmani.
Cardinalis cardinalis.
Guiraca caerulea.[B]
Cyanospiza cyanea.[B]
Piranga rubra.
Vireo olivaceus.
Vireo noveboracensis.[B]
Vireo flavifrons.
Mniotilta varia.
Protonotaria citrea.
Dendroica dominica albilora.[B]
Dendroica vigorsi.[B]
Geothlypis trichas brachidactyla.
Geothlypis formosa.
Icteria virens.
Wilsonia mitrata.[B]
Mimus polyglottos.
Galeoscoptes carolinensis.[B]
Thryothorus ludovicianus.
Sitta carolinensis.
Sitta pusilla.
Baeolophus bicolor.
Parus carolinensis agilis.[B]
Polioptila caerulea.
Hylocichla mustelina.
Sialia sialis.

A FEW OF THE LIZARDS AND SNAKES OF EASTERN TEXAS.

Lizards.

Anolis carolinensis.
Phrynosoma cornutum (local form).
Ophisaurus ventralis.[R]

Cnemidophorus sexlineatus.
Leiolopisma laterale.[R]
Eumeces quinquelineatus[R]

Snakes.

Opheodrys aestivus.
Callopeltis obsoletus.[R]
Lampropeltis getula holbrooki.
Natrix clarkii.[R]
Natrix fasciata transversa.[R]
Storeria dekayi.
Eutainia proxima.[R]

Tropidoclonium lineatum.
Tantilla gracilis.
Elaps fulvius.[R]
Agkistrodon piscivorus.
Agkistrodon contortrix.
Crotalus horridus.

PLANTS CHARACTERISTIC OF HUMID EASTERN IN DISTINCTION FROM ARID WESTERN
TEXAS.

Pinus taeda.	*Crataegus spathulata.*
Pinus palustris.	*Crataegus texana.*
Pinus echinata.	*Persea borbonia.*
Taxodium distichum.[P]	*Leitneria floridana.*
Juniperus virginiana.[P]	*Ilex opaca.*
Liquidambar styraciflua.	*Ilex decidua.*
Nyssa sylvatica.	*Ilex vomitoria.*
Nyssa aquatica.	*Ilex lucida.*
Platanus occidentalis.	*Morus rubra.*
Magnolia faetida.[P]	*Gleditsia tricanthos.*[P]
Magnolia virginiana.	*Gleditsia aquatica.*
Tilia leptophylla.	*Fagara clavaherculis.*[P]
Acer drummondi.	*Aralia spinosa.*
Acer rubrum.	*Viburnum rufotomentosum.*[P]
Hicoria ovata.[P]	*Viburnum molle.*
Hicoria alba.[P]	*Viburnum (nudum?).*
Hicoria glabra.[P]	*Callicarpa americana.*
Hicoria aquatica.[P]	*Cyrilla racemiflora.*
Juglans nigra.	*Vaccinium* sp.——?
Castanea pumila.	*Morella crispa.*
Carpinus caroliniana.	*Azalea* sp.——?
Ostrya virginiana.	*Schmaltzia lanceolata*
Betula nigra.	*Schmaltzia copallina.*
Quercus phellos.	*Rhus radicans.*[P]
Quercus nigra.	*Cephalanthus occidentalis.*
Quercus marylandica.	*Rhamnus caroliniana.*
Quercus digitata.	*Hamamelis virginiana.*
Quercus rubra.	*Vitis* sp.——?
Quercus virginiana.	*Smilax laurifolia.*
Quercus acuminata.[P]	*Smilax (renifolia?).*[P]
Quercus macrocarpa.[P]	*Smilax pumila.*
Quercus lyrata.	*Gelsemium sempervirens.*
Quercus minor.	*Bignonia crucigera.*[P]
Quercus alba.	*Campsis radicans.*
Populus deltoides.	*Bradleia* (wisteria).
Salix (nigra?).	*Passiflora incarnata.*[P]
Ulmus americana.	*Rubus (trivialis?).*
Ulmus fulva.[P]	*Rubus (procumbens?).*[P]
Ulmus alata.	*Yucca louisianensis.*[P]
Toxylon pomiferum.[P]	*Yucca arkansana.*
Celtis mississippiensis.[P]	*Sabal adiantinum.*
Asimina triloba.	*Arundinaria macrosperma.*[P]
Diospyros virginiana.	*Dendropogon usneoides.*[P]
Sassafras sassafras.[P]	*Mitchella repens.*
Cynoxylon floridum.	*Sphagnum* sp.——?

For crops of the Austroriparian faunal area of the United States see Life Zones and Crop Zones, page 46, under the headings 'Cereals,' 'Fruits,' 'Nuts,' and 'Miscellaneous.' Only a part of the crops listed are adapted to the east Texas region, however, while other varieties have been introduced since the preparation of these lists.

LOWER SONORAN OF WESTERN TEXAS.

In Texas the annual rainfall decreases gradually from about 50 inches in the eastern part of the State to about 10 inches in the extreme western part. While the extremes are so great and there is no abrupt change from eastern humid to western arid, there is still a well-defined division between the two regions, approximately where the annual rainfall diminishes to below 30 inches, or near the ninety-eighth meridian. By combining the limits of range of eastern and western species of mammals, birds, reptiles, and plants an average line of change can be traced across the State, beginning on the north at the ninety-eighth meridian, just east of Henrietta, and running south to Lampasas, Austin, Cuero, and Port Lavaca. This line conforms in a general way to the eastern limit of the mesquite, which more nearly than any other tree or shrub fills the whole of the arid Lower Sonoran zone. While scattering outlying mesquite trees are found farther east, the line is intended to mark the eastern edge of their abundance, or the transition from eastern prairie and timber country to the region dominated by the mesquite and associated plants.[a]

West of this line the region may be again subdivided into semiarid, or region of mesquite and abundant grass, stretching west to the Pecos Valley and from the northern Panhandle to the mouth of the Rio Grande, and extreme arid, or region of creosote bush and scanty grass, lying mainly between the Pecos and Rio Grande.

SEMIARID LOWER SONORAN.

The semiarid region is largely mesquite plains, varying from open grassy plains with scattered mesquite bushes to a miniature forest of mesquite trees, in places densely filled in with other thorny bushes and cactus, as along its southern stream valleys and over much of the plains of the Lower Rio Grande. Scattered oaks and other scrubby timber growth characterize the higher, rougher parts of the region, and narrow strips of tall timber are found along some of its streams. Toward the coast, flower-strewn grassy prairies extend irregularly nearly across the southern part of the State, forming a broken westerly extension of the more continuous eastern coast prairie. West of Matagorda Bay this prairie is mainly crowded

---

[a]The Mesophytic plant region of eastern Texas and the Xerophytic of western Texas of Coulter and Bray. (See Plant Relations, by John M. Coulter, pp. 168, 193, 230, 1899, and Ecological Relations of Vegetation of Western Texas, by William L. Bray, Botanical Gazette, XXXII, p. 111, 1901.)

back from the coast by dense thickets, consisting of mesquite, huisache, and numerous thorny shrubs mixed with cactus, or of miles of live-oak brush, in places only knee high; again, in dense jungles 10 or 20 feet high, in patches, strips, or isolated oak 'motts.' In Cameron County the oak motts occur as widely scattered islands on the prairie, and are usually made up of a few gnarled old trees. Along the stream bottoms and on the low coast flats the chaparral is especially dense and in places almost impenetrable from the abundance of cactus and thorny branches that interlace over the trails. The bulk of this chaparral is composed of common arid Lower Sonoran shrubs, such as *Momesia pallida,* *Zizyphus obtusifolia, Condalia obovata,* *Koeberlinia spinosa, Opuntia engelmanni,* *O. lepticaulis,* and other associated species, which in this semiarid region of rich soil grow with unusual vigor. Many other widely distributed species, such as *Parkinsonia aculeata, Vachellia farnesiana, Tillandsia recurvata,* and *Manfreda maculosa,* range through it, while a few others are peculiar to it or barely extend into it from farther south.

As Padre Island lies within this semiarid division, and is sufficiently large and isolated to provide a habitat for a few species of mammals, the following brief description by William Lloyd, who traveled its whole length in November, 1891, is of interest:

Padre Island is about 90 miles long, and at the south end runs out to a point, the last 10 miles of which is not over a mile wide, while for the last 5 miles it is only 300 or 400 yards wide. Its central and greater breadth is nearly 4 miles, including about two-thirds of the distance a muddy flat so soft that one sinks in it over 3 inches. From here it tapers again to its north extremity, which is about 300 yards wide. It is divided from the mainland by the Laguna Madre, which is only about a mile wide from Point Isabel and 2 miles wide opposite Arroyo Coloral. Here, however, the water is 8 to 10 feet deep in the channels. Farther north at the noted wagon crossing, about 15 miles south of Corpus Christi, near the north end of the island, the channel is 7 miles wide, with the water $4\frac{1}{2}$ to 5 feet deep at its ordinary elevation, although south winds raise it very rapidly so as to be impassable. The main island is surrounded by a network of smaller islands, with Mustang Island at the north end separated from it by a channel a mile wide. The drift or wrack and floating timbers on the Gulf side are rapidly embedded in the restless sand and form a nucleus for the sand dunes which stretch along the beach and form the backbone of the island. Beyond them are smaller mounds with some little vegetation, and at their feet lie sandy fields of grass, broken by numerous salt-water lakes where the sea has washed in from time to time.

The island has no arborescent growth worth noticing, with the exception of a shin-oak, which extends from the north end for about a mile and continues on sandy hills on the lagoon side for 5 or 6 miles farther. This is usually 6 inches to 18 inches high, but there are trees, perhaps a different species, 6 to 8 feet high. As this oak is always loaded with acorns, even now it is the favorite wintering ground of birds such as wood ibis, whooping and sand-hill cranes. Wild celery abounds also in the lagoon and attracts great numbers of ducks of various species.

A few willows, presumably *Salix nigra,* grow at the settlement and at one point north of it, and a few patches of buttonbush, *Cephalanthus occidentalis,* were observed, also a few stunted 'huisache,' *Acacia farnesiana,*[P] and crab grass, cockleburr, and wild grapes. These are all on the north and center of the island, south of which grow salt grass and various waxy and creeping plants.

Strange to say, neither hackberry, mesquite, nor Mexican persimmon, though abundant on the adjacent mainland, have succeeded in obtaining a footing anywhere, and two straggling prickly pears (*Opuntia engelmanni*)[P] were the sole representatives of the cactus family. Although palmetto and banana stumps wash ashore in great numbers, none were seen growing.

Gales cover the Gulf side of the island with debris that must come from the districts of Tampico or Vera Cruz. An iguana was taken a short time since on the island, and at least three species of snakes, including the rattlesnake, occur there. Deer and coyotes have been seen by several parties swimming or wading across to and from the island and mainland.

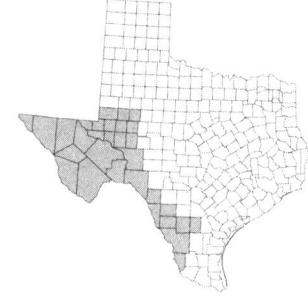

[3] The scientific name for the creosote bush is now *Larrea tridentata* and the current distribution in Texas is depicted in Figure 34.

Figure 34. The current distribution of creosote bush (*Larrea tridentata*) in Texas.

FIG. 2.—Distribution area of creosote bush (*Covillea tridentata*) **3**

EXTREME ARID LOWER SONORAN.

The extreme arid section of the arid Lower Sonoran zone of Texas includes the Pecos Valley and the Rio Grande Valley south to about Eagle Pass and all the country between the two valleys except the several mountain masses that rise as somewhat less arid Upper Sonoran and Transition zone islands. It has an irregular annual rainfall of 10 to 20 inches, and a half-barren soil, rich and mellow in the valleys, stony and baked on the mesas. It is subject to long, scorching drought, but after a single heavy rainfall bursts into verdure and bloom with a sudden brilliancy seen only in the desert. Its most characteristic shrub is the evergreen creosote bush, the range of which defines its extent better than any other plant, but its most conspicuous vegetation consists of yuccas, agaves, sotol, cactus, fouquiera, allthorn, and mesquite. Its mammals are mainly the species of the whole arid Lower Sonoran, but a few of these extend farther west without extending farther east than the Pecos Valley, among which are the following species:

Odocoileus hemionus canus.[M]
Ammospermophilus interpres.
Citellus spilosoma arens.[M]
Onychomys torridus.
Peromyscus leucopus texanus.
Peromyscus sonoriensis blandus.[M]
Peromyscus eremicus.
Perognathus penicillatus eremicus.[M]
Perognathus intermedius.[M]
Perognathus nelsoni[M]
Perognathus nelsoni canescens.[M]
Perognathus flavus.
Perognathus merriami gilvus.

Perodipus ordi.[M]
Dipodomys merriami.
Dipodomys merriami ambiguus.
Geomys arenarius.
Thomomys aureus lachuguilla.[M]
Canis mearnsi.[M]
Vulpes macrotis neomexicanus.[M]
Myotis californicus.
Myotis yumanensis.
Pipistrellus hesperus.[M]
Corynorhinus macrotis pallescens.
Antrozous pallidus.
Promops californicus.[M]

Including these somewhat mixed elements of semiarid, half open plains, strips of low prairie, dense cactus, thorny chaparral, and the more barren region of extreme aridity, under the heading of "Lower Sonoran Zone," we have in Texas an area which covers a little more than half of the State, and includes by far the largest number of species of mammals, birds, reptiles, and plants common to any subdivision in the State. It is characterized by the following species, some of which fill the subdivision and are restricted to it, while many more are restricted to definite areas within its limits, and still others range beyond through one or more of the other zones. Few of the species, however, extend through both arid and humid divisions of the zone without undergoing at least a subspecific change.

MAMMALS OF LOWER SONORAN OF WESTERN TEXAS.

Tatu novemcinctum texanum[M]
Didelphis marsupialis texensis.[M]
Tayassu angulatum.[M]
Odocoileus virginianus texanus.[M]
Odocoileus hemionus canus.[M]
Sciurus ludovicianus limitis.[M]
Ammospermophilus interpres.

Citellus variegatus couchi.[M]
Citellus buckleyi.[M]
Citellus mexicanus parvidens.[M]
Citellus spilosoma major.[M]
Citellus s. arens.[M]
Citellus s. annectens[M]
Onychomys torridus.

Onychomys longipes.[M]
Peromyscus leucopus texanus.
Peromyscus leucopus mearnsi.
Peromyscus michiganensis pallescens.[M]
Peromyscus sonoriensis blandus.[M]
Peromyscus eremicus.
Peromyscus attwateri.
Peromyscus taylori.[M]
Oryzomys aquaticus.[M]
Reithrodontomys intermedius.[M]
Reithrodontomys megalotis.
Reithrodontomys griseus.[M]
Neotoma micropus.
Sigmodon hispidus berlandieri.
Fiber zibethicus ripensis.[M]
Castor canadensis frondator.[M]
Liomys texensis.[M]
Perognathus hispidus.[M]
Perognathus penicillatus eremicus.[M]
Perognathus intermedius.[M]
Perognathus nelsoni[M]
Perognathus nelsoni canescens.[M]
Perognathus flavus.
Perognathus merriami.
Perognathus merriami gilvus.
Perodipus ordi.[M]
Perodipus sennetti.[M]
Perodipus compactus.[M]
Dipodomys spectabilis.
Dipodomys elator.[M]
Dipodomys merriami.
Dipodomys merriami ambiguus.
Geomys breviceps attwateri.[M]
Geomys breviceps llanensis.[M]
Geomys arenarius.
Geomys texensis.
Geomys personatus.
Geomys personatus fallax.
Cratogeomys castanops.
Thomomys aureus lachuguilla.[M]

Thomomys perditus.[M]
Lepus merriami.[M]
Lepus texianus.[M]
Lepus arizonae minor.[M]
Lepus floridanus chapmani.[M]
Felis onca hernandezi.[M]
Felis hippolestes aztecus.[M]
Felis pardalis limitis.[M]
Felis cacomitli.
Lynx texensis.[M]
Canis rufus.
Canis nebracensis texensis.[M]
Canis microdon.[M]
Canis mearnsi.[M]
Vulpes macrotis neomexicanus.[M]
Urocyon cinereoargenteus scotti.
Bassariscus astutus flavus.
Taxidea taxus berlandieri.
Procyon lotor mexicanus.[M]
Nasua narica (yucatanica?).[M]
Putorius frenatus.[M]
Putorius neomexicanus.[M]
Spilogale leucoparia.[M]
Mephitis mesomelas varians.[M]
Conepatus mesoleucus mearnsi.[M]
Conepatus leuconotus texensis.[M]
Scalopus texensis.[M]
Notiosorex crawfordi.
Blarina berlandieri.[M]
Myotis velifer.
Myotis californicus.
Myotis incautus.[M]
Myotis yumanensis.
Pipistrellus hesperus.[M]
Dasypterus intermedius.
Antrozous pallidus.
Corynorhinus macrotis pallescens.[M]
Nyctinomus mexicanus.[M]
Promops californicus.[M]
Mormoops megalophylla senicula.[M]

BREEDING BIRDS OF LOWER SONORAN OF WESTERN TEXAS.

Colinus virginianus texanus.
Callipepla squamata.
Callipepla squamata castanogastris.
Lophortyx gambeli.[B]
Meleagris gallopavo intermedia.
Leptotila fulviventris brachyptera.[B]
Melopelia leucoptera.[B]
Columbigallina passerina pallescens.[B]
Scardafella inca.[B]
Elanus leucurus.
Parabuteo unicinctus harrisi.
Buteo borealis calurus.[B]

Buteo abbreviatus.[B]
Buteo albicaudatus sennetti.[B]
Buteo swainsoni.
Urubitinga anthracina.[B]
Falco mexicanus.
Falco fusco-caerulescens.[B]
Falco sparverius phalaena.[B]
Polyborus cheriway.[B]
Syrnium varium helveolum.[B]
Megascops asio mccalli.
Bubo virginianus pallescens.
Speotyto cunicularia hypogaea.[B]

Micropallas whitneyi.[B]
Crotophaga sulcirostris.
Geococcyx californianus.
Coccyzus americanus occidentalis.
Ceryle americana septentrionalis.[B]
Dryobates scalaris bairdi.[B]
Centurus aurifrons.[B]
Phalaenoptilus nuttalli.
Nyctidromus albicollis merrilli.
Chordeiles acutipennis texensis.
Amizilis cerviniventris chalconota.[B]
Tyrannus vociferans.
Myiarchus cinerascens.
Sayornis saya.
Sayornis nigricans.
Pyrocephalus rubineus mexicanus.[B]
Xanthoura luxuosa glaucescens.[B]
Corvus corax sinuatus.
Corvus cryptoleucus.
Molothrus ater obscurus.
Tangavius aeneus involucratus.[B]
Sturnella magna hoopesi.
Icterus auduboni.[B]
Icterus cucullatus sennetti.
Icterus parisorum.
Icterus bullocki.[B]
Megaquiscalus major macrourus.[B]
Carpodacus mexicanus frontalis.[B]
Astragalinus psaltria.[B]
Amphispiza bilineata.
Amphispiza b. deserticola.[B]

Peucaea cassini.
Aimophila ruficeps eremoeca.
Arremonops rufivirgata.[B]
Cardinalis cardinalis canicaudus.[B]
Pyrrhuloxia sinuata.[B]
Pyrrhuloxia s. texana.[B]
Guiraca caerulea lazula.[B]
Cyanospiza versicolor[B]
Piranga rubra cooperi.[B]
Phainopepla nitens.
Lanius ludovicianus excubitorides.
Vireo atricapillus.[B]
Vireo belli medius.[B]
Vireo b. arizonae.[B]
Vireo noveboracensis micrus.[B]
Dendroica aestiva sonorana.[B]
Dendroica chrysoparia.[B]
Icteria virens longicauda.[B]
Mimus polyglottos leucopterus.[B]
Toxostoma longirostre sennetti.
Toxostoma curvirostre.
Heleodytes brunneicapillus couesi.[B]
Salpinctes obsoletus.
Catherpes mexicanus albifrons.
Thryomanes bewicki cryptus.[B]
Thryomanes b. leucogaster.[B]
Baeolophus atricristatus.
Auriparus flaviceps.
Polioptila caerulea obscura.[B]
Polioptila plumbea.

REPTILES OF LOWER SONORAN.

Lizards.

Crotaphytus reticulatus.
Crotaphytus wislizenii.[R]
Holbrookia texana.[R]
Holbrookia propinqua.
Holbrookia maculata.
Holbrookia m. lacerata.[R]
Sceloporus clarkii.[R]
Sceloporus spinosus floridanus.[R]
Sceloporus consobrinus.

Sceloporus dispar.[R]
Sceloporus merriami.
Phrynosoma cornutum.
Phrynosoma modestum.
Coleonyx brevis.
Ophisaurus ventralis.[R]
Cnemidophorus tessellatus.
Cnemidophorus perplexus.[R]
Cnemidophorus gularis.

Snakes.

Diadophis regalis.[R]
Heterodon nasicus.
Bascanion flagellum.[R]
Bascanion ornatum.[R]
Drymobius margaritiferus.
Callopeltis obsoletus.
Drymarchon corais melanurus.[R]
Rhinocheilus leconti.[R]

Natrix fasciata transversa.[R]
Eutainia elegans marciana.[R]
Eutainia proxima.[R]
Tantilla gracilis.
Elaps fulvius.[R]
Agkistrodon piscivorus.
Crotalus atrox.

CONSPICUOUS PLANTS OF LOWER SONORAN.

Prosopis glandulosa.
Prosopis pubescens.
Acacia constricta.[P]
Acacia tortuosa.[P]
Acacia roemeriana.[P]
Acacia schottii.[P]
Acacia wrightii.[P]
Acacia amentacea.[P]
Acacia berlandieri.[P]
Vachellia farnesiana.[P]
Leucaena retusa.
Mimosa emoryana.
Mimosa lindheimeri.
Mimosa borealis.
Mimosa fragrans.[P]
Parkinsonia aculeata.
Cercidium floridanum.[P]
Cercidium texanum.[P]
Eysenhardtia amorphoides.[P]
Sophora secundiflora.[P]
Parosela frutescens.[P]
Parosela formosa.[P]
Juglans rupestris.[P]
Celtis helleri.[P]
Momesia pallida.[P]
Chilopsis linearis.
Ehretia eliptica.[P]
Koeberlinia spinosa.
Adelia angustifolia.
Adelia neomexicana.
Fraxinus greggii.
Porlieria angustifolia.
Covillea tridentata.[P]
Schmaltzia microphylla.
Schmaltzia mexicana.
Schmaltzia virens.
Nicotiana glauca.
Brayodendron texanum.[P]
Berberis trifoliata.[P]
Zizyphus obtusifolia.
Zizyphus lycioides.[P]
Condalia obovata.[P]

Condalia spathulata.
Lycium berlandieri.
Lycium pallidum.
Leucophyllum texanum.[P]
Leucophyllum minus.
Krameria canescens.[P]
Fouquiera splendens.[P]
Aloysia ligustrina.[P]
Tecoma stans.
Ephedra antisyphilitica.
Ephedra trifurcata.
Croton torreyanus.[P]
Bernardia myricaefolia
Euphorbia antisyphilitica.
Mozinna spathulata.[P]
Baccharis (salicina?).
Baccharis (glutinosa?).[P]
Flourensia cernua.
Agave lecheguilla.
Hechtia texensis.
Tillandsia recurvata.
Tillandsia baileyi.
Yucca macrocarpa.
Yucca treculeana.
Yucca radiosa.[P]
Yucca rostrata.
Yucca rupicola.
Samuela faxoniana.
Samuela carnerosana.[P]
Hesperaloe parviflora.
Opuntia lindheimeri.[P]
Opuntia engelmanni.[P]
Opuntia leptocaulis.[P]
Cereus paucispinus.[P]
Cereus enneacanthus.[P]
Cereus stramineus.[P]
Echinocactus horizonthalonius.
Echinocactus hamatocanthus.[P]
Echinocactus wislizeni.[P]
Echinocactus wrighti.[P]
Cactus heyderi.[P]

For crops adapted to Lower Sonoran, see Life Zones and Crop Zones, pages 42—45, under heading "Crops of the Lower Sonoran Faunal Area," where under "Cereals," "Fruits," "Nuts," and "Miscellaneous" are listed the varieties that have proved a success in other parts of the area. Although many of these have not been tested in the Texas region, and while varieties other than those listed have proved successful, the list will be found helpful in selecting varieties for experiments.

[4] The current distribution of *Agave lechuguilla* is depicted in Figure 35. Large-scale production of agave fibers was never attempted in Texas because by the 1920s and 1930s synthetic fibers had become widely available, replacing natural fibers for most uses. Agave is still harvested and cultivated in Mexico for production of fiber, various beverages (including tequila), and other products.

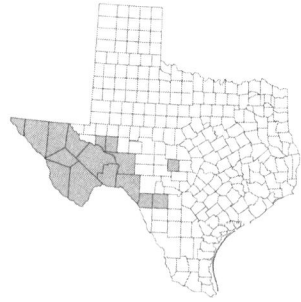

Figure 35. The current distribution of lechuguilla (*Agave lechuguilla*) in Texas.

30   NORTH AMERICAN FAUNA.   [No. 25.

Some practical suggestions may be derived also from the native species of plants, as in the case of *Schmaltzia mexicana,*[P] variously known as *Rhus mexicana* and *Pistacia mexicana,* and related to the *Pistacia vera* of the Mediterranean region, from which the pistachio of commerce is obtained. In places in the canyons of the Rio Grande this large shrub grows in profusion, suggesting that the real *pistachio* also might succeed here.

One of the conspicuous plants often dominant over much of the

FIG. 3.—Distribution area of lecheguilla (*Agave lecheguilla*). **4**

extreme arid Lower Sonoran zone of western Texas is a little century plant (*Agave lecheguilla*), best known by its Mexican name of 'lecheguilla.' Its rigid leaves are about a foot long, well armed with marginal hooks and stout terminal spines, which effectually protect them from the attacks of grazing animals. Even the hardy burros and hungry goats refrain from eating them, and pick their way cautiously among their dagger points. But within each leaf is a bundle of smooth, strong fibers suitable for the manufacture of brushes, matting, coarse twine, and rope. These plants grow in greatest abundance over limestone and lava mesas and steep rocky slopes that can never be irrigated and are often too steep and rough for grazing, even if the scanty grass were not crowded out by the cactus and agaves. Here, over thousands of square miles of the most worthless part of the desert, is a crop, not only offering in its leaf fibers a profitable industry awaiting development, but also suggesting that other species of agaves, yielding fiber of still more valuable quality, can be successfully introduced into this region—a region that now lies unimproved and almost uninhabited while hundreds of thousands of dollars worth of agave fibers are annually imported from Mexico.

In the Davis and Chisos and Guadalupe mountains the large *Agave wislizeni*[p] and *applanata,* the mescal plants of the Mescalero Apaches, offer a nutritious food that might well find place on our tables as a delicacy. They grow over the barest and roughest slopes, not only yielding in the starchy caudex a rich store of food, but in the beautiful flowers a quantity of delicious honey equaled by few other plants. A single plant during its flowering period of about a month bears from one to two thousand flowers, each yielding nearly half a teaspoonful of honey. That the country is well adapted to bees is evidenced by numerous and extensive bee caves in the rocks, by bee trees, and by the success of domestic swarms. The numerous leguminous shrubs—acacias, mimosas, and mesquites, several species of the 'bee bush' (*Lippia* and *Goniostachyum*)[p] and the abundant flowers of numerous species of Compositae—all yield rich stores of honey. In semiarid gulches where the native black-fruited Texan persimmon (*Brayodendron texanum*) bears an abundance of its almost worthless fruit it is probable that varieties of the delicious Japanese persimmon would thrive.

Other plants besides grass and cactus are important as food for stock or are of service to man. The sotol (*Dasylirion texanum*)[p] with its double-edged saw-bladed leaves and stout caudex, when split open so that the inner starchy heart can be reached, yields a large amount of hearty food for stock. The plant is widely distributed over most of the region west of the Pecos and Devil rivers, and is most abundant over the barest and stoniest slopes. Like most desert plants, it is of slow growth, and its greatest value has been in tiding stock over periods of scarcity. Sotol cutting becomes an important business with sheep and cattle men when a dry summer is followed by a winter of bare pastures.

The value of the mesquite and screw bean (*Prosopis glandulosa* and *pubescens*) to stockmen and ranchers of western Texas can hardly be overestimated. Over much of the arid and semiarid region of the State they yield fuel, fence posts, and building material for the ranch, and also shade, shelter, and food for stock. The common

[5] Both the black and yellow persimmons currently are included in the genus *Diospyros* (*Diospyros texana* and *Diospyros virginiana*). The current distributions for the two species are depicted in Figure 36.

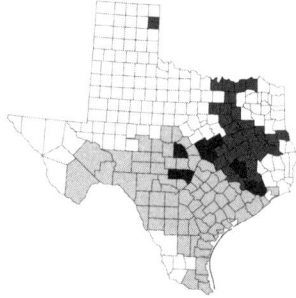

Figure 36. The current distribution of Texas persimmon (*Diospyros texana*, light shaded area) and common persimmon (*Diospyros virginiana*, medium shaded areas) in Texas. The darkest shaded areas represent regions of overlap or sympatry between the two species.

32   NORTH AMERICAN FAUNA.   [No. 25.

mesquite, though barely reaching the dignity of a tree and often dwarfed to a mere shrub, is the only available timber over thousands of square miles. The wood is heavy, strong, and durable. The feathery foliage, while so thin that grass grows under the trees, affords a welcome shade to man and beast. The fragrant, honey-laden, catkinlike flowers blossom quickly, and in warm weather after a good rain a crop of long bean pods will mature and ripen with

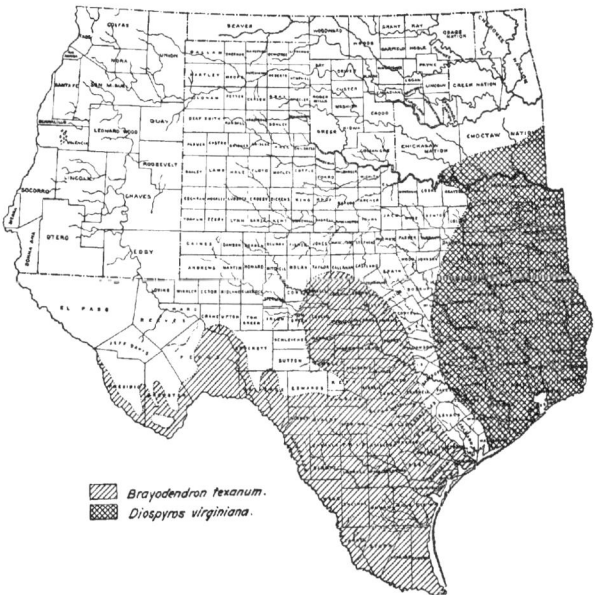

FIG. 4.—Distribution area of the black and the yellow persimmons (*Brayodendron texanum* and *Diospyros virginiana*). **5**

little regard to season. Often two crops a year mature if rains come at proper intervals. The sugary pods serve to fatten cattle, horses, mules, burros, sheep, and goats. The small, hard beans pass through animals and seed new ground, so that the spread and increase of the mesquite has been a notable result of stock raising. The use of the sweet, nutritious pods as food by both Indians and early settlers seems to have been mainly given up, but the actual food value of the pods needs no better demonstration than is afforded by the condition of animals feeding on them.

A much neglected product of the mesquite is the gum which exudes from the branches and can be gathered in large quantities. Apparently, it has all the qualities of gum arabic, the gum of closely related Old World acacias, and needs only introduction to a market to become of commercial value.

The seeds and pods of other leguminous shrubs, the acorns of several species of oaks, and the sugary berries of the alligator-barked juniper also are of considerable value in special areas as feed for stock or poultry.

UPPER AUSTRAL ZONE, UPPER SONORAN DIVISION.

East of the Pecos Valley, Upper Sonoran zone covers most of the Panhandle, the Staked Plains and the narrower secondary plain, or Edwards Plateau, running south as far as Rock Springs, as well as the tops and cold slopes of the ridges and bottoms of shaded gulches breaking down from the edge of these plains. West of the Pecos Valley it covers the foothills and lower slopes of the mountains, extending on southwest slopes nearly or quite to the tops of most of the peaks, but on the northeast slopes of the Guadalupe, Davis, and Chisos mountains giving place to Transition zone at about 6,000 feet. On such steep, arid slopes as these mountains present to the sun's rays the difference of zone level on opposite sides is often 2,000 or 3,000 feet, increasing with the steepness and barrenness of the slope. Over the mountains and rough country the zone is marked by a scattered growth of nut pines, junipers, and oaks, but over the plains, where short grass is the principal vegetation, its limits are often best determined by the absence of mesquite and other shrubs of the surrounding Lower Sonoran zone. Some of its most characteristic plants in the mountain region are *Pinus edulis* and *cembroides, Juniperus pachyphloea,*[P] *monosperma,* and *flaccida, Quercus grisea* and *emoryi, Adolphia infesta, Nolina texana, Mimosa biuncifera,*[P] *Cercocarpus parvifolius, Garrya lindheimeri,*[P] *Fallugia paradoxa, Yucca baccata, Agave wislizeni*[P]*and applanata,* while those on the plains, aside from grasses, are *Asclepias latifolia* and *speciosa, Laciniaria punctata,* several species of *Psoralea*[P]*and Astragalus, Polygala alba, Yucca glauca,* and *Opuntia cymochila.*[P]

In the mountains and rough country Upper Sonoran zone is especially characterized by the occurrence in the breeding season of birds such as *Cyrtonyx mearnsi,*[B] *Coeligena clemenciae,*[B] *Calothorax lucifer, Aphelocoma couchi, cyanotis,*[B]*and texana,*[B] *Pipilo mesoleucus,*[B] *Vireo plumbeus* and *stephensi,*[B] and *Psaltriparus plumbeus*[B]*and lloydi,*[B] and on the plains by such breeding species as *Podasocys montanus,*[B] *Numenius longirostris,*[B] *Chordeiles henryi,*[B] *Pooecetes confinis,*[B] and *Otocoris leucolaema.*[B]

In the mountains and rough country some of the most characteristic mammals of Upper Sonoran zone are *Ovis mexicana*,[M] *Odocoileus couesi*[M] and *canus*,[M] *Citellus grammurus*[M] and *couchi*,[M] *Peromyscus rowleyi*,[M] *attwateri*, and *laceyi*,[M] *Neotoma attwateri*[M] and *albigula*,[M] and on the plains *Antilocapra americana, Odocoileus macrourus*,[M] *Cynomys ludovicianus, Citellus pallidus*,[M] *Onychomys pallescens*,[M] *Perognathus paradoxus*[M] and *copei*,[M] *Perodipus richardsoni*,[M] *Lepus melanotis*,[M] *Vulpes velox*,[M] *Putorius nigripes*.[M]

Including both plains and mountain slopes, the Upper Sonoran zone in Texas is characterized by the following species:

MAMMALS OF UPPER SONORAN.

Ovis mexicanus.[M]
Antilocapra americana.
Odocoileus couesi.[M]
Odocoileus virginianus macrourus.[M]
Citellus variegatus grammurus.[M]
Citellus v. couchi.[M]
Citellus tridecemlineatus pallidus.[M]
Citellus spilosoma marginatus.[M]
Cynomys ludovicianus.
Onychomys leucogaster pallescens.[M]
Peromyscus sonoriensis.[M]
Peromyscus rowleyi.[M]
Peromyscus attwateri.
Peromyscus boylei laceyi.[M]
Neotoma attwateri.[M]
Neotoma albigula.[M]
Thomomys baileyi.[M]
Perognathus flavescens copei.

Perognathus hispidus paradoxus.[M]
Perodipus richardsoni.[M]
Lepus texianus melanotis.[M]
Lepus arizonae minor[M] (mainly Lower Sonoran).
Lepus pinetis robustus[M] (mainly Transition).
Felis hippolestes aztecus.[M]
Lynx baileyi.[M]
Canis griseus.[M]
Canis nebracensis.[M]
Vulpes velox.[M]
Urocyon cinereoargenteus scotti (also Lower Sonoran).
Putorius nigripes.[M]
Spilogale interrupta.[M]
Mephitis mesomelas varians.[M]
Taxidea taxus berlandieri.

BIRDS OF UPPER SONORAN.

Numenius longirostris.[B]
Podasocys montanus.[B]
Cyrtonyx montezumae mearnsi.
Accipiter cooperi.
Chordeiles virginianus henryi.[B]
Calothorax lucifer.
Coeligena clemenciae.[B]
Trochilus alexandri.[B]
Phalaenoptilus nuttalli.
Aeronautes melanoleucus.[B]
Tyrannus verticalis.
Otocoris alpestris leucolaema.[B]
Aphelocoma woodhousei.[B]
Aphelocoma cyanotis.[B]
Aphelocoma texana.[B]
Aphelocoma sieberi couchi.[B]

Cyanocephalus cyanocephalus.[B]
Icterus bullocki.[B]
Sturnella magna neglecta.[B]
Carpodacus mexicanus frontalis.
Spizella socialis arizonae.
Aimophila ruficeps scotti.
Pipilo fuscus mesoleucus.[B]
Cyanospiza amoena.[B]
Zamelodia melanocephala.[B]
Ampelis cedrorum.[B]
Vireo solitarius plumbeus.[B]
Vireo gilvus swainsoni.[B]
Troglodytes aëdon aztecus.[B]
Baeolophus inornatus griseus.[B]
Psaltriparus plumbeus.[B]
Psaltriparus melanotis lloydi.[B]

LIZARDS AND SNAKES OF UPPER SONORAN.

Lizards.

Crotaphytus collaris.
Crotaphytus c. baileyi.[R]
Uta ornata.[R]
Sceloporus torquatus poinsettii.[R]
Sceloporus consobrinus.

Phrynosoma hernandesi.
Gerrhonotus liocephalus infernalis.[R]
Eumeces guttulatus.[R]
Eumeces obsoletus.
Eumeces brevilineatus.[R]

Snakes.

Diadophis regalis. [R]
Heterodon nasicus.
Liopeltis vernalis. [R]
Bascanion flagellum. [R]
Pituophis sayi. [R]

Chionactis episcopus isozonus.[R]
Eutainia cyrtopsis.[R]
Crotalus molossus.
Crotalus lepidus.
Crotalus confluentis.[R]

PLANTS OF UPPER SONORAN PLAINS.

Asclepias latifolia.
Asclepias tuberosa.
Polygala alba.
Laciniaria punctata.[P]
Yucca glauca.
Yucca stricta.[P]
Opuntia davisi.[P]
Opuntia macrorhiza.
Cactus missouriensis.
Artemisia filifolia.
Ratibida columnaris.[P]
Helianthus annuus.
Helianthus petiolaris.
Gutierrezia sarothrae.

Mentzelia nuda.
Astragalus molissimus.[P]
Astragalus caryocarpus.[P]
Psoralea linearifolia. [P]
Psoralea digitata.[P]
Parosela enneandra.[P]
Acuan illinoensis.[P]
Amorpha canescens.
Hoffmanseggia jamesi.[P]
Petalostemon purpureus.[P]
Ipomea leptophylla. [P]
Merolix intermedia.[P]
Linum rigidum.
Verbena stricta.

PLANTS OF UPPER SONORAN MOUNTAINS AND FOOTHILLS.

Pinus edulis.
Pinus cembroides.
Juniperus pachyphloea.[P]
Juniperus flaccida.
Juniperus monosperma.
Juniperus sabinoides.[P]
Quercus grisea.
Quereus emoryi.
Quereus undulata.
Quercus texana.
Celtis reticulata.
Morus microphylla[P]
Adolphia infesta.

Mimosa biuncifera.[P]
Cercocarpus parvifolius.
Garrya lindheimeri.[P]
Garrya wrightii.
Philadelphus microphyllus.
Schmaltzia trilobata.
Arbutus xalapensis.
Cercis occidentalis. [P]
Fallugia paradoxa.
Agave wislizeni. [P]
Agave applanata.
Yucca baccata.
Nolina microcarpa.

The two long strips of Upper Sonoran zone lying east and west of the Pecos Valley at the present time are largely devoted to grazing, to which they are peculiarly adapted, but the time will come when they will be in part reclaimed for agriculture or horticulture, and the advantage of their position in adaptation to crops not grown in surrounding Lower Sonoran zone will be recognized. While grazing will long continue to be the chief industry, the introduction of successful crops will be of the greatest advantage to the stockmen. Most of the region is semiarid, with only sufficient rainfall for a good stand of native grasses, but by intelligent methods of handling the soil, deep plowing, dust mulch, and a system of cross-furrowing to utilize all the water that falls on sloping areas many kinds of fruits and other crops will thrive without irrigation. Where irrigation is possible, however, as it is in many places along streams or by means of water storage, artesian wells, or pumping, the returns will, of course, be far more certain and abundant.

For lists of cereals, fruits, nuts, and miscellaneous crops adapted to Upper Sonoran zone see Life Zones and Crop Zones, pages 37–40. This is preeminently the zone of standard varieties of apples and of many other fruits and grains. It is the only zone of any extent in Texas adapted to the sugar beet for the manufacture of sugar.

TRANSITION ZONE.

The Transition is the most restricted and broken of any zone within the State, being confined to the Chisos, Davis, and Guadalupe mountains from about 6,000 feet on northeast slopes to the tops of the ranges, while between these mountains it is divided by wide strips of Upper Sonoran zone. It is well marked in each of these ranges by its most characteristic tree—the yellow pine (*Pinus ponderosa*)—but in each is characterized by a different combination of plants and animals. Although the home of the wild potato (*Solanum t. boreale*), it is too rough for extensive agriculture and is important mainly for its timber and for the hidden sources of streams that break out at lower levels.

In the Guadalupe Mountains the Transition zone mammals are:

Odocoileus (hemionus?).	*Thomomys fulvus.*[M]
Eutamias cinereicollis canipes.[M]	*Lepus pinetis robustus.* [M]
Neotoma mexicana.	*Ursus (americanus?).*
Microtus mexicanus guadalupensis.	

In the Davis Mountains:

Odocoileus sp.——?	*Lepus pinetis robustus.*[M]
Neotoma mexicana.	*Ursus americanus ambliceps.*
Thomomys fulvus texensis.[M]	*Ursus horribilis horriaeus.*[M]
Erethizon sp.——?	*Vespertilio fuscus.*[M]

In the Chisos Mountains:

Odocoileus couesi.[M]	*Lepus pinetis robustus.* [M]
Neotoma (mexicana?).	*Ursus a. ambliceps.*
Sigmodon ochrognathus.	*Vespertilio fuscus.*[M]

The lists of breeding birds of the Transition zone show little variation in the Guadalupe, Davis, and Chisos mountains, and more thorough collecting high up in these ranges would doubtless show a still closer similarity of species.

The following species are common to all three ranges:

Cyrtonyx montezumae mearnsi.[B]   *Pipilo maculatus megalonyx.*
Columba fasciata.[B]   *Vireo solitarius plumbeus.*[B]
Melanerpes formicivorus.   *Piranga hepatica.*[B]
Trochilus alexandri.[B]   *Sitta carolinensis nelsoni.*
Selasphorus platycercus.

The following species occur and probably breed in the Guadalupe Mountain Transition:

Meleagris gallopavo merriami.   *Helminthophila celata orestera.*[B]
Syrnium occidentale.[B]   *Dendroica auduboni.*[B]
Megascops flammeolus.[B]   *Dendroica graciae.*[B]
Dryobates villosus hyloscopus.[B]   *Sitta pygmaea.*
Selasphorus rufus.   *Hylocichla guttata auduboni?*[B]
Aimophila ruficeps scotti.   *Merula migratoria propinqua.*[B]
Junco dorsalis.[B]

The following occur in both the Guadalupe and Davis Mountain Transition:

Colaptes cafer collaris.[B]   *Parus gambeli.*[B]
Cyanocitta stelleri diademata.[B]   *Sialia mexicana bairdi.*
Piranga ludoviciana.

The following species occur in both the Guadalupe and Chisos mountains, but were not observed in the Davis Mountains:

Antrostomus macromystax.[B]   *Empidonax difficilis.*
Nuttallornis borealis.[B]   *Wilsonia pusilla pileolata.*[B]
Contopus richardsoni.[B]

The following were found in the Chisos Mountains only:

Aphelocoma sieberi couchi.[B]   *Oreospiza chlorura.*[B]
Loxia curvirostra stricklandi.   *Vireo huttoni stephensi.*

In the Davis Mountains a single *Asyndesmas torquatus*[B] was seen. Most of these birds are Rocky Mountain forms, some of which reach their southern breeding limits in one of these groups of mountains; others range south into Mexico along this chain of Transition zone islands, while still others are more particularly southern and western forms that in one or more of the ranges reach approximately their northeastern limits.

The Transition zone plants of the Guadalupe Mountains are:

Pinus ponderosa.   *Prunus (serotina?).*
Pinus flexilis.[P]   *Amelanchier alnifolia.*
Pseudotsuga mucronata.[P]   *Rhamnus purshiana.*
Acer grandidentatum.   *Ceanothus greggii.*
Ostrya baileyi.[P]   *Robinia neomexicana.*
Quercus acuminata.[P]   *Berberis repens.*[P]
Quercus novomexicana.[P]   *Symphoricarpos (longiflorus?).*
Quercus grisea (dwarf form).   *Solanum tuberosum boreale.*
Quercus fendleri.   *Linum perenne.*
Quercus undulata.[P]   *Oxalis violacea* or var.[P]

Of the Davis Mountains:

Pinus ponderosa.	*Quercus emoryi* (dwarf form).
Pinus flexilis.[P]	*Prunus serotina acutifolia.*
Acer grandidentatum.	*Rhamnus purshiana.*
Quercus leucophylla.	*Symphoricarpos (longiflorus?).*
Quercus novomexicana.[P]	*Solanum tuberosum boreale.*
Quercus grisea (dwarf form).	

Of the Chisos Mountains:

Pinus ponderosa.	*Quercus emoryi.*
Pseudotsuga mucronata.[P]	*Quercus texana.*
Cupressus arizonica.[P]	*Prunus s. acutifolia.*
Acer grandidentum.[P]	*Rhamnus purshiana.*
Quercus grisea.	*Symphoricarpos (longiflorus?).*

The Transition area is so restricted in Texas as to be of comparatively little agricultural importance, especially as it lies within the arid section of the zone. Its greatest value, aside from its native timber, will be found in its adaptation to the culture of northern fruits, varieties of apples, pears, cherries, plums, grapes, and berries that will never prove successful elsewhere in the State.

For list of fruits that have been grown successfully in arid Transition zone in western Montana, eastern Washington and Oregon, and in parts of Idaho and Utah, see Life Zones and Crop Zones, pages 25–27.

CANADIAN ZONE.

In the Davis Mountains a thicket of *Populus tremuloides* along the northeast base of a high cliff near Livermore Peak indicates a mere trace of Canadian zone, while a single specimen of the hoary bat, shot as it came down one of the gulches near the northeast base of Livermore Peak, July 10, strongly suggests that this Canadian zone species was on its breeding ground. *Nuttallornis borealis*[B] and *Loxia curvirostra stricklandi* seen in the Chisos Mountains in June were probably breeding there, while *Nuttallornis borealis*[B] and immature *Junco dorsalis,* observed in the Guadalupe Mountains in August, may or may not have been migrants.

REPORT ON THE BIOLOGICAL SURVEY COLLECTION OF LIZARDS AND SNAKES FROM TEXAS.

The Biological Survey collection of reptiles in the United States National Museum contains 353 specimens from Texas, including 102 specimens and 31 species of snakes from 81 localities, 252 specimens and 32 species of lizards from 167 localities. No attempt has been made to identify or include the turtles and batrachians. The material has been gathered by the field assistants of the Survey as opportunity offered in connection with other work, but none of the collectors has made a specialty of reptiles. Until a systematic field study of these groups is taken up, we can not expect to know much of the distribution and habits of the species; but so much that is vague, erroneous, and misleading has been published, especially in regard to the Texas region, that it seems doubly important to put on record all definite localities from which specimens have been positively identified. Every specimen in the present collection is fully labeled with exact locality, including altitude in the case of most specimens from the mountains, date, name of collector, and in many cases notes on habitat. Only a knowledge of the country is requisite to enable each specimen to be referred to its proper zone. Of some species there are specimens from enough localities to determine with considerable accuracy their range and zonal position in the State, but of many others the few records will be useful mainly in future works of broader scope.

But few of the collectors' field notes on reptiles, when unaccompanied by specimens, have been made use of owing to the danger of confusing closely related species, and when such notes are used they are carefully distinguished. In a few cases the records of specimens in the United States National Museum collection from localities of peculiar importance are given separately. References to published records usually are avoided.

To Dr. Leonhard Stejneger I am greatly indebted for identification of the specimens. With the aid of his assistant, Mr. Richard G. Paine, I was able to simplify his task in many cases by making preliminary determinations.[a]

Anolis carolinensis Cuvier.   Carolina Chameleon.[R]

6   This little chameleon-like lizard is represented in the collection by specimens from Waskom, Joaquin, Sour Lake, and Columbia, and I have seen it in abundance at Jefferson and Timpson, but never in the western half of Texas.

Crotaphytus collaris (Say).   Ring-necked Lizard.[R]

7   Specimens of this beautiful lizard from Wichita Falls, Henrietta, Miami, Gail, Castle Mountains, Fort Lancaster, Fredericksburg, and Rock Springs, Tex., and from Roswell and Santa Rosa, N. Mex., carry the range of the species over the middle plains region of Texas, the region lying between the Pecos River and the eastern timbered

---

[a]Additional specimens identified by Mr. Paine and myself during Doctor Stejneger's absence are as follows: From Sour Lake, *Crotalus horridus, Callopeltis obsoletus, Agkistrodon contortrix, Storeria dekayi, Sceloporus consobrinus, Anolis carolinensis;* from Hempstead, *Tropidoclonium lineatum;* from Seguin, *Eutainia elegans marciana* and *Sceloporus spinosus floridanus;* from Washburn, *Heterodon nasicus* and *Liopeltis vernalis,* and from Cleveland, *Elaps fulvius.*

[6] *Anolis carolinensis* is now referred to as the northern green anole. This species is monotypic in Texas and is common throughout the eastern half of the state.

[7] The eastern collared lizard, *Crotaphytus collaris,* occurs throughout the northern and western portions of the state where it appears to be common.

[8] Taxonomists now recognize two subspecies of collared lizard in Texas with specimens of *Crotaphytus collaris baileyi* being placed in synonymy of *Crotaphytus collaris collaris*, which occupies the western two-thirds of the state (Ingram and Tanner, 1971). The other subspecies, *Crotaphytus collaris fuscus*, is confined to El Paso County.

[9] The common name for *Crotaphytus reticulatus* is the reticulate collared lizard. This monotypic species remains rare and occurs only in a few southern counties of the state. Some populations appear to be threatened and will require careful monitoring.

[10] Current taxonomy assigns specimens of *Crotaphytus wislizenii* to *Gambelia wislizenii*, the long-nosed leopard lizard (Smith, 1946). Specimens of this monotypic species are rare and have been documented from only a few counties in the Trans-Pecos region.

[11] Current taxonomy regards *Holbrookia texana* as a junior synonym of *Cophosaurus texanus*, the greater earless lizard. Two subspecies are recognized in Texas, *Cophosaurus texanus texanus* across the central and southern parts of the state and *C. t. scitulus* in the Trans-Pecos region.

country. The above localities lie near the junction of Upper and Lower Sonoran zones, but this lizard inhabits at least a part of both zones. Fourteen out of the 15 specimens referred to *collaris* have the single row of interorbital plates. One specimen from Miami has two full rows of interorbitals, but with the large supraoculars and blunt nose of *collaris*.

Crotaphytus collaris baileyi[R] Stejneger.

8 Nine specimens of *Crotaphytus* from eight localities are referred by Doctor Stejneger to *baileyi*. They are from Comstock, Alpine, Paisano, Chisos Mountains (west base), Davis Mountains (east base), Toyah, 70 miles north of Toyah, Tex., and one from the east base of Guadalupe Mountains, west of Carlsbad, N. Mex. These localities are near the junction of Upper and Lower Sonoran zones. Five of the nine specimens from Comstock, Paisano, and Chisos Mountains and two from the Davis Mountains are typical *baileyi*, with two full rows of interocular plates, small supraoculars, and relatively narrow muzzle. The specimen from Alpine has the interoculars joined in a single row, but otherwise possesses the characters of *baileyi*. The specimen from the Guadalupe Mountains and the two from Toyah and 70 miles north of Toyah have the interoculars joined in a single row and other characters intermediate between *baileyi* and *collaris*. Considering the close relationship and evident intergradation of the two forms, it seems best to follow Witmer Stone in placing *baileyi* as a subspecies of *collaris*.[a]

Crotaphytus reticulatus Baird.

9 Lloyd collected a specimen of this rare and apparently very locally distributed lizard at Rio Grande City, Tex., May 28, 1891.

Crotaphytus wislizenii[R] Baird & Girard. Leopard Lizard.[R]

10 A fine, large individual of this big, spotted lizard was shot near Boquillas, in the Great Bend of the Rio Grande, by McClure Surber, and Cary and Hollister each collected a specimen near Toyahvale, in the Pecos Valley. The species is not common and occurs only in the low, hot valleys of extreme arid Lower Sonoran zone.

Holbrookia texana[R] Troschel.

11 This most brilliantly colored of the Texas lizards is represented by 11 specimens from the following nine localities: Fort Stockton, Adams, Toyahvale, Pecos River (5 miles west of Sheffield), Davis Mountains (east base), Boquillas, McKinney Spring (60 miles south of Marathon), Comstock, and Benbrook. It is a common and conspicuous species over all the hot, bare Lower Sonoran desert of western Texas and as far up the Pecos Valley as Santa Rosa, N. Mex.

[a] Proc. Acad. Nat. Sci. Phila., 1903, 30.

12   This species is so similar in both general appearance and habits to *Callisaurus draconoides,* the 'gridiron-tailed lizard'[R] of the Death Valley country, that I never noticed the difference between them until Doctor Merriam pointed it out in a beautiful colored study of *Holbrookia texana*[R] made by Fuertes in western Texas. Few animals possess more wonderful protective markings than these bar-tailed lizards. As they dash away, well up on their legs, with tail curled over the back, exposing their brightly colored sides and the black and white barred lower surface of the tail, they are strikingly conspicuous, until, stopping suddenly, they flatten themselves on the ground, when the speckled back blends into the earth colors and the lizards are lost to view.

Holbrookia propinqua B. & G.   Long-tailed Holbrookia.[R]

13   There are 15 specimens of this slender-tailed Holbrookia from five localities in southern Texas: Brownsville, Sauz Ranch, Santa Rosa Ranch and Padre Island in Cameron County, and King's Ranch in Nueces County.

Holbrookia maculata Girard.   Spotted-sided Holbrookia.[R]

14   The collection contains 19 specimens of this little, short-tailed lizard from the following eight localities in Texas: Mouth of Devils River, Fort Stockton, Fort Davis, Alpine, Paisano, Dimmitt (45 miles south), Henrietta, and Amarillo, places lying some in Upper and some in Lower Sonoran zones of the arid and the semiarid regions. Apparently the species has not been taken south of Devils River and San Antonio.

Holbrookia maculata lacerata[R] Cope.   Spotted-tailed Holbrookia.[R]

15   One specimen of the spotted-tailed lizard from Cotulla, two from 15 miles west of Japonica, and one from 25 miles southwest of Sherwood considerably extend the southern and western range of the species. Apparently it belongs to Lower Sonoran zone.

Uta stansburiana B. & G.

16   Specimens of this little lizard from El Paso, Pecos City, and Fort Stockton carry its range across the extremely arid part of Lower Sonoran zone in western Texas and mark the eastern limit of a widely distributed species.

Uta ornata[R] B. & G.

17   Twelve specimens of this little lizard are from the following localities in western Texas: Mouth of Pecos, Langtry, Ingram, Chisos Mountains, Altuda, Paisano, and Fort Davis. The Chisos Mountain specimen was taken at 6,000 feet, and the four Fort Davis specimens at approximately 5,700 feet, in the midst of Upper Sonoran zone. Altuda, Paisano, and Ingram are at the lower edge of the zone, while

[12] *Callisaurus draconoides* is now referred to as the western zebra-tailed lizard, and this species does not occur in Texas.

[13] *Holbrookia propinqua* is now referred to as the keeled earless lizard. The distribution of this monotypic species is restricted to the southern part of the state.

[14] *Holbrookia maculata* is now called the common lesser earless lizard. Three subspecies are now recognized in Texas: *Holbrookia maculata approximans* in the Trans-Pecos region, *Holbrookia maculata maculata* in northwestern Texas, and *Holbrookia maculata perspiculata* in portions of central Texas. This species appears to be common throughout its range.

[15] Taxonomists (Smith and Taylor, 1950) have returned *Holbrookia maculata lacerata*, the spot-tailed earless lizard, to its original status as *Holbrookia lacerata*. Two subspecies are recognized in Texas, *Holbrookia lacerata lacerata* in the Edwards Plateau area and *Holbrookia lacerata subcaudalis* in southern Texas (Axtell, 1956).

[16] The monotypic species *Uta stansburiana*, the common side-blotched earless lizard, has a spotty distribution across the western one-third of the state, where it appears to be common.

[17] Taxonomists (Mittleman, 1942) have returned *Uta ornata*, the ornate tree lizard, to *Urosaurus ornatus* as described by Baird and Girard (1852). Two subspecies are recognized in Texas, *Urosaurus ornatus ornatus* in central Texas and along the southern portions of the Rio Grande and *Urosaurus ornatus schmidti* in the Trans-Pecos region. Both are common throughout their range.

[18] Taxonomists (Smith, 1936) now recognize *Sceloporus torquatus poinsettii* as *Sceloporus poinsettii poinsettia*, the crevice spiny lizard. This monotypic species occurs in the west-central portions of the state and populations appear to be stable.

[19] Texas specimens formerly assigned to *Sceloporus clarkii* by Bailey (1905) are now regarded as *Sceloporus bimaculosus*, the twin-spotted spiny lizard, which occurs only in the Trans-Pecos region. The species is monotypic and apparently rare in the state.

[20] Taxonomists (Smith, 1934) now regard *Sceloporus spinosus floridanus* to be *Sceloporus olivaceus*, the Texas spiny lizard, which occurs throughout the state except for the extreme western areas.

[21] *Sceloporus consobrinus*, the prairie lizard, occurs throughout Texas except for the extreme western areas, and it appears to be common throughout its range.

[22] *Sceloporus dispar* is now considered to be *Sceloporus grammicus microlepidotus*, the graphic spiny lizard (Sites and Dixon, 1981). It is found in only a few counties in southern Texas where it appears to be rare.

[23] *Sceloporus merriami*, the canyon lizard, includes three subspecies from Texas (*Sceloporus merriami annulatus, Sceloporus merriami longipunctatus*, and *Sceloporus merriami merriami*), all with extremely small distributions in the Trans-Pecos region. This species appears to be rare in the state.

Langtry and mouth of Pecos, as well as the type locality of the species, Devils River, are just below the edge in Lower Sonoran. However, enough Upper Sonoran species of plants cling to the cold walls of side canyons in the Langtry, Pecos, and Devils River country to account for the presence of such rock-dwelling species, and I am inclined to consider the range of this lizard as strictly Upper Sonoran, at least in Texas. If all the southern Arizona and California records of the species are correct, it is certainly Lower Sonoran in that region.

Sceloporus torquatus poinsettii [R] (B. & G.).

18   This splendid, big, scaly rock lizard is represented in the collection by 14 specimens from the following localities in western Texas: Japonica, East Painted Cave, Marathon (50 miles south), Chisos Mountains (6,000 feet), Paisano, Davis Mountains (5,700 feet), Fort Stockton, Castle Mountains, near Toyah, and Guadalupe Mountains (south end of Dog Canyon, at about 6,700 feet). The species ranges throughout the width of Upper Sonoran zone, but in many places comes well into Lower Sonoran, as at Toyah and Fort Stockton.

Sceloporus clarkii [R] B. & G.

19   The collection contains but 4 Texas specimens of this big scaly lizard—3 from Boquillas and 1 from Langtry. Both localities are on the Rio Grande, in extremely arid Lower Sonoran zone.

Sceloporus spinosus floridanus [R] Baird).

20   Eighteen specimens of this medium-sized *Sceloporus* from southern Texas come from the following localities: Seguin, Ingram, Brownsville, Rio Grande City, Lomita Ranch (6 miles north of Hidalgo), Devils River, Langtry, and Pecos River (50 miles from mouth), which carry its range across the State in Lower Sonoran zone.

Sceloporus consobrinus B. & G.

21   Six localities in Texas are represented by the 8 specimens of this medium-sized *Sceloporus:* Joaquin, Sour Lake, Kerrville (Lacey's Ranch), Santa Rosa (Cameron County), Langtry, and Fort Davis. The range of the species apparently covers nearly the whole State where there are trees, in both Upper and Lower Sonoran zones.

Sceloporus dispar [R] B. & G.

22   Lloyd collected 5 specimens of this little slender *Sceloporus* at Lomita Ranch, 6 miles north of Hidalgo, in June, 1891.

Sceloporus merriami Stejneger.

23   This beautiful little *Sceloporus*, which Doctor Stejneger has just described as new,[a] is represented by 5 specimens, 2 from the East

---

[a] Proc. Biol. Soc. Washington, XVIII, p. 17, Feb. 5, 1904.

Painted Cave, near the Rio Grande, a mile below the mouth of the Pecos; 1 from Comstock; 1 from the Pecos River Canyon, 55 miles northwest of Comstock, and 1 from Boquillas, in the Great Bend of the Rio Grande. Apparently the species is confined to the rocky walls of the Canyons of the Rio Grande and Pecos rivers.

Phrynosoma cornutum (Harlan).   Horned Toad.[R]

24   This commonest and longest-horned species of the Texas horned toads is represented by specimens from El Paso, Grand Canyon of Rio Grande, Alpine, Altuda, Valentine, Davis Mountains (east base), Toyahvale (20 miles southeast), Fort Stockton, Fort Lancaster, Painted Caves, Carrizo, Roma, Rio Grande City, Sauz Ranch (Cameron County), King's Ranch (Nueces County), Corpus Christi, Center Point, Llano, Dimmitt, Henrietta, and Tascosa, and from Antioch and Virginia Point in eastern Texas, a series of localities covering at least the whole arid Lower Sonoran zone of the State and extending irregularly into the humid eastern division. Considerable variation however, appears within this range. The three specimens from Virginia Point and one from Antioch are much darker than any of the others, with sharply marked face bars and gray, thickly spotted bellies, while those of the upper Rio Grande Valley, El Paso, and Rio Grande Canyon are the lightest and brightest colored of all, with narrow face bars and white bellies. The transition from these extremes is gradual across the State, but the dark individuals from the eastern localities, the gray ones from the grassy plains, the strongly marked ones from the Pecos Valley, and the paler specimens from the Rio Grande Valley, indicate color forms comparable with the subspecies of horned larks found breeding in regions of corresponding differentiation.

Phrynosoma hernandesi (Girard).

25   This rusty-brown horned toad with short, stubby horns is represented in the collection by a single specimen from Texas, collected at about 7,000 feet altitude in the southern end of the Guadalupe Mountains. Farther north it is abundant in Upper Sonoran and Transition zones.

Phrynosoma modestum Girard.

26   *Anota modesta* of Cope and others.

Specimens of this little gray, short-horned horned toad from the west base of the Davis Mountains, 40 miles south of Alpine, 20 miles northwest of Toyah, Salt Valley, at west base of Guadalupe Mountains, and Big Springs carry the range of the species well over the arid region of western Texas in a series of localities where Upper and Lower Sonoran species are more or less mixed. Apparently the species belongs to Lower Sonoran zone and extends to its extreme upper limit.

[24] The Texas horned lizard, *Phrynosoma cornutum*, occurs nearly statewide with the exception of the extreme regions of the state. Populations are declining and biologists with TPWD are making efforts to conserve all three species that occur in the state (see Annotations 25 and 26, below, and Chapter 6).

[25] *Phrynosoma hernandesi*, the greater short-horned lizard, is known from only four counties of the Trans-Pecos region.

[26] *Phrynosoma modestum,* the round-tailed horned lizard, occurs throughout much of the western portions and a few counties in the southern region of the state.

[27] *Coleonyx brevis*, the Texas banded gecko, occurs in the Trans-Pecos and southern regions of the state where it appears to have stable populations.

[28] Taxonomists refer Texas populations of *Ophisaurus ventralis* to *Ophisaurus attenuates*, the slender glass snake (McConkey, 1954). This species occurs throughout the eastern regions of the state and in a few isolated counties of the Panhandle where its populations appear to be stable.

[29] Taxonomists consider *Gerrhonotus liocephalus infernalis* to be *Gerrhonotus infernalis*, the Texas alligator lizard (Good, 1994). This species has a restricted distribution in the central and Trans-Pecos portions of the state and because of declining populations should be monitored in the future.

[30] Although some authorities are now using *Aspidoscelis* (Reeder et al., 2002), we have followed Dixon (2013) in continuing to use the generic designation *Cnemidophorus* for whip-tailed lizards. *Cnemidophorus tessellatus*, the common checkered whiptail, is common with a restricted distribution in the Panhandle and Trans-Pecos regions.

[31] Taxonomists recognize *Cnemidophorus perplexus* as *Cnemidophorus neomexicanus*, the New Mexico whiptail (Good, 1994). See Annotation 30 regarding our use of the generic name *Cnemidophorus*. This species is found in only four counties in the Trans-Pecos region and populations should be carefully monitored.

A specimen from the west base of the Davis Mountains is rusty brown instead of ashy gray, like those from other localities, a peculiarity of coloration agreeing with the brown *Crotalus lepidus* from this lava soil region.

Coleonyx brevis Stejneger.　Gecko.[R]

27　A single specimen of this odd little brown and yellow lizard was collected by Merritt Cary at Sheffield, August 9, 1902.

Ophisaurus ventralis[R] Linn.　Glass Snake.[R]

28　The glass snake is represented by two specimens from Texas, one collected by Lloyd near Santa Rosa, Cameron County, and the other by H. P. Attwater and W. H. Rawson, 3 miles north of Kerrville. These localities help to fix a western limit for this legless lizard of the Lower Sonoran zone of the Southeastern States.

Gerrhonotus liocephalus infernalis[R] (Baird).

29　A single specimen of this glassy smooth lizard was collected in the Chisos Mountains at approximately 6,000 feet altitude in Upper Sonoran zone. It was nosing about in the dry leaves under scrub oaks on the mountain side in the manner peculiar to the individuals of the genus.

Cnemidophorus tessellatus[R] (Say).　Whip-tailed Lizard.[R]

30　This species is represented by specimens from Boquillas, Langtry, Fort Stockton (35 and 45 miles west), Castle Mountains, Monahans, and Van Horn, all of which localities lie in the extremely arid Lower Sonoran zone of western Texas. They mark the eastern limit of the known range of the species.

Cnemidophorus perplexus[R] B. & G.

31　Two specimens of this species were collected by Cary, one at Pecos and one 4 miles west of Adams, in Pecos County. Both localities are in Lower Sonoran zone.

Cnemidophorus sexlineatus[R] (Linn).

32　The collection contains specimens of this eastern species of the whip-tailed lizard from Waskom, Long Lake, Nacogdoches, Henrietta, Canadian, and Padre Island.

Cnemidophorus gularis[R] B. & G.

33　This seems to be the commonest and most widely distributed of the whip-tailed lizards in Texas. There are specimens from Brownsville, Lomita Ranch (Hidalgo County), Rio Grande City, Roma, Carrizo, Cotulla, San Diego, near Alice, Corpus Christi, Cuero, Kerrville, Fort Lancaster, Fort Stockton, Devils River (near mouth), Comstock, Painted Caves, Paisano, and Marfa, and up the Pecos Valley in New Mexico to Santa Rosa and Ribera. Among these localities Ribera is the only one fairly out of Lower Sonoran zone.

---

[32] Three subspecies of *Cnemidophorus sexlineatus*, the six-lined racerunner, are now recognized in Texas: *Cnemidophorus sexlineatus sexlineatus* in the eastern regions, *Cnemidophorus sexlineatus stephensae* in extreme southern portions of Texas, and *Cnemidophorus sexlineatus virdis* in the northwestern parts of the state. See Annotation 30 regarding our use of the generic name *Cnemidophorus*. The species appears to be common throughout its range.

[33] *Cnemidophorus gularis*, the common spotted whiptail, is common throughout most of the state. See Annotation 30 regarding our use of the generic name *Cnemidophorus*.

Leiolopisma laterale[R](Say).

34 Specimens of this little slender Eumeces-like lizard from Waskom, in northeastern Texas, and from Velasco, near Columbia and near the mouth of Navidad River in southeastern Texas, do not extend the range of the species, which apparently covers all the eastern part of the State.

Eumeces quinquelineatus[R](Linn).

35 Hollister obtained a specimen of this species at Joaquin, near the eastern boundary of Texas.

Eumeces guttulatus[R](Hallowell). Skink[R]

36 Cary collected a specimen of this little skink at the east base of the Davis Mountains, 20 miles southwest of Toyahvale, at about 5,000 feet, and I took one at 6,800 feet in the south end of the Guadalupe Mountains, well toward the upper edge of Upper Sonoran zone.

Eumeces obsoletus[R](B. & G.). Skink[R]

37 A specimen of this larger skink was collected in the southern end of the Guadalupe Mountains, at 6,800 feet, in the same locality with the little *guttulatus*.

Eumeces brevilineatus[R]Cope.

38 Lloyd collected a single specimen at Paisano "in a damp fernery at 5,300 feet." Cope records it from Helotes Creek (the type locality), on the front line of hills 20 miles northwest of San Antonio, and from near Fort Concho (across the river from San Angelo).[a] While these localities lie at the upper edge of Lower Sonoran zone the species, if an inhabitant of damp gulches, may well belong to Upper Sonoran.

Diadophis regalis[R]B. & G. Ring Snake.[R]

39 A specimen of this little spotted-bellied snake was collected in the Chisos Mountains at 5,000 feet, June 3, 1901. The two previously recorded localities for Texas—Fort Davis and Eagle Springs[b]—are both close to 5,000 feet and, like the Chisos Mountain locality, at the edge of Upper and Lower Sonoran zones.

Heterodon nasicus[R]B. & G. Hog-nosed Snake.[R]

40 Specimens of the hog-nosed snake from Cameron County (El Haboncillo), Sycamore Creek, North Llano River, Washburn, and Amarillo carry its range over the whole north and south length of the State through Lower and Upper Sonoran zones in the semiarid region. It is not a common or conspicuous species, and with the exception of the specimens collected, I have not seen it in Texas.

[a]Ann. Rep. U. S. Nat. Mus. for 1898 (1900), p. 665.
[b]Ibid., p. 745

[40] Specimens formerly assigned to *Heterodon nasicus* are now composed of two species, *H. nasicus*, the plains hog-nosed snake, and *Heterodon kennerlyi*, the Mexican hog-nosed snake (Smith et al., 2003). *Heterodon nasicus* is common throughout much of the northern two-thirds of the state, and *Heterodon kennerlyi* is restricted to the Rio Grande region where it has been suggested that populations should be monitored.

[34] Taxonomists refer *Leiolopisma laterale* to *Scincella lateralis*, the little brown skink (Mittleman, 1950). The species is common throughout the eastern two-thirds of the state.

[35] Most taxonomists (Taylor, 1936) refer *Eumeces quinquelineatus* to *Eumeces fasciatus*, although a few researchers (Brandley et al., 2005) have arranged *Eumeces* in the genus *Plestiodon*. However, we have followed Dixon (2013) in continuing to use *Eumeces* for these skinks. *Eumeces fasciatus*, the common five-lined skink, is found throughout the eastern one-third of the state where it appears to be common.

[36] Taxonomists now assign *Eumeces guttulatus* to *Eumeces obsoletus* (Taylor, 1936). See Annotation 35 regarding our use of the generic name *Eumeces*.

[37] *Eumeces obsoletus*, the Great Plains skink, is common throughout the western two-thirds of the state. See Annotation 35 regarding our use of the generic name *Eumeces*.

[38] Two subspecies of *Eumeces tetragrammus*, the four-lined skink, are now recognized in Texas, *Eumeces tetragrammus brevilineatus* in the central and Trans-Pecos regions and *Eumeces tetragrammus tetragrammus* in the exteme southern part of the state. See Annotation 35 regarding our use of the generic name *Eumeces*.

[39] *Diadophis regalis* is now regarded a subspecies of *Diadophis punctatus*, the ring-necked snake (Mecham, 1956). Three subspecies are recognized in Texas: *Diadophis punctatus arnyi* in the central and northwestern regions, *Diadophis punctatus regalis* in the extreme western and Trans-Pecos areas, and *Diadophis punctatus strictogenys* in the eastern part of the state.

[41] *Heterodon platirhinos*, the eastern hog-nosed snake, occurs throughout the eastern half of the state and portions of the Panhandle where it appears to be common.

[42] *Opheodrys aestivus*, the rough green snake, occurs throughout the eastern half of the state where it appears to be common.

[43] *Liopeltis vernalis*, the smooth green snake, is now placed in the genus *Opheodrys* (Oldham and Smith, 1991). *Opheodrys vernalis* occurs only in a few counties in the coastal marsh areas, and it has not been observed for more than 30 years.

[44] The genus *Bascanion* is now referred to as *Masticophis*, although some taxonomists (e.g., Nagy et al., 2004) prefer the generic designation *Coluber*. However, we have followed Dixon (2013) in retaining *Masticophis*. Two subspecies of *Masticophis flagellum*, the coachwhip, are recognized in Texas, *Masticophis flagellum flagellum* in the eastern one-fourth and *M. f. testaceus* occupying the remainder of the state. The species is common throughout its range.

[45] *Masticophis taenatus*, the striped whipsnake, occurs only in the Edwards Plateau and Trans-Pecos regions where it appears to be common. See Annotation 44 regarding our use of the generic name *Masticophis*.

[46] *Drymobius margaritiferus*, the speckled racer, occurs only in a few counties in extreme southern Texas where monitoring has been recommended.

Heterodon platirhinos Latreille. Blow Snake[R].

41 A single immature specimen collected by Lloyd, at Matagorda, does not add much to our knowledge of the range of the species, which apparently covers eastern Texas and reaches west to the Pecos and Devils rivers. All the Texas records are from Lower Sonoran zone, but farther east the species also ranges across at least Upper Sonoran.

Opheodrys aestivus (Linn.). Rough Green Snake[R].

42 The little green snake is represented by 3 specimens from Corpus Christi, Kerrville, and Rock Springs. From a range over the whole eastern or humid division of Lower Sonoran zone the species apparently reaches in Texas its western limit. It does not enter the plains region nor the arid region except where brushy gulches enable it to cross the rough country south of the lower arm of the Staked Plains. The specimens from near Kerrville and Rock Springs were taken in gulches with such vegetation as pecan, sycamore, elm, black cherry, oak, and abundant underbrush.

Liopeltis vernalis De Kay. Smooth Green Snake.

43 A specimen of the little smooth green snake was collected at Washburn by Gaut in July, 1904.

Bascanion flagellum[R]Shaw. Whip Snake[R].

44 The coach whip or whip snake is represented in the collection by specimens from Matagorda, Matagorda Peninsula, Padre Island, and the Chisos Mountains, and by a flat skin from Devils River. While apparently distributed over the whole State, the species is most abundant, or at least most frequently seen, in the half brushy, half open arid region, where it is one of the commonest snakes.

Bascanion ornatum[R]B. & G.

45 Two specimens of this rare species were collected in Texas, one at the head of Devils River, by Cary, the other on the Rio Grande, 8 miles south of Comstock, by Hollister, both in July, 1902. Hollister wrote on the back of the label of his specimen, "Up in bushes."

Cope records the species from two localities only, "Western Texas" and "Howard Springs, Texas."[a]

Drymobius margaritiferus (Schlegel).

46 Lloyd collected a single specimen of this species at Brownsville on July 17, 1891.

Callopeltis obsoletus[R](Say). Pilot Snake; Mountain Black Snake[R].

47 A specimen was collected by Lloyd, near the mouth of the Nueces, November 21, 1891, and another by Gaut, near Sour Lake, April 1, 1905.

<hr>

[a] Ann. Rep. U. S. Nat. Mus. for 1898 (1900), p. 814.

---

[47] The genus *Callopeltis* has been placed in the genus *Elaphe* by Burt (1935), although some taxonomists (Utiger et al., 2002) now consider *Elaphe* to be synonymous with *Pantherophis*. However, we have followed Dixon (2013) in retaining *Elaphe*. *Elaphe obsolete*, the Texas ratsnake, occurs throughout most of the eastern two-thirds of the state where it appears to be common.

Drymarchon corais melanurus[R] Dum. & Bibr.

48　Lloyd collected a specimen of this Mexican black snake at Brownsville, July 6, 1891.

Pituophis sayi[R] (Schlegel).　Prairie Bull Snake.[R]

49　The prairie bull snake is represented in the collection by 3 specimens from Gail, Comstock (20 miles north), and Paisano, and by a flat skin from the head of Dog Canyon, in the southern Guadalupe Mountains. It is common over at least middle and northern Texas in Lower and Upper Sonoran zones. In a prairie-dog town near Gail I killed an unusually large individual, measuring 7 feet 8 inches in length. It was about 3 inches in diameter—large enough to have readily swallowed a full-grown prairie dog. Near Rock Springs a smaller individual was found in the act of swallowing a freshly killed squirrel (*Citellus m. parvidens*).

Lampropeltis getula holbrooki Stejneger.　King Snake.[R]

50　Lloyd collected two specimens at Matagorda in January and February, 1892, and I collected one at Arthur in June of the same year.

Rhinocheilus lecontei B. & G.

51　A single specimen of this beautiful yellow and black ringed snake was collected about 30 miles west of Rock Springs, where a tongue of Lower Sonoran runs up into the Upper Sonoran plains. If the published records can be trusted, the species ranges over the whole arid Lower Sonoran zone of Texas.

Chionactis episcopus isozonus[R] Cope.

52　A single specimen of this little, bright-colored, pink and black ringed snake was collected in the Chisos Mountains at about 6,000 feet altitude, in Upper Sonoran zone.

Natrix fasciata transversa[R] (Hallowell).　Water Snake.[R]

53　Specimens from Lipscomb, the Nueces River (near mouth), mouth of Devils River, the Pecos River (55 miles northwest of Comstock), and from Carlsbad, N. Mex., together with previously published records, indicate for the species a range in most of the rivers of the western half of Texas, excepting the Rio Grande. Lipscomb is apparently the only locality where the species has been found outside of pure Lower Sonoran, and this is just at the edge of the zone.

Natrix clarkii[R] B. & G.).　Striped Water Snake.[R]

54　A specimen of the striped water snake[R] was collected by J. D. Mitchel at Carancahua Bay, Calhoun County, Tex., in January, 1892, and the National Museum collection contains 5 specimens from Indianola, including the type, and one specimen from Galveston.

[48] Wüster et al. (2001) recently elevated *Drymarchon corais melanurus*, the Central American indigo snake, to specific status (*Drymarchon melanurus*). The species occurs throughout most of the southern two-fourths of the state, and there is concern that it is declining and in need of monitoring.

[49] Taxonomists now consider the gopher snake, *Pituophis sayi*, to be a subspecies of *Pituophis catenifer* (Sweet and Parker, 1990). Two subspecies are recognized in Texas, *Pituophis catenifer affinis* in the Trans-Pecos region and *Pituophis catenifer sayi* in the central two-thirds of the state. The species appears to be common throughout its range.

[50] Two subspecies of *Lampropeltis getula*, the common kingsnake, are now recognized in Texas, *Lampropeltis getula holbrooki* in the eastern one-fourth and *Lampropeltis getula splendida* throughout the remainder of the state. The species may be declining due to commercial collecting.

[51] *Rhinocheilus lecontei*, the long-nosed snake, occurs throughout most of the western two-thirds of the state where it appears to be common.

[52] The subspecies mentioned by Bailey, *Chionactis episcopus isozonus*, is now regarded by taxonomists as *Sonora semiannulata semiannulata*, the western groundsnake (Parker, 1982). Two subspecies are recognized in Texas, *Sonora semiannulata semiannulata* in the southeast and *Sonora semiannulata taylori* in the central and western regions. The species appears to be common throughout its range.

[53] Current taxonomists assign *Natrix fasciata transversa* to *Nerodia erythrogaster transversa*, the plain-bellied watersnake (Conant and Collins, 1991). Two subspecies are recognized in Texas, *Nerodia erythrogaster flavigaster* in the Piney Woods and *Nerodia erythrogaster transversa* in the extreme western and southern regions. The species appears to be common throughout its range.

[54] Modern taxonomists (Conant and Collins, 1991) now refer *Natrix clarkii* to *Nerodia clarkii*, the salt-marsh watersnake. The species occurs in a few counties along the Coastal Marsh area where populations appear to be stable but in need of monitoring.

[55] Two subspecies of Dekay's brownsnake, *Storeria dekayi*, are recognized in Texas, *Storeria dekayi limnetes*, restricted to a few counties along the Coastal Marsh area, and *Storeria dekayi texana* throughout the eastern half of the state and a few counties in the Panhandle. The species appears to be common throughout its range.

[56] Taxonomists now refer *Eutainia elegans marciana* to *Thamnophis marcianus*. The subspecies *Thamnophis elegans marcianus* has now been removed from *T. elegans* and treated as a separate species (McLain, 1899; Ruthven, 1907), *Thamnophis marcianus*, the checkered gartersnake. This species is common statewide except for the Piney Woods.

[57] Current taxonomy refers *Eutainia proxima* to *Thamnophis proximus*, the western ribbonsnake (see Annotation 56). Four subspecies are recognized in Texas: *Thamnophis proximus diabolicus* in the Trans-Pecos, *Thamnophis proximus orarius* in the southeast, *Thamnophis proximus proximus* in the northeast, and *Thamnophis proximus rubrilineatus* in the central region. This species appears to be common throughout its range.

[58] *Eutainia cyrtopsis* is now referred to *Thamnophis cyrtopsis*, the black-necked gartersnake (see Annotation 56). Two subspecies are recognized, *Thamnophis cyrtopsis cyrtopsis* in the Trans-Pecos and *Thamnophis cyrtopsis ocellatus* throughout the Edwards Plateau region. The species appears to be common throughout its distribution.

The following letter from Mr. Mitchel accompanied the specimen and 8 of the well-developed embryos:

> This snake was captured with 3 others in a salt marsh on Carrancahua Bay, Calhoun County. It took to the salt water freely. One had 4 small mullets in its stomach, another some fiddler crabs. The other inclosure is part of the womb of one of the snakes, showing embryos. She had 4 on one side and 10 on the other, 14 in all.

Storeria dekayi Holbrook.

55 Lloyd collected a specimen of this little brown snake at Barnard Creek, west of Columbia, March 4, 1892, and Gaut collected one at Sour Lake, March 14, 1905.

Eutainia elegans marciana[R](B. & G.).   Garter Snake.[R]

56 There are specimens of this plain little striped snake from Brownsville, Santa Rosa Ranch (Cameron County), Corpus Christi, Victoria, Seguin, Sycamore Creek, Devils River, Paisano, and Boquillas. It is the common garter snake of the whole arid Lower Sonoran zone of western Texas, apparently reaching its eastern limit at Victoria.

Eutainia proxima[R](Say).   Spotted Garter Snake.[R]

57 The collection contains 12 specimens of this garter snake from Brownsville, Lomita Ranch (Hidalgo County), Sycamore Creek, Corpus Christi, and San Antonio River, near San Antonio. The species apparently has a wide range, including most of Texas.

Eutainia cyrtopsis Kennicott.

58 One specimen of this beautiful garter snake with black nuchal spots was taken in the Davis Mountains, July 12, 1901, at about 5,700 feet, in Upper Sonoran zone.

Tropidoclonium lineatum Hallowell.

59 Gaut collected a specimen of this little striped snake at Hempstead, February 28, 1905.

Tantilla gracilis B. & G.

60 I collected a specimen of this tiny brown snake at Lacey's Ranch, near Kerrville, May 5, 1899. The species has been recorded from various localities in eastern and southern Texas in Lower Sonoran zone.

Elaps fulvius (Linn.).   Coral Snake.[R]

61 Four specimens of the coral snake from Brownsville, Corpus Christi, Cleveland, and Kerrville nearly double the list of definite localities for the State. Cope records the species from Indianola, San Diego, Fort Clark, and Hempstead, but his records for Cameron County, Rio Grande, and Rio Pecos are too indefinite for practical purposes in outlining distribution. A live specimen sent to Doctor Stejneger from Beaumont, December 4, 1903, carries the range of the

[59] Three subspecies of the lined snake, *Tropidoclonium lineatum,* occur in Texas: *Tropidoclonium lineatum annectens* in the northeastern region, *Tropidoclonium lineatum lineatum* in the Panhandle, and *Tropidoclonium lineatum texanum* in the central region. The species appears to be common throughout its range.

[60] *Tantilla gracilis*, the flat-headed snake, occurs throughout much of the eastern two-thirds of the state where it appears to be common.

[61] Current taxonomy refers *Elaps fulvius* to *Micrurus tener*, the Texas coral snake (Collins, 1991). The species occurs throughout much of the eastern and southern portions of the state where it appears to be common.

species nearly across the State in both humid and arid Lower Sonoran zones, but the Cleveland and Beaumont specimens are much darker and richer in coloration than those from farther west.

Fortunately this is not a common species in Texas. I have found it but once in the State. From its beautiful colors and harmless appearance it is likely to be handled carelessly, and its bite is dangerous.

Agkistrodon piscivorus (Lacépède).   Cottonmouth; Water Moccasin.

62   A single specimen of the water moccasin, collected by William Lloyd at the mouth of Devils River, September 24, 1890, slightly extends the known range of the species. (Indianola and Eagle Pass are the westernmost records given by Cope.[a]) A specimen collected in the Big Thicket of Liberty County was lost, but I could not have been mistaken in its identity.

Agkistrodon contortrix (Linn.)   Copperhead.

63   A specimen of the copperhead collected by H. P. Attwater near San Antonio and three others collected near Kerrville, Arthur, and Sour Lake help to fill out the range of the species in the State. Cope records it from Cook County, Sabinal, and between Indianola and San Antonio.[b] All of these localities are in Lower Austral zone.

Sistrurus catenatus consors[R](B. & G.).   Massasauga.

64   There is a single specimen of the massasauga in the Biological Survey collection, from Santa Rosa, Cameron County, Tex., collected by Lloyd in 1891.

Crotalus horridus Linn.   Eastern Rattlesnake.[R]

65   Gaut collected a specimen of this rattlesnake in the Big Thicket, 8 miles northeast of Sour Lake, April 1, 1905.

Crotalus atrox B. & G.   Western Diamond Rattler.[R]

66   This, the largest of the Texas rattlesnakes, ranges throughout at least the arid part of Lower Sonoran zone of Texas. In the collection there are specimens from Corpus Christi, Japonica, Devils River, Comstock, Sycamore Creek, Eagle Pass, Langtry, and Boquillas, and from as far up the Pecos Valley as Santa Rosa, N. Mex. Mr. Cary also saved a flat skin from Pecos, Tex., and Doctor Fisher one from Colorado, Tex. Specimens recorded by Cope[c] from Indianola, San Antonio, and Brazos River apparently mark the eastern border of the known range of the species. I have never found it in eastern Texas.

Throughout its Texas range this is the commonest rattlesnake.

---

[a]Ann. Rep. U. S. Nat. Mus. for 1898 (1900), p. 1135.
[b]Ibid., pp. 1137 and 1138.
[c]Ibid., p. 1166.

[62] The cottonmouth, *Agkistrodon piscivorus*, occurs throughout the eastern half of the state where its populations appear to be stable.

[63] Three subspecies of copperhead are recognized in Texas: *Agkistrodon contortix contortix* in the extreme eastern counties, *Agkistrodon contortix laticinctus* in the central region, and *Agkistrodon contortix pictigaster* in the Trans-Pecos and northwestern region. The species appears to be common throughout its range.

[64] Current taxonomy (Kubatko et al., 2011) considers the South Texas form of the massasauga to be *Sistrurus catenatus edwardsii* instead of *Sistrurus catenatus consors* as reported in Bailey (1905). Two subspecies are recognized in Texas, *Sistrurus catenatus edwardsii* in the western and southern regions of the state and *Sistrurus catenatus tergeminus* in the north-central and coastal regions. The species appears to be common throughout its range.

[65] *Crotalus horridus*, the timber rattlesnake, occurs throughout the eastern one-third of the state. This species may be declining due to loss of habitat to the growing human population in the eastern part of the state.

[66] The western diamond-backed rattlesnake, *Crotalus atrox*, occurs throughout the western four-fifths of the state where it appears to be common, even though there is considerable mortality associated with rattlesnake roundups (see Chapter 6).

[67] Taxonomists consider specimens formerly assigned to *Crotalus confluentus* to be *Crotalus viridis*, the prairie rattlesnake (Klauber, 1936). This species occurs throughout the western one-third of the state where it appears to be common.

[68] *Crotalus molossus,* the black-tailed rattlesnake, occurs throughout the Edwards Plateau and Trans-Pecos regions of the state where it appears to be common.

50   NORTH AMERICAN FAUNA.   [No. 25.

While it is often reported as excessively abundant, I have never found more than a dozen individuals in a season's field work of four or five months in its favorite haunts. Over most of the range we do not see more than an average of one or two in a month's field work. On an 18-days' camping and collecting trip—April 24 to May 11, 1900—from Corpus Christi to Brownsville and return, we saw only five rattlesnakes, where we had been led to expect hundreds; and in this region of dense cactus beds and thorny thickets they find perfect protection and probably reach their maximum abundance. Specimens from extreme southern Texas are reported by the residents as reaching a length of 11 to 13 feet; but the largest specimen I have seen alive measured only 50 inches in length.

Crotalus confluentus Say.   Plains Rattlesnake.[R]

67   This small, dull-colored rattlesnake is represented in the Biological Survey collection from Texas by one specimen from Amarillo, on the Staked Plains, in Upper Sonoran zone. I am familiar with the species on Upper Sonoran plains of New Mexico and Nebraska, but have not found it elsewhere in Texas. The specimen collected by Captain Pope (No. 4962, U. S. N. M.), and labeled "Pecos River, Texas," has not even a date by which to locate the place where obtained; but as it is well known that Captain Pope's specimens from the top of the Staked Plains on the east and from the Guadalupe Mountains on the west were labeled "Pecos River," this record can have no zonal significance. The specimen recorded by Cope[a] from San Antonio, Tex., is entered in the Museum catalogue as collected "between San Antonio and El Paso," and the one recorded from Rio San Pedro (Devils River) is catalogued from "between Ash Creek and Rio San Pedro." Mr. Brown's record for Pecos[b], based on a specimen collected by Meyenberg, can not, as Mr. Brown admits, be used for "minute zone work," as Meyenberg's specimens, while said to have been collected somewhere "within a day's journey by team of Pecos," may have come from much farther away. In 1902, at a place 75 miles northwest of Pecos, I met one of his men bringing in a wagonload of live animals, among them numerous snakes and lizards which had been collected along the base of the Guadalupe Mountains, mainly in Upper Sonoran zone, but probably also in Transition. As these specimens apparently were sent out as collected at Pecos, in Lower Sonoran zone, the difficulty of using Meyenberg's material for zonal work is apparent.

Crotalus molossus B. & G.

68   This is the common rattlesnake of the Guadalupe Mountains in Upper Sonoran zone on both sides of the Texas and New Mexico line.

---

[a] Ann. Rep. U. S. Nat. Mus. for 1898 (1900), p. 1172.
[b] Proc. Acad. Nat. Sci. Phila., 1903, p. 551.

Specimens were collected near the edge of Transition zone on the east and west slopes of the mountains at 6,300 and 6,800 feet, but I assume that the species belongs to Upper Sonoran. A flat skin collected by Cary at a point 25 miles west of Sheffield is apparently this species.

We found this snake in August, 1901, in the gulches high up on the range. It is pugnacious, quick to sound its rattle and throw itself on the defensive. Because of its prevailing color of olive green, we always referred to it in the field as the "green rattlesnake."

Crotalus lepidus Kennicott.

69   Five specimens of this little rattlesnake from western Texas, the mouth of the Pecos, Paisano, Chisos Mountains, and Davis Mountains, indicate a range confined to the most arid and rocky part of the State in both Upper and Lower Sonoran zones. The specimen from the mouth of the Pecos was in Lower Sonoran zone on the gray limestone ledges at the east end of the High Bridge. In color it is pale ashen gray, with the pattern faintly indicated in dusky lines. The color has not changed materially in the alcoholic specimen from that of the live animal collected May 20, 1900, when I called it the 'white rattlesnake' to distinguish it from anything previously known to me. Mr. Cary reports several seen on limestone ledges along the Pecos Canyon at Howard Creek and Sheffield, all of which were whitish in color like the lime rock on which they were found. One specimen from the Chisos Mountains at about 6,000 feet, one from Paisano, and two from 5,700 feet in the Davis Mountains, were all taken in Upper Sonoran zone. Others were seen in both the Chisos and Davis mountains at similar altitudes, but none lower. All of these specimens in life or when freshly killed, and the others seen but not collected, were dark rusty brown with a pinkish tinge, heavily marked with velvety black crossbars. The brown has faded considerably in the alcoholic specimens, as shown by comparison with a careful color study of the fresh snake made in the field by L. A. Fuertes.

These little brown rattlers were fairly common about our camp near the head of Limpia Creek, in the Davis Mountains. On the dark brown lava soil they were very inconspicuous, and they had a way of suddenly springing up between our feet that made us slightly nervous.

REPORT ON THE MAMMALS OF TEXAS.

The following report on the mammals of Texas is based mainly on my own field notes and those of the other members of the Biological Survey who have worked in the State, supplemented by records from local naturalists and ranchmen.

[69] Two subspecies of *Crotalus lepidus,* the rock rattlesnake, occur in Texas, *Crotalus lepidus klauberi* in the Trans-Pecos and southern parts of the Edwards Plateau, and *Crotalus lepidus lepidus* in the two most western counties of the state. The species appears to be common throughout its restricted distribution.

[70] *Tatu novemcinctum texanum* is now referred to as *Dasypus novemcinctus mexicanus*. Although Bailey classified *Tatu novemcinctum texanum* as distinct from *Tatu novemcinctum mexicanum*, the two are now considered synonymous, and *D. n. mexicanus* is the proper scientific name for the nine-banded armadillo (see McBee and Baker, 1982). The range of the armadillo has expanded northward and eastward considerably since the time of the Biological Survey and now occupies nearly all counties in the state, with the exception of the extreme western regions (for a complete discussion of the status of this species in Texas, see Chapter 6). Populations seem to be declining statewide, presumably due to recent droughty periods and the reduction of soil invertebrates. Although not known in Bailey's time, reproduction in the armadillo is marked by the unusual phenomenon of specific polyembryony, which results in the formation of identical quadruplets.

52   NORTH AMERICAN FAUNA.   [No. 25.

Tatu novemcinctum texanum [M] subsp. nov.   Texas Armadillo.[M]

70   Type from Brownsville, Tex., No. $\frac{34352}{46438}$ ♂ ad., U. S. Nat. Mus., Biological Survey Coll. Collected by F. B. Armstrong, June 10, 1902; original No. 4.

General characters.—Similar to *mexicanum*[M] from Colima, but with relatively heavier dentition, larger and more acutely triangular lachrymal bone, and with larger epidermal plates on forehead and wrists.

Measurements of type.—Total length, 800; tail, 370; hind foot, 100. Skull of type: Basal length, 81; occipito-nasal length, 100; nasals, 36; greatest zygomatic breadth, 43; mastoid breadth, 29; alveolar length of upper molar series, 27; of lower molar series, 27.

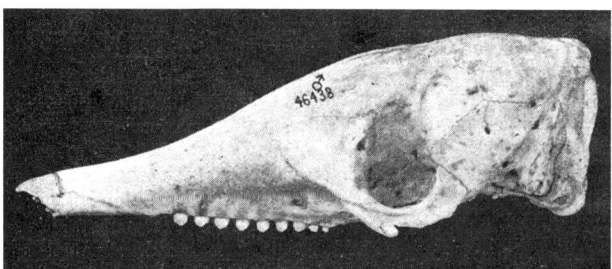

FIG. 5.—Skull of armadillo (*Tatu novemcinctum texanum*)[M] from Brownsville, Texas. (Natural size.)

Specimens examined.—Brownsville, 12; El Blanco, near Hidalgo, 1; Corpus Christi, 2; Nueces Bay, 1 skull; Kerrville, 2 skulls.[a]

Armadillos are common in southern Texas from the Lower Rio Grande to Matagorda Bay, the mouth of the Pecos, and north to Llano. A few are reported farther up the Pecos Valley, at old Fort Lancaster, Grand Falls, and Loving County, and one from 22 miles north of Stanton, while a specimen in the National Museum is labeled "Breckenridge." They have been taken in Burnet County and at Austin and Elgin, and reported from Inez, Seguin, Columbus, Navasota, and as far east as Antioch on Nevils Prairie. The belief is

[a] A series of 9 specimens, collected at Colima, Mex., by Nelson and Goldman, agrees in respect to the small quadrate lachrymal and light dentition with the excellent figure of a skull of *Tatusia mexicana* (Pl. II, fig. 3) in Gray's Hand List of the Edentate, Thick-Skinned, and Ruminant Mammals in the British Museum. As no locality more definite than Mexico is assigned to either the type of *Dasypus novemcinctus* var. *mexicanus* Peters or the specimen figured by Gray, the type locality may be fixed by considering the Colima specimens typical. The exact relationship between *mexicanum* and *novemcinctum*, of Brazil, remains to be worked out.

general that they are spreading eastward and northward, but whether this belief is founded on a real extension of range or on an increase in numbers throughout an established range is not entirely settled by the data at hand.

In 1890 Streator reported armadillos as rare on Raglans Ranch, 32 miles southeast of Eagle Pass, where two had been taken within ten years. In 1891 Lloyd reported them as common north of Brownsville and "much sought after for eating purposes." One taken at La Hacienda, 10 miles southeast of Hidalgo, he says, "was very tender, without any gamy smell," and he adds, "they eat small coleoptera and ants, greater quantities of the latter." He says that a cowboy saw an armadillo near the center of Padre Island, and at Nueces Bay he reported finding the remains of one and traces where others had been digging.

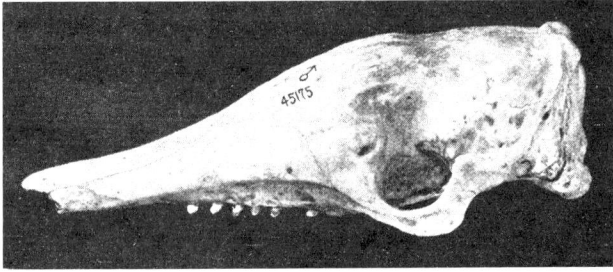

FIG. 6.—Skull of armadillo (*Tatu novemcinctum mexicanum*) from Colima, Mexico. (Natural size.)

In 1900 Oberholser reported them at Port Lavaca as not common, though occurring on the rivers at the head of Matagorda Bay and in the timber along small creeks west of there; at O'Connorport as not found except several miles back in the country, where quite rare; at Beeville as common at a little distance from town and frequently brought in by the Mexicans; at San Diego as common, living chiefly in timber along creeks and in chaparral about ponds; at Laredo as rare immediately south of town, but said to be common in places toward the north; at Cotulla as abundant, inhabiting principally the timber along the Nueces; at Uvalde as occasionally found along the Nueces; at Rock Springs as tolerably common in places. In 1901 he reported them at Fort Lancaster as said on good authority to occur on Independence Creek, 25 miles down the Pecos; at Langtry and the Pecos High Bridge as rare, but said to be occasionally found along the river; at Del Rio as reported to be common along the Rio Grande; at Comstock as rare. In 1902 he reported armadillos of occasional occurrence at Austin, where one was taken a few years ago, and as rare at Elgin, where there are well-authenticated instances of their capture along some of the creeks.

Near Antioch, Houston County, in 1902, Hollister obtained two records of the capture of armadillos on Nevils Prairie, the last about 1899, where one wandered into a smokehouse and was caught and kept alive for some time. Cary in his reports for 1902 records the capture of an armadillo by John Hutto, a sheep man living 22 miles north of Stanton, in 1892. Mr. Finnegan, the hotel proprietor at Stanton, saw the animal at the time. At Monahans Cary was told by Landlord Holman that armadillo shells were rarely found in the sandhills, but that he had seen a specimen killed at Grand Falls in 1899. In February, 1902, Mr. Royal H. Wright, of Carlsbad, N. Mex., wrote me that he had picked up an old armadillo shell in Loving County, Tex., close to the New Mexico line. At Llano, in 1899, I was told that armadillos were frequently killed around there or brought into town alive. In 1904 a few armadillos were reported at Seguin, and Mr. Samuel Neel told me that a few years ago he had found one in the garden under the vines of his cowpeas. At Columbus Mr. Henry Mathee told me of two armadillos killed there during the fall of 1904, and Mr. J. F. Leyendeker wrote me that one was taken near Frelsburg. At Navasota Mr. Charles Hardesty told me of one caught near there during the summer of 1904. In a letter of June 4, 1904, Mr. H. P. Attwater furnishes the following note:

> When in Port Lavaca last week I obtained some notes in regard to armadillos which may be of interest. Mr. J. M. Boquet, a very intelligent and reliable ranchman, says that these animals were first noticed in Calhoun County in 1886 or about that year; that they are now very common, and that he has no doubt there are hundreds of them in the county to-day. He says their favorite resort on the prairie ranches is in the long Cherokee-rose hedges, which have been grown in many parts of that and adjoining counties as a wind-break in winter for cattle. During the last few years since armadillos have become so common in the southwest and south central Texas I see baskets made of armadillo skins or shells in the curio stores at San Antonio and other places. The legs are cut off and the tail fastened to the mouth, forming the handle of the basket. At a curio shop in San Antonio two or three days ago I was informed that they sold for about $1.50 to $2 each.

The armadillos are strictly Lower Sonoran, but in the rough country between Rock Springs and Kerrville they range fairly into the edge of Upper Sonoran Zone. As a rule they do not extend east of the semiarid or mesquite region, nor to any extent into the extremely arid region west of the Pecos, but occupy approximately the semiarid Lower Sonoran region of Texas, north to near latitude 33°. They are partial to low, dense cover of coarse grass, thorny thickets, cactus patches, and scrub oaks, under which they make numerous burrows and trails, or root about in the leaves and mold, where they enjoy comparative safety under the double protection of leafy screen and armor plate. But they thrive best in a rocky country, especially where limestone ledges offer numerous caves and crevices of various sizes, from which they can select strongholds that will admit no larger animal. Almost every rock-walled gulch along the head-waters of Guadalupe River has one or more dens with smoothly worn doorways from which much traveled trails lead away through the bushes or to little muddy springs, where tiny hoof-like tracks and the corrugated washboard prints of ridged armor suggest that the arma-

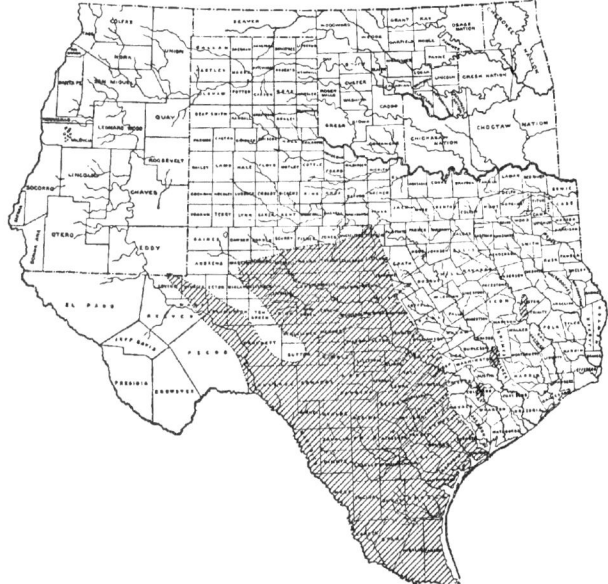

Fig. 7.—Distribution area of armadillo (*Tatu novemcinctum texanum*).[M]

dillos not only dig and nose about in the soft ooze for their insect food, but, pig-like, enjoy also a cooling mud bath. Other trails lead along rocky shelves, up the sides of gulches, and away from thicket to thicket, and are easily followed sometimes for half a mile till they branch and scatter or connect with cattle trails, where the rope-like prints of dragging, horny tails are visible among the dusty cow tracks. Late in the afternoon one occasionally meets an armadillo trotting vigorously along a trail on his stumpy little feet, his tail

[71] Historically, there were three subspecies of the Virginia opossum, *Didelphis virginiana*, in Texas: *Didelphis virginiana pigra* in the southeast, *Didelphis virginiana virginiana* in northern and central Texas, and *Didelphis virginiana californica* in the Trans-Pecos and Rio Grande Valley region. However, taxonomists now recognize all Virginia opossums in the eastern portion of Texas as *D. v. virginiana* (see Schmidly, 1983) and *D. v. pigra* is no longer a valid name for Texas opossums. The range of the opossum has expanded westward considerably since the time of the survey, and collecting records (Hollander and Hogan, 1992) show that its range now encompasses all of the state including much of the xeric areas of the Trans-Pecos, where specimens are assignable to *D. v. californica*. Range extensions have occurred primarily along streams and rivers, where woody vegetation has permitted the opossum to penetrate the otherwise treeless grasslands and deserts of western Texas.

dragging after him in a useless sort of way as he hurries nervously across the open spaces and stops in the thickets to nose about under the leaves in search of dainties from the fragrant soil. At such times the long, pointed nose seems to be the keenest organ of sense. The little eyes, half the time buried in rustling leaves, rarely detect an object not close by and in motion. I have followed one of these preoccupied little animals for half an hour, often within 20 or 30 feet, moving only when it was rustling in the leaves, and watching its motions without being discovered or creating alarm. Hunters say that if you stand still the armadillos will sometimes bump against your feet without discovering you, so short sighted are they and so intent on their own business. But when alarmed, they get over the ground with a rush that is surprisingly rapid considering their turtle-like build. If the first rush does not carry them to cover and an enemy overtakes them, they curl up in an ironclad ball that is not easily uncurled. In autumn, during the deer hunting months, when the young of the year are full grown, they are especially numerous and particularly obnoxious to the still hunters, who repeatedly mistake their rustling in the leaves for the noise of feet of bigger game. Where a dozen or twenty armadillos are met in a day's hunting, as sometimes happens, and possibly no deer are seen, the nervous strain and disappointment on the part of the hunter sometimes result in serious consequences to the innocent armadillo.

The excrement of the armadillos found scattered along the trails in the form of clay marbles and with the texture of baked mud gives some clue to the food habits of the animals. Careful examination shows only the remains of insects, mainly ants and a few small beetles, embedded in a heavy matrix of earthy matter.

Didelphis virginiana Kerr. Virginia Opossum.

71 Specimens of opossums examined from northern and middle Texas, Gainesville, Vernon, Brazos, Mason, and Kerrville, have the light-gray coat, white ear tips, and comparatively short tail of *virginiana*. To the west the species does not extend much beyond the one hundredth meridian, except along some of the stream valleys, up which it reaches as far as San Angelo, Colorado, and Tascosa. I have seen no specimens from extreme eastern Texas, except from near the coast, where they are referable to *pigra*.[M]

The Virginia opossums are more or less abundant throughout their range, and live mainly in the woods and brush along streams. In the daytime they sleep in hollow trees or logs, in holes in the ground, or merely curled up in the brush or weeds or sometimes on a large branch of a tree, and if disturbed appear stupid and dazed. At night they prowl about in search of food, and, not being epicureans, usually find it in abundance. They are especially fond of chickens and eggs and do considerable mischief in the henhouse. They will eat any kind of meat, even when it is old and stale, and often persist in getting into traps baited for more desirable game. Through the summer they feed extensively on fruit, and are usually lean and rangy. In the fall they become very fat and by many are then considered a great delicacy. Their importance as food and game animals and the value of their fur make up for the inconsiderable losses they now and then occasion poultry raisers. Winter skins in prime fur are quoted at 55 to 60 cents, and they usually constitute a large share of the fur harvest of local trappers.

Didelphis v. pigra[M]Bangs.   Florida Opossum.[M]

Specimens of the opossum from the coast region of Texas east of Matagorda Bay are generally a little darker than typical *virginiana,* with more dusky about the face. While not typical, they are nearer to *pigra*[M]than to *virginiana,*[a] from which there is no sharp line of separation. They are merely the darker southern form inhabiting Florida and the South Atlantic and Gulf coast region, with habits modified by local conditions and environment. On the coast prairies of southeastern Texas they live much in the open, wandering along the margins of ponds and bayous, sleeping under fallen grass or low bushes and feeding extensively on crawfish and other small crustaceans. In the stomach of one taken near Galveston I found a horned toad and bird's feathers, besides the meat used for trap bait.

Gaut found them abundant in the Big Thicket northeast of Sour Lake, where he caught them in a line of traps set along Black Creek in the timber. He reports two females, caught March 18, carrying in the pouches young apparently four or five days old, one with five young, the other with six. He found the stomachs of two individuals filled with crawfish, one full of carrion of a dead hog, and in another traces of maggots and carrion.

Didelphis marsupialis texensis[M]Allen.   Texas Opossum.[M]

72   This subspecies of a widely distributed Mexican form is easily distinguished from both *virginiana* and *pigra*[M]by its longer and blacker tail, the wholly black ears of adults, and by its dichromatism, about half of the individuals being entirely black instead of light gray. It inhabits the southern part of Texas, from Brownsville to Nueces Bay and San Antonio, and up the Rio Grande to Del Rio and the mouth of the Pecos. Doctor Allen, in his monograph of the genus *Didelphis,* considers the species distinct from *virginiana* and records specimens of both from San Antonio.[b]

---

[a]Doctor Allen refers these coast specimens rather doubtfully to *virginiana* (Bul. Am. Mus. Nat. Hist., XIV, 166, 1901), but all, with possibly one exception, seem to me nearer to *pigra*.[M]

[b] Bul. Am. Mus. Nat. Hist., Vol. XIV, 149—188, June 15, 1901.

[72] The name *Didelphis marsupialis texensis* has been changed to *Didelphis virginiana californica,* and its application is restricted to opossums occurring in the Trans-Pecos and Rio Grande Valley regions of Texas (see Gardner, 1973).

[73] The name *Tayassu angulatum* has been changed to *Pecari tajacu* (Woodburne, 1968), and the subspecies in Texas is *Pecari tajacu angulatus*. Interestingly, Bailey and the federal agents did not collect or observe javelina (more appropriately referred to as peccaries) in the Big Bend or the Davis Mountains, places where they are common today. It appears the range of this species has contracted in the east and north and expanded to the west during the last century, with the most recent information suggesting that this species is declining statewide (Schmidly and Bradley, 2016). An introduced population occurs along the Red River in Wilbarger and surrounding counties (Dalquest and Horner, 1984). Figure 37 depicts the current range of the collared peccary in Texas.

Figure 37. The current distribution of the collared peccary (*Pecari tajacu*) in Texas. Light shaded area represents natural distribution; dark shaded area represents counties where the species has been introduced.

58   NORTH AMERICAN FAUNA.   [No. 25.

The habits of this opossum are peculiar only in so far as they have been modified to adapt it to the country in which it lives, a more or less open region of mesquite, brush, and cactus, with few hollow logs or stumps. For home and shelter the animals depend largely on burrows, which apparently they dig for themselves. In setting traps where fresh earth was being brought out of the burrow or out of several of a group of burrows each night, hoping to get a badger or armadillo, I have on several occasions been disappointed to find in my trap next morning only an old black opossum.

A female caught at Del Rio January 30, 1890, had nine tiny young in her pouch, each clinging to one of the slender teats and grasping the moist, crinkled, brown hair of the pouch lining with all four of its little hands. If forcibly pulled loose they would immediately regain their hold, the only instinct of their embryonic life being to hold on tight and get nourishment. They were too small to noticeably distend the pouch, and I discovered them only in preparing the specimen on the following day. I then noticed that the nine teats, arranged in two semicircular rows with one in the center, were not the full set. The anterior pair were functionless, but as the mother was not fully grown these probably would have developed with the next and larger litter of young. While in the trap this female showed no disposition to fight or defend herself, but an old male caught a day or two later fought viciously, growling and biting anything that came within reach, actually cutting deep gashes in the hard-wood stock of my gun. Another female, caught by James H. Gaut at the mouth of Sycamore Creek, 12 miles east of Del Rio, June 1, 1903, was carrying six very small young in the pouch.

Tayassu angulatum[M](Cope).   Texas Peccary; Javeline; Musk Hog.[M]

73   Peccaries are still more or less common in southern Texas and along the Rio Grande to above the mouth of the Pecos, thence up the east side of the Pecos Valley into the unsettled sandhill region of southeastern New Mexico, and east along the broken edge of the plains to San Angelo and Kerrville, and along the coast to Corpus Christi. A few may remain here and there still farther east and north, where they once ranged, but they have been pretty thoroughly driven out by the settlement of the country, and are now merely clinging to existence in regions of deep rocky canyons or dense thorny cactus and chaparral and in an uninhabited waste of sand dunes. They are extremely wary, depending for protection mainly on caves in the rocks and impenetrable cover, and so may be able to hold their own for a few years longer. They are usually hunted with dogs and horses, as it is almost impossible to discover or get near them in any other way. The cowboys occasionally rope one, but claim that many good horses are ruined by being ridden over boars, which never fail to cut and gash the horses' legs in a dangerous manner.

In October, 1904, I was told that a half-grown peccary in the San Pedro Park, at San Antonio, had been captured in Nueces County within a month.

The following reports by Merritt Cary were made in September, 1902:

The peccary is common in Castle Mountains and on sand ridges northwest of there. Along Castle Gap I went into several caves beneath the rim rock

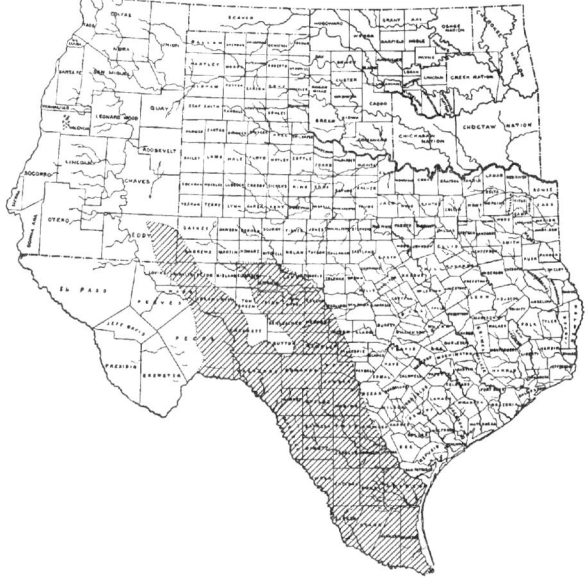

FIG. 8.—Distribution area of peccary (*Tayassu angulatum*).[M]

and found the ground all tramped up by them, where they had evidently made their dens.

A peccary was killed 2 miles northeast of Odessa about September 1, 1902, but was very thin and evidently a wanderer. The animals are said to be common in the western portion of Gaines County.

Peccaries range throughout the sand along the Texas Pacific Railroad west to about Quito station and east rarely as far as Odessa. North they follow the sand belt well up into New Mexico, according to report, while to the south their range is continuous through the Castle Mountains and down the valley of the Pecos. From what I could learn, their center of abundance in the sand belt is some 10 to 15 miles north of Monahans, where the 'shinrick' is densest, and their principal food, the acorns, most plentiful. They are said to hide

[74] Modern taxonomists classified Merriam's elk (now referred to as western elk or wapiti) as a subspecies of *Cervus elaphus* and then moved it to the same status within *Cervus canadensis* (see Groves, 2003). Although elk were extinct in Texas by the time of the Biological Survey, over the last 70 years there have been several reintroductions; however, these have not been of the native subspecies *Cervus canadensis merriami*, which is presumed extinct. Current populations may be the result of reintroductions from *Cervus canadensis canadensis* and *Cervus canadensis nelsoni* from various western states, although a recent study (Dunn et al., 2017) suggests that native elk from nearby populations in New Mexico may have naturally repopulated portions of the Trans-Pecos region. If these findings are accurate, then it is possible that individuals of *C. c. nelsoni* may have naturally repopulated Texas. For a discussion, see Chapter 5.

[75] The spelling of *Odocoileus virginianus texanus* has been changed to *Odocoileus virginianus texana*. Historically, four subspecies of white-tailed deer (*Odocoileus virginianus carminis, Odocoileus virginianus macroura, Odocoileus virginianus mcilhennyi,* and *Odocoileus virginianus texana*) have been recognized in Texas (see distribution of *Dama virginiana* by Hall, 1981). The subspecies *O. v. texana* occurs naturally throughout most of the state (except for the extreme southern portions of the Trans-Pecos and extreme eastern and coastal regions of the state). They have been stocked in the northeastern region of Texas where *O. v. macroura* once occurred

---

60   NORTH AMERICAN FAUNA.   [No. 25.

in the 'cats-claw' (*Acacia*) during the day and range out into the sand hills for acorns at night or in cloudy weather. One was killed while we were at Hawkins's ranch, but the hide and skull were spoiled before we could secure them. Several peccaries have been killed at San Angelo in years past.

Cervus merriami [M]Nelson.   Merriam Elk.[M]

74   There are no wild elk to-day in the State of Texas, but years ago, as several old ranchmen have told me, they ranged south to the southern part of the Guadalupe Mountains, across the Texas line. I could not get an actual record of one killed in Texas, or nearer than 6 or 8 miles north of the line, but as they were common to within a few years in the Sacramento Mountains, only 75 miles farther north, I am inclined to credit the rather indefinite reports of their former occurrence in this part of Texas. Specimens of horns and a part of a skull from the Sacramento Mountains indicate that the species was very similar to and probably identical with the Arizona elk described by E. W. Nelson who has aided me in making the comparisons.

In his field report for May, 1904, from the Wichita Mountains, Oklahoma, Gaut says:

Mr. A. T. Hopkins, of Lawton, killed an elk in 1881 on Rainy Mountain, about 40 miles west of Lawton. This apparently is the last specimen recorded from that region. Several antlers have been picked up within the last few years on Rainy Mountain, a high ridge about 12 miles west of Mount Scott, and Mr. O. F. Morrisey, the forest ranger, informed me that he frequently runs across elk antlers while on his rides through the reserve.

In 1852 elk were reported[a] by Captain Marcy from the Wichita Mountains, Indian Territory, and if they ever were common there they would naturally at times have strayed across the interval of less than 50 miles to the border of northern Texas. This report may have led to the inclusion of 'moose' with the game of Texas in Mary Austin Holley's History of Texas, 1836, but it is by no means certain. The particular form of elk which inhabited the Wichita Mountains will probably never be known, although it may have been referable to *C. canadensis.*

That northern Texas is well adapted to elk is shown by the perfect condition in 1902 of a herd of nine owned by Mr. Charles Goodnight, of Goodnight, Tex.

Odocoileus virginianus texanus[M]Mearns.   Texas White-tailed Deer.[M]

75   Specimens of the white-tailed deer from Corpus Christi, Kerr County, Rock Springs, and Langtry agree with the type and topotypes of *texanus*[M] from Fort Clark, and indicate for this form a wide range over the semiarid part of southern and middle Texas, but do not in any way define the limits of its range. In the open and arid

---

[a] Explorations of Red River of Louisiana, p. 186, 1854.

---

and along the Gulf coast where *O. v. mcilhennyi* once existed; consequently, it is difficult to determine if *O. v. macroura* and *O. v. mcilhennyi* remain in these regions. With the exception of *O. v. carminis*, the Carmen Mountain whitetail, which occurs in the higher elevations of the isolated mountain tops in the Big Bend region, it is virtually impossible to distinguish subspecies of white-tailed deer in Texas because of the extensive introductions and movements of animals from location to location across the state during

the latter half of this century. The specimen from El Paso County that was mentioned by Bailey likely was brought in by hunters or other persons traveling through the area.

region west of the Pecos white-tailed deer are rare. The skull of an old buck from San Elizario, on the Rio Grande, just below El Paso, has abnormally large molars and large audital bullae, and can only provisionally be referred to *texanus*.[M] The few once inhabiting the Davis and Guadalupe Mountains have been almost exterminated, and they may have been the little *couesi* instead of *texanus*.[M] Excepting a part of the Trans-Pecos region and, possibly, the open top of the Staked Plains, the whole of Texas is or has been occupied by some form of the white-tailed deer.[a]

A few imperfect specimens from Liberty, Hardin, and Jasper counties in extreme eastern Texas are apparently *texanus*,[M] but more and better specimens from this region may show that they are nearer to *virginianus*. They certainly are not the large, dark *louisianae*,[M] which, from geographical considerations alone, they might be supposed to be. In the Big Thicket region of Liberty and Hardin counties deer are still common, owing to the dense forest and the tangle of vines, briers, palmettoes, and canes which afford them almost impenetrable cover. They are now hunted mainly with hounds, but formerly when more abundant night hunting with a headlight was the favorite method, and deer were wantonly slaughtered in great numbers. One hunter told me that he had no idea how many deer he had killed for their skins, but that in the fall of 1886 he remembered selling 69 skins of deer, the carcasses of which were left in the woods. Even at the present time the law in this region is rarely enforced, and deer are killed without regard to season. With the natural advantages offered by extensive tracts of unoccupied forest and swamp land, dense cover, and abundance of acorns and other food, deer with some slight protection would soon become abundant again.

A collection of about 350 pairs of deer horns at Rock Springs is especially interesting, as all the horns came from deer killed in the vicinity, which is near the type locality of *texanus*,[M] and consequently show the local variation. As usual in a large collection, there are some abnormal sets, and a number of very small sets which seem to be of either young or imperfectly developed individuals and not of the little Sonora deer. Through the courtesy of Mr. Fleischer I obtained photographs of a large number of the horns. It is to be hoped that the collection will eventually find its way to some museum, where it would be of considerable scientific interest.

At the edge of the little town of Rock Springs Doctor Richardson had six tame deer in a small inclosure—a 3-year-old buck, a 2-year-old doe, and four yearling does. They were all in the red coat when

---

[a] The group of white-tailed deer is sadly in need of revision, but the material at hand is too scanty for final conclusions in regard to the several described forms.

examined July 15, 1902, and the nearly grown horns of the buck were in full soft velvet. All these deer were perfectly tame, and would push and crowd the Doctor as he fed them bran from a basin. Bran and hay constituted their main food. Though in good health and spirits they were thin, and I urged the Doctor to give them some of their natural food—the leaves of live-oak brush, acorns, mesquite and other bean pods.

On many of the large ranches between Corpus Christi and Brownsville, where the oak and mesquite thickets are interspersed with prairie and grassy openings, deer find ideal conditions, with abundant food and cover. The nature of the ground is such as to protect them from wolves and other natural enemies; but it is well suited to either hunting on horseback or still hunting, which, if freely allowed, would soon exterminate them. With protection, however, they increase rapidly, and in many places are abundant. Fresh tracks were common in the trails and even along the stage road in places, and the ranchmen usually know where to find a deer when needed. Similar conditions are reported over most of southern Texas, although varying greatly on the different ranches. From Kerrville west to Devils River and the Rio Grande deer are more or less abundant, both in the half-open mesquite valleys and over the rough juniper and oak-covered ridges. On certain large ranches they are still numerous, while on others they have become extremely scarce and would be entirely exterminated but for the recruits from surrounding and better protected ranches. Some of the ranchmen do not consider it worth while to protect the deer, while others leave the matter to indifferent foremen or else allow so much hunting by their friends that few of the animals escape. But such indifference is unusual. Almost every ranch gate we passed through bore the sign "Posted."

In spite of the protection of State laws and ranch owners there are still remote sections of rough, uncontrolled range where every year hunters kill wagonloads of deer for the market, or worse, kill the deer for the hides only, leaving the carcasses to rot. I was told that in the winter of 1901-2 hundreds of deer skins were brought out of the country west of Kerrville.

No part of the United States affords more perfect conditions for deer than southern Texas, and all that is required for their maintenance and rapid increase is efficient protection. In the past this has not been provided by the game laws; and the fact that the deer have not been wantonly destroyed over the entire region is due to the practical, business-like methods of the large ranch owners, who control the hunting on their ranges, and would as soon think of depleting their herds of cattle as the game under their control. On some of the larger ranches mounted rangers are regularly employed to ride over the country and protect both stock and game, and to see that fences are kept up and that there is no hunting. But usually this is an important part of the business of the regular cowboys. As a practical business proposition the protection of deer can not be urged too strongly. Their presence on the range does not interfere with the cattle and horses. The deer rarely, if ever, eat grass or any forage plant eaten by horses and cattle, but live on the leaves and twigs of bushes, seeds, pods, and flowers of a great variety of plants, including acorns and the pods of the numerous kinds of bean bushes.

Opposition to private control of game was never more groundless than in this open country where without such control the deer would long ago have been exterminated over extensive areas where they are now common.

Odocoileus macrourus?[M](virginianus) (Rafinesque). Plains White-
  tailed Deer.

76 Two specimens of white-tailed deer from the sandhills 20 miles north of Monahans, and 3 from Beaver Creek where it crosses the Texas and Oklahoma line at the north edge of the Panhandle, can not be referred to *texanus*[M]or *virginianus*. On geographic grounds they should be *macrourus*,[M]described from the "plains of the Kansas River," so provisionally, at least, I refer them to this form.[a]

The two specimens from north of Monahans are fully adult. A doe, collected September 18, is in the pale yellow summer coat with traces of the fall 'blue coat' showing through, and a large buck taken November 12 is in full, fresh winter pelage. These specimens are larger than corresponding sexes of typical *texanus*,[M]with relatively heavier, wider skulls. The doe is apparently lighter and brighter colored, with no trace of black on the tips of the ears, and the buck is much lighter colored around the face and ears than strictly comparable specimens of *texanus*.[M] The three specimens in the United States National Museum collection from Beaver Creek, collected by Hornaday in 1889, are in faded, late winter pelage, very pale and yellowish. The three imperfect skulls without horns from the same locality agree in a general way with the Monahans skulls.

In September, 1902, Merritt Cary reported the white-tailed deer as common in the sandhill region south and north of Monahans, and as feeding principally on the acorns of the little shin oak which covers this region. At Canadian, in the northeast corner of the Panhandle, in July, 1903, A. H. Howell reported white-tailed deer as occurring

---

[a] At present I do not know of a specimen of the white-tail deer from the Plains near enough to the type locality of *macrourus*[M]to be safely assumed to be typical of that form, and until typical specimens are obtained the status of the form must remain somewhat in doubt.

[76] The spelling of *Odocoileus virginianus macrourus* has been changed to *Odocoileus virginianus macroura*. This subspecies once occurred in the extreme northeastern corner of the state (see Annotation 75). Taxonomists now refer specimens from Monahans and the High Plains to the subspecies *Odocoileus virginianus texana* and restrict the name *O. v. macroura* to deer occurring in the Big Thicket in extreme southeastern Texas (see distribution of *Dama virginiana* by Hall, 1981).

[77] Taxonomists now restrict the use of the name *Odocoileus virginianus couesi* to populations that occur in Arizona, western New Mexico, and Mexico (see distribution of *Dama virginiana* by Hall, 1981). The specimens Bailey described from the Chisos Mountains are now recognized as *Odocoileus virginianus carminis*. This subspecies is represented by a small, isolated population in the Chisos Mountains, where it is well protected within the borders of Big Bend National Park, and by populations from the Rosillos, Christmas, Chinati, and Davis mountains in the Trans-Pecos (Krausman et al., 1978).

in small numbers in the brushy bottoms; and in July, 1904, Gaut reported them as common in the region about Mobeetie.

Odocoileus couesi[M] Coues and Yarrow.   Sonora Deer.[M]

77   This little deer is the smallest of the white-tailed group found in the United States, an old buck rarely being estimated at over 100 pounds, while the does are variously estimated at from 50 to 75 pounds. The horns are small and closely curved in, with usually three or four points to a beam; the ears are considerably larger than in specimens of *texanus*[M] weighing nearly twice as much. The young, after losing the spots, are light yellowish brown until after the change to the gray coat, which apparently takes place with the fall molt of the third year. The adults, after about two and a half years old, are light gray at all seasons, without black on ears or tail.

The species is widely distributed through the desert mountains of southern Arizona and northern Mexico and probably reaches its eastern limit in the Chisos Mountains of western Texas. Here these little deer range from 5,000 feet at the upper edge of Lower Sonoran zone through Upper Sonoran and Transition to the top of the mountains at 9,000 feet. They are closely associated with the oaks, junipers, and nut pines, and depend much on the cover of brush and timber. During the day they are usually found lying under a low, branching juniper tree or in a thicket of oak brush, and when started are more often heard bounding over the rocks than seen in the open. They are most numerous on the plateau top of the mountains, at 8,500 feet, where a steep 3,000-foot slope protects them from most hunters and where the sweet acorns of the little gray oak are abundant. Between rains the only water on this plateau is held in the rock basins, but it is usually ample for the needs of the deer. A few springs around the base of the mountains are permanent and always accessible in case of drought. The deer on the plateau are so little disturbed that they are often seen feeding or wandering about during the day. While sweeping the slopes with a field glass I often located deer and watched them without arousing their suspicion. At 11 o'clock one warm day I watched three does come down from an open sunny hillside and select cool beds in the shade of bushes along a deep gulch. At another time I watched a doe and two yearling fawns feed until they were satisfied and then scatter to make their beds under different trees on an open grassy slope. On several occasions, by moonlight, the flash of their white tails at close quarters was seen with startling effect, the gray bodies being quite invisible.

The food of the deer in June consisted mainly of leaves, flowers, green seeds, and capsules or pods of a great variety, of shrubs and plants, including the leaves of the little gray oak (*Quercus grisea*),

Fig. 1.—Cypress Swamp, near Jefferson.

Fig. 2.—Mixed Swamp Timber near Jefferson, Eastern Texas.

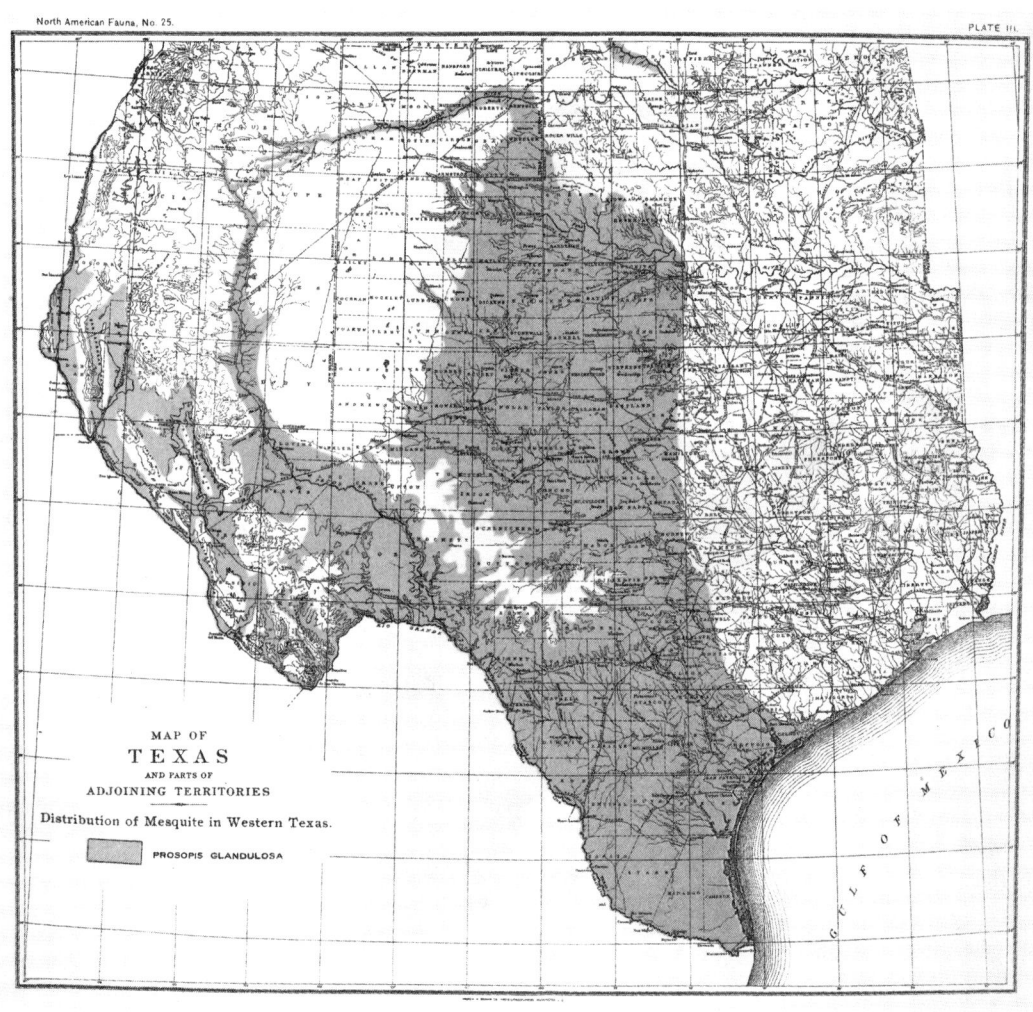

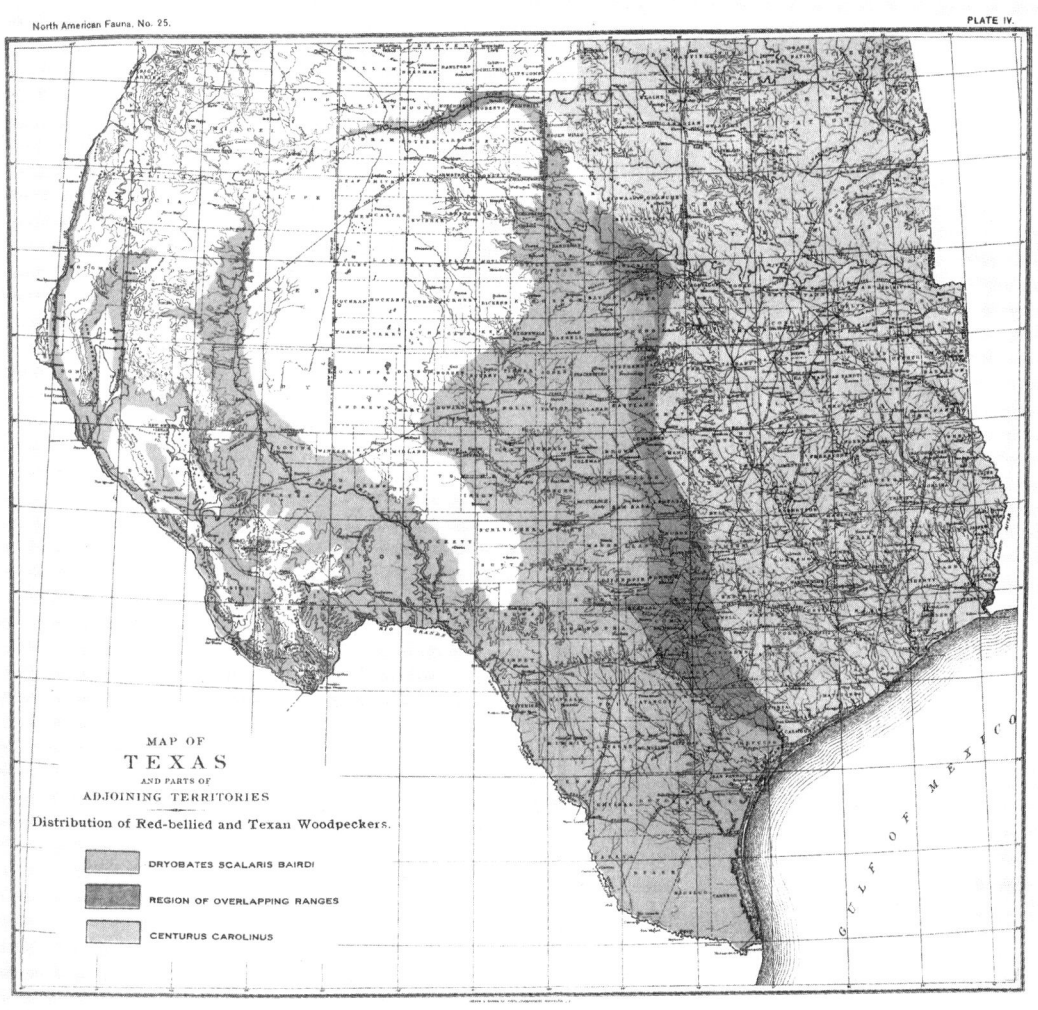

North American Fauna, No. 25.   Plate V.

Fig. 1.—Ocotillo (Fouquiera splendens) and Creosote Bush
(Covillea tridentata)[P].

Fig. 2.—Desert Vegetation of Great Bend Region.

Fig. 1.—Lecheguilla with Flowers and Fruit.

Fig. 2.—Lecheguilla (Agave lecheguilla) new Boquillas, Great Bend of Rio Grande.

North American Fauna, No. 25.   Plate VII.

Agave wislinzeni[P]in Flower, Davis Mountains. Hummingbird
at Flower Cluster on Left.

North American Fauna, No. 25.   Plate VIII.

Fig. 1.—Sotol on Mesa near Comstock.

Fig. 2.—Sotol After the Leaves are Burned Off.

North American Fauna, No. 25.   Plate IX.

Views on Staked Plains near Hereford and Dimmitt.

Fig. 1.—Transition Zone Timber of Guadalupe Mountains.

Fig. 2.—Head of Dog Canyon, Guadalupe Mountains.

North American Fauna, No. 25.

Plate XI.

Transition Zone Timber of CHisos Mountains (Approximately 6,500 Feet).

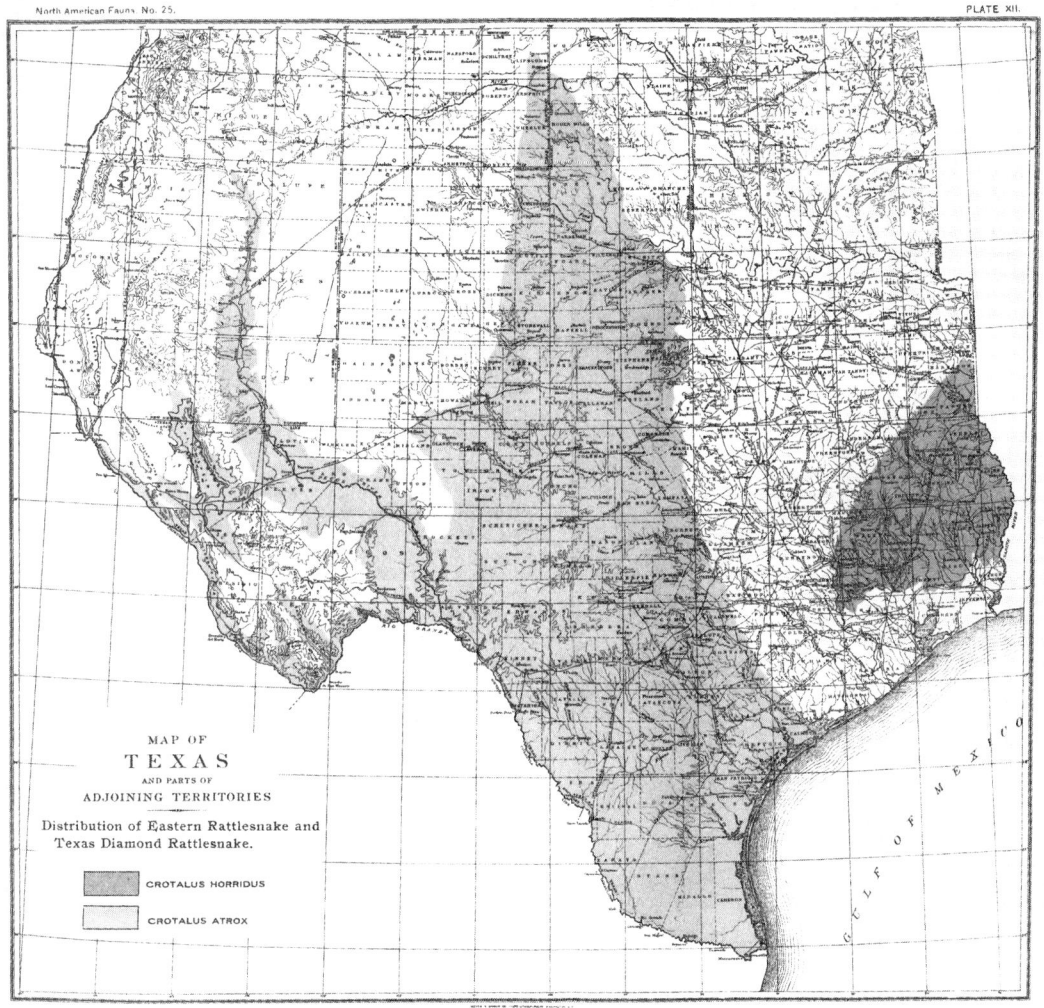

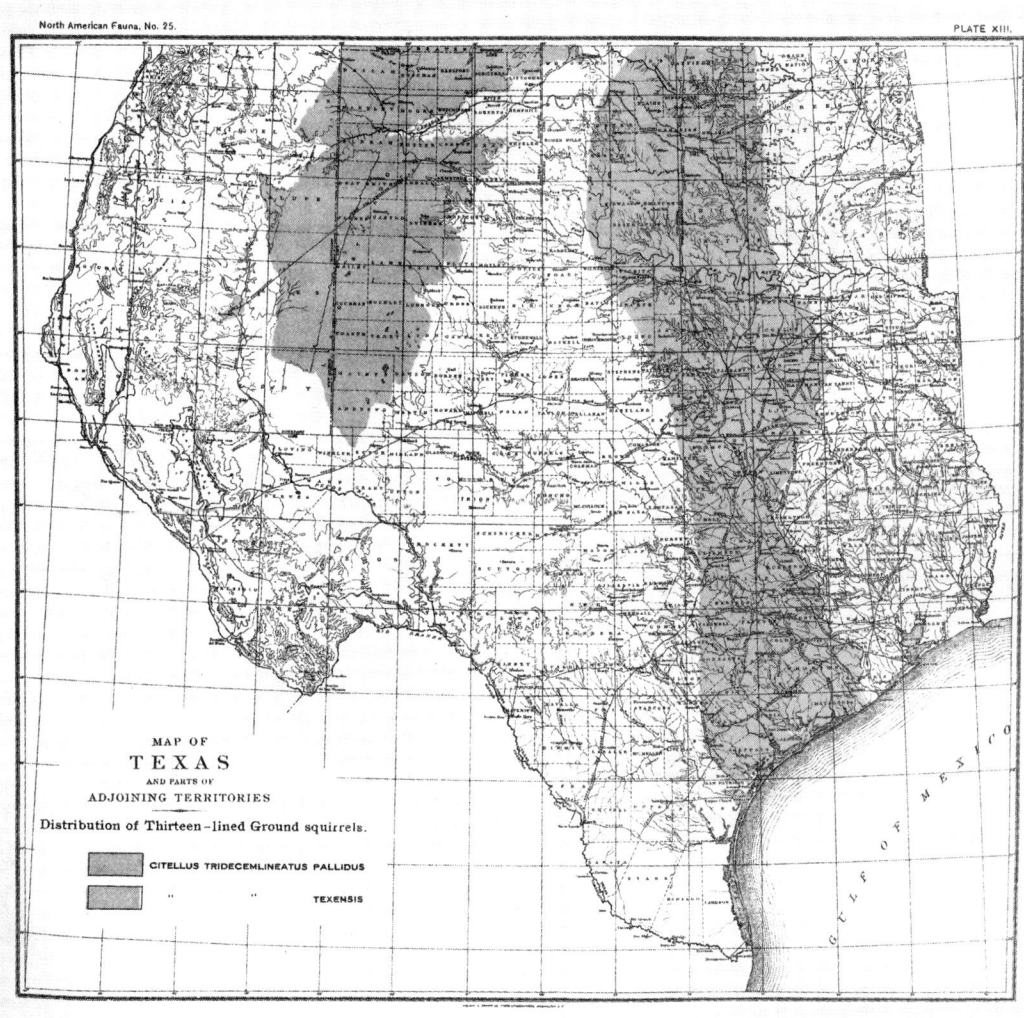

North American Fauna, No. 25.

PLATE XIII.

MAP OF
TEXAS
AND PARTS OF
ADJOINING TERRITORIES

Distribution of Thirteen-lined Ground squirrels.

CITELLUS TRIDECEMLINEATUS PALLIDUS

"   "   TEXENSIS

GULF OF MEXICO

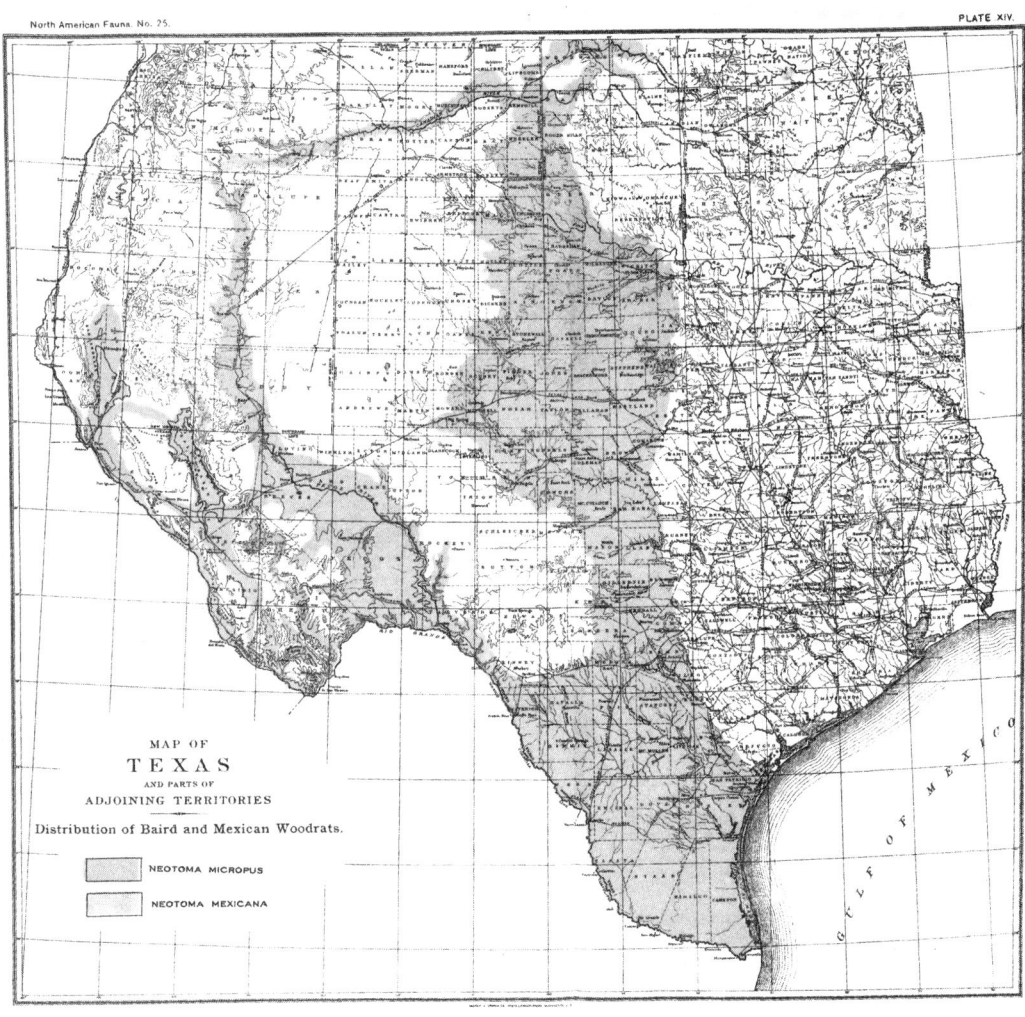

North American Fauna. No. 25.

PLATE XIV.

MAP OF
TEXAS
AND PARTS OF
ADJOINING TERRITORIES

Distribution of Baird and Mexican Woodrats.

NEOTOMA MICROPUS

NEOTOMA MEXICANA

GULF OF MEXICO

North American Fauna, No. 25.   Plate XV.

Head of Plateau Wild-cat [M] (*Lynx baileyi*). [M]

Civet Cat; ^MBassariscus (*Bassariscus Astutus flavus*).

[This page intentionally left blank]

leaves and wide, flat pods of several bean bushes (*Acacia roemeriana*[P] and others), leaves and berries of sumac (*Schmaltzia microphylla*), leaves and capsules of a large *Pentstemon*, and flowers and stems of bear grass (*Nolina lindheimeriana*). The prints of the deer's teeth were often found on the half-eaten green stalks of the century plant (*Agave wislizeni*).[P] Much-used trails and abundance of winter 'sign' among the oaks showed that acorns were the great attraction during fall and winter. No trace of grass could be found in any of the three stomachs examined.

Odocoileus hemionus canus [M]Merriam. Gray Mule-deer.[M]

78 Two skins (with skulls) of mule-deer and 8 skulls (or horns with parts of skulls) from the region about Samuels, Langtry, and the

FIG. 9.—Yearling buck of gray mule-deer,[M] Langtry, Texas.

mouth of the Pecos River, a head and horns from Alpine, and a few old horns from the Chisos Mountains agree in a general way with the type and topotypes of *canus*.[M] The skins show the same light gray color, and the skulls are small, the forehead flat, and the horns usually low and widespreading. I have seen no specimens from the northern end of the Staked Plains, but should expect the deer in this region to be *hemionus*. A skull of an old buck from the Guadalupe Mountains north of the Texas line in the collection of Royal H. Wright is not of the *canus*[M] type; and a very large buck that I saw at 8,500 feet on the side of Guadalupe Peak was of the full size of *hemionus*. In the outlying desert ridges west and south of these mountains, where the deer are most abundant, the country is typical of the Lower Sonoran desert region inhabited by *canus*.[M]

a   b

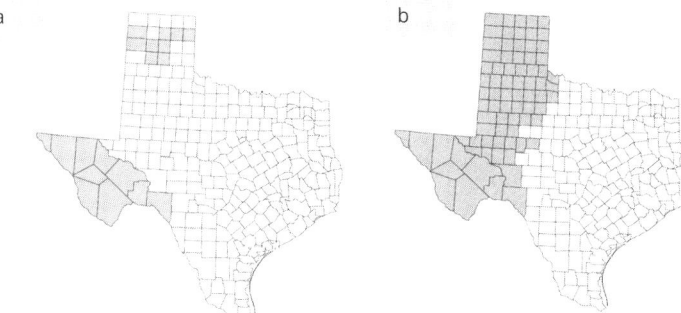

[78] The name *Odocoileus hemionus canus* was changed to *Odocoileus hemionus crooki*, one of two sub-species (*O. h. crooki* and *Odocoileus hemionus hemionus*) of mule deer historically recognized in Texas. *Odocoileus h. crooki* was thought to occur throughout West Texas and the Trans-Pecos region, and *O. h. hemionus* being restricted to the extreme Texas Panhandle. However, Heffelfinger (2000) determined that *O. h. crooki* was actually a hybrid between *O. hemionus* and *O. virginianus couesi*; consequently, *O. h crooki* is an invalid name and all mule deer formally assigned to *crooki* should now be referred to as *O. h. eremicus*. Although mule deer are not nearly as numerous in Texas as white-tailed deer, they do represent an important big game resource in the Trans-Pecos and Panhandle areas. Populations are now fairly stable (expanding in the Panhandle) but they experience periodic declines due to natural mortality factors such as predation, disease, and weather. Population numbers fluctuate in Texas from about 150,000 animals during dry years to about 250,000 during wetter years. It is interesting to note that Vernon Bailey's 1890 mammal report for the Davis Mountains includes a comment that white-tailed deer and mule deer may be producing hybrids in this area. It is now common knowledge that these two species do in fact inter-breed and produce hybrid offspring (Carr et al., 1986; Bradley et al., 2003), although the production of first-generation hybrids appears to be a rare event (Ballinger et al., 1992). A relatively high proportion of backcross individuals have been documented (Cathey et al., 1998). Figure 38a-b depicts the historic and current distribution of mule deer in Texas.

Figure 38a-b. The historic (a) and current (b) distribution of the mule deer (*Odocoileus hemionus*) in Texas.

The mule-deer is still more or less common in many parts of western Texas, in the Guadalupe, Diablo, Franklin, Davis, Santiago, and Chisos mountains, and eastward to Devils River and the rough country along the east side of the Pecos River as far as Fort Lancaster and the Castle Mountains. A few are found in the deep canyons and gulches cutting into the edges of the northern part of the Staked Plains as far east as Washburn and Mobeetie, but it is much more probable that these range from the Rocky Mountains through the extremely rough country along the south side of the Canadian River than that they have a continuous range southward to meet those of the lower Pecos country. Still, it is not improbable that they range, or have ranged in the past, all along the east escarpment of the Staked Plains. None were found in the timbered part of the Chisos Mountains, but they were common in the barren foothills and outlying desert ranges at long distances from any known water. In the Guadalupe Mountains the same distribution was conspicuous, the mule-deer being far more numerous in the barren foothills on the west side of the range where no water has been found than in the high central timbered part. The deer apparently can go for a long time without water, getting an occasional supply from the rock basins after each rain, or, in cases of long drought, possibly making journeys of 20 or 30 miles to permanent springs. It is commonly believed by ranchmen and hunters, and on good grounds, that these deer can live indefinitely without water, getting all the moisture required from juicy plants. They eat the green stalks of the big century plants (*Agave wislizeni* and *applanata*) and paw open the cabbage-like caudex of the sotol (*Dasylirion texanum*) for its starchy and juicy center. Sheep are often herded on green feed for from three days to a month without water, and where there is snow, dew, or rain, for a much longer time. It is not strange that in a country where most of the springs are utilized for ranch use the deer should adapt themselves to desert conditions, especially as they offer them the greatest possible protection. In this open country, however, they are entirely at the mercy of hunters and unless protected by laws strictly enforced will be exterminated as soon as the country is settled.

At Langtry, in March and April, 1903, Gaut reported deer as "very plentiful a few years ago," and said:

I visited the localities where old hunters claimed to have seen large numbers a few years back, and it is safe to say the large numbers are not there now. A young buck and a very large doe were seen. The heads of small rough canyons seem to be their favorite feeding grounds, and at this time of year they seem to feed to a great extent on the blossoms of *Yucca macrocarpa* and *Dasylirion texanum.*

At Mobeetie in 1904 Gaut was told by Mr. Long, a thirty-two-year resident of the locality, that the only mule-deer he could remember having seen in that country was killed in 1896. In the Franklin Mountains in February, 1903, Gaut reported a few mule-deer, but said they were scarce and very wild. On February 12 he saw the track of a small buck at 4,500 feet altitude on the east slope of the mountains about 10 miles north of El Paso.

Antilocapra americana (Ord). Antelope.[M]

79 In traveling by wagon from Ringgold Barracks to Corpus Christi in December, 1852, Bartlett found abundance of antelope on the plains of southern Texas. On January 1, 1853, he says of the prairie:

> Thousands of deer and antelope were scattered over it. Never before had we seen such numbers. Droves of mustangs also appeared. The deer and antelope were usually grazing in herds of from ten to fifty, and as we approached they leisurely trotted off to a short distance and again stopped. We shot none, for I was desirous of reaching Corpus Christi before night.[a]

A few antelope still remain, scattered over the plains of western Texas, mainly west of the one hundredth meridian.

In 1899 they were frequently seen along the stage road 30 miles south of Colorado City and 10 miles north of Sterling, and were said to be common near Gail. Thirty miles north of Gail I saw three antelope and the tracks of others, and along the road to Lubbock the next day saw several more small bunches and many tracks. A few were reported near Tascosa, while from the train I counted 32, singly or in little bunches, scattered over the prairie from Canyon, Tex., to Portales, N. Mex.

In 1900 ranchmen told me that a few still remained on the prairies west of Alice, where they were once numerous; a few were reported to Oberholser 40 miles northwest of San Diego, a few 20 miles west and a few 25 miles southwest of Cotulla, a small herd 30 to 40 miles northwest of Rock Springs, and another small herd 35 miles northwest and another 50 miles southwest of Henrietta; while they were said to have entirely disappeared within a few years from the country about Laredo and from the big valley in which Alpine is located.

In 1901 a bunch of about a dozen antelope was reported near Bone Spring, 50 miles south of Marathon. Oberholser reported them as fairly common in bunches of 3 to 6 between Sherwood and Fort Lancaster, as occasionally seen in the open country a little north of Cornstock and Langtry, and as common on the plains about Hereford and Mobeetie. In 1890 they were common about Sierra Blanca and Marfa and the base of the Davis Mountains, but in 1901 these bands mainly had disappeared, as they had also from most of the open, uncontrolled stock range in that section.

---

[a] Bartlett's Personal Narrative, Vol. II, 526, 1854.

[79] Two subspecies of pronghorn occur in Texas today, *Antilocapra americana americana* in the Panhandle and *Antilocapra americana mexicana* in western and central Texas, although reintroductions, beginning in the late 1930s, to augment a declining population may have altered this situation. In some areas of the Texas Panhandle, pronghorn have adapted to feeding in agricultural areas and seem to be doing quite well, whereas populations in the more xeric west and Trans-Pecos areas have been declining. Currently, pronghorns are estimated to number about 13,000 in the Panhandle and 6,500 in the Trans-Pecos (Woodward, 2020). Figure 39a-b depicts the historic and current distribution of pronghorn in Texas. See Chapter 6 for a discussion of the decline and recovery of the pronghorn (formerly referred to as antelope) in Texas.

a

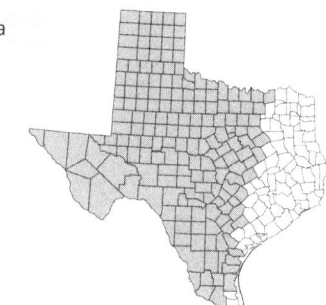

b

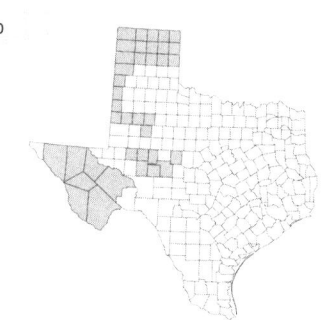

Figure 39a-b. The historic (a) and current (b) distribution of the pronghorn (*Antilocapra americana*) in Texas.

[80] The name *Bison bison* has reverted to the original name, *Bos bison*. The subspecies is *Bos bison bison*. Most of the bison had been exterminated in Texas by the time the Biological Survey began in the late 1880s (see Chapter 5 for a discussion). Descendants of the Charles Goodnight herd (maintained on Goodnight's JA ranch in the Palo Duro Canyon area) were established as the Texas State Bison Herd in 1997 at Caprock Canyons State Park. The herd is kept at a carrying capacity of just under 150 individuals.

In 1902 a few were said to be still found on the open plains north of Rock Springs, and a few on the plains between Valentine and the Davis Mountains, where Mr. John Finley reported 3 in his pasture. At Van Horn within a few years they had disappeared from the valley near the station, but a few were still found in the region farther back from the railroad. At Sierra Blanca, August 3, I saw 3 young antelope which had just been brought in and picketed near the station. Late in August, Hollister and I saw a fine old buck and tracks of a few others on the high plateau at the south end of the Guadalupe Mountains, and on October 2, Hollister saw two small bunches near the railroad between Dalhart and Texline. In crossing the Staked Plains in September I saw from the train 5 antelope near Hereford and 9 near Canyon City, and Mr. Goodnight told me that about 30, which until the previous year had been protected in his pasture, had escaped to neighboring ranches. Cary obtained reports from resident ranchmen of a few on the mesas 15 miles east of Sheffield, of others a short distance to the northeast of Ozona, of a few to the west of Fort Stockton and in the vicinity of Grand Falls, and of others on the east and west sides of the Castle Mountains. He reported them as occasionally seen in the vicinity of Odessa, as common 25 to 50 miles north of Stanton and in smaller numbers 10 or 20 miles south of that place, and as occurring 15 miles north of Abilene, in Jones County. He reported also a bunch of 10 or 12 on the plains 10 miles northwest of Clyde, and a number in the vicinity of Pecos City, but mainly in fenced pastures, where the stockmen protected them and strictly prohibited hunting.

In August, 1903, Howell saw a bunch of 8 or 9 antelope on the prairie 15 miles west of Texline. In February of the same year Gaut reported a few in the valley east of the Franklin Mountains and was informed by Mr. Thomas Robinson, foreman of a large cattle ranch, that these antelope were being protected in one of the ranch pastures.

In 1904 Gaut reported antelope as still common on the plains near Washburn.

It is greatly to be hoped that the Texas State law prohibiting the killing of antelope for a period of five years will be extended indefinitely, as without it both the antelope in the open range and those on the big fenced ranches will soon be exterminated. In no other part of America, with the exception of the Yellowstone and a few private parks, can antelope be expected to last many years, and it is to be hoped for the credit of the State and nation that Texas will protect them for all time.

Bison bison (Linn.). Buffalo; American Bison.

80   Buffalo once ranged over almost the whole of the present State of Texas, and were exceedingly numerous from the coast prairies north over the prairies and plains of the middle part of the State. They were slowly driven back until in 1870 their range, as defined by Doctor Allen,[a] was limited to the plains of the northwestern part of the State. In the next five years they were mostly killed or driven back to the top of the Llano Estacado, where a few remained in the northwest corner of the Pan Handle until 1889, when W. T. Hornaday estimated their number at 25.[b] In 1901 Oberholser was informed by local ranchmen that the last buffalo were seen in 1889 in the Devils River country, and that a few were seen 20 miles north of the mouth of the Pecos the same year. In 1903 Gaut found well-preserved skulls and skeletons of two bulls and a cow in a shallow cave about 10 miles east of Langtry, and saved the skulls for specimens.

In 1894 numerous reports were published in the local Texas papers and copied in Forest and Stream and other journals, describing a herd of buffalo variously estimated at from 30 to 60 head in Valverde County, near the Rio Grande. Later this herd was supposed to have crossed the river and disappeared in Mexico. These reports were shown by Mr. H. P. Attwater to be wholly fictitious.[c] Again, in 1897, a herd of about 80 was reported from Presidio County and the Great Bend of the Rio Grande. In 1901 I could find no one in the Great Bend country who had ever heard of buffalo in that region, nor could I find any evidence to indicate that they ever inhabited the extremely rough and arid country along that part of the Rio Grande Valley.

In 1902 Cary made the following report from Monahans, in the sand-hill region east of Pecos:

Landlord Holman, of the Monahan Hotel, who is an old-timer here, informs me that the last buffalo in the sand-hill region was killed in the winter of 1885 by a professional hunter, George Cansey, who is credited with having killed more buffalo than any other man in Texas. In the fall and summer of 1884 Cansey killed several near the southeast corner of New Mexico, and finally, in January, 1885, while riding to Midland, came up with the last two remaining animals, a cow and calf, near the Water Holes. Cansey shot the cow and roped the calf, which he finally turned over to Mr. C. C. Slaughter, of Fort Worth, who eventually had it killed for a large barbecue. From the same source I learned that the last bull buffalo in the San Angelo region was killed in the fall of 1883, in the southern part of Green County, by a Mr. Mertz, of San Angelo.

At Stanton, Cary says:

I heard from a number of reliable sources that Will Work, who lived at Marienfeld (now Stanton) in the early eighties, killed several buffalo near the New Mexico line, in the western part of Gaines County, Tex., in the winter of 1885.

---

[a] The American Bison, Living and Extinct; J. A. Allen. Geol. Survey of Kentucky, Vol. I, pt. 2, 1876.

[b] Extermination of the American Bison. W. T. Hornaday, Rept. U. S. Nat. Mus. 1889.

[c] Dr. J. A. Allen in Bul. Am. Mus. Nat. Hist., VIII, p. 53, April 22, 1896.

[81] The bighorn or mountain sheep of Texas was once recognized as *Ovis mexicanus*, but was later relegated to the subspecies *Ovis canadensis mexicana*. This subspecies, however, was presumed to have been extirpated by the late 1950s and subsequent reintroductions have been of other subspecies (*Ovis canadensis canadensis* and *Ovis canadensis nelsoni*) from Utah and Nevada. Bighorn sheep were still wide-ranging over many parts of Trans-Pecos Texas during the time of the Biological Survey. Captive breeding and reintroduction programs have been successful and bighorn populations have reached 1,500 individuals and now occur on 11 mountain ranges in Texas. For a discussion of their demise and reintroductions, see Chapter 5. Figure 40a-b depicts the historic and current distribution of the bighorn sheep in Texas.

70   NORTH AMERICAN FAUNA.   [No.25.

In 1899, while crossing the top of the Staked Plains from Gail to Amarillo and Tascosa, I found a few old, much-weathered buffalo horns, but the bones had mostly disappeared. In places the old deeply worn trails leading to water holes were a conspicuous feature of the plains, but where not kept open by range cattle they were heavily sodded over. Farther west on the slope, toward the Pecos River, the outcropping layers of soft limestone are deeply furrowed by hundreds of parallel trails trending toward the river valley. These are the last traces of the wild buffalo in Texas. The well-known herd of Mr. Charles Goodnight, at Goodnight, Tex., numbered in September, 1902, about 50 full-blooded buffalo and 70 crosses of various grades with polled angus cattle. The buffalo are in good condition, quiet and contented, breed freely, and are very hardy. The cows bear only full-blooded calves, and the crosses are made from buffalo bulls to polled angus cows, and then from these half-bloods to three-quarters and seven-eighths buffalo and to three-quarter polled angus, which last cross Mr. Goodnight believes gives promise of establishing a very superior grade of cattle.

Ovis mexicanus[M] Merriam.   Mexican Bighorn; Mountain Sheep.[M]

81   Two 5-year old rams and one 4-year-old from the southern end of the Guadalupe Mountains, Texas, and one 7-year-old ram from the mountains north of Van Horn agree in almost every detail of character with the type and topotypes of *Ovis mexicanus*[a][M] from Santa Maria, Chihuahua. They are older and a little larger than the type, and serve to accentuate some of the characters of the species.

Mountain sheep inhabit the Upper Sonoran and Transition zones of the desert ranges of extreme western Texas. They are found in the Guadalupe Mountains. A few have been killed in the Eagle and Corozones mountains and on the northwest side of the Chisos Mountains. They come into the Grand Canyon of the Rio Grande mainly from the Mexican side. Mr. R. T. Hill reports specimens killed in the Diablo Mountains, 25 miles north of Van Horn. The sheep are by no means confined to isolated mountain ranges. In several valleys I saw tracks where they had crossed from one range to another through open Lower Sonoran country. In this way they easily wander from range to range over a wide expanse of country in western Texas, and might be considered to have an almost or quite continuous distribution between the Guadalupe Mountains and the desert ranges of Chihuahua. Most of the ranges are steep, extremely rugged, and barren, with deep canyons and high cliffs. Here the sheep find ideal homes on the open slopes of terraced lime rock or jagged crests of old lava dikes, and, thanks to the arid and inaccessible nature of the country, they have held their own against the few hunters of the

[a] *Ovis mexicanus* Merriam, Proc. Biol. Soc. Wash., XIV, 29, Apr. 5, 1901.

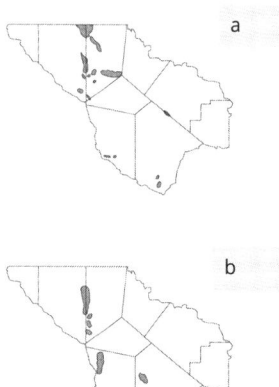

Figure 40a-b. The historic (a) and current (b) distribution of the bighorn sheep (*Ovis canadensis*) in Texas. All current populations are of a different subspecies and were introduced into the state. No descendants of native bighorns exist in Texas today.

region. An old resident of one of the canyons, who has supplied his table with wild mutton for many years, considers them fully as numerous now as fifteen years ago. He has seen as many as 30 in a herd, but says they usually go in small bunches of 3 to 10, sometimes all rams and sometimes all ewes and lambs, but usually in mixed bunches. They come down the sides of the canyon in sight of the ranch, and are shot only when needed.

While sweeping the slopes with the glass one evening near our camp in one of the big canyons opening into the Guadalupe Mountains, I located three sheep halfway up the face of the rocky slope, 1,000 feet above me. To the unaided eye they were invisible among the ledges and broken rocks, whose colors they matched to perfection, but through the glass they were conspicuous as they moved about feeding and climbing over the rocks. There were an old ram, a young ram, and a ewe. It was too near dark to make the long roundabout climb necessary to reach them, so I returned to camp and early the following morning started my camp man up the slope to the spot, while I went back up the canyon to get beyond them if they should run up the ridge. As I swept the slopes with the glass I heard a shot up where the sheep had been the evening before, and soon locating the hunter, watched him shoot two of them, while three others which were above climbed the cliff and finally disappeared over the crest of the canyon wall. The three that escaped were not much alarmed by the shooting. They jumped from rock to rock, pausing to look and listen, and turned back in one place to find a better way of retreat. They made some long leaps to reach the ledges above, but made no mistakes in their footing. Their motions were deliberate, and there was a moment's pause before each bound. I was amazed at the strength of the old ram, as, slowly lifting his massive horns, he flung himself with apparent ease to the rock above. The two lighter animals followed more nimbly, but with less show of power and without the splendid bearing of their leader, who often paused with head high in the air to watch the hunter below or to plan his way up the next cliff. While from below they seemed to be mounting the face of a steep cliff, I found later that it was not difficult to follow where they had gone.

It was interesting to note that these sheep had remained almost exactly where they were seen the night before. The two others may have joined them during the night, but more likely were all the time somewhere near, either lying down or hidden by the rocks.

The stomachs of the two sheep killed were full of freshly eaten and half-chewed vegetation, and most of the plants composing the contents were easily recognized by the stems, leaves, and fruit. The leaves, twigs, and carpels of *Cercocarpus parvifolius* formed a large part of the contents, while the leaves, twigs, and seed pods of *Phila-*

delphus microphyllus were present in less abundance. The seeds, stems, and leaves of the common wild onion of the mountain slopes were abundant and conspicuous in the mass, giving it a strong odor, while the black onion seeds, still unbroken and often in the capsules, were especially noticeable. A few bits of stems and leaves of grass were found in each of the stomachs, but they formed probably not over 2 per cent of the total mass.

Both of these sheep were in good condition, and the meat was tender, juicy, and delicious, with no strong or unpleasant taste. While it lacked the peculiar gamy flavor of venison, it came as near equaling it in quality as the meat of any game I know.

On August 22, in another range in which the bighorns were reported, I left the ranch accompanied by an old resident hunter. Riding hard up one gulch and down another we were soon 10 miles back in the mountains in a canyon with steep terraced walls rising from 1,000 to 2,000 feet above the open bottom. As we crossed the bottom a band of 12 or 15 mountain sheep bounded from the farther edge and started up the rocky slope in a long line of conspicuous bobbing white rumps led by three magnificent old rams. They had a quarter of a mile start, but in a very short time our hard-hoofed little horses had covered the stony gulch bottom and landed us at the base of the rocky slope within 250 yards of the sheep, which, having gained a point of sharp rocks above and feeling more secure, stopped to look down. As the king of the bunch suddenly paused on a sharp point and with a ponderous swing of his heavy horns turned to face us, my little 32-20 sounded weak and ineffective and only served to make him seek a higher ledge. But at the more spirited crack of the old ranchman's 30-30 the next in line, a buck with almost as heavy horns, rolled off the cliff with a broken neck and came sliding and tumbling to the base of the rocks a hundred feet below. The rest had scampered around the point of rocks, and as they came out again farther up and climbed cliff after cliff that from our base level seemed smooth and sheer a few more shots were wasted at long range. The herd divided and passed around both sides of the high peak. Following both trails for a mile or so to see if any of the sheep had been wounded, I found that I could go wherever they had gone. The cliffs were not so steep or so smooth as they had looked from below. In one place the animals had followed a narrow shelf above a sheer drop of 300 feet. Although they had jumped from point to point, striking their feet within an inch of the edge, I could not resist the impulse to lean close to the wall and keep my feet as far from the edge as the narrow shelf, which in places was not a foot wide, would allow. But some of the rocks crossed sloped at a steep angle, and the sheep had made daring jumps from rocky point to sloping surface, where their lives depended on their sure-footedness. The farther I followed the more I admired their skill and nerve. I asked my companion if he had ever known sheep to go where a man could not. He said he thought that they would sometimes make longer leaps down a sheer ledge than a man could attempt with safety, but that otherwise a man could go where they could.

I was especially interested in examining the feet of the old ram we had secured, and was struck first of all by the difference between the front and the hind feet—the front being fully twice as large as the hind, much squarer in form, with deeper, heavier cushioned heels, and lighter and less worn dewclaws. As the hind quarters of the sheep are light and fully two-thirds of the animal's weight comes over the front feet, this difference in size is not surprising. The greater wear of the hind dewclaws is easily accounted for by their constant use in holding back as the sheep goes down hill. While the points and edges of the hoofs are of the hardest horn, the deep, rounded heels are soft and elastic—veritable rubber heels—with a semihorny covering over a copious mass of tough, elastic, almost bloodless and nerveless tissue. While fresh, before the specimen is dried, these cushioned heels may be indented slightly with the thumb. It is easy to see how they would fit and cling to the smooth surface of a sloping rock where wholly hard hoofs like those of a horse would slip, just as you can turn your back to a steep slope of glacier-polished granite and walk up it on the palms of your hands where you can not take one step with the roughest hobnailed shoes. The dewclaws are also heavily cushioned beneath, but have fairly hard, horny points—mere movable, boneless knots. Among other peculiarities noticed in the fresh specimen were the pads of the breast and the knee, where the skin had developed to an almost cartilaginous shield over a quarter of an inch thick and so hard that it was not easily cut through with a sharp knife. The whole sternum and front of the knees were thus protected, and for very evident reasons. The beds where the sheep had been lying were found on rocky or stony shelves, usually above a sharp cliff and below a high wall of rocks sometimes on a bare surface of rock and almost always with at least a foundation of rough stones. If possible the sheep paw out a slight hollow, but they do this apparently more to make an approximately level bed than for the sake of the softness of the little loose dust they can scrape up among the stones. The hair is worn short over the knee and the breast pads, but the skin is unscratched either by rocks or thorns.

The legs of the sheep secured were filled, especially below the knee, with cactus and agave thorns that had gone through the skin and broken off in spikes half an inch to an inch long and lodged against the bone or the inner surface of the skin. A large share of these thorns were the terminal spikes from the leaf blades of *Agave lecheguilla,* which grows in great abundance over the hot slopes of the mountains, and which the horses avoid with even greater care than they do the numerous species of cactus.

The glandular disks under the eyes of this ram were more conspicuous than in any other specimen I have ever examined, probably on account of his mature age, which his horns showed to be 6 or 7 years. The gland is an elevated rim of thickened, black, scantily haired skin, with a depressed center, and measures about an inch across. It stands out prominently on the surface, and appears from the flesh side of the skin as well as from the front as a round thick pad. It has an oily or waxy secretion and a rank, sheepy odor.

In color the old rams were decidedly darker than the ewes and younger members of the herd, but all blended with astonishing harmony into the browned, rusty, old, weathered limestone of their chosen hillsides. Even the soiled white rump patches were just the color of freshly broken faces on the rocks seen here and there over the slopes. As the band of sheep sprang away up the slope the white rump patches were so conspicuous that I could not believe at first that the animals were not antelope; but higher up, as they stopped among the rocks to face us, they could easily have been mistaken for a group of rocks. As they appeared again farther away on the ridge beyond the gulch, the bobbing, white rump patches were conspicuous signal marks so long as the animals were running away from us, but when they turned their forms were completely lost in the background.

These sheep did not appear to run very fast, but probably few animals save the panther can catch them in a race over the rocks. A few days later, while hunting panther in these same hills, it was demonstrated that deerhounds can not catch nor tire out the sheep over their own trails, although my companion claimed that they were not very swift runners on open ground.

The meat of our 7-year-old ram was rather tough and dry, but without any bad flavor. The people at the ranch where I was staying, who had eaten young sheep, considered the meat superior to venison. Although shot at 4 p.m., our sheep had a full stomach and must have been feeding for an hour or two. His teeth were imperfect. One or two molars were missing in the lower jaw, and, as a result, the contents of his stomach were rather coarse, and many of the plants were easily recognized. Over half of the contents was composed of the green stems of *Ephedra trifurcata,* which I at first mistook for grass, but which could not be mistaken on careful examination. The stems, leaves, and flowers of *Tecoma stans,* a beautiful yellow-flowered bush, were conspicuous, as also were the leaves, stems, and berries of *Garrya wrighti.* A few twigs with leaves and fruit pods of *Pentstemon* were found, and a quantity of ripe fruit of *Opuntia engelmanni,* [b] including the chewed-up pulp and seeds of at least half a dozen of the large pear-shaped berries. Some other leaves and stems were found that I could not recognize, but a careful search failed to reveal a trace of grass in the stomach. Part of these plants are Lower and part Upper Sonoran species, and the sheep seem to inhabit the two zones freely. The cold slopes and upper benches of the mountains are Upper Sonoran, however, and probably are to be considered the animals' real home. Transition zone does not occur in this range.

It is with some hesitation that I make public these facts as to the abundance, distribution, and habits of mountain sheep in western Texas, and only in the hope that a full knowledge of the conditions and the importance of protective measures may result in the salvation instead of extermination of the species. It would not be difficult for a single persistent hunter to kill every mountain sheep in western Texas if unrestrained. Not only should the animals be protected by law, but the law should be made effective by an appreciation on the part of residents of the country of the importance of preserving for all time these splendid animals.

Sciurus ludovicianus[a] Custis.   Western Fox Squirrel; Louisiana Fox Squirrel.[M]

82 In eastern Texas the fox squirrels are large and richly colored like those of Louisiana, and a small proportion of their numbers are melanistic. Of 7 specimens taken at Arthur one was almost black, and the hunter with me said that among 14 squirrels killed on a previous hunt 4 were black or very dark. Hollister saw a black squirrel at Antioch and reported many at Rockland. A few black individuals among many of the others were reported at Tarkington. To the west the animals grade without any abrupt change into the smaller, paler colored *limitis.* Specimens from Gainesville and Matagorda county, while intermediate, are in size and color nearer to *ludovicianus* than to typical *limitis.*

Fox squirrels are reported by Loring, Oberholser, and Hollister as more or less common at Texarkana, Waskom, Joaquin, Antioch, Long Lake, Troup, Milano, Brenham, Rockland, Conroe, Jasper, near Beaumont and Sour Lake; and I have found them at Tarkington, Lib-

[82] Current taxonomy refers all fox squirrels (formerly western fox squirrel or Louisiana fox squirrel) in Texas and the United States to a single species, *Sciurus niger* (eastern fox squirrel); consequently, *Sciurus ludovicianus* is no longer considered to be a valid name, having been relegated to subspecific status, *Sciurus niger ludovicianus.* Three subspecies occur in Texas: *Sciurus niger limitus* in most of the western part of the state, *S. n. ludovicianus* in the east, and *Sciurus niger rufiventer* from the Canadian River drainage and adjacent areas of northwestern and extreme north-central Texas. See Chapter 5 for a discussion of the status of this species in Texas.

---

[a] If a type locality can be established for *Sciurus rufiventer*[M] E. Geoffroy (Cat. Mus. Hist. Nat., 1803, p. 174) within the range of the form known since 1800 as *Sciurus ludovicianus* Custis, or if the type specimen sent to Geoffroy by Michaux from America can be identified as the Louisiana form, it will become necessary to revert to the name *rufiventer.*[M] Meanwhile I prefer to use a long-established name in preference to one three years older, the application of which is still open to question.

erty, Richmond, Cuero, Jefferson, Gainesville, and Arthur. Others
reported from Elgin, Austin, Decatur, Brazos, and Wichita Falls are
probably intermediate between *ludovicianus* and *limitis*.

At Arthur in northeastern Texas, and in the Big Thicket region
of southeastern Texas, they inhabit the hickory and oak covered
ridges, and leave the dense river bottoms and swamps entirely to the
gray squirrels; but farther west in the more open country they inhabit
both the timbered river bottoms and the oak ridges. They live

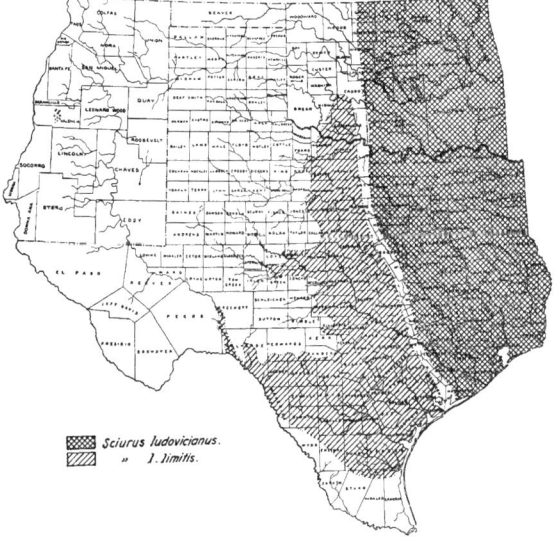

Fig. 10—Distribution areas of fox squirrels (*Sciurus ludovicianus* [M] and *S. l. limitis*). [M]

mainly in hollow trees, but also make bulky nests of leaves and
twigs out on the branches. When alarmed these squirrels run to the
nearest hollow tree or up the first tree with branches leading to
one, and are soon safely hidden inside, but if they do not reach some
safe retreat they are so skillful at hiding that they often escape the
hunter by keeping on the farther side of trunk and branch. Their
food consists mainly of nuts and acorns, but fruit, berries, and
lichens also are eaten. When feeding on nuts their flesh has a delicious
nutty flavor.

Sciurus ludovicianus limitis[M] Baird.   Texas Fox Squirrel.[M]

Sciurus texianus Allen, Bul. Am. Mus. Nat. Hist., XVI, p. 166, 1902. (Not of Bachman, 1838.)[a]

83 The Texas fox squirrel differs from the Louisiana animal in smaller size and paler coloration. So far as I can learn, it is never black. It inhabits the semiarid part of the Lower Sonoran zone, on the west reaching the canyon of the Pecos, the Rio Grande at the mouth of Devils River and at Del Rio, and farther south, extending across into Coahuila and Nuevo Leon, Mexico. To the east it grades into *ludovicianus,* but specimens from the mouth of the Nueces River, San Antonio, Seguin, Brownwood, and Henrietta are referable to *limitis.* There are specimens from Devils River, Del Rio, Rock Springs, Japonica, Ingram, Kerr County, San Antonio, San Antonio River in Victoria County, near the mouth of Nueces River, Cotulla, Mason, San Angelo, Brownwood, Henrietta, and Vernon; and Oberholser reports a few 12 miles north of San Diego and in the Pecos Valley near Fort Lancaster.

In this half-forested mesquite region the little fox squirrels inhabit the timber along the streams, where the pecan, hickory, oak, and little walnut trees furnish their favorite food and a few hollow trees afford protection, but nowhere within their range do they get the deep shade of the forests farther east. Wherever the pecan tree is found along the streams from Kerrville to the Rio Grande they are abundant. Specimens were collected on the Guadalupe at Ingram and Japonica, on the Hackberry near Rock Springs, and on Devils River. A few were seen on ridges between rivers, but they keep mainly to the bottoms. They are closely associated with the pecan tree, in the branches or hollow trunks of which they build their nests, living mainly on its nuts, and rarely wandering away from its shade. Along the Devils River, where these magnificent old trees reach their greatest perfection and form a miniature forest overarching the river with their spreading branches and shading its cool banks for miles, the little fox squirrels abound. Their leafy stick nests are common among the branches, but their safe retreats are the numerous hollows in the gnarled old trunks, the openings of which have been

---

[a] In using the name *texianus*[M] in place of *limitis* for the little pale west Texas fox squirrel Doctor Allen seems to ignore Bachman's excellent description (P. Z. S., 1838, 87) and to base his decision on the fact that one of the specimens mentioned by Bachman was said to have come from Mexico, one from Texas, and one from southwestern Louisiana. It is necessary only to read Bachman's description, with specimens of both species in hand, to be convinced that it applies strictly to the large dark-colored *ludovicianus* and not to the little pale *limitis.* His measurements are the maximum for *ludovicianus.* I find nothing to indicate that Bachman had ever seen *limitis,* unless it be his statement that a specimen of an apparently undescribed species seen in the Museum at Paris was said to have been received from Mexico.

[83] This subspecies is now classified as *Sciurus niger limitis.* See Annotation 82.

[84] The common name of *Sciurus carolinensis* is now the eastern gray squirrel. This species is restricted to the riparian habitats in the eastern third of the state, although an isolated population, presumably the result of an introduction, exists in the Lubbock area (see Chapter 6 for a discussion). Eastern gray squirrels, or cat squirrels as they are sometimes called, once represented an important game animal in eastern Texas. Prior to 1910, these squirrels were abundant over the entire region where suitable habitat was found. However, during the last century there was a drastic reduction of suitable habitat as a result of detrimental land-use practices and their population numbers have declined. Where gray and fox squirrels occur together in eastern Texas, gray squirrels prefer the poorly drained types of forest cover in the hardwood timber along the larger creeks and rivers. Fox squirrels prefer the upland creeks and well-drained bottomlands generally found along the smaller creeks and in the upland pine and hardwood timber. There is evidence to suggest that fox squirrels may be increasing in abundance at the expense of gray squirrels. The drainage of lowland bottomlands seems to result in a reduction of the number of gray squirrels and an increase in the number of fox squirrels. Gray squirrels have been introduced in many places in Texas outside of their natural range.

worn smooth by ages of use as doorways. Usually, however, no protection is needed beyond their quick ear for detecting an approaching footstep, their natural skill at hiding on the farther side of a trunk or branch, and their rapid retreat among the branches from tree to tree.

Late in July we found the squirrels beginning to cut off many of the green pecan nuts, apparently just to test if they were nearly ripe.

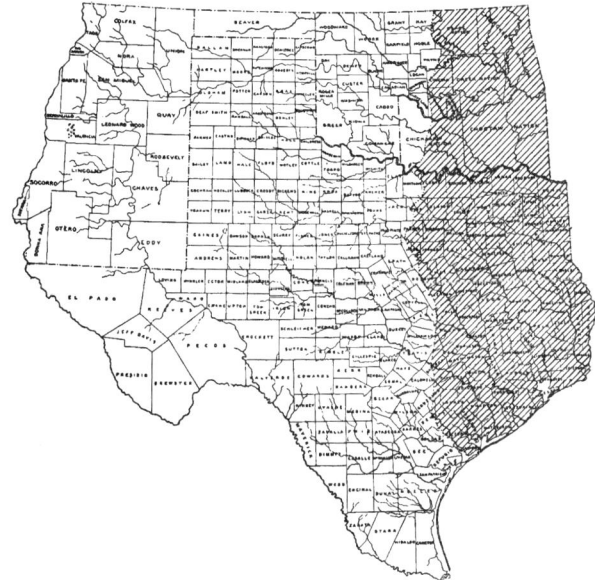

FIG. 11.—Distribution area of gray squirrel (*Sciurus carolinensis*).

The last year's crop of nuts was probably exhausted, as the squirrels were feeding on various other things. Along the Guadalupe River, July 4 to 7, they were eating seeds of the cypress cones, and had their hands and lips covered with pitch, while the ground was strewn with half-eaten cones. It was then too early for them to begin barking much, but a few soft barks of warning were heard near our camp on Devils River late in July.

Sciurus carolinensis Gmelin.   Gray Squirrel.[M]

84   Gray squirrels inhabit the timbered region of eastern Texas as far west as the mouth of the Colorado, Cuero, Austin, and Brazos.

Specimens examined from the mouth of the Colorado, Sour Lake, Liberty, Long Lake, Jasper, Troup, Arthur, and Joaquin are almost typical *carolinensis,* which seems to have a continuous range from the Atlantic coast west through Lower Sonoran zone to its extreme western limit in central Texas. Gray squirrels are reported from Texarkana, Jefferson, Waskom, Antioch, Long Lake, Jasper, Conroe, Rockland, Tarkington, Saratoga, near Beaumont, Brenham, Aledo, and Benbrook, and except along the western edge of their range are usually said to be common or abundant.

They seem to prefer the tall timber of the river bottoms and not to extend west into the lower and more open woods. At Arthur I found them abundant on the flats of the Red River, but found none on the upland ridges, where the fox squirrels were common. The two species seemed to keep entirely apart, and old hunters claim that the gray squirrels choose their ground and keep the fox squirrels away from it. In the Big Thicket of Hardin and Liberty counties, in November and December of 1904, the grays were numerous throughout the heavy timber and dense swamps of the bottoms, while the few fox squirrels were found in scattered groves along the edge of Tarkington and Liberty prairies. Acorns and nuts furnish abundance of food and countless hollow trees offer safe retreats. The squirrels also build numerous branch nests of twigs or Spanish moss or a mixture of the two. The perfect blending of the pelage of a gray squirrel with the gray moss which loads the branches of the trees saves many a squirrel from the hunter.

Sciurus fremonti lychnuchus[M]Stone & Rehn.   Pine Squirrel.[M]

85 Several people in the Guadalupe Mountains claimed to have seen a small, dark-colored tree squirrel, which they said was very rare. I failed to find any traces of it, however, although the timber and country are well adapted to squirrels. Pine squirrels are common in the Sacramento Mountains, a little farther north, and it is not improbable that a few may find their way south along the crest of the range and across the Texas line.

Sciuropterus volans querceti[M]Bangs.   Florida Flying Squirrel.[M]

86 Texas specimens from Texarkana, Gainesville, Troup, and Tarkington agree perfectly with the Florida subspecies and differ from typical *volans* in slightly darker coloration, dusky instead of whitish toes of the hind feet, and in slenderer nasals and muzzle and larger audital bullae.

Flying squirrels are reported from numerous localities over eastern Texas, where they seem to be fairly common and to have a continuous distribution throughout the forested region. The westernmost records are from Elgin, where Oberholser reported the species

[85] Current taxonomy refers the red squirrel (formerly pine squirrel) from New Mexico to *Tamiasciurus hudsonicus lychnuchus* instead of *Sciurus fremonti lychnuchus*. Despite the speculation by Bailey, this squirrel has never been documented in the Guadalupe Mountains or any other place in Texas.

[86] Current taxonomy refers *Sciuropterus volans querceti* to *Glaucomys volans texensis* and restricts the former name to a subspecies in Florida and southeastern Georgia. The subspecies of southern flying squirrel that occurs in Texas, *G. v. texensis*, was described by A. H. Howell (1915) on the basis of specimens collected during the Biological Survey from 7 miles (11 km) northeast of Sour Lake, Hardin County, in the Big Thicket (Howell, 1918). The status of this species is difficult to predict as we extend into the twenty-first century. It depends on old hardwood timber for nesting cavities, and logging practices could negatively impact populations.

as tolerably common; and from Aledo and Benbrook (just west of Fort Worth), where Cary saw a stuffed specimen which came from that place, and was told of a family of 8 taken in 1901 from a hole in an elm a mile west of Aledo. They have been reported from Guadalupe River, Richmond, Brenham, Long Lake, Antioch, Rockland, Saratoga, Sour Lake, Conroe, Jasper, Waskom, Jefferson, and Texarkana.

Flying squirrels are among the most strictly nocturnal of mammals and are rarely noted except by timber cutters, who see them

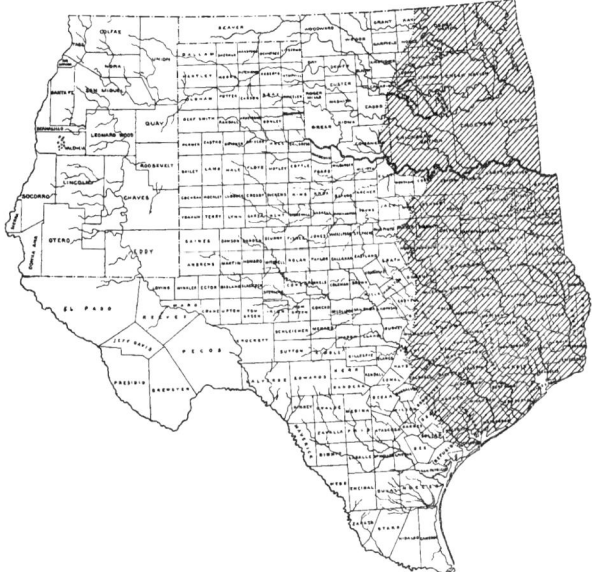

FIG. 12.—Distribution area of flying squirrel (*Sciuropterus volucella querceti*).

flying from their nests in falling trees. While every wood chopper in the east Texas region is familiar with them, it is difficult to get specimens. They are not easily trapped and often live in hollows in the large trees, where pounding with an ax does not start them from their nests. While hunting in the Big Thicket of Liberty and Hardin counties I often heard their fine, whistling squeak from the branches over my head at night, and occasionally the rustle of their feet on the bark of a tree close to the trail I was following.

At Mike Griffin's place, 8 miles northeast of Sour Lake, Gaut was shown a dead pine out in the field where flying squirrels were said to live in a deserted woodpecker's hole. By pounding on the base of the tree two flying squirrels were driven out and secured. A few days later two more were driven out of the same tree and secured, and again a few days later two more, making in all six specimens from one woodpecker's nest.

Eutamias cinereicollis canipes[M]Bailey.   Gray-footed Chipmunk.

87 The gray-footed chipmunks are common in Transition zone throughout the Guadalupe Mountains, from 7,000 feet in Dog Canyon and 6,000 feet in Timber Canyon up to at least 8,500 feet and probably to the top of the peaks at 9,500 feet, at which altitude they are common in the Sacramento Mountains a little farther north. While none were found in the lower part of the range, between the Guadalupes and the Sacramentos, they seem to be identical in the two ranges and may easily have a continuous distribution between. In the Sacramento Mountains they occupy the whole width of both the Transition and Canadian zones. In the Guadalupe Mountains they range from the lower edge of the Transition zone upward with the yellow pine and Douglas spruce, but in September they are more closely associated with the shrubby oaks, several species of which are abundant over the upper slopes of the mountains. They were occasionally seen in the densest timber, but more often in the open oak scrub, gathering the little sweet acorns in the tops of the bushes, or sitting on logs or rocks eating them. Both logs and rocks were covered with acorn shells. Occasionally these chipmunks were seen in the lower branches of a tree, but when alarmed they always ran to the ground and disappeared among rocks, logs, or brush. They were very shy, and in the thick cover it was difficult to get specimens. Their light 'chipper' was often heard from the bushes, and on a few occasions I heard their low 'chuck-chuck-chuck,' repeated slowly from a log or rock or the low branch of a tree, but it always ceased as soon as danger was suspected.

Ammospermophilus interpres (Merriam).   Texas Antelope Squirrel.

88 The Texas antelope squirrel is common along the Rio Grande from El Paso to the mouth of the Pecos, but less common up the Pecos Valley to the Castle Mountains and in the country between the Rio Grande and Pecos valleys in Texas. Specimens collected at El Paso, Boquillas, Pecos High Bridge, Fort Lancaster, Castle Mountains, south end of Guadalupe Mountains, and Sierra Blanca carry the range of the species over the extremely rough and arid Lower Sonoran region of western Texas, but indicate a very irregular range along the course of canyons and the foothills of barren, desert moun-

[87] *Eutamias cinereicollis canipes* is now classified as *Tamias canipes canipes* (Nadler et al., 1977; Levenson et al., 1985). Bailey described this taxon on the basis of specimens collected from 7,000 feet (2,133 m) in Dog Canyon in the Guadalupe Mountains on 24 August 1902. These chipmunks are most numerous in coniferous forests where fallen logs, in which they often build their nests, are common. In a survey of the mammals of Guadalupe Mountains National Park, Hugh Genoways and associates (1979) obtained nine specimens of the gray-footed chipmunk near The Bowl and in Upper Dog Canyon at the higher elevations of the park. The current status of the species appears to be good within the protected confines of the national park. Subsequent to the Biological Survey, populations of the gray-footed chipmunk were documented from the Sierra Diablo Mountains in southwestern Culberson County, but this population appears to have been extirpated.

[88] Bailey is the first author credited with the application of the name *Ammospermophilus interpres* for the Texas antelope squirrel which was described as *Tamias interpres* by Merriam (1890) on the basis of specimens obtained by Bailey at El Paso in 1889. This species is more widely distributed in western Texas than depicted by Bailey's map. It occurs throughout the Trans-Pecos region (not just adjacent to the drainages of the Rio Grande and Pecos rivers), extending eastward to Reagan, Crockett, and Kinney counties and as far north as Gaines County (Fig. 41). These ground squirrels inhabit rocky foothill terrain, characterized by shrub desert habitat. Despite the speculation by Bailey, it is now known that these squirrels do not hibernate and are active throughout the year. In fact, they are one of the few small mammals active during the day in the hot summer months of the Big Bend.

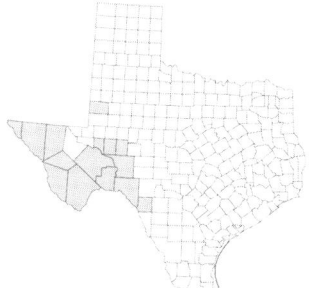

Figure 41. The current distribution of the Texas antelope squirrel (*Ammospermophilus interpres*) in Texas.

tains. In fact the presence of canyons, bare cliffs, and rocks, with which the species is closely associated, seems to be the determining factor of its range within its zone.

Near El Paso and in the Great Bend of the Rio Grande, near Boquillas, these little squirrels live along the steep banks of the river or in the narrow side gulches that cut back into the barren mesas. Along the Pecos Canyon they are found on the rock shelves of the canyon walls; and around the Castle and Guadalupe mountains, and

FIG. 13.—Distribution area of Texas antelope squirrel (*Ammospermophilus interpres*).

at Sierra Blanca they occur in rocky gulches or along low cliffs. They burrow under the edge of a bowlder or around the base of a bunch of bushes or cactus, and are usually seen either running from bush to bush, sitting on a point of rock, or running over the rocks with their short, bushy tails curled tight over their rumps. Sometimes they climb to the top of a cactus or low bush, apparently in search of food, but at the first alarm they rush for a burrow or the nearest rocks.

Near Boquillas in May the half-grown young were out with the others getting their own food from the various seeds and fruits, and climbing the acacia and mesquite bushes to secure the ripening bean pods, which were found scattered in abundance about their burrows. The stomach of one shot in September in the Castle Mountains by Gordon Donald was full of the fruit of *Opuntia engelmanni*,[P] which Cary, who examined the specimen, thinks must have been the squirrel's steady diet for some time, as its flesh was tinted throughout with the purple color of the fruit.

In autumn these little fellows become very fat and probably hibernate during the coldest weather. At El Paso in December, 1889, I found them out on warm days, although very lazy and sluggish. They were then feeding on various seeds, including those of the creosote bush. They were in the beautiful long silky winter fur, very different from the short, harsh summer coat. Along the east base of the Franklin Mountains, in February, 1903, Gaut found them running about in a drizzling rain when the temperature was close to freezing.

Citellus variegatus couchi[M] (Baird).   Couch Rock Squirrel.[M]

89   These black-headed, or often entirely black, rock squirrels are common throughout the Chisos and Davis mountains and along the canyons of the Rio Grande, Pecos, and Devils rivers. While varying in color in the gray phase from entirely dark gray to the usual gray back and black head or crown, no specimens I have seen show the combination of black back and gray rump of *buckleyi*,[M] nor the light gray head and shoulders of *grammurus*. A specimen collected at Boquillas and one on the Rio Grande near Comstock are entirely black, exactly like Baird's type of *couchi*.[M] Several other entirely black individuals have been seen along the Rio Grande and near the month of the Pecos in company with the gray ones, while seven entirely black specimens collected at Santa Catarina, Mexico, the type locality of *couchi*,[M] by Nelson and Goldman, seem to prove complete dichromatism for the species.

In the Davis and Chisos mountains the rock squirrels range with the oaks and junipers in canyons and over rocky slopes throughout the Upper Sonoran zone; while along the river canyons they are confined to the Lower Sonoran zone with the modifying influence of canyon walls and narrow gulches. Along the canyons they are usually found sitting on the prominent points of rocks, and their loud whistle often reverberates from side to side. When alarmed they disappear among the rocks or climb to the tops of the tallest cliffs. In the mountains they live mainly among the rocks, cliffs, and ledges, but range out among the oaks and junipers for food. They climb the trees for acorns and berries, but when surprised in the branches they always rush to the ground and scamper away to the nearest rock pile or burrow. During early summer they feed ex-

[89] For several decades, taxonomists referred all of the ground squirrels to the genus *Citellus*, until Bryant (1945) provided the justification for the application of the name *Spermophilus*. More recently, Helgen et al. (2009) split *Spermophilus* into multiple genera, including three in Texas: *Ictidomys*, *Otospermophilus*, and *Xerospermophilus*. Consequently, *Citellus variegatus* was changed to *Otospermophilus variegatus*. The subspecies in Texas are *Otospermophilus variegatus grammurus* from the Trans-Pecos region and Stockton Plateau (Hall, 1981; Goetze, 1998) and *Otospermophilus variegatus variegatus* in the Edwards Plateau and Hill Country.

[90] Under the arrangement proposed by Helgen et al. (2009; see Annotation 89), *Citellus variegatus buckleyi* has been changed to *Otospermophilus variegatus buckleyi* for populations in the Edwards Plateau and central Texas. Rock squirrels show considerable variation in color, ranging from specimens that are light gray to dark gray to entirely black. Specimens of the subspecies *O. v. buckleyi* from the Edwards Plateau show a large preponderance of entirely black individuals, which is why Bailey applied the common name "black-backed rock squirrel" to this taxon. Rock squirrels tend to be found in the lower life zones from the pine-oak forests down into the desert. They occupy broken, usually rocky terrain and are common in rocky hillsides and along arroyos. They are quite common and have adapted well to human encroachment. The taxonomy of rock squirrels is badly in need of revision, with many taxonomists suggesting that all rock squirrels in West Texas should be referred to a single subspecies, *Otospermophilus variegatus grammurus* (*Spermophilus variegatus grammurus* of Jones and Jones, 1992).

tensively on the old juniper berries and acorns of the previous year, digging for them under the trees and in many places keeping the ground well stirred. By the middle of July they begin on the nearly matured acorns of one of the black oaks (*Quercus emoryi*) and also on the new crop of juniper berries (*Juniperus pachyphlœa*).[P] Some of those shot were feeding largely on green foliage, the leaves of clover and various plants, and along the Rio Grande mainly on the juicy fruit of *Opuntia engelmanni*. None of those taken in summer were very fat, but in January, 1890, in the Davis Mountains, I found them excessively so. They were then keeping very quiet and came out of their rocky dens only on warm days.

Citellus variegatus buckleyi[M] (Slack). Black-backed Rock Squirrel.[M]

90 This, the handsomest of the rock squirrels, with glossy black head and shoulders, inhabits a restricted area in the rough and semiarid mesquite country along the eastern slope of the southern arm of the Staked Plains, from Mason and Llano to a little west of Austin and San Antonio, and again west to Kerrville and the head of the Nueces River. Specimens examined from Mason, Llano, near Austin, near Kerrville, Japonica, and Rock Springs (39 in all) do not vary to any great extent, except that in a few the black extends over the back to base of tail.

Along the upper branches of the Guadalupe and Nueces rivers these squirrels are common in rocky places. I saw them near Ingram and collected specimens near Japonica and on Hackberry Creek near Rock Springs, while the ranchmen reported them as common in all rocky gulches throughout this strip of rough country. West of Rock Springs we did not find any trace of rock squirrels, there being no suitable country, until *couchi*[M] was found in the lower part of Devils River Canyon. Apparently the open divide between the headwaters of the Nueces and the headwaters of the streams flowing into the Rio Grande separates the ranges of *buckleyi*[M] and *couchi*[M] with a neutral strip in which neither occurs. Near Camp Verde, in Kerr County, Cary found them common in the rocky cliffs, where he secured specimens, and was told by the ranchmen that the squirrels had a habit of appearing in considerable numbers on the cliff just before a storm. They did this with such regularity that the ranchmen depended on it as a sure sign of rain.

Mr. J. H. Tallichet, of Austin, sent a specimen from Bull Creek, Travis County, and wrote, under date of September 18, 1893:

> I send to you by this mail a specimen of the spermophile which occurs in this part of the State. * * * The specimen is an immature male which I killed while camping last year. His cheek pouches were filled with corn and melon seeds. These rock squirrels live in the debris at the foot of the canyon walls and are very wary. Full-grown specimens are nearly as large as tree squirrels and are eaten by the country people.

In habits *buckleyi*[M] is a true rock squirrel, and is never seen at any great distance from cliffs or broken ledges. At Llano I found one pair near a cliff living in a hollow oak tree which they entered by holes in the branches 15 or 20 feet from the ground. They climbed the tree and disappeared—as quickly as any tree squirrel could have done—and did not show themselves at the openings for half an hour. Generally, however, the squirrels are found sitting on the rocks doing picket duty, ready at the slightest alarm to slide noiselessly over the edge of a rock into a burrow, under a bowlder, or into a break in the cliff. They are exceedingly shy and have to be stalked as carefully as an antelope. By the middle of May the half-grown young are out caring for themselves and feeding in the same manner as the adults.

Piles of acorn shells near the burrows indicate that acorns, when obtainable, are the principal food of the squirrels, which in summer, however, feed mainly on flowers, fruit, and green vegetation. The stomach of one examined contained mostly pulp of green cactus fruit (*Opuntia engelmanni*), together with parts of the big yellow cactus flower, while several of these flowers with the green berry attached were found on the rocks where the squirrels were in the habit of sitting. The stomach of another was filled with the white starchy pulp from the base of young leaves of *Yucca stricta.*[P] Most of the yucca plants near the dens of the squirrels had part of their leaves cut out, and on examination I found the base of these leaves tender, sweet, and starchy, with a rather pleasant flavor. Another individual had the stems and leaves of a little stonecrop in its pouches. Flowers seemed to be a rather common food, and the contents of the stomachs often showed spots of red, yellow, and blue from the various species eaten. A squirrel shot on Hackberry Creek at the edge of a little corn field July 14 had its cheeks stuffed full of green corn, and the field showed many ragged ears.

Most of the squirrels collected in May were lean and muscular, but one that happened to be in good condition proved as good eating as any tree squirrel, while the young of the year were always tender and delicious.

Citellus variegatus grammurus[M] (Say).   Rock Squirrel.[M]

91   The rock squirrels of the southern Guadalupe Mountains and the Franklin Mountains near El Paso, Tex., are typical *grammurus,* with light gray head and shoulders. In the Guadalupe Mountains they are common, together with the junipers and oaks, from 4,000 to 7,000 feet throughout the Upper Sonoran zone. They usually live along the rocky canyons, but are sometimes seen in the open woods, where they climb the trees for the sweet berries of *Juniperus pachyphlœa*[P] and the little acorns of the gray oak, or dig acorns of the previous year from the ground under the trees. Down in the foothill canyons

[91] Under the arrangement proposed by Helgen et al. (2009; see Annotation 89), *Citellus variegatus grammurus* has been changed to *Otospermophilus variegatus grammurus.*

[92] Under the arrangement proposed by Helgen et al. (2009; see Annotation 89), *Citellus mexicanus parvidens* has been changed to *Ictidomys parvidens*, the Rio Grande ground squirrel. Edgar Mearns (1896), who led the biological survey of the United States Mexican boundary area (see Chapter 1), described this subspecies on the basis of specimens collected at Fort Clark in Kinney County in 1893.

[93] Under the arrangement proposed by Helgen et al. (2009; see Annotation 89), *Citellus tridecemlineatus texensis* has been changed to *Ictidomys tridecemlineatus texensis*. This subspecies has become rare in the prairies extending through the middle part of the state as those areas have become encroached by brush and chaparral. Merriam described this subspecies on the basis of specimens sent to the Biological Survey by a Mr. G. H. Ragsdale from Gainesville in Cooke County (Merriam, 1898a). Ragsdale, who obtained the specimens on 15 April 1886, wrote newspaper articles about natural history in Cooke County.

we found them feeding on cactus fruit (*Opuntia engelmanni* [P] and *Cereus stramineus*) [P] and walnuts (*Juglans rupestris*). [P] One specimen shot in Dark Canyon had thirteen of these little walnuts of the size of small cherries in its cheeks and a lot of cactus fruit in its stomach. They are shy and usually silent, but when danger threatens, their loud, vibrant whistle rings back and forth from the canyon walls.

Citellus mexicanus parvidens [M] (Mearns).   Rio Grande Ground Squirrel.

92   The Rio Grande ground squirrels show no important geographic variation over a wide range in western Texas. Specimens from Brownsville are a little larger than typical individuals, and those from Altuda are of minimum size. They inhabit approximately the whole mesquite region or arid Lower Sonoran zone of Texas; are common at Brownsville, Rockport, Mason, Colorado, and Gail, in the Pecos Valley north to Roswell, and westward to the Rio Grande and beyond. Wherever the scrubby mesquite tree grows their burrows are sure to be found under its shade, or, if in the open, near enough to it for them to feed on the sweet pods, pieces of which are often seen scattered around their holes. They are strictly 'ground squirrels,' and climb only into low bushes for seeds and fruit, and depend entirely on their burrows for protection. Like most of the smaller ground squirrels of the arid regions, they usually burrow under the edge of a cactus or some low, thorny bush, where they obtain shade and the protection of thorny cover. They apparently do not hibernate, but during the cold weather have the unsquirrel-like habit of closing their burrows and remaining inside. I have caught them in these closed burrows at Del Rio in January and at Dryden on the 9th of May, when my traps were set, as I supposed, for pocket gophers or moles. Also near Rock Springs in July I found closed burrows that I attributed to this species. The habit of closing the entrance of the burrow is unusual in the squirrel family, but may probably be accounted for as a protection against enemies, and especially snakes. Near Rock Springs I took a half-grown squirrel from a bull snake which had killed and just begun to swallow it.

Like other members of the genus, these ground squirrels feed on seeds, grain, fruit, green foliage, lizards, and numerous insects, and often gather around gardens and grain fields, where they do considerable damage in spring by digging up corn, melons, beans, and various sprouting seeds, and, in summer and fall, by feeding on the ripening grain. Specimens examined at Roswell, N. Mex., in June were feeding on about equal proportions of seeds and insects.

Citellus tridecemlineatus texensis [M] (Merriam).   Texas Ground Squirrel. [M]

93   This southernmost form of the 13-striped ground squirrel occupies a narrow strip of half prairie country through the middle part of

Texas, where the timber and plains intermingle—from Gainesville and Vernon on the north to Richmond and Port Lavaca on the south. Apparently its range is more or less broken and scattered, although the animal is common in places. A little colony was found at Richmond, and Oberholser saw a mounted specimen at Port Lavaca, said to have been killed near the town, where they were reported as occurring.[a]

In habits, voice, and general appearance they do not differ much from *tridecemlineatus.* They live in the open grassy prairies or around fields and depend on their burrows for shelter and their striped brown coats for protection. They feed largely on grasshoppers and other insects, together with seeds, grain, fruit, green herbage, and flowers.

Citellus tridecemlineatus pallidus[M] (Allen).　Pale Ground Squirrel.[M]

94 The little, pale striped ground squirrel is common in Upper Sonoran zone over the top of the Staked Plains, where it is often seen running through the short grass or standing erect and stake-like at the edge of its burrow. A number of specimens collected in August at Washburn had been feeding mainly on grasshoppers, which were abundant over the plains. A few other insects were noted in the stomachs examined, and one of the spermophiles had been eating the fruit of the small prickly pear (*Opuntia macrorhiza?*), the seeds of which were stored away in his pouches.

Citellus spilosoma major[M] (Merriam).　Large Spotted Ground Squirrel.[M]

95 The spotted spermophiles from Lipscomb, Canadian, Miami, Mobeetie, Colorado, Pecos City, and Monahans, Tex., and Carlsbad, Roswell, and Santa Rosa, N. Mex., agree in their large size and coarse, indistinct spotting with *major* from Albuquerque, N. Mex., but show slight variation with almost every change of soil and surroundings. The foregoing localities, which completely surround the Staked Plains, lie near the junction of Upper and Lower Sonoran zones, but as the species ranges north to Las Animas and Greeley, Colo., it apparently belongs to Upper Sonoran.

These quiet, shy, inconspicuous little ground squirrels live in burrows under the edge of clumps of bushes or on open, grassy plains. Their fine, trilling whistle is often heard from behind a bush or weed patch. I have found their stomachs full of grasshoppers and beetles and their pouches full of seeds of sand bur (*Cenchrus tribuloides*),[P] and have seen little heaps of the empty bur shells scattered about

[94] Under the arrangement proposed by Helgen et al. (2009; see Annotation 89), *Citellus tridecemlineatus,* the thirteen-lined ground squirrel, has been changed to *Ictidomys tridecemlineatus.* Consequently, the specimens recorded by Bailey in the Panhandle are recognized today as *Ictidomys tridecemlineatus arenicola.* The ranges of the two subspecies, *I. t. arenicola* and *Ictidomys tridecemlineatus texensis,* are not as disjunct as indicated by Bailey (Fig. 42). Specimens of *I. t. texensis* have been recorded from Hardeman and Foard counties along the Red River, and *I. t. arenicola* is known from most of the counties in the Panhandle. This is a widely occurring and common ground squirrel throughout the Texas Panhandle (Jones, J K., Jr., et al., 1988b).

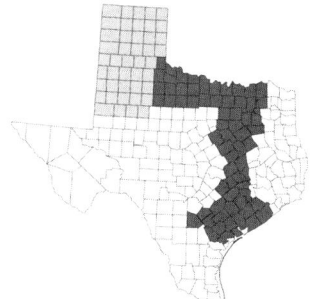

Figure 42. The current distribution of the thirteen-lined ground squirrel (*Ictidomys tridecemlineatus*) in Texas. Light shaded areas represent *I. t. arenicola* and the dark shaded areas represent *I. t. texensis.*

---

[a] Two specimens recorded by Doctor Allen from Bee County, Tex. (Bul. Am. Mus. Nat. Hist., III, p. 223, 1890), as *Spermophilus tridecemlineatus,*[M] eight years before *texensis* was described, I assume to be referable to this form.

[95] Under the arrangement proposed by Helgen et al. (2009; see Annotation 89), the specimens identified by Bailey as *Citellus spilosoma major* have been changed to *Xerospermophilus spilosoma major.* Although Bailey recognized four subspecies of spotted ground squirrels in the state, only three are considered valid today: *Xerospermophilus spilosoma annectens* in the southern part of the state, *Xerospermophilus spilosoma canescens* in far western Trans-Pecos, and *Xerospermophilus spilosoma marginatus* in the remainder of the range. *Xerospermophilus s. major* is now placed in synonymy of *X. s. marginatus.* This species is common throughout its range in Texas.

[96] Under the arrangement proposed by Helgen et al. (2009; see Annotation 89), the name *Citellus spilosoma marginatus* has been changed to *Xerospermophilus spilosoma marginatus.*

[97] The name *Citellus spilosoma arens* was changed to *Spermophilus spilosoma arens* and later placed into synonymy with *Spermophilus spilosoma canescens* by Bailey (1932). However, under the arrangement proposed by Helgen et al. (2009; see Annotaton 89), the currently recognized name is *Xerospermophilus spilosoma canescens.*

[98] *Citellus spilosoma annectens* was described by C. Hart Merriam on the basis of specimens collected by William Lloyd on Padre Island (Merriam, 1893). However, under the arrangement proposed by Helgen et al. (2009; see Annotation 89), the currently recognized name is *Xerospermophilus spilosoma annectens.* This subspecies is common throughout its range in Texas, and populations remain common on the island today.

their burrows. Usually they are not sufficiently numerous in agricultural regions to do serious damage in the grain fields.

Citellus spilosoma marginatus[M] (Bailey).   Brown Ground Squirrel.[M]

96 This little brown, sharply spotted ground squirrel is apparently an Upper Sonoran form, living on the dark lava soil of the Davis Mountain plateau. The type was caught in the open valley near Alpine, and others were seen on the mesa at Fort Davis and along the east base of the mountains. Specimens from Valentine, Presidio County, Van Horn, and Toyahvale are referred to the species, though not all are typical. The Toyahvale specimens show some of the characters of *major.*[M]

In no part of their range have we found these spermophiles common, but like other members of the group they are inconspicuous, shy little fellows, rarely heard or seen. They burrow in the open or under the edge of a bush or cactus and usually keep close to their homes. They often live under the dense, spinescent bushes of *Microrhamnus,* which is common in this region.

Citellus spilosoma arens[M] (Bailey).   Spotted Sand Squirrel.[M]

97 These little sand-colored ground squirrels are common in the open part of the valley bottom below the town of El Paso, where they make their burrows in the sand banks among scattered bushes of *Atriplex, Suaeda,* and mesquite, with little protection from the glaring light and scorching heat of summer. Their coloration is wonderfully protective, and being shy little animals and not very abundant they are rarely seen unless located by their fine bird-like whistle. They seem sensitive to a slight degree of cold and apparently hibernate early in winter, for I could find no trace of them in December about the same holes where I had caught them the previous July. Doctor Fisher found them common in May, but says that a windy day kept them in their burrows.

Citellus spilosoma annectens[M] (Merriam).   Padre Island Ground Squirrel.[M]

98 A number of specimens, including the type of *annectens,* were taken by William Lloyd near the two ends of Padre Island, and others by H. P. Attwater on Mustang Island.[a] In August, 1891, Lloyd says they were abundant, but in November, when the island was again visited, only one was seen. Apparently they were keeping mainly in their burrows. He reported them also from the mouth of the Rio Grande and at Rio Grande City, but secured only one specimen on the mainland—on the sandy beach at the mouth of the Rio Grande. He says that they seem to live in the crab burrows and are very shy, but their call note, similar to that of a grass finch, is occasionally heard.

---

[a] Bul. Am. Mus. Nat. Hist., VI, p. 182, 1894.

Cynomys ludovicianus (Ord).   Prairie Dog.[M]

99 The prairie dogs inhabit an area comprising more than one-third of the State of Texas. Their range extends from Henrietta, Fort Belknap, Baird, and Mason west almost to the Rio Grande, north over the Staked Plains and the Pan Handle region, and south to the head draws of Devils River, to 10 miles south of Marathon and 25 miles south of Marfa. While to the northward inhabiting mainly Upper Sonoran zone, in Texas they extend well into the upper edge

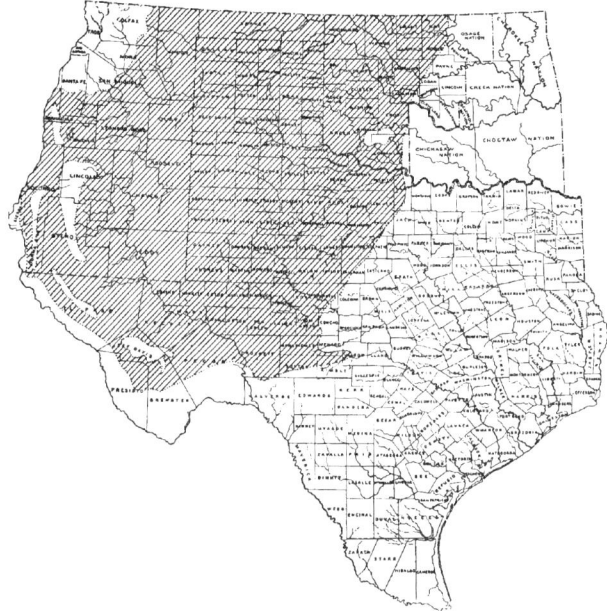

Fig. 14.—Distribution area of prairie dog (*Cynomys ludovicianus*).

of Lower Sonoran. So far as I can learn, they are not found in the immediate valley of the Rio Grande or nearer to the river than Sierra Blanca, except one little colony 2 miles east of Fort Bliss, nor do they occur elsewhere in the Lower Sonoran zone much beyond the scattered traces of Upper Sonoran species of plants. Normally they belong to Upper Sonoran zone, but their strong tendency to expansion carries them slightly beyond its bounds. In the Davis Mountains they range up to 5,800 feet in an open valley on Limpia Creek at a point where the first yellow pines appear, while on the main ridge

[99] Bailey did not refer Texas populations of the prairie dog (now referred to as the black-tailed prairie dog) to subspecies. Today, however, two subspecies are known in the state: *Cynomys ludovicianus arizonensis* in the Trans-Pecos and *Cynomys ludovicianus ludovicianus* elsewhere (Hall, 1981). The large concentrations of prairie dogs described by Bailey, such as between San Angelo and Clarendon, are a thing of the past in Texas. Land conversion to agriculture and the extensive use of poisons to kill animals have considerably reduced their former range and numbers. Past state law, since repealed, required a landowner to destroy all prairie dogs on his property. The result is a scarcity of large prairie dog towns today. In the northern Panhandle, prairie dogs have been observed at numerous localities on rangelands, but these colonies rarely number more than a few hundred individuals. The decline in prairie dogs has been so dramatic during the past few decades that in 1998 the National Wildlife Federation petitioned the USFWS to list the species as threatened or endangered (see Chapter 5).

of the Guadalupe Mountains and in Dog Canyon, which is named for them, they straggle up to 6,900 feet, or to the very upper limit of the Upper Sonoran zone. Usually they are found in scattered colonies or 'dog towns,' varying in extent from a few acres to a few square miles, but over an extensive area lying just east of the Staked Plains they cover the whole country in an almost continuous and thickly inhabited dog town, extending from San Angelo north to Clarendon in a strip approximately 100 miles wide by 250 miles long. Adding to this area of about 25,000 square miles the other areas occupied by them, they cover approximately 90,000 square miles of the State, wholly within the grazing district. It has been roughly estimated that the 25,000 square mile colony contains 400,000,000 prairie dogs.[a] If the remaining 65,000 square miles of their scattered range in the State contains, as seems probable, an equal number, the State of Texas supports 800,000,000 prairie dogs. According to the formula for determining the relative amount of food consumed by animals of different sizes (Yearbook Department of Agriculture, 1901, p. 258), this number of prairie dogs would require as much grass as 3,125,000 cattle.

In many places the prairie dogs are increasing and spreading over new territory, but on most of the ranches they are kept down by the use of poison or bisulphid of carbon, or, better, by a combination of the two. As a Texas cattle ranch usually covers from 10,000 to 100,000 acres, the expense of destroying the prairie dogs in the most economical manner often means an outlay of several thousand dollars to begin with and a considerable sum each year to keep them down.[b]

The increase of prairie dogs is plainly due to the destruction of their natural enemies, badgers, coyotes, foxes, ferrets, hawks, eagles, owls, and snakes, many of which are destroyed wantonly.

The prairie dog is a plump, short-eared, short-legged, short-tailed little animal of the squirrel family, cleanly in habits, good-natured, and eminently social in disposition. If there are only a dozen in a big valley they will be located on an acre of ground where they can visit back and forth among the burrows, play or fight, and take turns in standing guard. If there are thousands of them their burrows will be found close together over the plain to the edge of the 'dog town,' beyond which none will be seen for perhaps 10 or 20 miles. On a trip from San Angelo north over the Staked Plains we were with them for weeks, both in the region of their continuous range and among

---

[a] Yearbook U. S. Department of Agriculture, 1901. p. 258.
[b] For methods of destroying prairie dogs see 'The Prairie Dog of the Great Plains,' by C. Hart Merriam (Yearbook U. S. Department of Agriculture, 1901, pp. 257-270), and 'Destroying Prairie Dogs and Pocket Gophers,' by D. E. Lantz (Bul. 116, Experiment Station, Kansas State Agricultural College, Manhattan, Kans.).

scattered colonies. In places they were comparatively tame, and would sometimes let us drive within 20 feet, and even then they would not go entirely down their burrows. From a distance they could be seen watching us. A few were always sitting on top of their mounds barking an alarm, but on our nearer approach all scampered for the nearest burrows, while those farther ahead took up the alarm. When once half within the funnel-shaped entrance of the burrow the courage of the prairie dog revives, and with hands braced across the doorway, and with erect, flipping tail, the animal keeps up a steady barking at the intruder, sinking lower and lower, until finally with a quick dive, a shrill chatter, and a farewell twinkle of the tail, it vanishes down the hole. Frequently when you reach the burrow the animal can still be heard sputtering and chuckling deep down in the earth, and when once driven into its hole it does not soon reappear. It takes no little patience to await an hour or more the reappearance of the little black eye that cautiously peeps over the rim to see if the coast is clear.

Promptly with the rising sun the prairie dogs come out for their breakfasts, at which time a 'dog town' is as animated as any metropolis, but with the setting sun they retire to their burrows. Breakfast lasts for a good share of the day, with intermissions for work and play and a good long midday nap. There are always burrows to be dug deeper or new ones to be started, rims to be built higher, and in damp weather the crater-like mounds to be molded. Immediately after a shower, often before the last drops have fallen, the prairie dogs are out scraping up damp earth on the rim around the burrow and pressing the funnel-shaped inside into proper form with their stubby noses. In this way an effectual dike, sometimes a foot or two in height, is formed around the entrance of the burrow. But during a cloud-burst I have stood in a dog town and seen all of the burrows with rims not over 6 inches high fill with water; and in the track of an unusually violent downpour have seen the bodies of dozens of drowned prairie dogs scattered along the gulches.

The food of the prairie dog is mainly grass. Not only are the leaves and stems eaten, but the roots are dug up until the circles of bare ground around the burrows become wider and wider. Many other plants, some seeds, and a few insects are eaten, but to a less extent than grass. After a long season of drought or a succession of dry years it often happens that every green thing is exterminated in a prairie-dog town and the animals are forced to move on to new pastures. In a dry season I have ridden over long stretches of barren and deserted dog towns; and, again, after a year of abundant rain, have found this same ground growing up to worthless weeds, or, if to grass, only to the equally worthless foxtail.

[100] Although not native to Texas, the house mouse (*Mus musculus*) has become widespread and occurs either as a commensal—living in buildings and farm structures—or feral animal throughout the State. We have learned much about this introduced rodent since the time of Vernon Bailey. Commensal populations differ from feral ones in having longer tails and darker coloration, and there are also differences in social behavior between the two types. Commensal house mice live in small groups dominated by an adult male that does most of the mating with females in the group. The group occupies a defended home area from which casual immigrants are excluded. Wild or feral house mice do not occur in rigidly structured colonies, but instead live within a home range in a manner that is typical of native rodents.

[101] The scientific name of the wharf or brown rat (now commonly known as the Norway rat), *Mus norvegicus*, has been changed to *Rattus norvegicus*. As noted by Bailey, the Norway rat is common in Texas.

In autumn the prairie dogs become fat, but in Texas they do not regularly hibernate as they do to some extent in the North. If their fur should become fashionable, or roast prairie dog an epicurean dish, the problem of keeping them in check would be settled, and there is no reason, save their name, for not counting them, properly prepared and cooked, a delicacy. While owing their name to a chirping or 'barking' note of warning, they are in reality a big, plump, burrowing squirrel of irreproachable habits as regards food and cleanliness. An old stage driver expressed the idea in graphic words one day: "If them things was called by their right name there would not be one left in this country. They are just as good as squirrel and I don't believe they are any relation to dogs."

Mus musculus Linn. House Mouse.

100 Common house mice are found practically over all the settled part of Texas, even at most of the isolated ranches at a distance from railroads and towns. They were caught at deserted adobe cabins in the Great Bend of the Rio Grande, 100 miles south of the Southern Pacific Railroad. They are by no means confined to houses and outbuildings, but over much of the country have become established in the fields, meadows, hedgerows, and weed patches, from which they collect in the stacks of hay and grain, and are ready to attack each crop as it matures.

Mus norvegicus[M] Erxl. Wharf Rat.[a] Brown Rat.[M]

101 Wharf rats are common in most of the towns of Texas and on some of the ranches, but they are not so generally distributed in thinly settled regions as the house mouse, nor do they take so readily to the fields and country. In the years 1889 and 1890 there were reports of swarms of rats overrunning parts of the State, but the species are uncertain, nor is it known whether the wharf rat was one of them. At Seguin, Guadalupe County, in November, 1904, I found wharf rats in great abundance around farm buildings and along fences and weedy borders of fields, wherever sufficient cover was offered. Their runways and burrows resembled those of the cotton rats, but were larger and did not extend so far out from cover. Along the edge of a cornfield numerous cobs were scattered under the fence, where the corn had been eaten off. In Mr. Neel's tomato patch the ripe tomatoes were being rapidly devoured, and I caught a rat in the midst of the patch by using a ripe tomato for bait.

---

[a] *Mus rattus*[M] Linn. Black Rat. A black rat collected by Lloyd at Brownsville proves to be a melanistic *Mus norvegicus*,[M] closely resembling *M. rattus*[M] in color. I find no specimens or records of the latter species from Texas, but as it is found farther east and west it undoubtedly will be taken in the State.

Mus alexandrinus[M]Geoffroy.   Roof Rat.[M]

102 Two specimens of the roof rat, caught in July, 1902, on the Guadalupe River, at Ingram, Kerr County, constitute, so far as I can learn, the second record of the species for the State. It seemed strange to find this exotic mammal, which is usually found near the coast, so far in the interior of a thinly settled country, but the explanation is simple. The Guadalupe River is subject to violent floods, sometimes rising suddenly to 50 feet above low water. The enormous heaps of drift rubbish deposited along the bottoms and in the branches of trees have evidently furnished a highway for the distribution of the rats from the coast up the river. The two individuals secured were living in these drift heaps and were caught in traps set for *Neotoma attwateri*.[M]One was caught on the ground at the edge of a drift heap; the other on a pole reaching across from one heap to another. A specimen reported by H. P. Attwater in 1894 was caught on a boat that made trips between St. Charles Peninsula and Rockport, and was said to have been on the boat about a year.[a]

Onychomys longipes[M]Merriam.   Texas Grasshopper Mouse.[M]

103 This large dull-colored form of the grasshopper mouse occupies the semiarid Lower Sonoran zone of southern Texas, and, so far as known at present, reaches its eastern limit at Rockport, its northern limit at San Angelo, and its western limit at Comstock and Sycamore Creek, and extends south of Brownsville into Mexico. As it occupies so much of the brushy, half-open cactus and mesquite country, its apparent absence from the region of San Antonio and Austin and north to the Red River on the east side of the Staked Plains is probably due to the fact that this strip of country has not been thoroughly worked. Unlike most species of *Onychomys*, *longipes*[M]inhabits weedy, grassy, brushy land, and specimens are found in the woods as well as the open. It is strictly nocturnal, and its shrill little whistle is often heard not far from our camp fires in the evening.

Onychomys leucogaster pallescens[M]Merriam.   Pale Grasshopper Mouse.[M]

104 Throughout most of its range this pale, plains form of the grasshopper mouse is found in the Upper Sonoran zone and crowding into the edge of both Transition and Lower Sonoran. In Texas it extends over the Staked Plains, meeting or overlapping the range of *torridus* in the Pecos Valley at Monahans and Fort Lancaster, Tex., and Carlsbad, N. Mex. Specimens examined from Lipscomb, Texline, Miami, Mobeetie, Washburn, Amarillo, and Hereford are fairly typical *pallescens*,[M]and one specimen from Fort Lancaster, not fully adult, is referable to *pallescens*[M]rather than *longipes*.[M]

At Texline, Howell caught a series of 12 specimens in the valley of a small dry creek, where he found that they preferred sandy soil

[a] Bul. Am. Mus. Nat. Hist., VI, p. 174, 1894.

[102] The scientific name for the roof rat, or black rat, is now *Rattus rattus* instead of *Mus alexandrinus* as listed by Bailey. Roof rats are largely commensal and live in close association with humans. They seldom establish feral populations as do Norway rats. As noted by Bailey, the wharf or Norway rat seems to be more common in Texas. Black rats are darker in color than Norway rats and have a tail that is longer than the head and body rather than shorter. A more recent invader to the United States, the Norway rat is larger and more aggressive and appears to be supplanting the roof rat in many parts of the country.

[103] *Onychomys longipes*, referred to by Bailey as the Texas grasshopper mouse, was described by Merriam (1889) on the basis of a specimen collected in Concho County in 1887 by William Lloyd. Current taxonomy follows Riddle (1999) and treats *Onychomys longipes* as a subspecies of the more wide-ranging species, *Onychomys leucogaster*, the northern grasshopper mouse. Three subspecies are known from Texas: *Onychomys leucogaster ruidosae* in El Paso and Hudspeth counties; *Onychomys leucogaster articeps* in the Panhandle, eastward to Wichita County and south to Pecos County; and *Onychomys leucogaster leucogaster* from Tom Green and Terrell counties southward to the Rio Grande and southeastward to Refugio County. Their conservation status is considered stable throughout their range in Texas.

[104] The name *Onychomys leucogaster pallescens*, which Bailey applied to grasshopper mice from the Panhandle and northern Texas, applies now to populations occurring in New Mexico, Arizona, Colorado, and Utah; Texas specimens are referred to *Onychomys leucogaster articeps*.

[105] The scientific name *Onychomys torridus*, as used by Bailey, is no longer applicable to populations of grasshopper mice from western Texas. Recent genetic work has revealed that two "cryptic" species are included under this name, and the populations occurring in the Trans-Pecos region of Texas are now classified as *Onychomys arenicola*. The name *Onychomys torridus* (now referred to as the Chihuahuan grasshopper mouse) applies to populations occurring from southwestern New Mexico westward (Hinesley, 1979). The type specimen of *O. arenicola* was collected by E. A. Mearns along the Rio Grande about six miles (9.6 km) above El Paso, El Paso County, in 1892; hence the common name, Mearns' grasshopper mouse. This species is common throughout its range in Texas.

[106] Deer mice of the genus *Peromyscus* are the most taxonomically diverse and complex group of rodents in Texas. Not until Osgood (1909) published his classic taxonomic revision of the group was it possible to make much sense of the confusion surrounding this widespread genus of rodents. At the time Bailey wrote his accounts of deer mice, there were 130 nominal species and more than 60 subspecies ascribed to the genus. Bailey recognized seven species and nine subspecies of deer mice in Texas. Osgood (1909) reduced the number of valid species to 43 and increased the number of subspecies to 100, including seven species and 10 subspecies in Texas. The application of scientific names has changed substantially from those used by Bailey. Today, nine species and 16 subspecies of deer mice occur in Texas (see Schmidly and Bradley, 2016), and all but four were recorded by the Biological Survey field agents.

with a good growth of sagebrush (*Artemisia filifolia*). They make few holes, though two were taken at the mouths of small burrows. A turtle ate the head of one specimen and a rattlesnake tried to swallow another, but was prevented by the trap.

Throughout a wide range these little animals live on the short-grass plains or in the sagebrush country, and are caught at all sorts of burrows, an old badger, prairie-dog, or spermophile hole being a favorite resort, probably on account of the insects to be found within. They are strictly nocturnal, and while never seen by daylight, their long-drawn, fine whistle is often heard in the grass between dusk and early dawn. The morning round of a line of traps usually reveals one or more specimens that have been attracted by the oatmeal bait, and just as often shows some half-eaten *Perognathus*, *Peromyscus*, kangaroo rat, or other small rodent that happened to be in the trap when this forager came along. The *Onychomys* stomachs usually contain, besides finely chewed seeds and grain, an interesting assortment of grasshoppers, crickets, beetles, scorpions, and small insects, and occasionally parts of a lizard or mouse.

Onychomys torridus[M] (Coues). Arizona Grasshopper Mouse.[M]

Onychomys torridus arenicola Mearns. Proc. U. S. Nat. Mus., XIX, Advance Sheet, May 25, 1896, p. 3. Type from El Paso.[a]

105 These little long-tailed grasshopper mice occupy the arid Lower Sonoran zone of western Texas from El Paso to near the mouth of the Pecos, and up the Pecos Valley to old Fort Lancaster and Monahans, and Carlsbad, N. Mex. They are found on the open, barren mesas among stones and cactus and the characteristic desert vegetation, or in the sandy mesquite bottoms of the Rio Grande and Pecos rivers. Like all the genus they are strictly nocturnal, and while prowling about at night get into traps set at the burrows of various other mammals. About all that we know of their habits is gained from examination of their stomachs, which usually contain, besides a small portion of seeds or grain, a larger share of scorpions, grasshoppers, crickets, beetles, and various other insects.

Peromyscus leucopus (Rafinesque). White-footed Mouse.[M]

106 The dark-colored *Peromyscus* from the coast region of southeastern Texas, while not quite typical *leucopus*, seems to be nearer to it than to *mearnsi*.[M] There are specimens from near Alvin, near Galveston, Velasco, Elliott, Arcadia, Matagorda, Deming Station, and east Carancahua Creek. To the west apparently it grades into

[a] Specimens in the Biological Survey collection from El Paso, Sierra Blanca, Marfa, and Alpine do not differ so far as I can see from the type of *torridus*[M] and from specimens taken around the type locality, when corresponding pelages are compared. It is a little, dark, richly colored species, becoming pale in late winter and spring. Neither in the dimensions nor in the skulls can I find any character by which to recognize the subspecies *arenicola*.

mearnsi,[M] while immature specimens from Gainesville, Decatur, and Benbrook suggest intergradation with the same form.

Lloyd reports these mice at Deming Bridge, Matagorda County, "as found only where a quantity of brush had been cut down to fill a gap in the road." Near Matagorda he says "they live in trees, both in nests in the moss and in hollows in the roots." At Velasco he records one from "edge of creek" and another from "edge of old field." At Austin Bayou, near Alvin, he collected an old female containing 4 fully grown embryos, March 17, 1892. Beyond these fragmentary notes by Lloyd nothing is known of the habits of the species in this region, where apparently it inhabits the timbered and brushy bottoms with the palmetto and Spanish moss.

Peromyscus leucopus texanus (Woodhouse).   Texas White-footed Mouse.[M]

> *Peromyscus tornillo* Mearns. Proc. U. S. Nat. Mus., Vol. XVIII, Advance Sheet, March 25, 1896, p. 3. Type from Rio Grande 6 miles above El Paso, Tex.

107 This is a common species over the arid Lower Sonoran zone of western Texas from El Paso to Del Rio, Rock Springs, and Fort Lancaster. To the south it grades into *mearnsi.* Specimens from Del Rio, Rock Springs, and San Antonio are not typical of either animal, but combine enough of the characters of both to be considered fairly intermediate. A series of 11 specimens from Lipscomb, 4 from Canadian, 2 from Miami, and 3 from Mobeetie, on the plains of the northern Panhandle, and one from Henrietta, while not typical *texanus,* can be referred to it better than to *leucopus.*

In this arid region these mice take the place of *leucopus* and other members of the *leucopus* group, to which they belong. They have the general habits of the group and in places live among rocks, but more often on the weedy and brushy bottoms under rubbish or dense vegetation, where they are often the most abundant mammal. At El Paso and Juarez, Loring says: "Common on both sides of the river. They were caught in traps set at holes and in the brush along irrigation ditches and baited with oatmeal and small pieces of meat." At Sierra Blanca I found them only in an old *Neotoma* house in a bunch of yuccas in the open valley, while *eremicus* occupied the nearest cliffs and the little *blandus*[M] lived out on the open plain. At Fort Lancaster Oberholser reported them as "abundant in the chaparral," and at Langtry as "not common." The two specimens taken at Langtry were caught under logs and among dead leaves and rubbish near water in a deep side canyon. At Del Rio I found them common in holes in the creek bank, under thick brush, and in old houses; and Gaut collected two specimens "in high grass along the main irrigation ditch west of town."

At Lipscomb Howell took specimens only "in brushy places along

[107] The white-footed deermouse, *Peromyscus leucopus*, has a state-wide distribution, and it is the most common and widespread of all the species of deer mice in Texas. Three subspecies are recognized: *Peromyscus leucopus leucopus* in the eastern one-fourth of the state, *Peromyscus leucopus texanus* in central Texas (west to Brewster, Terrell, and Winkler counties), and *Peromyscus leucopus tornillo* in the Panhandle and much of the Trans-Pecos. The subspecies *Peromyscus leucopus mearnsi* mentioned by Bailey was described by J. A. Allen in 1891 on the basis of specimens from Brownsville in Cameron County and placed in synonymy of *P. l. texanus* by Osgood (1909).

Two chromosomal races of *P. leucopus*, designated southwestern and northeastern, have been identified in the south-central United States (Baker et al., 1983). The two genetically distinct races divide in the ecotone between the Great Plains Grassland and Eastern Deciduous Forest biomes in Oklahoma (Stangl, 1986; Nelson et al., 1987). Interestingly, the ranges of the two races do not correspond to the classical subspecific boundaries for *P. leucopus*. So far, only one of the races, the southwestern one, has been documented in Texas, although no specimens have been examined from the woodlands of extreme eastern Texas, where the northeastern race might be expected to occur (Stangl and Baker, 1984).

[108] Specimens considered by Bailey to be *Peromyscus leucopus mearnsi* are now referred to *Peromyscus leucopus texanus* (Osgood, 1909).

[109] *Peromyscus michiganensis pallescens* is currently classified as *Peromyscus maniculatus pallescens* (Osgood, 1909). The loss of native grassland habitat in the central part of Texas presents a major concern for the future of this subspecies. We have attempted without success to collect this species at several localities from which it was previously recorded. Recently, taxonomists recognized five subspecies of *Peromyscus maniculatus* in Texas: *Peromyscus maniculatus luteus* and *Peromyscus maniculatus nebracensis* in the Panhandle counties; *Peromyscus maniculatus ozarkiarum* in Cooke and Grayson counties; *Peromyscus maniculatus blandus* in central and western regions; and *Peromyscus maniculatus pallescens* in the eastern one-third of the state. However, genetic evidence (Bradley et al., 2019) suggests that *P. maniculatus* is composed of multiple species and that the subspecies *P. m. luteus*, *P. m. nebracensis*, *P. m. ozarkiarum*, and *P. m. pallescens* should be assigned to a newly recognized taxon, *Peromyscus sonoriensis* (*Peromyscus sonoriensis luteus*, *Peromyscus sonoriensis nebracensis*, *Peromyscus sonoriensis ozarkiarum*, and *Peromyscus sonoriensis pallescens*, respectively), whereas *P. m. blandus* should be assigned to a separate taxon, *Peromyscus labecula* (*Peromyscus labecula blandus*).

the creek bottoms." At Canadian he caught one "in the grass along an irrigation ditch" and another "in a deserted cabin." At Miami he caught two "in the rocky bluffs near town and one on the sandy bottoms," and at Mobeetie others "along the sandy creek bottoms in traps set for *Perognathus* and *Perodipus*."[M]

Like other members of the genus, they are strictly nocturnal, and during the day keep safely within their burrows in the ground or in some other dark retreat. As a result they are almost never seen alive except when they get their tails instead of their necks in our traps. Very little is known of the habits of this species.

Peromyscus leucopus mearnsi[M] (Allen).   Mearns White-footed Mouse.[M]

Peromyscus canus Mearns. Proc. U. S. Nat. Mus., Vol. XVIII, Advance Sheet, March 25, 1896, p. 3. Type from Fort Clark, Tex.

108 The Mearns white-footed mouse ranges over southern Texas from Brownsville north to Eagle Pass, Fort Clark, and San Antonio, and east to Rockport, grading into *leucopus* on the east and *texanus* along its northern boundary. There is certainly not room for an intermediate form between *texanus* and *mearnsi*,[M] which *canus*[M] proves to be.

The region inhabited by *mearnsi*[M] is semiarid chaparral and cactus plains in Lower Sonoran zone. At Brownsville, the type locality of the species, Lloyd collected a large series of specimens and reported the species as "very common out to the sand belt." He also caught a few on Padre Island and around Nueces Bay. Another series was collected at Brownsville by Loring, who says of the mice: "Quite common. Found in the willows along the river bank and under logs and brush near the overflows. Several were caught in traps baited with meat." At Hidalgo, Loring says: "They were taken in traps baited with oatmeal and set by hedge fences, cactus beds, and underbrush." In writing of specimens obtained at San Lorenzo Creek and Santa Tomas, Lloyd says: "They prefer an arid region where grass is scant among the cactus." At Corpus Christi the mice were very scarce, and I caught but one in a long line of traps. It was at a hole in the bank just back of the beach. At Beeville Oberholser reports them as "evidently not very common, as all my trapping failed to reveal more than a single individual." At San Diego he says: "This animal does not appear to be more than tolerably common. It lives only in the damp thickets bordering the ponds and water holes in the chaparral; at least my trapping in all kinds of situations failed to reveal its presence anywhere else."

Peromyscus michiganensis pallescens[M] Allen.   Little Pale Peromyscus.[M]

109 This pale little mouse is represented in the collection by 9 specimens collected at San Antonio and one from near Alice. The one from Alice was collected by Lloyd 12 miles southwest of the town on open prairie. San Antonio specimens were caught by H. P. Attwater in traps set for harvest mice around brush piles.[a]

Peromyscus sonoriensis (Le Conte).   Sonoran Peromyscus.[M]

110 A few specimens of this little *Peromyscus* from Washburn, Tex., seem to be nearer to typical *sonoriensis* than to any of the subspecies in the group. One specimen was caught at a tiny burrow on the short-grass plains, miles from any cover that would conceal even a mouse, and others were caught on the prairie at the edges of fields.

Peromyscus sonoriensis blandus Osgood.   Frosted Peromyscus.[M]

111 This pale, silky-haired little mouse is common in western Texas over the rough and arid region between the Pecos and Rio Grande valleys. There are specimens from the Franklin Mountains (15 miles north of El Paso), Sierra Blanca, Valentine, Onion Creek (Presidio County), and Bone Spring (53 miles south of Marathon). All the above localities are in rough country near the junction of Upper and Lower Sonoran zones, where more or less mixture of the two occurs, so that the zonal range of the species is not perfectly determined by them.

Although in a rough country, broadly speaking, these white-footed mice inhabit the smooth spots in the bottoms of open valleys. At Sierra Blanca they were on the broad flats southeast of the station, where the principal vegetation was low, scattered, composite shrubs (*Gutierrezia microcephala?* and *Crassina grandiflora*), among which they burrowed in the mellow soil, and the seeds of which furnished in winter a large share of their food. So far as their own genus was concerned, they held this ground by themselves, the larger *P. texanus* [M] being caught in an old woodrat's nest under a yucca and the long-tailed *P. eremicus* in the nearest cliff of tilted rocks. At Onion Creek they were found living in holes in the soft, level ground of the creek valley, where none of the other species of *Peromyscus* were taken. At Valentine, out in the middle of a big open valley, I caught one of these little fellows under the doorstep of the house where I boarded, on the edge of town, and near Bone Spring, 50 miles south of Marathon, I caught another under a mesquite out in the open valley.

Peromyscus gossypinus (Le Conte).   Pine Woods Peromyscus.[M]

112 Specimens of this large, dark-colored *Peromyscus* from Texarkana, Jefferson, Long Lake, Joaquin, Jasper, and Sour Lake, indicate an

[a] Bul. Am. Mus. Nat. Hist., VIII,, p. 64, 1896.

[110] *Peromyscus sonoriensis* was relegated to a subspecific level, *Peromyscus maniculatus sonoriensis*, but this subspecies did not occur in Texas; consequently, populations of deer mice from the Texas Pan-handle were referred to *Peromyscus maniculatus luteus* (Judd, 1970) and *Peromyscus maniculatus nebracensis* (Manning et al., 2008). However, recent genetic evidence (Bradley et al., 2019; see Annotation 109) suggests that *P. maniculatus* is com-posed of multiple species and that the subspecies *P. m. luteus* and *P. m. nebracensis* should be recognized as subspecies of the newly recog-nized taxon, *Peromyscus sonoriensis* (*Peromyscus sonoriensis luteus* and *Peromyscus sonoriensis nebracensis*, respectively).

[111] *Peromyscus sonoriensis blandus* was reclassified as *Peromyscus maniculatus blandus* (Osgood, 1909); however, recent genetic evidence (Bradley et al., 2019; see Annotation 109) suggests that *P. maniculatus* is composed of multiple species and that the subspecies *P. m. blandus* is affiliated more with the Mexican form than with the Texas forms and consequently, it should be assigned to the newly recognized taxon, *Peromyscus labecula*. This taxon is still common throughout its range in Texas, which encompasses the Trans-Pecos and areas immediately to the east of the Pecos River.

[112] The cotton deermouse, *Peromyscus gossypinus*, is the most common rodent in the Big Thicket region of southeastern Texas. The subspecies in Texas is *Peromyscus gossypinus megacephalus*.

[113] The spelling of *boylei* has been changed to *boylii*. Bailey listed *Peomyscus boylii penicillatus* from El Paso as a synonym of *Peromyscus boylii rowleyi* (now referred to as the brush deermouse), and it remained as such until Diersing (1976) demonstrated that the type specimen of *Peromyscus boylii penicillatus* was actually a representative of another species of deermouse, *Peromyscus difficilis*. Diersing and Hoffmeister (1974) also documented that a specimen collected by Bailey on 22 August 1901 at McKittrick Canyon in the Guadalupe Mountains was a representative of the species *P. difficilis. Peromyscus difficilis* is now considered to be a composite of two cryptic species, with the northern populations, including those from Texas, referable to *Peromyscus nasutus* (Carleton, 1989). *Peromyscus nasutus* has a spotty distribution in far western Texas with specimens known from the Chinati, Chisos, Davis, Guadalupe, and Franklin mountains (Bradley et al., 1999b). Two subspecies are recognized: *Peromyscus nasutus nasutus* in Culberson, Jeff Davis, Presidio, and Brewster counties and *Peromyscus nasutus penicillatus* in El Paso County. The conservation status of *P. nasutus* in Texas is uncertain.

extensive range for the species over the timbered region of eastern Texas.

Apparently it is not an abundant species anywhere in this region, and much trapping is necessary to procure a few specimens. At Jefferson Hollister caught two in the woods near Big Cypress Creek, and at Joaquin one under a log on heavily timbered creek bottoms. Oberholser caught one in a canebrake along the Red River at Texarkana, one in heavy woods on the edge of McCracken Lake, near Long Lake, in Anderson County, and several in a cabin and one along a stream in the woods north of Jasper. In the heart of the Big Thicket, 7 miles northeast of Sour Lake, I caught several in and around old tumbledown buildings, and Gaut caught them around old logs and stumps in the woods.

Peromyscus boylei rowleyi[M](Allen).   Rowley Peromyscus.[M]

    Peromyscus boylii penicillatus Mearns. Proc. U. S. Nat Mus., Vol. XIX, Advance Sheet, May 25, 1896, p. 2. Type from El Paso.

113   A large series of this big, long-tailed *Peromyscus* from the Franklin and Organ mountains is typical *rowleyi*, as apparently are also six other specimens taken in Dog and McKittrick canyons in the Guadalupe Mountains. In this region and farther north they range throughout Upper Sonoran zone, being closely associated with junipers and nut pines, as well as with rocks and cliffs. In places they follow the cliffs slightly below the junipers, but only where canyon walls offer especially favorable haunts. In the Guadalupe Mountains they range to the upper limits of junipers, where the yellow pines begin on dry, hot slopes at 7,800 feet, and down in the northeast gulches near Carlsbad at the east base of the mountain slope at 3,100 feet. While usually found along cliffs or among rocks, they are often common among junipers, nut pines, and oaks at considerable distance from any rocks. In such places they live in hollow trees or logs or take advantage of any convenient cover. I have occasionally found them curled up in a soft nest in a hollow tree, and have often found a nest that I attributed to this species in a knothole or under a loose layer of bark. At one of our camps on top of the Guadalupe Mountains, in a beautiful orchard-like park of junipers, one took possession of the camp wagon and made its nest among boxes and sacks.

The food of these mice consists largely of juniper berries, or at least the seeds of juniper berries, of which there is usually an abundant supply at all times of the year, but acorns and pine nuts are eaten while they last. The empty shells of seeds and nuts and acorns show the favorite feeding grounds to be under the hollow base or low spreading branches of a juniper.

Peromyscus boylei laceyi[a]M subsp. nov.   Lacey Peromyscus.M

Type from Turtle Creek, Kerr County, Tex. Adult male, No. 92746, U. S. National Museum, Biological Survey Coll. Collected by H. P. Attwater, Dec. 4, 1897. No. 1372, X Catalogue.

114 *General characters.*—Size and proportions about as in *rowleyi,* to which it is most nearly related. Color decidedly darker; under surface of tail more grayish.

Color.—Adults in winter pelage dull, dark ochraceous, brightening on sides; ankle and upper surface of long hairy tail blackish; lower surface of tail, dusky gray; belly and feet, pure white. Summer pelage, brighter in the rufescent phase; paler in the gray phase; lower surface of tail less grayish.

Skull with interpterygoid fossa generally narrower than in either *boylei* or *rowleyi.*

Measurements.— Type not measured in the flesh, but the hind foot measures 24 when dry. Average of four topotypes: Total length, 188; tail vertebrae, 97; hind foot, 23.2.

Skull of type.—Total length, 28; basioccipital length, 23.3; nasals, 10; zygomatic breadth, 14; width of braincase, 13; mastoid breadth, 12; alveolar length of upper molar series, 4.

This big rufescent species of the group of long-tailed *Peromyscus* inhabits the Upper Sonoran, rocky, juniper-covered hills of Kerr and Edwards counties, the Davis Mountains, and the Chinati Mountains. Specimens from Turtle Creek and Ingram, Rock Springs, the Davis Mountains, Paisano, and the Chinati Mountains show some variation, but may all be included under one name. Three specimens from Ozona and one from Big Springs, at the edge of the plains, are very pale. From *attwateri,* the only similar species with which they are associated, *laceyi*M may be easily distinguished by larger size, darker color, black ankle and heel, and gray underside of tail, as also by good cranial characters.

At Lacey's ranch, near Kerrville, I caught them in cliffs and gulches with *attwateri,* without noting any difference in habits or habitat of the two species. At Rock Springs and in the Davis Mountains also they occur with *attwateri,* and apparently have very similar habits. They are largely cliff dwellers, but live also in open woods, on oak and juniper covered ridges, and in brushy gulches. In the central part of the Davis Mountains they were the only species of *Peromyscus* that we found at 5,500 to 6,500 feet in the basalt cliffs and over the timbered slopes, but down near the east base of the mountains Cary caught one specimen in the same cliff with an *attwateri.* Lloyd caught them among the rocks at Paisano and in the

[a] Named for Mr. Howard Lacey, at whose ranch the specimens were taken.

[114] Taxonomists now refer *Peromyscus boylei laceyi* to *Peromyscus attwateri.* Although Bailey classified specimens of this mouse as a new subspecies, *Peromyscus boylei laceyi,* this assignment proved to be an error. In a paper he published in 1906, he wrote, "I gave the name *laceyi* to a mouse of the genus *Peromyscus* occurring in central Texas. Through a most unfortunate misconception the name was applied to the wrong one of the two species found together at the type locality, to the larger, darker colored form previously named *attwateri* by Dr. J.A. Allen. The smaller, paler animal is now for the first time described under the name *laceianus* as a subspecies of *pectoralis,* its nearest relative." J. A. Allen, a contemporary mammalogist of Bailey who worked at the American Museum of Natural History in New York City, described *Peromyscus attwateri* in 1895 on the basis of specimens from Turtle Creek in Kerr County. Thus, the two species that Bailey collected at Lacey's Ranch on Turtle Creek were *P. attwateri* and *Peromyscus pectoralis* (but see below). Bailey (1906) subsequently arranged *P. attwateri* as a subspecies of *Peromyscus boylii* (*Peromyscus boylii attwateri*), an arrangement that Osgood followed when he revised the genus in 1909. On the basis of chromosomal and morphological differences, Schmidly (1973a) showed that *P. b. attwateri* was a distinct species from both *P. boylii* and *P. pectoralis. Peromyscus attwateri,* now commonly referred to as the Texas deermouse, occurs in the central part of the state where it is one of the most common rodents in the juniper-breaks habitat. Specimens of *P. b. laceyi* (*sensu* Bailey 1905) from the Davis

Mountains, Paisano, and the Chinati Mountains are now referred to *Peromyscus boylii rowleyi.* Specimens from Turtle Creek, Ingram, Rock Springs, Ozona, and Big Springs now are referred to either *P. attwateri* or *P. laceianus.* Bradley et al. (2015) showed that populations of *P. laceianus,* previously regarded as a subspecies of *P. pectoralis,* were genetically and specifically distinct from Mexican populations of *P. pectoralis.* Thus, two species, *P. attwateri* and *P. laceianus,* occur sympatrically throughout the Hill Country of Texas, and they are among the most common rodents in the region (Goetze, 1998).

[115] Cactus deermice, *Peromyscus eremicus*, are most commonly found on the bajadas and canyon bottoms of lowland desert areas. This species is common throughout the Trans-Pecos and western Rio Grande regions in Texas. An analysis of mitochondrial DNA in populations of *P. eremicus* from the Chihuahuan and Sonoran Deserts indicate these populations likely represent recently diverged "cryptic" species (Walpole et al., 1997). If this proves to be the case, specimens from the Chihuahuan Desert in Texas would be classified as *Peromyscus arenarius* Mearns (1896).

Chinati Mountains, 35 miles south of Marfa. At Ozona and Big Springs they were taken in cliffs in the comparatively open country.

Peromyscus attwateri Allen.   Attwater Peromyscus.[M]

The Attwater peromyscus inhabits the cliffs around the edges of the lower arm of the Staked Plains from Big Springs, Llano, Austin, and Kerrville, west to Comstock and Langtry, and up the Pecos Valley to Fort Lancaster and Sheffield, the cliff and canyon country along the Rio Grande, and at least the lower slopes of the Davis and Chisos mountains, and extends westward into Mexico. Most of these localities are so near the edge of Upper and Lower Sonoran zones that the species might belong to either, except that among the cliffs and canyons it probably gets the cooler temperature of the higher zone.

At Howard Lacey's ranch, near Kerrville, where the type of the species was collected in 1895 by H. P. Attwater, I found these animals abundant in 1899, and caught them in crevices along the cliffs, under logs in the woods, and under fallen grass and weeds on the creek bank in the bottom of the gulch. At Ingram, also near Kerrville, Cary and I caught them in the rocks along the bluffs and under the heaps of flood drift on the river bottoms. At Camp Verde Cary caught a few under rocks and logs. One was taken on the crest of a juniper ridge near Rock Springs; Gordon Donald took one in the cliffs near Devils River Station, and N. Hollister caught a series in a little canyon near Comstock and one in the cliffs of the Rio Grande Canyon 8 miles south of Comstock. Lloyd caught one at the Painted Caves and Gaut collected 12 specimens in the canyons around Langtry. Oberholser caught 5 among the rocks at Fort Lancaster. Cary and Hollister found them in the cliffs along the Pecos Canyon as far up as Sheffield, and Cary took one individual of this species, together with *laceyi*,[M] in a canyon at the east base of the Davis Mountains. In the Chisos Mountains a few were taken in Upper Sonoran zone from 6,000 to 7,000 feet. A small series taken among the rocks at Llano is typical *attwateri,* but a series of 13 specimens collected in the cliffs near Austin, where Oberholser reported them as the commonest small mammal of the locality, is not typical.

Peromyscus eremicus (Baird).   Desert Peromyscus.[M]

    Peromyscus eremicus arenarius Mearns. Proc. U. S. Nat. Mus., XIX, Advance Sheet, May 25, 1896, p. 2. Type from El Paso, Tex.

115 This wide-ranging desert species inhabits the arid Lower Sonoran zone of western Texas from the Pecos Valley to El Paso. Specimens from Comstock and vicinity and from Carlsbad (Eddy), N. Mex., mark the eastern limit of its known range. There are specimens in the Biological Survey collection also from El Paso, Franklin Mountains, Sierra Blanca, Presidio County, Boquillas, Terlingua, 20 miles south of Marathon, Langtry, Painted Caves, and 65 miles northwest of Toyah. The slight variation in specimens from Texas localities does not warrant separation from typical *eremicus* of Fort Yuma, Cal.

At El Paso the species is common in the cliffs just back of town, and at Sierra Blanca two were caught in a cliff near the station. At Lloyd's ranch, in Presidio County, 30 miles south of Marfa, it was common in the cliffs, and a few were taken in cliffs at Terlingua, Boquillas, 20 miles south of Marathon, at Comstock, and along the canyons of the Rio Grande at Langtry and Painted Caves. At Carlsbad, N. Mex., it occupied a limestone cliff near the river with *rowleyi,* which belongs to the zone above, and in this same cliff I caught both *Neotoma micropus* and *albigula*,[M] belonging, respectively, to Lower and Upper Sonoran zones. Of the habits of the desert peromyscus little is known save what our traps reveal of its choice of homes on the dusty rock shelves of cliffs and caverns, where lines of tiny footprints lead to and from cracks and small openings in the rocks. I have never known of its being found away from rocks, and this peculiar habitat may have some connection with the wholly naked sole of the foot.

Peromyscus taylori[M] (Thomas).   Taylor Baiomys.[M]

116 Specimens of this tiny, short-tailed *Peromyscus*[M] from Brownsville, Beeville, and San Antonio do not show any appreciable variation, and are assumed to be typical *taylori,* as they come from both north and south of San Diego, the type locality of the species.

At Brownsville Loring reported them as "common in weeds and brush and along fences in meadows and a few in small willows near the river;" but Lloyd found them "only in open fields and meadows, where they have very small, round holes." At Beeville Oberholser caught one "at the edge of a clump of *Opuntia engelmanni*[M] in the chaparral." H. P. Attwater collected a series of specimens at Watson's ranch, 15 miles south of San Antonio, and furnished the following interesting notes on their habits:

> The specimens sent were taken under a pile of dry weeds and rubbish in an orchard, where the two nests sent were also found. There were several others with them which escaped. The two specimens taken in March were kept alive till May 29. They were fed on sugar-cane seed, oats, corn, and bran. They used to drink water when I put it in the cage, but appeared to do just as well without it. • • •
>
> One of the nests sent was found by Mr. Watson while digging up a small pecan tree in the river bottom near his ranch. The nest was about a foot below the surface of the ground, among the roots of the tree, and several passages led

[116] The subgenus *Baiomys* was elevated to generic status by Miller (1912) and confirmed by Packard (1960), so *Peromyscus taylori* is now recognized as *Baiomys taylori* (see Annotation 1). Consequently, modern taxonomists now refer to *Peromyscus taylori subater* as *Baiomys taylori subater.* Two subspecies of *Baiomys taylori* are recognized, *B. taylori taylori* throughout most of the state and *B. t. subater* in the Navasota River Valley and coastal marsh region. The northern pygmy mouse has been expanding its range northward in recent years (Choate et al., 1990; Schmidly and Bradley, 2016), possibly in response to dispersal along highway right-of-ways, an increase in CRP acreage, transportation of hay, and climate change.

[117] *Peromyscus taylori subater* is now recognized as *Baiomys taylori subater* (see Annotations 1 and 116).

down into the ground below the nest. In one of these holes a number of pecan nuts were found. The nest contained an old female and three half-grown young.[a]

Peromyscus taylori subater[M] subsp. nov.   Dusky Baiomys.[M]

Type from Bernard Creek near Columbia, Brazoria County, Tex. No. $\frac{32616}{44539}$ ♀ ad., U. S. Survey Nat. Mus., Biological Coll. Collected by Wm. Lloyd Feb. 25, 1892. Original No. 1122.

117 *Characters.*—Size and proportions of *P. taylori*,[M] but much darker colored. Upper parts blackish or sooty gray, belly buffy.

Measurements of type.—Total length, 91; tail, 37; hind foot, 15. Average of 7 topotypes: Total length, 95; tail, 39; hind foot, 14.8.

Skull of type.—Basal length, 14.8; nasals, 6.3; zygomatic breadth, 10; mastoid breadth, 8.4; alveolar length of upper molar series, 3.

FIG. 15.—Distribution areas of the two forms of the subgenus *Baiomys* (*Peromyscus taylori*[M] and *subater*).[M]

This dusky form of the little *Baiomys* inhabits the coast prairies of Texas east of Matagorda Bay. Specimens from Matagorda, Matagorda Peninsula, Bernard Creek (12 miles west of Columbia), Richmond,

[a] Allen, Mammals of Bexar County, Tex. Bul. Am. Mus. Nat. Hist., VIII, p. 66, 1896.

Virginia Point, Alvin, and Sour Lake are referred to it, although the Matagorda specimens are a little grayer and evidently tend toward *taylori.*

At Virginia Point on the mainland opposite Galveston, I caught two of these little mice in a grassy orchard at a ranch on the broad prairie. They were trapped in the grass-covered runways of sigmodons. At Richmond these mice were fairly common under the rich carpet of grass on the open prairie. Their tiny runways, leading from one little burrow to another, wound about over the surface of the ground among the plant stems and indicated habits so similar to those of *Microtus* that at first I thought I had discovered traces of a diminutive species of that genus. At Sour Lake Hollister collected one specimen "on the open prairie." On Matagorda Peninsula Lloyd found these mice living under logs near the Gulf shore where he collected both old and young. One young, about two weeks old, was found in a nest under a log February 11. In another nest two young were found, and an old female taken the same day contained two fully developed embryos. On the mainland near Matagorda Lloyd "caught them in the long grass skirting the edges of fields," and a nest containing three young was plowed up in the field February 2.

Oryzomys palustris[M](Harlan).   Rice Rat.[M]

  Oryzomys palustris texensis Allen, Bul. Am. Mus. Nat. Hist., VI, p. 177, May 31,
     1894. Type from Rockport, Aransas County, Texas.

118 The rice rats inhabit the coast marshes of Texas as far south as Corpus Christi and a little reef near the north end of Padre Island. Apparently they have not been found farther from the coast than Wharton County, some 40 miles up the Colorado River, and they seem to be common only in the salt marshes. At Port Lavaca, Oberholser says, "they are common in the tall grass bordering the bayous and are apparently confined to such places. The ground where they live is quite wet, but still out of reach of ordinary tides, though the whole area was flooded during part of my stay. The runways are not covered and not plain, though there are usually fresh signs at intervals." On Matagorda Island he says the rice rats are "tolerably common in the tufts of coarse grass bordering bayous, making conspicuous covered runways where the grass is thickly matted, but are not found more than a short distance back from the bayous." At Matagorda Bay, Lloyd says, "they occur along the shore of the bay and also on Selkirk Island and Peninsula, where they were found in the high, rank grass near the shore;" and at Nueces Bay, he says, "they are common out in the low grass on the marshes, where they take to water readily. Several were found drowned while held down by my traps. On a small island reef about 100 yards off the north

[118] Based on molecular data (DNA sequences), Hanson et al. (2010) recommended that populations of *Oryzomys palustris* in Texas be assigned to *Oryzomys texensis*. Bailey described the distribution of *Oryzomys texensis* (now referred to as the Texas marsh rice rat) as confined to the coastal marshes from Corpus Christi to Galveston and no further inland than forty miles (64 km). Today, the species has been recorded throughout the eastern part of Texas, west to Denton County, north to the Red River counties, and then southward to Cameron County. Rice rats are common on the dikes and levees of the coastal marshes. Inland, they prefer wetlands, such as marshes and moist meadows, but occasionally they live in forested areas. Allen (1894) described the subspecies *Oryzomys palustris texensis* on the basis of specimens collected by Mr. George B. Sennett, a well-known amateur ornithologist, who made extensive biological collections in south Texas during the 1880s and 1890s. This species is common over most of its range in Texas although the continued loss of wetlands habitat could place it in jeopardy in the future.

[119] Coues's rice rat is a seemingly rare, subtropical rodent of the Texas-Tamaulipas, Mexico borderlands. It is classified as a subspecies, *Oryzomys couesi aquaticus*, and it has been recorded from four counties in extreme southmost Texas (Benson and Gehlbach, 1979). It is listed as threatened by TPWD due to loss of habitat. The resaca environment is declining in southern Texas, largely because of drainage for irrigated agriculture. Resacas bordered by cattail-bulrush marsh and subtropical woodland are confined primarily to Cameron, Kenedy, and Hidalgo counties.

[120] The taxonomy of harvest mice (genus *Reithrodontomys*) in Texas has changed dramatically since Bailey's time. Bailey recognized five species from Texas, but today only four are considered valid. *Reithrodontomys griseus* has been relegated to a subspecies of *Reithrodontomys montanus*, and most of Bailey's other taxa have been subsumed under *Reithrodontomys fulvescens*. *Reithrodontomys intermedius*, *Reithrodontomys aurantius*, and *Reithrodontomys laceyi* are now classified as subspecies of the wide-ranging species, *R. fulvescens*. *Reithrodontomys fulvescens* occurs in eastern and central Texas (west to Armstrong, Briscoe, and Floyd counties in the north) and in parts of the Trans-Pecos region. This species prefers weedy or grassy habitats intermixed with shrubs, vines, and bushes. It has fared well since the turn of the century as mesquite and other brush have covered many areas of the state that were formerly prairie or grassland. Four subspecies are recognized: *Reithrodontomys fulvescens aurantius* in the eastern part of the

104 NORTH AMERICAN FAUNA. [No. 25.

end of Padre Island they were found in patches of marsh 'cranberry.' Two of their round cup-shaped nests, composed of fine rootlets, were found under old boards." At Virginia Point, opposite Galveston, I caught them in runways under the grass and rushes along the edge of the salt marsh. At that time, in April, they were rather scarce, but the people say that occasionally they become very numerous, especially in and around the rice fields.

Oryzomys aquaticus[M]Allen.   Rio Grande Rice Rat.[M]

119   This species is known only from the vicinity of Brownsville, near the mouth of the Rio Grande.

Loring reported it as common in grassy spots in the mesquite brush.

Reithrodontomys intermedius[M]Allen.   Rio Grande Harvest Mouse.[M]

Reithrodontomys laceyi[a] Allen, Bul. Am. Mus. Nat. Hist., VIII. p. 235, Nov. 21, 1896. Type from Watson's ranch, 15 miles south of San Antonio, Tex.

120   This long-tailed harvest mouse inhabits the Lower Sonoran zone of southern Texas from Brownsville to Corpus Christi, San Antonio, Kerr County, and Del Rio, and extends south into Mexico.

At Brownsville Loring reported that he caught several specimens of this species in traps baited with meat and set among small willows, weeds, and high grass near the river. At Del Rio I caught them at little burrows on the brushy flats, and near Kerrville found them common around fields and in weedy places generally and caught them at burrows and runways under the fallen grass. Lloyd reported one found on Padre Island in an old cow's horn, and two dead ones in an old barrel. Between Laredo and Rio Grande City he reported two as living in old nests of the cactus wren, and near Corpus Christi he found one in a nest in a catsclaw bush. In April, 1900, I found what looked like an old verdin's nest in a bush of *Momesia pallida*[b] near Corpus Christi. The nest was about 4 feet from the ground, a globular structure of grass, lichen, and short gray moss (*Tillandsia recurvata*), with a small opening at one side. As I touched the side, two black eyes appeared at the doorway, but after watching me for a moment were withdrawn. At a slight shake of the

---

[a] Specimens of *Reithrodontomys laceyi*[M] from San Antonio and Kerr County agree perfectly with specimens of comparable age and pelage from Brownsville, Matamoras, and Santa Tomas, and I see no way but to consider them typical *intermedius*.[M] The slightly smaller and grayer specimens are evidently young of the year. The difference in size indicated by Mr. Attwater's measurements does not appear in comparison of skulls or hind feet and may be due to a slight difference in the methods of measurement. My own measurements of Kerrville specimens and Goldman's of his Matamoras specimens agree almost to a millimeter. While all of the small series of topotypes of *laceyi*[M] can be matched from the large series of *intermedius*[M] from Brownsville, there are none in as bright summer pelage as some specimens in the Brownsville series.

---

state; *Reithrodontomys fulvescens canus* in the southern Trans-Pecos; *Reithrodontomys fulvescens intermedius* on the Rio Grande Plain and in adjacent areas of south Texas; and *Reithrodontomys fulvescens laceyi* in the central and northwestern parts of the state. Much of the taxonomic revisionary work discussed above was published by Howell (1914), who assisted Bailey with the Texas survey. The distribution of all taxa of *Reithrodontomys* in Texas differs from the description provided by Bailey. For example, the federal biologists during the

Texas survey did not collect *R. fulvescens* in the Big Bend region or in west-central Texas, nor did they record *R. megalotis* from the High Plains and Panhandle (Fig. 43). *Reithrodontomys fulvescens*, *R. megalotis*, and *R. montanus* have now been documented in the Big Bend region of Texas, and the former two species have been taken sympatrically at Big Bend Ranch State Park (Yancey et al., 1995a).

bush, out popped a trim little long-tailed harvest mouse, which sat undecided on the branch for a moment and then ran gracefully along branches and stems from one bush to another and finally down to the ground, where it disappeared in the tall grass. On examining the nest I found a firm base, evidently an old bird's nest that had been arched over with a substantial roof which left an opening at the side only large enough for my finger. It was neither a verdin's nor a cactus wren's nest, and evidently had been built or remodeled by the present tenant. When I returned the next day, the mouse was at home, but so sleepy that I merely disturbed him enough to make him come out and sit for a moment on the branch, after which I withdrew and let him go back to finish his nap. Further search revealed two more similar but old and unoccupied nests in the bushes near by, but no trace of runway, burrow, or other signs of the mice on the damp sticky soil beneath. A good line of traps set among the bushes and under the adjoining prairie grass remained untouched until the bait grew moldy. Even at the base of the bush under the occupied nest nothing was caught in several days' trapping and after a trip of two weeks I returned to find the little fellow still occupying his nest.

Along the Medina River 15 miles south of San Antonio, Mr. H. P. Attwater says he occasionally came across these mice in 1889 and 1890 while hunting for birds' nests. He says they were found singly in the daytime in little round nests made of grass and placed in the lower branches of small trees.[a]

Reithrodontomys aurantius[M] Allen. Louisiana Harvest Mouse.[M]

121 This largest and richest colored of the Texas harvest mice inhabits eastern Texas, and extends along the coast region as far west as Matagorda Bay, and in the interior north to Hempstead, Nacogdoches, Joaquin, and Texarkana. There is every reason to suppose that it inhabits the whole of eastern humid Texas, as usually it is not an abundant or easily captured species and is often overlooked by collectors. At Texarkana Oberholser caught one and reported the species as rare about thickets on the edge of cleared ground. Hollister caught several at Joaquin on grassy ground along the railroad and at the edge of a cotton field, and at Sour Lake a few in tall grass at the edge of the woods. At Hempstead Gaut caught them in brushy woods between cultivated fields. In southern Louisiana I found them in runways among weeds and tall grass on low ground at Iowa Station, and caught one in a trap where an *Oryzomys* was caught the preceding night. In Matagorda County, Lloyd reported

---

[a] Mammals of Bexar County, Tex., J. A. Allen, Bul. Am. Mus. Nat. Hist., VIII, pp. 66-67, 1896. For further interesting notes on habits of this harvest mouse in Bexar County by H. P. Attwater, see also page 236, same volume.

[121] *Reithrodontomys aurantius* is now classified as a subspecies of *Reithrodontomys fulvescens* (see Annotation 120 and Howell, 1914). It occurs in the eastern part of the state.

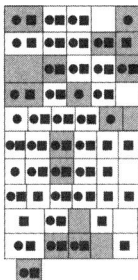

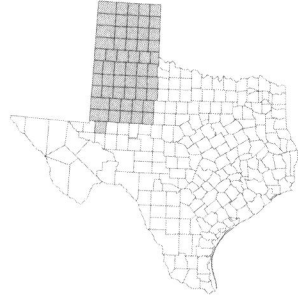

Figure 43. The current distributions of the western harvest mouse (*Reithrodontomys megalotis,* closed circles) and the plains harvest mouse (*Reithrodontomys montanus,* closed squares) in the Panhandle region of Texas (shaded area of state map) according to recent county specimen records. Shaded counties indicate those that were surveyed by federal agents from 1889 to1905; neither species was documented in the region at that time.

[122] *Reithrodontomys megalotis,* the western harvest mouse, occurs in western Texas from the Panhandle southward to the Trans-Pecos region (Choate, 1997). The subspecies are *Reithrodontomys megalotis aztecus* in the northern part of the range and *Reithrodontomys megalotis megalotis* to the south. Surprisingly, Bailey did not record *R. megalotis* from the Panhandle (Fig. 43), a region where it is known to occur today (Choate, 1997).

These mice prefer grassy or weedy areas. They appear to have stable populations over most of their range, although they are seldom captured in large numbers. Yancey and Jones (1997) studied the dispersal of *Reithrodontomys fulvescens* and *R. megalotis* between the High Plains and Rolling Plains of Texas and found these species have dispersed back and forth between the regions by making extensive use of the grassy habitats associated with a "railtrail" created when TPWD converted about 100 km of former railway into hiking and horseback trail. This example of using manmade dispersal routes may be representative of the way other small mammals have dispersed across different vegetative regions during this century.

[123] Taxonomists now regard *Reithrodontomys merriami* as a subspecies of *Reithrodontomys humulis,* the eastern harvest mouse. *Reithrodontomys humulis merriami* occurs in the eastern part of Texas, and there is no evidence that it intergrades into the more westernly distributed *Reithrodontomys montanus* as speculated by Bailey. The distribution of this species is considerably more

106   NORTH AMERICAN FAUNA.   [No. 25.

them under old logs and in low brush, where numerous nests were seen with holes leading into the ground beside them. He called this species the "tree mouse," but does not speak of any nests in bushes or anywhere except on the ground. At Hempstead, Gaut caught a few in traps set at the bases of trees in a brier thicket between two cultivated fields, and in the Big Thicket, northeast of Sour Lake, he reports it as the most abundant mouse, living under the dead grass wherever there was dry ground.

Reithrodontomys megalotis (Baird).   Big-eared Harvest Mouse.[M]

122 This pale, desert harvest mouse comes into western Texas between the Rio Grande and the Pecos, as shown by specimens from Fort Stockton, Pecos City, Alpine, and the southern parts of the Guadalupe and the Franklin mountains.

Cary secured a single specimen on the grassy plain 25 miles west of Fort Stockton, and another under matted grass near a flowing well at Pecos City, but was unable to catch any more in either locality. Gaut caught one in a patch of high grass about two miles north of Alpine, and another on a grassy flat in the foothills of the Franklin Mountains, 15 miles north of El Paso, at 4,400 feet. One was caught in a *Microtus* runway at 8,400 feet altitude on top of the ridge at the head of Dog Canyon in the Guadalupe Mountains. It was among grass, shin oak, and other low brush, and in a most unexpected locality for a *Reithrodontomys.* No other specimens were secured, although considerable trapping was done in the vicinity.

Reithrodontomys merriami[M]Allen.   Merriam Harvest Mouse.[M]

123 This little dusky harvest mouse, the smallest species in the State, inhabits the coast prairies of southeastern Texas west to Richmond, but apparently is nowhere common. Near Richmond I caught two under the grass on the open prairie in the same runways where *Peromyscus taylori subater*[M]was caught. Both of these specimens while in the traps were eaten by some other mouse, so that only the skull of one and the ragged skin of the other could be saved. At Austin Bayou, Lloyd caught the species in "rank grass on the prairie," and at Lafayette, La., R. J. Thompson caught one in "tall meadow grass on the prairie."

Reithrodontomys griseus[M]sp. nov.   Little Gray Harvest Mouse.[M]

124 Type from San Antonio, Tex., No. 87852, ♂ ad., U. S. Nat. Mus., Biological Survey Coll. Collected March 4, 1897, by H. P. Attwater. Collector's number 1068 (X Catalogue No. 371).

General characters.—Size small, tail short and sharply bicolor; color buffy gray with indistinct dorsal streak of dusky; brain case short and wide.

Color.—Upper parts dark buffy gray, darkened especially along the

extensive than depicted by Bailey, but nowhere is it common.

[124] Modern taxonomists now regard *Reithrodontomys griseus* as a subspecies of *Reithrodontomys montanus. Reithrodontomys montanus griseus* occurs in the northwestern and central parts of Texas and *Reithrodontomys montanus montanus* occurs in the Trans-Pecos region. There is no evidence that this species intergrades into the more easterly distributed

Reithrodontomys humulis, as speculated by Bailey. The distribution of this species is considerably more extensive than depicted by Bailey.

As with *Reithrodontomys megalotis* (see Annotation 122), Bailey did not record *R. montanus* from the Panhandle region of Texas, a region where it is known to occur today (Choate, 1997). See Figure 43.

dorsal line with black tipped hairs; ear with a large black spot on upper outer surface and another on lower inner; feet and whole lower parts white; tail white with a narrow blackish line above.

Cranial characters.—Compared with that of *merriami*,[M] the geographically nearest neighbor in the group, the skull is larger with relatively lower, shorter, wider brain case, and flattened instead of circular foramen magnum, smaller bullae, and wider basioccipital. From *albescens*[M] it differs as from *merriami*[M] in relatively shorter, wider brain case.

Measurements.—Type: Total length, 120; tail vertebrae, 56; hind foot, 14.5 (15 measured dry). Average of six adult males from type locality measured by H. P. Attwater: Total length, 114; tail vertebrae, 55; hind foot, 14.6.

Skull of type.—Occipitonasal length, 19.2; basal length, 16; nasals, 7; zygomatic breadth, 10.4; mastoid breadth, 9; greatest breadth of brain case, 9.8; interorbital constriction, 3.

Distribution.—Specimens examined from San Antonio, Mason, San Angelo, Clyde, and Gainesville, Tex., indicate a rather unusual distribution along the eastern edge of the plains. At San Angelo, Oberholser caught one at a hole in the grassy margin of a cultivated field; at Clyde, Cary caught one in a patch of sand burs in the corner of a sandy cotton field. Another specimen was taken at Gainesville on open prairie, but in all of these localities they seemed to be extremely scarce.

Remarks.—The present species holds its characters with surprisingly little variation over an extensive area from San Antonio north to southeastern Nebraska, where, if it grades into *albescens*[M] as seems probable, it must do so entirely between London and Neligh in that State. The smaller, darker *merriami*[M] shows no variation throughout a wide range over the coast prairies of Texas and Louisiana, and if it grades into *griseus*[M] the complete transition must occur between Richmond and San Antonio.

Neotoma floridana rubida Bangs.   Swamp Wood Rat.[M]

125 The common wood rats throughout the Big Thicket of eastern Texas are typical *rubida* of southern Louisiana, while a specimen from Texarkana possibly indicates a shading toward *baileyi*.[M] The Big Thicket is a continuation of southern Louisiana swamp country, extending into Texas from the lower Sabine west to the San Jacinto and marking the western limit of range of many species. Wood rats are well known to settlers throughout its extent. They are reported from near Cleveland and Tarkington in Liberty County and at Bragg and Saratoga in Hardin County, and I found them common in the thickest woods and around old deserted buildings near Dan Griffin's place, 7 miles northeast of Sour Lake. The first

[125] *Neotoma floridana rubida* remains a valid subspecies. The nominal species, *Neotoma floridana* (eastern woodrat), has been recorded from the eastern part of Texas, south to Victoria County, and westward to Edwards and Kerr counties (Fig. 44). Three subspecies are known from Texas, the same number recognized by Bailey, but different trinomial combinations apply to these populations today. *Neotoma floridana illinoensis*, described by Howell (1910), is known from three counties along the Louisiana border; the other subspecies are *Neotoma floridana attwateri*, which is common in the northern and western parts of the range in the state, and *Neotoma floridana rubida*, in the southeast. *Neotoma f. rubida* is the second most common rodent, aside from *Peromyscus gossypinus*, in the Big Thicket region of southeastern Texas. It is especially abundant in flatland hardwood, flatland hardwood pine, and lower slope hardwood pine forests.

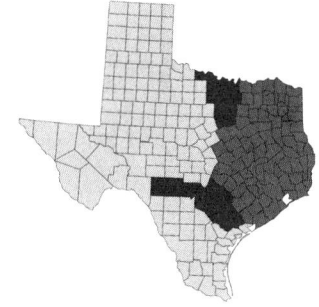

Figure 44. The current distributions of the Southern Plains woodrat (*Neotoma micropus*, light shaded area) and the eastern woodrat (*Neotoma floridana*, medium shaded area) in Texas. Darkest shaded areas represent overlap between the two species.

one secured was in a house of its own building at the base of an old dead pine. It had piled up pine bark and pieces of rotten wood around the base of the tree to a height of 2 feet, and in the cavities in this mound had made several soft nests of grass and bark fiber. There was a nest also in an old hollow log close by and several holes under a rotten stump not far away. As I tore the house to pieces in search of its builder a gray squirrel ran out of the first nest of grass and bark near the top and rushed up the old dead pine. As I uncovered deeper chambers one was found well filled with white-oak acorns and berries of the cat brier, and a cache of green leaves was safely stored away under a shelf of pine bark. The rat was found in a chamber deeper down near the bottom of the house. When finally uncovered it ran to the hollow log near by, then to the holes under the stump, then back to the house before I got a shot at it. It proved to be an old female, as were two others caught the next night under an old log in an equally dense part of the thicket. No trace of the rats was found except under the protecting cover of dense timber, brush, or vine tangle, or in hollow logs, trees, or old buildings. An old log house where hay was stored was apparently well stocked with them, judging from the stick piles under the floor, tracks in the ashes of the old fireplace, piles of characteristic pellets in the corners, and a familiar wood-rat odor pervading the air. More or less evidence of their presence was noticed in other old buildings.

In the thicket near Saratoga the Flower boys told me that a wild cat (*Lynx*) killed a short time before had been opened and its stomach found to be full of wood rats. The abundance of wild cats and barred owls throughout the Big Thicket probably accounts for the habit of the wood rats of choosing the most impenetrable cover.

At Houma, La., near the type locality of the species, I found these wood rats common in the woods and swamps. Some of the houses were built at the bases of hollow trees, over old logs, or under thick brush mats, but just as commonly they were placed in the lower branches of trees or in vines 10 to 30 feet from the ground. Those in the branches were usually in a fork or on a large limb close to the body of a tree, or in a thick tangle of branchlets and connected with the ground by numerous vines, while those suspended in the vines were globular stick masses from 1 to 4 feet in diameter, worked in among a lot of ascending vine stems or into a snarl of vine branches and resembling magpies' nests. Slender sticks, twigs, and pieces of bark and gray moss formed the main body of these elevated houses, while a hole at one side afforded entrance to the soft nest of bark fiber and moss within. By shaking and jerking the vines I drove the rat out of one of these houses and watched him climb up the vines and branches to near the top of the medium-sized tree, probably 60 feet from the ground. He climbed readily, but not with squirrel-like freedom and speed, and avoided the trunk of the tree. Another that I shook out of a house at the base of a small tree climbed up the vines to the top of the tree, some 20 feet from the ground, but I have never seen one climb the trunk of a large tree. No doubt, however, they could climb a rough-barked trunk. Several of the houses located on the ground were examined and in each was found at least one nest of fine bark or moss in a chamber near the ground. No holes could be found entering the ground below the houses, probably owing to the dampness of the soil, which may also account for the elevated houses in this region. Some stick piles and nests were found in hollow logs, and on the ground inside the shell of an old hollow sycamore stub, that measured $10\frac{1}{2}$ feet across, the rats had built a good-sized house against the wall. Several holes entered the sides of this house, and superficial examination located one snug nest in a back corner. Well-marked trails sometimes were found leading through grass and weeds from one house to another or from a house to the nearest log, tree, or brush heap.

Neotoma floridana baileyi [M]Merriam.   Nebraska Wood Rat.[M]

126 This northernmost form of the *floridana* group of wood rats barely gets into northern Texas. Two specimens in the Merriam collection from Gainesville, Cook County, are best referred to it, although they are a shade darker in color and in this respect intermediate between *baileyi*[M]and *rubida*. As a larger series of specimens from across the line in the Wichita Mountains, Oklahoma, is more nearly typical *baileyi*,[M]it seems necessary to refer the Gainesville specimens also to this species.

Neotoma baileyi[M]is a large, pale, bicolor-tailed form of the *floridana* group, extending up the wooded river valleys across the plains country from Texas to northern Nebraska. At Gainesville Mr. G. H. Ragsdale secured a few of the wood rats in wooded ravines, but said they were very scarce. In the Wichita Mountains Gaut found them common from the bases to the tops of the ridges. In the timber along Medicine Creek he occasionally found them in hollow logs or about the overhanging roots of a tree at the edge of a steep creek bank, in houses made of sticks, leaves, bones, and cow chips. Up the steeper slopes of the ridges they were more numerous among the rocks and in crevices of the bluffs. At Valentine, Nebr., where the timber is restricted to the canyons, these rats inhabit the cliffs and caves along the canyon walls, and forage in the brush and timber along the sides and bottoms of the canyons. In fact, over most of the range of the species cliffs, caves, and cut banks furnish the favorite homes. At Marble Cave, Stone County, Mo., I found their tracks in the deepest recesses of the great cave, but found the animals and their stick

[126] *Neotoma floridana baileyi* should be applied to woodrat populations from South Dakota and Nebraska (Hall, 1981). The specimens from Gainesville, discussed by Bailey, are now classified as *Neotoma floridana attwateri.*

[127] Modern taxonomists refer woodrats from the southern portions of the Staked Plains to the species *Neotoma micropus,* not *Neotoma floridana attwateri* (Choate, 1997). The range of *N. f. attwateri* extends no farther west than a line extending from Wichita to Edwards counties in the central part of the state.

houses more common under the shelving limestone ledges along the sides of the ravines. Three or four of those collected were cooked at the ranch where I was staying, and we all pronounced them better than gray squirrels. The meat was very tender and of good flavor, with no trace of the external musky odor peculiar to wood rats.

Neotoma floridana attwateri Mearns.   Attwater Wood Rat.[M]

127 On the juniper ridges of the southern arm of the Staked Plains this big buffy-brown wood rat, which appears to be an Upper Sonoran form of the *floridana* group, lives in a rocky, half-forested region. It makes its house sometimes among the rocks, piling up its rubbish in a broken cliff, rock pile, or old stone wall, and sometimes in the woods at the base of a tree, under a brush pile, in some old cabin, or along the river in heaps of flood drift.

In company with Mr. Howard Lacey, on his ranch in Kerr County, I uncovered one of the houses in the corner of an old log cabin where the rats had built up a pile of rubbish among the fallen logs and boards. As the material was removed the rat ran out of the nest into a hollow log, where he was easily caught. The nest on the ground under the rubbish pile was a bulky mass of soft juniper bark, with an opening at the side. Above it the spaces between logs and boards were filled with several bushels of rubbish, including a large number of cactus thorns. A quantity of green leaves of walnut and some pieces of green cactus stems were found near the nest, while scattered acorn and walnut shells, juniper berries, and cactus capsules showed part of the menu of the occupant.

Mr. H. P. Attwater, who first collected this species, tore down numbers of the houses and found nests in underground burrows as well as in the rubbish piles. He says:

> In one of the underground passages at the nest on the oak ridge were found stored away about three dozen bunches of wild grapes; also many acorns and black haws. In another nest in the cedar brake were about two dozen small mushrooms, partly dry and shriveled. All the heaps in the cedar brakes contained large stores of cedar berries, most of them with the outside pulp eaten off and the seeds eaten out. When the very small size of the seed is taken into consideration, it is surprising what an immense amount of work is necessary before enough can be obtained for a meal, as probably a thousand would be required. One nest contained shells of nuts of the Mexican buckeye (*Ungnadia speciosa*), although these nuts are reputed to be poisonous.[a]

Near Ingram, in the valley of the Guadalupe River, a few of these wood rats were caught in the cliffs and rocks bordering the river valley, but they were more common under the great heaps of driftwood and rubbish along the river bottoms. The Guadalupe, like many of the Texas rivers, is subject to floods, and in a sudden rise of sometimes 50 feet great quantities of driftwood are washed into the

---

[a] Bul. Am. Mus. Nat. Hist., 1896, pp. 61-62.

bottoms and left wherever the trees are close enough to hold it. In places among the old cypress trees tons of this driftwood lie in heaps like haystacks, and in and under these the wood rats find ideal homes. They make holes and runways through the heaps, and hollow out cavities for their nests inside. Often instead of making runways they traverse the logs from one heap to another. A favorite place for a nest is in the drift lodged in vines and branches of trees and reached by means of the vines or rough bark. The presence of the rats in these drift piles is easily detected by their peculiar musky odor. In spite of the odor, which apparently comes from the large gland along the skin of the belly, the flesh of the animals is delicious, of good flavor, white, tender, and more delicate than that of the squirrel.

Neotoma micropus Baird.   Baird Wood Rat.[a][M]

128 This large, light slaty-gray *Neotoma* inhabits the arid mesquite country of the western half of Texas and the adjoining parts of Mexico, and extends north up the Pecos Valley in New Mexico to at least Santa Rosa, and from central Texas northward across western Oklahoma. Specimens examined from Rockport, San Antonio, Brazos, Seymour, Henrietta, Mobeetie, and Lipscomb mark approximately the eastern limit of the species in the State, as at present known. Judging by the characteristic houses which I found abundant near Wichita Falls, the species ranges east to the western edge of the Upper Cross Timbers, while a few old houses at Tascosa and Logan indicate a continuous range with the low mesquite up the Canadian River and across to Santa Rosa on the Pecos, thus completely encircling the Staked Plains. It is the most abundant and widely distributed of the Texas wood rats. It lives mainly in the half open country and builds houses under mesquites, acacias, zizyphus, allthorn, yuccas, cactus, or anything else sufficiently thorny to prove an effectual protection against its enemies. Rarely it lives among rocks. The favorite building site, however, is in and around a bunch of the big flat blades of the prickly pear (*Opuntia engelmanni*), where the stack of rubbish—cow chips, sticks, bark, leaves, stones, bones, pieces of metal, dishes, leather, rags, or any other available material, well salted with bits of cactus and other thorny things—is often built into a dome 4 or 5 feet high. An allthorn bush is another choice building site, and when the house is largely composed of its rigid angular thorns, well mixed with cactus, a more

[128] Bailey's distribution map for *Neotoma micropus* shows the Southern Plains woodrat to be absent from the Llano Estacado and from the region extending southeastward to the western reaches of the Edwards Plateau. However, recent collecting has shown the species is actually common in these areas (Choate, 1997; Goetze, 1998; Schmidly and Bradley, 2016) and that it has a much wider distribution than conceived by Bailey. Both *N. micropus* and *Neotoma floridana* occur in the western region of the Edwards Plateau and there is some evidence that the two species may hybridize with each other in this area (Birney, 1973), but this has not been proven conclusively. Recent genetic evidence (produced by RDB and his students) suggests that the two species also may hybridize along the Red River near Burkburnett. Again, refer to Figure 44, which depicts the distribution and area of range overlap of *N. micropus* and *N. floridana*. Two subspecies of *Neotoma micropus* are recognized, *Neotoma micropus canescens* in the western portions of the state and *Neotoma micropus micropus* in the central region. *Neotoma micropus* likely has been excluded from, or its numbers greatly reduced in, areas of intensive agricultural activity because regular plowing, use of defoliants, and the absence of habitat in some areas have reduced populations. Bailey described this species as a serious pest in many places and this is true even today.

---

[a] The name of black wood rat applied to this species by Professor Baird is as much of a misnomer as its specific name *micropus*. As the species is one of the palest of the genus, I have thought best to change its name to Baird Wood rat. The name 'rat' leads many people to associate with the wharf rats— filthy animals introduced from the Old World and naturalized around our stables and cellars—the wood rats, which belong to a different genus, are natives of America, and animals of exemplary and extremely interesting habits.

bristling and formidable combination can hardly be imagined. Most of the houses, wherever located, are so well protected with thorns that they are rarely molested by the larger mammals, not even by the tough-hided badger. But how the rats can run over these houses and along the trails strewn with cactus spines and never show a scratch on the bare, pink and white soles of their feet is a mystery. One or more nests placed in cavities of the house or in the ground beneath, and entered by openings through the sides or under the edges of the mass of rubbish, are well protected not only from outside enemies, but from occasional violent storms and the glaring heat of the sun. Usually these nests are slight structures of leaves and grass, always kept neat and clean when in use, and quite free from scattered remains of food and excrement. Well-worn trails lead under the brush from one house to another, or away to feeding grounds, or to neighboring rock piles, for the rats seem to be of a social disposition, several usually living together and apparently doing much visiting.

Their food consists of a great variety of green vegetation, especially the juicy flesh of cactus, but mainly of seeds, nuts, and fruit. Cactus fruit and the sweet pods of the mesquite bean are extensively eaten; also acorns, nuts, and any kind of grain within their reach.

At times the wood rats become exceedingly numerous, and their houses appear in every nook and corner of brush, thicket, and cactus patch, while the animals crowd into fields and about ranch buildings, and do some mischief even in a thinly settled stock country. But at such times they attract great numbers of hawks, owls, and other enemies, and after a year or two of unusual abundance they decrease to, and sometimes below, their normal numbers. I have seen a *Parabuteo u. harrisi* come out from under the mesquite with one of this species in its claws, and have found the skulls of large numbers of the rats in and around the nest of this hawk, as well as their flesh and fur in the crop of the bird. The skulls are among the commonest bones recognized in pellets under the cliffs where the great horned owls roost. Being both diurnal and nocturnal, these rats are subject to the attacks of both hawks and owls. Coyotes, foxes, and wild cats catch them whenever opportunity offers, and especially when they are numerous enough to be frequently encountered away from their houses. Snakes are apparently still more deadly enemies, as they enter the holes and houses of the rats and swallow the occupants. Rattlesnakes, bullsnakes, blacksnakes, and whipsnakes are often found in and around the rat houses, and at Comstock, Lloyd opened a rattlesnake and found a wood rat in its stomach. Under ordinary conditions these wood rats are of little economic importance, and will never prove to be a serious pest unless as a result of the destruction of their natural enemies.

Cary found this species abundant at Monahans, and says:

They usually have their nests in mesquite or zizyphus thickets, but frequently take up their abode in the abandoned burrows of *Dipodomys spectabilis*, where thorny branches in the mouth of the burrow give notice of their presence. Their stores usually consist of mesquite beans. They proved a veritable nuisance by continually getting into our small traps and running off with them. They took a number also of our traps home. So commonly was this done that on missing a trap Donald and I would go to the nearest rat house, where we were almost certain to find it. The people at Monahans, and in fact throughout the region, call them chicken rats on account of their supposed fondness for young chickens.

Neotoma albigula [M] Hartley.   White-throated Wood Rat. [M]

129 This wood rat extends into Texas from the west, reaching its eastern limit along the eastern edge of the Staked Plains at Llano, near

FIG. 16.—Distribution area of white-throated wood rat *(Neotoma albigula)*. [M]

Colorado, and in a canyon near Washburn. It apparently belongs to Upper Sonoran zone, but along cliffs and rocky gulches extends into the upper edge of Lower Sonoran, and so slightly overlaps

[129] Edwards et al. (2001) used molecular genetics to demonstrate a distinct taxonomic division between populations of woodrats located east and west of the Rio Grande in New Mexico, with populations east of the Rio Grande representing a different species, *Neotoma leucodon* (white-toothed woodrat), from those west of the river, *Neotoma albigula*. Using this logic, all Texas populations would be assigned to the species *N. leucodon*, including *N. a. robusta* which had been subsumed into *N. a. albigula* by Rogers and Schimdly (1981). If the genetic data summarized in Edwards et al. (2001) remain valid, then two subspecies occur in Texas, *Neotoma leucodon warreni* from the extreme northern part of the Texas Panhandle and *Neotoma leucodon robusta* from the remainder of the range (Blair, 1954a). The distribution of the white-toothed woodrat has changed little since the time of the Biological Survey, although populations recently have been reported in extreme northern Texas, north-central Texas (Cottle, Hardeman, Foard, and Baylor counties), and over most of the Edwards Plateau. As described by Bailey, this woodrat occurs in a variety of habitats in arid regions, but it is almost always associated with slopes and other rocky areas.

[130] *Neotoma mexicana* is a small woodrat, which occupies rimrocks, canyon walls, and other rocky areas at mid to high elevations. It is known only from Trans-Pecos Texas (Brewster, Culberson, Hudspeth, Jeff Davis, and Presidio counties). Although he did not collect it there, Bailey speculated it would occur in the Chisos Mountains, and indeed that is the case. The subspecies in Texas is *Neotoma mexicana mexicana*, and it is common throughout its range in Texas.

[131] Cotton rats (*Sigmodon hispidus*) are more broadly distributed in Texas than Bailey envisioned. They are statewide in distribution, probably occurring in every county. The two subspecies are the ones recognized by Bailey, *Sigmodon hispidus texianus* in the eastern and central parts of the state and *Sigmodon hispidus berlandieri*, which occurs from the Panhandle southward to the Trans-Pecos and the Rio Grande Plain. Recent studies by Peppers and Bradley (2000) and Phillips et al. (2007) reported levels of genetic distinction between eastern populations and those to the west that approach levels observed between other species of rodents; therefore, additional studies are needed to further resolve this taxonomic issue.

Populations of cotton rats are cyclical and subject to extreme fluctuations in density. Incredible densities of this rat have been documented following several successive wet, rainy years and mild winters. H. P. Attwater referred to the huge outbreak of cotton rats around San Antonio in 1889 (Allen, 1896) and many other episodes have been

the range of the larger and grayer *micropus*. Both species occur side by side at El Paso, Sierra Blanca, Kent, Stanton, and Colorado, and in Presidio County, Tex., and at Carlsbad, N. Mex., but each retains its distinctive characters and habits, *micropus* living mainly in its stick houses in the brush, and *albigula*[M] always keeping among the rocks along cliffs and gulches. In a few cases I have caught *micropus* in the rocks, but have never found *albigula*[M] away from them. Being a cliff dweller, its houses are largely provided by nature, and a few sticks, chips, and stones piled among the rocks in addition are often all that seems to be required, but sometimes these accumulations of rubbish in a favorite and long-inhabited den amount to 20 or 30 bushels. The doorways are usually plainly indicated by scattered remains of food and various unmistakable signs. A strong musky odor characteristic of the genus is the usual indication that the dens are inhabited.

I have never known this species to become very abundant or very troublesome. It sometimes enters houses and barns located near the rocks and does a little mischief, but is easily caught in traps. Along its native cliffs and canyon walls it is the especial prey of *Lynx*, *Urocyon*, and *Bassariscus*, which, with the owls, keep its ranks thinned until in many places few are left.

The remains of food scattered about the dens show a varied taste for fruit, seeds, and green things, and usually include pieces of cactus stems and fruit, mesquite, acacia, and other leguminous pods, juniper berries, acorns, and various seeds, green foliage, and flowers.

Neotoma mexicana Baird.   Mexican Wood Rat.[M]

130 This little dark-colored wood rat is the smallest of the species occurring in Texas, and, being mainly a Transition zone animal, has but a limited distribution in the State. It is common in the upper parts of the Davis and Guadalupe mountains, and probably occurs also high up in the Chisos Mountains, where we found old signs but failed to get specimens. In the Davis and Guadalupe mountains it lives in the rocks and cliffs where the junipers and yellow pines are mixed, and also ranges to the very tops of the mountains, where Transition species predominate. In habits it does not differ materially from *albigula*[M] or any of the rock-dwelling species. Its food seems to be largely acorns and the sweet berries of *Juniperus pachyphloea*.[P]

Sigmodon hispidus texianus (Aud. & Bach.).   Texas Cotton Rat.[M]

131 The cotton rats of the eastern half of Texas, while lacking the rich brown color of true *hispidus* of the Atlantic coast, are distinctly darker and more brownish gray than those of western Texas. Specimens from Gainesville, Vernon, Richmond (on the west bank of the

documented this century. Records reveal a severe outbreak in 1919 (Davis and Schmidly, 1994), in McLennan County in 1928 (Strecker, 1929), statewide in the late 1930s and 1940s (Davis and Schmidly, 1994), throughout east Texas from 1958 to 1960 (Haines, 1963, 1971), and near Wichita Falls in 1961 (Dalquest and Horner, 1984). One of us (RDB) witnessed a similar outbreak near Dickens, Texas, in the winter of 2015 and spring of 2016. A variety of factors have been postulated by mammalogists to account for these cyclic

fluctuations, including rainfall cycles, parasitism, and protracted periods of temperature extremes.

Brazos), Sour Lake, Port Lavaca, Seguin, and San Antonio are fairly typical, although they become slightly paler at San Antonio. Along the Gulf coast and lower Rio Grande they become still paler without reaching the extreme of the light gray *berlandieri*.

Although not often seen the cotton rats are usually common, and at times they become excessively numerous, living under cover of tall grass and weeds, in meadows, around the edges of fields, and along the banks of streams and ditches. They live in bulky nests of grass

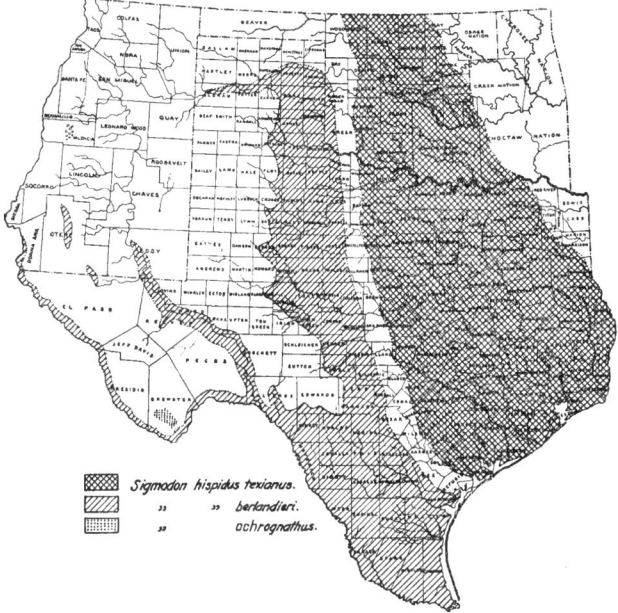

FIG. 17.—Distribution areas of cotton rats (genus *Sigmodon*).

on the surface or in underground burrows, and make numerous long runways under cover of fallen grass and dense vegetation. Apparently they breed rapidly. Gaut records a female containing 8 embryos. For food they cut the green stems of grass and various plants along their runways—eating stems, leaves, and seeds—and along the edges of grain fields they gather to feed on both green and ripening grain. The amount of damage they do depends on their abundance and the kind of crop attacked.

Near Seguin, in November, 1904, I found the cotton rats numerous around fields, in grass patches, under brush heaps and fallen weeds, in the mesquite woods, and in fact everywhere that any cover or concealment could be found. Thickets of thorny chaparral and bunches of cactus offered the most perfect protection, and even at midday the animals often were seen running about under the prickly pear. A network of their runways covers the surface of the ground and connects the numerous burrows wherever protecting cover is offered. Along the edges of cotton fields they are especially numerous, and the runways opening into the fields are often fairly lined with cotton that has been pulled from the bolls and dragged under cover where the seeds can be eaten in safety. Some cotton minus its seeds is also found scattered over the ground near the edges of fields where the animals are abundant, and a smaller amount is carried away for nests. The loss of cotton is not great in any one field, but considered over the entire range of this group of cotton rats, which coincides in a general way with the cotton-producing area of the United States and Mexico, it is considerable.

A simple and effective remedy would be to clean out the borders of fields by burning the weeds, grass, and rubbish accumulating along the fences year after year as a harbor for various rodent and insect pests and a perennial source of supply of weed seeds. If these borders were burned yearly, mowed and raked, treated with oil or chemicals to prevent weed growth, closely pastured, or thoroughly cultivated, the hawks and owls would quickly dispose of the rodents, which would then have no protecting cover. Marsh hawks are abundant and constantly skim over the fields, frequently diving into the grass. Harris hawks sit on the mesquite trees and telegraph poles watching the ground below; sparrow hawks sit on the fence posts, and barred owls are heard hooting every evening from the moss laden live oaks. There is no lack of enemies eager to prey on the rodents, and no simpler way of reducing the number of such pests than by the aid of their natural enemies.

As is the case with rabbits, and many other species of rodents, the abundance of the cotton rats varies greatly with the different localities during the same year, and with different years in the same locality. At times they are extremely scarce over extensive areas, and again so numerous as to suggest the plagues of voles that from time to time have overrun parts of Europe. Mr. H. P. Attwater describes one of their invasions, and the enemies that attacked them, as follows:

In the year 1889, Sigmodons appeared suddenly in this [Bexar] county in great numbers, and were known as "tramp rats." Where they came from, or from which direction, I have been unable to find out. Thousands first appeared about the 1st of May, and were heard from in all the region for many miles around San Antonio. They were most numerous in the high, dry parts of the country, and were not noticed in the lowlands along the rivers. They were very numerous all through the "chaparral," and made their nests with the wood rats (*Neotoma*) in the bunches of *Opuntia*, with a network of runways leading in every direction, through which they were often seen running in the daytime. They seemed to agree with the wood rats, but in the oat stacks and around the ranch buildings the common brown rats fought, killed, and ate them. Mr. Watson's boys killed over 100 in one afternoon in a brush fence, and for several months their cat used to bring in from 6 to 12 every night. He says that on one occasion, when the rats were thickest, they counted 38 which this cat in one night had piled up in the wood box for the amusement of her kittens.

The "tramp rats" played particular havoc with all kinds of grain crops, and corn in particular, but they were not good climbers, and consequently the ears on leaning stalks suffered most. Some farmers lost half their corn crop, and in some instances small patches were entirely destroyed.

During the winter of 1889 and 1890 marsh hawks were very numerous, no doubt attracted by the rats. The hawks were seen skimming over the fields in the daytime chasing the "tramps." In 1890 and 1891 short-eared owls, on their way north in the month of March, stopped over to attend to the Sigmodons; in other years I have not noticed these owls during migration. Weasels and little striped skunks were much more common than usual in 1890 and 1891, which I attribute to the same cause. Rattlesnakes and other snakes were seldom seen abroad, and when disturbed in their retreats were found gorged with cotton rats. The large skunks and coyotes hunted them, and dogs, generally in the habit of killing rats and mice and shaking them, also ate them.

The bulk of these rats stayed for about eighteen months. After the crops were gathered in 1890 they began to get scarce, and gradually disappeared during 1891. Whether they died out or "tramped" out I am unable to say, but I am inclined to think many of them migrated. Old settlers say they remember a similar invasion about the year 1854.[a]

Sigmodon hispidus berlandieri Baird.   Berlandier Cotton Rat.[M]

Sigmodon hispidus pallidus Mearns, Proc. U. S. Nat. Mus., XX. Advance Sheet, March 15, 1897, p. 4. Type from 6 miles above El Paso, Tex.

132 This pale-gray form of the cotton rat inhabits the desert region of eastern Mexico and western Texas along the Rio Grande and Pecos valleys. East of the Pecos Valley it grades into *texianus* so gradually that no dividing line can be drawn.

The habits of this species do not differ from those of *texianus*, except in so far as modified by the character of the arid desert country in which it lives. The rats find suitable food and cover mainly along the more fertile stream valleys or in the irrigated sections, where they usually live under the fallen grass, canes, weeds, or brush, and they eagerly gather in fields of growing grain or alfalfa. Their burrows often perforate the banks of creeks and irrigation ditches, but their nests are found also on the surface of the ground, scattered through the fields and over the level bottoms.

[a] Quoted by J. A. Allen, Mammals of Bexar County, Tex.: Bul. Am. Mus. Nat Hist, Vol. VIII, pp. 62-64, 1896.

[132] The subspecies *Sigmodon hispidus berlandieri*, hispid cotton rat, is much more broadly distributed today than indicated by Bailey. It occurs in virtually every county in West Texas, most likely having spread there along railroad and highway rights-of-way that provide suitable habitat for dispersal.

[133] *Sigmodon ochrognathus*, the yellow-nosed cotton rat, is similar in appearance to *Sigmodon hispidus* but it differs in having a paler overall coloration and a distinctly orange or rusty snout. As discussed by Bailey, it is common at the higher elevations of the Chisos Mountains, and it has now been recorded from the Davis Mountains in Jeff Davis County, from Big Bend Ranch State Park in southern Presidio County (Yancey and Jones, 1996), the Elephant Mountain Wildlife Management area in Brewster County (Heaney et al., 1998), the Sierra Vieja range in western Presidio County, and from the Guadalupe Mountains in Culberson County. Until recently, it was thought to be rare and restricted to montane habitat. But recent collecting shows it to occupy a number of non-montane habitats in the Trans-Pecos, albeit in small numbers.

118   NORTH AMERICAN FAUNA.   [No.25.

An old female, taken at Carlsbad, N. Mex., September 9, 1901, contained 11 nearly matured embryos, which is probably an unusual number, as the old one had but 10 mammae—inguinal $\frac{1}{1}$, abdominal $\frac{2}{2}$, pectoral $\frac{2}{2}$. The front pair were between the arms, almost on the throat. Another specimen had 8 mammae.

FIG. 18.—Cotton rat (dead) and nest in Johnson grass, Pecos Valley.

Besides grass, grain, and alfalfa, a few grasshoppers were found in the stomachs of the specimens examined at Carlsbad.

Sigmodon ochrognathus Bailey.   Chisos Mountain Cotton Rat.[M]

133   These little yellow-nosed sigmodons are abundant in grassy parks among the oaks, nut pines, and junipers over the top of the Chisos Mountain plateau at 8,000 feet altitude. They live in numerous burrows and runways under short grass and feed on the stems of grass and various small plants. They are mainly diurnal, and we often saw them running along their little roadways in the daytime, while our traps were rarely disturbed at night. On June 13, 1901, besides young of several ages, two females were caught, one of which contained four small and the other four large embryos. Some old grass nests were found on the surface of the ground, but these apparently were winter nests. The runways all led to fresh burrows in the ground, which were at least the summer homes of the sigmodons.

As the country around the Chisos Mountains is a hot, Lower Sonoran desert, the species seems to be entirely isolated on top of the mountains. Its nearest relatives are on similar isolated ranges in Mexico.

Microtus mexicanus guadalupensis[M]Bailey.   Guadalupe Vole.[M]

134   These little, short-tailed, snuff-brown voles are common over the brushy or grassy slopes of the Guadalupe Mountains from 7,800 to 8,500 feet in Transition zone. Unlike most species of *Microtus,* neither the presence of water nor moist nor grassy ground is required for their homes. In the head of McKittrick Canyon they live in the dry grassy parks and open places in the woods, where their runways, burrows, and old winter nests are abundant under the tall grass and weeds. Higher up on open ridges their runways wind about among stones and leaves under the shin oak and other low bushes of the driest mountain slopes, and sometimes well into the edge of the woods. The runways are distinct, well-worn little roads leading from burrows to feeding grounds or to other burrows. The summer homes seem to be entirely under ground, but unused grass nests found here and there on the surface appear to have been built for winter use under the snow. Green vegetation seems to be the principal food of this vole, and little, clean-cut sections of grass and various plant stems are found scattered along the runways on the feeding grounds.

Several old females, caught late in August, contained embryos, and at the same time young of various ages were caught in the traps.

Microtus ludovicianus[M]Bailey.   Louisiana Vole.[M]

135   At Sour Lake, in southeastern Texas, Hollister secured a single specimen of this little vole, previously known only from Calcasieu Parish, La. It was caught in a brush patch at the edge of the prairie in company with the cotton rat. The prairie about Sour Lake is very similar to that just east of Lake Charles, La., where I found these little voles fairly numerous, living in the peculiar, flat mounds that are scattered over the low, damp prairie, and making their runways through the grass from one to another. Some of the mounds were perforated with a dozen or more of the little round holes, from each of which a smooth trail led away. A colony of a dozen or less of the voles, in some cases all adults, in others both adults and half-grown young, was usually occupying a mound. One female taken April 8 contained three well-developed embryos, and several others taken on the same date were giving milk. As usual in the females of this subgenus (*Pedomys*) the mammae were uniformly inguinal $\frac{2}{2}$, pectoral $\frac{1}{1}$. A few winter nests of grass were found on the surface of the ground where the standing grass had burned off, but the breeding nests apparently were all in the burrows below the surface.

Fire had recently run over most of the prairie and left the burrows exposed and the trails sharply defined over the blackened ground, but as the animals were caught as readily over the burned area as in the standing grass, the burrows are evidently a safe retreat in case of fire.

[134] During his professional career, Vernon Bailey was considered a leading authority on the taxonomy and life history of voles belonging to the genus *Microtus*, having written a taxonomic revision of the genus in 1900. Bailey (1902) described *Microtus mexicanus guadalupensis*, although current taxonomy refers this subspecies to *Microtus mogollonensis guadalupensis*, the Mogollon vole. This is a relict species with Rocky Mountain affinities, and it represents one of the unique taxa in Guadalupe Mountains National Park, the only place in Texas where it occurs. Collections of mammals made in the Guadalupe Mountains subsequent to the Biological Survey (Davis, 1940; Davis and Robertson, 1944; Genoways et al., 1979) show this to be one of the most common small mammals at the highest elevations of the park (above 6,300 feet [1,920 m] elevation). The species is stable within the confines of the national park.

[135] *Microtus ludovicianus*, the prairie vole, is now regarded as a subspecies of a wide-ranging species, *Microtus ochrogaster*, and it is thought to be extinct throughout its range in Texas and Louisiana (for a discussion, see Chapter 6). J. K. Jones Jr. et al. (1988b) reported eight specimens of another subspecies of *M. ochrogaster, Microtus ochrogaster taylori*, from three counties (Armstrong, Hansford, and Lipscomb) in the northern Panhandle. A. H. Howell worked extensively in Lipscomb County in 1903, collecting twenty species of small mammals, none of which were voles. More recently, Roberts et al. (2015) reported a specimen of *M. ochrogaster* from Lubbock County, suggesting that this subspecies is expanding its range. The occurrence of *M. o. taylori* in Texas is almost certainly an invasion that occurred in the twentieth and twenty-first centuries.

[136] Although Bailey only recorded *Microtus pinetorum* (referred to as the woodland vole) from two counties in Texas (Bowie County in the far northeastern corner of the state and Gillespie County in the Hill Country), it subsequently has been reported from numerous places in the eastern and central parts of the state, but nowhere does it appear to be common within its range in Texas. Two subspecies are known from Texas, both of which were described by Bailey on the basis of specimens collected in other states, namely *Microtus pinetorum auricularis* (type locality in Mississippi) in the southern part of the state and *Microtus pinetorum nemoralis* (type locality in Oklahoma) to the north. Although Bailey assigned both of his specimens to the subspecies *M. p. auricularis*, modern taxonomists recognize both subspecies as part of the Texas fauna. Continued degradation of grassland habitats could have a major impact on this species in Texas.

[137] The name *Fiber zibethicus* has been changed to *Ondatra zibethicus*. Bailey and the other field agents found muskrats to be abundant at the beginning of the century in the Canadian River drainage. Similarly, Blair (1954b) reported a dense population in the tule marshes of Moore and Bugby creeks in Hutchinson County. J. K. Jones Jr. et al. (1988b) conducted an extensive survey of the mammals of the northern Texas Panhandle during the 1980s and did not report any evidence of muskrats at the sites where they had been previously reported. In fact, it appears that most of the creeks are now dry and the tule marshes are greatly reduced (late Clyde Jones,

120　　　　　　NORTH AMERICAN FAUNA.　　　　　[No. 25.

The stomachs of those caught contained only green vegetation, and along the runways grass and various small plants had been cut for food. As rice is the principal crop over these low prairies and as the ground is flooded while the rice is growing, this little vole is not likely to do serious damage.

Microtus pinetorum auricularis[M]Bailey.　Bluegrass Vole.[M]

136 Two specimens from Jefferson, in northeastern Texas, prove to be nearest to this form of the subgenus *Pitymys,* although differing slightly in the more elongated skull and larger bullae. They were caught about a mile south of town at the edge of a swampy run under a tangle of old grass and blackberry bushes. Most of their numerous runways, nests, and burrows were unused at the time the specimens were taken (June 12, 1902), which would indicate that previously the occupants had been much more numerous. There were none of the surface ridges which are usually found marking the tunnels of *pinetorum* and allied species, probably owing to the ample cover of vegetation which hid their runways. Neat, little grass nests were found here and there on the surface of the ground under the leaves along the trails, and burrows entered the ground at frequent intervals. A few bits of grass and tender plant stems were the only traces of food noticed along the runways.

A flat skin and smashed skull, apparently of this subspecies, sent in 1895 to the Department from Baron Springs, near Fredericksburg, Tex., by Fritz Grosse, formed the only previous record of the subgenus from Texas, although it ranges over the southeastern United States and reappears in Vera Cruz, Mexico.

Fiber zibethicus (Linn.).　Muskrat.[M]

137 Nine specimens of the muskrat from Lipscomb and three from Canadian fail to show any cranial characters that will separate them from typical *zibethicus,* assuming that New York, Massachusetts, and Minnesota specimens are typical; but size and cranial characters separate them widely from their near neighbors, *ripensis,* of the Pecos Valley. The pelage of these 12 specimens, which were collected June 25 to July 16, is worn, faded, and very pale, while the more northern specimens, collected in fall, winter, and spring, are comparatively fresh and dark. I have seen, however, equally pale summer specimens at Elk River, Minnesota.

Lipscomb and Canadian are practically at the junction of Upper and Lower Sonoran zones, and apparently mark the extreme southern limit of range of *zibethicus.* At Canadian, Howell reports muskrats as "numerous at Clear Creek, living in the fish ponds and irrigation ditches, where they cause considerable trouble by tunneling into the banks and thus releasing the water." At Lipscomb he says:

They are found in small numbers in nearly all the small grassy creeks throughout this region. I secured two on Cottonwood Creek, 5 miles east of

personal communication). One of us (DJS) is aware of populations north of Pampa, Texas, along the Canadian River, and muskrats have been reported from several localities in southwestern Oklahoma, just outside of the boundary with Texas.

Bailey did not attempt a subspecific assignment of the specimens from the Texas Panhandle, but he did assign the specimens from the Rio Grande and the Trans-Pecos to the subspecies *Ondatra zibethicus*

ripensis, which he had previously described in 1902. A recent study by Falcone et al. (2019) examined genetic variation in *O. z. ripensis* and suggested it was closely aligned with populations in New Mexico. In addition to *O. z. ripensis*, the subspecies recognized today are *Ondatra zibethicus cinnamominus* along the Canadian River drainage and *Ondatra zibethicus rivalicius* on the Gulf Coastal Plain. See Chapter 5 for a discussion of the conservation status of muskrats in Texas.

[138] The name *Fiber zibethicus ripensis* has been changed to *Ondatra zibethicus ripensis.*

here, and a man who went fishing there a few days later saw three more. He approached near enough to one, which was feeding on the bank, to hit it with his fishing pole, and after it had retreated into a hole in the bank he prodded it until it came out and swam away. I set traps at this place later, but caught nothing. In a creek known as First Creek, flowing into Wolf Creek from the north 15 miles west of Lipscomb, I found the muskrats really abundant, the local conditions being peculiarly favorable for them. This stream consists of a series of wide and deep holes, with abundance of marsh grass growing on their borders, and partially filled with a flowering water plant (*Batrachium divaricatum*) upon which the muskrats feed. Their trails could be seen leading in every direction through this mass of floating vegetation, and one could hardly walk a half mile along the creek at any time of day without seeing one or more of the rats. Their favorite feeding times are about sundown and sunrise, and at these times I sometimes saw eight or ten in a short distance. They swim out from the bank into the water plant, and rest quietly on the surface while they feed. Several which I shot had the flowers of this plant in their mouths. These rats do not build nests, as their eastern cousins do, but live entirely in holes in the banks, entering either below or just at the surface of the water. When alarmed they dive and take refuge in one of these hidden retreats. When I first began to hunt them they were much less wary than after several had been killed, and if one were to sit quietly on the bank they would feed and move about unconcernedly. I secured seven in two evenings' hunting, besides wounding several which got away. I failed to catch any in traps, except one, which got away with the trap. I was told that they are common for miles up this stream, and, if so, there must be hundreds of them. (June 19 to July 10, 1903.)

Fiber zibethicus ripensis[M] Bailey.   Pecos River Muskrat.[M]

138 This small, dull-colored muskrat lives apparently in suitable places along the whole length of the Pecos River and on some of its tributaries, and along the Rio Grande near the mouth of the Pecos. In 1890 I found a few unmistakable muskrat tracks and signs on the banks of the Rio Grande near Del Rio, and ten years later again found their signs in the Pecos Canyon above the High Bridge. In 1902 Cary and Hollister collected a series of specimens at Fort Stockton, where they were common in the rushes along the banks of Comanche Creek, and Gaut collected a few higher up on the Pecos at Santa Rosa, N. Mex. They are common near Carlsbad (Eddy), N. Mex., in the river and irrigation canals, where their burrows enter the banks below the surface of the water and are high enough up for a dry nest chamber, often at a considerable distance from the brink. Grassy or tule-fringed banks are chosen, if possible, with the double advantage of cover and a supply of food close at hand. The muskrats are largely nocturnal, but usually come out of their burrows before dark and are sometimes seen swimming at midday. They bring up roots and stems of grass, sedges, and various aquatic plants, and after eating them on little shelves or niches in the bank, leave rejected and scattered parts behind that show the nature of the food. At the slightest alarm they dive with a splash and are seen

[139] By the 1850s, beavers (*Castor canadensis*) were reduced to very low population numbers over a considerable part of Texas because of excessive annual harvests by trappers. However, these animals were able to survive on the remote streams of the upper and western Hill Country, along the Devils River, along the edge of the Panhandle, and around El Paso (Weniger, 1997). By the early 1900s, the federal agents found them to be decreasing in numbers and shortly thereafter strict harvest regulations were imposed and restocking of depleted populations became common practice. Today, beavers are found over most of the state where suitable aquatic habitat prevails. Two subspecies are recognized, *Castor canadensis texensis* in the northern four-fifths of the state and *Castor canadensis mexicanus* along the Rio Grande.

no more, either coming up at some distant point or hiding under the banks or in their nests.

In several places their burrows were found in the banks of the large irrigation canals, where no doubt they cause some of the mysterious breaks that occur in the ditches.

Castor canadensis texensis subsp. nov.　Texas Beaver.[M]

139 Type from Cummings Creek, Colorado County, Tex., No. 135744, U. S. Nat. Mus., Biological Survey Coll. Original number, 5139, X Catalogue. Made over from a mounted specimen purchased of A. Hambold, New Ulm, Tex. Caught in Cummings Creek by Florence Brune, Dec. 25, 1900, and kept alive until Jan. 10, 1901. Sex not indicated. Old and large.

Characters.—Coloration pale, as in *frondator*, possibly due in part to fading.

Skull.—Sagittal crest short and lateral ridges lyrate or spreading even in extreme old age; supraoccipital crest doubly curved, nasals long, spatulate, and tapering to narrow point posteriorly.

Measurements.—Type: Hind foot, measured dry, 174; naked portion of tail, measured dry, 265 long, 113 wide.

Skull of type.—Basal length, 136; nasals, 57; breadth of nasals, 30; zygomatic breadth, 107; interorbital breadth, 29; mastoid breadth, 67; alveolar length of upper molar series, 32.

Specimens examined.—Type, skin, and skull, and two skulls from Cypress Mills, Blanco County, farther up the Colorado River.

Remarks.—The characters shown by these three specimens are so well marked and uniform as to justify describing the subspecies, even on so scanty material. Whether the beaver of other streams north and south of the Colorado Valley of eastern Texas are the same can be settled only by specimens; but I have grouped the scattered notes and records for all but the Rio Grande and Pecos valleys of Texas under this form.

Beaver are still found in many of the streams of eastern Texas, especially in the larger rivers, where deep water and steep banks afford protection against relentless trapping. In 1892, at Arthur, in northeastern Texas, I was informed that they were fairly common along the Red River and that trappers caught a few each year. In 1902, at Texarkana, Oberholser was told that a few were still found in the Red River, and in 1901, at Mobeetie, was informed that they were common in Sweetwater Creek, a branch of the North Fork of the Red River. In 1903 Howell reported them as still common in the Sweetwater and Gage creeks not far from Mobeetie; also in the Wichita and Canadian rivers not far from Canadian. In the Colorado River a few were reported in 1892, by B. H. Dutcher, about 10 miles below Colorado City; again, in 1902, they were reported by Oberholser as rare near Austin and Elgin, while in 1892 Lloyd found trees girdled by them along the Colorado in Matagorda County. In 1901 Oberholser saw a skin that was brought into San Angelo, and was probably taken near there on the Concho, a branch of the Colorado. In 1902 he reported beaver in the Brazos as rare in the region of Brenham; in the Trinity River as occurring at Long Lake; in the Neches as occurring rarely in the river and bayous in the region of Beaumont, and as occurring in some of the larger streams about Jasper (probably branches of the Neches or Sabine). In 1899, at Lake Charles, La., I was told that trappers came down the Sabine River every winter, and among other furs brought some beaver. In the Big Thicket, in 1904, Dan Griffin told me that beaver were abundant a few years before in Village Creek, Polk County. In 1900 Oberholser was told of a colony of beaver 40 or 45 miles northwest of Uvalde, which would place them on the headwaters of the Nueces. In 1902 Mr. Gething, of Rock Springs, told me of a fine beaver skin that he bought the previous winter, which was obtained on the headwaters of the Rio Frio.

Through the kindness of Mr. Attwater, I am able to give the following interesting notes from his correspondence with Mr. J. F. Leyendeker, who writes from Frelsburg, Tex., under date of June 6, 1904:

I have your favor of the 3d instant and will cheerfully give you all the information at my command in regard to beavers in this section of Texas. I have heard of beavers and seen them in the Colorado River and Cummings Creek, a tributary of the Colorado River, which has its source near Giddings in Lee County, and empties into the Colorado River in the big bend about 2 miles nearly north of the town of Columbus. It is quite a large stream, with many deep-water holes or pools, sometimes over half a mile long and from a few to 10 or 12 feet deep.

The first beaver I ever saw was a very large male, weighing over 40 pounds, killed by my brother in said creek, in February or March, 1866. But few were noticed until after the big overflow of the Colorado River in 1869 and 1870, after which they were more numerous, especially where the creek passed through Mr. F. A. Brune's plantation, about 7 miles nearly north of Columbus. In this place there was quite a colony of the beavers, in fact so many that they did considerable damage to Mr. Brune's growing corn crop by cutting off the stalks, and I suppose using the ears as food. About six or seven years ago they constructed a dam across the creek, 40 or 50 feet long, in Mr. Brune's field, using blood weeds mostly and some other material for that purpose. This dam was perhaps a foot to 15 inches high, and strong and compact, but of course the first rise in the creek washed it away.

Mr. Emil Brune, a son of F. A. Brune, was here yesterday, and, after questioning him in regard to beavers, he said that he trapped six or seven, among these the one sent to San Antonio in January or February two years ago—he did not recollect the exact date. He also stated that while fishing he broke through into a beaver cave and there found four young beavers, which he carried home, but they soon died. I have been informed that there are still some beavers in Cummings Creek, near Mr. Justin Stein's place, a few miles nearly west of Frelsburg. It is also said that there are still some beavers in

[140] The subspecies of beaver that occurs along the Rio Grande is now classified as *Castor canadensis mexicanus*. Bailey (1913b) described this subspecies, making the following comments: "In my report on the mammals of Texas in 1905 I referred the beavers of the Rio Grande and Pecos rivers to *Castor canadensis frondater* Mearns. Since that time specimens have been collected at additional localities along the Rio Grande and its tributaries, and in working over the material in the Biological Survey Collection from New Mexico I find that the beavers of the Rio Grande drainage differ so markedly and constantly from those of the Colorado drainage that it becomes necessary to provide a name for them."

the Colorado River, near Mr. William Schulenburg's place, about 4 miles above the town of Columbus.

In 1872, while surveying land 18 miles above Fredericksburg, I found the beaver quite abundant in the Perdinales and White Oak creeks, and I have no doubt that some may be found there yet.

Castor canadensis frondator[M]Mearns.   Broad-tailed Beaver.[M]

140   Beaver are still found in many places along the Rio Grande, Pecos, and Devils rivers. In 1891 Lloyd reported them as common on the Mexican side of the Rio Grande 12 miles below Matamoras, and in 1900 at Brownsville I was told that a good many beaver were caught in the river above there every winter. In 1902 a fine specimen was taken by Goldman at Camargo, on the Mexican side of the river, and Mr. F. B. Armstrong told Mr. Nelson that the live beaver sent to the New York Zoological Gardens were caught in the Lower Rio Grande within 8 miles of the mouth. In the summer of 1901 we found fresh beaver 'sign' near Boquillas, in the Great Bend, and in the following winter trappers reported a good many beaver caught in the Rio Grande, Pecos, and Devils rivers, but stated that their numbers were rapidly decreasing. Still, one of these trappers assured me that he expected to make $500 on a trapping trip down the Rio Grande from Langtry to Brownsville the next winter, and was counting on getting $5 for each of his beaver skins.

In the winter of 1902-3 one trapper was reported to have caught 200 beaver on the Rio Grande between the Grand Canyon and Del Rio.

In the summer of 1902 I visited a beaver pond in the Pecos River Canyon, where apparently a good-sized family of beavers was living. This pond was a natural reservoir in a deep, sheer, walled side canyon, and was filled from the river in times of flood until it formed a deep lake a hundred yards wide and half a mile long, held in at the narrow outlet by a dam not over 30 feet long at the time of my visit only 2 or 3 feet high. This pond—or lake, as it is called—with steep earth banks on one side, overhung by willow trees, with deep holes, big bowlders, and little islands, is an ideal spot for a beaver home. The willows furnish the principal food of the beaver, and have for ages, as shown by the old stumps and fallen timber along the shore, together with the freshly cut trees and gnawed bark and branches. The beaver often cut a tree so that it falls into the water, leaving the base anchored to the stump, and then at their leisure gnaw off the bark and cut the branches. Many trees fall inland, however, and in that case are abandoned, or else are well trimmed of branches and bark or cut into sections and carried away. The banks offer such good retreats that apparently no houses have been built around the lake, and at the time of my visit the river was high and the top of the dam was 2 feet under water. The photographs taken of the pond and its surroundings showed some of the cut trees, though little else of the beaver's work.

A beaver house near the head of Devils River was built on the bank of a deep rock-bottomed pond, where the clear, blue water spread out into a quiet little lake full of fish and margined in places with lily pads and willows. The house was placed on a rocky bank just above deep water and was mainly composed of old beaver cuttings—willow stems and branches cut to a convenient length for transportation. These were simply piled up in a mound some 8 or 10 feet wide and 3 or 4 feet high without mud or other filling, but when I tried to open a doorway to the nest I found them interlaced in a snarl that was not easily broken through. The house had the appearance of a big brush heap or a pile of driftwood on the bank, and might have been passed unnoticed save for its position and the gnawed ends of the sticks. Apparently it was either new or merely the summer house of one old beaver, and consequently was small and not substantially built. Its walls were so thin that as my shoes touched the rocky ledge at the back I distinctly heard the beaver get up and slide out of his nest into the water. As he left the house I caught a glimpse of him deep under the water, and for some time followed his course of travel by the line of bubbles that came to the surface as he swam up and down the lake or came back near the house to watch for a chance to return and finish his nap. At no time did he show himself at the surface, and the glimpses I had of him were at a depth of 6 or 8 feet, where he looked like a great fish dashing along with the speed of a racing boat. Quietly withdrawing, I returned at sundown to watch for his appearance. Just before reaching the house I saw a big head with short stubby ears rise quietly from the water near the middle of the lake and lie motionless for a few minutes and then move toward the bank a few rods below the house and disappear just before reaching it. A moment's stealthy creeping put me in the bushes close to the house, where I could watch the water, and after a few minutes the beaver again came to the surface with a stick in his mouth, apparently a willow root from under the bank. He swam leisurely around a big bowlder and then came directly toward me. When about a rod from shore his head went down and his round back rolled up as he dived to his submarine doorway. A moment later I heard him enter the house beside me. For fifteen minutes I could hear his big chisel teeth crunch, crunch through the wood and bark as he munched his evening meal. When the munching stopped there was another stir inside, followed by a gurgling of water from below. Then a line of bubbles spread out along the surface of the water for several rods from shore, and soon the familiar head rose to the surface. After remaining quiet for about a minute the beaver started back to the same feeding spot at the bank and again dived at the base of the willow tree. For about three minutes he remained below, and then came up again with food and started for the return trip to the house. As dusk was now deepening and as I fully realized the importance of securing a specimen from a river where no beaver had ever been collected, I dared not wait longer, but decided to shoot him with buckshot, as the light was far too dim for rifle sights. For fear of injuring his skull I aimed for his neck, which was deeper under water than I counted on. At the report a thundering splash told that he was not dead. A second later he leaped from the water close to my feet and at a single dash crossed a narrow point of land at the edge of his house and disappeared in the deep water, followed by a line of bubbles that shot up the pond. Such strength, such powerful bounds, and racehorse speed I had never dreamed of in the clumsy looking beaver. I had emptied my pockets of notebook and cash, to be ready to dive for the prize in case he sank, but I would as soon have jumped on a grizzly bear in his native gulch as this live beaver in the water. A little later a loud slap of his tail on the water far up the pond sounded like a "come on," and the old trappers tell me that this is really a fighting challenge. I waited until after dark without further developments, and then picked my way over the rocks for the long 2 miles back to camp. In the morning the old moon was still shining, and I was at the beaver house before day began to break, but there was no beaver either in the house or outside. He had moved, and probably had not returned to that part of the river since his fright. All that was left for me to do was to examine and photograph his house. With a good deal of difficulty I forced an opening through the stick wall so that I could put my arm in and feel the damp walls of the chamber, the big round hole where the water came just to the edge of his bed, and the bed of grass and weeds scattered over with peeled branchlets and roots of willow. No trace of other food was found. It was evident from the size of the house and the nest chamber that this was the bachelor quarters of an old beaver. Carefully closing the opening, I left the house as nearly as possible as I found it.

In talking with John Seawel, an old beaver trapper, I asked him why it would not pay to protect the beaver in a pond like that above the Pecos Bridge and let them multiply. The idea was not new to him, for he had talked it over with other trappers and all agreed that it was not worth trying, because they considered the beaver naturally ferocious, to a great extent solitary, and a slow breeder. Seawel says that two old beavers rarely live together in one house or even in one small pond; that they fight and chase away any newcomers; that if a family grows up and is undisturbed in a pond or a deep bend of the river, its members keep all others of the species away, and that they attack and kill any one of their number that is found in a trap or is sick or crippled. While he thinks that systematic breeding for fur is out of the question, he admits that the beaver should be protected all over the country, until the few that remain increase and restock the rivers. There are probably more beaver in the Rio Grande and Colorado rivers than in any other southern streams, and it is important that Mexico should cooperate with the United States in protection of the mammal that has played so important a part in the history of the development of the country.

Liomys texensis Merriam.   Spiny Pocket Rat.[M]

Heteromys alleni Allen, Bul. Am. Mus. Nat. Hist., III, 1891, p. 268 (in part, specimens from Brownsville).

141 A large series of these little spiny pocket rats has been collected in the region about the mouth of the Rio Grande, at Brownsville, Matamoras, and Lomita.

Loring reports them at Brownsville as "common in the timber under logs and the roots of trees;" and Lloyd says they are "found at Lomita in the densest brush on the ridges forming the old banks of the river, and around old corrals." He adds:

> Their habit of throwing out a white clayey mound like the gophers attracts attention, and, although the mound may be a month old, by cleaning out a hole and putting a trap in it you will in time capture the occupant. The ordinary outlets are generally covered up by fallen leaves, which in some instances seem to have been placed there by the occupants. They are strictly nocturnal in their habits, and feed on the seeds of hackberry, mesquite, and various other shrubs. Young and old inhabit the burrows together.

Geomys breviceps Baird.   Louisiana Gopher.[M]

142 These little, dark-colored pocket gophers, usually known throughout their range as 'salamanders,' extend from Louisiana into eastern Texas, and, with considerable variation, westward to Navasota, Brenham, Milano, Peoria, Decatur, and Gainesville, or a little beyond the ninety-seventh meridian, thus inhabiting most of the eastern humid area of Texas, to the edge of the semiarid mesquite country, where they grade into a larger, paler form.

Their range is broken and irregular. Across sandy ridges their hills abound for miles, and then across miles of occasionally flooded bottom lands or wide stretches of black wax-land prairie they are entirely wanting. They live impartially in timber and open country, and are rarely found on clay or hard soil, but are most abundant on the sandiest and mellowest land. At the edge of flood lands they burrow mainly in the large flat mounds so characteristic of the region, and if not responsible for the construction of a certain class of these mounds, at least constantly add to them the earth brought up from below. Owing to the small size of these gophers, their scattered distribution and choice of poor, sandy soil for their most active

[141] These mice are now classified as *Liomys irroratus texensis*, the Mexican spiny pocket mouse. This species is widely distributed across much of Mexico, but it occurs in the United States only in extreme southern Texas. In addition to the specimens reported by Bailey from Brownsville in Cameron County, the species has now been collected in six other counties (Zapata, Jim Hogg, Starr, Hidalgo, Kenedy, and Willacy) in the lower Rio Grande Valley. This species is relatively abundant where it occurs in Texas.

[142] The systematics of pocket gophers of the genus *Geomys* have changed drastically since Bailey's 1905 publication. Each of the major changes are discussed in annotations 142–150, and see Chapter 7 for a contemporary distribution map of the *Geomys* species. Baird's pocket gopher (*Geomys breviceps*) initially was recognized as a distinct species by Baird (1855) on the basis of specimens from Louisiana, but Baker and Glass (1951) placed it as a subspecies of *Geomys bursarius*. Bohlin and Zimmerman (1982) demonstrated the distinctness of *G. breviceps*.

[143] Whereas Bailey referred pocket gophers from the coastal prairies around the western edge of Galveston Bay to *Geomys breviceps sagittalis,* Honeycutt and Schmidly (1979) demonstrated that all populations of *Geomys* from eastern Texas should be referred to this subspecies. Bailey reported this gopher as common all the way from Virginia Point to Houston. Apparently, this is no longer true. Land clearing and land use at the turn of the century have reduced much of the coastal prairie; consequently, this species is not as abundant as it once was. Hice and Schmidly (1999) found sign of gophers at Virginia Point, and they even collected a specimen near Hitchcock, but these burrowing rodents have completely disappeared around Clear Creek and toward Houston. Sudman et al. (2006) demonstrated a significant level of genetic divergence among populations of Baird's pocket gopher, and it may prove that *G. b. sagittalis* from Texas is distinct relative to gophers from further east in Arkansas and Louisiana.

work, there is comparatively little complaint of their mischief. In the heavier, better soil of cultivated fields they are not so common, and they throw out fewer and smaller mounds, but in pastures, potato fields, gardens, and orchards they sometimes do serious damage, besides leaving unsightly mounds over lawns and parks. They are easily trapped and there is no excuse for allowing them to injure crops or trees. A field once cleaned out will not be repopulated to any extent for several years, as the animals rarely travel except by extending their underground tunnels.

Like all species of the genus, they are strictly vegetarian in diet and cleanly in habits. Their flesh is sweeter, better flavored, and more delicate than that of squirrel or rabbit, and their small size is the only objection to their use as a table delicacy.

Geomys breviceps sagittalis Merriam. White-throated Pocket Gopher.[M]

143 This white-throated form of the *breviceps* group seems to have a very local distribution on the coast prairie west of Galveston Bay. There are specimens from Clear Creek, Arcadia, and Virginia Point. I failed to find any trace of this gopher on Galveston Island or the point east of the bay at Bolivar. Along the Santa Fe Railroad from Virginia Point to Houston they are common most of the way over the prairie, where low mounds furnish favorite burrowing places. In certain localities they are numerous, and there are many complaints of the mischief they do, especially to orchards.

On the ranch of Mr. Lee Dick, at Virginia Point, they had entirely destroyed an orchard of 200 six-year old fig trees in bearing. Most of the dead trees had been piled up over the fence, where I examined them and found that all the small roots had been cut off, and in many cases the tap root where it was 2 or 3 inches in diameter. A few dead trees that were still standing were tipped over and the roots found in the same condition—all bearing the unmistakable marks of the teeth of the gopher. Five hundred dollars would be a small cash value to place on this lot of trees, and probably a dozen gophers had done the mischief. Mr. Dick had wasted a good deal of time trying to shoot them, but he had given up and said the people might as well move out and let the gophers have the country. I set nine No. 0 steel traps in this orchard patch, and a few hours later took out of them seven gophers. Not more than two or three remained in the field. The owner then acknowledged that with half a dozen traps a few hours' work might have freed his orchard of gophers, and that the loss of his trees was wholly unnecessary. His claim, moreover, that other gophers would soon come in from the surrounding prairie is true only to a very limited extent, and the immigration could be entirely prevented by trapping in the immediate vicinity of the field.

Geomys breviceps attwateri[M] Merriam.   Attwater Pocket Gopher.[M]
144 This pocket gopher inhabits the islands and coast prairie between the mouth of the Colorado River and Nueces Bay and extends inland nearly to San Antonio. It is larger and lighter colored than typical *breviceps* and inhabits a decidedly more arid region.

Mr. H. P. Attwater has furnished the following interesting notes on their habits at Rockport:[a]

> The animals are very abundant all over the peninsula in Aransas County wherever the soil is sandy. There is hardly a foot of land that has not been 'plowed' several times over by gophers, and I believe the fertility of some sections has been greatly improved by them, by bringing the poorer soil up to the top. I have noticed that the richer the land the richer the gophers. Of course they do considerable damage to vegetable crops, especially to young fruit trees and cuttings just rooting. The samples sent you of mulberry trees cut by gophers were from the Faulkners' ranch, on St. Charles peninsula, in the eastern part of the county. Mr. Samuel Walker, the manager of the ranch, told me that he killed over 250 gophers in his young pear orchard between the 1st of March and April 15, 1893. This orchard was set out where sweet potatoes had grown the year before, and they came up again and covered the ground, and I think the potatoes attracted the gophers in the first place more than the pear trees.

Geomys breviceps llanensis[M] subsp. nov.   Mesquite Plains Gopher.[M]

Type from Llano, Tex., No. 97086, ♂ ad., U. S. Nat. Mus., Biological Survey Coll., May 15, 1899. Vernon Bailey. Original No. 6912.

145 *General characters.*—Similar to *breviceps,* but larger and lighter colored with more arched skull.

Color.—Upper parts light liver brown, in three of the females much darker, with dusky over the back; lower parts creamy or buffy white.

Skull.—Long and slender, with very narrow braincase and rostrum and small bullae as in *breviceps,* but with narrower and arched instead of convex interorbital region, nasals not sharply emarginate or abruptly constricted posteriorly; occiput sloping instead of abruptly truncate.

Measurements.—Type: Total length, 270; tail, 88; hind foot, 32. Adult male topotype: Total length, 270; tail, 82; hind foot, 32. Adult female: Total length, 230; tail, 74; hind foot, 30.

Skull of type.—Basal length, 44.3; zygomatic breadth, 29.6; mastoid breadth, 25; interorbital breadth, 6.3; breadth of muzzle at root of zygoma, 9; alveolar length of upper molar series, 8.5.

Remarks.—While closely resembling *texensis*[M] externally and while the ranges of the two almost or quite meet it needs but a cursory examination of the skulls to show that this form has no connection with that species. It is a large, light-colored plains form of *brevi-*

---

[a] Merriam, N. Am. Fauna No. 8, p. 136, 1895.

[144] *Geomys attwateri*, Attwater's pocket gopher, is now regarded as a distinct species from *Geomys breviceps* (Tucker and Schmidly, 1981). The dividing line between them is approximately along the Brazos River from Waco to the coast. *Geomys breviceps* occurs east of the river and *G. attwateri* to the west. However, the Brazos River occasionally changes course, resulting in land being transferred from one side of the river to the other. These changes have produced small zones of range contact and overlap between *G. breviceps* and *G. attwateri*. Merriam (1895) described *G. attwateri* on the basis of specimens collected at Rockport in Aransas County and named it in honor of H. P. Attwater for his numerous contributions to early Texas mammalogy. The species *G. attwateri* is monotypic and subspecies are not recognized. This gopher is locally common throughout its range in Texas.

[145] *Geomys breviceps llanensis* is now regarded as conspecific with *Geomys texensis*, the Llano pocket gopher, and has been synonymized with that species. This conclusion is based on detailed genetic analysis conducted by Block and Zimmerman (1991) and Sudman et al. (2006). Some authorities treat *Geomys texensis llanensis* as a valid subspecies (Block and Zimmerman, 1991), whereas other sources consider it as a synonym of *Geomys texensis texensis* (McAliley and Sudman, 2005). Modern taxonomists refer the specimens from Colorado, Stanton, Childress, Vernon, Newlin, Canadian, Lipscomb, and Tascosa to *Geomys bursarius major*. Specimens from Brazos County are assigned to the taxa *Geomys breviceps sagittalis*.

[146] Merriam described *Geomys texensis*, the Llano pocket gopher, on the basis of specimens from Mason in Mason County (Merriam, 1895). The subspecies *Geomys texensis texensis* has been reported from McCulloch, San Saba, Kimble, Mason, Gillespie, Lampasas, and Llano counties (Pitts et al., 1999). Smolen et al. (1993) described a new subspecies of *Geomys texensis*, *Geomys texensis bakeri*, from the drainages of the Frio River in Uvalde, Zavala, and Medina counties. Populations of *G. texensis* from Sycamore Creek and the Rio Grande are presumably extinct due to the flooding of their habitat when the Amistad Reservoir was created.

[147] *Geomys arenarius*, the desert pocket gopher, is known only from El Paso and Hudspeth counties in far western Texas. The specimen Bailey refers to from Ward County is now classified as *Geomys knoxjonesi* (Baker and Genoways, 1975). *Geomys arenarius* and *G. knoxjonesi* are "cryptic" species, but research on mitochondrial DNA, ribosomal DNA, chromosomes, and allozymes indicates that they are distantly related species (Baker et al., 1989; Bradley et al., 1991; Jolley et al., 2000; Sudman et al., 2006). *G. arenarius* remains common in El Paso County as well as several localities where it has been obtained in New Mexico, and *G. knoxjonesi* is locally common throughout its range as well. Where the range of *G. knoxjonesi* contacts that of *Geomys bursarius major* in eastern New Mexico, the two taxa hybridize in a narrow hybrid zone (Pembleton and Baker, 1978; Baker et al., 1989). In an extensive study of this zone, Baker et al. (1989) found that *G. b. major* and *G. knoxjonesi*

differed in chromosome number, mitochondrial and ribosomal DNA, and allozyme systems. Baker et al. (1989) and Bradley et al. (1991) determined that both premating and postmating isolating mechanisms reduced gene exchange and resulted in hybrids that were either sterile or had reduced fertility, thus indicating the two taxa were behaving as good biological species. It is virtually impossible to distinguish the two species on the basis of overall appearance and morphology.

ceps which follows up the river valleys from eastern Texas and becomes differentiated as it enters the open country. Specimens from Colorado, Stanton, Brazos, Childress, Vernon, Newlin, Canadian, Lipscomb, and Tascosa, Tex., are referable to it. Two females from Brazos are clearly intermediate between the present form and *breviceps*. In general contour of skull and especially in slender rostrum it resembles *phalax*[M], but in the slender audital bullae and small mastoids and consequent narrow base of skull it differs widely from that species.

So far as known at present the range of the form in Texas extends mainly along strips of sandy soil in the Llano, Colorado, Brazos, Red, and Canadian river valleys, in a region of scattered mesquite bushes, but does not reach the Staked Plains and rarely extends over the hard-soiled ridges between stream valleys. Gaut caught one gopher two miles south of Washburn, but could find no other trace of them in the country around there. At Lipscomb Howell says "they are plentiful both on the prairie and in the sandy bottoms. Their burrows are very difficult to open, as they are usually closed for a distance of about 18 inches below the surface, at which depth they take a horizontal direction."

Owing to their scattered distribution over a sparsely settled stock country, these gophers are at present of little economic importance, but as irrigation reclaims the mellow soil of these semiarid bottom lands they will constitute one of the problems to be dealt with by the farmers.

Geomys texensis Merriam.   Texas Pocket Gopher.[M]
146 This little, brown-backed, white-bellied gopher inhabits a few spots in central and western Texas. A series of 28 specimens in the Merriam collection from Mason, the type locality, indicates its abundance there, while a single specimen from each of two sandy patches along the Rio Grande, at Del Rio and at the mouth of Sycamore Creek, suggests a scattered distribution along this part of the Rio Grande Valley and a probable former extension of range up the Devils River and across to the head of the Llano as far as Mason. The country immediately north and south of its range has been pretty thoroughly worked without disclosing any species of *Geomys*. We succeeded in catching only *Cratogeomys* and *Thomomys* along Devils River, so at the present time the Mason and Rio Grande colonies seem to be widely isolated.

Geomys arenarius Merriam.   Desert Pocket Gopher.
147 This gopher is common on both sides of the river at El Paso. Specimens have been taken at Las Cruces and Deming, N. Mex., and in 1902 Cary caught one that is almost typical *arenarius*[M] in the sand hills near Monahans, Tex., at the east edge of the Pecos Valley.

This last locality can hardly be considered a part of the general range of the species, but probably marks a long isolated colony.

At El Paso gophers are common on the sandy river bottoms just below the town, where they throw up numerous and very large mounds of the mellow sand. I have never been able to find one in the irrigated orchards and fields, for there the water fills their burrows and drowns or drives them out. Loring reports them as especially abundant in railroad grades and banks of irrigation ditches at El Paso, and he caught seven in one day in the railroad grade a few miles north of Las Cruces, N. Mex. He says: "When pulled from their holes they hissed violently and when two were placed together they fought like bulldogs."

Geomys personatus True.   Padre Island Pocket Gopher.^M

148 This large, light-colored pocket gopher inhabits the central and northern part of Padre Island, a sandy belt along the mainland in Cameron County, and a sandy area near Carrizo, on the Rio Grande. Apparently it does not inhabit the lower Rio Grande Valley, as Lloyd did not find any trace of it between Carrizo and Brownsville nor between Brownsville and Sauz. On a trip from Corpus Christi to Brownsville I found its hills abundant across the sandy country between Olmos Creek and Sauz Ranch, but entirely wanting in the baked clay soil outside of these limits. At Carrizo Lloyd found them in only one patch of sandy soil, and there is nothing to show that they have a continuous range across from this point to Cameron County. On the light sandy soil and drifting dunes where these gophers abound there are no crops to be injured.

On Padre Island, Lloyd says:

> Their habits are in some respects peculiar, owing, perhaps, to the soft sand, that caves in on them, for they fill up their tunnels after throwing out the earth to a distance of 1 and sometimes 2 yards. They can not go very deep in the flats or they would reach water; in fact, the water filled some of the tunnels for about a foot until they curved upward.

Geomys personatus fallax Merriam.   Nueces Pocket Gopher.^M

149 Since this relatively small and dark subspecies of *personatus* was described I have been over its range pretty thoroughly and am convinced that it is an isolated and very local form, inhabiting the sandy strips near the coast between Nueces Bay and the Salt Lagoon at the mouth of San Fernando Creek and extending a short distance up the south side of the Nueces River. Except for very limited sandy strips along the coast and some of the stream shores the country is characterized by a tenacious black clay soil so sticky when wet and so hard when dry that no burrowing rodents inhabit it. From Corpus Christi west to San Fernando Creek we did not see any signs of gophers nor any soil that they could live in. Nueces Bay and the Nueces River, with its flood bottoms, cut off this range entirely from

[148] *Geomys personatus*, the Texas pocket gopher, has a much more extensive range and complicated taxonomy than originally presented by Bailey. Williams and Genoways (1981) recognized seven subspecies in Texas. However, more recent studies (Jolley et al., 2000; Sudman et al., 2006) suggest that *Geomys personatus streckeri* should be recognized as a distinct species, *Geomys streckeri*, leaving only six subspecies (*Geomys personatus davisi*, *Geomys personatus fallax*, *Geomys personatus fuscus*, *Geomys personatus maritimus*, *Geomys personatus megapotamus*, and *Geomys personatus personatus*) as valid. The subspecies *G. p. personatus* is restricted to Mustang and Padre Islands in Kleberg and Nueces counties, and it does not extend more than halfway down Padre Island. It is common throughout this area.

[149] The specimens from Laredo and Carrizo, referred to by Bailey as *Geomys personatus fallax*, are now considered to be *Geomys personatus davisi*, which was described by Williams and Genoways (1981) from western Webb and Zapata counties. The remaining specimens available to Bailey (from Nueces, Bee, Karnes, and Goliad counties) are now classified as *G. p. fallax*, which was described by Merriam (1895) on the basis of material from the south side of Nueces Bay in 1891. To further complicate matters, *Geomys personatus maritimus* is now known from Nueces and Kleberg counties adjacent to Nueces Bay and consequently next to *G. p. fallax*. The subspecies *Geomys personatus megapotamus* occurs from La Salle County southeastward to Starr and Cameron counties. The remaining subspecies, *Geomys personatus fuscus,* is restricted to Val Verde and Kinney counties, although many attempts to collect this subspecies in recent years have failed to produce any evidence that it is extant.

[150] Jones, J. K, Jr., et al. (1988b) referred the specimens collected by A. H. Howell from 15 miles (24 km) east of Texline to the subspecies *Geomys bursarius jugossicularis*. Subsequently, Genoways et al. (2008) elevated *G. b. jugossicularis* to species status (*Geomys jugossicularis*, Hall's pocket gopher), and molecular data have confirmed that populations of *Geomys* in the western portion of the Texas Panhandle should be recognized as a distinct species (Schmidly and Bradley, 2016). This gopher is common in the northern Panhandle of Texas.

[151] Collections made by mammalogists since the Biological Survey have revealed that the yellow-faced pocket gopher, *Cratogeomys castanops*, has a broader distribution and many more subspecies than recognized by Bailey. Hollander (1990) conducted a taxonomic review of this species in the United States and recognized seven subspecies (*Cratogeomys castanops angusticeps, Cratogeomys castanops clarkii, Cratogeomys castanops dalquesti, Cratogeomys castanops lacrimalis, Cratogeomys castanops parviceps, Cratogeomys castanops perplanus,* and *Cratogeomys castanops tamaulipensis*) in Texas. More recently, Hafner et al. (2008) subsumed all seven of these subspecies into *C. c. castanops*; however, in their study they did not include samples of all of the Texas subspecies. Clearly, the taxonomy of this species needs to be revisited in more detail.

Although Bailey commented about a conspicuous absence of gophers from the Lower Rio Grande Valley, Cleveland (1977) reported *C. castanops* from near the Rio Grande in

132   NORTH AMERICAN FAUNA.   [No. 25.

that of *attwateri* on the north, while the Laguna Madre, Salt Lagoon, and streams radiating from them separate as effectually the range from that of *personatus* on the south. Two females from Laredo agree more nearly with *fallax* than with any other form, but probably the range of this colony has no connection with that of *fallax* of the Nueces Bay region.

In the region of Corpus Christi the sandy soil is especially desirable for growing early vegetables, and the presence of the gophers is a source of much annoyance and considerable loss to the farmers.

Geomys lutescens[M] Merriam.   Yellow Pocket Gopher.[M]

150 Two specimens of barely adult females from near Texline agree with *lutescens*[M] in external characters, but possess cranial characters that suggest the possibility of a local subspecies. Howell reported numerous burrows in a range of sand hills 15 miles east of Texline, where the two specimens were caught, but elsewhere in the region none were seen.

Cratogeomys castanops (Baird).   Chestnut-faced Pocket Gopher.[M]

151 This, the largest of the Texas pocket gophers, with the single-grooved upper incisor, is common in Lower Sonoran zone and the edge of Upper Sonoran of western Texas from Eagle Pass, the headwaters of Devils River, Fort Lancaster, Big Springs, Hail Center, and Tascosa westward. A few are scattered here and there over the Staked Plains, but generally they inhabit valleys with fertile and mellow soil lower down, becoming very numerous and troublesome in some of the cultivated land. Their concentration on the best soil, together with the large size of their burrows and mounds, makes them one of the most injurious of the gopher family.

In habits they do not differ materially from the various species of *Geomys,* except in being more alert and possibly more diurnal. During the day they are often seen at the mouths of their burrows pushing out earth, at which times their comparatively large eyes are conspicuous, bright, and alert. They see a person much more quickly and at a greater distance than do most species of *Geomys* or *Thomomys,* and hence move about somewhat more freely at the entrance of the burrow. Still no protective measures are neglected, and the burrows are always promptly closed and packed with earth, sometimes for a distance of 2 or 3 feet back from the main tunnel. The mounds of these gophers often contain a bushel or more of earth, and when located in a meadow or alfalfa field they cover and destroy much of the crop, besides interfering with machinery in harvesting. The greatest damage caused by the gophers, however, is in cutting off roots, especially in such crops as alfalfa and garden vegetables, but most of all in the case of fruit trees. In many instances small orchards have been almost destroyed by a few gophers that could

Cameron County just across the river from Tamaulipas, Mexico. Cleveland observed numerous burrows southeast of Brownsville in 1976 but stated that no burrows had been observed at the same site in 1972. These observations suggest the possibility of a recent invasion by these gophers from the south side of the river. The population from Brownsville is assigned to the subspecies *C. c. tamaulipensis*. Bailey's specimens from Eagle Pass are now assigned to the subspecies *C. c. angusticeps*; those from Val Verde County to *C. c.*

clarkii; and those from Big Spring, Stanton, and Tascosa to *C. c. parviceps*. Other recognized subspecies include *C. c. dalquesti* from west-central Texas to the Edwards Plateau area; *C. c. lacrimalis* in Loving, Ward, and Winkler counties; and *C. c. perplanus* in El Paso, Culberson, and Hudspeth counties. This remains one of the most common pocket gophers throughout its range in Texas and may be increasing its range due to its preference for chirty soils often associated with roadbed construction.

have been trapped with little trouble. They are so easily caught in steel traps that it would hardly pay to poison them except on a large ranch, though undoubtedly they could be poisoned in the same way as other gophers by dropping raisins, prunes, soaked corn, or small potatoes containing strychnine into the burrow and then closing the opening from above. On a cattle ranch in the foothills of the Davis Mountains I found where a couple of the gophers were working in dangerous proximity to the roots of a half dozen flourishing and fruit-laden peach trees growing near the windmill reservoir, while in the 3-acre patch of alfalfa just below, the hills of the animals were numerous. Under a neighboring cliff where a pair of horned owls had raised their young the same year I counted 20 skulls of *Cratogeomys* among bones of other rodents, but for fear these owls would catch the chickens one had been killed by the ranchmen and the others driven away.

Thomomys fulvus[M](Woodhouse).   Fulvous Pocket Gopher.[M]

152 The pocket gophers from the Transition zone summit of the Guadalupe Mountains, while differing slightly from typical *fulvus*,[M] do not seem to require separation from that wide ranging species. They are abundant over the timbered slopes of these mountains in Transition zone and often in places where the yellow pines are mixed with nut pines and junipers. They were common in the head of Dog Canyon, at 7,000 feet, the head of McKittrick Canyon, at 8,000 feet, and on top of the ridges, from 7,000 to 9,000 feet, and probably to the highest peaks, at 9,500 feet. There, as elsewhere, they inhabit partly forested slopes covered with abundant vegetation. They make endless tunnels and throw up numerous hills, often working among the rocks and constantly bringing to the surface the rich, mellow soil. In walking over the mountain slopes one's feet break into the burrows that honeycomb the soil beneath. The long, rope-like ridges of dry earth on the surface of the ground show where the gophers have worked in winter under the snow and filled snow tunnels with the earth brought up from below.

In mountain districts the gophers can do no possible harm, and besides their beneficial effect on the soil their underground tunnels catch and carry into the ground much of the water that would otherwise run off the surface and be lost.

Thomomys fulvus texensis[M]Bailey.   Davis Mountain Pocket Gopher.[M]

153 This little, dark-brown gopher inhabits the timbered part of the Davis Mountains in Transition zone and in at least the upper edge of Upper Sonoran, ranging from about 5,000 feet up through the juniper and yellow-pine belts to the highest part of the mountains. The highest point where their mounds were seen was on the main

[152] Vernon Bailey authored the first serious revision of this genus in 1915. Bailey recognized four species of *Thomomys* in Texas: the wide-ranging *Thomomys fulvus* from the Guadalupe and Davis Mountains, *Thomomys baileyi* from Sierra Blanca in Hudspeth County, *Thomomys aureus* from El Paso to the Big Bend Country, and *Thomomys perditus* from Comstock and the head of the Devils River to the vicinity of Rock Springs. Modern taxonomists have grouped all of these taxa into a single species, *Thomomys bottae*, which occurs over much of the Trans-Pecos eastward across the Edwards Plateau, with seven recognized subspecies (see Beauchamp-Martin et al., 2019): *Thomomys bottae baileyi* restricted to Sierra Blanca in Hudspeth County and immediate surrounding areas, *Thomomys bottae confinalis* from Tom Green to Mason and Edwards counties, *Thomomys bottae lechugilla* along the Rio Grande from El Paso County to Brewster County, *Thomomys bottae limpiae* from lower elevations along Limpia Creek in Jeff Davis County, *Thomomys bottae robertbakeri* (new subspecies, Beachamp-Martin et al., 2019) from Crane to Irion counties and southward to Terrell and Val Verde counties, *Thomomys bottae spatuosus* from northern Brewster County, and *Thomomys bottae texensis* from the higher elevations of the Guadalupe Mountains southward to the Davis Mountains. Specimens from the Guadalupe Mountains, referred to *T. fulvus* by Bailey 1905, had been recognized as a distinct subspecies, *Thomomys bottae guadalupensis,* but were subsumed into *T. b. texensis* by Beauchamp-Martin et al. (2019).

[153] In 1902, Bailey described a small pocket gopher from the head of Limpia Creek at 5,500 feet (1,676 m) in the Davis Mountains in Jeff Davis County, under the name *Thomomys fulvus texensis* (see Annotation 152). Nelson and Goldman (1934) provided evidence that *Thomomys fulvus* and *Thomomys bottae* intergrade and placed all forms formerly recognized as races of *T. fulvus* in the species *T. bottae*. The Davis Mountains gopher then became known as *Thomomys bottae texensis*. Blair (1939) described a new race, *Thomomys*

bottae limpiae, from the mouth of Limpia Canyon, one mile north of Fort Davis. Thus, two subspecies are now recognized from the Limpia Creek drainage in the Davis Mountains, *T. b. texensis* at the higher elevations and *T. b. limpiae* in the foothills (Davis and Buecher, 1946; Beauchamp-Martin et al., 2019). Given the isolated nature of these subspecies, efforts should be made to monitor and conserve these populations.

There is some indication that *Cratogeomys* may be replacing *Thomomys* in this region as a result of climate changes involving increased aridity. Reichman and Baker (1972) studied the distributional relationships of *Cratogeomys* and *Thomomys* from 1968 to 1970 along Limpia Creek in the Davis Mountains. As the area became drier, *Thomomys*, which occurred from near the streambed to the foot of the rocky bluffs lining the canyon, moved closer to the stream, and *Cratogeomys* spread into the vacated areas. These distributional changes may be linked to a decrease in soil moisture and a subsequent increase in xerophytic plants, both of which would favor *Cratogeomys* over *Thomomys*. Recent fieldwork by mammalogists at Texas Tech University in 2017-2019 support the contention that *Cratogeomys* is replacing *Thomomys* throughout much of the Trans-Pecos.

[154] Modern taxonomists regard *Thomomys baileyi* as a subspecies of *Thomomys bottae* (see Annotation 151). Despite several attempts, this taxon has not been collected in several decades and it may now be extinct in Texas (see Chapter 6 for a discussion).

[155] The name *Thomomys aureus lachuguilla* has been changed to *Thomomys bottae lachuguilla* (see Annotation 152).

134   NORTH AMERICAN FAUNA.   [No. 25.

ridge of Mount Livermore, at about 8,200 feet. In the gulches they come down nearly to Fort Davis. In habits as well as general appearance and zonal position they are much like *fulvus*,[M] living in a region of abundant vegetation and considerable rainfall and burrowing in the rich mold on stony mountain slopes or in open grassy parks. For a part of each year they live under the snow. The sides of the mountains are plowed over by them, and the mellow earth that is brought up from between the stones is washed down by the rains and deposited in the gulches below, where other gophers, with their endless underground tunnels, are mixing and stirring the soil and steadily improving it for cultivation. The service to man thus performed by these little animals is not to be lightly estimated.

In this region of extensive stock ranges and very limited agriculture the gopher will never prove a serious pest. The few that get into gardens and orchards are easily caught in traps, while those outside go on cultivating the soil without harming anything. With the larger *Cratogeomys* of the lower country, before mentioned, the case is different.

Thomomys baileyi[M] Merriam.   Sierra Blanca Pocket Gopher.[M]

154 This unique little gopher is known only from the specimens collected at Sierra Blanca on the open arid plain at the junction of Upper and Lower Sonoran zones. It is probably an Upper Sonoran species of the open country, as no trace of any *Thomomys* has been found in the big valley to the south and east, while its hills are common over the mesas and low mountains northeast of Sierra Blanca and north of Van Horn. Its characters do not suggest relationship with its nearest neighbor *lachuguilla* from the Lower Sonoran, Rio Grande Valley, or with any other of the surrounding species. It probably represents a long-isolated colony of very limited distribution.

Thomomys aureus lachuguilla[M] Bailey.   Lachuguilla Gopher.[M]

155 This little gopher inhabits the hottest and most arid part of western Texas. It lives on the barren mesas along the east side of the Rio Grande Valley, from El Paso to the Great Bend country, where the principal vegetation consists of scattered desert shrubs, cactus, yuccas, and agaves. Its little mounds are distributed over the baked and stony mesas, sometimes in long lines across barren strips, but usually grouped around the base of a bunch of cactus or a group of yuccas or agaves, the roots of which furnish it with both food and drink. The roots, stems, and leaves of apparently every plant encountered are eaten, but the favorite and principal food of the species is the tender, starchy caudex of the little *Agave lecheguilla,* a plant protected by sharp hooks and rigid spines from every outside attack, but wholly unprotected from below. The gophers burrow under and eat out the whole pineapple-like heart of the stem until the leaves and flower stalk dry up and topple over, while they burrow along to the next plant in their way, often leaving a long trail of dead agaves to mark their course. As the agave is extremely abundant and generally is considered a nuisance, the gophers are given credit for good work in destroying it, but if its fiber proves of value, as seems probable, the verdict in favor of the gopher must be reversed.

Thomomys perditus[M] Merriam.   Little Gray Pocket Gopher.[M]

156   These little gray gophers are scattered sparingly over the high, stony mesa from Comstock to the Pecos High Bridge and Langtry, and still more sparingly to the head of Devils River. Farther east they do not seem to have a continuous range, but their hills were seen at points east and west of Rock Springs, and a specimen was taken on the high plain 35 miles east of Rock Springs, and another in the Castle Mountains in Crockett County.

The animals are not only scarce, but difficult to catch, as they live in scanty, stony soil where their little mounds are often mainly composed of stones instead of earth, while their tunnels become blocked by stones and are soon abandoned. Sometimes the doorways are left open apparently for lack of soil to close them, or because the gopher has abandoned the burrow in the hope of finding more favorable conditions elsewhere; sometimes they are merely blocked by two or three stones, but usually they are closed to a slight depth. The burrows do not extend far and the hills thrown up are few and small. Sometimes the old ones are almost obliterated before a fresh one is thrown up, and I have caught the gophers where the nearest hill appeared to be a month old.

Most of the food of the gopher is procured under ground from various roots, largely of yucca and sotol, or from the inside fleshy parts of cactus, *Cereus*,[P] *Echinocactus*,[P] and *Cactus*,[P] which they burrow into and eat out from below. The roots and starchy base of a yucca or sotol will furnish food for an individual apparently for a week or more.

Perognathus hispidus[M] Baird.   Hispid Pocket Mouse.

157   This big pocket mouse is common in the more or less brushy part of the Lower Sonoran zone over southern Texas and the adjoining part of Mexico. In Texas it ranges from Brownsville north to O'Connorport, Cuero, Seguin, Llano, and probably, judging by immature specimens, to Brazos and Henrietta on the east, and to Del Rio on the west. It is less partial to open ground than most species of the genus, and is often caught in brushy or grassy places among the mesquite, at the edge of a thicket, along the fence at the edge of a field, on a weedy sand flat, or even in the midst of a corn or cotton field.

[156] The name *Thomomys perditus* has been changed to *Thomomys bottae perditus* and applies to gophers that occur in Mexico. The specimens listed in the survey by Bailey (1905) were classified as *Thomomys bottae limitaris* (Goldman, 1936), but a recent review by Beauchamp-Martin et al. (2019) placed *T. b. limitaris* as a synonym of *Thomomys bottae lachuguilla.*

[157] Hafner and Hafner (1983) demonstrated that the hispid pocket mouse, along with all spiny-rumped pocket mice in Texas (formerly included in the genus *Perognathus*), should be placed in the genus *Chaetodipus*, including the hispid pocket mouse (*Chaetodipus hispidus*), desert pocket mouse (*Chaetodipus penicillatus*), rock pocket mouse (*Chaetodipus intermedius*), and Nelson's pocket mouse (*Chaetodipus nelsoni*; now *Chaetodipus collis* following Neiswenter et al,. 2019). Bailey recognized three subspecies of *C. hispidus* in Texas, although their distribution in the state has been rearranged by modern taxonomists (Fig. 45; Schmidly and Bradley, 2016): *Chaetodipus hispidus hispidus* in the eastern two thirds of the state, *Chaetodipus hispidus paradoxus* in the western one-third, and *Chaetodipus hispidus spilotis* in a limited area along the Red River (type locality at Gainesville, Cooke County). The conservation status of the subspecies seems to be satisfactory at the present time, but this is another species that could be impacted by the continued degradation of grassland habitats in Texas.

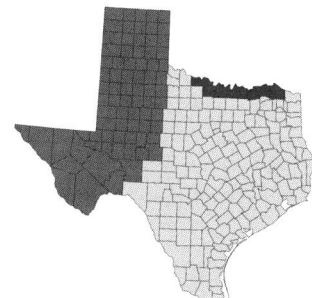

Figure 45. The current distribution of the hispid pocket mouse (*Chaetodipus hispidus*) in Texas. The light shaded area represents the distribution of *C. h. hispidus*, the medium shaded area represents *C. h. spilotus*, and the darkest shaded area represents *C h. paradoxus.*

Some of the burrows suggest inch auger holes bored straight down into the ground, with no trace of earth that has been brought out; others are closed flush with the surface of the ground so as to be almost invisible, while others are closed 2 or 3 inches below the surface. At almost every den, however, there is a mound of earth that has been brought out of the burrows and heaped up, sometimes to the size of a gopher or mole hill, over the closed main entrance. This fact probably accounts for the absence of earth at other burrows that have been opened out from the main tunnel.

FIG. 19.—Pocket mice (*Perognathus hispidus*)caught in traps at Seguin, Texas.

At Seguin, in Guadalupe County, in November, 1904, I found these big pocket mice unusually abundant. Their characteristic inch auger holes and gopher-like mounds were found mainly along the edges of sandy fields, but also frequently in the middle of corn and cotton fields that had been thoroughly cultivated. Some of the mounds were 6 inches high and contained 6 or 8 quarts of earth, and were distinguished from gopher hills only by being solitary instead of in a series. By opening the burrow under these mounds I could catch the occupant at any time of day, but most of my specimens were caught at night in traps set at the open doorways or in artificial runways scraped with my foot along the ground near by. Sorghum seed proved the most attractive of the several kinds of bait tried.

Mr. Neel, the market gardener with whom I stayed, complained of great trouble in raising cantaloupes and green peas, because something dug up the seeds as fast as he could plant them. In the midst of his cantaloupe patch, where only four or five plants had survived, I found traces of these mice and caught two of the animals. A few others lived around the edge of the field, but a few nights' trapping would have cleared them all out and no doubt would have prevented further trouble.

Like others of the genus, these mice are mainly nocturnal and are rarely seen alive. The little that is known of their habits has been gathered by trapping them for specimens. At times they are readily caught in traps baited with rolled oats or various grains, and again they obstinately refuse to touch any kind of bait or to come near the traps. When caught they often have their cheek pockets stuffed full of the trap bait or of wild seeds.

They are active all winter and apparently never become very fat or show signs of hibernating.

Perognathus hispidus paradoxus[M] Merriam.   Kansas Pocket Mouse.[M]

158   This large, pale subspecies of the *Perognathus hispidus*[M] group ranges over the open plains and desert country from South Dakota to Arizona, including northern and western Texas, south to Presidio County and Comstock and east to Rock Springs, Colorado, Mobeetie, and Lipscomb, mainly in Upper Sonoran zone. From the smaller and darker-colored *hispidus*[M] on the south and from *spilotus*[M] on the east the shading off is so gradual that no sharp line can be drawn between the ranges. Immature specimens from Brazos, Henrietta, and Tebo can not be positively referred to one rather than another of the three forms, and the Lipscomb specimens shade toward *spilotus*[M].

The habits of *paradoxus*[M] do not differ from those of other forms of the group except as they have been modified to meet the conditions of plains and desert. In the Guadalupe Mountains I caught one in the head of Dog Canyon at 6,800 feet altitude, just below the edge of Transition zone, and at Amarillo, on the top of the Staked Plains, I caught one and found part of the skin of another at the entrance of a burrowing owl's nest in a prairie dog hole. The subspecies is common in the Pecos Valley, but apparently does not occur along the Rio Grande. Loring took one at Henrietta by the stone foundation of a bridge, and Oberholser another under a mesquite tree. At Mobeetie Oberholser caught one in the stone foundation of an old house. At Brazos Cary caught one in a cane field and another with corn in its pockets in a field of Johnson grass. One taken in January in Presidio County had its pockets full of *Convolvulus* seeds, and

[158] The name *Perognathus hispidus paradoxus* has been changed to *Chaetodipus hispidus paradoxus* (see Annotation 157).

[159] The name *Perognathus hispidus spilotus* has been changed to *Chaetodipus hispidus spilotus* (see Annotation 157). This subspecies is widely distributed in eastern Oklahoma, Kansas, and Nebraska but occurs in Texas only along the Red River on the Texas-Oklahoma border. The specimens from Jefferson referred by Bailey as *Perognathus hispidus spilotus* are today regarded as *Chaetodipus hispidus hispidus*.

[160] The name *Perognathus penicillatus eremicus* has been changed to *Chaetodipus penicillatus eremicus* (see Annotation 157). In a recent study of speciation in the desert pocket mouse, Lee et al. (1996) determined that *Chaetodipus penicillatus* should be divided into two species (*C. penicillatus*, a Sonoran Desert form, and *Chaetodipus eremicus*, a Chihuahuan Desert form) on the basis of studies of allozymes, chromosomes, and mitochondrial DNA sequences. Thus, Texas specimens of this species are now classified as *C. eremicus*, the Chihuahuan Desert pocket mouse. This taxon was described by E. A. Mearns in 1898 on the basis of specimens from Fort Hancock in Hudspeth County. *Chaetodipus eremicus*, is a small heteromyid rodent with a widespread distribution in the low desert areas of Trans-Pecos Texas where it occupies the sandy or soft alluvial soils along stream bottoms, desert washes, and valleys (Porter, 2011). It is seldom found on gravelly soils or among rocks, a habitat preferred by the externally similar species, *Chaetodipus intermedius* and *Chaetodipus nelsoni*. Yancey (1997) recorded all three of these species at Big Bend Ranch State Park in southern Presidio County.

another at Smithville, S. Dak., in June had its pockets full of *Cymopterus* seeds. In the Castle Mountains Cary took one from the stomach of a rattlesnake, and near Texline Howell also found one in one of these snakes.

Perognathus hispidus spilotus Merriam.   Black-eared Pocket Mouse.[M]

159 A specimen collected by Hollister at Jefferson not only extends the range of this group of pocket mice eastward almost across the State, but exhibits in an accentuated degree the characters of *spilotus*,[M] described from Gainesville specimens. This record, with a few others, gives an extensive and logical range to what seems to be a fairly well-marked subspecies, which differs from typical *hispidus*[M] in slightly darker and richer coloration, with more of a tendency to suffusion of yellow over the belly and along the top of the foot and the under surface of the tail, in larger and blacker spot on upper edge of ear, and in the extension of the nasals back to or beyond the posterior tips of premaxillae. Besides the Gainesville and Jefferson specimens, I should refer to the subspecies a richly colored flat skin from Long Point in the National Museum collection, a good skull from Saginaw, a few specimens from Ponca and Orlando, Okla., Red Fork, Ind. T., and an immature specimen from Garden Plain, Kans.

At Gainesville I caught two specimens on the edge of a pasture in rather tall prairie grass, but they, as well as all other rodents, were scarce on the black, hard soil of that region. At Jefferson they were fairly common, and Mr. Richard Crain told me that he often plowed them out, and that his cat frequently brought them to the house. I found a number of their characteristic burrows with fresh tracks around them, but could not coax the animals into my traps with any kind of bait. Hollister caught one in a trap set in a path running between a cotton field and the woods, but at Antioch he could not catch them, although the farmers there described the species accurately and said that at times they were common. One of the Gainesville specimens had seeds of a little Mimosa in its pockets, and another at Ponca had its pockets full of *Petalostemon*[P] seeds.

Perognathus penicillatus eremicus[M] Mearns.   Desert Brush-tailed Pocket Mouse.[M]

160 This desert pocket mouse inhabits the Lower Sonoran zone of extreme western Texas, ranging from El Paso east to Monahans, south to Boquillas, and westward into Mexico. There are Texas specimens from El Paso, Boquillas, east base of Chisos Mountains, 35 miles south of Marathon, Toyahvale, Pecos City, and Monahans.

At El Paso, in 1889, I caught this soft-haired species in the sandy bottoms below town and supposed that it had a different range from the spiny-rumped *intermedius*[M] caught at the same time in the rocks above town; but later Doctor Fisher caught one on the gravelly mesa near El Paso, Cary caught one among the mesquites at Pecos City, and a number among the mesquites on a hard, limy ridge at Monahans. In the Boquillas and Great Bend region Oberholser and I found the species associated with *nelsoni*[M] on sandy bottoms and among rocks of the cliffs bordering the Rio Grande, and around old stone cabins. While they are evidently partial to valley bottoms, the one essential for their burrows is a bit of mellow soil which may be found among broken bowlders or between thin strata of limestone, as well as on the sandy flats, or in the soft mesa soil that collects around the base of desert bushes. The little mounds that usually cover the entrances of their closed burrows are easily distinguishable from the work of any other species of the region except *intermedius*[M] or *nelsoni*.[M] They are often elongated or fan shaped, and stretch away to a distance of a foot or so from the point where the earth was brought up, as if pushed or kicked out, much like the mound or strip of dirt thrown out in front of the burrows of the smaller species of *Dipodomys*. The entrance of the burrow is usually but lightly closed and can be easily broken into with the finger. By breaking the crust above it the burrow may be followed for a considerable distance where it runs near the surface. As usual with pocket mice and kangaroo rats, there are several openings and radiating tunnels from the central cavities of the subterranean den, and while part of these are closed, there are generally concealed openings or some means of ready escape.

These mice, like the whole family, are mainly nocturnal, but can be caught in the daytime by opening a closed burrow and setting a trap inside, sometimes in a very short time after placing the trap. Usually they take rolled oats readily, and are easily caught in traps set around their burrows or in long furrows drawn in the sand, which they almost invariably follow till a trap is reached. When caught their cheek pockets are often full of rolled oats from the trap bait, or partly filled with seeds of various plants.

Perognathus intermedius[M] Merriam.   Intermediate Pocket Mouse.[M]

161 A large series of specimens from the El Paso region are almost typical *intermedius*.[M] From *eremicus*,[M] with which these pocket mice are associated, they are easily distinguished by the spinescent hairs of the rump and apparently by a difference of habitat. At El Paso, in 1889, I caught them only in the rocks in the foothills of the Franklin Mountains, and *eremicus*[M] only on the sandy flats below town. In 1903 Gaut collected them in the eastern foothills of the Franklin Mountains up to 4,800 feet, and reported them as living around the rock slides and cliffs. While in other localities *eremicus*[M] also has been taken among rocks, *intermedius*[M] throughout its range is closely

[161] The name *Perognathus intermedius* has been changed to *Chaetodipus intermedius*, the rock pocket mouse (see Annotation 157). It occupies the western Trans-Pecos region and prefers rocky habitats in close association with cliffs, canyons, rocky gulches, or the edges of boulders in deserts and lower grasslands. The eastern limits of the range of *C. intermedius* roughly follow the western boundary of the range of *Chaetodipus nelsoni*, which also occupies areas characterized by large rocks and boulders in the eastern two-thirds of the Trans-Pecos. *Chaetodipus intermedius* is common throughout its range in Texas.

[162] The name *Perognathus nelsoni* was changed to *Chaetodipus nelsoni,* Nelson's pocket mouse (see Annotation 157). It is distinguished from *Chaetodipus eremicus* and *Chaetodipus intermedius* in having numerous elongate, black-tipped spiny hairs on the rump that overreach the normal guard hairs. The three species are difficult to distinguish even by trained taxonomists (Wilkins and Schmidly, 1979; Manning et al., 1996). See Porter (2011) for the most recent information concerning the ecology and natural history of Nelson's pocket mouse. More recently, Neiswenter et al. (2019) split *C. nelsoni* into at least two species with populations along the Rio Grande and Guadalupe Mountains being assigned to *Chaetodipus collis*.

[163] All specimens in Texas are referrable to the subspecies *Chaetodipus nelsoni canescens,* which is common throughout its range in the state; however, see Annotation 162 and the possible recognition of *Chaetodipus collis*.

[164] Silky pocket mice of the genus *Perognathus* are very small pocket mice characterized by tan, buff, or salmon-colored upperparts contrasting with white underparts, by conspicuous buff-colored patches behind the ears, and by short tails that lack tufts of long hairs on their ends. There are three species of "silky pocket mice" in Texas, *Perognathus flavus, Perognathus merriami,* and *Perognathus flavescens,* all of which are common in the state. Using karyology, allozyme, and DNA studies, Lee and Engstrom (1991) and Coyner et al. (2010) have shown that although *P. flavus* and *P. merriami* are highly similar

associated with cliffs, canyons, rocky gulches, stone walls, or the edges of bowlders.

Perognathus nelsoni[M] Merriam.   Nelson Pocket Mouse[M]

162   Specimens of this dark-colored form of brush-tailed pocket mouse from Boquillas, east base of Chisos Mountains, Alpine, and east base of Davis Mountains, carry the range of *nelsoni*[M] from Mexico well into western Texas, where it overlaps the range of the superficially similar but quite distinct *eremicus.*[M] At Boquillas and the east base of the Chisos Mountains, Oberholser and I caught the two species together along the cliffs, on sandy flats, and about old stone cabins. While the freshly caught animals were readily distinguished by the spinescent rump and dusky soles of *nelsoni,*[M] no constant difference was found in habits or habitat of the two species. Near Alpine Lloyd caught one specimen at the base of a cliff; and at the east base of the Davis Mountains, at approximately 5,000 feet, Cary caught one under a pile of rocks.

Perognathus nelsoni canescens[M] Merriam.   Gray Brush-tailed Pocket Mouse.[M]

163   The gray pocket mouse is represented from Texas by 5 specimens from Comstock, 4 from Langtry, and 1 from Sheffield. Except for the slightly larger and more angular interparietal, they seem to be typical *canescens,*[M] which was previously known only from the type locality, Jaral, Coahuila, Mexico.

Hollister caught one among the rocks in a small canyon near Comstock, several others along the edge of the Rio Grande Canyon a few miles south of there, and still another on a steep rocky slope near Sheffield; Gaut caught four in the vicinity of Langtry in small caves among the rocks of the river canyons.

Nothing is known of the habits of this pocket mouse save what can be gathered from the character of its habitat, an extremely hot and barren region with light-colored soil and gray limestone cliffs.

Perognathus flavus Baird.   Baird Pocket Mouse.[M]

164   In Texas the Baird pocket mouse is common at El Paso, Sierra Blanca, Valentine, Alpine; and probably in the Pecos Valley, and in the northwest corner of the Panhandle, since it occurs just beyond the Texas line at Carlsbad (Eddy), N. Mex., and at Beaver River, Okla.

At El Paso these little yellow pocket mice were common in December, 1889, along the edges of the sandy valley bottom 2 miles below town, where little sand drifts were heaped up around the base of *Atriplex* and *Suaeda* bushes. Their burrows were usually in groups of three or four, under the edges of the bushes. The occupied ones were closed, and were discovered only by following the lines of tiny footprints across the bare patches of sand from bush to bush till

morphologically, they are genetically distinct and do not appear to interbreed in areas of sympatry. The distribution of *P. flavus* is primarily to the west of Texas, resulting in a discontinuous range that includes the Trans-Pecos and Panhandle regions and an absence of the species between the northern and southern populations within the state.

they disappeared at little mounds of fresh earth that served as doors and blinds to the underground houses. By scraping away the earth a burrow big enough to admit my little finger was disclosed under each tiny mound. Traps baited with rolled oats set near the burrows and along the lines of tracks soon yielded a series of 8 specimens in the rich satiny winter coats—the daintiest, most exquisite of the rodents commonly classed as 'mice.' On chilly nights they did not move about much, but on mornings following a warm night their lines of tracks were abundant, and radiated from the burrows to the nearest patches of wild sunflower and pigweed, whose seeds seemed to furnish their favorite food. One specimen caught December 15 was apparently nursing young, or lately had been, as the teats contained milk.

At Valentine in August, 1902, I turned over a flat stone in the hotel yard and caught one of these little pocket mice as he jumped out of his burrow, and at Sierra Blanca in December, 1889, caught one at a hole in the mellow soil of the railroad bank. At Alpine Gaut caught one in an old gopher mound about 3 miles east of town.

Perognathus merriami Allen.   Merriam Pocket Mouse.[M]
165 This little dusky and yellow pocket mouse ranges over southern Texas from Padre Island and Brownsville north to Devils River, Austin, Mason, and southward into Mexico. Specimens from Devils River and Comstock are fairly intermediate between *merriami* and *gilvus,* as apparently are two specimens from Washburn, which would indicate that *merriami* ranges well up along the east side of the Staked Plains.

The species is common on sandy or mellow soil, more often among weeds and brush than in the open. Their little mounds of earth thrown out on two or three sides of a cactus, bunch of bushes, or flat rock mark the main entrances to their dens. These doorways are always closed during the day if the den is occupied, and when opened from without are usually promptly closed again from within. A careful search near the mounds will generally disclose several little round holes standing open, with no trace of earth thrown out, but with the openings often concealed under bushes or leaves. If you dig into the main burrow or stamp on the ground, a *Perognathus* will often dart out of one of these openings, or more often break through a thin crust of earth that covered a concealed exit and after a leap or two will sit trembling and blinking in the dazzling light of day. It is then so easily caught in the hands that many of our specimens are secured in this way. Most of these are young of the year, however, as apparently the adults are not so readily driven from their dens. When caught they do not offer to bite, but sometimes utter a fine squeak, and if held gently for a while soon cease struggling

[165] Allen (1892) described *Perognathus merriami* on the basis of specimens from the vicinity of Brownsville. He compared this material with specimens of *Perognathus flavus* from El Paso and concluded that they represented a distinct species. Mammalogists followed this taxonomic arrangement for most of the twentieth century (see Osgood, 1900). More recent genetic analyses (see Annotation 164) have confirmed that two species, *P. merriami* and *P. flavus*, are indeed represented. In the central Trans-Pecos region and perhaps in the extreme northern Panhandle region, these nearly identical species of pocket mice occur together but are reproductively isolated from each other. The distribution of the two species is essentially as described by Bailey (Fig. 46). Although Bailey comments that *P. merriami* does not hibernate, they are known to become torpid when stressed by low temperatures or by lack of food (Findley, 1987). See Porter (2011) for the most recent information concerning the ecology and natural history of Merriam's pocket mouse.

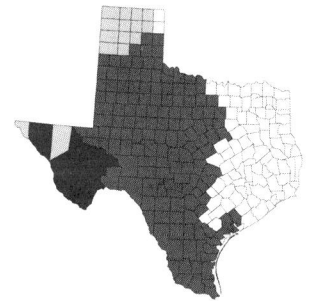

Figure 46. The current distributions of the silky pocket mouse (*Perognathus flavus*) and Merriam's pocket mouse (*Perognathus merriami*) in Texas. The light shaded area represents the distribution of *P. flavus,* the medium shaded area represents *P. merriami,* and the darkest shaded area represents overlap between the two species.

[166] Osgood named a new subspecies of Merriam's pocket mouse (*Perognathus merriami gilvus*) based on three specimens from West Texas and four from New Mexico, and he noted the intermediacy of this subspecies to both *Perognathus flavus* and *P. merriami*. Don Wilson (1973) used a detailed statistical analysis of skull measurements to confirm this interpretation, namely that the two taxa are conspecific, and he combined both under the name *P. flavus*. However, genetic analyses have shown that two species are indeed represented (see Annotation 164); consequently, *Perognathus merriami gilvus* is the appropriate name.

and seem to lose all fear. The light evidently hurts their eyes, and after blinking for a while they soon close them if held quietly in the hands or placed in an undisturbed position on the ground. While often abundant, these little mice are not easily caught in traps, and usually seem indifferent to any bait we use, frequently pushing the traps out of the way or turning them over when set near their burrows or in places where they run. Sometimes they can be caught by placing the trap where they have to step in it in going out of or into their burrows. Near Kerrville a number were caught in this way, while only one out of five had filled his pockets with the rolled oats used for trap bait. A couple were caught in traps baited with juniper berries, which seemed to be a favorite food. In a number of burrows I found juniper seeds or the empty shells from which the kernel had been eaten out through a little hole in one end. In some cases these berries must have been brought from a distance of 10 or 20 rods. In one den under a flat rock, where three tunnels, a foot to a foot and a half long, met in a nest chamber the size of my fist, there was a handful of fresh juniper seeds carefully cleaned of the outer pulp. As this was in May, and the occupant of the burrow was not a full-grown animal, this store was probably laid up for a rainy day rather than for a winter supply. At another burrow a lot of old moldy corn and bits of rubbish mixed with fresh earth were brought out, a little each night, as if in a general house cleaning, indicating that various seeds and grains are stored up in times of abundance. As the mice do not hibernate and as seeds of one kind or another are usually abundant, there is no need of laying up large stores of food.

Perognathus merriami gilvus Osgood.   Dutcher Pocket Mouse.[M]

166 The Dutcher pocket mouse inhabits the Pecos Valley from Carlsbad (Eddy), N. Mex., south to Langtry and the Painted Caves, eastward to Big Springs, and 20 miles east of Rock Springs, and westward to Van Horn and Presidio County; in other words its range coincides approximately with that of the creosote bush in all but the western corner of the extremely arid Lower Sonoran zone of western Texas. It overlaps the range of *flavus*,[M] occurring with it at Carlsbad, and apparently overlaps also the range of *copei*[M] in the country north of Monahans, from both of which it is quite distinct and easily distinguished. From *merriami*,[M] of which it is a larger, lighter yellow subspecies, it shades off along the southern edge of the open and extremely arid region. Specimens from near Rock Springs, along Devils River, and near the mouth of the Pecos are more or less intermediate between the two forms.

In habits these pocket mice do not differ from *merriami*[M] except in so far as they have become adapted to a more open and arid region.

At Langtry Oberholser found them common on the stony mesa, and at Fort Lancaster in the chaparral of the bottom of the Pecos Valley. At Monahans, Cary reported them as abundant in September throughout the sand dunes and as feeding extensively on the seeds of a low, shrubby *Baccharis*. In the dry and barren valley 6 miles south of Marathon I caught them in the baked soil among the scattered mesquite bushes and cactus, and 10 miles farther south found them fairly common in the still more arid and stony valley of Maravillas Creek. Their characteristic little burrows were found around the edges of stones, under bushes and cactus, and occasionally in open spots of bare ground, but the occupants refused to enter my traps or touch any bait. A few were dug out of their burrows and caught in our hands, and from these Mr. Fuertes was able to make some extremely lifelike studies. When first caught the little fellows were greatly frightened and struggled to escape, but never offered to use their teeth. After being held gently for a few minutes they seemed to forget their fear and would sit quietly on the open hand for a minute at a time, blinking sleepily in the unfamiliar glare of daylight. At a sudden motion they would bound away in long leaps, but soon stop, under a weed or bush. While sitting motionless with panting sides they could be easily recaptured by approaching cautiously and covering them quickly with the open hand.

Perognathus flavescens copei Rhoads.   Cope Pocket Mouse.

167 Three specimens taken by Gaut in July, 1904, at Mobeetie, the type locality of *copei*, possess characters which enable this form to be recognized as a bright-colored subspecies of *flavescens*. Two others taken by Cary in the sand hills 20 miles north of Monahans show slightly accentuated characters and, so far as known, mark the limit of its southern range. They suggest also that its range near the southeastern corner of New Mexico probably overlaps that of both *gilvus* and *flavus*.

The 3 Mobeetie specimens, 1 adult female and 2 young of the year, were caught at a den on the edge of a millet field in traps set by the closed entrances of two burrows on opposite sides of a sun-baked furrow. The millet in the field was about ready for harvesting and each of the animals had millet seed in its cheek pouches. A long line of traps yielded no more specimens, and as no other traces of the animal were found it is evident that the species is very scarce in this locality. The failure of several other collectors to procure topotypes of the species is a further compliment to the prowess of the rattlesnake from the stomach of which the type was taken by Professor Cope. (*Cf*. Proc. Acad. Nat. Sci., Phila., 1893, 405.)

[167] As discussed by Bailey, *Perognathus flavescens copei* occupies sandy soils with sparse vegetation in the Great Plains region of northwestern Texas. Its current distribution includes all of the High Plains east to Wilbarger County and south to Midland and Ward counties. Jones and Lee (1962) captured a silky pocket mouse near El Paso, which they identified as *Perognathus apache*, a species thought to be closely related to *Perognathus flavescens*. However, Williams (1978) demonstrated that *P. apache* and *P. flavescens* were conspecific, and the El Paso specimen is now considered to be a representative of *Perognathus flavescens melanotis*. There are no records of *P. flavescens* in the region of western Texas between El Paso and Ward counties.

[168] Bailey followed the conventional wisdom of the time in recognizing two genera of kangaroo rats: *Perodipus* for those taxa with five toes on the hind foot and *Dipodomys* for those species with four toes. However, Grinnell (1919) showed that some of the four-toed *Dipodomys* had five toes on one hind foot and four on the other, and, consequently, he arranged *Perodipus* as a synonym of *Dipodomys*, which now applies to all of the kangaroo rats in Texas.

Dipodomys ordii, Ord's kangaroo rat, occurs throughout most of the western one-third of the state, but its subspecies have changed radically since Bailey published the *Biological Survey* (Baumgardner and Schmidly, 1981). The subspecies now recognized in Texas are: *Dipodomys ordii medius* from the central Llano Estacado southward east of the Pecos River to Crane, Crockett, and Upton counties, and east to Jones County (identified as *Perodipus montanus richardsoni* by Bailey); *Dipodomys ordii obscurus* along the southern parts of the Rio Grande Plain and in the southern Big Bend area (not recorded from these areas by Bailey); *Dipodomys ordii ordii* in the western Trans-Pecos region; and *Dipodomys ordii richardsoni* from the Panhandle and adjacent areas southward at least to Floyd County and east to Montague County (identified as *P. m. richardsoni* by Bailey).

[169] Taxonomists now refer to this taxon as *Dipodomys ordii richardsoni* (see Annotation 168).

Perodipus ordi[M](Woodhouse).   Ord Kangaroo Rat.[M]

168 This little five-toed kangaroo rat is common in the Rio Grande Valley at El Paso and Fort Hancock, and a few specimens have been taken in a tributary valley of the Rio Grande at points 6 to 20 miles south of Marathon. A specimen from Kent, one from Toyahvale, and an imperfect one from Pecos seem to be almost typical *ordi*, while the larger, brighter-colored *richardsoni*[M]is almost typical at Monahans, only 37 miles east of Pecos. The specimens at hand do not clearly prove intergradation in this region and the two forms have well-defined ranges which conform closely to Upper and Lower Sonoran zone limits.

Perodipus ordi[M]is one of the few Lower Sonoran species of this mainly Upper Sonoran genus. In the extremely hot and arid valleys of western Texas it ranges over much of the same ground as *Dipodomys ambiguus,*[M]which it closely resembles in habits as well as appearance. At El Paso I caught specimens on the sandy flats below the town under brush and cactus on the same ground and even at the same holes with *Dipodomys ambiguus.*[M]At Deming, N. Mex., they were common in the sandy strips along the dry valley of the Rio Mimbres, where, in patches of scattered brush and weeds, they were feeding on seeds of wild sunflowers, *Parosela,*[P] and other wild beans. Of ten adult females caught November 29 to December 6, four were giving milk. At the same time numbers of nearly full-grown young were caught, which would indicate either that two litters of young are raised in a season or that the breeding season is very irregular.

Perodipus montanus richardsoni[M](Allen).   Richardson Kangaroo Rat.[M]

169 This largest and brightest colored of the four species of five-toed kangaroo rats inhabiting Texas comes into the State from the Upper Sonoran plains to the north, but instead of keeping to the hard-soiled, Upper Sonoran part of the Staked Plains it completely encircles them. It lives in the sandy stream valleys in the upper edge of the Lower Sonoran zone, but nowhere extends far enough down to be out of reach of Upper Sonoran plants. There are specimens in the Biological Survey collection from Texline, Lipscomb, Tascosa, Canadian, Mobeetie, Newlin, Vernon, Colorado, Stanton, and Monahans in Texas, and from Carlsbad, Roswell, Fort Sumner, and Santa Rosa, in the Pecos Valley, New Mexico. At Carlsbad and Monahans it meets the range of and occupies the same ground with *Dipodomys merriami.*

Throughout its range this species shows a marked partiality for sand, and from Nebraska to Texas fairly revels in the mellow soil of the yellow, shifting, naked drifts and dunes that the wind piles up along the edges of most of the river valleys. It digs an apparently unnecessary number of burrows, which it abandons to other less energetic rodents or uses only as convenient resorts in case of sudden danger. It scampers over the smooth surface with the apparent enjoyment of rabbits on a crusted snow or boys on a skating pond, and paired tracks of the long hind feet are found in the morning in zigzag lines over the drifts, sometimes registering hops of a few inches, again flying leaps of 4 to 6 feet, only to be wiped out each day by the drifting sand and re-registered each night in varying form. Through the weeds and grass of a sandy prairie or the standing grain or scattered stubble of a wheat field the kangaroo rats make little roads, either from burrow to burrow or radiating from burrows to the feeding grounds, and always keep a clear track for retreat to doors that usually are left wide open day and night. Many of the burrows are single, but generally the home den has several openings, with trails leading away from each. For the size of the animal the burrows are large, and in a mound or slope they go back horizontally, so that in case of a hard rain the water runs out of instead of into them. Even on level ground the holes enter as nearly horizontally as possible, and sometimes run along for 10 or 15 feet without going down a foot below the surface. If no sand bank offers the proper angle, the burrow is usually placed under a bunch of cactus, a clump of mesquite bushes, or under some shrub that affords protection as well as a slight eminence to burrow into.

The food of this, as of other species of the genus, is almost entirely seeds, including those of many grasses, various native plants, and any of the small grains. These seeds are neatly shelled out and eaten on the spot or carried in the ample cheek pouches to the dens to be eaten at leisure. No matter how small the seed the shell is always removed, and the contents of the very small stomach of the little animal are always clean and free from indigestible particles. Often the bottom of the burrow is covered with the shells of seeds, but in the several dens examined I never found stores of seeds or grain. Occasionally a little ripe grain is eaten, and a small amount of seed wheat or other grain is dug up; but unless the animals become far more numerous than usual the loss from their depredations is too insignificant for serious consideration.

Perodipus sennetti[M](Allen).   Sennett Kangaroo Rat[M].

170   The type of *Perodipus sennetti*[M] was labeled "near Brownsville," Cameron County, Tex., but the efforts of several collectors to procure topotypes have not resulted in specimens from nearer than the Rio Coloral, 35 miles north of Brownsville. In reply to a letter asking just where he collected the type of *Perodipus sennetti*,[M] Mr. Priour writes under date of February 22, 1903, that it was taken at

[170]   Bailey's conclusion that *Dipodomys sennetti* and *Dipodomys compactus* were separate species and distinct from *Dipodomys ordii* was followed until Davis (1942) observed resemblances in external proportions and cranial features between *D. ordii* and *D. sennetti* and concluded that they were only subspecies of one species (Davis, 1942). Davis further observed that the difference between *D. compactus* and *D. sennetti* was of approximately the same degree as that between *sennetti* and *ordii*, and consequently he treated both *D. compactus* and *D. sennetti* as subspecies of *D. ordii*. However, Schmidly and Hendricks (1976) and Baumgardner and Schmidly (1981) used karyotypes and features of the skull, analyzed with sophisticated statistical techniques, to show that *D. compactus* and *D. ordii* were distinct from one another and that they occurred at the same locality without evidence of hybridization. Thus, there are two species of kangaroo rats recognized today in south Texas: *Dipodomys compactus* with two subspecies, *Dipodomys compactus compactus* on the barrier islands and *Dipodomys compactus sennetti* on the adjacent mainland; and *Dipodomys ordii* with a single subspecies, *Dipodomys ordii obscurus*, from the western two-thirds of the south Texas mainland. Both species are common throughout their ranges in south Texas. With the exception of three specimens collected in 1900 by Oberholser at Cotulla, all of the specimens obtained during the Biological Survey were of the species *D. compactus*. As noted by Bailey, island populations of *compactus* exhibit intrapopulational variation in color, characterized by distinct red and gray color phases.

Although not mentioned by Bailey, William Lloyd made a remarkable observation about the feeding behavior of kangaroo rats, which to our knowledge has not been observed or noticed by other mammalogists. To quote from Lloyd's field itinerary about Padre Island: "But what did they [kangaroo rats] eat was the question! Dissection showed in some cases some animal matter, others were the sagebrush seeds. We went out each night with a lantern and could see them making strange antics, when at last we saw what they were doing—jumping up in the air and catching minute *Coleoptera*. We watched them several nights, having wondered why they always kept in the open sand and away from the brush but their madness had method in it."

[171] *Perodipus compactus* is now recognized as *Dipodomys compactus* (see Annotation 168). The Gulf Coast kangaroo rat is restricted to the southern portions of the state where two subspecies are recognized, *Dipodomys compactus compactus* on the barrier islands and *Dipodomys compactus sennetti* on the adjacent mainland.

Santa Rosa stage station, 85 miles southwest of Corpus Christi. This is on the Alice and Brownsville stage road, near the northwest corner of Cameron County, 145 miles from Brownsville. From this point north to Santa Rosa, across 60 miles of mainly sandy prairie, the species is abundant, and a series of specimens from Sauz Ranch and Santa Rosa shows no variation in characters. A specimen recorded by Mr. Oldfield Thomas[a] from San Diego is shown by the skull measurements to be of this species, but whether collected by Mr. Taylor at San Diego or from the sandy country farther south is not stated. Oberholser found no trace of any kangaroo rat at San Diego. Mr. H. P. Attwater collected 5 specimens 18 miles south of San Antonio. Through the kindness of Dr. J. A. Allen I have examined two of these specimens, now in the American Museum of Natural History, and agree with him that they are typical *sennetti*.[M] Mr. Attwater reported a female taken August 23, containing "two small embryos," and says: "These beautiful little animals appear to be quite common in the sandy black-oak region south of the Medina River in Bexar County. Their burrows seem to be most numerous in the poorest, sandy soil."[b]

Along the Alice and Brownsville stage road the burrows of the Sennett kangaroo rat are common in the yellow sand, sometimes remaining open during the day and sometimes being securely closed with earth. William Lloyd, who camped in this region, says:

> In the deep sand around the stage stations they soon learn what corn and oats are and become great robbers. They seem to enjoy the moonlight nights, skipping about, and on several occasions coming close up to my bed. A motion and they are ten steps away, crouched against the sand; then, if not noticed, they rise and continue their rambles. A lighted lantern seems to puzzle them, and leaving one on the ground to attract them I have caught two of the animals in my hands. At Santa Rosa, while out with the lantern, I saw one starting a burrow. It tried two or three places, presumably to find one sufficiently soft, and at last, apparently suited, pushed its nose in, and drawing its hind feet up close to its jaws, scratched vigorously and soon had made a good beginning to a burrow, when I caught it in my hands.

Perodipus compactus[M] (True).   Padre Island Kangaroo Rat.[M]

171 While closely resembling *sennetti*[M] in cranial characters, *compactus*,[M] even in its darkest color phase, differs from all the mainland forms of Texas in its light coloration, white-margined ears, usually white soles of feet, and mainly white under surface of tail. In the light phase it is unique in having the upper parts a pale ashy gray. In a series of 23 specimens there are 9 of the dark phase and 14 of the light, caught on the same ground by William Lloyd, who said that

---

[a] Proc. Zool. Soc. London. 1888. p. 446.
[b] Allen, Mammals of Bexar County, Tex., Bul. Am. Mus. Nat. Hist., VIII, 57, 1896.

while he found no difference in their habits he could tell them apart even by moonlight.

These kangaroo rats are probably common over the whole length of the 100-mile sand reef known as Padre Island, as Lloyd found them at both the north and south ends. He reported them as sometimes found in the level soil, but usually living in the sand dunes and always on the side away from the prevailing wind. He says:

> At the north end of the island, where most abundant, they close their burrows before daylight, throwing out several quarts of sand in a little mound like a small gopher hill and opening them again after dark. Their object in thus closing their doors is not very evident, as snakes and crabs are too few to bother them. I believe it must be to keep out the black carrion beetles that occupy every disused hole, and a species of pale sand grasshopper that lives in similar situations. Traps set in their burrows were usually covered up with sand, and most of the specimens were caught in the runways, where the prints of their two hind feet and the swish of their tails made unmistakable signs. They feed on the seed of a small sand plant like purslane, and take oatmeal readily as trap bait. After a violent storm their bodies are common objects among the wreckage along the shore, attracting the attention of the boatmen, who call them white rats.

Dipodomys spectabilis Merriam.   Large Kangaroo Rat.[a][M]

172 This beautiful, big kangaroo rat is common in the upper edge of the arid Lower Sonoran zone of extreme western Texas, east to the eastern edge of the Pecos Valley at Monahans and Odessa, and north and south along the Pecos Valley from Adams, Tex., to Santa Rosa, N. Mex. It apparently does not inhabit the lower half of the zone, as it extends neither into the Rio Grande Valley of Texas nor the Gila Valley of Arizona. I have not found it nearer to El Paso or the Rio Grande than Sierra Blanca, Tex., and Jarilla, N. Mex., on the east, and Deming, N. Mex., on the west. While ranging to the extreme upper edge of the zone, it does not enter Upper Sonoran to any extent. In Texas it is common at Sierra Blanca, Van Horn, Valentine, Kent, Toyah, Toyahvale, Adams, Pecos, Grand Falls, Castle Mountains (west base), Monahans, and Odessa; and Gaut collected one specimen at the east base of the Franklin Mountains, 10 miles north of El Paso.

Although strictly nocturnal animals and rarely seen alive, these kangaroo rats usually make their presence evident by conspicuous mounds scattered here and there over the barest and hardest of gravelly mesas, mounds as characteristic and unmistakable as muskrat houses or beaver dams, and as carefully planned and built for as definite a purpose—home and shelter. An old mound that has been inhabited for years is often 3 or 4 feet high and 10 or 12 feet wide, a

---

[a] It is to be regretted that the name 'kangaroo rat' has become firmly fixed to this group of beautiful Jerboa-like rodents, which are as unratlike as they are widely removed from the Marsupials.

[172] The banner-tailed kangaroo rat has now been recorded from several counties in the southern part of the Llano Estacado and northern Trans-Pecos (Schmidly and Bradley, 2016). At the time of the Biological Survey, Texas specimens were not assigned to a subspecies, but Goldman (1923) assigned them to the subspecies *Dipodomys spectabilis baileyi,* named in honor of Vernon Bailey. This taxonomic assignment is still relevant today. Banner-tailed kangaroo rats remain common throughout their range in Texas, but further degradation or loss of grassland habitat could severely reduce populations.

[173] The only specimens of the Texas kangaroo rat, *Dipodomys elator*, captured during the Biological Survey were from Bellevue (by Loring in 1894) and Henrietta (by Oberholser in 1900) in Clay County. Subsequently, the species has been reported from Montague County to the east and as far west in north-central Texas as Motley and Cottle counties, including Archer, Baylor, Foard, Hardeman, Wichita, and Wilbarger counties. This species has been declining in numbers, and it has been proposed for listing under the Endangered Species Act (see Chapter 7 for a discussion of its status).

148　　　　　　NORTH AMERICAN FAUNA.　　　　　[No. 25.

dome-shaped pile of earth entered from the top and sides by a half dozen, or sometimes a dozen, big burrows that would easily accommodate a cottontail rabbit. Well-beaten paths lead away from each of these doorways to others or to neighboring mounds. Usually one or more of the doorways are closed each morning with earth behind the retiring inmates, probably to keep out rattlesnakes and other unwelcome guests. At night these earth doors are opened for use, and the best place to set a trap for the animal is in front of a closed, rather than an open, door. While all of the holes are used more or less at night, apparently only the closed ones are occupied in the daytime. All the fresh earth brought out of the burrows and much that is dug up outside is scraped back on to the mound, so that its size slowly increases with age. Inside, the burrows widen out into roomy chambers, some of which are close to the surface, while others are deep and at the ends or sides of winding burrows. In trying to walk over these mounds one is almost sure to break through knee deep into the chamber below.

　　While the kangaroo rats do not hibernate or store up great quantities of food, they carry considerable food into the burrows to be eaten probably during the day, as shown by deposits in their chambers, by the shells of seeds and grain brought outside during house cleaning or found scattered over their chamber floors, and by the presence of seeds in the fur-lined cheek pouches of individuals caught in traps. That the inmates of these mounds are not always asleep during the daytime can be proved by tapping or scratching at the entrance of the burrow and then listening for a response. A low drumming sound can usually be heard from deep underground, sometimes from two or three points. Apparently it is made as the similar drumming of the wood rats is known to be, by beating the soles of the hind feet rapidly on the ground, which produces a tiny, vibrating roar, and is used as a signal of alarm, call note, or challenge. The animals are social. Often three or four are caught in a mound, and the trails lead from one mound to another. The paired prints of the two long hind feet are fresh every morning in the trails and dusty roads, but I have never seen a print of the tiny hands which apparently are never used in locomotion. When caught in traps or in the hands, the animals struggle violently, but never make a sound or offer to bite. Like rabbits, they are gentle and timid and depend on flight and upon their burrows for protection.

Dipodomys elator Merriam.　　Loring Kangaroo Rat.[M]

173 Specimens of these kangaroo rats are known only from near Henrietta and a point 10 miles to the southwest, and from Chattanooga, Okla. Oberholser says:

　　They are not common in the immediate vicinity of Henrietta, but seem to be of frequent occurrence from 20 to 30 miles to the southwest and most abundant between 2 and 13 miles in this direction. The approximate limits of their range are from Henrietta about 4 miles north, 5 miles east, 22 miles south, 8 miles west, and about 43 miles southwest. They live, so far as determined, almost exclusively among the mesquites and make their holes around the roots of the mesquites and bunches of *Opuntia*. One of the specimens caught was found in the throat of a large rattlesnake that had swallowed it as far as the trap would permit.

Loring, who first caught these kangaroo rats at Henrietta, says:

At one set of holes the main entrance was closed every morning with dirt from the inside, and my traps were not touched. The hole was so small that I thought it might be a *Perognathus*, so got a pick and shovel and dug it out. The burrow branched from below and opened out at four different points. One of the rats was caught in a muddy pocket the size of my fist at the end of the main burrow, the other was covered with dirt in a sharp bend of the burrow, but escaped into another hole near by. The deepest and longest burrow ran about 3 feet underground. I did not find any grass or seeds in any of the burrows. Taking the rat that I had caught to a large field, I turned it loose. It sat for a minute, dazed by the sun, but when I poked it scampered off at such a lively rate that I could hardly keep up and could not see whether it used its fore feet or not. It was very quick and graceful. While jumping its tail was slightly curved up and was not used in any way to aid in its progress.

Near Chattanooga, Okla., some 50 miles northwest of Henrietta, Tex., Prof. D. E. Lantz collected a specimen and reported on the species as follows:

While not numerous, they seem to be well distributed in the vicinity of Chattanooga. Nearly all of the settlers with whom I talked were acquainted with them and informed me that they lived about the premises of their homes. Several were confident that they could capture one or more specimens for me, but only one was secured. This was killed by a farmer as he was walking across the prairie on a dark night with a lantern. It had been foraging in a Kafir corn field, and I found its pouches widely distended with grain. They contained 100 seeds of Kafir corn and 65 seeds of *Solanum rostratum*. In the vicinity of Chattanooga the animals are found on hard clay soils, and they seem to prefer the vicinity of houses, living under houses and outbuildings and in caves made for storing vegetables and other household supplies. They seem to be attracted by lanterns or other lights carried on dark nights.

Mr. Laurie, living in Chattanooga, has a cave back of his hardware store in which a pair of kangaroo rats had taken up their winter quarters. He purchased a couple of bushels of wheat to feed to his poultry and placed it in the cave. Some time later, when he wished to begin to use it, he found that it had all disappeared. Last spring he removed some boards which lined the lower part of the cave on the inside and found all of the wheat carefully stored away behind the boards by the kangaroo rats.

Dipodomys merriami Mearns.   Merriam Kangaroo Rat.[M]

174 This little dull-colored kangaroo rat of the four-toed group ranges over most of the extremely arid Lower Sonoran zone of western Texas, except where it gives place to *ambiguus*[M] in the immediate valley of the Rio Grande. Specimens from near Langtry and 6 miles south of Marathon, Fort Stockton, Toyahvale, Pecos, Monahans,

[174] Merriam's kangaroo rat, *Dipodomys merriami*, is one of the most common kangaroo rats in the state. Federal field agents during the time of the Biological Survey did not collect this species on the Llano Estacado or in south Texas, but mammalogists have now collected it from the southwest sector of the Llano (Choate, 1997) and as far south as Dimmit County in south Texas.

[175] Bailey assigned specimens of *Dipodomys merriami* to two subspecies in Texas, *Dipodomys merriami ambiguus* along the Rio Grande from El Paso to Boquillas in the Big Bend Country and *Dipodomys merriami merriami* over the remainder of the species range. Today, taxonomists recognize only a single subspecies, *D. m. ambiguus*, in the state (Lidicker, 1960).

[176] *Erethizon epixantham* is now recognized as *Erethizon dorsatum* (Miller, 1912). Bailey found evidence of the North American porcupine in Texas only from the Panhandle (Tascosa) and Trans-Pecos (Davis Mountains) regions. Today, the species has been documented from the entire western one-half of the state, east to Van Zandt County and south to Hidalgo County. This is another example of a mammal that has expanded its range in Texas since the turn of the century (see Chapter 6 for a discussion). Systematists now recognize three subspecies in Texas: *Erethizon dorsatum epixantham* along the extreme western edges of the Panhandle, *Erethizon dorsatum bruneri* in the Panhandle and along the Red River, and *Erethizon dorsatum couesi* over the remainder of the distribution (Fig. 47).

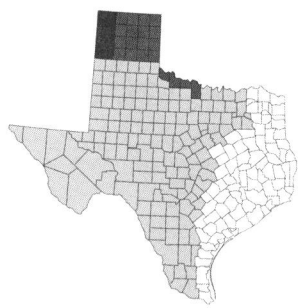

Figure 47. The current distribution of the porcupine (*Erethizon dorsatum*) in Texas. The light shaded area represents the distribution of *E. d. couesi*, the medium shaded area represents *E. d. bruneri*, and the darkest shaded area represents *E. d. epixantham*.

150   NORTH AMERICAN FAUNA.   [No. 25.

Kent, and Sierra Blanca, Tex., and from Carlsbad and Tularosa, N. Mex., are almost typical *merriami*,[M] differing slightly in duller and darker coloration, the opposite extreme from *ambiguus*.[M] The cranial characters throughout the group are extremely uniform.

The habitat of this species is mainly dry, half-barren mesas or open desert valleys, where the animals make their homes in baked and stony soil or less frequently in sandy patches. At Monahans Cary found them only on the hard soil of the valley and never among the sand dunes with *Perodipus richardsoni*,[M] though at Marathon, Kent, Stockton, and Sierra Blanca they were caught in the same ground with *Perodipus ordi*.[M] Except in choice of higher, rougher ground they seem not to differ in habits from *ambiguus*[M]

Dipodomys merriami ambiguus Merriam. El Paso Kangaroo Rat.[M]

175 This little four-toed kangaroo rat, differing from *merriami* in its brighter, more golden color, seems to be of very local distribution along the sandy bottom of the Rio Grande Valley from El Paso and Juarez south to Boquillas in the Great Bend country. A series of specimens from Sierra Blanca is intermediate and can be referred in part to this and in part to *merriami*.

In the Rio Grande Valley it does not differ much in range or habits from *Perodipus ordi*,[M] which it resembles so closely externally that specimens can not be safely named without reference to the toes on the hind feet. In the flesh the two can be distinguished at a glance by the much slenderer feet and tail of the *Dipodomys*. On the sandy river bottoms just below El Paso, where I caught many specimens of the two genera on the same ground and sometimes at the same burrow, I could find no difference in habits or local habitat.

Their little burrows usually are under a bunch of mesquite, acacia, or creosote bushes or cactus, evidently for the sake of protection from enemies, or in order to get a little shade from the fierce heat of the sun's rays. Often their burrows enter from several sides of the bunch of bushes or cactus, and converge toward a common center, where they apparently meet below. Some of their doorways are usually closed during the day, while others are left open. Like all the genus, they are strictly nocturnal and at night feed on the ripe seeds of various plants or carry them into the burrows to be eaten at leisure. Often when caught in traps their cheek pouches are stuffed with seeds or with the rolled oats used for trap bait. Occasionally a bit of green leaf is found in the pockets, but I have seen only the fine white pulp of ripe seeds in their stomachs—no green foliage or anything that would seem to furnish moisture.

Erethizon (epixanthum?)[M] Brandt. Yellow-haired Porcupine.[M]

176 The only specimen of porcupine I have seen from Texas was a badly stuffed skin and a fragment of skull brought me at Tascosa in

1899 by a ranchman who had killed the animal there the previous year. At Alpine in 1900 the ranchmen told me that porcupines were occasionally found there, and in the Davis Mountains in 1901 I found a bushel of their unmistakable signs in a cavity under the rocks, where a porcupine had evidently lived for a good part of the previous winter. As the Finleys, who have lived for many years in the Davis Mountains, had never seen nor heard of these animals, they are evidently not common there.

Lepus merriami Mearns.   Black-naped Jack Rabbit.[M]

177 The black-naped jack rabbits are common over southern Texas, from Brownsville north to the mouth of the Devils River, Fort Clark, and San Antonio, and east to Cuero, Port Lavaca, and Matagorda. They occur with *texianus*[M] in the eastern part of their range and meet or overlap its range in the Devils River country.

In April, 1900, I found them common all the way from Corpus Christi to Brownsville, both on the prairies and in the mesquite and chaparral; and in April of the previous year, between Victoria and Port Lavaca, I counted from the train six that were so close that the black necks showed conspicuously and served to distinguish them from *texianus*,[M] which also was seen along the road. In February, 1894, Loring reported them as common from Alice to Brownsville, as many as ten being seen in a bunch. In March, 1900, Oberholser reported them as numerous on the Thomas ranch, near Port Lavaca, inhabiting chiefly the open prairie, where they found cover under tall bunches of grass. In April and May of the same year he reported them at O'Connorport as only fairly common and very wild; at Beeville as common, living principally on the prairie and in the more open areas in the chaparral; in the immediate vicinity of town as very wild, but farther away, where not often hunted, as much less so; at San Diego as abundant in the more open portions of the chaparral; at Laredo as abundant all through the chaparral; at Cotulla as abundant in the chaparral and very tame, apparently living in the thick brush, although late in the afternoon frequenting the more open, grassy places, where sometimes as many as six or eight were seen together; at Uvalde as abundant and very tame, inhabiting the more open parts of the chaparral, particularly the area between town and the base of the hills to the northward; and at Rock Springs as common on some of the more open areas. On May 29 and 30, 1903, Gaut collected an old male and a 2-weeks-old young one near Del Rio, where he found jack rabbits scarce, although they were said to have been very numerous a short time previous. Only one other individual was seen in three weeks, although considerable time was devoted to hunting for them. Lloyd reported

[177] *Lepus merriami* is now classified as *Lepus californicus,* the black-tailed jackrabbit. Three subspecies are recognized in Texas: *Lepus californicus melanotis* in the north, *Lepus californicus merriami* in the south and southeast, and *Lepus californicus texianus* from the western Edwards Plateau and the Trans-Pecos. Populations in the southeastern part of the state have declined dramatically this century as a result of diminishing habitat, especially the loss of native grasslands.

[178] Black-tailed jackrabbits from the western Edwards Plateau, Trans-Pecos, and southern edge of the Llano Estacado are now classified as *Lepus californicus texianus*. This subspecies is common throughout its range in Texas. Jackrabbit abundance varies in a cyclical pattern, with years of abundance followed by years of scarcity. One period of particular abundance in Texas occurred in the 1920s, spurring ranchers and farmers to mount large campaigns to round up and destroy the jackrabbits. The estimate of five jackrabbits consuming as much grass as one sheep is an exaggeration. It is generally believed today that it takes 18 jackrabbits to consume the same amount of range vegetation as one sheep, or 128 jackrabbits to consume the same vegetation as one cow. Jackrabbits tend to concentrate on pastures that are overgrazed by livestock, and an overabundance of jackrabbits is a good indication of overstocking. Attempts to eliminate predators, ostensibly to protect livestock, often result in an increase in jackrabbits and the subsequent damage to the range may be more detrimental than the occasional loss of a calf to a coyote. Jackrabbits are no longer considered a desirable food resource because they often carry disease and parasites.

them, in 1891, as common on Padre Island and as generally found crouched in the short grass on the open sand.

Over most of their range the black-naped jack rabbits are sufficiently numerous to do considerable damage on farms and truck gardens, as they are fond of many cultivated plants. At Laredo, in April, Oberholser reported that a field of cantaloupes with vines 6 inches high was entirely ruined by them, and a similar field of watermelons was extensively damaged. As raising early vegetables and fruits has become an important industry in southern Texas, the abundance of the rabbits is a serious matter. Their consumption of range grasses in a region that is mainly devoted to stock raising is also a matter of considerable importance.

In general characters this species does not differ much from *texianus*,[M] but in life the difference appears much greater than when specimens are compared. As the rabbit sits up at a distance or, with ears erect, runs across the prairie, the black nape contrasts sharply with the snowy white backs of the ears, and the black tail and rump stripe with the almost white hams and flanks. Near Cuero one jumped from beneath a low mesquite bush close to me and, after a few long leaps that were like flashlights of black and white, suddenly stopped and crouched, lowering the dull gray ears until their white surface rested on and wholly concealed the black neck; the tail was curled up till its black upper surface concealed the black rump stripe and left only the gray lower sides exposed, while at the same time the white hams and flanks close to the ground served to cut out shadow and obliterate form, so that the whole animal was transformed into a part of the great prairie.

This same rabbit, now made into a specimen, does not seem very different from *texianus*, except when the ears are raised to show the black neck; but alive and running, its specific characters might have been recognized at a distance of 40 rods.

Lepus texianus[M] Waterhouse.[a]   Texas Jack Rabbit.[M]

Lepus texianus griseus Mearns, Proc. U. S. Nat. Mus., XVIII, p. 562, 1896. Type from Fort Hancock, Tex.

178   The jack rabbits, except from extreme northern and southern Texas, can be referred to this form with gray or whitish nape and

---

[a] I find no grounds whatever for following Doctor Mearns (Bul. Am. Mus. Nat. Hist., II, p. 296, 1890) and Doctor Allen (Bul. Am. Mus. Nat. Hist., VI, p. 347, 1894) in restricting the name *texianus*[M] to the Arizona jack rabbit. In Waterhouse's original description (Natural History of the Mammalia, II, p. 136, 1848) the only real characters mentioned which distinguish the Texas and the Arizona forms are in the fifth and sixth lines of the description on page 136, "throat and abdomen white; haunches and outer surface of legs gray; tarsus nearly white"—all of which applies to the Texas animal rather than to the Arizona form. The measurements, while evidently from a mounted specimen and of no real value, indicate the slightly smaller size of the Texas form.

light-gray flanks. They are common over the arid plains country of central and western Texas, south to Rock Springs and San Antonio, and from Austin and Brazos westward to Langtry and El Paso; and less common in the strips or islands of prairie country eastward nearly or quite across the State. A specimen collected by Hollister at Antioch, Houston County, is nearly typical *texianus.*[M] A few jack rabbits, probably of the same species, are reported by Oberholser from the prairie near Boston, in the northeastern corner of the State, and a few from the coast prairie west of Beaumont. From residents of the country I have obtained reports of a few from Calcasieu Parish, in the southwest corner of Louisiana, and from the Texas prairies near Virginia Point and Richmond; and on the prairie near Houston, Cuero, and Port Lavaca have myself seen the rabbits close enough to be sure that they were not *merriami.*[M] Specimens from San Antonio, Rockport, and Colorado City show a tendency toward *melanotis,*[M] while others from Vernon and Henrietta can be referred to that subspecies, as can also those of the Panhandle country. In the Davis Mountains the jack rabbits ascend to the edge of the yellow pines, or completely through Upper Sonoran zone, where I have found them common in both July and January. In the Guadalupe Mountains they were common in August up to 7,000 feet on the open ridges, but the main part of their range in Texas lies in Lower Sonoran zone, in the arid part of which they are most abundant.

The abundance of the jack rabbits varies with different seasons and localities, but seems to have a wave-like sequence. After increasing for a few years until extremely numerous, they disappear rather suddenly, are unusually scarce for a few years, and then gradually increase again. This periodic change does not affect the whole country simultaneously, however, for at the same season the rabbits may fairly swarm in one valley and be scarce in another. In January, 1890, on a 30-mile trip from Marfa south to a ranch on Onion Creek, there was hardly a moment when jack rabbits were not in sight— sitting by the road or scurrying through the scattered brush of the desert. In places as many as 20 could be counted, and during December and February of the same winter they were almost as numerous about El Paso and Del Rio. I did not visit the Rio Grande Valley again for ten years, and then could not find one in the region about El Paso.

At Llano in May, 1899, they were numerous in spite of the 5-cent bounty that had been paid on 5,600 of them that year in the county. I often saw a dozen as I made the morning round to my traps, and many of these were limping about with great lumps on their backs and sides where the tapeworm larvae had developed under the skin. Along the wagon road from San Angelo to Colorado City, thence northwest to Gail and the eastern escarpment of the Staked Plains, they were numerous, and in places where no rain had fallen that year and the vegetation was scant and dried up the jack rabbits had seconded the prairie dogs in eating the bark from the small mesquite bushes, from *Opuntia arborescens,*[P] and a large part of the fleshy pads of *Opuntia engelmanni.*[P]

The food of the jack rabbit consists mainly of grass and green vegetation, of which growing grain of all kinds, clover, and alfalfa are especial favorites. It has been estimated that five jack rabbits eat as much grass as one sheep.[a] Allowing one rabbit to the acre, which surely would not be overestimating their maximum abundance, the rabbits on a 1,000-acre ranch would consume as much grass as 200 sheep. That the rabbits are a serious drain on the grass supply of the stock range, especially in the more arid parts of the State, can not be denied. It is a question if they are not even more injurious than the prairie dog, as they cover about twice the area in the State that the prairie dog does, and instead of being in colonies and keeping to a definite locality they travel about freely, seeking cultivated fields, meadows, gardens, orchards, and the best pastures. They are as independent of water supply as any of the desert mammals, and in many of the valleys must go for months without water save what is obtained from their food.

For protection from their enemies the jack rabbits depend on protective coloration, the keenness of their ears and eyes, and the length of their legs, and all they ask of a coyote is a fair start and an open field. A greyhound will pick up one on a straight run, however, and foxhounds will often tire them out if there is moisture enough for good tracking. Coyotes, foxes, and wild cats catch them apparently with a quick bound in brushy places, leaving only patches of scattered fur and a few tracks to mark the spot next morning. Hawks, owls, and eagles prey extensively upon them. Their bones were among the commonest of those scattered over the ground under a great horned owl's nest on a cliff at the edge of the Davis Mountains and under a golden eagle's nest on a cliff near Marathon.

During the morning and evening hours jack rabbits may be seen loping along the trails to their feeding grounds, nibbling grass on the green patches, standing with ears erect, on the *qui vive,* or scurrying in alarm from real or fancied enemies. In the twilight they become almost invisible, and their highly protective coloration probably serves them better by night than by day, as they are then most active. During most of the day they sit in their forms, or merely crouch close to the ground under the edge of a bush or weed, or even in the open without other protection than the blending of their gray coats with

---

[a] 'Jack Rabbits of the United States,' by T. S. Palmer, Biological Survey, Bul. No. 8, p. 30, 1897.

the gray desert vegetation. When they bound away from the bare ground or short grass close to your feet, the surprise is greater than when they start from under a fuzzy-topped weed, though in both cases they may have been in plain view all the time. The only home they can claim for themselves and their young is the form, a slight depression scratched in the ground, usually under the shady side of a bush or weed. They can not endure the heat of the midday sun, and in hot weather always seek some shade. I have never known one to enter a burrow, though they could easily go down badger holes. The young are hidden in grass and weeds until large enough to escape their enemies by running, and they are such experts at hiding that they are rarely discovered.

Extermination of jack rabbits, even if practicable, is not desirable, as they have considerable value for game and food purposes, aside from the interest and pleasure of maintaining a reasonable number of our native animals. When in good condition their flesh is excellent, though usually not so tender as that of the cottontail or Belgian hare. The common prejudice against using them as food has been shown by Dr. T. S. Palmer to be entirely without foundation.[a]

Lepus texianus melanotis[M] Mearns.   Kansas Jack Rabbit.[M]

179 Jack rabbits from Texline, Lipscomb, Canadian, Washburn, Henrietta, and Vernon, and apparently also an immature specimen from Saginaw, are almost typical *melanotis*.[M] They are scarce over the northern part of the Staked Plains, and I have seen but a single specimen from that region and only a few individuals in life. One seen close to the train near Washburn September 22, 1902, had every appearance of being the brown-backed *melanotis* instead of the paler gray *texianus*. A few others reported from Tascosa, Hereford, Washburn, and Gainesville are probably *melanotis*.[M] More specimens from eastern Texas may show closer affinities with *melanotis*[M] than with *texianus*,[M] as the few examined are to some extent intermediate between the two forms.

The habits of *melanotis*[M] do not differ much from those of *texianus*,[M] of which it is the plains and prairie representative. Instead of depending on low desert bushes for shade and concealment, the former usually hide in tall prairie grass, which habit may in some way account for their slightly shorter ears and smaller audital bullae and the browner coloration that constitute their principal subspecific characters.

They are generally less numerous than *texianus*[M] or *merriami*,[M] and consequently of less serious economic importance. They are also freer from parasites, and therefore more acceptable as game.

---

[a] See 'Jack Rabbits as Game,' in Jack Rabbits of the United States, by T. S. Palmer, Biol. Surv., Bul. 8, p. 71, 1897.

[179] The name *Lepus texianus melanotis* has been changed to *Lepus californicus melanotis*. This subspecies occurs in the northern Panhandle of Texas where it is very abundant and extends eastward to the Piney Woods where it is rare.

[180] The name *Lepus floridanus alacer* has been changed to *Sylvilagus floridanus alacer*, the eastern cottontail (Lyon 1904). This rabbit occurs statewide, with three subspecies recognized: *Sylvilagus floridanus alacer* in eastern Texas; *Sylvilagus floridanus chapmani* in the central, southern, and western parts of the state; and *Sylvilagus floridanus llanensis* in the Llano Estacado and Panhandle regions. The species is common throughout its range.

[181] The name *Lepus floridanus chapmani* has been changed to *Sylvilagus floridanus chapmani* (Lyon, 1904).

Lepus floridanus alacer[M] Bangs.   Bangs Cottontail.[M]

180 The cottontails of eastern Texas as far west as Port Lavaca and Gainesville are readily distinguished from *floridanus*[M] by the small audital bullae, and from *chapmani*[M] by the darker colors, in both of which characters they agree with *alacer*[M] described by Outram Bangs, from Stilwell, Ind. T. The darkest specimens are from extreme eastern Texas, but to the westward the transition into the lighter, grayer *chapmani*[M] takes place mainly along the line where timber and thick grass prairie change to mesquite plains.

The cottontails are common over practically all of eastern Texas, living in the densest timber and brush patches, in the open woods, in the rich prairie grass, or about fields and buildings. Where there are no dogs to chase them, their favorite home is under a house or other building. In the woods an old log, tree top, or brush heap usually protects them, though they are often found crouched in their forms under a bunch of briers, weeds, or bushes. On the prairie they often jump from under a tuft of overhanging grass and run to the nearest brush or weed patch for cover. They rarely find burrows to make use of and apparently never dig them.

Nowhere have I found them more than moderately common or in any way a serious pest. Their value as a food and game animal probably compensates for what little mischief they do in cutting off young fruit and forest trees, and for the small amount of grain and vegetables they injure in fields and gardens. If in places they become troublesome, it is easy to thin them out by hunting them with dogs, but usually the hawks and owls keep their numbers sufficiently reduced.

Lepus floridanus chapmani[M] Allen.   Chapman Cottontail.[M]

Lepus floridanus caniclunis Miller, Proc. Acad. Nat. Sci. Phila., p. 388, Oct. 5, 1899. From Fort Clark, Tex.

Lepus simplicicanus Miller, Proc. Biol. Soc. Wash.. XV, p. 81, Apr. 25, 1902. From Brownsville, Tex.

181 This small-eared cottontail ranges from Rockport, Brazos, Henrietta, Mobeetie, Canadian, and Lipscomb, westward to Stanton and Comstock, south to Brownsville in Texas, and across into Mexico. It is a small, pale-gray form of the *floridanus*[M] group, amply covered by the name *chapmani*,[M] given by Doctor Allen to Corpus Christi specimens. It is quite distinct from the long-eared cottontail (*Lepus arizonae minor*),[M] with which it occurs at Stanton, Comstock, Del Rio, Fort Clark, and along a wide strip of country where the ranges of the two overlap. It inhabits the semiarid mesquite country of Lower Sonoran zone and usually is abundant throughout its range.

At Corpus Christi, and thence to Brownsville and Del Rio, these little rabbits live among the big bunches of prickly pear and in the thickets of mesquite and catsclaw, finding in the thorny cover the same protection that the wood rats and many other mammals do, and seeming to ignore the presence of thorns in and along their trails. One of their favorite resorts for a midday nap is in or among the big flat pads of a prickly pear, where they will stick to their form until fairly forced out. In the still more dense and thorny thickets of *Zizyphus* and *Momesia pallida*[P] it is impossible to force them out. In the evening and morning hours they may be seen hopping around the edges of these thickets, where they are often comparatively tame, so confident are they of being able to dodge quickly into a safe retreat. In the country about Kerrville, and westward to Devils River, they are less common, and usually are found in the oak thickets or among junipers and scrub oaks, but the country does not seem to suit them as well as the more open mesquite region farther north at Llano, where they are abundant in the thickets of mesquite and *Zizyphus*. At Mobeetie, Miami, and Lipscomb, Howell found them inhabiting the brushy creek bottoms, and at Canadian both the brushy bottoms and the plum thickets over the sand hills. He says they were very wild, and that none were seen on the open country where the long-eared *minor*[M] ranged.

I have never found these rabbits making use of burrows or of openings in the rocks.

Like all the cottontails, they are excellent food, and are usually free from grubs and other parasites. The young are especially delicious, and as white and tender as quail. Complaints are rarely made of the harm they occasionally do in orchards and gardens, as this seems to be compensated by their value as game.

Lepus arizonae minor[M] Mearns.   Desert Cottontail.[M]

182 The long-eared desert cottontail can be distinguished from the short-eared *chapmani*[M] by its much larger audital bullae with even more certainty than by its longer ears, and, as the two occur together over a wide stretch of country, this distinction is important. It is the common cottontail of western Texas, and extends from El Paso and along the Rio Grande east to Wichita Falls, Tebo, Colorado, San Angelo, Fort Clark, Cotulla, and San Diego, south to Rio Grande City, and north to Tascosa and Lipscomb. The eastern edge of its range overlaps the western edge of the range of *chapmani*[M] in places for a distance of a hundred miles or more, where the two occur commonly together. While distributed mainly over the arid Lower Sonoran zone, it ranges into Upper Sonoran in the Davis and Guadalupe Mountains and on the Staked Plains, but perhaps not farther than a rabbit would wander in a few warm months.

Unlike *Lepus chapmani*,[M] these cottontails are largely inhabitants of the plains and open country, caring little for cover when they can

[182] The name *Lepus arizonae minor* has been changed to *Sylvilagus audubonii minor* (Nelson, 1909). *Sylvilagus audubonii* occupies upland habitats in the western one-half of the state where it is abundant throughout the region. Three subspecies are now recognized: *Sylvilagus audubonii minor* from the southern Trans-Pecos eastward to Val Verde County, *Sylvilagus audubonii parvulus* in south Texas, and *Sylvilagus audubonii neomexicanus* in the Panhandle.

find prairie dog or badger holes for safe retreats. They are often most conspicuous and abundant in a half-deserted prairie-dog town in a barren valley, where they sit under a bush or weed until alarmed, when they rush to the nearest burrow and disappear, or stop, perhaps, at the edge to see if they are pursued.[a] Where there are no prairie-dog holes, badger holes are usually common throughout the range of this rabbit, and a large kangaroo rat burrow is often made use of in an emergency. Openings in and under rocks also are favorite retreats, and rabbits are usually common along the base of cliffs and in canyons and gulches where, besides the natural cavities among the rocks, they make use of burrows where skunks and badgers

FIG. 20.—Long-eared cottontail (*Lepus a. minor*) at badger hole under mesquite, Pecos Valley.

have dug out smaller rodents or made dens under big bowlders. Dense tangles of brush and impenetrable cactus patches also are resorted to for cover, but nowhere within the range of the species is there anything more nearly approaching real woods than scattered mesquite and junipers, with the exception of willows and cottonwoods on some of the river bottoms, which the cottontails seem to avoid. Like jack rabbits, they seem to feel more secure in the open country, where safety depends on keenness of sight and hearing and speed of foot.

---

[a] At Lipscomb and Canadian, Howell found them inhabiting old prairie-dog holes to such an extent that they were called by the ranchmen "prairie-dog rabbits" or "dog rabbits."

Over much of their range they are usually very abundant. While on the train going from Wichita Falls to Seymour, a distance of about 50 miles, I counted 16 of these rabbits, and from Wichita Falls to Childress, a distance of about 100 miles, I noted over 30. At Sycamore Creek in half an hour Lloyd counted 18, and in many places we found them equally numerous. Around ranches they are generally shot for food or chased away by dogs, so that there is little complaint of injury to crops. The amount of range grass consumed by them under ordinary circumstances can not be very great, but without natural enemies they would soon become so numerous as to be a serious pest.

They breed rapidly, but are preyed upon constantly by coyotes, foxes, wildcats, hawks, eagles, and owls. Under the nesting cliff of a great horned owl at the west base of the Davis Mountains parts of fully 100 skulls of this cottontail were found among other bones in the owl pellets.

As food these cottontails are equal to any rabbit and when young they are especially delicious. In camp they are often the only available fresh meat. As other game becomes scarce their importance as food and game will be greatly increased.

Lepus arizonae baileyi[M] Merriam.   Plains Cottontail[M]

183 Two specimens of cottontail from Texline and one from Buffalo Springs, 20 miles to the northeast of Texline, can be referred to *baileyi*[M] better than to *minor*[M] or *arizonae*,[M] although not typical of either form. It is a question if the Lipscomb and Tascosa specimens referred to *minor*[M] do not shade also toward *baileyi*,[M] which appears to be an Upper Sonoran plains form of the *arizonae*[M] group.

At Texline Howell says that these rabbits are numerous in the sagebrush draws near town. When started from the sagebrush (*Artemisia filifolia*), they usually make for the nearest rocks, or else run into a burrow. They are very wild, and if no cover offers quickly run out of sight.

Lepus pinetis robustus[M] subsp. nov.   Mountain Cottontail.[M]

Type from Davis Mountains, Texas, 6,000 feet altitude. No. $\frac{18262}{25165}$ ♀ ad., U.S. Nat. Mus., Biological Survey Coll. Collected Jan. 6, 1890, by Vernon Bailey. Original No. 873.

184 *General characters.*—Similar to *Lepus pinetis holzneri*,[M] but larger, with relatively narrower braincase and conspicuously wider, more prominent postorbital processes.

Color.—Winter pelage: Crown and rump brownish gray, sides and rump light ashy gray, nape and exposed part of legs bright fawn color, lower parts white with buffy gray throat patch. The short summer pelage is not known.

[183] The name *Lepus arizonae baileyi* has been changed to *Sylvilagus audubonii baileyi* (Lantz, 1908), but the application of this name is restricted to rabbits occurring in the northern plains states. The specimens from Texas, discussed by Bailey, are today classified as *Sylvilagus audubonii neomexicanus*, which occurs in the Panhandle of Texas, southward to Brewster and Terrell counties. A third subspecies, *Sylvilagus audubonii parvulus*, occurs from Llano County southward in south-central Texas to the Rio Grande.

[184] Originally described by Bailey as a subspecies of *Lepus pinetis*, Hall and Kelson (1959) arranged *robustus* as a subspecies (*Sylvilagus floridanus robustus*) of the widespread eastern cottontail, *Sylvilagus floridanus*. Ruedas (1998) presented evidence that *Sylvilagus robustus* was a species based on distinct characteristics of the teeth, and Lee et al. (2010) used molecular data to confirm this position. Its known range includes the Guadalupe Mountains of Texas and New Mexico, the Chisos and Davis mountains of Texas, and the Sierra de la Madera of Coahuila, Mexico. The Guadalupe and Chisos Mountains populations have been severely reduced and no specimens had been seen or collected from these areas for a period of nearly 30 years. However, Nalls et al. (2012) collected *S. robustus* in the Chisos Mountains in 2003–2004. Recently, mammalogists from Texas Tech University have obtained several specimens of *S. robustus* from the Davis Mountains, and it appears that a healthy population remains in that area. The species currently is not listed as threatened or endangered by the USFWS or TPWD. Prior to 1996, it was listed as a potential Category 2 taxon.

[185] *Lepus aquaticus* is now classified as *Sylvilagus aquaticus* (Nelson,
1909).

Cranial characters.—Skull larger than in *pinetis* [M] or *holzneri,* [M] with relatively narrower braincase, slightly larger bullae, and conspicuously wider, more prominent postorbital processes.

Measurements.—Type specimen ♀ ad.: Total length, 460; tail vertebrae, 55; hind foot, 104; ear from notch (measured dry), 67. Average of 5 adults from western Texas: Total length, 458; tail vertebrae, 59; hind foot, 103.6; ear from notch, 68.

Skull of type.—Basal length, 60; nasals, 32; zygomatic breadth, 34; greatest breadth of braincase, 26.5; mastoid breadth, 25.5; interorbital breadth, 19.

Remarks.—This large brush rabbit needs comparison only with *pinetis* [M] and *holzneri,* [M] from both of which it differs enough to form a good subspecies, and from the range of which it is apparently entirely cut off by intervening valley country. From *arizonae* [M] and its subspecies it is entirely distinct, differing widely in size and cranial characters and occupying the same ground in the lower part of its range. It is a Transition zone species, ranging from 6,000 to 8,000 feet in the Davis and Chisos mountains, rarely coming down the brushy slopes into Upper Sonoran zone. A specimen collected in midwinter in Presidio County was in the Upper Sonoran zone near the edge of the Chinati Mountains, at about 4,200 feet, where it may have wandered down along the brushy creek from a higher level.

It lives in brushy and timbered country and makes runways through the thickets, which, when started, it follows at a lively speed and with much noise. It is almost as large and heavy as the varying hare, and needs only to be seen or heard running to be distinguished from the light, slender *minor.* [M] While many were seen or heard in the Davis and Chisos mountains, but few specimens were collected, owing to the difficulty of getting shots at them in the thickets. They seem to be entirely free from grubs and other parasites and are fine eating.

Lepus aquaticus [M] Bachman.   Swamp Rabbit.

185   The swamp rabbits from near the coast of southeastern Texas agree in general appearance with *aquaticus,* in referring them to which species I follow Doctor Allen, although more specimens from this region, as well as more of typical *aquaticus,* are necessary to a final decision. Specimens examined: Selkirk Island, Matagorda County, 1; Bernard Creek (12 miles west of Columbia), 2; Austin Bayou (near Alvin), 1. Oberholser reported them as common in the moist woods near Beaumont, and Hollister found them "exceedingly numerous at Sour Lake, especially about the wooded islands." Lloyd reported them from Selkirk Island at the mouth of the Colorado River and in the salt marshes near Matagorda. My own acquaintance with the species has been in the Big Thicket of Hardin County, Tex., and in southern Louisiana, where they live in swamps, marshes, and low brushy woods near the bayous, making trails that often lead through shallow water. They usually jump from under old logs or tangles of briers and underbrush and go dashing off with a heavy thumping run, but usually with speed enough to escape the dogs. Fires are said sometimes to drive them out of the swamps and marshes by hundreds. In the Big Thicket in December, 1904, they were especially abundant under the dense growth of palmettoes and tangle of vines. At this season the ground was dry, but the quantity of large flattened pellets covering the tops of old logs suggested that during wet weather the rabbits spent much of their time on the logs.

Late in the following March Gaut found them abundant in this region during high water, and was informed by Mike Griffin, a hunter living on Black Creek, that they were great swimmers, and when chased by the dogs would invariably swim back and forth across the creeks. One female examined contained five embryos and two others were nursing young.

Lepus aquaticus attwateri [M] Allen.　　Attwater Swamp Rabbit. [M]

186 Swamp rabbits are common along the streams of eastern Texas as far west as Port Lavaca, San Antonio, Austin, and Gainesville. Specimens from Richmond, Antioch, Joaquin, Troup, and Gainesville are large and gray like typical *attwateri*, [M] and can be referred to nothing else. They are reported as common in the swamps or bottom lands at Arthur, Texarkana, Jefferson, Waskom, Rockland, Brenham, and near Elgin. Those reported from Conroe and Jasper are probably nearer to *aquaticus.*

In habits these rabbits are similar to *aquaticus,* living in the timbered bottom lands along the rivers, often among the palmettoes, or in wet, half-swampy places in the woods. On the Brazos bottoms, near Richmond, I found them under old logs and brush in the densest woods, and at Troup, Loring reported them as hiding in fallen tree tops or under roots of trees and brush piles in low, swampy places. H. P. Attwater says:

> When frightened from their hiding places and chased by dogs they take refuge in hollow trees and in holes in the river bluffs. The dogs seem to have more difficulty in trailing them than they do the cottontails and jack rabbits, the swamp rabbits often eluding the hounds by taking to water. I have seen them on several occasions swimming across the river while the dogs were hunting for them on the other side.[a]

---

[a] Bul. Am. Mus. Nat. Hist., VII, p. 328, 1895.

[186] The swamp rabbit, *Sylvilagus aquaticus*, occurs in the eastern third of the state, from the Red River counties west to Brown and Bandera counties, then south to Refugio County along the coast. Optimum habitat in much of its range is shrinking with drainage of wetlands and clearing of hardwood forests, and consequently there is much concern about the future conservation status of this rabbit in Texas. In the Hill Country, they are threatened by habitat fragmentation.

[187] *Felis hippolestes aztecus* is now classified as *Puma concolor couguar* and generally refers to all mountain lions from the US (Kitchener et al., 2017). Previously, mountain lions in Texas were classified as *Puma concolor stanleyana* except for populations in El Paso County, which were assigned to *P. concolor azteca* (Culver et al., 2000). Mountain lions occur in almost every kind of habitat throughout the state. Years of predator control efforts by livestock producers, however, forced them into the more remote, thinly populated areas. In recent years, they have increased in number and numerous sightings have been reported around the state (see Chapter 5 for a discussion of their status today).

Felis hippolestes aztecus[M]Merriam.   Mexican Cougar; Mountain Lion; Panther.

187 The specimens of mountain lion from Texas available for determination of the species are few, and their status is unsatisfactory; but they indicate that at least the western part of the State is inhabited by *aztecus.*[M] Two skulls of females from the Davis Mountains and one from Brownsville do not possess important specific characters, but they and a flat skin from the Davis Mountains and one from near Boquillas agree with *aztecus*[M]more nearly than with any other species. A fine male seen in the San Pedro Park, at San Antonio, October 30, 1904, said to have come from Langtry, was in the light-gray coat of *aztecus,*[M]and a nearly perfect skull of an old male from 20 miles north of Comstock shows the best-marked characters of that species.

In the rough and sparsely settled western part of the State mountain lions are still fairly common in certain sections, where they often lay a heavy tribute on colts, calves, and sheep. At Langtry Gaut reported them in 1903 as "quite common a few years ago, but now scarce," and adds: "One was shot in a pasture about half a mile from the station last winter, and an old hunter (Mr. E. B. Billings) at Samuels informs me that the stomach of one that he killed near Langtry a few years ago contained part of the foot of a raccoon and also some of the remains of a gray fox." Gaut reported a few panthers in the Franklin Mountains the same year, and said that he was shown a mule killed by one. The mule's neck showed deep gashes which had been cut by teeth and claws. Large numbers of colts were said to be killed every year in these mountains by panthers. Near Oakville, Live Oak County, Mr. F. A. Lockhart reported that a horse and two colts had been killed by a panther July 25, 1895, and that a hound was killed the next day by the same animal.

The rough desert ranges, full of canyons, cliffs, and caves, are the favorite haunts of the panthers, and will be their last strongholds, not only because of the advantages they offer for foraging but because of the protection they afford from hounds and hunters. In the desert mountains just north of Van Horn in August, 1902, a panther and I were mutually surprised at meeting in a narrow gulch, he evidently expecting a venison supper, and I, in my search for rock squirrels, discovering his big, round, yellow face between the rocks above me. I drew my sight a little too fine and caught the rock just under his chin with no more damage than to fill his eyes with rock dust and cause a quick retreat behind the crest of the ridge. I was scarcely more disappointed than was the ranchman in the next valley whose colts had been disappearing at frequent intervals. In the Davis Mountains these cougars have been hunted with hounds until scarce, but in the Santiago range, in the Chisos Mountains, and along the canyons of the Rio Grande and Pecos they are still common. A few are found even on the edges of the Staked Plains. On a ranch 22 miles north of Monahans Cary saw the skin of one that had been roped by a cowboy in July, 1902. It was a very large female, and was said to have measured 11 feet in total length. The size strongly suggests *hippolestes*,[M] which it ought to be from geographic considerations.

The form inhabiting the timber and swamps of eastern Texas undoubtedly is different from either *aztecus*[M] or *hippolestes*,[M] but whether it is the Florida panther (*coryi*),[M] the Adirondack panther (*couguar*), or something else, will remain doubtful until specimens are procured from the region. Baird speaks of the redness of a skin collected by Captain Marcy on the Brazos River (Pac. R. R. Rep., VIII, 84, 1857), but the skin can not now be found. An old female panther, which died in the National Zoological Park January 19, 1900, was caught August 12, 1892, when about two or three weeks old, near Memphis, Tex., in the Red River Valley, east of the Staked Plains. The skin of this animal, now in the National Museum collection, agrees fairly well with skins of *hippolestes*,[M] but the skull does not agree with any of the skulls in the National Museum and shows peculiarities probably due to life-long confinement.

In most of eastern Texas panthers are reported as formerly common, but now as very rare or entirely extinct. Individuals have been killed, however, within a few years in the swamps not far from Jefferson in the northeastern part and Sour Lake in the southeastern part of the State. At Tarkington Prairie Mr. A. W. Carter says there were a few panthers when he was a boy in 1860, but he has not seen one since. In the Big Thicket of Hardin County a few panthers have been killed in past years, and Dan Griffin, who lives 7 miles northeast of Sour Lake, says a very large one occasionally passes his place. He saw its tracks in the winter of 1903-4.

Felis onca hernandezi[M] (Gray).   Jaguar.

Felis onca Baird, Mamm. N. Am., p. 86, 1857 (in part).

188 The jaguar, the largest of North American cats, once reported as common over southern Texas and as occupying nearly the whole of

[a] An adult male jaguar killed near Center City, Mills County, Tex., September 3, 1903, agrees very closely in color and markings with a skin of *Felis hernandezi*[M] from near Mazatlan, Mexico. The ground color in the Texas skin is a shade yellower, and the spotting slightly coarser, but the difference is too slight for any important significance. Unfortunately there is no skull with the Mazatlan skin, but the Texas skull is scarcely distinguishable from comparable skulls of typical *onca*[M] from Brazil.

Baird's detailed description of a skin from the Brazos River also agrees in a general way with this topotype skin from Mazatlan.

[188] The name *Felis onca hernandezi* has been changed to *Panthera onca hernandesii* and refers to jaguars located in western Mexico (Pocock, 1939). The jaguar that once ranged throughout the southern one-third of Texas is now classified as *Panthera onca veraecrucis* and is now extinct in Texas (see Chapter 5 for a discussion).

the eastern part of the State to Louisiana and north to the Red River, is now extremely rare. Occasionally there is a report that one has been killed, but in very few cases have the reports been substantiated by specimens. A skin from the Brazos River, Texas, without a date, but entered in the National Museum Catalogue in 1853 and described in detail by Professor Baird in the Mexican Boundary Survey (Vol. II, part 2, p. 6), seems to have disappeared. This specimen was obtained from J. M. Stanley, but no more definite locality was given than 'Brazos River.' The following note from H. P. Attwater was published by Doctor Allen in his list of mammals of Aransas County, Tex.[a] "Captain Bailey says he formerly owned a fine skin of a jaguar killed on the point of Live Oak Peninsula by J. J. Wealder and A. Reeves in 1858, but has not heard of any in this neighborhood since."

In reply to a request for detailed information relating to a mounted specimen of a jaguar mentioned in a previous letter, Mr. H. P. Attwater writes under date of June 4, 1904, as follows:

> Since writing I have been in San Antonio, and while there hunted up my friend Mr. Frank Toudouze, and from him obtained some information about the jaguar referred to in previous communication. Mr. Frank Toudouze, who is now living in San Antonio, remembers the circumstances very well, and tells me that the jaguar was killed by his brother, Henry Toudouze, and party of hunters, in 1879, about 10 miles south of Carrizo Springs, in Dimmit County, Tex. He tells me that it was a male, and even at that time considered a rare animal in that part of Texas. Mr. Henry Toudouze died a few years ago, but I have heard him tell about the killing of this particular animal many times, as well as hearing his father and brothers speak of it. When I came to Texas it was in Mr. Gustave Toudouze's (the old gentleman) collection, and he and I took it with other specimens to the New Orleans Exposition in 1884, as a part of the Texas Natural History Exhibit of which we had charge. At the close of the New Orleans Exposition it was brought with our collection back to San Antonio, and subsequently taken to Mexico by Mr. Frank Toudouze, who tells me that he sold it with the rest of the collection to the officials of the State Museum at Saltillo, State of Coahuila, and he says he has no doubt that the specimen is still there.

In 1902 Oberholser heard of a jaguar that was killed south of Jasper a few years before, and also obtained reports of the former occurrence of the species along the Neches River near Beaumont and in the timber south of Conroe. There have been several reports from different sources of one killed near the mouth of the Pecos in 1889, or near that date, and in 1901 Oberholser got a record of one killed south of Comstock "some years ago," but without a definite date. At Camp Verde, Cary was told by a Mr. Bonnell of a jaguar killed in 1880 at the head springs of the Nueces River, but this may have been the Toudouze specimen from Carrizo Springs.

---

[a] Bul. Am. Mus. Nat. Hist., VI, 198, 1894.

The skin and skull of a fine old male jaguar killed near Center City, Mills County, Tex., September 3, 1903, through the efforts of Mr. H. P. Attwater, the enthusiastic naturalist of Houston, Tex., have been secured and safely lodged in the National Museum. Through the courtesy of Mr. Gerrit S. Miller, Jr., assistant curator of mammals in the U.S. National Museum, the correspondence relating to the capture of the animal and the securing of the skin and skull for the museum has been placed at my disposal. The following extracts from this correspondence are of special interest.

In a letter of March 21, 1904, Mr. Attwater wrote to Mr. Miller:

> Last fall I heard that a jaguar had been killed near Goldthwaite, in Mills County, north of the Colorado River, in west central Texas, and wrote to parties in that section for particulars, but with poor satisfaction, so made up my mind to go there the first opportunity for the purpose of getting at the facts. I was so much engaged with my work that I was unable to spare time to do this until several weeks ago, but I am glad to say that I am now able to report very satisfactory results, and that I found the skull and hide still there, also ordered photographs (from the negative taken at the time), which have just come to hand. I take pleasure in sending you one with this letter, and later on will send you full particulars, with date of killing, etc., the most important of which I already have.

*   *   *   *   *   *   *

> In regard to the killing of the jaguar, I understood from Mr. Hudson, who skinned him, and from others, that they found it accidentally, and that they were hunting wildcats at the time they ran across him. I was told while I was there that another jaguar had been reported in the same locality after this one was killed, and it was supposed that there was a pair of them, but as far as I could find out nothing had been heard or seen of the other one for some time past. * * * In regard to how the jaguar came there, my idea is that it strayed there probably with its mate from the Rio Grande region, which it could easily have done by the route indicated on the inclosed map. The character of the country all along this route from the Rio Grande to Mills County is similar and not thickly inhabited, and I am inclined to think the animal made its way up the San Saba River and across the Colorado into Mills County. I took particular note of the country around Goldthwaite and in that part where the animal was killed it is rough with rocky ridges which they call 'mountains,' running parallel with the creeks and rivers, with uneven valley lands between the streams and the mountains. There is no tall timber, but the entire country is covered with a thick brush or chaparral, consisting chiefly of shin oak thickets known as the 'shinnery,' also sumac thickets and Spanish oak clumps with live oak trees scattered among them. On the lower flats there are considerable mesquite trees.

Later Mr. Attwater wrote as follows:

> I send you by express box containing the skin and skull of jaguar. * * * Miss Julia Kemp, the photographer in Goldthwaite, very kindly promised to write to the parties who killed the animal to get the data and other particulars for me. I herewith inclose you the correspondence, together with a letter she sent, received from Homer Brown, one of the parties in the "fight."

[189] The name *Felis pardalis limitis* has been changed to *Leopardus pardalis albescens* (Allen, 1919). The ocelot was still common over much of the Rio Grande and coastal marsh regions and extended as far north as Rock Springs and Kerrville during the time of the Biological Survey, but its range and numbers have been reduced significantly. Although Bailey reported a single specimen from the Trans-Pecos region that had been killed near the Alamo de Cesarae [Cesario] Ranch in Brewster County between Marfa and Terlingua in 1903, it appears that ocelots were never a common resident of the western portions of the state. The ocelot was still common over much of the Rio Grande and coastal marsh regions of Texas during the time of the Biological Survey, but its range and numbers have now been greatly reduced in the state. It is listed as "Endangered" by the USFWS and TPWD, and it is now restricted to small, isolated patches of suitable habitat in two counties of the Rio Grande Plains. Mike Tewes, of the Caeser Kleberg Wildlife Research Institute in Kingsville, Texas, has been studying the ecology and conservation of ocelots in Texas. He has documented aspects of their biology and natural history, and he has developed a habitat conservation plan for this species (Tewes, 2019).

The following letter from Homer Brown was addressed to Miss Julia Kemp, March 20, 1904:

> Yours of 15th at hand. In regard to the jaguar, we killed him Thursday night, September 3. I will give some of the particulars. Henry Morris came to go hunting with me that night. I had a boy staying with me by the name of Johnnie Walton. We three took supper at my home and then started for the mountains, 3 miles southwest of Center City, where we started the jaguar just at dark. We ran him about 3 miles and treed him in a small Spanish oak. I shot him in the body with a Colt .45. He fell out of the tree and the hounds ran him about half a mile and bayed him. I stayed with him while Morris went to Center City after guns and ammunition. In about an hour and a half he came back and brought several men with him, so then the fight commenced. We had to ride into the shinnery and drive him out, and we got him killed just at 12 o'clock that night. We commenced the fight with ten hounds, but when we got him killed there were three dogs with him, and one of them wounded. He killed one dog and very nearly killed several others. He got hold of Bill Morris's horse and bit it so bad it died from the wounds. * * * The men in the chase were three of the house boys, Al and Joe Tangford, George Morris, Bill Morris, Thad Carter, Claud Scott, Henry Morris, Johnnie Walton, and myself. The jaguar measured 6 1/2 feet from tip to tip, 36 inches around chest, 26 inches around head, 21 inches around forearm, 9 inches across the bottom of foot; weight, 140 pounds.

Felis pardalis limitis[M] Mearns. Ocelot; Leopard Cat.

> *Felis pardalis* Baird, Mamm. N. Am., 87, 1857 (in part).
>
> *Felis limitis* Mearns, Proc. Biol. Soc. Washington, XIV, p. 146, 1901.

189 The ocelots are still found in brushy or timbered country over southern Texas, as far north as Rock Springs and Kerrville, and up the Pecos Valley to the region of Fort Lancaster. One killed near the Alamo de Cesarae Ranch, in Brewster County, between Marfa and Terlingua, in 1903, was reported by Mr. G. K. Gilbert, and later its beautiful light-gray skin was purchased from Mrs. M. A. Bishop, of Valentine. This seems to be the westernmost record for the State. Farther east ocelots are still reported as very rare about Beaumont and Jasper, near the eastern line of the State, and farther north, near Waskom and Long Lake. Early records carried their range across into Louisiana and Arkansas, but it is doubtful if at the present time they are to be found in the United States beyond the limits of Texas. Most of the records are from hunters, ranchmen, or residents of the country, who know the animal by the name of ocelot or leopard cat, or describe it as a long-tailed, spotted cat the size of the lynx. In 1902 at Sour Lake Hollister reported "several so-called leopard cats killed near there," and says: "They are described as about the size of the wildcat but of a different build, spotted and with a long tail." Near Beaumont Oberholser reported them as occasionally killed in the woods along the Neches River. In Kerr County Mr. Moore, the sheriff, told me that he saw a beautiful skin of a large, long-tailed, spotted cat that was killed 10 miles south of Kerrville the latter part of June, 1902. At Rock Springs in July of the same year Mr. Gething told me that each year a few ocelot skins were brought into the store for sale. In his report from Sheffield Cary says: "I am informed that leopard cats are fairly common in the cedar brakes along the Pecos southeast of here." In 1899 Mr. Howard Lacey, the well-known naturalist of Kerr County, told me that he occasionally caught an ocelot while hunting with dogs for other game, and in January, 1903, he wrote of the species as follows:

> The few that I have seen have all been found by the hounds, usually when we were hunting bear, and always in just the kind of country a bear would choose—the roughest, rockiest part of a dense cedar brake. Once on the head of the Frio River in November the hounds struck a hot trail and were just beginning to get off well together on it when a splendid male ocelot sprang into a large cedar close to us. Thinking the hounds might be on a bear trail I shot the cat at once, put him behind me on the saddle, and made after the hounds, that were getting off at a good pace. They ran about 2 miles and then treed a female ocelot in the bottom of a steep canyon. This we also shot and I think the two were together when we started them, and that they often go in pairs. They are not common here, but I fancy that they often rest in the trees and so escape the dogs.
>
> They are heavier and more muscular than the bobcat, and our hounds, that always make short work of a bobcat, find the leopard cat 'a tough proposition.' Unlike the bobcat, they have the strong odor peculiar to the larger felines, and I never killed one without being reminded of the lion house at the London Zoo.
>
> I have never had the luck to find any kittens, but a friend of mine ran a female into a cave with his hounds and killed her; then the dogs went into the cave and killed and brought out two kittens a few weeks old. This was in November. On another occasion he killed a female that in the course of a few days would have brought forth two kittens. Another of my neighbors killed a female and two kittens in a cave near here. This was also in November, and the kittens had not yet got their eyes open.
>
> These cats do much damage to the stockmen, being especially fond of young pigs, kids, and lambs. They probably also kill fawns and turkeys, and, like many other cats, often hide what they can not eat under a heap of leaves.

Felis cacomitli [M] Baird (Berlandier MS.). Red and Gray Cat. [M]

1857. *Felis yaguarundi* Baird, Mamm. N. Am., 88, 1857. From Lower Rio Grande region; not *Felis yagouaroundi* Geoffroy, 1803, from Guiana. Gray phase.

1857. *Felis eyra* Baird, Mamm. N. Am., 88, 1857. From Lower Rio Grande region; not *Felis eyra* Fisch., 1815, from Paraguay. Red phase.

1859. *Felis cacomitli* Baird (Berlandier MS.), Report Mex. Boundary Survey, II, 12, 1859. From Matamoras, Mexico. Gray phase.

1901. *Felis apache* Mearns, Proc. Biol. Soc., Washington, XIV, 150, 1901. From Matamoras, Mexico. Red phase.

1902. *Felis cacomitli* Mearns, Proc. U. S. Nat. Museum, XXIV, 207, 1902. Gray phase.

190 A study of the specimens in the Biological Survey and U. S. National Museum collections, including five skins and skulls of the red cats and six of the gray from southern Texas and eastern Mexico,

[190] *Felis cacomitli* is now classified as *Puma yagouroundi*, and populations that once occurred in Texas belonged to the subspecies *Puma yagouroundi cacomitli* (Culver et al., 2000). Although it is now extirpated from Texas, at the time of the Biological Survey the jaguarundi was still common in the brush country of extreme South Texas. As with the jaguar and ocelot, predator control and habitat destruction associated with clearing brushlands in the Rio Grande Valley were responsible for its demise (see Chapter 5 for a discussion).

reveals no constant difference in cranial or external characters other than color. The striking coincidence of range and similarity of habits, as well as structure, of the red and gray cats strengthen the evidence tending to show that these supposedly distinct species present only another case of dichromatism, comparable to the black and cinnamon bear and the red and gray phases of the screech owl. A wide range of individual variation in size, shades of color, and in cranial characters is shown in the series of specimens examined. The type skull of *apache*[M] shows the widest departure in characters, and especially in dwarfed size; but as the animal was captured when very young and kept in confinement throughout the rest of its life without becoming wholly domesticated, this may account for abnormal development.

Owing to lack of enough Central and South American specimens to show the relationship of *cacomitli*,[M] through the several intervening forms, with typical *yagouaroundi*[M] and *eyra*,[M] it seems best for the present to treat this northernmost and relatively light-colored form as a species. When the relationships of the group are fully worked out it will doubtless stand as a subspecies of *yagouaroundi*.[M]

The Biological Survey collection contains four specimens of the gray and one of the red cat from Brownsville, collected by F. B. Armstrong in 1891 and 1892, and a young one of the gray form collected by Lloyd, August 9, 1891. This quarter-grown young was reported as one of a litter of four caught by a boy and a dog in a 'resaca' near Brownsville. Lloyd also reported seeing one fresh and several dry hides of the gray cat in Brownsville, and mentioned two "ancient mounted specimens" of the red cat in Armstrong's collection there, but did not say where they originally came from. Since then Armstrong has sent to the National Zoological Park at Washington four of the red and two of the gray cats alive from the Lower Rio Grande region.

In a letter from Brownsville to Dr. Frank Baker, superintendent of the National Zoological Park, Armstrong writes:

Eyra and yaguarundi cats inhabit the densest thickets where the timber (mesquite) is not very high, but the underbrush—catsclaw and granjeno—is very thick and impenetrable for any large-sized animal. Their food is mice, rats, birds, and rabbits. Their slender bodies and agile movements enable them to capture their prey in the thickest of places. They climb trees, as I have shot them out of trees at night by 'shining their eyes' while deer hunting. I capture them by burying traps at intervals along the trails that run through these thick places. I don't think they have any regular time for breeding, as I have seen young in both summer and winter, born probably in August and March. They move around a good deal in daytime, as I have often seen them come down to a pond to drink at midday, and often see them dart through the brush in daytime. They are exceedingly hard to tame. Their habitat is from the Rio Grande, 40 miles north of here, as far as Tampico, Mexico. Beyond that I don't know.

A 'long-tailed yellow lynx' reported by John M. Priour from west of Corpus Christi in December, 1902, may have been this species. Mr. Priour thought it might be a partially albino ocelot. Apparently the same animal was seen there two years before by Dr. Adolph Huff, of San Antonio, who thought it might be a young panther.

Lynx rufus texensis Allen.   Texan Lynx.[M]

191 The large, dark, and usually much spotted and lined lynx of southern and eastern Texas ranges in Lower Sonoran zone north to at least Montague and Cooke counties and west to Kinney County. An immature specimen from Antioch and three skulls and six skins from Hardin and Liberty counties carry its range to near the eastern part of the State. More material may show that the form inhabits the whole Lower Sonoran zone of Texas, including the Pecos and Rio Grande valleys, and grades into *baileyi*[M] or overlaps it in range in the Davis Mountains country. It is common over southern and eastern Texas and especially abundant in the dense chaparral of cactus and mesquite along brushy stream bottoms and in the timbered gulches where the lower arm of the Staked Plains breaks down into the low country, and in the swamp country farther east. At Port Lavaca Oberholser reports: "The wildcats are common in places away from town where there is sufficient cover, such as live-oak thickets and the great rose hedges. In the thickets they are not so difficult to hunt, but in the hedges they have almost impenetrable cover, and it is well-nigh impossible to reach them except by trapping." He says that at O'Connorport "a good many wildcats inhabit the thicker part of the oak brush, where they can be hunted only with dogs;" on Matagorda Island "they occur in the little chaparral that grows on the island;" and at Beeville "they are common in the denser portions of the chaparral, where specimens are frequently secured not far from town." From Corpus Christi to Brownsville in 1900 wildcats were common along streams and in the chaparral, where their tracks were abundant in the dusty trails and on the muddy margins of streams and pools, but where the cover was generally too dense and thorny to admit dogs or to allow any method of hunting save by traps or poison. Lloyd states in his report that along the lower Rio Grande and in Cameron County "most of the ranchmen will not allow the wildcats to be killed for fear their ranches will be overrun with wood rats, mice, and rabbits." Not only in this region, but farther north and east this fear has been realized many times in swarms of wood rats, cotton rats, and rabbits, but the services of such predatory mammals as wildcats, foxes, and coyotes are not always recognized by the ranchmen. I have found this lynx common at Uvalde, Devils River, Kerr County, and farther east at Seguin, but in no other locality so abundant as in the Big Thicket of

[191] The bobcat across much of the state, excluding the Trans-Pecos and western Panhandle, is now regarded as *Lynx rufus rufus* (Kitchener et al., 2017). Previously, all bobcats in Texas belonged to the subspecies *Lynx rufus texensis* (Schmidly and Read, 1986). The bobcat is distributed statewide and prefers rocky habitats or thickets, depending on availability in the region. Unlike the other wild cats of Texas, the bobcat is highly adaptable and, in most areas, has coped well with the encroachment of humans.

[192] The bobcat of the Trans-Pecos and western Panhandle is now regarded as *Lynx rufus fasciatus* (Kitchener et al., 2017).

Liberty and Hardin counties. Here its tracks were seen in every muddy spot in roads and trails, and on damp mornings the dogs started one about as soon as they got into the thicket. The cat would rarely tree, but usually, rabbitlike, would run round and round in a limited circle in the thickest part of the swamp, depending on out-running or dodging the dogs. Cat hunting is a favorite sport in this region, and the hunters usually take stands in open spots and wait for the dogs to drive the game within shot. In one case I shot the cat in front of the hounds as it passed me for the third time. It did not seem tired or much alarmed, but easily kept out of sight of the dogs.

The stomach of this individual was full of venison that had not been perfectly fresh when eaten, probably from a deer that had been wounded by some hunters a week before. The hunter with me said he had examined the stomach of one not long before that was full of wood rats, and Gaut found wood rats in the stomach of one examined at Sour Lake. The food of this, like other species of lynx, consists mainly of rodents, rabbits, wood rats, ground squirrels, gophers, and mice, with a few birds, and occasionally some poultry. There are a few complaints of their killing sheep, young goats, and pigs.

Lynx baileyi[M] Merriam.   Plateau Wildcat.[M]

192 The lynx of the mountains and Staked Plains regions of western Texas, as shown by specimens from the Davis Mountains, and from near Alpine and Van Horn, and flat skins from Stanton and Odessa, is indistinguishable from *baileyi*,[M] which seems to occupy at least the Upper Sonoran zone of Texas. An immature specimen from Presidio County, a flat skin from the east base of the Chisos Mountains, and a flat skin labeled El Paso [?], are referred somewhat doubtfully to this species, but good material from the Rio Grande Valley may change this decision.

The country occupied by this plateau wildcat is mainly open, arid, and rocky. Canyons, gulches, and cliffs are its favorite haunts and hunting grounds, while caves and clefts in the rocks furnish dens and safe retreats from which hunting excursions are made into the valleys and even to the edge of the plains. Fresh tracks are frequently seen where the cats have followed the lines of the cliffs, crept along narrow shelves of rock from one wood rat's den to another, or walked noiselessly in the dust under and around the great bowlders and broken talus at the base of a cliff where the cottontails hide. Most of the wildcat's hunting is done at night, but occasionally one is surprised at midday crossing a valley to another cliff or found toward evening getting an early supper. One shot among the rocks near Alpine just before sundown had already caught and eaten a wood rat, which made a good beginning for a meal. On the head of Onion

Creek, Presidio County, in January, 1890, while watching the hawks come into the cottonwoods to roost one evening at sundown, I saw a pair of bright eyes among the branches overhead and slowly traced out the almost invisible form of a wildcat flattened along a rough gray branch. I needed the specimen, so did not wait to see if hawks were the object of his hunt, but an empty stomach showed that he had met with no success.

Here and there in some rocky corners of the cliffs one finds elongated pellets of bones and fur, some freshly deposited, others old and bleached, and these throw important light on the food habits of the animal. Bits of fur, teeth, and jaws serve to identify many of the mammals that have been eaten, and usually disclose a great preponderance of rabbits and wood rats. Traces of many smaller rodents and a few bird feathers and bones are found, but no remains of food other than animal. The ranchmen complain of some poultry's being killed, and, still worse, a few sheep. This, with a few quail and other birds, is about all that stands against the account of the wildcat, with a much larger amount on the credit side.

Wildcats are not readily trapped, as they rarely follow the same trail twice or touch any kind of bait. A few are shot, and the cowboys occasionally rope one in the open, but they are most successfully hunted with dogs at night or early in the morning. When started, they quickly take to a tree or to the rocks, and are shot or driven out of the tree, or sometimes smoked out of the rocks.

Canis griseus[M] Sabine.   Gray Wolf; Loafer; Lobo.

193 The big, light-gray wolf, 'loafer,' or 'lobo' is still common over most of the plains and mountain country of western Texas, mainly west of the one hundredth meridian. As its range seems to extend into Lower Sonoran zone no farther than a wolf would naturally wander in a few nights, the animal seems to be restricted approximately to the Upper Sonoran and Transition zones in the State. The only Texas specimens which I have for comparison are a skull from the top of the Guadalupe Mountains, just south of the New Mexico line, and one from Monahans, east of the Pecos Valley, both of which agree with skulls of the Colorado and Wyoming animals, the skins examined by Merritt Cary from Monahans and 50 miles north of Stanton, on the southern end of the Staked Plains, and by myself from the Pecos Valley, and a live animal seen at Portales, N. Mex., all agree essentially with the fine series of Colorado, Wyoming, and Montana skins in the Biological Survey collection. Moreover, descriptions by the ranchmen over this region apply in every instance to the large, light-gray wolf, while along the southern edge of the plains almost all of the ranchmen distinguish between the red wolf or big coyote of the rough country and the larger, lighter-

[193] At the time Bailey wrote the *Biological Survey*, the classification of dog-like carnivores (genus *Canis*) was almost totally based on a typological species concept. Virtually every local population, including color variants, was regarded as a distinct species. Today, with the application of the modern biological species concept, which uses the criterion of reproductive connectedness among populations, the number of recognized species has been reduced dramatically. Thus, many of the species names used by Bailey for canids are now placed in synonymy.

To our knowledge, *Canis griseus* has never been applied to the gray wolf. Perhaps Bailey mistakenly used *griseus* in combination with *Canis*. The name for the gray wolf is *Canis lupus* (see Wozencraft, 2005). Predator control and habitat destruction led to the extirpation of the gray wolf from Texas. The elimination of the wolf and other large predators over most of its former range released predator pressure on big game such as deer, which in part created a serious problem of overpopulation of deer in several localities in Texas. Under the Trump administration, the gray wolf was removed from the Endangered Species list. However, various environmental groups are suing the USFWS over this decision.

colored 'loafer' of the plains to the north. At Comstock, where special bounties are paid by sheep owners for the coyote and the common red wolf, the 'loafer' is unknown. A specimen killed 20 miles north of there on the higher plains in 1901 excited especial comment and raised the question whether or not the range of the gray wolf is being extended southward.

These wolves are most abundant in and about the Davis and Guadalupe mountains and over the Staked Plains and open country east of the Pecos River. Whether they are residents in the Pecos Valley or merely wanderers between the plains and the mountains is not easily determined, but I have no record of their breeding in the low part of the valley, while they are known to breed commonly in the high country on both sides. The present abundance of the species in any given place is not easily determined, as inferences are mainly drawn from the numbers killed, rather than the numbers left alive. Personally I have known of six or eight that were killed in 1901 and 1902 in the Davis Mountains, and a few in the Guadalupe Mountains and on the Staked Plains that were poisoned or dug out of their burrows. While my own observations have been limited, they aid in determining the accuracy of numerous other reports from resident hunters and ranchmen. These reports indicate that the wolves are not decreasing in numbers rapidly, if at all, in spite of those killed by ranchmen and by professional wolf hunters. On many of the large ranches a special bounty of $10, $20, or sometimes $50, is paid for every wolf killed. Several smaller ranches often combine to offer a large bounty in addition to that paid by the county, so that wolf hunting becomes a profitable business. In such cases there is a strong temptation for the hunters to save the breeding females and dig out the young each year for the bounty, thus making their business not only profitable but permanent. The hunters also bring wolves from a distance to the ranch paying the highest bounty. The bounty system offers dangerous temptations and has never proved effectual or even highly beneficial over any large area.[a]

To protect themselves from fraud and their stock from wolves many of the large ranch owners employ wolf hunters by the month and pay them well to keep the wolves and other noxious animals from their range. On the whole, when skilled hunters can be procured, this seems by far the most economical and satisfactory method.

When opportunity offers, the 'loafer' not only kills sheep but often kills a large number, apparently for the pleasure of killing. His regular and most serious depredations, however, are on the scattered

---

[a] Extermination of Noxious Animals by Bounties, T. S. Palmer, Yearbook U. S. Department of Agriculture, 1896, p. 55.

and unguarded cattle of the range. Two or three wolves usually hunt together and sometimes pull down a steer, but most of their meat is procured from yearlings or cows. Occasionally a colt is killed, but not often. Where two or three wolves take up their residence on a ranch and kill one or more head of cattle almost every day, the ranchmen become so seriously alarmed that they frequently offer a reward of $50 or $100 apiece for the scalps. In his report from Monahans, Merritt Cary writes:

I secured a skull of a very large female lobo wolf, which was killed on Hawkins's ranch in March, 1902, by Hugh Campbell. The skin when stretched on the side of the house is said to have measured 8 feet 4 inches from nose to end of tail, and was turned in to the Stockmen's Association, which paid Campbell $50 bounty on the animal. This female wolf was the mate to 'Big Foot,' a famous wolf throughout the region, whose track is always recognized by an extremely large right forefoot. On the second day of my stay at Hawkins's ranch Campbell and I got on the trail of 'Big Foot' and another wolf, which had crossed our own trail within two hours. Although on the trail for four hours we got no sight of them, nor did we find where they had killed any calves. There is a standing reward of $75 for 'Big Foot' by the Stockmen's Association; but although persistently hunted and trapped for a half dozen years, and thoroughly known to every cowboy in the region, the wily old wolf still retains his freedom, spurning poisoned baits, even disdaining to touch any meat not freshly killed by himself.

From Lipscomb, July, 1903, Howell reports: "Gray wolves occur in small numbers in this county, and a few cattle have recently been killed by them."

In disposition the 'loafer' is quite different from the coyote, lacking its cunning and assurance in the vicinity of man, and showing greater intelligence in the wild state and a better disposition when tamed. A half-grown 'loafer' that I found playing about the hotel at Portales, a little town on the edge of the Staked Plains, was like a big, good-natured puppy, full of fun and play, but soon became fighting angry if roughly handled. Although running at liberty over the town, he had never tried his puppy teeth on the chickens and pigs around him. He was the only survivor of a litter of seven, dug out of a burrow before their eyes were open. The others died, but 'Sampson' was nursed on a bottle for seventeen days—until his eyes opened. When I saw him in June he already gave promise of becoming a good-sized 'loafer.' He had a powerful voice and always responded to music with a doleful howl.

Canis (ater?)[M] Richardson. Black Wolf.[M]

194 The black wolf is reported from a few localities in the timbered region of eastern Texas, but in most cases as "common years ago, now very rare or quite extinct." The more numerous reports of a "large gray wolf" or "timber wolf" in the same region merely indicate

[194] *Canis ater* is now considered to have been a melanistic form of the gray wolf and consequently has been placed in synonymy with *Canis lupus*.

[195] The historic distribution of the red wolf, *Canis rufus*, as described by Bailey, was not completely accurate according to modern accounts. The Laredo (Webb County) and Matamoros, Mexico, specimens are erroneous because this species did not occur in southern Texas except along the coast as far south as Kenedy County, nor did it occur as far west as the Pecos River. Red wolves occurred primarily in the eastern half of Texas (Fig. 48), and their numbers and range quickly declined under pressure of intensive land use in the region (see Chapter 6 for a discussion of their status).

The taxonomic status of the red wolf has been disputed for some time, with certain experts calling it a species and others suggesting that it should be considered a subspecies of the gray wolf. Morphological and molecular genetic studies of captive animals and museum specimens raised the possibility that the red wolf group represents a hybrid between gray wolf subspecies and coyotes (O'Brien and Mayr, 1991). A 2016 genomic study (von Holdt et al., 2016) supported the hypothesis that the red wolf is a population of hybrids, formed primarily since the 1800s, from gray wolves and coyotes (von Holdt et al., 2016). However, Heppenheimer et al. (2018) suggested that the red wolf is a distinct species, but that hybridization with coyotes led to a mixing of genomes, such that some red wolf genes have been retained in modern coyotes. See Chapter 6 for a more detailed discussion.

[196] *Canis nebracensis* is now classified as *Canis latrans* (Young, 1951; Beckoff, 1977). All coyotes in Texas, once considered to be many different species, now belong to three subspecies: *Canis latrans latrans* in the Panhandle region, *Canis latrans frustrator* in the eastern third of state, and *Canis latrans texensis* in the western half. Coyotes occur statewide and are undoubtedly the most adaptable of all North American predators. Until the twentieth century, coyotes were rare in East Texas where red wolves were more common, but with the eradication of the latter species, coyotes expanded their range to include that part of the state.

174   NORTH AMERICAN FAUNA.   [No. 25.

variation in color, and show that only a minority of the individuals are entirely black. Presumably they all are of the same species. Apparently there is not extant a Texas skin or skull of this wolf to show whether or not it is the same species as the one in Florida, and it is greatly to be hoped that specimens will find their way to the National Museum before the species becomes entirely extinct.

Audubon, who had more experience with these wolves in their wild state and original abundance than any naturalist will ever have again, considered the black wolf of eastern Texas, Louisiana, southern Missouri, Kentucky, North Carolina, and Florida as one species, and carefully distinguished it from the "red wolf" of southern Texas and the white or gray wolf of the plains.[a]

Canis rufus[M]Aud. and Bach.   Texan Red Wolf.[M]

195   Since his work on the coyotes in 1897, Doctor Merriam has made special effort to procure specimens of the large coyote or small wolf of southern Texas. As a result there are at the present time fourteen skulls and four skins of this wolf in the Survey collection from Columbus, Corpus Christi, O'Connorport, Port Lavaca, Kerr County, Edwards County, and Laredo, in addition to two skulls in the National Museum, one from Fort Richardson, Jack County, Tex., and one from Matamoros, Mexico. Based on these specimens and the field reports of the Biological Survey a definite range can be assigned the species, covering the whole of southern Texas north to the mouth of the Pecos and the mouth of the Colorado, and still farther north along the strip of mesquite country east of the plains, approximately covering the semiarid part of the Lower Sonoran zone. As yet there are no specimens to show whether these wolves extend into the more arid region west of the Pecos. While apparently nowhere overlapping the range of the larger, lighter-colored 'lobo' or 'loafer' of the plains, they take its place to the south and east as soon as the plains break down and the scrub oak and mesquite country begins, but their whole range is shared with the coyote. The ranchmen invariably distinguish between them and coyotes, and with good reason, for the wolves kill young cattle, goats, and colts with as much regularity as the coyotes kill sheep. While paying a bounty of $1 or $2 for coyotes, the ranchmen usually pay $10 or $20 for red wolves.

Canis nebracensis[M]Merriam.   Plains Coyote.[M]

196   Five coyote skulls from Canadian and three from Sherwood, Tex., and three from Clayton and two from 30 miles southeast of Carlsbad, N. Mex., agree with typical *nebracensis*[M] skulls from Johnstown, Nebr.; while a flat skin from Monahans is as pale as the type of

[a] Aud. and Bach., Quad. N. Am., II, pp. 130-131 and 243, 1851.

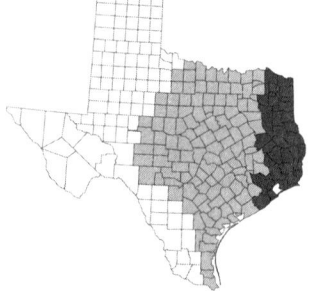

Figure 48. The historic distribution of the red wolf (*Canis rufus*) in Texas. The light shading represents *C. r. rufus* and the medium shading represents *C. r. gregoryi.*

nebracensis.[M] This gives to the species a perfectly logical range over the Panhandle and Llano Estacado, or the open Upper Sonoran plains of Texas, but specimens from many more localities are needed before its full range can be accurately outlined. At Lipscomb, in the northeast corner of the Panhandle, Howell reported coyotes July 10, 1903, as "common at some seasons." At Canadian, where five old skulls were secured at Simpson's ranch on Clear Creek, he reported them as "killed here in winter in some numbers;" and at Texline he stated that they were "fairly common in this region," and added that "two were seen during my stay (August 1-8), and another was killed at Buffalo Springs."

In crossing the summit of the Staked Plains I have often seen the coyotes, both from the train and from our camp wagon, and night after night from our camp fires have heard their long quavering howls. But when seen they were always just out of rifle range. They were not afraid, and in this open, level country have little reason for fear.

Canis nebracensis texensis[a][M] subsp. nov.　Texas Coyote.[M]

> Type from 45 miles southwest of Corpus Christi, Tex., ♂ young adult, No. 116277, U. S. Nat. Mus., Biological Survey Coll. Collected by J. M. Priour, Dec. 14, 1901. Original No. 3478, X catalogue.

General characters.—Similar to *C. nebracensis*,[M] but darker and brighter colored and with lighter dentition. Smaller, brighter, and more fulvous than *latrans;* almost as richly colored as *ochropus*,[M] but without the large ears of that species. Not in the same group as *microdon*,[M] *mearnsi*,[M] and *estor*.[M]

Color.—Fresh winter pelage buffy gray, heavily clouded with black, becoming clear, bright, fulvous on legs, ears, and nose, and whitish on throat and belly; a strong line of black down front of foreleg. Summer pelage duller and darker.

Skull.—Slightly slenderer than in *nebracensis*,[M] with conspicuously lighter dentition, narrower molars and carnassials.

Measurements of type.—Total length, 1,143; tail vertebrae, 355 (measured by collector); hind foot, 180 (measured from dry skin). Skull of type: Basal length, 169; greatest length of nasals, 67; zygomatic breadth, 94; mastoid breadth, 61; interorbital breadth, 30; length of crown of upper carnassial tooth, 19.8.

The Texas coyote is more or less common over at least middle and

---

[a] In his Revision of the Coyotes, published in the Proc. Biol. Soc. Wash., XI, 26, 1897, Doctor Merriam referred this coyote provisionally to *frustror*,[M] of which the half-grown type was then the only available specimen. A series of topotypes of *frustror* [M] secured since at Red Fork, Ind. T., shows it to be a widely different species, more nearly related to *Canis rufus*. The coyote of southern Texas is thus left without a name, and its nearest relative proves to be the pale *nebracensis* [M] of the more northern plains.

[197] Current taxonomy classifies *Canis nebracensis texensis* as *Canis latrans texensis*.

southern Texas and apparently eastward on strips of prairie as far as Gainesville and Richmond. There are vague reports of a small wolf occurring farther east on the coast prairie even to the border of Louisiana, but specimens are needed before these reports can be associated with definite species. East of the semiarid mesquite region coyotes are rare and probably mere stragglers. True to their name of prairie wolf, they do not enter the timbered country to any extent, although at home in the scrub oak, juniper, mesquite, and chaparral, as well as over the open prairies of the southern part of the State. In the extreme southern part of the State their range is slightly overlapped by that of the little *microdon*,[M] and in the extreme western part by that of *mearnsi*,[M] while specimens from the northern Pan Handle country and Staked Plains are nearer to *nebracensis*.[M]

In spite of the enmity of man, in spite of traps, poison, gun, and dogs, the coyote over most of his old range fairly holds his own. Combining with the cunning and suspicion of the fox a speed and endurance that almost insures his safety from ordinary hounds, he has little to fear except an occasional long-range shot or the traps and poison of the professional coyote hunter.

On many of the large ranches men are employed by the month to kill the coyotes, lobos, and panthers, and some of these men have attained such skill as to be able almost to extirpate the coyotes over a considerable area. But the coyotes are wanderers, and while they soon gather where food is abundant and easily procured, they quickly leave an inhospitable region for better hunting grounds. Civilization has little terror for them. I have heard them howling near many of the little towns and ranches, where they were attracted by the smell of freshly killed beef or by carcasses that were far from fresh, and near a ranch corral have found many dead coyotes poisoned at the carcass of a cow. After dark they show little fear of the ranch dogs, and sometimes seem even to invite a chase. In fact they not infrequently cross with the ranch dogs and produce hybrids with erect ears and wolfish appearance. I have seen several of these hybrids with characters that substantiated the statement that they were half coyote. At San Pedro Park, San Antonio, I was shown a 6-months-old cross between a coyote and shepherd dog, bred and born in the zoo. Except for being nearly black it had the general appearance of a coyote. It was kept chained in the open and was on friendly terms with the keeper.

About our camps the coyotes on rare occasions are surprisingly familiar, coming close to the camp wagon, especially if there is fresh meat in it, though usually paying their visits after dark. Sometimes the first man up in the morning gets a glimpse of one sneaking away or on rare occasions gets a good shot within easy range. Except during the breeding season, when they are very quiet, their frequent serenades are our regular camp music.

Within certain limits the credit and debit sheets of the coyote are well balanced. On the one hand, he kills many sheep and a few goats, some poultry, and considerable game. On the other hand, the bulk of his food the year round consists of rabbits, prairie dogs, ground squirrels, gophers, wood rats, mice, and all the small rodents that come in his way. An unusual increase of jack rabbits in any region is always followed by a corresponding influx of coyotes, which probably accounts in part for the often observed fact that in the years following their maximum abundance jack rabbits are unusually scarce.

At times the food of the coyote consists largely of fruit, including that of several species of cactus, juniper and forestiera berries, persimmons, and the sugary pods of mesquite; but in times of scarcity a piece of rawhide garnished with a few horned toads, lizards, and some horse manure suffices for a meal.

Canis mearnsi[M] Merriam.   Mearns Coyote.[M]

198 Four good specimens, skins and skulls, of coyotes collected near El Paso late in February, 1903, by James H. Gaut are mearnsi[M] in slightly worn and faded pelage. One skull from near the Texas and New Mexico line, in Salt Valley, at the west base of the Guadalupe Mountains, a good skin and skull from the same valley a little farther north, several specimens from the edge of Tularosa Valley, three old skulls from Sanderson, and two from Samuels, near the mouth of the Pecos, a skull from Grand Falls, in the Pecos Valley, and one from 30 miles southeast of Carlsbad, N. Mex., all belong to this slender, bright-colored desert form of the small-toothed coyotes. From the locality 30 miles southeast of Carlsbad, with the skull of an old male that is unmistakably mearnsi,[M] were collected two skulls of nebracensis,[M] while from Sanderson and Salt Valley there are skulls that I can refer only to texensis.[M]

There is not yet material enough to show whether mearnsi[M] grades into microdon[M] farther south or into estor[M] farther north, but it evidently overlaps the range of both nebracensis[M] and texensis.[M] Nor are there any specimens from the Davis Mountain plateau to show what form or forms occur there. Canis mearnsi,[M] so far as known, is Lower Sonoran in range.

Coyotes are common throughout the extremely arid valleys of western Texas, including the Pecos and Rio Grande valleys south to their junction. Distance from water seems to have no effect on their

[198] Canis mearnsi is now classified as Canis latrans mearnsi (Young, 1951; Beckoff, 1977) and it refers to a subspecies that ranges from New Mexico westward. The specimens from Texas are today classified as Canis latrans texensis.

[199] *Canis microdon* has been changed to *Canis latrans microdon*. This subspecies occurs in Mexico, just south of the Texas border.

[200] The red fox of Texas is today recognized as a single subspecies, *Vulpes vulpes fulva*. This introduced canid has done remarkably well in Texas, having expanded to cover most of the state except for the extreme western and southern regions.

abundance, although in this region they can hardly find a spot more than an easy night's journey, 20 or 30 miles, from open water. We find their tracks along every road and trail, and often see one of the animals loping across the valley or watching us from a ridge, and frequently hear them from our evening camp fires. At El Paso in 1889 I jumped one from under a creosote bush, where it was sleeping at midday, within rifle shot of the town, and at another time saw four together on the mesa half a mile out from the railroad station. At Fort Hancock Gordon Donald reported them in 1902 as very abundant, and said: "I heard them calling in the evenings, and the Mexicans had several young ones that they had caught in the vicinity. A ranchman told me that in the low foothills where his ranch was situated he saw two or three coyotes every day."

Canis microdon[M]Merriam.   Small-toothed Coyote.[M]

199 This little dark-colored coyote of the lower Rio Grande Valley overlaps the range of *texensis*[M]in southern Texas. Specimens from Brownsville, Roma, and Alice show all that we know of its range in the State. These localities indicate that it is a chaparral rather than a prairie species, but there is nothing to prove that its habits are different from those of *texensis.*[M]

Vulpes fulvus[M](Desm.).   Red Fox.

200 Apparently the red foxes are not natives of Texas,[a] but since their introduction they are becoming locally common, especially over the eastern half of the State. Oberholser obtained reports of their occurrence at Texarkana, Jasper, and Austin; Hollister, at Antioch, Rockland, and Sour Lake; and Cary, from Kerr County and along Howard Creek and the Pecos River. The following extract from a letter from Mr. T. H. Brown to Mr. H. P. Attwater is, as Doctor Allen says, a document of historic interest:[b]

I was the first to introduce 'red foxes' into this part of the State. We had exchanged our old-time native hounds, or, as they are usually called, 'pot lickers,' for the Walker dogs from Kentucky, and the gray foxes proved themselves no match for these dogs, only being able to run from twenty to forty-five minutes ahead of them. Having the dogs, it became necessary to get game that would give them a respectable race. Accordingly, in 1891, I imported from Kentucky and Tennessee 10 red foxes and placed them among the Bosque brakes, about 4 miles above where it empties into the Brazos River. They gradually scattered over a large area of country. The next spring (1892) I again brought in 23 more reds from the older States, planting 13 of them again among the Bosque brakes and 10 of them on White Rock Creek, on the east side of the Brazos River. These foxes afforded us some fine sport; but they, too,

[a] See Aud. & Bach., Quad. N. Am., II, 271, 1851.
[b] Extract of letter from T. H. Brown, Waco, Tex., in Bul. Am. Mus. Nat. Hist., VIII, p. 77, 1896.

gradually scattered, only a few remaining in the neighborhood of their adopted home, some wandering off through Bosque and Erath counties. The next spring I only succeeded in getting 2 reds from the East and planted these on the Bosque, and they remained and are still affording fine races. In the spring of 1895 I again planted 5 reds on the river near Lovers Leap, where the waters of all the Bosques mingle with the waters of the Brazos. Some of the bluffs here are 300 feet high and have a great many caves in them, and these last foxes seem well satisfied with their new home. Occasionally I hear of a red fox in various parts of this (McLennan) county, and I am satisfied that within a few years they will be as numerous here as in the old States.

I understand that Messrs. Eli and James Rosborough and Capt. T. H. Craig, all of Marshall, Harrison County, some ten or fifteen years since planted quite a number of reds in that, the eastern, part of the State, and occasionally they find them where they have located off some 20 or 30 miles from where originally turned loose.

Dr. John D. Rogers has, I think, during the spring of 1895, planted some 6 or 8 on his Brazos bottom farms in Brazos and Washington counties. I should suppose that in all there have been at least 100 red foxes imported and planted in the State.

Vulpes velox (Say).   Swift; Kit Fox.[M]

201 So far as known the swift in Texas ranges only over the Upper Sonoran Staked Plains. It is reported at Tascosa and Washburn on the northern end of the plains and near Stanton and Midland at the southern end. In 1902 Cary secured five flat skins at Stanton, but says the ranchmen reported the swifts as scarce there in comparison with their numbers in former years. Most of these skins were secured by poison put out in winter, when the swifts were said to come to the poisoned bait generally the first night after it was put out, while the coyotes usually waited until later.

Vulpes macrotis neomexicanus[M] Merriam.   New Mexico Desert Fox.[M]

202 The little desert fox has been taken in the Rio Grande, Tularosa, and Pecos valleys just north of the Texas line, and one specimen was taken by James H. Gaut in Texas 10 miles north of El Paso. It is reported from as far south as the mouth of the Pecos. A flat skin brought in to the store at Sierra Blanca in December, 1889, had the characteristic large ear of the group, the ear measuring 78 mm. from crown. Apparently the range of the species corresponds in this region to that of *Dipodomys spectabilis* in the open desert valleys of the Lower Sonoran zone. It is by no means common in the region, and many of the ranchmen have never seen it, or else have never distinguished it from the common and much larger and darker-colored gray fox of the genus *Urocyon.*

[201] For most of the last century, arid-land foxes were regarded as comprising two similar but separate species, the swift fox (*Vulpes velox*) and the kit fox (*Vulpes macrotis*). This was the arrangement used by Bailey. A study by Dragoo et al. (1990) concluded that these taxa were not sufficiently distinct to warrant separate species status; however, a few years later Mercure et al. (1993) concluded that sufficient genetic differences existed to warrant the recognition of two species. Consequently, these foxes are now treated as separate species. The Swift Fox, *V. velox*, occurs in the Panhandle and adjacent areas of Texas; the Kit Fox, *V. macrotis*, occurs in the Trans-Pecos (Wozencraft, 2005; Fig. 49). The study by Mercure et al. (1993) determined little genetic variation within each of the two species and suggested the abandonment of subspecific names. These foxes are notoriously susceptible to trapping and poisoning, which has greatly reduced or entirely eliminated them in areas where predator control campaigns have been carried out.

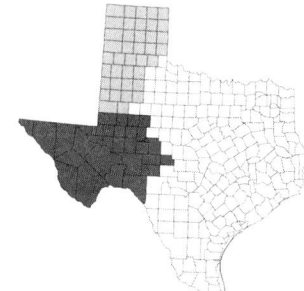

Figure 49. The current distributions of the swift fox (*Vulpes macrotis,* light shaded area) and the kit fox (*Vulpes velox,* dark shaded area) in Texas.

[202] *Vulpes macrotis neomexicanus* is no longer a valid name; the Kit Fox, *Vulpes macrotis,* is monotypic.

[203] The common gray fox, *Urocyon cinereoargenteus* , is widely distributed throughout Texas except in the northern Panhandle region. Two subspecies are recognized, *Urocyon cinereoargenteus scottii* in the western two-thirds of the state and *Urocyon cinereoargenteus floridanus* in the eastern one-third.

180   NORTH AMERICAN FAUNA.   [No. 25.

Urocyon cinereoargenteus scotti [M]**Mearns.   Gray Fox.**[M]

> *Urocyon cinereoargenteus texensis* Mearns,[a] Proc. U. S. Nat. Mus., XX, Advance Sheet, January 12, 1897, p. 2.

203 The gray fox is common over all the western half of Texas, except on the open plains. It is mainly an inhabitant of the timbered or brushy country, living in hollow trees or logs, but preferably in dens among the rocks. It lacks the cunning and swiftness of the red fox, is easily caught in traps, and quickly overtaken by the hounds, except where it can keep in dense cover. Often after a short run, and sometimes at the very start, it trees or takes to its rock den, where it is safe from the dogs; but if no such protection offers there

FIG. 21.—Gray fox (*Urocyon c. scotti*)[M] in trap, Langtry, Texas.
(Photographed by Oberholser.)

is little hope for the fox. Even over rocks and in the brush I have seen the hounds catch one in a 200-yard dash. With a good start,

---

[a] The original label on the type of *Urocyon c. texensis*[M] reads: "Rio Bravo and San Pedro. 1851. A. Schott." As is well known, Rio Bravo is synonymous with Rio Grande, and at that time the Devils River was commonly known as the San Pedro. (See Baird's Mammals of N. Am., p. 713, and Pacific R. R. Rept., Vol. I, p. 110. Also see query after Eagle Pass in Mammals, Mex. Boundary Survey, Vol. II, pt. 2, p. 17.) This seems to necessitate changing the type locality of *texensis*[M] from 'near Eagle Pass' to the junction of the Devils River with the Rio Grande, which, however, has no important bearing on the validity of the species. In comparing the type of *texensis*[M] and other specimens from near the mouth of Devils River, Painted Caves, Langtry, San Diego, and the Davis Mountains, in western Texas, with the type of *scotti*[M] and with specimens from all around the type locality—near Tucson, Fort Huachuca, Fort Bowie, Chiricahua Mountains, and Fort Verde—I am unable to find any constant difference, either cranial or external, on which to recognize *texensis*[M].

however, one will lead the hounds a long chase over the roughest ground it can find, and if it does not make the mistake of climbing a tree, instead of taking to the rocks, it is pretty safe. Strange as it may seem, these foxes go up the trunk of a tree with almost cat-like ease. I have found them looking down at the dogs from 20 to 40 feet up in the branches of nut pines and live oaks, and have known of their climbing a yellow pine (*Pinus ponderosa*) where 20 feet of straight trunk over a foot in diameter intervened between the ground and the first branch. More often they take to a live oak or juniper, where the lower branches can be reached at a bound, and then, squirrel-like, hide in the swaying topmost branches. On the approach of the hunter they become anxious and seem to doubt the security of their position, sometimes making a flying leap to the ground. Stones and clubs will usually dislodge them from the tree top, but as they still have a good chance to escape the dogs and take to the rocks, it is a common and heartless practice to shoot them so as to break a leg and make escape impossible.

With his smaller but laterally flattened tail the gray fox certainly equals, if he does not surpass, the red fox in quickness of motion and skill at dodging the dogs. If uninjured, he will often strike the ground in the midst of the hounds and escape by a few quick bounds to right and left. Apparently it is only his small size that puts him at a disadvantage in a test of speed with the hounds or with his larger cousin, the red fox.

In choice of food the gray foxes are almost as omnivorous as the coon. Various fruits form the bulk of their food in summer and part of it in winter, while a great variety of small game, beetles, grasshoppers, maggots, mammals, birds, and some poultry fall a prey to them during the year. In June they were feeding extensively on berries of *Zizyphus obtusifolia* and *Adelia angustifolia* along the Rio Grande near Boquillas, while around the Davis Mountains in early August they were feeding mainly on the ripe fruit of *Opuntia engelmanni*.[P] In December in the Davis Mountains and in September in the Guadalupe Mountains they were eating the sweet pulpy berries of *Juniperus pachyphloea*,[P] which grow in great abundance in these ranges and in the Chisos Mountains. Mice, wood rats, ground squirrels, rabbits, and various other small rodents are eaten when obtainable, and, much to our annoyance, are often taken from our traps or carried away, trap and all. At Langtry, Gaut examined several stomachs, and in one found part of a mocking bird and in another a *Perognathus*. At most of the ranches there are enough dogs to keep the foxes at a respectful distance from the poultry; but they have a keen relish for chickens, and are often complained of in vigorous terms. Without data for positive statements it seems probable that

[204] The subspecies *Urocyon cine-reoargenteus floridanus* occurs in the eastern one-third of the state, where the species is more common than in the remainder of its range in Texas.

[205] The distribution of the ringtail, *Bassariscus astutus flavus,* is nearly statewide, although it apparently is uncommon in the lower Rio Grande and Coastal Plains of southern Texas. As with many of Texas' fur-bearing species, the ringtail histori-cally was an economically important species on the Edwards Plateau.

182   NORTH AMERICAN FAUNA.   [No. 25.

the good done in destroying small rodents equals, if it does not exceed, the mischief done among poultry.

As a game animal this fox is holding its ground better than many more important species, and even from the sportsman's point of view needs little protection. The skin is of so little value for fur that it is rarely saved when the fox is killed.

Urocyon cinereoargenteus floridanus[M]Rhoads.   Florida Gray Fox.[M]

204 A nearly adult male gray fox from the Big Thicket, near Sour Lake, Tex., agrees with the Florida specimens in dark color, dusky legs, feet, and face, and in most of the cranial characters. The shorter, heavier muzzle is evidently due to slight immaturity. A flat skin from Tarkington Prairie is less dusky, and while probably shading toward *scotti*,[M]more nearly resembles *ocythous*.[M]A skull in the National Museum from Washington County, a little farther west, also shows some of the characters of *ocythous*,[M] but is not typical of any form. While I have no hesitation in referring the Big Thicket *Urocyon* to *floridanus*,[M]it is probable that this is not the only form inhabiting the eastern part of the State. Before final conclusions can be reached more specimens are needed, especially from farther north.

To show how generally the gray fox is distributed over eastern Texas the following localities are given from which it is reported as more or less common: Henrietta, Gainesville, Arthur, Texarkana, Waskom, Rockland, Jasper, Sour Lake, Tarkington Prairie, Ever-green, Hempstead, Matagorda, Washington, Antioch, and Long Lake. Those from Rockport and Brazos are likely to be nearer to *scotti*.[M] My information in regard to the habits of the animal in this region has been received mainly from residents, who say that the foxes keep in the brush and timber, especially along the river bottoms, where the thickest growth is found. They are said to climb trees, and com-plaints of their killing poultry are more frequent than in the more open country farther west.

Near Sour Lake Gaut reports them as found mainly in the pine woods at the edge of the thicket, but as occasionally straying down into the densest part of the thicket, where he caught one on Black Creek, near Mike Griffin's place. The stomach of this individual contained a mass of crayfish.

Bassariscus astutus flavus Rhoads.   Civet Cat; Cacomistle.[M]

205 The civet cat is common all over Texas except the open plains country of the western half from Brownsville, Corpus Christi, Seguin, Austin, Brownwood, and Grady westward. It has been reported east to Matagorda County, near the coast, and a specimen in the U. S. National Museum is labeled "Red River." One from Grady, Fisher

County, seems to be the northernmost authentic record for the State, but the species undoubtedly continues along the canyons and cliffs of the eastern edge of the Staked Plains to the Red and Canadian rivers.[a]

Although preeminently inhabitants of rocks, cliffs, and canyon walls, civet cats are common over the chaparral, mesquite, and cactus plains of southern Texas down to the very coast, a peculiarity of distribution shared by a number of other mammals which find in the thorny cover of dense patches of cactus and tangled thickets of chaparral ample protection and a greater abundance of small game than in the rocky haunts of the higher country. In habits they are catlike, mainly nocturnal and carnivorous. At night they prowl along the ledges of cliffs from cave to cave, leaving the prints of their little, round, catlike feet in the dry dust of the darkest corners, and helping themselves to a liberal share of the *Peromyscus* and *Neotoma* found in the traps of careless collectors. Usually, however, the small rodents are extremely scarce where the civet cats are at all common, and the wise collector scatters his small traps out over the valley until his steel traps have cleaned the cliffs of carnivorous species.

Owing to their nocturnal habits and the fastness of their rock dens, the civet cats are rarely seen in the wild state, but when tamed the ranchmen say they make affectionate pets and are better mousers than the domestic cat. A pair was caught in traps in one of the canyons of the Rio Grande and the male fought and screamed viciously as we approached, but the female was quiet and gentle. Even in the traps the animation and brightness of their faces were wonderful. The large ears, when directed forward, were in constant motion. The long, black, vibrating moustache, the striking black and light face markings, and, most of all, the big, soft, expressive eyes give a facial expression of unusual beauty and intelligence. L. A. Fuertes, who was with me when these two were caught, made a careful color study of the head of the male, which loses but little of its excellence in the black and white reproduction.

An old female caught near Boquillas May 27 contained three nearly

---

[a] The range of *Bassariscus* has been supposed to extend eastward to Arkansas (see Baird, Mammals of North America, p. 147), and a skin in the U. S. National Museum is labeled "Red River, Ark." On the remaining fragment of the original label of this specimen is only "Red River, Capt. Marcy." There is no date on the label, but the skin was entered in the Museum catalogue March 31, 1853. In 1852 Captain Marcy explored the headwaters of the Red River and in his report records *Bassariscus* from the "Cross Timbers," probably this same specimen. (See Exploration of the Red River of Louisiana, p. 186, 1854, by Capt. Randolph B. Marcy. Also, for route of Captain Marcy, see map opposite p. 36 of Annual Report of Wheeler Survey, Rept. Chief of Engineers for 1876, App. JJ.)

[206] *Taxidea taxus berlandieri*, the American badger, is distributed throughout much of the state. Bailey commented on its absence in eastern Texas, but recent records suggest it has been extending its range eastward as a result of land-clearing operations. Although the range and numbers of prairie dogs have been greatly reduced since the time of the Biological Survey, badgers apparently tolerated this reduced food source well by relying more on ground squirrels and other rodents. The fact that their range is expanding supports the belief that badgers are quite adaptable, and in some areas their densities appear to be increasing.

184   NORTH AMERICAN FAUNA.   [No. 25.

developed fetuses, which, with a litter of four young recorded by Mr. Clark from San Pedro River, would indicate small families.[a]

Most of the stomachs of *Bassariscus* examined have been found to contain the bones and hair of small rodents, which make up also most of the excrement found along ledges and in caves where the animals live. Fragments of a large centipede were found in the stomach of one caught by Gordon Donald on Devils River; and in other localities they have been reported as eating fruit. At Langtry, Gaut caught several in traps baited with meat.

Taxidea taxus berlandieri Baird.   Mexican Badger.[M]

206   The badger is generally distributed over the western half of Texas, but apparently is unknown in the eastern part of the State. Its eastern limit corresponds, in a general way, with the eastern edge of the mesquite country. Specimens have been taken as far east as Corpus Christi, San Antonio, and Mason, and there are records from Clyde, Henrietta, and Mobeetie. A significant fact is that the badger's eastern limit of range agrees closely with the eastern limits of the prairie dog and the Mexican ground squirrel. Its abundance depends mainly on food supply, reaching a maximum on the open plains in the prairie-dog country and decreasing slightly in the southern part of the State and in the mountains and rocky country of the extreme western part. But in speaking of badgers, abundance may mean one to a square mile, while with prairie dogs it may mean 10,000 to a square mile.

When food is scarce the badgers become great wanderers. Their short legs are fully compensated by their unusual strength and by their capacity for digging and fighting, that enable them to escape from most enemies. But with such abundance of food as is found in a populous prairie dog town, they waste little time in travel. They become fat and lazy; but as food grows scarce they start off again on their travels, sinking a house in the earth wherever sleeping time overtakes them.

The badger feeds mainly on small rodents, varied with grasshoppers, beetles, scorpions, lizards, or some larger animal found dead. It is accused of killing poultry, but the accusation is so rarely substantiated that it may well be ignored. Pocket gophers, kangaroo rats, wood rats, and various kinds of mice are always acceptable, but the badger lives mainly on prairie dogs and ground squirrels, which fall an easy prey. He often digs a dozen holes along the interminable tunnel of a pocket gopher and then gives up in disgust, but a fat spermophile or prairie dog at the bottom of its simple burrow is entirely at his disposal, nor does he have much trouble in digging it

[a] Baird, Mammals of North America, p. 147, 1857.

out. A few minutes' work with his powerful claws will unearth the spermophile, while by merely enlarging the prairie-dog hole about two diameters he enters its deepest chambers and is sure of a good square meal at the end. On a ranch in the Pecos Valley I found a badger living in an alfalfa field that had been overrun with prairie dogs. Every morning there was at least one new hole that he had enlarged, and while he may have secured two or three prairie dogs in some of the burrows he was evidently destroying at least one a day. This badger was needed for a specimen, and at the earnest solicitation of the ranch people, who were afraid he would kill their

FIG. 22.—Prairie-dog burrow enlarged by badger, Pecos valley.

poultry, I finally shot him as he came out about 4 o'clock one afternoon to get his supper. He had begun on a Swainson hawk that had been shot the day before. Otherwise his stomach was empty, but the lower part of his alimentary tract was full of wads of prairie-dog fur from his meal of the previous night. He was fat and had evidently been working all summer in that 20-acre field. The people had no reason to believe that he had ever killed any of their poultry, but they were afraid that he would. There were already two badger skins hanging in the tool house on this ranch, while a 20-acre field of alfalfa was rendered almost worthless by prairie dogs. When I tried to convince the owners that every badger on the ranch was

[207] The black bear, *Ursus americanus*, once ranged across most of Texas except for the southernmost counties. By the early 1900s, it was virtually extirpated from the state except for the mountainous areas in the Trans-Pecos region and a few individuals in the Big Thicket region. In recent years, populations have slowly increased in number and black bears once again are being documented throughout the Trans-Pecos, Big Thicket, in northeastern Texas, and occasionally in the Edwards Plateau and Panhandle regions. On the basis of geographic probability, Bailey referred central Texas bears to *Ursus americanus americanus*, Trans-Pecos bears to *Ursus americanus amblyceps*, and east Texas bears to a separate species, *Ursus luteolus* (later relegated to a subspecies of *Ursus americanus*). Current taxonomy recognizes four subspecies in Texas: *U. a. americanus* in northeastern Texas along the Red River counties, *Ursus americanus luteolus* in southeastern Texas along the Louisiana border, *U. a. amblyceps* in the Trans-Pecos and Panhandle regions, and *Ursus americanus eremicus* in the western portions of the Hill Country (Hall, 1981).

worth $100 to them they only laughed. Some of the ranchmen, however, appreciate the services of the animal, but even then the temptation to try a shot at one at long range or to let the dogs catch one for a fight is often too great to be resisted. Dead badgers are frequently seen by the roadside with smashed skulls or bullet holes through them, and this most often in the heart of the prairie-dog country. When taken to task for their folly in destroying these valuable animals the ranchmen have usually stoutly denied the charge, saying that most of them were killed by emigrants and other 'tenderfeet.'

The cowboys, however, have a real grievance against the badgers, especially those who have been thrown from running horses that had inadvertently stepped in old and half-concealed holes. Such accidents are by no means rare, and sometimes they are fatal to both horse and rider. It is hardly surprising, therefore, that the cowboys look upon the badger as a legitimate target for their six-shooters. In a prairie-dog country, however, this is not a fair excuse, for prairie-dog holes are just as dangerous, and each badger helps to reduce the total number of pitfalls.

The rapid increase in the abundance of prairie dogs in certain parts of the State and their constant extension of range is unquestionably due in great measure, if not mainly, to the destruction of badgers. It seems unaccountable that the intelligent observations of ranch people should not result in a strong sentiment in favor of protecting badgers, but it must be remembered that without the support of protective laws nothing can be done to prevent the destruction of the animals by uninterested and irresponsible people.

Ursus americanus Pallas.   Black Bear; Cinnamon Bear.[a]

207 Specimens of the black bears collected in the Wichita Mountains, Oklahoma, prove to be *americanus,* and the bears reported from Mobeetie and near Washburn were undoubtedly the same. Others reported farther south from west of Austin and even to Kerrville may have been the same, also the bears from the Guadalupe Mountains, but as no specimens from these Texas localities have been seen the species can be admitted to the State list only provisionally.

At Washburn in 1892 I was told that there were a few black bears south of there in the canyon of the Prairie Dog Fork, and at Mobeetie in 1901 Oberholser reported them as "formerly common, now extinct." In 1902 Oberholser obtained a rather indefinite report at Austin that "a few bears were still to be found in the rough country west of there," and the same year at Kerrville I was told that bears

---

[a] As is well known, the black and cinnamon bears are merely dichromatic forms, or color phases of the same species, one cub of a litter often being black and another brown.

were becoming very scarce, but that one had been killed the previous year only 7 miles from there.

Ursus americanus amblyceps Baird.   New Mexico Black Bear.[M]

208   Black bears are still found in the timbered mountains of western Texas, where in a few restricted areas they are fairly common. A few specimens examined from the Davis and Chisos mountains can best be referred to *amblyceps,* but there are no specimens from the Guadalupe Mountains or from middle Texas to show where this form gives place to *americanus* on the north or to *luteolus*[M] of the eastern part of the State. The records from Kerrville, west of Austin, Prairie Dog Fork, (near Washburn), and Mobeetie I am inclined to refer provisionally to *americanus.*

In July of 1902 a young black bear was caught by the section men on the railroad near Comstock, and a few were reported from the Pecos Canyon and vicinity. Bears were formerly abundant in this region, but apparently no specimen has been preserved to show what form ranged in the Pecos, Devils River, and Rio Grande country. On Onion Creek, 30 miles south of Marfa, in January, 1890, I picked up a skull from one of three bears killed near there in 1887. In June of 1901 black bears were common in the upper canyons of the Chisos Mountains, where fresh tracks of old and young were frequently seen and where there was an abundance of old 'sign' and turned-over stones. The old excrement was made up largely of acorns, juniper berries, and pine nuts, while the seeds of cactus fruit were noticed in the fresher deposits.

In the Davis Mountains black bears hold their own surprisingly well against unusual odds. In July, 1901, I found abundant 'sign,' fresh tracks, and turned-over stones along the crest of the higher ridges on the east slope of Mount Livermore, and again in August, 1902, found 'sign' equally abundant in the canyons on the west slopes. In following up a deep canyon west of the main peak on August 13 after a heavy rain of the previous day, I saw fresh tracks of bears of at least three different sizes—cubs, yearlings, and adults—and found numerous little diggings in the black mellow soil where roots or beetles had been unearthed, and many stones freshly turned over for the ants and beetles beneath them. In a side gulch a large buckthorn bush (*Rhamnus purshiana*) had been freshly torn up and half stripped of its ripening berries, while close by was a lot of fresh bear 'sign,' made up entirely of the skins and seeds of these berries. In other places on the east slope I found fresh 'sign,' composed mainly of the sugary berries of the checker-barked juniper (*J. pachyphloea*),[P] and some that was older, largely composed of acorn shells.

In the southern part of the Guadalupe Mountains, on the upper slopes of almost inaccessible canyons, black and brown bears were

[208] Black bears (*Ursus americanus amblyceps*) are being sighted more frequently in the Trans-Pecos region of Texas, especially in the Chisos Mountains and other portions of Big Bend National Park, and recent sightings suggest that there is a resident, breeding population in the park. Black bears also have been increasing in number in the Davis Mountains around Alpine, and even in the Panhandle. Apparently, black bear populations have rebounded in northern Mexico and northeastern New Mexico and are beginning to cross over the border and disperse into Texas (see Chapter 6).

[209] The black bear of eastern Texas is considered a subspecies, *Ursus americanus luteolus,* of the more widespread continental species, *Ursus americanus*. Black bears are increasing in number in the wooded areas of eastern Texas, with resident populations being established (see Chapter 6).

common in 1901. In the head of McKittrick Canyon they had well-worn trails leading to and from their feeding grounds on the oak and juniper ridges and down the canyon to the upper water holes. In places along the sides of narrow, bowlder-strewn gulches the trails were series of big, deep tracks, where for ages each bear had stepped in the footprints of his predecessor. On the open slopes the trails spread out and were lost. On some of these slopes almost every loose stone had been turned over by bears in their search for insects, but at the time of my visit, in August, they were feeding mainly on the sweet acorns of several species of shin oak, berries of the checker-barked juniper, and, to a less extent, berries of *Berberis fremonti*. Some of the previous year's excrement contained shells of pine nuts (*Pinus edulis*), but this was the off year, when the nut pines did not bear.

Near one of the trails in the head of Dog Canyon in the Gaudalupe Mountains a Douglas spruce a foot in diameter had served for many years as a gnawing tree, while farther up the gulch a larger yellow pine was well blazed and deeply scarred by many old and a few new gashes of powerful teeth. In the Davis Mountains, on the ridge just north of Livermore, a yellow pine a foot and a half through had served as a bear register for apparently ten or twenty years. It was deeply scored on all sides from 4 to 6 feet from the ground, but on one side from 5 to 6 feet up, the bark had long been cut away and the dry weathered wood was splintered and gashed with deep grooves of various ages. Two fresh sets of tooth prints showed on opposite sides of the tree near the top of the ring, and one little bear had lately tried his teeth in the green bark about 4 feet from the ground. At the head of a gulch on the east side of Limpia Creek stood another big yellow pine that had been similarly treated, and on it, as on the others, the upper limit of reach was about 6 feet from the ground. Apparently the bear at each visit to one of these register trees had given but a single bite, leaving the marks of an opposing pair of canines.

In January of 1890 I learned that ten or twelve bears had been killed in the Davis Mountains the fall before, and the annual bear hunt of the ranchmen has become as firmly established an institution there as the annual camp meeting. In November a large crowd gathers with camp wagons, hounds, and saddle horses for a week's bear hunting. In 1900 ten bears were killed by the party, and in 1902 four were killed. Others are killed each year by local hunters.

At present the black bears do no serious damage to stock, and it is greatly to be hoped that their numbers will not be materially reduced.

Ursus luteolus[M] Griffith.   Louisiana Bear.[M]

209 The Louisiana bear formerly ranged over most of eastern Texas, and still is found in considerable numbers in the more extensive swamps and thickets. Skulls examined from Kountze,[a] Sour Lake, Tarkington, and Wharton have the long, low brain case and very large molars characterizing the species, while the skins are indistinguishable from those of *americanus* in the black phase.

The following reports of field naturalists for 1902 from scattered localities will give an idea of the present status of the bears over eastern Texas:

Texarkana: Now very rare; one killed a few years ago.

Waskom: Formerly common; now very rare.

Jefferson: Very scarce; one killed near here a few years ago.

Antioch: Formerly common; now extinct.

Rockland: Now very rare or quite extinct.

Conroe: A few still found in the 'big thicket' 15 miles south of here.

Beaumont: A few still found in the forest northwest of here.

Brenham: Formerly common along the Brazos; now extinct.

Elgin: Formerly common; now rare or extinct.

Sour Lake: Still common in the swamps near here; a few killed every year. An old one and two cubs seen during July.

At Richmond in 1899 I was informed that bears were still fairly common in the timbered bottoms along the Bernard River, 18 miles to the southwest, where in the fall one old trapper made a business of trapping them. At Seguin, in 1904, they were said to have been exterminated years ago, though formerly common.

The following reports, made in 1900 by Oberholser, probably also relate to this species:

Beeville: Bears are still found on the Nueces, 20 miles west of here.

San Diego: One was seen a few years ago some 12 miles northwest of here.

Uvalde: A few are still found in the canyon of the Nueces.

At Wharton in November, 1904, I secured the skull of a bear killed the previous year by a negro who said there were still a good many in the thicket near there. Mr. W. O. Victor also told me that he knew where several bears were living in the thicket, and that he hoped to kill some of them later in the season when they became fat. Mr. Victor has an apiary with a large number of hives located at several points in the dense woods and thickets bordering the Colorado River below Wharton, and the bears have caused him much trouble and considerable loss through their fondness for honey. During the past ten years he has killed eight or nine bears, mainly for the protection of his bees. Some of these were killed with set guns, some by trapping, and others in the hunt. One was shot at night by Mr. Victor

---

[a] The Biological Survey is indebted to Mr. J. B. Hooks, of Kountze, for the loan of one skull of this species and the presentation of another.

and two companions who were watching for it in the bee stand. When the men approached the bear after he was located, they could hear him whining and sniffling as if the bees were making it hot for him. This probably accounts for his letting them come near enough by moonlight for a fatal shot. This bee stand was about 3 miles from town and back from any settlement or ranch, and the bear had been feasting on honey for several nights before the mischief was discovered. Mr. Victor says about fifty swarms were destroyed, the hives turned over, part of the honey scooped out, and the bees scattered. In many cases the bear apparently became enraged at the stings and smashed the hives in retaliation. A photograph of this bear, taken the following morning, shows him stretched out among the overturned hives and gives some idea of the mischief he had done.

Mr. Victor says the bears in that region 'den up' for a little while during the coldest part of winter, or at least keep quiet in the densest thickets. He says they are invariably black, and he thinks the nose also is black.

In November, 1904, an old bear hunter, Ab Carter, living on the west edge of Tarkington Prairie, in Liberty County, told me that there were no bears at that time in Liberty County west of the Trinity River, but the active part taken by Mr. Carter in exterminating the bears in that locality makes his statements of peculiar interest. Forty-nine years ago he was born on the ranch he now owns, and his principal occupation, like that of his father, has been keeping hogs and killing bears. To a man with several hundred hogs running in the woods, bear killing was the most important part of the season's work, but it was not until about 1883 that the extermination of the bears began in earnest. At that time Mr. Carter and a neighbor each got a pack of good bear hounds and in the following two years they killed 182 bears, mainly within a radius of 10 miles from the ranches. This reduced the number of bears so that later not more than ten to twenty were killed annually up to 1900, when Mr. Carter killed the last two of the vicinity. Two years ago he killed the last of his bear dogs, and now keeps only hog and wolf dogs, while his hogs eat acorns in safety over 100 square miles of magnificent forest and dense thicket.

The number of hogs killed in a year could be only approximately estimated, but Mr. Carter thinks the bears sometimes got nearly half of the pigs and many of the hogs. Pigs were their favorite prey, and were easily caught, but the bears took anything they could get. One large 4-year-old boar was killed and partly eaten only a mile from the house.

As soon as acorns began to fall the bears would feed on them and let the hogs alone for a while, but during spring and summer pork was their principal food. The first berry to ripen in summer, Mr. Carter says, is on the 'grandaddy graybeard' bush (apparently *Ame-*

lanchier), of which the bears are very fond. Blackberries and huck-leberries are abundant summer food for bears. Later the sour gum (*Nyssa sylvatica*) is a favorite food, and nearly every sour-gum tree in the woods has its top branches bent and twisted and its bark well clawed.

Mr. Carter went with me to an old pine 'measuring tree' in the woods that he said had been bitten deep into the wood about as high up as he could reach, but when we found the tree it was only a charred stump. Fire had destroyed all trace of the bear marks. Another small pine that we found had grown well out around the old bites that still showed plainly in the dead wood. Mr. Carter says cypress trees are sometimes bitten in the same way by bears, but less commonly than pines.

In the Big Thicket of Hardin County black bears were common in many parts of the thicket in December, 1904, but not so abundant as they were a few years ago. I had no trouble in starting one almost every day, but could not get a pack of dogs that would hold one till I could get to it. I had five good bear hounds, but each of the several bears that we started escaped. The bears in this region rarely tree for dogs, and unless the dogs keep one fighting on the ground he travels faster than a man can run through the jungle of palmettoes, brush, and vines. Horses are useless in the thicket.

While hunting I found numerous bear beds and old and fresh 'sign,' some composed of acorns, some of sour gum and other berries. We also saw half chewed acorns where the bears had been feeding. During summer the bears feed extensively on pigs belonging to the settlers, but in December both pigs and bears were rapidly fattening on the abundant acorn crop.

In several places in the heart of the Thicket I found cypress trees gnawed by the bears as high as they could reach, 6 to 7 feet from the ground, and I photographed two of these trees. One, which was about a foot in diameter, had been bitten lately and at different times previously for at least eight or ten years. Several large spots of wood were dead and bare of bark and full of old tooth prints. The other tree, over 2 feet in diameter, had been bitten for a longer time, probably fifty years, and the old dead wood was sunken 4 or 5 inches deep in the surrounding growth. The fresher bites were on new spots and some were made apparently the day before, as fresh mud had been rubbed against the trunk as high up as 4 feet. One old-field pine about 14 inches in diameter had been well bitten at the usual height, but in this region cypress seems to be the favorite biting tree, or 'measuring tree,' as called by the hunters. Several magnolia trees showed deep claw marks in the smooth, gray bark, and the rough bark of the sour gum is often clawed extensively, although the marks are indistinct. The bears are said to feed to some extent

[210] The grizzly or brown bear (*Ursus horribilis horriaeus* of Bailey, 1905) is now recognized as a single species, *Ursus arctos*, and it is listed as Threatened by the USFWS. The subspecies in Texas was *Ursus arctos horribilis*, and the grizzly bear reported by Bailey from the Davis Mountains remains the only available specimen from the state. The date when it was obtained remains uncertain. Bailey said it was taken in October 1890, but the specimen record in the National Museum lists the date of collection as 2 November 1899. C. O. Finley, who shot the bear, reported it was killed on 3 November 1900 (see Chapter 1).

[211] The subspecies of white-nosed coati, *Nasua narica yucatanica*, reported by Bailey does not occur in Texas; *Nasua narica molaris* is the recognized subspecies in Texas. Because of its erratic distribution and concern about its habitat, the coati is listed as Endangered by TPWD. Available evidence suggests that it was never as abundant in Texas as it is in southwestern Arizona. Schmidly et al. (2016a) report evidence they may be repopulating Big Bend National Park and the Devils River regions, although there is no evidence that breeding populations have been established.

[212] Three subspecies of the northern raccoon, *Procyon lotor*, are recognized in Texas: *Procyon lotor hirtus* in the Panhandle north of the Canadian River, *Procyon lotor mexicanus* in the western part of the Trans-Pecos, and *Procyon lotor fuscipes* throughout the remainder of the state. Thus, the specimens identified here by Bailey as *Procyon lotor lotor* would today be classified as *P. l. fuscipes*. Raccoons are common throughout the state, but they are seldom found far from water, which has an important influence upon their distribution.

on magnolia berries and very extensively on the berries of the sour gum.

I have inquired of many hunters and find none who have ever seen a brown bear in this region. The nose is said to be brown in some and entirely black in others. The large old male, of which I secured the skull and incomplete skin, was said to have had a brown nose, as did the perfect skin of the female sent with it. Dan Griffin, who killed it, says it was the largest bear he ever saw. He thinks it would have weighed 400 pounds, although poor, and says that two men while skinning it had hard work to turn it over.

Ursus horribilis horriaeus[M]Baird.　Sonora Grizzly.[M]

210　The only specimen of grizzly bear that I have seen or heard of from Texas was killed in the Davis Mountains in October, 1890, by C. O. Finley and John Z. Means. The skull, which Mr. Finley has kindly sent me for comparison, proves to be that of a large and very old male of the Sonora grizzly, agreeing in all essential characters with Baird's type of *horriaeus*[Mc]from southwestern New Mexico. The measurements of the skull are: Greatest length, 370; basal length, 310; zygomatic breadth, 220; mastoid breadth, 157; interorbital breadth, 71; postorbital breadth, 69. The claws on the front feet, Mr. Finley says, were about 3 1/2 inches long, and the color of the bear was brown with gray tips to the hairs. Its weight was estimated at 1,100 pounds, 'if it had been fat.' Mr. Finley says that this bear had killed a cow and eaten most of it in a gulch near the head of Limpia Creek, where the dogs took the trail. Out of a pack of fifty-two hounds only a few would follow the trail, although most of them were used to hunting black bear. These few followed rather reluctantly, and after a run of about 5 miles over rough country stopped the bear, which killed one of them before it was quieted by the rifles of Finley and Means. It took four men to put the skin, with head and feet attached, upon a horse for the return to camp.

Nasua narica yucatanica[M]Allen.　Nasua; Coati.[M]

211　A specimen of this long-nosed, long-tailed, coon-like animal in the National Museum, collected in 1877 at Brownsville by the late Dr. J. C. Merrill, furnishes apparently the only record for the State. As nasuas occur over most of Mexico up to near the border of the United States, other records along the Rio Grande may be expected.

Procyon lotor (Linn.).　Raccoon; Coon.[M]

212　The raccoon of eastern Texas, as represented by specimens from the coast region as far west as Matagorda and in the interior from Tex-

---

[a] Dr. J. A. Allen, in Bul. Am. Mus. Nat. Hist., XX, 53, 1904, identifies the Brownsville specimen as *Nasua narica yucatanica*[M]Allen; it is possible, therefore, that this specimen may have been an imported animal that escaped from captivity.

arkana west to Kerrville and Mason, differs but little from typical *lotor* of the northeastern United States. The slightly larger size, wider muzzle, and usually heavier dentition show a tendency toward *mexicanus,* into which it grades to the west. The high frontals of specimens from the coast marshes of southeastern Texas suggest an approach to *elucus,* the Florida form, but in the light of the present material these coast specimens can best be referred to *lotor.*

Coons are abundant along the margins of streams, lakes, and bays, along the coast, in marshes, or around water holes, adapting their habits to almost any condition save that of dryness. In the timbered country hollow trees, hollow logs, cavities under old logs, or upturned roots provide them temporary homes in which to spend the day, and on the great salt marshes of the coast country masses of fallen grass and rushes provide dark cover, or hollow banks and windrows of drift stuff afford safe retreats, while the broken walls of rocky canyons and gulches toward the headwaters of the streams furnish the favorite, because the safest, dens. It is not uncommon for coons to leave the stream where they have been hunting and travel half a mile or a mile to dens in a cliff, though otherwise they are rarely found so far from water. They are mainly nocturnal, and every morning their unmistakable plantigrade tracks mark the shores of the streams, following the trails, logs, or mud flats, now in, now out of the water, often disappearing where the animals swam from point to point or from one side to the other of the stream in search of food. Often the coons follow the same line of travel again and again, until well-worn trails are formed along the margins of the streams or through the marsh grass. Along these trails scattered remains of food tell half the story of the coon's life. In places along the Guadalupe River, in Kerr County, almost every little point and island has its pile of mussel shells from which the mussels have been eaten, and every morning a few shells freshly scooped out are found on the piles until sometimes a bushel is accumulated. On the coast marshes the shells of crawfish are found scattered along the coon trails, while the excrement deposited here and there in well-chosen spots is made up largely of the indigestible parts of crustaceans mixed with a few scales and bones of fish and occasional traces of frogs and small mammals. As these marshes swarm with crawfish and small crabs, the coons have a perennial feast and naturally become numerous. On Matagorda Peninsula Lloyd reported them feeding on oysters as well as crabs and crawfish, and in the stomach of one caught near the mouth of the Colorado River he reported finding a meadow lark. In their selection of food coons are quite as omnivorous as bears, seeming to relish almost any kind of flesh, fruit, grain, nuts, and acorns. At Brazos they were reported by B. H. Dutcher as feeding on melons.

[213] The subspecies *Procyon lotor mexicanus* is restricted to the western part of the Trans-Pecos (see Annotation 212).

Their nightly raids on fields of green corn are too well known to need comment, and small fields of corn planted in or near the woods are sometimes almost destroyed, the ears being torn open and the corn eaten from the cob from the time of the early milk stage until ripe, and even after being cut and shocked.

In the Big Thicket coons are numerous along every stream and bayou, as shown by fresh tracks along roads and trails and in the muddy margins of ponds and water ways, and by skins drying under the sheds of almost every ranch. Their fur is the principal catch of most of the trappers and their abundance makes trapping fairly profitable in this region. During November and December they were feeding mainly on acorns, but were still eating crawfish, while the old shells of mussels, including the enormous pearl-bearing species (*Quadrula heros*) and the smaller thick-shelled *Quadrula forsheyi*, piled here and there along the banks of bayous, apparently marked the remains of summer feasts.

While watching for fox squirrels one morning in the heavily timbered bottoms I heard a scratching sound from an old cypress in the edge of the swamp near by, followed by a loud splash. A young coon less than half grown had fallen from the tree into the water. At the sound the old coon and two more young ones came out of a hollow some 30 feet up in the trunk and climbed down to near the bottom of the tree. They came down the tree slowly but steadily, head first, as a squirrel would have done, with the hind feet reversed and slightly divergent. When the old coon saw the young one climb out of the water upon the tree trunk she turned about and ascended the trunk, followed by the three young. The one that had fallen, besides being very wet, was slightly hurt, and climbed with difficulty. When halfway up he stopped on a limb to rest and began whimpering and crying. The mother had already reached the hole, but on hearing his cries turned about and climbed down to him. Taking a good hold of the back of his neck and placing him between her fore legs so that he, too, could climb she marched him up the tree and into the hollow.

Procyon lotor mexicanus Baird.   Mexican Raccoon[M]

213 Raccoons are common along every stream in Texas, and especially common along the coast and on the islands. Specimens from the Rio Grande, Pecos, and Devils River country are large and pale; they have a long tail and the more quadrate molars of *mexicanus* to which subspecies they are referred, although differing in having the narrower basioccipital and yellow nape of *lotor*.[a] [M]

[a] It has been customary to refer specimens from western Texas to *hernandezi,* but a number of specimens of that species from the type region in Mexico, collected by Nelson and Goldman, prove to be quite different from the Texas animal.

In the northern part of the State the range of *mexicanus* is partly cut off from that of the smaller, darker coon of eastern Texas by the plains; but near the coast, where there is no break in the ranges, only an arbitrary division can be made between the two forms. Specimens from as far east as Corpus Christi can safely be referred to *mexicanus* and others from as far west as Matagorda to *lotor*,[M] while specimens between, from Port Lavaca and Aransas County, can be referred as well to one as the other. Assuming, as seems necessary, that Baird's redescribing and correctly naming "*Procyon lotor*[M] variete mexicaine" of St. Hilaire[a] fixes the type locality at Mazatlan, Mexico, the name *mexicanus* becomes available for the coon of western Texas, which, though not typical, is certainly nearer to this form in general characters, as well as in geographic position, than to any other.

In western Texas coons are closely restricted to the streams, and consequently are rare over the wide intervals of dry desert country between. Along the Rio Grande, Pecos, and Devils River valleys they are especially abundant, and their dens are almost invariably located in the broken walls of cliffs and canyons. In the low country toward the coast of southern Texas, where dense chaparral, cactus patches, and the tall grass of the salt marshes offer ample shelter and streams are not infrequent, they have a more continuous distribution. From Corpus Christi to Brownsville their tracks were seen along the shores of every stream and pond and were especially numerous near the coast, where the animals apparently lived on the little fiddler crabs (*Gelasimus pugilator?*), always found in abundance on the low, sandy soil. Lloyd reported hackberries (probably *Momesia pallida*)[P] in the stomach of one caught at Corpus Christi. In the Pecos and Devils River canyons the heavy shells of one of the pearl-bearing mussels (*Lampsilis berlandieri*) are often found in piles along the banks of the streams, but the ripe fruits of the prickly pear (*Opuntia engelmanni*)[P] and of the black persimmon (*Brayodendron texanum*)[P] were their principal food in July and August. The sweet pods of mesquite were also eaten, and apparently some of the insipid berries of *Zizyphus, Condalia,*[P] *Adelia, Lycium,* and *Momesia.*[P]

Lutra (canadensis?)[M] (Schreber).   Otter.[M]

214 Otters are not uncommon in the streams of eastern Texas, but, being unable to procure a specimen from any part of the State, I can only provisionally refer the species to *canadensis*. The only specimen that throws any light on the question is a fine old male collected at Tallulah, Madison County, in northeastern Louisiana, by W. E. Forbes and N. Hollister, which agrees in most characters with *canadensis* and shows no tendency toward intergradation with the Florida otter, *L. c. vaga*.[M]

---

[a] Voyage de la Venus, Zoologie, p. 125, 1855.

[214] *Lutra canadensis* is now classified as *Lontra canadensis* (van Zyll de Jong, 1972; Larivière and Walton 1998), and the subspecies in Texas is *Lontra canadensis lataxina*. The otter currently is known from about the eastern and southern one-fourth of the state in major watersheds but may be expanding into the Panhandle regions (Schmidly and Bradley, 2016). For a discussion of its status, see Chapter 6.

[215] The name of the American mink has been changed from *Lutreola lutreocephala* to *Vison vison* (Harding and Smith, 2009), and the subspecies in Texas is *Vison vison mink*. This species occurs in the eastern one-half of the state and Hansford County in the Panhandle. In all places, it requires habitats near permanent water. It has declined in abundance this century with the decline in natural surface water and does not appear to be common anywhere in the state.

In the Big Thicket of Liberty and Hardin counties otters are common, and a few are caught each year by the local trappers. During low water the black pools of the half-dry bayous, swarming with landlocked fish, are their favorite haunts. Oberholser obtained reports of otters at Mobeetie and along the Red River at Texarkana, and along the Neches and San Jacinto rivers near Beaumont and Conroe. Lloyd reported them from Palacio Creek, Matagorda County, and John M. Priour writes that they are found on the Colorado River in the region of Austin. None of our field men have ever heard of them along the Rio Grande or Pecos rivers, however, while several old trappers, long familiar with the Rio Grande, Pecos, and Devils rivers, have assured me that otters were never found along these streams. In addition to this evidence, Mr. W. H. Dodd, of Langtry, has told me that for many years in buying fur of the local trappers no skins or even reports of otters had come to his notice in that region. Along Big Cypress Bayou, below Jefferson, in northeastern Texas, Mr. Richard Crane told me in 1902 that otters were fairly common, and in fifteen years' hunting and fishing along this stream he had killed eight or ten, most of which he shot. One that came up near his boat and then dived, leaving its tail temptingly above water for a second, he caught by the tail, whereupon it promptly curled up and severely bit his legs and hands before he could kill it. He says $50 would not tempt him to catch another otter by the tail.

Lutreola lutreocephala[M](Harlan).   Large Brown Mink; Southeastern Mink.[M]

215 Minks are common over approximately the eastern half of Texas, but apparently are unknown in the western part of the State.[a] The western limit of their range is roughly indicated by specimens from Gainesville, Brazos, and Mason, and by reports of occurrences near Austin and on the lower Guadalupe River.

I have examined specimens from Gainesville, Brazos, Mason, Navasota, Harris County, Matagorda, Tarkington Prairie, Rockland, Antioch, and Texarkana, but find no characters, cranial or external, by which to separate them from typical *lutreocephala*[M] from Maryland and the District of Columbia.

Along most of the streams and bayous of the timbered country of eastern Texas, minks are so common as to form an important item in the catch of the local trappers. In fall and winter a few of their skins are usually found among the more numerous coon, opossum, and

---

[a] Cary obtained an indefinite report of 'mink' at Fort Stockton, in the Pecos Valley, but there is a possibility that the name may have been applied to some other animal.

skunk skins at trappers' camps or cabins, or in general merchandise or fur stores of the town. While usually closely associated with stream courses, where much of their food consists of fish, frogs, crustaceans, birds, and mice, minks are perfectly at home in the dry parts of woods and swamp, and even on the open prairie. At Navasota I caught one in woods near the river and another in a trap set in a cut bank gulch in the middle of a wide field. Both were attracted by bodies of birds that had been shot for specimens, and while in the traps had gorged themselves with the bait. At Tarkington Prairie minks are said to be much less common in the timber than on the open prairie, where myriads of birds roost at night in the long prairie grass, and crawfish chimneys thickly dot the margins of shallow ponds. Along the coast marshes the minks follow the shores of bayous and ditches, where their tracks usually may be found in the mud and sand, or range back over the wide expanse of marsh and prairie, where tall grass and drift heaps furnish ample cover. Over these marshes they feed extensively on crawfish and minnows, as shown by their excrement. One caught by Lloyd on Matagorda Peninsula had a freshly eaten cotton rat in its stomach.

The occasional losses from the raids of minks on the poultry yard in most cases can be prevented by a little care on the part of the farmer in providing roosting places out of reach of the prowling minks, if necessary, with tin-covered uprights. Minks are good climbers, and will sometimes climb to the top of a tall tree to escape the dogs, but they seem to hunt almost entirely on the ground. An ordinary poultry fence with fine wire mesh affords perfect protection, not only from minks, but from many other troublesome 'varmints.' The value of the mink's fur makes the animal of considerable economic importance, especially as it has proved its ability to hold its own in thickly settled districts. Its value as a destroyer of small rodents compensates in part, if not fully, for its depredations.

Putorius nigripes[MA]Aud. and Bach.   Black-footed Ferret.

216   The black-footed ferret has been reported from a number of localities in the prairie-dog country of Texas east and south of the Staked Plains. A very large weasel, described by B. H. Dutcher in 1893 at Stanton, may or may not have been of this species. Merritt Cary learned of one that was killed in 1894 at Seymour. J. A. Loring found an almost perfect skull of a fine adult at Childress on the house of a wood rat. A flat skin in the U. S. National Museum, labeled "Gainesville, Texas," probably came from some point west of there, as it is merely a rough hunter's skin, evidently not prepared by G. H. Ragsdale, whose name is on the label. If this were a bona fide record for Gainesville it would be the first from any point far out of the range of the prairie dog.

[216] *Putorius nigripes* is now classified as *Mustela nigripes*, the black-footed ferret (Abramov, 2000). Its historic range was roughly the northwestern third of the state including the Panhandle, much of the Trans-Pecos, and a considerable part of the Rolling Plains region, which corresponds with the distribution of the black-tailed prairie dog, the ferret's principal prey. The black-footed ferret was extirpated primarily as a result of the destruction of the prairie dog towns (see Chapter 5).

[217] The name *Putorius frenatus* has been changed to *Mustela frenata*, the long-tailed weasel (Abramov, 2000). This species occurs statewide, although it is scarce in most areas, especially in western and northern Texas. The reduction in natural surface water in Texas may have prompted this decline, as the absence of drinking water is a factor limiting their distribution (Hall, 1951). Five subspecies are now recognized in Texas: *Mustela frenata neomexicana* west of the 100th meridian, *Mustela frenata texensis* in the north-central part of the state, *Mustela frenata primulina* in the extreme northeastern part of the state, *Mustela frenata arthuri* in east-central and southeastern areas, and *Mustela frenata frenata* along the southern Gulf Coast (Hall, 1951).

[218] The name *Putorius frenatus neomexicanus* has been changed to *Mustela frenata neomexicana* (Abramov, 2000).

[219] In his revision of the genus *Spilogale*, Van Gelder (1959) suggested that the only differences between the eastern spotted skunk (*Spilogale putorius*) and the western spotted skunk (*Spilogale gracilis*) were size and color patterns. He therefore relegated *S. gracilis* as a subspecies of *S. putorius*, and treated spotted skunks as a single, wide-ranging species with four subspecies: *Spilogale putorius gracilis*, *Spilogale putorius leucoparia*, *Spilogale putorius interrupta*, and *Spilogale putorius putorius*. However, Mead (1968a,b) used reproductive data to demonstrate that the eastern and western spotted skunks did behave as species because the

198   NORTH AMERICAN FAUNA.   [No. 25.

In September, 1902, Cary writes:

> A number of black-footed ferrets are said to have been caught at the dog town south of the Stanton stock yard in past years, and every person questioned was familiar with the animal and could give a good description of it. Doctor Vance, living just north of town, saw one about a week before I arrived there and set a rude box trap at the hole in an attempt to capture the animal alive, but without success. A Mr. Williams, living at Fort Stockton, kept for a year or more a black-footed ferret which a Mexican caught in a trap set at an old adobe house on the edge of a dog town just north of the Pecos River at Grand Falls. It was described to me as built like a mink, with dark-brown feet and a bar across the face.

At Lipscomb, in July, 1903, A. H. Howell "saw the hide of one killed there the previous summer and was told of a den of them located near First Creek."

Putorius frenatus[M](Lichtenstein).   Bridled Weasel.[M]

217 While never common, the bridled weasel seems to be generally distributed over the low country of southern Texas. There are specimens in the Biological Survey collection from Brownsville and near Hidalgo. Oberholser examined mounted specimens at San Diego, Beeville, and Port Lavaca. Lloyd reported the species from Corpus Christi, and Attwater from San Antonio.

Putorius frenatus neomexicanus[M]Barber and Cockerell.   New Mexico Bridled Weasel.[M]

218 This species, so far as I know, is not positively known to occur in the State of Texas, but in the winter of 1889 I found the tracks of a weasel winding in and out of the *Dipodomys* and *Perodipus*[M]holes in the sandy bottoms just below El Paso. A record of a weasel taken several years ago at Langtry (reported to Oberholser by W. H. Dodd, of that place) may have been of this species, and suggests a continuous range from the country of *frenatus*[M]up the Rio Grande to the type locality of *neomexicanus*[M]at Mesilla Valley, N. Mex.

Spilogale leucoparia[M]Merriam.   Rio Grande Spotted Skunk.[M]

219 This beautiful little spotted skunk, with broad white stripes, occupies the rough country bordering the southern arm of the Staked Plains from Mason and Waring to Langtry, Comstock, and Eagle Pass and farther south into Mexico. It is probably the form occupying also the rough country east and west of the Pecos Valley. In the Davis Mountains the ranchmen report a spotted skunk as common, and say that it climbs trees as readily as a squirrel. It is often treed by the dogs at night and shot from the branches by the hunters. Under the nest of a great horned owl in the face of a cliff at the west base of the Davis Mountains I found several jaws of these little skunks in the owl pellets. Throughout most of its known range it inhabits rocky gulches, cliffs, and canyons, or the brushy bottoms usual in such places.

western form exhibited a period of delayed implantation that was absent in the eastern form, which meant their breeding seasons were separate and they were reproductively isolated. A molecular study of spotted skunks by Dragoo et al. (1993) has corroborated these differences. So, in Texas today, there are two species of spotted skunks, *Spilogale gracilis,* subspecies *leucoparia*, in the western part of the state, and *S. putorius,* subspecies *interrupta,* in the eastern part (Fig. 50). Populations of both species appear to be in decline, and there are serious concerns about their future in Texas. The name *S. leucoparia*, as used by Bailey, has been changed to *S. g. leucoparia*.

Spilogale interrupta[M](Rafinesque).   Prairie Spotted Skunk.[M]

220 This dark form of the spotted skunk, or spilogale, with the narrow white stripes, comes into Texas from the more northern plains, and is represented by specimens from Canadian, Gainesville, and Brazos. Beyond these localities there are no specimens to show the limits of its range in the State or to indicate whether it grades into the neighboring forms to the south. Though the little "spotted skunks," "hydrophobia cats," or "phoby cats" are reported from

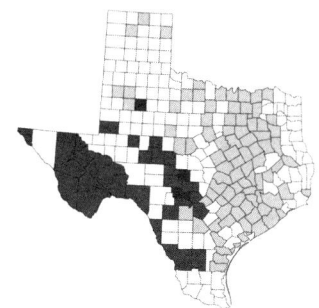

Figure 50. The current distribution of the western spotted skunk (*Spilogale gracilis*) and the eastern spotted skunk (*Spilogale putorius*). The light shaded area represents the distribution of *S. putorius*, the medium shaded area represents *S. gracilis*, and the darkest shaded area represents overlap between the two species.

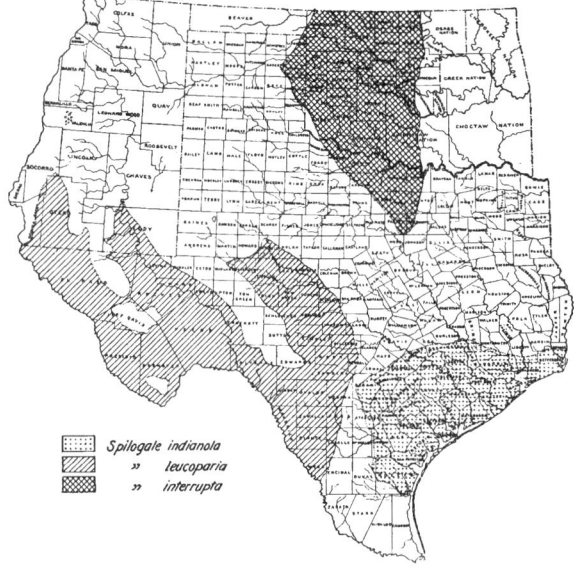

Spilogale indianola
    "   leucoparia
    "   interrupta

FIG. 23.—Distribution areas of spotted skunks (genus *Spilogale*).

[220] *Spilogale interrupta* is now recognized as *Spilogale putorius interrupta* (see Annotation 219). This subspecies was listed as Category 2 by the USFWS prior to 1996 and is now being considered for listing under the Endangered Species Act.

almost every part of Texas, even including the top of the Staked Plains, there is still much to be learned of the range and relationships of the several forms inhabiting the State.

Although, broadly speaking, plains animals, these spilogales, like most species of the genus, take advantage of any cover in the way of bushes, tall grass, stream banks, or old buildings that the country offers. In Kansas I have caught them in burrows in the sandy soil, but whether the burrows were of their own digging or borrowed from spermophiles or other burrowing mammals I could not tell. At Canadian, Tex., one was caught in a No. 0 steel trap set in an old

[221] *Spilogale indianola* is now classified as *Spilogale putorius interrupta* (Van Gelder, 1959).

[222] *Mephitis mesomelas* is now classified as *Mephitis mephitis*, the striped skunk (Wade-Smith and Verts, 1982). Striped skunks in the eastern part of Texas are recognized as *Mephitis mephitis mesomelas* and those in the western part as *Mephitis mephitis varians*. The combined range of the two subspecies covers the entire state, and this is unquestionably the most common skunk in Texas.

Another species of *Mephitis*, not obtained during the Survey, is the hooded skunk, *Mephitis macroura milleri*. This skunk is superficially similar to the striped skunk, but has longer, softer fur and a distinct ruff of longer hair on the upper neck. This is primarily a Mexican species that occurs in Texas in the Big Bend region and adjacent parts of the central Trans-Pecos, where it is relatively rare. Its status is unknown but there have been few sightings or specimens collected in recent years. Efforts should be made to monitor the status of the hooded skunk in the future.

200   NORTH AMERICAN FAUNA.   [No. 25.

tumbledown shed in the corner of a field and baited with the bodies of birds that had been skinned for specimens. I had with me a bottle of bisulphid of carbon for experiments on prairie dogs. Thinking to try a new experiment, I scraped a hollow about 8 inches deep in the sand, and with a stick gently loosened the trap chain and slowly drew the little skunk to the hole. He tumbled in, thinking he had escaped, and curled up in the bottom. I then poured a couple of ounces of bisulphid on a bunch of grass and threw it into the hole, and after waiting five minutes found the skunk dead and perfectly free from unpleasant odor. This method of killing any of the skunks, when it becomes necessary to trap them around buildings, can safely be recommended.

Spilogale indianola [M] Merriam.   Gulf Spotted Skunk. [M]

221   This little spotted skunk inhabits the coast region of Texas from Corpus Christi to southwestern Louisiana and extends inland as far as Beeville, San Antonio, and Navasota. So far as known, it is mainly an animal of the cactus and chaparral patches of the open country. In the Big Thicket region I could get no reports of it east of Conroe, but at Navasota I found it common and caught two in traps in brushy places. At Beeville Oberholser caught one in a trap set in the runway of a wood rat. Of two specimens secured by Lloyd in Matagorda County one was taken in a group of burrows in a thicket on the prairie and the other in an old cotton gin. In the stomach of the former were found parts of a *Perognathus hispidus* [M] and some crawfish. Near Corpus Christi I caught the animals in bunches of prickly pear and in wood-rat houses under the mesquites. At Virginia Point, on the prairie near Galveston, I shot and trapped them in the big bunches of cactus (*Opuntia engelmanni* [P] found here and there on the prairie. Such confidence had they in the protection of these thorny masses that one came out repeatedly, thrusting its head between the cactus blades to watch me with its keen little eyes, first at one window then at another, moving about freely among the thorns and refusing to enter its burrow even when I approached to within a few yards. Its motions were quick and alert, and its expression bright and weasel-like rather than skunk-like, which, added to its beautiful markings, made it a most attractive little animal. The burrows under the cactus and thorny huisache bushes were apparently dug by skunks, as no other burrowing animal near their size occurs there. The stomach and intestines of the specimens taken contained only shells and legs of a large brown beetle which swarmed about the houses at night. A few legs and wings of grasshoppers were found in the lower intestines of one individual.

Mephitis mesomelas [M] Lichtenstein.   Louisiana Skunk. [M]

222   The Louisiana skunk is common over the whole of eastern Texas and about as far west as Wichita Falls and Matagorda Bay. Specimens from O'Connorport are clearly intermediate between *mesomelas*[M] and *varians,*[M] as are also specimens from Wichita Falls. There is apparently no locality in Texas where skunks are not more or less common, and the transition from *mesomelas*[M] to *varians,*[M] while not abrupt, seems to follow approximately the line of transition from humid forest and prairie country to semiarid mesquite plains.

Skunks are generally less common over eastern than western Texas, owing probably to the more thickly inhabited country to the eastward, to the number of dogs kept at every little farm or cabin, and to the popular superstition that all skunks convey hydrophobia and should be destroyed whenever possible. There are undoubtedly authentic cases of rabies in skunks, as well as in other animals that have been inoculated with the disease, but there is no reason to suppose that they are any more subject to it than dogs or cats nor more dangerous to human beings when they do have it. On the other hand, they are among the most useful of the predatory mammals, destroying great numbers of small rodents, grasshoppers, beetles, and larvae, and should be protected, except in rare cases of mischief. There are a few complaints of their destroying poultry, but in most cases this mischief can be easily prevented.

At Virginia Point, on the prairie opposite Galveston, I trapped a skunk one morning in a bunch of cactus and by a bungling shot allowed it to discharge its odorous fluid. Being anxious to save the skin in spite of its odor, I sat down on a patch of dry sand to skin it, and in a few minutes a black shadow passed me on the ground. Looking up I saw not less than 50 turkey buzzards and black vultures beating up the wind in a long line straight toward me. They were flying low and keenly scanning the ground. Many came within 20 feet, apparently, before seeing me, and soon I was the center, though not the object of attraction, of the constantly increasing flock. As my work ended and I moved away they pounced on the carcass, and soon there was nothing but the large scent gland and its odor to mark the spot. Even the bones had mostly disappeared. This is but one of many similar instances in which turkey buzzards and vultures have quickly responded to the smell of a freshly killed skunk, although they usually leave a cleanly picked skeleton as well as the scent gland.

Mephitis mesomelas varians[M] Gray.   Long-tailed Texas Skunk.[M]

223   The long-tailed skunk ranges over western Texas from Brownsville to El Paso and east to Rockport, San Antonio, Mason, Brazos, Canadian, and Lipscomb, or approximately over the mesquite region and plains of Texas in both Upper and Lower Sonoran zones. Although generally distributed even over the top of the Staked Plains, these skunks are most abundant in the chaparral or brushy

[223] *Mephitis mesomelas varians* is now classified as *Mephitis mephitis,* the striped skunk, and the subspecies is *M. m. varians* (Wade-Smith and Verts, 1982).

[224] All Texas hog-nosed skunks are now recognized as *Conepatus leuconotus*, and the three subspecies mentioned by Bailey are now placed under this species. Consequently, *Conepatus mesoleucus mearnsi* is now *Conepatus leuconotus leuconotus* (Dragoo et al., 2003). This subspecies remains relatively common across southwestern, central, and southern Texas (Fig. 51).

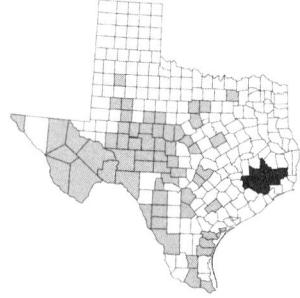

Figure 51. The distribution of the hog-nosed skunk (*Conepatus leuconotus*) in Texas. The light shaded area represents the subspecies *C. l. leuconotus* and the dark shaded area represents *C. l. telmalestes* (assumed to now be extinct).

202   NORTH AMERICAN FAUNA.   [No. 25.

country, especially along bushy-bottomed, rocky walled gulches and in canyons, where to an abundance of food are added ample cover and the protection of numerous safe retreats. The sandy bottoms and dusty trails are almost invariably marked with their tiny, bear-like tracks, and they are frequently met with morning or evening racking along the trail on their way home or abroad. At night they often come into camp, and leave tracks in the ashes of the campfire or around the 'grub box,' but in years of camp life where they are common I have never know them, when unprovoked, to be discourteous or disagreeable. One morning in the Davis Mountains we noticed tracks and numerous little holes dug in search of beetles around our beds and among the frying pans and kettles. We had evidently camped on the favorite digging ground of this particular skunk and he had quietly put up with the inconvenience of our presence.

The skunks often acquire the habit of coming to camp for the discarded bodies of birds and mammals that have been skinned for specimens, but if their favorite foods—grasshoppers, cicadas, beetles, and grubworms—are abundant, it is difficult to entice them into traps with any kind of bait. Any small game that they can catch for themselves is welcome and they sometimes raid an unprotected chicken coop. I have found their stomachs filled with berries of ziziphus, and have noted the remains of cactus fruit, black persimmons, and small berries in the 'sign' along their favorite trails. But legs and shells of grasshoppers and beetles usually form the bulk of their 'sign.' One caught at Santa Tomas by Lloyd had just dined on a cotton rat, and in other places Lloyd reported them as feeding on wood rats. Occasionally they find our traps and eat the small rodents caught in them.

Conepatus mesoleucus mearnsi [M]Merriam.   Mearns Conepatus; Hognosed Skunk.

224   The white-backed or hognosed skunk is common over most of western Texas, from Kerrville, Mason, and Llano to the Rio Grande and beyond, and south to Dimmit County. Along Devils and Pecos rivers and the canyon country of the Rio Grande and in the Davis Mountains it is evidently the commonest skunk. It apparently has not been taken in the El Paso part of the Rio Grande Valley, but as it is found farther north, it undoubtedly occurs there also. Oberholser obtained a report of its rare occurrence at Austin, which is its easternmost record. Specimens reported by H. P. Attwater from San Antonio are probably of this species.[a]

The scarcity of specimens of *Conepatus* in collections is not due

---

[a] Allen, Mammals of Bexar County, Tex., in Bul. Am. Mus. Nat. Hist., VIII, 72, 1896.

entirely to the scarcity of the animals. In several localities where they were common and their long-clawed tracks and peculiar dig- gings were abundant and fresh every morning, I utterly failed to trap them, as they would not come near any kind of bait that I could offer; but in these localities their favorite food—a large brown beetle—was abundant. Near Boquillas, in the side canyons of the Rio Grande, the mellow sandy bottoms were pitted with little funnel-shaped holes about 2 inches deep where the animals had dug out the beetles, whose round holes perforated the ground on all sides like half-inch auger holes. One of the skunks shot by moonlight early in the evening on his digging ground had already filled his stomach with these crisp juicy beetles to the number of several hundred. In skinning him the next morning I was struck with the adaptability of his long naked nose to the work of probing the beetle holes. A sniff would probably show whether the beetle was at home and worth digging for or whether the hole was occupied by a taran- tula. This and two other specimens, which I failed to shoot in such a way as to break their backs and prevent the discharge of their scent gland, curled up with their last gasp and drenched their bodies from head to tail with the reeking fluid, which differs neither in quantity nor strength from that of *Mephitis*. The repetition of this act by the two individuals indicates a habit not shared with the com- mon skunk, which to its last breath tries to avoid soiling itself in using its weapon of defense. In general the habits of *Conepatus* and *Mephitis* are very similar even to a choice of the same brush patches and gulch bottoms for foraging ground. They must frequently meet, whether on friendly terms or otherwise.

Along Devils River in July *Conepatus* was common, but as usual was difficult to catch. One got into a trap set in a trail and another was shot by moonlight as it trotted through camp. They were feed- ing on beetles, grasshoppers, crickets, and the ripe fruit of the prickly pear (*Opuntia engelmanni*).[P] Near Langtry Gaut caught an old female, March 24, which contained a single embryo that he thought would have been born a week later.

Conepatus mesoleucus telmalestes[M] subsp. nov.   Swamp Conepatus; White-backed Skunk.[M]

Type from the Big Thicket, 7 miles northeast of Sour Lake, Tex., ♂ ad., No. 136551, U. S. Nat. Mus., Biological Survey Coll. Collected by James H. Gaut, March 17, 1905. Original No. 3485.

225 *General characters.*—Similar in general appearance to *Conepatus mesoleucus mearnsi*,[M] skull usually slenderer, dentition lighter.

Color.—Whole upper parts and tail white, the white extending for- ward on forehead nearly to eyes; lower parts, sides, legs, and face black.

[225] *Conepatus mesoleucus telma- lestes* is now *Conepatus leuconotus telmalestes* (see Annotation 224), which is thought to be extinct (see Chapter 5). Its historic distribution is shown in Figure 51.

204   NORTH AMERICAN FAUNA.   [No. 25.

Skull of type elongated, with slender muzzle, narrow interorbital region and prominent mastoid processes; upper molar relatively long and narrow, upper and lower carnassials strikingly smaller than in comparable specimens of *mearnsi.*[a][M]

Measurements of type.—Total length, 625; tail vertebrae, 257; hind foot, 78. Of two female topotypes: Total length, 610; tail vertebrae,

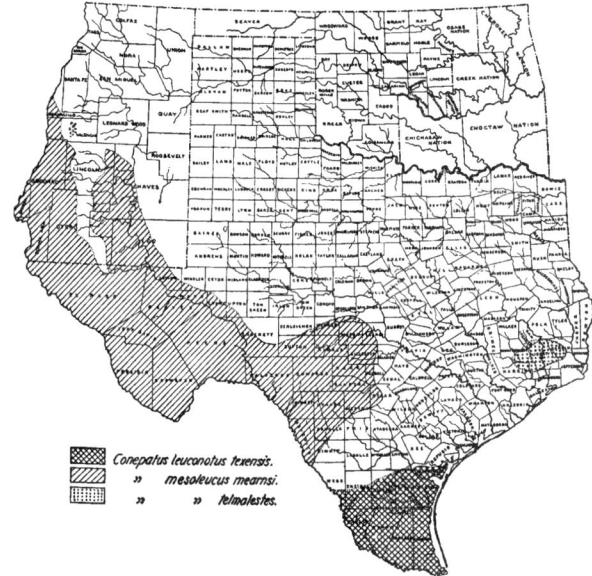

FIG. 24.—Distribution areas of white-backed skunks[M] (genus *Conepatus*).

265; hind foot, 67; and total length, 676; tail vertebrae, 304; hind foot, 74.

Skull of type.—Basal length, 65.2; zygomatic breadth, 44.3; interorbital breadth, 22.3; postorbital constriction, 20; mastoid breadth, 40.3; alveolar length of upper molar series, 16.3.

Three skins and four skulls have been examined from the Big Thicket, 7 to 10 miles northeast of Sour Lake, in Hardin County, and one skin and two skulls from Tarkington Prairie, in Liberty County.

---

[a] The skull of a large male from Tarkington lacks the slender rostrum and narrow interorbital region, but agrees with the others in tooth characters and spreading mastoid processes.

At Saratoga, Kountze, and Cleveland the white-backed skunk is said to be the commonest species, and under a trapper's shed at a ranch on Tarkington Prairie in November, 1904, I saw eight or ten of their skins hanging up to dry with a smaller number of skins of *Mephitis mesomelas.* They were valued at 40 cents each, or less than half as much as the blacker skins of *Mephitis.*

Apparently no *Conepatus* are found in the country west of Liberty County until the range of *mearnsi* is reached near Austin, or that of the more widely different *texensis* at Rockport. The extension of range of the genus is less surprising than that a local form of a group so generally associated over a wide area with arid desert regions should be found restricted to the most humid and densely timbered corner of the State of Texas.

The residents of the Big Thicket country are familiar with these animals, which they call "white-back skunks" to distinguish them from the black-backed or two-striped *Mephitis.* I could not learn of any difference in habits or habitat of the two species. Gaut reports two females taken in April as nursing young, and with one of these he found two small young about a week old in a hollow stump. He also reports that the stomachs of three adults were filled with ground up insects—mostly beetles—with a few grubworms, large brown flies, and grasshoppers.

Conepatus leuconotus texensis Merriam.   Texas Conepatus.

226 From Brownsville, on the lower Rio Grande, this larger form of the white-backed skunk extends up the coast as far as Rockport and up the Rio Grande Valley to Laredo. Lloyd, who collected specimens at Brownsville and Laredo, reported them as rare. He also reported them as occurring occasionally on Padre Island and at Nueces Bay.

Scalopus aquaticus (Linn.). Eastern Mole.

227 One specimen from Joaquin and fifteen specimens from the Big Thicket, 7 miles northeast of Sour Lake, show no distinguishing characters when compared with a large series of typical *aquaticus* from Virginia, Maryland, and the District of Columbia. The slightly lighter color and larger molars indicate a shading toward *texanus,* but in so slight a degree as to be merely a suggestion.

The sandy pine ridges of eastern Texas, and even the mellow soil of the river bottoms and the low mounds above flood level, are criss-crossed by innumerable mole ridges, and dotted here and there with little heaps of yellow sand pushed up through the carpet of fallen leaves and pine needles. The moles are abundant, and save for the barriers of rivers have an almost unobstructed range west to the black wax land prairies. Their work is most conspicuous on the lightest, sandiest soil, which is kept so well stirred and plowed that

[226] *Conepatus mesoleucus texensis* is now *Conepatus leuconotus leuconotus* (see Annotation 224). The hog-nosed skunk of southern Texas is extremely rare (see Chapter 6) and the population has declined drastically in recent years.

[227] Five subspecies of *Scalopus aquaticus* are currently recognized in Texas (Yates and Schmidly, 1977): *Scalopus aquaticus aereus* along the Louisiana border, *Scalopus aquaticus alleni* in the south-central region, *Scalopus aquaticus cryptus* in the north-central region, *Scalopus aquaticus inflatus* in extreme southern regions, and *Scalopus aquaticus texensis* in the northern and Panhandle regions. Their distribution encompasses the eastern two-thirds of the state, and in the northern Panhandle it extends to the New Mexico line along the Canadian River drainage. Moles remain very common in Texas wherever suitable soils exist.

[228] *Scalopus aquaticus texanus* is an enigmatic subspecies known only from Presidio County by a single specimen taken in 1887. The specimens Bailey discusses from southern and central Texas would probably be recognized today as *Scalopus aquaticus inflatus, Scalopus aquaticus alleni,* or possibly *Scalopus aquaticus cryptus.*

206   NORTH AMERICAN FAUNA.   [No. 25.

in walking over it the feet constantly sink into the network of old burrows. In fields the freshly raised ridges can be traced for long distances. The moles are commonly accused of eating sweet potatoes, cutting the roots of fruit trees, and of doing other mischief, for most of which the pocket gopher or 'salamander' is responsible. The food of the moles consists almost entirely of insects, earthworms, and various other inhabitants of the soil, in pursuit of which the animals sometimes are troublesome by disturbing the roots of young plants and by marring the surface of lawns and parks with ridges and little mounds of earth. But all things considered the mole is too valuable an ally of the farmer to be destroyed.

Scalopus aquaticus texanus Allen.   Texas Mole.

228   So far as known, *Scalopus texanus*[M] is found only in semiarid Lower Sonoran zone, from Cameron County north to Mason. Specimens examined from Rockport, Corpus Christi, Santa Rosa Ranch (near northwest corner of Cameron County), Padre Island (north end), and Mason, as apparently also two imperfect specimens from San Antonio and Long Point, while showing marked variation at every locality from which perfect specimens were secured, can all be referred to this form. While at each locality the specimens are surprisingly uniform in characters, and the variation is sufficient for recognition, a careful comparison of specimens indicates that the result of further subdivision would only be confusing. The physiography of the middle Gulf region of Texas tends to the isolation of all burrowing mammals. Some of the rivers with headwaters in sandstone and granite formations cut through wide plains of the most impervious, waxy soil, in which no mammal can burrow, and while some of these streams leave more or less continuous deposits of mellow, sandy soil along their courses, others carry their contributions to the coast, to be built into interrupted areas of sand flats, dunes, and islands, between which the rivers with their wide flood bottoms form as impassable barriers as the wide stretches of waxy prairie. In some cases the isolation is complete; in others, partial. While the general conditions are similar, locally they are more or less varied, and their effect on the burrowing mammals is analogous to that on mammals found on a series of oceanic islands.

In habits *texanus*[M] does not differ from other species of the genus. At Corpus Christi it is common on the scattered patches of sandy soil, and common also over the sandy prairie for a distance of 65 miles, from near Santa Rosa to Sauz on the Alice and Brownsville stage road. Lloyd reported it as abundant on Padre Island. On the half-naked sands near the coast mole ridges are usually conspicuous, and the mounds, while less numerous, are often as large as those of the pocket gopher.

Scalopus aquaticus intermedius[M]Elliot.   Plains Mole[M]

229 Two specimens of moles from Mobeetie and three from Lipscomb, while not typical *intermedius*,[M] are nearer to it than to any other species. Externally they agree with topotypes from Alva, Okla., but the slender skulls indicate a distant connection with *aquaticus* farther east. One of the specimens from Mobeetie is tinged all over with a delicate purple, evidently from the root juice of a *Lithospermum*.

Howell reports that these moles are more or less abundant at Mobeetie, Miami, Canadian, and Lipscomb, where their runways are especially numerous in cultivated fields, among the sand hills, or on sandy bottoms, while a few were found on wet bottoms and on ground that was flooded in times of high water. At Tascosa in 1899 I found mole ridges common over the sandy river bottoms.

[Sorex personatus[M]Geoffroy Saint Hilaire.   Common Eastern Shrew.[M]

230 A specimen of this shrew, recorded by Mr. Oldfield Thomas[a] as received with the William Taylor collection from San Diego, Tex., is apparently the only record of a *Sorex* for the State. As numerous collectors have failed to find the species in the State, or anywhere within the life zone including San Diego and most of Texas, it seems probable that this specimen originally came from some other part of the country.]

Notiosorex crawfordi (Coues) (Baird MS.).   Crawford Shrew. Eared Shrew.[M]

231 *Notiosorex* differs from *Blarina* in having 28 instead of 32 teeth. *N. crawfordi* is larger than *B. parva*[M] or *berlandieri*,[M] with more conspicuous ears, and with tail about $2\frac{1}{2}$ instead of $1\frac{1}{2}$ times as long as hind foot.

This shrew was described from specimens collected at old Fort Bliss, 2 miles above El Paso, and additional specimens have since been collected at San Diego, Corpus Christi, and San Antonio. It has a wide range in the arid Lower Sonoran zone of Mexico, southern California, and Arizona, and so far as we know reaches its eastern limit near Corpus Christi and at San Antonio.

Blarina brevicauda carolinensis (Bach.).   Carolina Short-tailed Shrew.[M]

232 A specimen of the Carolina short-tailed shrew from Joaquin and two from the Big Thicket, 8 miles northeast of Sour Lake, extend the range of this species from eastern Arkansas and western Mississippi into eastern Texas. Though these shrews are never abundant and are easily overlooked in collecting, they may yet be found over much of eastern Texas where the conditions are favorable. Hollister caught the Joaquin specimen in a runway under old grass on low ground at

---

[a] Proc. Zool. Soc. London, 1888, p. 443.

[229] *Scalopus aquaticus intermedius* is now classified as *Scalopus aquaticus aereus*, which is known from the Panhandle and extreme eastern Texas (Yates and Schmidly, 1977).

[230] The name *Sorex personatus* has been changed to *Sorex cinereus* (Demboski and Cook, 2003). As Bailey concluded, this species does not occur in Texas.

[231] *Notiosorex crawfordi* is a relatively rare small shrew that occurs throughout much of the western half of Texas. Bailey did not record this species from northern Texas, but it is now known from throughout the Panhandle region.

[232] *Blarina brevicauda carolinensis* has been elevated to specific status as *Blarina carolinensis*, which ranges over the eastern one-fourth of the state (Schmidly and Brown, 1979). Two subspecies are recognized, *Blarina carolinensis carolinensis* in the northeastern region and *Blarina carolinensis minima* in the southeastern part of the state. Another species of short-tailed shrew, *Blarina hylophaga*, has been discovered in Texas during the twentieth century (George et al., 1981). This species is known from Montague County in northern Texas and from disjunct populations in southeast Texas (Aransas County on the Gulf Coast and Bastrop County in the Lost Pines region). Both species have been taken in near sympatry in Bastrop State Park near Bastrop, Texas (Baumgardner et al., 1992). *Blarina carolinensis* and *B. hylophaga* are cryptic species that are distinguishable from one another only by subtle morphometric differences and features of the karyotype (Stangl and Carr, 1997). The conservation status of both species is unknown, although neither one is particularly common. There may be reason to have concern about the status of *B. hylophaga*, especially given the disjunct distribution of its two subspecies (*Blarina hylophaga hylophaga* from a single county along the Red River and *Blarina hylophaga plumbea* from two counties in south-central Texas).

[233] Bailey mistakenly placed the least shrew, *Cryptotis parva*, in the genus *Blarina* (see Whitaker, 1974). The range of this species covers the eastern one-half of the state and the Panhandle where it appears to be relatively common. Bailey did not capture the least shrew in the Panhandle, despite extensive trapping efforts by the agents in that region (Fig. 52).

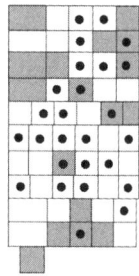

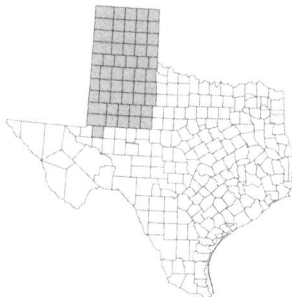

Figure 52. The current distribution of the least shrew (*Cryptotis parva*, closed circles) in the Panhandle region of Texas (shaded area of state map) according to recent county specimen records. Shaded counties indicate those that were surveyed by federal agents from 1889 to1905, but the species was not documented in the region at that time.

the edge of a cotton field about a mile east of town. It was half eaten while in the trap by some other animal, probably by one of its own species. Gaut caught the two Big Thicket specimens in traps set by old logs in the woods near Mike Griffin's place.

Blarina parva[M](Say).　Least Short-tailed Shrew.

233　This smallest of the United States species of short-tailed shrews has been taken at Gainesville, Hempstead, and Richmond. As throughout a wide range over the eastern United States it is a rare, or at least a rarely taken species, it may well be as common over a large part of eastern Texas as over the rest of its range.

The Gainesville specimens in the Merriam collection were taken by G. H. Ragsdale, but on the same ground in 1892 I was unable to find any trace of these animals save a few old runways under a carpet of fallen prairie grass. At Richmond in 1899, while trapping for *Sigmodon* on the big coast prairie, I caught one in its own little runway under the prairie grass. At Hempstead Gaut caught one in a trap set in the dry grass near a rain pool.

Blarina berlandieri[M]Baird.　Rio Grande Short-tailed Shrew.[M]

234　The Rio Grande *Blarina*[M]is slightly larger and paler than *parva*,[M] but very similar in general appearance. It was described from specimens collected at Matamoras, Mexico, and other specimens have been taken at Brownsville, San Diego, and Del Rio, Tex. Little is known of its habits, which apparently are similar to those of *parva*.[M]

At Del Rio in February, 1890, I caught one in a *Sigmodon* runway on grassy bottoms of San Felipe Creek a couple of miles from the point where the creek joins the Rio Grande.[a]

Myotis velifer (J. A. Allen).　Cave Bat.[M]

235　The four localities from which this little brown bat is known in Texas—the mouth of the Pecos, Langtry, New Braunfels, and San Antonio—when added to its wider range from Arizona to Missouri and south to southern Mexico, indicate that the species covers at least the western half of Texas. Specimens collected at mouth of Pecos by Lloyd, August 23 and September 4, 1890, and at Langtry by Gaut, March 29, 1893, indicate that it is a summer resident along the Rio Grande. Lloyd's specimens were "found in a cave tunnel," and Gaut's were taken in Pump Canyon, a deep box canyon near Langtry. I collected three adult males of this bat at Marble Cave, Mo., on June 28 and 30, 1892. One was caught in the cave 150 feet below the surface of the earth; the others were shot as they came out of the mouth

[a] This Del Rio specimen, which is typical *berlandieri*,[M]was by some accident referred by Doctor Merriam to *parva*, although he had previously written the name *berlandieri*[M]against it in the catalogue. (N. Am. Fauna, No. 10, p. 18, 1895—Revision of Shrews.)

[234] The least shrew is now placed in the genus *Cryptotis* (see Annotation 233). The subspecies on the Rio Grande Plain is *Cryptotis parva berlandieri* and populations in the northern two-thirds of the state are assigned to *Cryptotis parva parva*.

[235] The cave myotis, *Myotis velifer*, is a year-round resident of Texas, although it exhibits a varied seasonal distribution in the western two-thirds of the state. There are two subspecies in Texas, *Myotis* *velifer incautus* in the Trans-Pecos, Edwards Plateau, and South Texas Plains region, and *Myotis velifer magnamolaris* in the High Plains and Rolling Plains. Bailey's specimens were all *M. v. incautus*. This is one of the most abundant bats in Texas, roosting in caves and tunnels, rock fissures, manmade structures, and even in abandoned cliff swallow nests.

of the cave in the evening. If this bat is habitually a cave dweller, the distribution of caves probably accounts for its somewhat erratic range.

Myotis californicus (Aud. and Bach.). Little California Bat.[M]

236 This tiny bright brown bat comes into the desert country of western Texas, but evidently is not very common. A single specimen collected at Paisano by Lloyd on July 21, and another that I shot at Peña Coloral, 5 miles south of Marathon, May 14, and one on Terlingua Creek, July 1, seem to furnish the only records for the State. The species is common in New Mexico just north of the Texas line. Five specimens—three males and two females—collected by James H. Gaut in the foothills on the east slope of the San Andreas Mountains, New Mexico, January 19 and 20, 1903, indicate that the bats are not only resident, but are active during winter months. At Santa Rosa, N. Mex., I found them common in May, and a female shot on the 29th contained one small embryo. On the wing they are scarcely distinguishable from *Pipistrellus hesperus*,[M] but they are usually found in the open or among trees, while *hesperus*[M] keeps mainly to the canyons and cliffs.

Myotis incautus[M] (J. A. Allen). House Bat.[M]

237 Apparently the only known specimens of this bat are the five taken at San Antonio by Mr. H. P. Attwater, from which Doctor Allen described the species; seven collected by M. Cary and myself 15 miles west of Japonica, Kerr County; one collected at Langtry by James H. Gaut, and eight collected at Carlsbad, N. Mex. The San Antonio specimens were collected March 12 and October 10, which would suggest that they were migrants. The Japonica specimens were taken July 7 and 8, and were on their breeding ground, as probably were those taken at Carlsbad July 29 and September 17. A female collected in Pump Canyon, near Langtry, March 29, may have been either resident or migrant.

Little is known of the habits of this species. On the North Fork of the Guadalupe, west of Japonica, Cary and I found them early in the evening, flying up and down the rocky bed of the stream in great abundance, dipping to the water pools to drink and then zigzagging through the air in pursuit of insects. With a fairly good light, we secured seven of the bats after a few minutes' rapid shooting. The bats apparently came from the limestone cliffs both above and below the open space where we found them.

At the water tower 3 miles southwest of Carlsbad, where a large pool is formed from the pure mountain water pumped up to supply the town, these bats came in over the dry plain on the evening of July 29 from some limestone hills several miles away.

[236] The California myotis, *Myotis californicus*, is a year-round resident of the Trans-Pecos, where it occurs in desert, grassland, and wooded habitats. One specimen has been recorded from Canyon in Randall County near the breaks of the Llano Estacado (Choate and Killibrew, 1991). This is one of the most common bats in the western part of Texas and is represented by the subspecies *Myotis californicus californicus*.

[237] *Myotis incautus* is now regarded as a subspecies of *Myotis velifer* (Fitch et al., 1981). It is the second most common bat, following *Tadarida brasiliensis*, in the cave region of the Hill Country.

[238] The Yuma myotis, *Myotis yumanensis*, is restricted in Texas to the southern Trans-Pecos and Rio Grande Valley where it inhabits desert, lowland habitats. A single specimen obtained by the Texas Department of State Health Services indicates a record from Dallas County. Most specimens come from areas near the Rio Grande where the species appears to be common. Seven other species of *Myotis* have been recorded in Texas since the Biological Survey: *Myotis austroriparius*, *Myotis ciliolabrum*, *Myotis lucifugus*, *Myotis occultus*, *Myotis septentrionalis*, *Myotis thysanodes*, and *Myotis volans*. Most of these species are rare in Texas and have restricted distributions.

[239] Current taxonomy places the western pipistrelle, *Pipistrellus hesperus*, in the genus *Parastrellus* (Hoofer and Van Den Bussche, 2003; Hoofer et al., 2006). This species is widely distributed in suitable rocky habitats from the Trans-Pecos region northward along the eastern escarpment of the Llano Estacado. It is one of the most common bats of the desert southwest and the subspecies in Texas is *Parastrellus hesperus maximus*.

210   NORTH AMERICAN FAUNA.   [No. 25.

They were flying straight for the water pool without a crook or turn, and I shot four without missing, a rare occurrence in bat shooting. These were all females, but four taken on September 17 at the Bolles ranch, 6 miles south of Carlsbad, were all males. Three of these were shot in the evening as they flew about the house, and one was caught in the daytime in a corner of an outhouse.

In the original description of the species, based on a series of five specimens taken at San Antonio by Mr. Attwater, March 12 and October 10, Doctor Allen says: "It is a 'house' bat, all of the specimens having been taken in the house except one, which was caught in a barn."[a]

Myotis yumanensis (H. Allen). Yuma Bat.[M]

238 This little light-brown bat was not known from Texas until May 26, 1903, when Gaut found a breeding colony near Del Rio. He collected a series of eight adult females and one young, and says: "They were taken from a colony of bats, all the same species, in a shallow cave near the railroad about 10 miles west of Del Rio. When disturbed they flew about, the females each carrying a young one clinging to its breast. One of these young was obtained and prepared." It was very small, almost naked, and apparently its eyes were not yet open.

Pipistrellus hesperus[M] (H. Allen). Little Canyon Bat.[M]

239 These tiny gray bats are easily recognized by their jet-black ears, tail, and wings. They come into arid Lower Sonoran zone of western Texas as far east as the Pecos Valley. There are specimens from El Paso, Chinati Mountains, Grand Canyon of Rio Grande, Terlingua Creek, Boquillas, points 15, 20, and 80 miles south of Marathon, Alpine, Paisano, Davis Mountains (east base), Sanderson, Pecos River (at mouth), and farther up the Pecos Valley from near Carlsbad and Santa Rosa, N. Mex.

These bats are usually the most abundant of the species where they occur, and they are, more than any other species I know, strictly canyon or cliff dwellers. They often follow up the canyons to the extreme limits of Lower Sonoran zone on the warm slopes where the surrounding country is entirely Upper Sonoran or even Transition, and the hotter, dryer, and barer the canyon the thicker these midgets swarm. They fly early, sometimes coming out on the shady side of a canyon before the last trace of sunlight has disappeared, but even with a fair amount of light they are not easily shot. Their flight is rapid and crooked, and the collector wastes more ammunition on them than on almost any other bat.

The Texas records for this species are all for summer, May 10 to

---

[a] Bul. Am. Mus. Nat. Hist., VIII, 239, 1896.

August 26, and breeding specimens are found throughout their range. A female collected 20 miles south of Marathon, May 10, contained two half-developed embryos, and two collected near Boquillas, May 23 and 24, each contained one large embryo. Another taken at Santa Rosa, N. Mex., May 27, contained two small embryos.

In specimens of this bat shot only a few minutes (twenty or thirty at most) after they began to fly in the evening, I have invariably found the stomachs stuffed full of freshly eaten insects—a fact which speaks well for their skill as flycatchers.

Pipistrellus subflavus[M] (F. Cuvier).   Georgian Bat.[M]

240 Specimens of the Georgian bat from Clear Creek, in Galveston County, Long Lake, Brownsville, Devils River, Comstock, and Del Rio indicate a range over the eastern part of the State and as far west as the timber extends along streams in Lower Sonoran zone. The species has a wide range over the southeastern United States, and finds its western limits in Texas. The Brownsville specimen collected October 10, 1891, may have been a migrant, as may also have been those from Clear Creek, taken on March 28. One from Devils River, collected July 23, and three from Long Lake, procured July 19 and 20, were undoubtedly on their breeding grounds, as were probably those from Del Rio, collected May 21 and 22, and one from Comstock, collected May 3.

Vespertilio fuscus[M] Beauvois.   Large Brown Bat.[M]

241 Specimens of the brown bat from Jefferson, Sour Lake, the Brazos River, Grady, and the Davis and Chisos mountains carry the range of the species across Texas from east to west without defining any limits of range, but the species apparently has not been taken in the southern part of the State. The specimen collected by Hollister at Jefferson, June 14, may have been a late migrant, as the species is supposed not to breed in Lower Sonoran zone. One collected by Gaut near Sour Lake, March 17, was undoubtedly a migrant. The Brazos River specimen is an old alcoholic (No. 11217, U.S.N.M.), without date. The Davis and Chisos mountains specimens were in Transition zone, and probably on their breeding grounds. Both are males, and in both localities the species seemed to be common. The one from the Davis Mountains was shot by L. A. Fuertes on the evening of July 12 as it came down the gulch over our camp at 5,700 feet altitude. The Chisos Mountain specimen was shot by McClure Surber, June 9, at our camp in the gulch at 6,000 feet altitude, at the edge of Transition zone.

At Mr. C. O. Finley's ranch, at the west base of the Davis Mountains, I found two lower jaws of this bat among numerous other bones in pellets under the nest of a great horned owl.

[240] Current taxonomy places *Pipistrellus subflavus* in the genus *Perimyotis* (Hoofer and Van Den Bussche, 2003; Hoofer et al., 2006), and the common name for this bat is now the American perimyotis (previously it had been known as the eastern pipistrelle). It occurs over the entire state except for far West Texas, but recent records from Big Bend Ranch State Park (Yancey, 1997) and the Davis Mountains (the late Clyde Jones, personal communication) in the Trans-Pecos suggest it is expanding its range westward. Two subspecies are now recognized in Texas, *Pipistrellus subflavus subflavus* over most of the state (including Bailey's specimens from Clear Creek, Long Lake, and Brownsville), and *Pipistrellus subflavus clarus* from Val Verde County (Bailey's specimens from the Devils River, Comstock, and Del Rio). The two subspecies were not delineated at the time of the Texas biological survey.

[241] *Vespertilio fuscus* is now placed in the genus *Eptesicus* (see Simmons, 2005). The big brown bat, with two recognized subspecies (*Eptesicus fuscus pallidus* and *Eptesicus fuscus fuscus*), has a disjunct distribution in Texas. *Eptesicus f. fuscus* occurs in the eastern part of the state, and *E. f. pallidus* in West Texas. This species is a year-round resident and is common over most of its range in Texas.

[242] The eastern red bat, *Lasiurus borealis*, occurs statewide but it is much more common in the eastern part of the state, where it is a year-round resident, than in the Trans-Pecos, where it occurs only in the summer. Another closely related species, *Lasiurus blossevilli*, has been recorded in Texas in the Sierra Vieja Mountains of Presidio County (Genoways and Baker, 1988) and from Starr County in south Texas (Weaver et al., 2020), but none of the specimens secured by Bailey were representative of that species.

[243] *Lasiurus borealis seminolus* is now recognized as a distinct species, *Lasiurus seminolus* (Baker et al., 1988). The seminole bat is a nonmigratory species that occurs throughout the oak-hickory, pine-oak, and longleaf pine forest regions of east Texas where its preferred roosting sites are in Spanish Moss. This species is locally abundant and does not appear to have declined substantially. A recent record along the Devils River suggests it may be expanding its range.

212   NORTH AMERICAN FAUNA.   [No. 25.

Lasiurus borealis (Müller).   Red Bat.[M]

242 The red bat is common over eastern Texas and westward to the lower Rio Grande, Devils River, and Wichita Falls. Specimens have been taken at Jefferson, Clarksville, Arthur, Paris, Waco, Tarkington Prairie, Wichita Falls, Camp Verde, Ingram, Nueces Bay, Corpus Christi, Brownsville, Fort Clark, and Devils River. Its western limit in the State apparently corresponds to the limit of essentially treeless plains. Being a tree bat and partial to the deep shade of bottom-land forests, it follows the stream courses into the plains as far as they carry timber.

The dates on Texas specimens, covering a period from March 19 to November 30, do not indicate whether the species is migratory or resident, or whether, if resident, it hibernates or is active during the winter months, but there is abundant proof that it breeds throughout its Texas range. A female shot at Clarksville June 10 contained two fully developed fetuses, as also another, shot the next evening at Paris. Two females shot at Paris June 11, and 3 at Arthur June 16, were all nursing young. In a large series of specimens collected by F. B. Armstrong at Brownsville there are 39 young, ranging in size from tiny, almost naked individuals a few days old to almost full-grown animals, and bearing dates from May 19 to July 25. Most of the very small young were taken in May, but one of the smallest is dated July 18. Adults were collected at Brownsville by Lloyd as late as September 10, at Corpus Christi November 13, and at San Patricio, near the mouth of the Nueces, November 30.

These bats are among the least difficult to collect, as they come out early in the evening and their flight is comparatively slow. In leafy woods they often come out soon after sundown, while there is still light enough to distinguish the species by its color, form, and flight. At Wichita Falls I shot one as it was flying about in the woods near the river in bright sunlight at about 4 p.m. In the big pecan grove near the head of Devils River they were very numerous in July.

Lasiurus borealis seminolus[M](Rhoads).   Florida Red Bat.[M]

243 A single specimen of this rich mahogany-brown subspecies, recorded by Gerrit S. Miller, jr., from Brownsville[a] is the only record for Texas. It was killed September 8, 1891, and was probably a migrant from its usual summer range in the South Atlantic and Gulf States east of Texas. Upon all grounds of geographic distribution this bat should be a summer resident of eastern Texas, and it will probably be found there when this most neglected group of all our North American mammals becomes better known.

---

[a] N. Am. Fauna, No. 13, p. 109, 1897. Revision of N. Am. Bats of the Family Vespertilionidae, by Gerrit S. Miller, jr.

Lasiurus cinereus[M](Beauvois).　Hoary Bat.

244 The hoary bat probably migrates over the whole of Texas, but it is known in the State only from nine specimens collected at Brownsville and one from the Davis Mountains. The seven available Brownsville specimens are dated October 23, November 16, December 20, January, May 7, and May 23, and probably are migrants. The Davis Mountain specimen was shot by L. A. Fuertes, July 10, 1891, at 5,700 feet altitude, in a gulch northeast of Mount Livermore. It is an adult male, and was shot early in the evening as it came down the gulch from the east side of the mountain.

Dasypterus intermedius H. Allen.　Yellow Bat.[M]

245 Specimens of this large, yellow, short-eared bat from Brownsville and the south end of Padre Island, Texas, and Matamoras, Mexico, furnish all that is known of the range of the species, a range covering scarcely 30 miles near the Gulf coast in the semiarid cactus and mesquite country of Lower Sonoran zone.

　　None of the collectors of this bat have written anything on its habits, but a male collected by Lloyd on Padre Island August 26, and a series of 57 males, females, and young collected by Armstrong at Brownsville from May 12 to August 4, show that this region is the breeding ground of the species. Females collected May 12, 14, and 19 contained each two large fetuses, and seven young collected June 7 to 17 are about half grown, while one taken July 16 is but little larger.

Nycticeius humeralis Rafinesque.　Evening Bat.

246 This little, dark brown bat has been taken over eastern and southern Texas at Paris, Arthur, Texarkana, Jefferson, Jasper, Hidalgo, Lomita Ranch, and Brownsville, at dates ranging from May 8 to August 19. At Texarkana Oberholser took two nearly full-grown young, June 23, and at Jasper three not fully adult specimens on August 18 and 19. At Brownsville a series of fifteen less than half-grown young was collected by F. B. Armstrong, June 1 to 12, and two about half-grown young on June 17 and 24. Twelve adults taken May 8 to June 17 were all females, but an apparently adult male was taken by Lloyd on July 23. It was "found hanging on mesquite."

　　At Texarkana Oberholser reports this species as "the common bat of the bottoms," and at Jasper as abundant. Near Jefferson Hollister shot two on the evening of June 13 at our camp in the timber by Big Cypress Creek, where bats apparently of this species were numerous.

[244] Modern taxonomists place the hoary bat, *Lasiurus cinereus*, in the genus *Aeorestes* (Baird et al., 2015). *Aeorestes cinereus* is a spring–fall migratory species that has been recorded seasonally in all vegetative regions of the state. In the Trans-Pecos, it occupies mountainous, wooded areas and can be locally abundant.

[245] At one time the name *Dasypterus intermedius* was changed to *Lasiurus intermedius*, but recent evidence has returned it to the genus *Dasypterus* (Baird et al., 2015). The northern yellow bat occurs primarily along the Gulf Coast, with a single record from Dallas County. It does not appear to be common anywhere within its range. Texas specimens are referrable to two subspecies, *D. intermedius floridanus* from Bexar and Travis counties eastward and *Dasypterus intermedius intermedius* from Victoria County southward. A recent molecular genetics study (Decker and Ammerman, 2020) has revealed two genetically distinct lineages that correspond to the two subspecies' geographic distributions. The specimens obtained by Bailey are *D. i. intermedius*.

Two other "cryptic" species of *Dasypterus*, *Dasypterus ega* and *Dasypterus xanthinus*, have been recorded in Texas since the Biological Survey. *Dasypterus ega* is known from the coastal region of south Texas, and *D. xanthinus* is known from the Big Bend region and from a single record in Starr County in southern Texas. Both of these species are rare and little is known about their status in Texas (see Chapter 5).

[246] The evening bat, *Nycticeius humeralis*, is among the most common species of bats east of the 100th meridian in Texas. It is a year-round resident of the state, roosting in hollow trees and building attics. Captures of this species (Boyd et al., 1997; Dowler et al., 1999; Krejsa et al., 2020) west of the 100th meridian (see Schmidly and Bradley, 2016) suggests it may be expanding its range to the west, much like the American perimyotis (*Perimyotis subflavus*).

[247] The big-eared bat mentioned by Bailey, *Corynorhinus macrotis pallescens*, is now classified as *Corynorhinus townsendii* (Hoofer and Van Den Bussche, 2001). This species seems to prefer caves and mine tunnels, and two subspecies occur in Texas, *Corynorhinus townsendii australis* in the Trans-Pecos and *Corynorhinus townsendii pallescens* in the Panhandle. A second big-eared bat, *Corynorhinus rafinesquii macrotis*, occurs in extreme east Texas. Bailey was correct in predicting this species eventually would be found in the state. It was first recorded in 1965 (Michael and Birch, 1967), and specimens have now been recorded in 15 counties in East Texas (Horner and Maxey, 1998). It is among the rarest bats in Texas.

[248] The pallid bat, *Antrozous pallidus,* occupies the western one-half of the state. Martin and Schmidly (1982) initially assigned all specimens of *A. pallidus* to *Antrozous pallidus pallidus,* including specimens from the Texas Panhandle. Manning et al. (1988) were able to examine a larger sample of pallid bats from the Panhandle region and concluded that they were affiliated with specimens of *Antrozous pallidus bunkeri* distributed to the north and east (Oklahoma and Kansas). Consequently, two distinct subspecies are recognized in Texas, *Antrozous pallidus bunkeri* in the northern Panhandle and *Antrozous pallidus pallidus* in the western and southern regions. Bailey and the federal agents obtained representatives of both of these subspecies, which are among the most abundant bats in the Trans-Pecos.

Corynorhinus macrotis pallescens[M] Miller.   Long-eared Bat.[M]

247 A single specimen of this pale subspecies of the long-eared bat, collected by Lloyd in the "east" Painted Cave, September 5, 1890,[a] apparently forms the only record for Texas. From its wide range over Mexico and arid Lower Sonoran of southern California and Arizona, the species may be expected to inhabit at least a large part of western Texas. In the eastern part of the State its place would naturally be taken by the darker colored *macrotis*,[M] which has not yet been recorded from Texas, but which breeds abundantly in southern Louisiana.

Antrozous pallidus (Le Conte).   Pale Bat.[M]

248 This large, light-colored bat is common throughout the summer in arid Lower Sonoran zone of western Texas from Sycamore Creek, Devils River, and the Pecos Valley westward, and a single specimen was obtained at Tascosa, in the northwestern part of the Panhandle. The records cover a period from April 18 to October 11, but these limits are apparently dates of collectors' entering and leaving the region rather than of the migration of the bats. Still in the Rio Grande Valley enough winter work has been done to prove that the bats either migrate or hibernate during cold weather. That they breed in the region is amply proved by their remaining throughout the summer months, and by a female shot near Boquillas, May 28, and three females taken at Comstock May 11, each containing two large fetuses. Lloyd collected a half-grown young at Paisano, July 18.

During the day they hide in cracks of buildings, and probably also in cliffs, as they inhabit rocky country where there are no buildings. At Comstock, May 11, 1901, Oberholser found "eight or nine roosting behind the signboard of a store, and they were said to have been driven from a similar place at the railroad station." He secured five of these, which proved to be four females and one male. On July 24 and 25 of the following year Hollister caught seven more (four males and three females) from behind the signboard at the railroad station at Comstock. At Van Horn one came into my room in the evening of August 23, and was caught. Near Carlsbad, N. Mex., these bats were abundant around the house on the Bolles ranch in September, coming out of cracks in buildings early in the evening and flying softly around the house in the twilight before the smaller bats began to appear. During the day I often heard them squeaking behind the casings, and with a pair of forceps took five from behind a board. Of six specimens taken, three were males and three females. In this species I have never found a striking preponderance of either sex, probably because I have found them only in the breeding season.

---

[a] Probably the third cave, about a mile below the mouth of the Pecos.

Their flight is soft and noiseless, and, while rapid, it is not so quick and jerky as that of most bats. Their light color and large size render them unmistakable in the early evening; even the long, projecting ears can sometimes be distinguished as the bats fly over. An old female, previously mentioned as containing two fetuses, measured 345 mm. (approximately a foot) from tip to tip of wings while fresh.

Nyctinomus mexicanus[M] Saussure.   Free-tailed Bat.[M]

249 The free-tailed bat is the most abundant species over approximately the western half of Texas in arid Lower Sonoran zone. Its eastern limit of range, so far as known, agrees closely with the eastern limits of mesquite. There is a specimen in the U. S. National Museum collection labeled Indianola. The species is abundant at San Antonio, and I have examined specimens from Brazos, San Angelo, Kerrville, Ingram, Padre Island, Brownsville, Hidalgo, Eagle Pass, Del Rio, Comstock, mouth of Pecos, Langtry, Boquillas, Alpine, Davis Mountains, Fort Stockton, and up the Pecos Valley as far as Roswell, N. Mex. The abundant bats in the town of El Paso are probably of this species.

In at least a part of their Texas range these bats are not only resident, but active throughout the winter months. At Del Rio I found them abundant in January and February, 1889; Lloyd and Streator found them common at Eagle Pass in November, and Lloyd collected one on Padre Island November 11. At Brazos Cary found them as late as October 9, 1902, and says: " I shot twenty of these bats in a crack in the bridge where the Texas Pacific Railroad crosses the Brazos. The bats were in the cracks by hundreds." Most of his alcoholic specimens are very fat, which would suggest that later they might have hibernated. At San Angelo Oberholser reported the species April 2 to 4, 1901: "Abundant along the Concho, where one was taken. All the bats seen were apparently of this genus." At Fort Stockton, in August, Cary reports them as the "most abundant bat." At Alpine, July 5, they swarmed out of the adobe walls of empty houses in the evening until the town was full of them and their musky odor. A few were shot in the canyons of the Davis Mountains July 10, and their unmistakable odor was very noticeable among the old adobe walls at Fort Davis. At the ranch of Mr. Howard Lacey, near Kerrville, these bats were numerous May 1 to 7, 1899. Some were shot around the ranch buildings in the evening, and one of their roosting places was found in a crack under an overhanging rock of a high cliff. I heard them squeaking and apparently fighting in the crack. A few shots of the auxiliary brought sixteen of them to the ground, and examination showed these

[249] *Nyctinomus mexicanus* is now classified as *Tadarida brasiliensis* (Simmons, 2005). There are two subspecies in Texas, *Tadarida brasiliensis cynocephala* in the extreme eastern portions of the state and *Tadarida brasiliensis mexicana* over the remainder. This is the most common bat in Texas, reaching its greatest concentration in the caves of the Hill Country where summer populations may number up to 10–20 million bats in each cave. The subspecies *T. b. cynocephala* roosts primarily in human structures and does not occur in such huge concentrations. The specimens obtained by Bailey were all *T. b. mexicana*.

[250] *Promops californicus* is now classified as *Eumops perotis californicus* (Simmons, 2005). This is the largest bat, and one of the most rare, in the United States. Besides the specimen obtained by Gaut, individuals have been captured in Texas in the rugged, rocky canyon country adjacent to the Rio Grande in Presidio, Brewster, Terrell, and Val Verde counties, with a single record from Midland County. This is a species that bears monitoring in the future.

[251] *Mormoops megalophylla senicula*, now a synonym of *Mormoops megalophylla megalophylla* (Rezsutek and Cameron, 1993), has the most distinctive facial ornamentation of any bat in Texas. Specimens have now been obtained in the Trans-Pecos, the southern escarpment of the Edwards Plateau, and extreme south Texas. At Big Bend Ranch State Park in Presidio County, this was one of the most common bats collected (Yancey, 1997). It is probably not as rare within its geographic range as previously thought.

to be males and females in about equal numbers. The embryos in these females were just beginning to enlarge noticeably, but a female shot at Boquillas May 28 contained a half-developed embryo. On a hot May evening in San Antonio I have watched a stream of these bats fly from under the cornice of the old adobe hotel, making the hot air heavy with their odor. They are partial to towns and adobe houses. At Del Rio, in January and February, 1889, they were excessively numerous. At dusk the air seemed full of them, and several people told me that their houses were so infested with bats that no one would rent them. On visiting one of these vacant houses in the evening I found bats pouring out of cracks and holes in the boards that covered the adobe walls. There was an incessant squeaking and scratching as they climbed over the inner surface of the boards and fought and pushed each other at the narrow places of exit. The noise could be heard across the street. I stood at a knothole and caught them as they came out one at a time, until I had nine in my hands, but by the time I had dispatched these the others were all out and on the wing. My specimens were covered with lice and redolent with a peculiar rank, musky odor that did not leave my hands for a couple of days. The odor is like that of the house mouse, only much stronger, and it is often noticeable as you walk along the sidewalk past some of the bat-infested houses.

Promops californicus[M](Merriam).   Bonnet Bat.[M]

250   A single specimen of this large bat, collected at Langtry, Tex., March 8, 1903, by James H. Gaut, adds the species to the Texas fauna and extends its range from the southern parts of Arizona and California. Gaut says it was caught in the pump house at the bottom of Pump Canyon, near Langtry.

Mormoops megalophylla senicula[M]Rehn.   Rehn Bat.[M]

251   The only specimen of this Mexican and West Indian bat recorded from the United States was taken by Dr. E. A. Mearns at Fort Clark, Tex., December 3, 1897. Doctor Mearns says:

> A lady called me to her house to see a 'very remarkable bat' which had attached itself to the inner side of a door-screen. I found this bat very much alive, at a season when all other bats of the locality were dormant or had migrated. No other bats were seen until the following March, when the common *Nyctinomus* reappeared in the usual abundance.[a]

---

[a] Proc. Biol. Soc. Wash. XIII, p. 166, 1900.

Table 4. Scientific names of plant species as referenced in the *Biological Survey of Texas* (1905) and the current taxonomic designations for those species.

Old Name	Current Name	Common Name
Acacia	*Acacia, Senegalia,* and *Vachellia*	
Acacia amentacea	*Vachellia rigidula*	Blackbrush Acacia
Acacia berlandieri	*Senegalia berlandieri*	Guajillo
Acacia constricta	*Vachellia constricta*	Mescat Acacia
Acacia farnesiana	*Vachellia farnesiana*	Huisache
Acacia roemeriana	*Senegalia roemeriana*	Roemer Acacia
Acacia schottii	*Vachellia schottii*	Schott Acacia
Acacia tortuosa	*Vachellia tortuosa*	Twisted Acacia
Acacia wrightii	*Senegalia greggii*	Catclaw Acacia
Acer	*Acer*	
Acer drummondi	*Acer rubrum*	Red Maple
Acer grandidentatum	*Acer grandidentatum*	Bigtooth Maple
Acer grandidentum	*Acer grandidentatum*	Bigtooth Maple
Acer rubrum	*Acer rubrum*	Red Maple
Acuan	*Desmanthus*	
Acuan illinoensis	*Desmanthus illinoensis*	Illinois Bundleflower
Adelia	*Forestiera*	
Adelia angustifolia	*Forestiera angustifolia*	Narrowleaf Forestiera
Adelia neomexicana	*Forestiera pubescens*	Elbowbush
Adolphia	*Adolphia*	
Adolphia infesta	*Adolphia infesta*	Texas Adolphia
Agave	*Agave*	
Agave applanata	*Agave havardiana*	Havard Agave
Agave lecheguilla	*Agave lechuguilla*	Lechuguilla
Agave wislizeni	*Agave parrasana*	Cabbage Head Agave
Aloysia	*Mulguraea*	
Aloysia ligustrina	*Mulguraea ligustrina*	Common Beebush
Amelanchier	*Amelanchier*	
Amelanchier alnifolia	*Amelanchier alnifolia*	Serviceberry
Amorpha	*Amorpha*	
Amorpha canescens	*Amorpha canescens*	Leadplant
Amyris	*Amyris*	
Amyris parvifolia	*Amyris texana*	Texas Torchwood
Aralia	*Aralia*	
Aralia spinosa	*Aralia spinosa*	Angelica Tree
Arbutus	*Arbutus*	
Arbutus xalapensis	*Arbutus xalapensis*	Texas Madrone

Old Name	Current Name	Common Name
Artemisia	*Artemisia*	
Artemisia filifolia	*Artemisia filifolia*	Sagebrush
Arundinaria	*Arundinaria*	
Arundinaria macrosperma	*Arundinaria gigantea*	Giant Cane
Asclepias	*Asclepias*	
Asclepias latifolia	*Asclepias latifolia*	Broadleaf Milkweed
Asclepias speciosa	*Asclepias speciosa*	Showy Milkweed
Asclepias tuberosa	*Asclepias tuberosa*	Butterfly Milkweed
Ascyrum	*Ascyrum*	
Asimina	*Asimina*	
Asimina triloba	*Asimina triloba*	Common Pawpaw
Astragalus	*Astragalus*	
Astragalus bisulcatus	*Astragalus bisulcatus*	Two-grooved Milkvetch
Astragalus caryocarpus	*Astragalus crassicarpus*	Groundplum Milkvetch
Astragalus molissimus	*Astragalus mollissimus*	Wooly Locoweed
Atriplex	*Atriplex*	
Azalea	*Azalea*	
Azalea sp.	*Rhododendron sp.*	Azalea
Baccharis	*Baccharis*	
Baccharis glutinosa	*Baccharis salicifolia*	Seepwillow
Baccharis salicina	*Baccharis salicina*	Willow Baccharis
Baptisia	*Baptisia*	
Batrachium	*Batrachium*	
Batrachium divaricatum	*Batrachium divaricatum*	none available
Berberis	*Berberis* and *Mahonia*	
Berberis fremonti	*Berberis fremontii*	Fremont Barberry
Berberis repens	*Mahonia repens*	Creeping Barberry
Berberis trifoliata	*Mahonia trifoliata*	Agarita
Bernardia	*Bernardia*	
Bernardia myricaefolia	*Bernardia myricifolia*	Brush Myrtlecroton
Betula	*Betula*	
Betula nigra	*Betula nigra*	River Birch
Bignonia	*Bignonia*	
Bignonia crucigera	*Bignonia capreolata*	Crossvine
Bradleia	*Wisteria*	
Brayodendron	*Diospyros*	
Brayodendron texanum	*Diospyros texana*	Texas Persimmon
Cactus	*Cactus, Mammillaria,* and *Escobaria*	

Old Name	Current Name	Common Name
Cactus heyderi	*Mammillaria heyderi*	Flattened Mammillaria
Cactus missouriensis	*Escobaria missouriensis*	Missouri Foxtail Cactus
Callicarpa	*Callicarpa*	
Callicarpa americana	*Callicarpa americana*	American Beautyberry
Callirhoe	*Callirhoe*	
Campsis	*Campsis*	
Campsis radicans	*Campsis radicans*	Trumpet Creeper
Carpinus	*Carpinus*	
Carpinus caroliniana	*Carpinus caroliniana*	American Hornbeam
Castalia	*Nymphaea*	
Castalia elegans	*Nymphaea elegans*	Tropical Royalblue Waterlily
Castanea		
Castanea pumila	*Castanea pumila*	American Chinquapin
Castela	*Castela*	
Castela nicholsonii	*Castela erecta*	Goatbush
Ceanothus		
Ceanothus greggii	*Ceanothus greggii*	Desert Ceanothus
Celtis	*Celtis*	
Celtis helleri	*Celtis lindheimeri*	Lindheimer Hackberry
Celtis mississippiensis	*Celtis laevigata*	Sugar Hackberry
Celtis reticulata	*Celtis reticulata*	Netleaf Hackberry
Cenchrus	*Cenchrus*	
Cenchrus tribuloides	*Cenchrus tribuloides*	Dune Sandbur
Cephalanthus	*Cephalanthus*	
Cephalanthus occidentalis	*Cephalanthus occidentalis*	Common Buttonbush
Cercidium	*Parkinsonia*	
Cercidium floridanum	*Parkinsonia floridana*	Blue Palo Verde
Cercidium texanum	*Parkinsonia texana*	Texas Palo Verde
Cercis	*Cercis*	
Cercis occidentalis	*Cercis canadensis*	Eastern Redbud
Cercocarpus	*Cercocarpus*	
Cercocarpus parvifolius	*Cercocarpus montanus*	True Mountain Mahogany
Cereus	*Echinocereus*	
Cereus enneacanthus	*Echinocereus enneacanthus*	Pitaya
Cereus paucispinus	*Echinocereus triglochidiatus*	Claret-cup Echinocereus
Cereus stramineus	*Echinocereus stramineus*	Strawberry Cactus
Chilopsis	*Chilopsis*	
Chilopsis linearis	*Chilopsis linearis*	Desert Willow

303

Old Name	Current Name	Common Name
Condalia	*Condalia*	
Condalia obovata	*Condalia hookeri*	Brazilian Bluewood
Convolvulus	*Convolvulus*	
Coreopsis	*Coreopsis*	
Covillea	*Larrea*	
Covillea tridentata	*Larrea tridentata*	Creosote Bush
Crassina	*Crassina*	
Crassina grandiflora	*Crassina grandiflora*	Plains Zinnia
Crataegus	*Crataegus*	
Crataegus spathulata	*Crataegus spathulata*	Littlehip Hawthorn
Crataegus texana	*Crataegus texana*	Texas Hawthorn
Croton	*Croton*	
Croton torreyanus	*Croton incanus*	Torrey Croton
Cupressus	*Hesperocyparis*	
Cupressus arizonica	*Hesperocyparis arizonica*	Arizona Cypress
Cynoxylon	*Cynoxylon*	
Cynoxylon floridum	*Cornus florida*	Flowering Dogwood
Cyrilla	*Cyrilla*	
Cyrilla racemiflora	*Cyrilla racemiflora*	Swamp Cyrilla
Dasylirion	*Dasylirion*	
Dasylirion texanum	*Dasylirion texanum*	Texas Sotol
Daubentonia	*Sesbania*	
Daubentonia longifolia (Sesban cavanillesii)	*Sesbania drummondii*	Drummond's Rattlebox
Dendropogon	*Tillandsia*	
Dendropogon usneoides	*Tillandsia usneiodes*	Spanish Moss
Diospyros	*Diospyros*	
Diospyros virginiana	*Diospyros virginiana*	Common Persimmon
Echinocactus	*Echinocactus, Ferocactus* or *Glandulicactus*	
Echinocactus hamatocanthus	*Ferocactus hamatacanthus*	Turk's Head
Echinocactus horizonthalonius	*Echinocactus horizonthalonius*	Devilshead Cactus
Echinocactus wislizeni	*Ferocactus wislizeni*	Southwestern Barrel Cactus
Echinocactus wrighti	*Glandulicactus uncinatus*	Wright's Fishhook Cactus
Ehretia	*Ehretia*	
Ehretia eliptica	*Ehretia anacua*	Anacua
Ephedra	*Ephedra*	
Ephedra antisyphilitica	*Ephedra antisyphilitica*	Mormon Tea
Ephedra trifurcata	*Ephedra trifurca*	Longleaf Ephedra

Old Name	Current Name	Common Name
Euphorbia	*Euphorbia*	
Euphorbia antisyphilitica	*Euphorbia antisyphilitica*	Candelilla
Eustoma	*Eustoma*	
Eysenhardtia	*Eysenhardtia*	
Eysenhardtia amorphoides	*Eysenhardtia texana*	Kidneywood
Fagara	*Zanthoxylum*	
Fagara clavaherculis	*Zanthoxylum clava-herculis*	Hercules' Club
Fallugia	*Fallugia*	
Fallugia paradoxa	*Fallugia paradoxa*	Apache Plume
Flourensia	*Flourensia*	
Flourensia cernua	*Flourensia cernua*	Tarbush
Fouquiera	*Fouquieria*	
Fouquiera splendens	*Fouquieria splendens*	Ocotillo
Fraxinus	*Fraxinus*	
Fraxinus greggii	*Fraxinus greggii*	Gregg's Ash
Garrya	*Garrya*	
Garrya lindheimeri	*Garrya ovata* ssp. *lindheimeri*	Lindheimer Silktassel
Garrya wrightii	*Garrya wrightii*	Wright's Silktassel
Gelsemium	*Gelsemium*	
Gelsemium sempervirens	*Gelsemium sempervirens*	Carolina Jessamine
Gleditsia	*Gleditsia*	
Gleditsia aquatica	*Gleditsia aquatica*	Water Locust
Gleditsia tricanthos	*Gleditsia triacanthos*	Common Honey Locust
Goniostachyum	*Lippia*	
Grindelia	*Grindelia*	
Gutierrezia	*Gutierrezia*	
Gutierrezia microcephala	*Gutierrezia microcephala*	Sticky Snakeweed
Gutierrezia sarothrae	*Gutierrezia sarothrae*	Broom Snakeweed
Hamamelis	*Hamamelis*	
Hamamelis virginiana	*Hamamelis virginiana*	Common Witch-hazel
Hartmannia	*Hartmannia*	
Hechtia	*Hechtia*	
Hechtia texensis	*Hechtia texensis*	Texas False Agave
Helianthus	*Helianthus*	
Helianthus annuus	*Helianthus annuus*	Common Sunflower
Helianthus petiolaris	*Helianthus petiolaris*	Prairie Sunflower
Hesperaloe	*Hesperaloe*	
Hesperaloe parviflora	*Hesperaloe parviflora*	Red Yucca

Old Name	Current Name	Common Name
Hicoria	*Hicoria*	
Hicoria alba	*Carya tomentosa*	Mockernut Hickory
Hicoria aquatica	*Carya aquatica*	Water Hickory
Hicoria glabra	*Carya glabra*	Pignut Hickory
Hicoria ovata	*Carya ovata*	Shagbark Hickory
Hoffmanseggia	*Pomaria*	
Hoffmanseggia jamesi	*Pomaria jamesii*	James Rush-pea
Hymenocallis	*Hymenocallis*	
Ibervillea	*Ibervillea*	
Ibervillea lindheimeri	*Ibervillea lindheimeri*	Lindheimer Globeberry
Ilex	*Ilex*	
Ilex decidua	*Ilex decidua*	Possum-haw
Ilex lucida	*Ilex coriacea*	Baygall Holly
Ilex opaca	*Ilex opaca*	American Holly
Ilex vomitoria	*Ilex vomitoria*	Yaupon
Inodes	*Sabal*	
Inodes texana	*Sabal mexicana*	Texas Palmetto
Ipomea	*Ipomea*	
Ipomea leptophylla	*Ipomoea leptophylla*	Bush Morning Glory
Jatropha	*Jatropha*	
Jatropha macrorhiza	*Jatropha macrorhiza*	Ragged Nettlespurge
Jatropha multifida	*Jatropha multifida*	Coral Plant
Juglans	*Juglans*	
Juglans nigra	*Juglans nigra*	Black Walnut
Juglans rupestris	*Juglans microcarpa*	Little Walnut
Juniperus	*Juniperus*	
Juniperus flaccida	*Juniperus flaccida*	Weeping Juniper
Juniperus monosperma	*Juniperus monosperma*	One-seed Juniper
Juniperus pachyphloea	*Juniperus deppeana*	Alligator Juniper
Juniperus sabinoides	*Juniperus ashei*	Ashe Juniper
Juniperus virginiana	*Juniperus virginiana*	Eastern Red Cedar
Karwinskia humboldtiana	*Karwinskia humboldtiana*	Coyotillo
Koeberlinia	*Koeberlinia*	
Koeberlinia spinosa	*Koeberlinia spinosa*	Allthorn
Krameria	*Krameria*	
Krameria canescens	*Krameria grayi*	White Ratany
Laciniaria	*Liatris*	
Laciniaria punctata	*Liatris punctata*	Dotted Gayfeather

Old Name	Current Name	Common Name
Lantana	*Lantana*	
Lantana camara	*Lantana camara*	Common Lantana
Leitneria	*Leitneria*	
Leitneria floridana	*Leitneria floridana*	Corkwood
Leucaena	*Leucaena*	
Leucaena retusa	*Leucaena retusa*	Littleleaf Leadtree
Leucophyllum	*Leucophyllum*	
Leucophyllum minus	*Leucophyllum minus*	Big Bend Silverleaf
Leucophyllum texanum	*Leucophyllum frutescens*	Texas Silverleaf
Linum	*Linum*	
Linum perenne	*Linum lewisii*	Lewis Flax
Linum rigidum	*Linum rigidum*	Stiffstem Flax
Lippia	*Lippia*	
Liquidambar	*Liquidambar*	
Liquidambar styraciflua	*Liquidambar styraciflua*	Sweetgum
Lithospermum	*Lithospermum*	
Lycium	*Lycium*	
Lycium berlandieri	*Lycium berlandieri*	Berlandier Wolfberry
Lycium pallidum	*Lycium pallidum*	Pale Wolfberry
Magnolia	*Magnolia*	
Magnolia faetida	*Magnolia grandiflora*	Southern Magnolia
Magnolia virginiana	*Magnolia virginiana*	Southern Sweetbay
Malpighia	*Malpighia*	
Malpighia glabra	*Malpighia glabra*	Barbados Cherry
Manfreda	*Manfreda*	
Manfreda maculosa	*Manfreda maculosa*	Texas Tuberose
Mentzelia	*Mentzelia*	
Mentzelia nuda	*Mentzelia nuda*	Bractless Mentzelia
Meriolix	*Calylophus*	
Meriolix intermedia	*Calylophus serrulatus*	Yellow Evening Primrose
Microrhamnus	*Microrhamnus*	
Mimosa	*Mimosa*	
Mimosa biuncifera	*Mimosa aculeaticarpa*	Catclaw Mimosa
Mimosa borealis	*Mimosa borealis*	Fragrant Mimosa
Mimosa emoryana	*Mimosa emoryana*	Emory's Mimosa
Mimosa fragrans	*Mimosa borealis*	Fragrant Mimosa
Mimosa lindheimeri	*Mimosa aculeaticarpa*	Catclaw Mimosa
Mitchella	*Mitchella*	

Old Name	Current Name	Common Name
Mitchella repens	*Mitchella repens*	Partridgeberry
Momesia	*Celtis*	
Momesia pallida	*Celtis pallida*	Spiny Hackberry
Monarda	*Monarda*	
Morella	*Morella*	
Morella cerifera	*Morella cerifera*	Southern Wax Myrtle
Morella crispa	*Morella cerifera*	Southern Wax Myrtle
Morus	*Morus*	
Morus microphylla	*Morus celtidifolia*	Texas Mulberry
Morus rubra	*Morus rubra*	Red Mulberry
Mozinna	*Jatropha*	
Mozinna spathulata	*Jatropha dioica*	Leatherstem
Nicotiana	*Nicotiana*	
Nicotiana glauca	*Nicotiana glauca*	Tree Tobacco
Nolina	*Nolina*	
Nolina lindheimeriana	*Nolina lindheimeriana*	Lindheimer's Bear Grass
Nolina microcarpa	*Nolina microcarpa*	Small-seed Nolina
Nolina texana	*Nolina texana*	Sacahuista
Nyssa	*Nyssa*	
Nyssa aquatica	*Nyssa aquatica*	Water Tupelo
Nyssa sylvatica	*Nyssa sylvatica*	Black Tupelo
Oenothera	*Oenothera* or *Calylophus*	
Opuntia	*Cylindropuntia* or *Opuntia*	
Opuntia arborescens	*Cylindropuntia imbricata*	Walking Stick Cholla
Opuntia cymochila	*Opuntia macrorhiza*	Grassland Prickly Pear
Opuntia davisi	*Cylindropuntia davisii*	Davis Cholla
Opuntia engelmanni	*Opuntia engelmannii*	Engelmann Prickly Pear
Opuntia leptocaulis	*Cylindropuntia leptocaulis*	Pencil Cholla
Opuntia lindheimeri	*Opuntia engelmannii*	Engelmann Prickly Pear
Opuntia macrorhiza	*Opuntia macrorhiza*	Plains Prickly Pear
Ostrya	*Ostrya*	
Ostrya baileyi	*Ostrya knowltonii*	Western Hophornbeam
Ostrya virginiana	*Ostrya virginiana*	American Hophornbeam
Oxalis	*Oxalis*	
Oxalis violacea or *var.*	*Oxalis violacea*	Violet Woodsorrel
Parkinsonia	*Parkinsonia*	
Parkinsonia aculeata	*Parkinsonia aculeata*	Retama
Parosela	*Dalea*	

Old Name	Current Name	Common Name
Parosela enneandra	*Dalea enneandra*	Nine-anther Prairie Clover
Parosela formosa	*Dalea formosa*	Feather Dalea
Parosela frutescens	*Dalea frutescens*	Black Dalea
Passiflora	*Passiflora*	
Passiflora incarnata	*Passiflora incarnata*	Maypop Passion Flower
Pentstemon	*Pentstemon*	
Persea	*Persea*	
Persea borbonia	*Persea borbonia*	Red Bay
Petalostemon	*Dalea*	
Petalostemon purpureus	*Dalea purpurea*	Purple Prairie Clover
Philadelphus	*Philadelphus*	
Philadelphus microphyllus	*Philadelphus microphyllus*	Littleleaf Mockorange
Pinus	*Pinus*	
Pinus cembroides	*Pinus cembroides*	Mexican Pinyon
Pinus echinata	*Pinus echinata*	Shortleaf Pine
Pinus edulis	*Pinus edulis*	Pinyon Pine
Pinus flexilis	*Pinus strobiformis*	Southwestern White Pine
Pinus palustris	*Pinus palustris*	Longleaf Pine
Pinus ponderosa	*Pinus ponderosa*	Ponderosa Pine
Pinus taeda	*Pinus taeda*	Loblolly Pine
Pistacia	*Pistacia*	
Pistacia mexicana	*Pistacia mexicana*	Mexican Pistache
Pistacia vera	*Pistacia vera*	Pistachio
Platanus	*Platanus*	
Platanus occidentalis	*Platanus occidentalis*	American Sycamore
Polygala	*Polygala*	
Polygala alba	*Polygala alba*	White Milkwort
Populus	*Populus*	
Populus deltoides	*Populus deltoides*	Eastern Cottonwood
Populus tremuloides	*Populus tremuloides*	Quaking Aspen
Porlieria	*Porlieria*	
Porlieria angustifolia	*Guaiacum angustifolium*	Guayacan
Prosopis	*Prosopis*	
Prosopis glandulosa	*Prosopis glandulosa*	Honey Mesquite
Prosopis pubescens	*Prosopis pubescens*	Screwbean
Prunus	*Prunus*	
Prunus serotina	*Prunus serotina rufula*	Black Cherry
Prunus serotina acutifolia	*Prunus serotina rufula*	Black Cherry

Old Name	Current Name	Common Name
Pseudotsuga	*Pseudotsuga*	
Pseudotsuga mucronata	*Pseudotsuga menziesii*	Douglas Fir
Psoralea	*Pediomelum* or *Dalea*	
Psoralea digitata	*Pediomelum digitatum*	Palm-leaf Scurf-pea
Psoralea enneandra	*Dalea enneandra*	Nine-anther Prairie Clover
Psoralea linearifolia	*Pediomelum linearifolium*	Slim-leaf Scurf-pea
Quercus	*Quercus*	
Quercus acuminata	*Quercus muehlenbergii*	Chinkapin Oak
Quercus alba	*Quercus alba*	White Oak
Quercus digitata	*Quercus falcata*	Southern Red Oak
Quercus emoryi	*Quercus emoryi*	Emory Oak
Quercus emoryi (dwarf form)	*Quercus emoryi* (dwarf form)	Emory Oak
Quercus grisea	*Quercus grisea*	Gray Oak
Quercus grisea (dwarf form)	*Quercus grisea*	Gray Oak
Quercus leucophylla	*Quercus falcata*	Southern Red Oak
Quercus lyrata	*Quercus lyrata*	Overcup Oak
Quercus macrocarpa	*Quercus macrocarpa*	Bur Oak
Quercus marylandica	*Quercus marilandica*	Blackjack Oak
Quercus minor	*Quercus stellata*	Post Oak
Quercus nigra	*Quercus nigra*	Water Oak
Quercus novomexicana	*Quercus gambelii*	Gambel Oak
Quercus phellos	*Quercus phellos*	Willow Oak
Quercus rubra	*Quercus rubra*	Red Oak
Quercus texana	*Quercus texana*	Nuttall Oak
Quercus undulata	*Quercus undulata*	Wavyleaf Oak
Quercus virginiana	*Quercus virginiana*	Live Oak
Ratibida	*Ratibida*	
Ratibida columnaris	*Ratibida columnifera*	Upright Prairie Coneflower
Rhamnus	*Rhamnus*	
Rhamnus caroliniana	*Rhamnus caroliniana*	Carolina Buckthorn
Rhamnus purshiana	*Frangula purshiana*	Cascara
Rhus	*Pistacia* or *Toxicodendron*	
Rhus mexicana	*Pistacia mexicana*	Mexican Pistache
Rhus radicans	*Toxicodendron radicans*	Poison Ivy
Robinia	*Robinia*	
Robinia neomexicana	*Robinia neomexicana*	New Mexico Locust
Rubus	*Rubus*	
Rubus procumbens	*Rubus flagellaris*	Common Dewberry

Old Name	Current Name	Common Name
Rubus trivialis	*Rubus trivialis*	Rio Grande Dewberry
Sabal	*Sabal*	
Sabal adiantinum	*Sabal minor*	Dwarf Palmetto
Salix	*Salix*	
Salix nigra	*Salix nigra*	Black Willow
Samuela	*Yucca*	
Samuela carnerosana	*Yucca carnerosana*	Giant Spanish Dagger
Samuela faxoniana	*Yucca faxoniana*	Faxon Yucca
Sassafras	*Sassafras*	
Sassafras sassafras	*Sassafras albidum*	Sassafras
Schmaltzia	*Rhus*	
Schmaltzia copallina	*Rhus copallina*	Flameleaf Sumac
Schmaltzia lanceolata	*Rhus lanceolata*	Prairie Sumac
Schmaltzia mexicana	*Pistacia mexicana*	Mexican Pistache
Schmaltzia microphylla	*Rhus microphylla*	Littleleaf Sumac
Schmaltzia trilobata	*Rhus trilobata*	Skunkbush Sumac
Schmaltzia virens	*Rhus virens*	Evergreen Sumac
Smilax	*Smilax*	
Smilax laurifolia	*Smilax laurifolia*	Laurel Greenbrier
Smilax pumila	*Smilax pumila*	Sarsaparilla Vine
Smilax renifolia	*Smilax bona-nox*	Saw Greenbrier
Solanum	*Solanum*	
Solanum triquetrum	*Solanum triquetrum*	Texas Nightshade
Solanum tuberosum boreale	*Solanum tuberosum boreale*	Potato
Sophora	*Sophora*	
Sophora secundiflora	*Sophora secundiflora*	Mescal Bean
Sphagnum	*Sphagnum*	
Sphagnum sp.	*Sphagnum sp.*	Peat Moss
Suaeda	*Suaeda*	
Symphoricarpos	*Symphoricarpos*	
Symphoricarpos (longiflorus?)	*Symphoricarpos longiflorus*	Long-flower Snowberry
Taxodium	*Taxodium*	
Taxodium distichum	*Taxodium mucronatum*	Bald Cypress
Tecoma	*Tecoma*	
Tecoma stans	*Tecoma stans*	Trumpetflower
Tilia	*Tilia*	
Tilia leptophylla	*Tilia americana*	American Basswood
Tillandsia	*Tillandsia*	

Old Name	Current Name	Common Name
Tillandsia baileyi	*Tillandsia baileyi*	Bailey Ballmoss
Tillandsia recurvata	*Tillandsia recurvata*	Small Ballmoss
Toxylon	*Maclura*	
Toxylon pomiferum	*Maclura pomifera*	Osage Orange
Ulmus	*Ulmus*	
Ulmus alata	*Ulmus alata*	Winged Elm
Ulmus americana	*Ulmus americana*	American Elm
Ulmus fulva	*Ulmus rubra*	Slippery Elm
Ungnadia	*Ungnadia*	
Ungnadia speciosa	*Ungnadia speciosa*	Mexican Buckeye
Vaccinium	*Vaccinium*	
Vaccinium sp.	*Vaccinium sp.*	Blueberry
Vachellia	*Vachellia*	
Vachellia farnesiana	*Vachellia farnesiana*	Huisache
Verbena	*Verbena*	
Verbena stricta	*Verbena stricta*	Hoary Vervain
Viburnum	*Viburnum*	
Viburnum molle	*Viburnum molle*	Kentucky Viburnum
Viburnum nudum	*Viburnum nudum*	Possumhaw Viburnum
Viburnum rufotomentosum	*Viburnum rufidulum*	Rusty Blackhaw
Vitis	*Vitis*	
Vitis sp.	*Vitis sp.*	Wild Grape
Yucca	*Yucca*	
Yucca arkansana	*Yucca arkansana*	Arkansas Yucca
Yucca baccata	*Yucca baccata*	Datil Yucca
Yucca glauca	*Yucca glauca*	Soapweed Yucca
Yucca louisianensis	*Yucca louisianensis*	Louisiana Yucca
Yucca macrocarpa	*Yucca torreyi*	Torrey's Yucca
Yucca radiosa	*Yucca elata*	Soaptree Yucca
Yucca rostrata	*Yucca rostrata*	Beaked Yucca
Yucca rupicola	*Yucca rupicola*	Texas Yucca
Yucca stricta	*Yucca glauca*	Soapweed Yucca
Yucca treculeana	*Yucca treculeana*	Spanish Dagger
Ziziphus	*Ziziphus*	
Ziziphus lycioides	*Ziziphus obtusifolia*	Lotebush
Ziziphus obtusifolia	*Ziziphus obtusifolia*	Lotebush

Table 5. Scientific names of bird species as referenced in the *Biological Survey of Texas* (1905) and the current taxonomic designations for those species.

Old Name	Current Name	Common Name
Accipiter	*Accipiter*	
Accipiter cooperi	*Accipiter cooperi*	Cooper's Hawk
Aeronautes	*Aeronautes*	
Aeronautes melanoleucus	*Aeronautes saxatalis*	White-throated Swift
Agelaius	*Agelaius*	
Agelaius phoeniceus	*Agelaius phoeniceus*	Red-winged Blackbird
Agelaius phoeniceus floridanus	*Agelaius phoeniceus floridanus*	Red-winged Blackbird
Agelaius phoeniceus richmondi	*Agelaius phoeniceus littoralis*	Red-winged Blackbird
Agelaius phoeniceus sonorensis	*Agelaius phoeniceus sonorensis*	Red-winged Blackbird
Aimophila	*Aimophila*	
Aimophila ruficeps eremaeca	*Aimophila ruficeps eremoeca*	Rufous-crowned Sparrow
Aimophila ruficeps scotti	*Aimophila ruficeps scotti*	Rufous-crowned Sparrow
Ajaia	*Platalea*	
Ajaia ajaja	*Platalea ajaja*	Roseate Spoonbill
Amizilis	*Amazilia*	
Amizilis cerviniventris	*Amazilia yucatanensis*	Buff-bellied Hummingbird
Amizilis cerviniventris chalconota	*Amazilia yucatanensis chalconota*	Buff-bellied Hummingbird
Amizilis tzacatl	*Amazilia tzacatl*	Rufous-tailed Hummingbird
Ammodramus	*Ammodramus*	
Ammodramus maritimus sennetti	*Ammodramus maritimus sennetti*	Seaside Sparrow
Ampelis	*Bombycilla*	
Ampelis cedrorum	*Bombycilla cedrorum*	Cedar Waxwing
Amphispiza	*Amphispiza*	
Amphispiza bilineata deserticola	*Amphispiza bilineata bilineata*	Black-throated Sparrow
Amphispiza bilineata	*Amphispiza bilineata*	Black-throated Sparrow
Antrostomus	*Antrostomus*	
Antrostomus carolinensis	*Antrostomus carolinensis*	Chuck-will's-widow
Antrostomus macromystax	*Antrostomus arizonae*	Mexican Whip-poor-will
Aphelocoma	*Aphelocoma*	
Aphelocoma couchi	*Aphelocoma wollweberi*	Mexican Jay
Aphelocoma cyanotis	*Aphelocoma woodhouseii*	Woodhouse's Scrub Jay
Aphelocoma sieberi couchi	*Aphelocoma wollweberi couchi*	Mexican Jay
Aphelocoma texana	*Aphelocoma woodhouseii*	Woodhouse's Scrub Jay
Aphelocoma woodhousei	*Aphelocoma woodhouseii*	Woodhouse's Scrub Jay
Arremonops	*Arremonops*	
Arremonops rufivirgata	*Arremonops rufivirgatus*	Olive Sparrow

Old Name	Current Name	Common Name
Astragalinus	*Spinus*	
Astragalinus psaltria	*Spinus psaltria*	Lesser Goldfinch
Asyndesmus	*Melanerpes*	
Asyndesmus torquatus	*Melanerpes lewis*	Lewis's Woodpecker
Auriparus	*Auriparus*	
Auriparus flaviceps	*Auriparus flaviceps*	Verdin
Baeolophus	*Baeolophus*	
Baeolophus atricristatus	*Baeolophus atricristatus*	Black-crested Titmouse
Baeolophus bicolor	*Baeolophus bicolor*	Tufted Titmouse
Baeolophus inornatus griseus	*Baeolophus ridgwayi ridgwayi*	Juniper Titmouse
Bubo	*Bubo*	
Bubo virginianus	*Bubo virginianus*	Great Horned Owl
Bubo virginianus pallescens	*Bubo virginianus pallescens*	Great Horned Owl
Bubo virginianus subarcticus	*Bubo virginianus subarcticus*	Great Horned Owl
Buteo	*Buteo*	
Buteo abbreviatus	*Buteo albonotatus*	Zone-tailed Hawk
Buteo albicaudatus sennetti	*Geranoaetus albicaudatus hypspodius*	White-tailed Hawk
Buteo borealis calurus	*Buteo jamaicensis calurus*	Red-tailed Hawk
Buteo lineatus	*Buteo lineatus*	Red-shouldered Hawk
Buteo swainsoni	*Buteo swainsoni*	Swainson's Hawk
Caeligena	*Lampornis*	
Caeligena clemenciae	*Lampornis clemenciae*	Blue-throated Hummingbird
Callipepla	*Callipepla*	
Callipepla sqaumata	*Callipepla sqaumata*	Scaled Quail
Callipepla squamata castanogastris	*Callipepla squamata castanogastris*	Scaled Quail
Calothorax	*Calothorax*	
Calothorax 1ucifer	*Calothorax 1ucifer*	Lucifer Hummingbird
Campephilus	*Campephilus*	
Campephilus principalis	*Campephilus principalis*	Ivory-billed Woodpecker
Cardinalis	*Cardinalis*	
Cardinalis cardinalis	*Cardinalis cardinalis*	Northern Cardinal
Cardinalis cardinalis canicaudus	*Cardinalis cardinalis canicaudus*	Northern Cardinal
Cardinalis cardinalis canicaudus	*Cardinalis cardinalis magnirostris*	Northern Cardinal
Carpodacus	*Haemorhous*	
Carpodacus mexicanus frontalis	*Haemorhous mexicanus potosinus*	House Finch
Catherpes	*Catherpes*	
Catherpes mexicanus albifrons	*Catherpes mexicanus albifrons*	Canyon Wren
Centurus	*Melanerpes*	

Old Name	Current Name	Common Name
Centurus aurifrons	*Melanerpes aurifrons*	Golden-fronted Woodpecker
Centurus carolinus	*Melanerpes carolinus*	Red-bellied Woodpecker
Ceophloeus	*Dryocopus*	
Ceophloeus pileatus	*Dryocopus pileatus*	Pileated Woodpecker
Ceryle	*Chloroceryle* and *Megaceryle*	
Ceryle americana septentrionalis	*Chloroceryle americana septentrionalis*	Green Kingfisher
Ceryle torquata	*Megaceryle torquata*	Ringed Kingfisher
Chondestes		
Chordeiles	*Chordeiles*	
Chordeiles acutipennis texensis	*Chordeiles acutipennis texensis*	Lesser Nighthawk
Chordeiles henryi	*Chordeiles minor*	Common Nighthawk
Chordeiles virginianus	*Chordeiles minor*	Common Nighthawk
Chordeiles virginianus chapmani	*Chordeiles minor chapmani*	Common Nighthawk
Chordeiles virginianus henryi	*Chordeiles minor henryi*	Common Nighthawk
Coccyzus	*Coccyzus*	
Coccyzus americanus	*Coccyzus americanus*	Yellow-billed Cuckoo
Coccyzus americanus occidentalis	*Coccyzus americanus occidentalis*	Yellow-billed Cuckoo
Colaptes	*Colaptes*	
Colaptes auratus	*Colaptes auratus*	Northern Flicker
Colaptes cafer collaris	*Colaptes auratus collaris*	Northern Flicker
Colinus	*Colinus*	
Colinus virginianus	*Colinus virginianus*	Northern Bobwhite
Colinus virginianus texanus	*Colinus virginianus texanus*	Northern Bobwhite
Columba	*Patagioenas*	
Columba fasciata	*Patagioenas fasciata*	Band-tailed Pigeon
Columba flavirostris	*Patagioenas flavirostris*	Red-billed Pigeon
Columbigallina	*Columbina*	
Columbigallina passerina pallescens	*Columbina passerina pallescens*	Common Ground Dove
Colymbus	*Tachybaptus*	
Colymbus domincus brachypterus	*Tachybaptus dominicus brachypterus*	Least Grebe
Contopus	*Contopus*	
Contopus richardsoni	*Contopus sordidulus*	Western Wood Pewee
Contopus virens	*Contopus virens*	Eastern Wood Pewee
Corvus	*Corvus*	
Corvus corax sinuatus	*Corvus corax sinuatus*	Common Raven
Corvus cryptoleucus	*Corvus cryptoleucus*	Chihuahuan Raven
Coturniculus	*Ammodramus*	
Coturniculus savannarum bimaculatus	*Ammodramus savannarum bimaculatus*	Grasshopper Sparrow

Old Name	Current Name	Common Name
Crotophaga	*Crotophaga*	
Crotophaga sulcirostris	*Crotophaga sulcirostris*	Groove-billed Ani
Cyanocephalus	*Gymnorhinus*	
Cyanocephalus cyanocephalus	*Gymnorhinus cyanocephalus*	Pinyon Jay
Cyanocitta	*Cyanocitta*	
Cyanocitta cristata	*Cyanocitta cristata*	Blue Jay
Cyanocitta stelleri diademata	*Cyanocitta stelleri macrolopha*	Steller's Jay
Cyanospiza	*Passerina*	
Cyanospiza amoena	*Passerina amoena*	Lazuli Bunting
Cyanospiza cyanea	*Passerina cyanea*	Indigo Bunting
Cyanospiza versicolor	*Passerina versicolor*	Varied Bunting
Cyrtonyx	*Cyrtonyx*	
Cyrtonyx mearnsi	*Cyrtonyx montezumae*	Montezuma Quail
Cyrtonyx montezumae mearnsi	*Cyrtonyx montezumae mearnsi*	Montezuma Quail
Dendrocygna	*Dendrocygna*	
Dendrocygna autumnalis	*Dendrocygna autumnalis*	Black-bellied Whistling Duck
Dendroica	*Setophaga*	
Dendroica aestiva sonorana	*Setophaga petechia aestiva*	Yellow Warbler
Dendroica auduboni	*Setophaga coronata auduboni*	Yellow-rumped Warbler
Dendroica chrysoparia	*Setophaga chrysoparia*	Golden-cheeked Warbler
Dendroica dominica albilora	*Setophaga dominica albilora*	Yellow-throated Warbler
Dendroica graciae	*Setophaga graciae*	Grace's Warbler
Dendroica vigorsi	*Setophaga pinus*	Pine Warbler
Dryobates	*Picoides*	
Dryobates borealis	*Picoides borealis*	Red-cockaded Woodpecker
Dryobates pubescens	*Picoides pubescens*	Downy Woodpecker
Dryobates scalaris	*Picoides scalaris*	Ladder-backed Woodpecker
Dryobates scalaris bairdi	*Picoides scalaris cactophilus*	Ladder-backed Woodpecker
Dryobates villosus auduboni	*Picoides villosus auduboni*	Hairy Woodpecker
Dryobates villosus hyloscopus	*Picoides villosus hyloscopus*	Hairy Woodpecker
Elanoides	*Elanoides*	
Elanoides forficatus	*Elanoides forficatus*	Swallow-tailed Kite
Elanus	*Elanus*	
Elanus leucurus	*Elanus leucurus*	White-tailed Kite
Empidonax	*Empidonax*	
Empidonax difficilis	*Empidonax difficilis*	Pacific-slope Flycatcher
Empidonax virescens	*Empidonax virescens*	Acadian Flycatcher
Falco	*Falco*	

Old Name	Current Name	Common Name
Falco fusco-caerulescens	*Falco femoralis*	Aplomado Falcon
Falco mexicanus	*Falco mexicanus*	Prairie Falcon
Falco sparverius	*Falco sparverius*	American Kestrel
Falco sparverius phalaena	*Falco sparverius sparverius*	American Kestrel
Florida	*Egretta*	
Florida caerulea	*Egretta caerulea*	Little Blue Heron
Fregata	*Fregata*	
Fregata aquila	*Fregata magnificans*	Magnificent Frigatebird
Galeoscoptes	*Dumetella*	
Galeoscoptes carolinensis	*Dumetella carolinensis*	Gray Catbird
Geococcyx	*Geococcyx*	
Geococcyx californianus	*Geococcyx californianus*	Greater Roadrunner
Geothlypis	*Geothlypis*	
Geothlypis formosa	*Geothlypis formosa*	Kentucky Warbler
Geothlypis poliocephala	*Geothlypis poliocephala*	Gray-crowned Yellowthroat
Geothlypis trichas brachidactyla	*Geothlypis trichas trichas*	Common Yellow-throat
Glaucidium	*Glaucidium*	
Glaucidium phalaenoides	*Glaucidium brasilianum*	Ferruginous Pygmy-owl
Guara	*Eudocimus*	
Guara alba	*Eudocimus albus*	White Ibis
Guiraca	*Passerina*	
Guiraca caerulea	*Passerina caerulea*	Blue Grosbeak
Guiraca caerulea lazula	*Passerina caerulea caerulea*	Blue Grosbeak
Heleodytes	*Campylorhynchus*	
Heleodytes brunneicapillus couesi	*Campylorhynchus brunneicapillus guttatus*	Cactus Wren
Helminthophila	*Leiothlypis*	
Helminthophila celata orestera	*Leiothlypis celata orestera*	Orange-crowned Warbler
Hydranassa	*Egretta*	
Hydranassa tricolor ruficollis	*Egretta tricolor ruficollis*	Tricolored Heron
Hylocichla	*Hylocichla* and *Catharus*	
Hylocichla guttata auduboni	*Catharus guttatus auduboni*	Hermit Thrush
Hylocichla mustelina	*Hylocichla mustelina*	Wood Thrush
Icteria	*Icteria*	
Icteria virens	*Icteria virens*	Yellow-breasted Chat
Icteria virens longicauda	*Icteria virens virens*	Yellow-breasted Chat
Icterus	*Icterus*	
Icterus auduboni	*Icterus graduacauda adubonii*	Audubon's Oriole
Icterus bullocki	*Icterus bullockii*	Bullock's Oriole

Old Name	Current Name	Common Name
Icterus cucullatus sennetti	*Icterus cucullatus sennetti*	Hooded Oriole
Icterus galbula	*Icterus galbula*	Baltimore Oriole
Icterus parisorum	*Icterus parisorum*	Scott's Oriole
Jacana	*Jacana*	
Jacana spinosa	*Jacana spinosa*	Northern Jacana
Junco	*Junco*	
Junco dorsalis	*Junco hyemalis*	Dark-eyed Junco
Lanius	*Lanius*	
Lanius ludovicianus excubitorides	*Lanius ludovicianus excubitorides*	Loggerhead Shrike
Leptotila	*Leptotila*	
Leptotila fulviventris brachyptera	*Leptotila verreauxi angelica*	White-tipped Dove
Lophortyx	*Callipepla*	
Lophortyx gambeli	*Callipepla gambelii*	Gambel's Quail
Loxia	*Loxia*	
Loxia curvirostra stricklandi	*Loxia curvirostra stricklandi*	Red Crossbill
Megaquiscalus	*Quiscalus*	
Megaquiscalus macrourus	*Quiscalus mexicanus*	Great-tailed Grackle
Megaquiscalus major	*Quiscalus major*	Boat-tailed Grackle
Megaquiscalus major macrourus	*Quiscalus major major*	Boat-tailed Grackle
Megascops	*Megascops*	
Megascops asio	*Megascops asio*	Eastern Screech Owl
Megascops asio mccalli	*Megascops asio mccalli*	Eastern Screech Owl
Megascops flammeolus	*Psilioscops flammeolus*	Flammulated owl
Melanerpes	*Melanerpes*	
Melanerpes erythrocephalus	*Melanerpes erythrocephalus*	Red-headed Woodpecker
Melanerpes formicivorus	*Melanerpes formicivorus*	Acorn Woodpecker
Meleagris	*Meleagris*	
Meleagris gallopavo intermedia	*Meleagris gallopavo intermedia*	Wild Turkey
Meleagris gallopavo merriami	*Meleagris gallopavo merriami*	Wild Turkey
Meleagris gallopavo silvestris	*Meleagris gallopavo silvestris*	Wild Turkey
Melopelia	*Zenaida*	
Melopelia leucoptera	*Zenaida asiatica*	White-winged Dove
Merula	*Turdus*	
Merula migratoria propinqua	*Turdus migratorius propinquus*	American Robin
Micropallas	*Micrathene*	
Micropallas whitneyi	*Micrathene whitneyi*	Elf Owl
Mimus	*Mimus*	
Mimus polyglottos	*Mimus polyglottos*	Northern Mockingbird

Old Name	Current Name	Common Name
Mimus polyglottos leucopterus	*Mimus polyglottos polyglottos*	Northern Mockingbird
Mniotilta	*Mniotilta*	
Mniotilta varia	*Mniotilta varia*	Black-and-white Warbler
Molothrus	*Molothrus*	
Molothrus ater obscurus	*Molothrus ater obscurus*	Brown-headed Cowbird
Mycteria	*Mycteria*	
Mycteria americana	*Mycteria americana*	Wood Stork
Myiarchus	*Myiarchus*	
Myiarchus cinerascens	*Myiarchus cinerascens*	Ash-throated Flycatcher
Myiarchus crinitus	*Myiarchus crinitus*	Great Crested Flycatcher
Myiarchus mexicanus	*Myiarchus tyrannulus*	Brown-crested Flycatcher
Nomonyx	*Nomonyx*	
Nomonyx dominicus	*Nomonyx dominicus*	Masked Duck
Numenius	*Numenius*	
Numenius longirostris	*Numenius americanus*	Long-billed Curlew
Nuttallornis	*Contopus*	
Nuttallornis borealis	*Contopus cooperi*	Olive-sided Flycatcher
Nuttallornis richardsoni	*Contopus sordidulus*	Western Wood-Pewee
Nyctidromus	*Nyctidromus*	
Nyctidromus albicollis merrilli	*Nyctidromus albicollis merrilli*	Common Pauraque
Oreospiza	*Pipilo*	
Oreospiza chlorura	*Pipilo chlorurus*	Green-tailed Towhee
Ornithion	*Camptostoma*	
Ornithion imberbe	*Camptostoma imberbe*	Northern Beardless-Tyrannulet
Ortalis	*Ortalis*	
Ortalis vetula maccalli	*Ortalis vetula maccalli*	Plain Chachalaca
Otocoris	*Eremophila*	
Otocoris alpestris giraudi	*Eremophila alpestris giraudi*	Horned Lark
Otocoris alpestris leucolaema	*Eremophila alpestris leucolaema*	Horned Lark
Otocoris leucolaema	*Eremophila alpestris*	Horned Lark
Parabuteo	*Parabuteo*	
Parabuteo unicinctus harrisi	*Parabuteo unicinctus harrisi*	Harris' Hawk
Parus	*Baeolophus* and *Poecile*	
Parus atricristatus	*Baeolophus atricristatus*	Black-crested Titmouse
Parus carolinensis agilis	*Poecile carolinensis agilis*	Carolina Chickadee
Parus gambeli	*Poecile gambeli*	Mountain Chickadee
Peucaea	*Peucaea*	
Peucaea aestivalis bachmani	*Peucaea aestivalis bachmani*	Bachman's Sparrow

Old Name	Current Name	Common Name
Peucaea cassini	*Peucaea cassini*	Cassin's Sparrow
Phainopepla	*Phainopepla*	
Phainopepla nitens	*Phainopepla nitens*	Phainopepla
Phalacrocorax	*Phalacrocorax*	
Phalacrocorax vigua mexicanus	*Phalacrocorax brasilianus mexicanus*	Neotropic Cormorant
Phalaenoptilus	*Phalaenoptilus*	
Phalaenoptilus nuttallii	*Phalaenoptilus nuttallii*	Common Poorwill
Pipilo	*Pipilo* and *Melozone*	
Pipilo fuscus mesoleucus	*Melozone fusca mesoleuca*	Canyon Towhee
Pipilo maculatus megalonyx	*Pipilo maculatus megalonyx*	Spotted Towhee
Pipilo mesoleucus	*Melozone fusca*	Canyon Towhee
Piranga	*Piranga*	
Piranga hepatica	*Piranga hepatica*	Hepatic Tanager
Piranga ludoviciana	*Piranga ludoviciana*	Western Tanager
Piranga rubra	*Piranga rubra*	Summer Tanager
Piranga rubra cooperi	*Piranga rubra rubra*	Summer Tanager
Pitangus	*Pitangus*	
Pitangus derbianus	*Pitangus sulphuratus*	Great Kiskadee
Podasocys	*Charadrius*	
Podasocys montanus	*Charadrius montanus*	Mountain Plover
Polioptila	*Polioptila*	
Polioptila caerulea	*Polioptila caerulea*	Blue-gray Gnatcatcher
Polioptila caerulea obscura	*Polioptila caerulea caerulea*	Blue-gray Gnatcatcher
Polioptila plumbea	*Polioptila plumbea*	Tropical Gnatcatcher
Polyborus	*Caracara*	
Polyborus cheriway	*Caracara cheriway*	Northern Crested Caracara
Pooecetes	*Pooecetes*	
Pooecetes confinis	*Pooecetes gramineus*	Vesper Sparrow
Protonotaria	*Protonotaria*	
Protonotaria citrea	*Protonotaria citrea*	Prothonotary Warbler
Psaltriparus	*Psaltriparus*	
Psaltriparus lloydi	*Psaltriparus minimus*	American Bushtit
Psaltriparus melanotis lloydi	*Psaltriparus minimus dimorphicus*	American Bushtit
Psaltriparus plumbeus	*Psaltriparus minimus plumbeus*	American Bushtit
Pyrocephalus	*Pyrocephalus*	
Pyrocephalus rubineus mexicanus	*Pyrocephalus rubinus mexicanus*	Vermilion Flycatcher
Pyrrhuloxia	*Cardinalis*	
Pyrrhuloxia s. texana	*Cardinalis sinuatus sinuatus*	Pyrrhuloxia

Old Name	Current Name	Common Name
Pyrrhuloxia sinuata	*Cardinalis sinuatus*	Pyrrhuloxia
Quiscalus	*Quiscalus*	
Quiscalus quiscula aeneus	*Quiscalus quiscula aeneus*	Common Grackle
Salpinctes	*Salpinctes*	
Salpinctes obsoletus	*Salpinctes obsoletus*	Rock Wren
Sayornis	*Sayornis*	
Sayornis nigricans	*Sayornis nigricans*	Black Phoebe
Sayornis saya	*Sayornis saya*	Say's Phoebe
Scardafella	*Columbina*	
Scardafella inca	*Columbina inca*	Inca Dove
Selasphorus	*Selasphorus*	
Selasphorus platycercus	*Selasphorus platycercus*	Broad-tailed Hummingbird
Selasphorus rufus	*Selasphorus rufus*	Rufous Hummingbird
Sialia	*Sialia*	
Sialia mexicana bairdi	*Sialia mexicana bairdi*	Western Bluebird
Sialia sialis	*Sialia sialis*	Eastern Bluebird
Sitta	*Sitta*	
Sitta carolinensis	*Sitta carolinensis*	White-breasted Nuthatch
Sitta carolinensis nelsoni	*Sitta carolinensis nelsoni*	White-breasted Nuthatch
Sitta pusilla	*Sitta pusilla*	Brown-headed Nuthatch
Sitta pygmaea	*Sitta pygmaea*	Pygmy Nuthatch
Speotyto	*Athene*	
Speotyto cunicularia hypogaea	*Athene cunicularia hypugaea*	Burrowing Owl
Spizella	*Spizella*	
Spizella pusilla	*Spizella pusilla*	Field Sparrow
Spizella socialis	*Spizella passerina*	Chipping Sparrow
Spizella socialis arizonae	*Spizella passerina arizonae*	Chipping Sparrow
Sporophila	*Sporophila*	
Sporophila morelleti	*Sporophila morelleti*	Morellet's Seedeater
Sturnella	*Sturnella*	
Sturnella magna hoopesi	*Sturnella magna hoopesi*	Eastern Meadowlark
Sturnella magna neglecta	*Sturnella neglecta neglecta*	Western Meadowlark
Syrnium	*Strix*	
Syrnium occidentale	*Strix occidentalis*	Spotted Owl
Syrnium varium helveolum	*Strix varia helveola*	Barred Owl
Tangavius	*Molothrus*	
Tangavius aeneus involucratus	*Molothrus aeneus aeneus*	Bronzed Cowbird
Thryomanes	*Thryomanes*	

Old Name	Current Name	Common Name
Thryomanes bewickii bairdi	*Thryomanes bewickii bairdi*	Bewick's Wren
Thryomanes bewicki cryptus	*Thryomanes bewickii cryptus*	Bewick's Wren
Thryomanes bewickii leucogaster	*Thryomanes bewickii sadai*	Bewick's Wren
Thryothorus	*Thryothorus*	
Thryothorus ludovicianus	*Thryothorus ludovicianus*	Carolina Wren
Toxostoma	*Toxostoma*	
Toxostoma curvirostre	*Toxostoma curvirostre*	Curve-billed Thrasher
Toxostoma longirostre sennetti	*Toxostoma longirostre sennetti*	Long-billed Thrasher
Trochilus	*Archilochus*	
Trochilus alexandri	*Archilochus alexandri*	Black-chinned Hummingbird
Trochilus colubris	*Archilochus colubris*	Ruby-throated Hummingbird
Troglodytes	*Troglodytes*	
Troglodytes aedon aztecus	*Troglodytes aedon parkmanii*	House Wren
Tympanuchus	*Tympanuchus*	
Tympanuchus americanus	*Tympanuchus cupido*	Greater Prairie Chicken
Tympanuchus americanus attwateri	*Tympanuchus cupido attwateri*	Attwater's Prairie Chicken
Tympanuchus attwateri	*Tympanuchus cupido attwateri*	Attwater's Prairie Chicken
Tyrannus	*Tyrannus*	
Tyrannus melancholicus couchi	*Tyrannus couchi*	Couch's Kingbird
Tyrannus tyrannus	*Tyrannus tyrannus*	Eastern Kingbird
Tyrannus verticalis	*Tyrannus verticalis*	Western Kingbird
Tyrannus vociferans	*Tyrannus vociferans*	Cassin's Kingbird
Urubitinga	*Buteogallus*	
Urubitinga anthracina	*Buteogallus anthracinus*	Common Black Hawk
Vireo	*Vireo*	
Vireo atricapillus	*Vireo atricapilla*	Black-capped Vireo
Vireo bellii arizonae	*Vireo bellii bellii*	Bell's Vireo
Vireo bellii medius	*Vireo bellii bellii* and *Vireo bellii medius*	Bell's Vireo
Vireo flavifrons	*Vireo flavifrons*	Yellow-throated Vireo
Vireo flavoviridis	*Vireo flavoviridis*	Yellow-green Vireo
Vireo gilvus swainsoni	*Vireo swainsoni swainsoni*	Warbling Vireo
Vireo huttoni stephensi	*Vireo huttoni stephensi*	Hutton's Vireo
Vireo noveboracensis	*Vireo griseus*	White-eyed Vireo
Vireo noveboracensis micrus	*Vireo griseus micrus*	White-eyed Vireo
Vireo olivaceus	*Vireo olivaceus*	Red-eyed Vireo
Vireo plumbeus	*Vireo plumbeus*	Plumbeous Vireo
Vireo solitarius plumbeus	*Vireo solitarius solitarius*	Blue-headed Vireo
Vireo stephensi	*Vireo huttoni*	Hutton's Vireo

Old Name	Current Name	Common Name
Wilsonia	*Setophaga* and *Cardellina*	
Wilsonia mitrata	*Setophaga citrina*	Hooded Warbler
Wilsonia pusilla pileolata	*Cardellina pusilla pileolata*	Wilson's Warbler
Xanthoura	*Cyanocorax*	
Xanthoura luxuosa glaucescens	*Cyanocorax yncas luxuosus*	Green Jay
Zamelodia	*Pheucticus*	
Zamelodia melanocephala	*Pheucticus melanocephalus*	Black-headed Grosbeak

Table 6. Scientific names of reptile species as referenced in the *Biological Survey of Texas* (1905) and the current taxonomic designations for those species. Names listed in parentheses indicate recent proposed changes for genus names.

Old Name	Current Name	Common Name
Agkistrodon	*Agkistrodon*	
Agkistrodon contortrix	*Agkistrodon contortrix*	Copperhead
Agkistrodon piscivorus	*Agkistrodon piscivorus*	Cottonmouth
Anolis	*Anolis*	
Anolis carolinensis	*Anolis carolinensis*	Green Anole
Bascanion	*Masticophis* (*Coluber*)	
Bascanion flagellum	*Masticophis* (*Coluber*) *flagellum*	Coachwhip
Bascanion ornatum	*Masticophis* (*Coluber*) *taeniatus*	Striped Whipsnake
Callopeltis	*Elaphe* (*Pantherophis*)	
Callopeltis obsoletus	*Elaphe* (*Pantherophis*) *obsoleta*	Texas Rat Snake
Chionactis	*Sonora*	
Chionactis episcopus isozonus	*Sonora semiannulata semiannulata*	Western Ground Snake
Cnemidophorus	*Cnemidophorus* (*Aspidoscelis*)	
Cnemidophorus gularis	*Cnemidophorus* (*Aspidoscelis*) *gularis*	Common Spotted Whiptail
Cnemidophorus perplexus	*Cnemidophorus* (*Aspidoscelis*) *neomexicanus*	New Mexico Whiptail
Cnemidophorus sexlineatus	*Cnemidophorus* (*Aspidoscelis*) *sexlineatus*	Six-lined Racerunner
Cnemidophorus tessellatus	*Cnemidophorus* (*Aspidoscelis*) *tesselatus*	Common Checkered Whiptail
Coleonyx	*Coleonyx*	
Coleonyx brevis	*Coleonyx brevis*	Texas Banded Gecko
Crotalus	*Crotalus*	
Crotalus atrox	*Crotalus atrox*	Western Diamond-backed Rattlesnake
Crotalus confluentus	*Crotalus viridis*	Prairie Rattlesnake
Crotalus horridus	*Crotalus horridus*	Timber Rattlesnake
Crotalus lepidus	*Crotalus lepidus*	Rock Rattlesnake
Crotalus molossus	*Crotalus molossus*	Black-tailed Rattlesnake
Crotaphytus	*Crotaphytus* and *Gambelia*	
Crotaphytus collaris baileyi	*Crotalus collaris collaris* and *C. c. fuscus*	Eastern Collared Lizard
Crotaphytus collaris	*Crotaphytus collaris*	Eastern Collared Lizard
Crotaphytus reticulatus	*Crotaphytus reticulatus*	Reticulate Collared Lizard
Crotaphytus wislizenii	*Gambelia wislizenii*	Longnose Leopard Lizard
Diadophis	*Diadophis*	
Diadophis regalis	*Diadophis punctatus regalis*	Ring-necked Snake
Drymarchon	*Drymarchon*	
Drymarchon corais melanurus	*Drymarchon melanurus*	Central American Indigo Snake

Old Name	Current Name	Common Name
Drymobius	*Drymobius*	
Drymobius margaritiferus	*Drymobius margaritiferus*	Speckled Racer
Elaps	*Micrurus*	
Elaps fulvius	*Micrurus tener*	Texas Coral Snake
Eumeces	*Eumeces (Plestiodon)*	
Eumeces brevilineatus	*Eumeces (Plestiodon) tetragrammus brevilineatus*	Four-lined Skink
Eumeces guttulatus	*Eumeces (Plestiodon) obsoletus*	Great Plains Skink
Eumeces obsoletus	*Eumeces (Plestiodon) obsoletus*	Great Plains Skink
Eumeces quinquelineatus	*Eumeces (Plestiodon) fasciatus*	Common Five-lined Skink
Eutainia	*Thamnophis*	
Eutainia cyrtopsis	*Thamnophis cyrtopsis*	Black-necked Garter Snake
Eutainia elegans marciana	*Thamnophis marcianus*	Checkered Garter Snake
Eutainia proxima	*Thamnophis proximus*	Western Ribbon Snake
Gerrhonotus	*Gerrhonotus*	
Gerrhonotus liocephalus infernalis	*Gerrhonotus infernalis*	Texas Alligator Lizard
Heterodon	*Heterodon*	
Heterodon nasicus	*Heterodon kennerlyi*	Mexican Hog-nosed Snake
Heterodon nasicus	*Heterodon nasicus*	Western Hog-nosed Snake
Heterodon platirhinos	*Heterodon platirhinos*	Eastern Hog-nosed Snake
Holbrookia	*Holbrookia* and *Cophosaurus*	
Holbrookia maculata	*Holbrookia maculata*	Common Lesser Earless Lizard
Holbrookia maculata lacerata	*Holbrookia lacerata*	Spot-tailed Earless Lizard
Holbrookia propinqua	*Holbrookia propinqua*	Keeled Earless Lizard
Holbrookia texana	*Cophosaurus texanus*	Greater Earless Lizard
Lampropeltis	*Lampropeltis*	
Lampropeltis getula holbrooki	*Lampropeltis getula holbrooki*	Common Kingsnake
Leiolopisma	*Scincella*	
Leiolopisma laterale	*Scincella lateralis*	Little Brown Skink
Liopeltis	*Opheodrys*	
Liopeltis vernalis	*Opheodrys vernalis*	Smooth Greensnake
Natrix	*Nerodia*	
Natrix clarkii	*Nerodia clarkii*	Saltmarsh Watersnake
Natrix fasciata transversa	*Nerodia erythrogaster transversa*	Plain-bellied Watersnake
Opheodrys	*Opheodrys*	
Opheodrys aestivus	*Opheodrys aestivus*	Rough Greensnake
Ophisaurus	*Ophisaurus*	
Ophisaurus ventralis	*Ophisaurus attenuatus*	Slender Glass Lizard
Phrynosoma	*Phrynosoma*	

Old Name	Current Name	Common Name
Phrynosoma cornutum	*Phrynosoma cornutum*	Texas Horned Lizard
Phrynosoma hernandesi	*Phrynosoma hernandesi*	Greater Short-horned Lizard
Phrynosoma modestum	*Phrynosoma modestum*	Round-tailed Horned Lizard
Pituophis	*Pituophis*	
Pituophis sayi	*Pituophis catenifer sayi*	Gopher Snake
Rhinocheilus	*Rhinocheilus*	
Rhinocheilus lecontei	*Rhinocheilus lecontei*	Long-nosed Snake
Sceloporus	*Sceloporus*	
Sceloporus clarkii	*Sceloporus bimaculosus*	Twin-spotted Spiny Lizard
Sceloporus consobrinus	*Sceloporus consobrinus*	Prairie Lizard
Sceloporus dispar	*Sceloporus grammicus*	Graphic Spiny Lizard
Sceloporus merriami	*Sceloporus merriami*	Canyon Lizard
Sceloporus spinosus floridanus	*Sceloporus olivaceus*	Texas Spiny Lizard
Sceloporus torquatus poinsettii	*Sceloporus poinsettii poinsettii*	Crevice Spiny Lizard
Sistrurus	*Sistrurus*	
Sistrurus catenatus consors	*Sistrurus catenatus edwardsii*	Massasauga
Sistrurus catenatus consors	*Sistrurus catenatus tergeminus*	Massasauga
Storeria	*Storeria*	
Storeria dekayi	*Storeria dekayi*	Dekay's Brownsnake
Tantilla	*Tantilla*	
Tantilla gracilis	*Tantilla gracilis*	Flat-headed Snake
Tropidoclonium	*Tropidoclonium*	
Tropidoclonium lineatum	*Tropidoclonium lineatum*	Lined Snake
Uta	*Uta* and *Urosaurus*	
Uta ornata	*Urosaurus ornatus*	Ornate Tree Lizard
Uta stansburiana	*Uta stansburiana*	Common Side-blotched Lizard

Table 7. Scientific names of mammal species as referenced in the *Biological Survey of Texas* (1905) and the current taxonomic designations for those species.

Old Name	Current Name	Common Name
Ammospermophilus	*Ammospermophilus*	
Ammospermophilus interpres	*Ammospermophilus interpres*	Texas Antelope Squirrel
Antilocapra	*Antilocapra*	
Antilocapra americana	*Antilocapra americana*	Pronghorn
Antrozous	*Antrozous*	
Antrozous pallidus	*Antrozous pallidus*	Pallid Bat
Atalpha	*Lasiurus*	
Atalpha noveboracensis	*Lasiurus borealis*	Eastern Red Bat
Bassariscus	*Bassariscus*	
Bassariscus astutus	*Bassariscus astutus*	Ringtail
Bassariscus astutus flavus	*Bassariscus astutus flavus*	Ringtail
Blarina	*Cryptotis* and *Blarina*	
Blarina berlandieri	*Cryptotis parva*	Least Shrew
Blarina brevicauda	*Cryptotis parva*	Least Shrew
Blarina brevicauda carolinensis	*Blarina carolinensis*	Southern Short-tailed Shrew
Blarina brevicauda carolinensis	*Blarina carolinensis minima* (in part)	Southern Short-tailed Shrew
Blarina parva	*Cryptotis parva*	Least Shrew
Bison	*Bos*	
Bison bison	*Bos bison*	Bison
Canis	*Canis*	
Canis ater	*Canis lupus*	Gray Wolf
Canis mearnsi	*Canis latrans*	Coyote
Canis microdon	*Canis latrans*	Coyote
Canis nebracensis	*Canis latrans*	Coyote
Canis nebracensis texensis	*Canis latrans texensis*	Coyote
Canis rufus	*Canis rufus*	Red Wolf
Castor	*Castor*	
Castor canadensis	*Castor canadensis*	American Beaver
Castor canadensis frondator	*Castor canadensis mexicanus*	American Beaver
Castor canadensis texensis	*Castor canadensis texensis*	American Beaver
Cervus	*Cervus*	
Cervus canadensis	*Cervus elaphus canadensis*	Wapiti or Elk
Cervus merriami	*Cervus elaphus merriami*	Wapiti or Elk
Citellus	*Ictidomys*, *Otospermophilus*, and *Xerospermophilus*	
Citellus couchi	*Otospermophilus variegatus variegatus*	Rock Squirrel

327

Old Name	Current Name	Common Name
Citellus grammurus	*Otospermophilus variegatus grammurus*	Rock Squirrel
Citellus mexicanus	*Ictidomys parvidens*	Rio Grande Ground Squirrel
Citellus mexicanus parvidens	*Ictidomys parvidens*	Rio Grande Ground Squirrel
Citellus pallidus	*Ictidomys tridecemlineatus arenicola*	Thirteen-lined Ground Squirrel
Citellus pallidus	*Ictidomys tridecemlineatus pallidus*	Thirteen-lined Ground Squirrel
Citellus spilosoma	*Xerospermophilus spilosoma*	Spotted Ground Squirrel
Citellus spilosoma arens	*Xerospermophilus spilosoma canescens*	Spotted Ground Squirrel
Citellus spilosoma major	*Xerospermophilus spilosoma annectens*	Spotted Ground Squirrel
Citellus spilosoma marginatus	*Xerospermophilus spilosoma marginatus*	Spotted Ground Squirrel
Citellus texensis	*Ictidomys tridecemlineatus texensis*	Thirteen-lined Ground Squirrel
Citellus tridecemlineatus	*Ictidomys tridecemlineatus*	Thirteen-lined Ground Squirrel
Citellus tridecemlineatus pallidus	*Ictidomys tridecemlineatus pallidus*	Thirteen-lined Ground Squirrel
Citellus tridecemlineatus texensis	*Ictidomys tridecemlineatus texensis*	Thirteen-lined Ground Squirrel
Citellus variegatus	*Otospermophilus variegatus*	Rock Squirrel
Citellus variegatus couchi	*Otospermophilus variegatus grammurus*	Rock Squirrel
Citellus variegatus couchi	*Otospermophilus variegatus variegatus*	Rock Squirrel
Conepatus	*Conepatus*	
Conepatus leuconotus	*Conepatus leuconotus*	Hog-nosed Skunk
Conepatus leuconotus mearnsi	*Conepatus leuconotus leuconotus*	Hog-nosed Skunk
Conepatus leuconotus texensis	*Conepatus leuconotus leuconotus*	Hog-nosed Skunk
Conepatus mesoleucus	*Conepatus leuconotus*	Hog-nosed Skunk
Conepatus mesoleucus mearnsi	*Conepatus leuconotus leuconotus*	Hog-nosed Skunk
Conepatus mesoleucus telmalestes	*Conepatus leuconotus telmalestes*	Hog-nosed Skunk
Corynorhinus	*Corynorhinus*	
Corynorhinus macrotis	*Corynorhinus townsendii*	Townsend's Big-eared Bat
Corynorhinus macrotis pallescens	*Corynorhinus townsendii australis*	Townsend's Big-eared Bat
Corynorhinus macrotis pallescens	*Corynorhinus townsendii pallescens*	Townsend's Big-eared Bat
Cratogeomys	*Cratogeomys*	
Cratogeomys castanops	*Cratogeomys castanops*	Yellow-faced Pocket Gopher
Cryptotis	*Cryptotis*	
Cryptotis parva	*Cryptotis parva*	Least Shrew
Cryptotis parva berlandieri	*Cryptotis parva berlandieri*	Least Shrew
Cryptotis parva parva	*Cryptotis parva parva*	Least Shrew
Cynomys	*Cynomys*	
Cynomys ludovicianus	*Cynomys ludovicianus*	Black-tailed Prairie Dog
Dasypus	*Dasypus*	

Old Name	Current Name	Common Name
Dasypus novemcinctus	*Dasypus novemcinctus*	Nine-banded Armadillo
Dasypus novemcinctus mexicanus	*Dasypus novemcinctus mexicanus*	Nine-banded Armadillo
Dasypterus	*Dasypterus*	
Dasypterus intermedius	*Dasypterus intermedius*	Northern Yellow Bat
Didelphis	*Didelphis*	
Didelphis marsupialis	*Didelphis virginiana*	Virginia Opossum
Didelphis marsupialis texensis	*Didelphis virginiana californica*	Virginia Opossum
Didelphis virginiana	*Didelphis virginiana*	Virginia Opossum
Didelphis virginiana pigra	*Didelphis virginiana virginiana*	Virginia Opossum
Dipodomys	*Dipodomys*	
Dipodomys ambiguus	*Dipodomys merriami ambiguus*	Merriam's Kangaroo Rat
Dipodomys elator	*Dipodomys elator*	Texas Kangaroo Rat
Dipodomys merriami	*Dipodomys merriami*	Merriam's Kangaroo Rat
Dipodomys merriami ambiguus	*Dipodomys merriami ambiguus*	Merriam's Kangaroo Rat
Dipodomys spectabilis	*Dipodomys spectabilis*	Banner-tailed Kangaroo Rat
Dipodomys spectabilis baileyi	*Dipodomys spectabilis baileyi*	Banner-tailed Kangaroo Rat
Erethizon	*Erethizon*	
Erethizon (epixanthum)	*Erethizon dorsatum*	North American Porcupine
Erethizon sp.?	*Erethizon dorsatum*	North American Porcupine
Eutamis	*Tamias*	
Eutamis cinnereicollis	*Tamias canipes*	Gray-footed Chipmunk
Eutamis cinnereicollis canipes	*Tamias canipes canipes*	Gray-footed Chipmunk
Felis	*Puma, Leopardus,* and *Panthera*	
Felis (sp.?) (panther)	*Puma concolor*	Mountain Lion
Felis cacomitli	*Puma yagouaroundi cacomitli*	Jaguarundi
Felis hippolestes	*Puma concolor*	Mountain Lion
Felis hippolestes aztecus	*Puma concolor couguar*	Mountain Lion
Felis hippolestes aztecus	*Puma concolor couguar*	Mountain Lion
Felis onca	*Panthera onca*	Jaguar
Felis onca hernandezi	*Panthera onca veraecrucis*	Jaguar
Felis pardalis	*Leopardus pardalis*	Ocelot
Felis pardalis limitis	*Leopardus pardalis limitis*	Ocelot
Fiber	*Ondatra*	
Fiber zibethicus	*Ondatra zibethicus*	Common Muskrat
Fiber zibethicus ripensis	*Ondatra zibethicus ripensis*	Common Muskrat
Geomys	*Geomys*	
Geomys arenarius	*Geomys arenarius*	Desert Pocket Gopher
Geomys breviceps	*Geomys breviceps*	Baird's Pocket Gopher

Old Name	Current Name	Common Name
Geomys breviceps attwateri	*Geomys attwateri*	Attwater's Pocket Gopher
Geomys breviceps llanensis	*Geomys texensis llanensis*	Llano Pocket Gopher
Geomys breviceps sagittalis	*Geomys breviceps sagittalis*	Baird's Pocket Gopher
Geomys arenarius (in part)	*Geomys knoxjonesi*	Jones's Pocket Gopher
Geomys lutescens	*Geomys jugossicularis*	Hall's Pocket Gopher
Geomys personatus	*Geomys personatus personatus*	Texas Pocket Gopher
Geomys personatus	*Geomys personatus maritimus*	Texas Pocket Gopher
Geomys personatus	*Geomys personatus davisi*	Texas Pocket Gopher
Geomys personatus fallax	*Geomys personatus fallax*	Texas Pocket Gopher
Geomys sagittalis	*Geomys breviceps sagittalis*	Baird's Pocket Gopher
Geomys texensis	*Geomys texensis*	Llano Pocket Gopher
Lasiurus	*Lasiurus* and *Aeorestes*	
Lasiurus borealis	*Lasiurus borealis*	Eastern Red Bat
Lasiurus borealis seminolus	*Lasiurus seminolus*	Seminole Bat
Lasiurus cinereus	*Aeorestes cinereus*	Hoary Bat
Lepus	*Lepus* and *Sylvilagus*	
Lepus aquaticus	*Sylvilagus aquaticus*	Swamp Rabbit
Lepus aquaticus attwateri	*Sylvilagus aquaticus*	Swamp Rabbit
Lepus arizonae	*Sylvilagus audubonii*	Desert Cottontail
Lepus arizonae baileyi	*Sylvilagus audubonii neomexicanus*	Desert Cottontail
Lepus arizonae minor	*Sylvilagus audubonii parvulus*	Desert Cottontail
Lepus arizonae minor	*Sylvilagus audubonii minor*	Desert Cottontail
Lepus chapmani	*Sylvilagus floridanus*	Eastern Cottontail
Lepus floridanus	*Sylvilagus floridanus*	Eastern Cottontail
Lepus floridanus alacer	*Sylvilagus floridanus alacer*	Eastern Cottontail
Lepus floridanus chapmani	*Sylvilagus floridanus chapmani*	Eastern Cottontail
Lepus melanotis	*Lepus californicus melanotis*	Black-tailed Jackrabbit
Lepus merriami	*Lepus californicus merriami*	Black-tailed Jackrabbit
Lepus pinetis	*Sylvilagus floridanus*	Eastern Cottontail
Lepus pinetis holzneri	*Sylvilagus floridanus holzneri*	Eastern Cottontail
Lepus pinetis robustus	*Sylvilagus robustus*	Davis Mountains Cottontail
Lepus sylvaticus	*Sylvilagus floridanus*	Eastern Cottontail
Lepus texianus	*Lepus californicus*	Black-tailed Jackrabbit
Lepus texianus melanotis	*Lepus californicus melanotis*	Black-tailed Jackrabbit
Liomys	*Liomys*	
Liomys texensis	*Liomys irroratus texensis*	Mexican Spiny Pocket Mouse
Lutra	*Lontra*	
Lutra canadensis	*Lontra canadensis*	Northern River Otter

Old Name	Current Name	Common Name
Lutra canadensis vaga	*Lontra canadensis lataxina*	Northern River Otter
Lutreola	*Vison*	
Lutreola lutreocephala	*Vison vison*	American Mink
Lynx	*Lynx*	
Lynx baileyi	*Lynx rufus*	Bobcat
Lynx baileyi	*Lynx rufus fasciatus*	Bobcat
Lynx rufus	*Lynx rufus*	Bobcat
Lynx rufus texensis	*Lynx rufus rufus*	Bobcat
Lynx texensis	*Lynx rufus*	Bobcat
Mephitis	*Mephitis*	
Mephitis mesomelas	*Mephitis mephitis mesomelas*	Striped Skunk
Mephitis mesomelas varians	*Mephitis mephitis varians*	Striped Skunk
Microtus	*Microtus*	
Microtus ludovicianus	*Microtus ochrogaster ludovicianus*	Prairie Vole
Microtus mexicanus	*Microtus mogollonensis*	Mogollon Vole
Microtus mexicanus guadalupensis	*Microtus mogollonensis guadalupensis*	Mogollon Vole
Microtus pinetorum	*Microtus pinetorum*	Woodland Vole
Microtus pinetorum auricularis	*Microtus pinetorum auricularis*	Woodland Vole
Microtus pinetorum auricularis	*Microtus pinetorum nemoralis*	Woodland Vole
Mormoops	*Mormoops*	
Mormoops megalophylla	*Mormoops megalophylla*	Ghost-faced Bat
Mormoops megalophylla senicula	*Mormoops megalophylla megalophylla*	Ghost-faced Bat
Mus	*Mus* and *Rattus*	
Mus alexandrinus	*Rattus rattus*	Black Rat
Mus musculus	*Mus musculus*	House Mouse
Mus norvegicus	*Rattus norvegicus*	Norway or Brown Rat
Mus rattus	*Rattus rattus*	Black Rat
Mustela	*Mustela*	
Mustela frenata	*Mustela frenata*	Long-tailed Weasel
Mustela frenata arthuri	*Mustela frenata arthuri*	Long-tailed Weasel
Mustela frenata frenata	*Mustela frenata frenata*	Long-tailed Weasel
Mustela frenata neomexicanus	*Mustela frenata neomexicana*	Long-tailed Weasel
Mustela frenata primulina	*Mustela frenata primulina*	Long-tailed Weasel
Mustela frenata texensis	*Mustela frenata texensis*	Long-tailed Weasel
Myotis	*Myotis*	
Myotis californicus	*Myotis californicus*	California Myotis
Myotis incautus	*Myotis velifer incautus*	Cave Myotis
Myotis velifer	*Myotis velifer*	Cave Myotis

Old Name	Current Name	Common Name
Myotis yumanensis	*Myotis yumanensis*	Yuma Myotis
Nasua	*Nasua*	
Nasua narica	*Nasua narica*	White-nosed Coati
Nasua narica yucatanica	*Nasua narica molaris*	White-nosed Coati
Neotoma	*Neotoma*	
Neotoma albigula	*Neotoma leucodon*	White-toothed Woodrat
Neotoma attwateri	*Neotoma floridana attwateri*	Eastern Wood Rat
Neotoma floridana	*Neotoma floridana rubida*	Eastern Wood Rat
Neotoma floridana	*Neotoma floridana attwateri*	Eastern Wood Rat
Neotoma floridana	*Neotoma floridana illinoensis*	Eastern Wood Rat
Neotoma floridana attwateri	*Neotoma floridana attwateri*	Eastern Wood Rat
Neotoma floridana baileyi	*Neotoma floridana attwateri*	Eastern Wood Rat
Neotoma floridana rubida	*Neotoma floridana rubida*	Eastern Wood Rat
Neotoma mexicana	*Neotoma mexicana*	Mexican Woodrat
Neotoma micropus	*Neotoma micropus*	Southern Plains Woodrat
Notiosorex	*Notiosorex*	
Notiosorex crawfordi	*Notiosorex crawfordi*	Crawford's Desert Shrew
Nycticeius	*Nycticeius*	
Nycticeius humeralis	*Nycticeius humeralis*	Evening Bat
Nyctinomus	*Tadarida*	
Nyctinomus mexicanus	*Tadarida brasiliensis*	Brazilian Free-tailed Bat
Odocoileus	*Odocoileus*	
Odocoileus canus	*Odocoileus hemionus*	Mule Deer
Odocoileus hemionus canus	*Odocoileus hemionus eremicus*	Mule Deer
Odocoileus couesi	*Odocoileus virginianus carminis*	White-tailed Deer
Odocoileus louisianae	*Odocoileus virginianus*	White-tailed Deer
Odocoileus macrourus	*Odocoileus virginianus texana*	White-tailed Deer
Odocoileus virginianus	*Odocoileus virginianus*	White-tailed Deer
Odocoileus virginianus macrourus	*Odocoileus virginianus macroura*	White-tailed Deer
Odocoileus virginianus texanus	*Odocoileus virginianus texana*	White-tailed Deer
Ondatra	*Ondatra*	
Ondatra zibethicus	*Ondatra zibethicus*	Common Muskrat
Ondatra zibethicus cinnamonius	*Ondatra zibethicus cinnamonius*	Common Muskrat
Ondatra zibethicus ripensis	*Ondatra zibethicus ripensis*	Common Muskrat
Ondatra zibethicus rivaliclus	*Ondatra zibethicus rivaliclus*	Common Muskrat
Onychomys	*Onychomys*	
Onychomys leucogaster	*Onychomys leucogaster*	Northern Grasshopper Mouse
Onychomys leucogaster pallescens	*Onychomys leucogaster pallescens*	Northern Grasshopper Mouse

Old Name	Current Name	Common Name
Onychomys longipes	*Onychomys leucogaster longipes*	Northern Grasshopper Mouse
Onychomys pallescens	*Onychomys leucogaster pallescens*	Northern Grasshopper Mouse
Onychomys torridus	*Onychomys arenicola*	Mearns' Grasshopper Mouse
Onychomys torridus arenicola	*Onychomys arenicola arenicola*	Mearns' Grasshopper Mouse
Oryzomys	*Oryzomys*	
Oryzomys aquaticus	*Oryzomys couesi*	Coues' Rice Rat
Oryzomys palustris	*Oryzomys texensis*	Texas Marsh Rice Rat
Ovis	*Ovis*	
Ovis mexicanus	*Ovis canadensis*	Bighorn Sheep
Pedomys	*Microtus*	
Perodipus	*Dipodomys*	
Perodipus compactus	*Dipodomys compactus*	Gulf Coast Kangaroo Rat
Perodipus montanus	*Dipodomys ordii richardsoni*	Ord's Kangaroo Rat
Perodipus ordi	*Dipodomys ordii*	Ord's Kangaroo Rat
Perodipus richardsoni	*Dipodomys ordii richardsoni*	Ord's Kangaroo Rat
Perodipus sennettii	*Dipodomys compactus sennetti*	Gulf Coast Kangaroo Rat
Perognathus	*Perognathus* and *Chaetodipus*	
Perognathus copei	*Perognathus flavescens copei*	Plains Pocket Mouse
Perognathus flavescens	*Perognathus flavescens*	Plains Pocket Mouse
Perognathus flavescens copei	*Perognathus flavescens copei*	Plains Pocket Mouse
Perognathus flavus	*Perognathus flavus*	Silky Pocket Mouse
Perognathus hispidus	*Chaetodipus hispidus*	Hispid Pocket Mouse
Perognathus hispidus hispidus	*Chaetodipus hispidus hispidus*	Hispid Pocket Mouse
Perognathus hispidus paradoxus	*Chaetodipus hispidus paradoxus*	Hispid Pocket Mouse
Perognathus hispidus spilotus	*Chaetodipus hispidus spilotus*	Hispid Pocket Mouse
Perognathus intermedius	*Chaetodipus intermedius*	Rock Pocket Mouse
Perognathus merriami	*Perognathus merriami*	Merriam's Pocket Mouse
Perognathus merriami gilvis	*Perognathus merriami gilvis*	Merriam's Pocket Mouse
Perognathus merriami merriami	*Perognathus merriami merriami*	Merriam's Pocket Mouse
Perognathus nelsoni	*Chaetodipus nelsoni*	Nelson's Pocket Mouse
Perognathus nelsoni canescens	*Chaetodipus nelsoni canescens*	Nelson's Pocket Mouse
Perognathus paradoxus	*Chaetodipus hispidus paradoxus*	Hispid Pocket Mouse
Perognathus penicillatus	*Chaetodipus eremicus*	Chihuahuan Desert Pocket Mouse
Perognathus penicillatus eremicus	*Chaetodipus eremicus*	Chihuahuan Desert Pocket Mouse
Peromyscus	*Peromyscus* and *Baiomys*	
Peromyscus attwateri	*Peromyscus attwateri*	Texas Deermouse
Peromyscus boylei (in part)	*Peromyscus attwateri* and *Peromyscus boylii*	Texas Deermouse and Brush Deermouse

Old Name	Current Name	Common Name
Peromyscus boylei laceyi	*Peromyscus attwateri* and *Peromyscus laceianus*	Texas Deermouse and Lacey's White-ankled Deermouse
Peromyscus boylei pennicillatus	*Peromyscus nasutus pennicillatus*	Northern Rock Mouse
Peromyscus boylei rowleyi	*Peromyscus boylii rowleyi*	Brush Deermouse
Peromyscus canus	*Peromyscus leucopus texanus*	White-footed Deermouse
Peromyscus eremicus	*Peromyscus eremicus*	Cactus Deermouse
Peromyscus eremicus arenarius	*Peromyscus eremicus eremicus*	Cactus Deermouse
Peromyscus gossypinus	*Peromyscus gossypinus*	Cotton Deermouse
Peromyscus leucopus	*Peromyscus leucopus*	White-footed Deermouse
Peromyscus leucopus mearnsi	*Peromyscus leucopus texanus*	White-footed Deermouse
Peromyscus leucopus texanus	*Peromyscus leucopus texanus*	White-footed Deermouse
Peromyscus michiganensis	*Peromyscus maniculatus*	North American Deermouse
Peromyscus michiganensis pallescens	*Peromyscus maniculatus blandus*	North American Deermouse
Peromyscus rowleyi	*Peromyscus boylii rowleyi*	Brush Deermouse
Peromyscus sonoriensis	*Peromyscus maniculatus sonoriensis*	North American Deermouse
Peromyscus sonoriensis blandus	*Peromyscus maniculatus blandus*	North American Deermouse
Peromyscus subater	*Baiomys taylori subater*	Northern Pygmy Mouse
Peromyscus taylori	*Baiomys taylori taylori*	Northern Pygmy Mouse
Peromyscus texanus	*Peromyscus leucopus texanus*	White-footed Deermouse
Peromyscus tornillo	*Peromyscus leucopus tornillo*	White-footed Deermouse
Pipistrellus	*Parastrellus*	
Pipistrellus hesperus	*Parastrellus hesperus*	American Parastrellus
Pipistrellus subflavus	*Perimyotis subflavus*	American Perimyotis
Pitymys	*Microtus*	
Procyon	*Procyon*	
Procyon elucus	*Procyon lotor elucus*	Northern Raccoon
Procyon lotor	*Procyon lotor*	Northern Raccoon
Procyon lotor fuscipes	*Procyon lotor fuscipes*	Northern Raccoon
Procyon lotor hernandezii	*Procyon lotor hernandezii*	Northern Raccoon
Procyon lotor lotor	*Procyon lotor fuscipes*	Northern Raccoon
Procyon lotor mexicanus	*Procyon lotor fuscipes*	Northern Raccoon
Procyon lotor mexicanus	*Procyon lotor mexicanus*	Northern Raccoon
Promops	*Eumops*	
Promops californicus	*Eumops perotis californicus*	Western Bonneted Bat
Putorius	*Mustela*	
Putorius frenatus	*Mustela frenata*	Long-tailed Weasel
Putorius neomexicanus	*Mustela frenata*	Long-tailed Weasel
Putorius nigripes	*Mustela nigripes*	Black-footed Ferret

Old Name	Current Name	Common Name
Reithrodontomys	*Reithrodontomys*	
Reithrodontomys aurantinus	*Reithrodontomys fulvescens aurantius*	Fulvous Harvest Mouse
Reithrodontomys griseus	*Reithrodontomys montanus griseus*	Plains Harvest Mouse
Reithrodontomys intermedius	*Reithrodontomys fulvescens intermedius*	Fulvous Harvest Mouse
Reithrodontomys laceyi	*Reithrodontomys fulvescens laceyi*	Fulvous Harvest Mouse
Reithrodontomys megalotis	*Reithrodontomys megalotis*	Western Harvest Mouse
Reithrodontomys merriami	*Reithrodontomys humulis merriami*	Eastern Harvest Mouse
Reithrodontomys montanus	*Reithrodontomys montanus*	Plains Harvest Mouse
Reithrodontomys montanus albescens	*Reithrodontomys montanus albescens*	Plains Harvest Mouse
Scalopus	*Scalopus*	
Scalopus aquaticus	*Scalopus aquaticus*	Eastern Mole
Scalopus aquaticus aereus	*Scalopus aquaticus aereus*	Eastern Mole
Scalopus aquaticus alleni	*Scalopus aquaticus alleni*	Eastern Mole
Scalopus aquaticus cryptus	*Scalopus aquaticus cryptus*	Eastern Mole
Scalopus aquaticus inflatus	*Scalopus aquaticus inflatus*	Eastern Mole
Scalopus aquaticus intermedius	*Scalopus aquaticus intermedius*	Eastern Mole
Scalopus aquaticus texanus	*Scalopus aquaticus texanus*	Eastern Mole
Scalopus texensis	*Scalopus aquaticus*	Eastern Mole
Sciuropterus	*Glaucomys*	
Sciuropterus volans	*Glaucomys volans*	Southern Flying Squirrel
Sciuropterus volans querceti	*Glaucomys volans texensis*	Southern Flying Squirrel
Sciurus	*Sciurus* and *Tamiasciurus*	
Sciurus fremonti	*Tamiasciurus hudsonicus*	Red Squirrel
Sciurus fremonti lynchnuchus	*Tamiasciurus hudsonicus lychnuchus*	Red Squirrel
Sciurus limitis	*Sciurus niger limitis*	Eastern Fox Squirrel
Sciurus ludovicianus	*Sciurus niger ludovicianus*	Eastern Fox Squirrel
Sciurus ludovicianus limitis	*Sciurus niger limitis*	Eastern Fox Squirrel
Sciurus rufiventer	*Sciurus niger rufiventer*	Eastern Fox Squirrel
Sciurus texianus	*Sciurus niger ludovicianus*	Eastern Fox Squirrel
Sigmodon	*Sigmodon*	
Sigmodon hispidus	*Sigmodon hispidus*	Hispid Cotton Rat
Sigmodon hispidus berlandieri	*Sigmodon hispidus berlandieri*	Hispid Cotton Rat
Sigmodon hispidus pallidus	*Sigmodon hispidus berlandieri*	Hispid Cotton Rat
Sigmodon hispidus texianus	*Sigmodon hispidus texianus*	Hispid Cotton Rat
Sigmodon ochrognathus	*Sigmodon ochrognathus*	Yellow-nosed Cotton Rat
Spermophilus	*Otospermophilus*	
Spermophilus grammurus	*Otospermophilus variegatus grammurus*	Rock Squirrel

Old Name	Current Name	Common Name
Spilogale	Spilogale	
Spilogale indianola	Spilogale putorius interrupta	Eastern Spotted Skunk
Spilogale interrupta	Spilogale putorius interrupta	Eastern Spotted Skunk
Spilogale leucoparia	Spilogale gracilis	Western Spotted Skunk
Spilogale leucoparia	Spilogale gracilis leucoparia	Western Spotted Skunk
Tatu	Dasypus	
Tatu novemcinctus	Dasypus novemcinctus	Nine-banded Armadillo
Tatu novemcinctus mexicanum	Dasypus novemcinctus mexicanus	Nine-banded Armadillo
Tatu novemcinctus texanum	Dasypus novemcinctus mexicanus	Nine-banded Armadillo
Taxidea	Taxidea	
Taxidea taxus	Taxidea taxus	American Badger
Taxidea taxus berlandieri	Taxidea taxus berlandieri	American Badger
Tayassu	Pecari	
Tayassu angulatus	Pecari tajacu	Collared Peccary
Thomomys	Thomomys	
Thomomys aureus	Thomomys bottae	Botta's Pocket Gopher
Thomomys aureus lachuguilla	Thomomys bottae limitarius	Botta's Pocket Gopher
Thomomys baileyi	Thomomys bottae baileyi	Botta's Pocket Gopher
Thomomys fulvus	Thomomys bottae	Botta's Pocket Gopher
Thomomys fulvus guadalupensis	Thomomys bottae texensis	Botta's Pocket Gopher
Thomomys fulvus texensis	Thomomys bottae texensis	Botta's Pocket Gopher
Thomomys perditus	Thomomys bottae limitarius	Botta's Pocket Gopher
Urocyon	Urocyon	
Urocyon cinereoargenteus	Urocyon cinereoargenteus	Common Gray Fox
Urocyon cinereoargenteus floridanus	Urocyon cinereoargenteus floridanus	Common Gray Fox
Urocyon cinereoargenteus ocythous	Urocyon cinereoargenteus floridanus	Common Gray Fox
Urocyon cinereoargenteus scotti	Urocyon cinereoargenteus scottii	Common Gray Fox
Ursus	Ursus	
Ursus americanus	Ursus americanus	American Black Bear
Ursus americanus ambliceps	Ursus americanus ambliceps	American Black Bear
Ursus americanus americanus	Ursus americanus americanus	American Black Bear
Ursus americanus luteolus	Ursus americanus luteolus	American Black Bear
Ursus horribilis	Ursus arctos	Grizzly or Brown Bear
Ursus horribilis horriaeus	Ursus arctos horribilis	Grizzly or Brown Bear
Ursus luteolus	Ursus americanus	American Black Bear
Vespertilio	Eptesicus	
Vespertilio fuscus	Eptesicus fuscus	Big Brown Bat
Vulpes	Vulpes	

Old Name	Current Name	Common Name
Vulpes fulvus	*Vulpes vulpes fulva*	Red Fox
Vulpes macrotis	*Vulpes macrotis*	Kit Fox
Vulpes macrotis neomexicanus	*Vulpes macrotis*	Kit Fox
Vulpes velox	*Vulpes velox*	Swift Fox

Texas's Historical Landscapes and Land Uses

Human modification of native landscapes continues to pose one of the most severe threats to wildlife diversity in Texas. These photos depict two sources of anthropogenic influences facing Texas natural history in current times: urban sprawl and the "border wall." As urban sprawl transforms natural landscapes into residential and commercial developments, natural habitats are fragmented or eliminated, air and water pollution increase, and groundwater is depleted, creating negative impacts for both wildlife and humans. The border wall that is being constructed along the Texas side of the Rio Grande River will act as a physical barrier for many wildlife species, preventing gene flow between populations in Texas and northern Mexico. Construction of the wall also is degrading or destroying vital habitat, causing erosion along the riverbanks, and may increase downstream flooding. Urban sprawl, public domain photo. Border wall, courtesy National Butterfly Center.

THIS SECTION INCLUDES TWO CHAPTERS THAT ADDRESS MAJOR LAND-scape and land use changes that have impacted the Texas biota during the last century and a half. Chapter 3 describes the landscapes of Texas as encountered by the federal agents who worked on the biological survey and includes photographs taken by them while working in the state. Chapter 4 uses the most current information available to describe landscape and faunal changes in the twentieth and twenty-first centuries. These changes are highlighted by comparing the photographs of landscapes taken more than a century ago, in Chapter 3, to what these landscapes look like today.

These two chapters represent examples of archival natural history. Archival natural history addresses the important concept of change over time periods of 50 to 150 years as documented through historical notes and publications, archives, photographs, and interviews. As such, it is designed to bring more clarity to two important issues, closely related to personal perspective, that continue to perplex conservationists: these issues have been labeled "landscape amnesia" and "environmental generational amnesia." In his 2005 book, *Collapse: How Societies Choose to Fail or Succeed* (Diamond, 2005), Jared Diamond introduced the concept of "landscape amnesia" to describe gradual changes in landscapes that appear to be almost imperceptible over a period of 50 to 100 years because the change from year to year is so gradual. It refers to the way a major change can be accepted as a normal situation if it happens slowly, through unnoticeable increments of change. Such change would otherwise be regarded as objectionable if it took place in a single step or short period.

The similar concept of "environmental generational amnesia" (see Kahn, 2002), addresses the phenomenon in which people assume the natural environment they encounter during childhood is the norm against which they will measure environmental degradation later in their life. With each ensuing generation, the amount of environmental degradation increases, but each generation takes that degraded condition as the non-degraded condition, essentially as the normal experience. Consequently, the baseline of a modern person's perception of the environment becomes problematic because they experience very little, if any, pristine nature and the nature they do encounter daily is highly regulated. They gradually come to value structured nature over wild nature because it is what they know to be "natural." As a result, they often will not work for, or even care about, the protection of wild nature and biodiversity. This is particularly a problem for new generations of children as they grow into adolescence and adulthood.

Herein, we explore both concepts through the lens of what happened to the natural environments of Texas during the twentieth and early part of the twenty-first centuries. We address these perspectives through a broad overview of Texas's natural history, documenting change in its climate, landscapes and land cover, major habitat types, water, and biological diversity during this period. In addition, we have speculated on the potential impact of environmental pollutants, wildlife diseases, and climate change to both Texas wildlife and to its human occupants.

CHAPTER 3

Texas Landscapes, 1889-1905

T EXAS'S LANDSCAPES REPRESENT A MICROCOSM OF THE REMAINDER OF
the entire United States. Grasslands stretch all the way from Canada southward
onto the High Plains and terminate at the prairies, marshes, and sandy beaches
along our coastline. Coming in from the west and southwest is the southern extension of
the Rocky Mountains and the northeastern extent of the Chihuahuan Desert, bisected by
the Rio Grande. On the east, the Big Thicket and Piney Woods represent the westernmost
extension of the same forest that extends all the way to the Appalachians. To the south,
the subtropical Rio Grande Valley and the brush country of southern Texas represent the
northern limits of landscapes that extend through Mexico and Central America. These
different landscapes converge in central Texas where the rugged hill country includes a
mixture of grasslands, shrublands, and forests.

Bailey and the federal agents visited Texas during a time of intensive livestock use
and severe drought in the western United States. The drought period of the 1880s continued over most of the West until about 1905. Until the late 1800s, the open range system dominated the cattle industry in Texas. Longhorn and "scrub" (free-ranging or feral)
cattle roamed freely over the vast acres of rangeland in the western half of the state. By
the turn of the century, however, the open range system had changed to a stock farming
system, largely as a result of the introduction of barbed wire fencing and the utilization
of drilled wells and windmills to provide water for stock. As the stock farming system
gained prominence, longhorn and scrub cattle were replaced with registered breeds such
as Herefords. Cattle production in the state increased dramatically, as did the number of
sheep and goats. It wasn't long before the overgrazing of Texas's vast rangelands became
a serious problem (Fig. 53). Also, during this time Texas's grassland landscapes were subjected to rapid encroachment of cacti and woody plants, largely as a result of a decrease

Figure 53. Drought conditions and overgrazing by cattle and other livestock decimated the rangelands of Texas during the late nineteenth and early twentieth centuries. Courtesy the Southwest Collection, Texas Tech University.

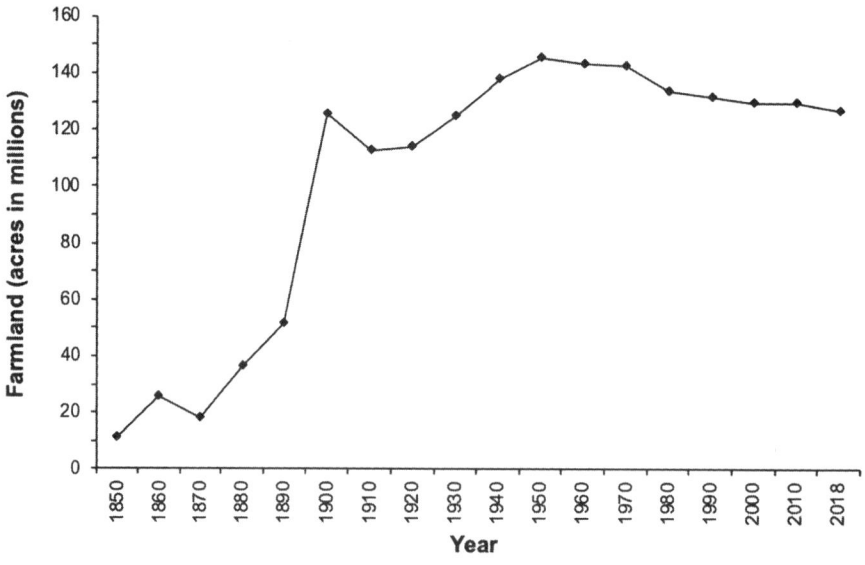

Figure 54. Acres of farmland in Texas, 1850-2018.

in prairie wildfires and the severe overstocking of rangelands (Lehman, 1969). A 1904 *Texas Almanac* article about the grasslands of Texas stated that what had been "one of the best stock ranges in . . . this country [is] so abused that today its capacity for sustaining livestock is very small in comparison with what it was 30 or even 20 years ago" (*Texas Almanac and State Industrial Guide*, 1904).

The turn of the century saw a change in Texas farming practices, moving from predominately cotton farming to the concept of diversified farming. This change coincided with an increase in immigration of farmers from northern, grain-producing states, which produced an enormous increase in the number of acres in farmland from 1880 to 1910 (see Fig. 54). Also, the efforts of the Texas Agricultural Experiment Station, established in 1887 to conduct research into every phase of the state's crop and livestock operations, helped to promote the idea of diversification. Although cotton farming continued to dominate agricultural production in the state, fruit, vegetable, and grain crops became profitable for landowners in many areas of the state less suitable for cotton production. From 1870 to 1900, cotton production increased 1,000 percent, while wheat production increased 5,200 percent, and oat production increased 3,700 percent. Rice growing became an important industry in the coastal regions, as irrigated land in this area increased from 8,700 acres (3,500 ha [hectares]) in 1899 to 235,000 acres (95,000 ha) in 1904. Rapid progress in irrigation was an important factor in the development of the arid and semi-arid regions of Texas, where irrigated land increased from 37,000 acres (15,000 ha) in 1899 to more than 100,000 acres (40,000 ha) in 1904.

Lumber production, primarily in the pine-oak forests of eastern Texas, was the leading industry in that region at the turn of the century and it was rapidly expanding, thus depleting the stands of virgin timber. The exploitation of these forests began in the early 1800s as settlers in the region viewed the trees as an impediment to settlement and agriculture. Logging became a dominant manufacturing concern in the region in the 1870s as the state's population began to grow and there was an increased demand for timber (Fig. 55). The ease of logging in the region, a climate conducive to year-round production, and technological improvements such as steam-powered sawmills and circular saws contributed to the growth of the lumber industry. At its peak in 1907, 2.25 billion board feet (5.3 million cubic meters) of lumber was produced in Texas. By 1926, at least 450 lumber mills were in operation in eastern Texas. The *Texas Almanac* for that year stated "at the present rate of lumber production, it is estimated that Texas within 9 or 10 years will have witnessed the passing of practically all of its virgin stand of timber and will be dependent largely upon timber products from the northwestern forests." The original area of Texas virgin forests had been estimated at 12 million acres (4.8 million ha). By 1926, 4.7 million acres (1.9 million ha) had been converted to agriculture, 3.9 million acres (1.6 million ha) had been cut with little or no forest renewal, 1.9 million acres (770,000 ha) were in second growth timber, and only 1.5 million acres (607,000 ha) of virgin timber remained.

During the time of the biological survey, most of Texas's three million people resided in the eastern and upper coastal prairies of the state. The far western and northwestern

Figure 55. Logging in East Texas, early 1900s. Courtesy East Texas Research Center, Steen Library, Stephen F. Austin State University.

Figure 56. Comanche Springs, Ft. Stockton, Pecos County, 1887. Courtesy Texas Parks and Wildlife Department.

regions still were sparsely populated and for the most part remained as ranching country. Farming was just beginning to expand in these areas. Most Texans lived in rural areas and not in cities. Texas's largest city, San Antonio, had fewer than 60,000 residents in 1900. Because people lived and earned a living on the land, natural resources were under intensive use during this period. There were no federal or state parks or preserves to protect scenic places, and there was little public concern about the need for this.

Texas Landscapes as Described by the Federal Agents

Bailey's 1905 publication (Bailey, 1905) included a description as well as a map of the life/crop zones of the state (see Chapter 2, Fig. 31, p. 65). The common faunal elements (primarily birds and mammals), together with the dominant plants and plant associations of each zone or important geographic region, were listed and augmented by a few photographs of landscapes and plant associations. For the most part, however, very few of the federal agents' extensive physiographic reports, landscape and habitat descriptions, or field trip itineraries were included in the 1905 publication. This chapter provides a description of Texas landscapes, selected from those reports and materials and supplemented with landscape photographs taken by the federal agents. These reports and photographs were selected to represent the ten ecological regions recognized by most biologists working in the state today (see Fig. 5, p. 16, and Table 2, p. 55, in Chapter 1). Thus, they provide a modern perspective of Texas's landscapes at the end of the nineteenth and beginning of the twentieth centuries. Biologists working in the state today do not use the life zone designations of Merriam and Bailey because of their limited utility (see Chapter 1).

Examination of the total portfolio of photographs reveals that profound changes to the landscape had already occurred prior to the time that Bailey and the other federal agents worked in Texas. With the exception of the High Plains and the northern parts of the Llano Estacado, most of the photos show remarkable impacts already by the beginning of the twentieth century. Vast expanses of native prairie that once covered parts of the state had already been turned under by the plow or overgrown by shrubs and woody plants. In fact, of all the community types represented in the photographs, the greatest adverse impacts appear to have been on grasslands.

Despite the fact that much of the survey was conducted during a period of prolonged drought in Texas, the federal agents encountered a considerable amount of natural, free-flowing surface water throughout their travels across the state. Almost every photograph the federal agents took of a stream or river showed an abundance of natural, free-flowing surface water, which is not true of most of those places today. The federal agents were aware of and collected specimens at many of the springs, and they liked to camp near springs because they provided a good source of drinking water. For example, Cary and Hollister in 1902 secured a series of Pecos River muskrats (*Ondatra zibethicus ripensis*) at Comanche Springs near Ft. Stockton (Fig. 56). Cary wrote the following description of the springs, which are dry today and no longer harbor muskrat, in 1902:

Comanche Creek, a beautiful stream of clear water running some thirty miles in a northeasterly direction, has its source in some extremely large springs at Stockton, and is bordered with a heavy growth of tules. This stream is utilized for irrigation purposes by several ranchmen and some fine cornfields and alfalfa meadows were noted. Fine grassy meadows along this creek afford fine pasturage.

The field agents conducted fieldwork in all ten of the ecological regions of the state, with the greatest effort in the South Texas Plains and Trans-Pecos Mountains and Basins (19 percent of all field days in each region), Edwards Plateau (15 percent), and Gulf Prairies and Marshes (15 percent). Other regions ranked as follows: Piney Woods (9 percent), Rolling Plains (6 percent), Post Oak Savannah (5 percent), Cross Timbers and Prairies (5 percent), Blackland Prairies (4 percent), and High Plains (3 percent). The following descriptions, with accompanying photographs, of the landscapes and land-cover in the ten regions have been taken directly from the accounts of various field agents.

Gulf Prairies and Marshes and Barrier Islands

Bailey and the federal agents found extensive marshlands and coastal prairies as they traversed the Gulf Coast of Texas at the beginning of the twentieth century (Figs. 57-65). Bailey visited Galveston in 1899, prior to the Great Storm (hurricane) in 1900 that killed more than 5,000 people, but he did not collect there. He spent most of his time collecting at Virginia Point just across the bay from the city of Galveston and wrote the following account of traveling by train from Galveston to Houston across vast stretches of coastal grasslands and bottomland forests:

> Virginia Point is the first station on the mainland north of Galveston. The station is only 5 miles from Galveston. The ranch where we stayed is 2 miles northeast of the station on open prairie only half a mile from the bay.
> The country is all flat and open. Level prairie stretches away to the west and, at least as far as Houston, to the north. Generally it comes to the edge of the bay and terminates in a beach but in places great salt marshes border the shores. There is practically no native timber until Buffalo Bayou is reached near Houston. There are no hills or high ground of any kind. There are some low, marshy places over the prairie but generally the surface is level and flat save for the mounds scattered uniformly over it. Shell ridges have been thrown up along the shore in places but these are rarely over 4 or 5 feet above the surface of the shore and merely mark the combined efforts of high tides and storms. On Galveston Island are extensive sand dunes but I have seen none on the mainland shore.

The federal agents visited many places along the coast during their work in Texas, describing the coastal prairie and marshy habitats. William Lloyd wrote the following description of the country around Matagorda Peninsula (Matagorda County) in 1892:

Figure 57. Shore of Galveston Bay, La Porte, Harris County, 1906. Courtesy National Archives, 222-WB-30-B9041.

Figure 58. Thicket of yaupon (*Ilex vomitorius*), High Island, Chambers County, 1907. Courtesy National Archives, 22-WB-40-B9908.

Figure 59. Mitchell's Point, Matagorda Peninsula, Calhoun County, 1900. Courtesy National Archives, 22-WB-40-B1257.

Figure 60. Open prairie near Port Lavaca, Calhoun County, 1900. Pools of water from recent rains. Courtesy National Archives, 22-WB-30-B1207.

Figure 61. Marsh by Big Chocolate Creek, near Port Lavaca, Calhoun County, 1900. Courtesy National Archives, 22-WB-30-1210.

Figure 62. Field of prickly pear (*Opuntia*) near Port Lavaca, Calhoun County, 1900. Courtesy National Archives, 22-WB-40-B1219.

Figure 63. Hedge of prairie rose (*Rosa setigera*) near Port Lavaca, Calhoun County, 1900. Courtesy National Archives, 22-WB-40-B1205.

Figure 64. Live oak chapparal and grassland near Port O'Connor, Calhoun County, 1900. Courtesy National Archives, 22-WB-41-B1300.

Figure 65. Clump of tall live oaks near Port O'Connor, Calhoun County, 1900. Courtesy National Archives, 22-WB-41-B1298.

The country east of the Nueces River along the coast and for 70 miles inland is generally a plain, bare of the slightest timber except along the river bottoms as far as the western edge of Jackson County, where commence copses of Live Oak usually in sharply defined areas. Further east one passes a Post Oak country around the two Caranchua Creeks and arrive at an interesting stream, the Tres Palacios, the western border of Matagorda County.

From here wide prairies stretch all over the county, low and often under water, a black loam, hog-wallow land it is generally termed, which is all of the same character except along the Bay shore where it is mixed with sand and intersected by pools of salt water.

The Colorado River passes through the center of the district and is blocked up by a raft 9 or 10 miles long that reaches up to near Elliott's crossing and is increasing every year. As the river has many tributaries it is frequently swollen and yearly the water backs in increasing volume from against the raft, spreading over the surrounding low country for miles and will probably in time seek an outlet through Caney Creek which is supposed to be an old mouth of the Colorado.

The river itself is deep, far more than the Rio Grande at Brownsville, and at Elliott's about 100 feet wide and red-colored as its name implies, and carries so much silt with it, that, on account of the overflows, all the trees in the bottom are girdled to a height of about 10 feet choked and dead. Below the raft however it is different. The river is studded with islands some of large size but covered with so much dense undergrowth that they cannot be kept cleared.

The Peninsula, now an island or rather two, was separated from the mainland during a great storm in 1876, which also washed away Indianola. At that time it was well settled, over a hundred lives being lost on the Peninsula alone, the storm washing away two-thirds of the houses and cutting channels through it in four places two of which have since filled up, and covering the land from shore to shore with immense amount of drift; hundreds of logs over a hundred feet long and 2 to 5 feet in diameter. A few people stayed on when in 1886 came a second storm wiping off both the last vestige of a dwelling in Indianola and everything—people, stock, houses— from the peninsula, leaving a few ruined houses standing, and leaving the place to the mammals, which flourished untouched so that some outfits with dogs captured this year 117 raccoons but also leaving the place without a vestige of fresh water.

Many of the former settlers, the remnant of which now live in Matagorda, state it had 3 rows of sand hills some 40 to 50 feet in height but that the storms leveled them. They are now only to be seen as one small continuous elevation of shifting sand about five feet high.

Live Oak Creek, as suggested by its name, near the east border of the County and Brazoria County, is also the east boundary of the Matagorda plain, and is the last body of open land in Texas (except Houston prairie), all east of the creek being timbered densely.

Mankind has modified the topography of the County in one respect. It, the County, is well settled and has been so for seventy years and all of the river land has been cut into small pastures about a mile square in each, so wild animals were quickly exterminated, the finishing touch being the Scalp Law that came into effect last year.

Along the coast at Port Lavaca (Calhoun County), Harry Oberholser wrote the following description of the countryside in March of 1900:

Port Lavaca is situated on the west shore of the northwest area of Matagorda Bay, some ten or twelve miles south of the head of this bay, and about 25 miles from the Gulf of Mexico. The town lies on a bluff some 18 or 19 feet above the sea, and, excepting towards the bay, is on all sides surrounded by almost perfectly level grassy prairies which extend for many miles, and in which the principal interruptions are two small creeks some few miles south and west of Port Lavaca. These creeks have cut down several feet from the level of the country, forming rather broad shallow valleys; the water, however, runs only in the wet season. There are a few small ponds on the prairie, most of which are dry in the summer, but while they contain water, particularly during spring and fall, are favorite resorts for water birds and waders. The entire region is so nearly flat that large areas are flooded after a heavy fall of rain. There is comparatively little marshy land along the shore of the bay near Port Lavaca, since the bluffs extend, though less in height, nearly all the way from some distance north of the town, south to O'Connorport; and the only marshes lie at the entrances to a few small creeks and bayous. The opposite side of this part of Matagorda Bay is much the same in general appearance as is the vicinity of Port Lavaca. The environs of Port Lavaca are evidently part of an extensive alluvial plain, and the surface soil is, for some distance about the town, a very heavy black clay which in the hot sun bakes hard and cracks; farther to the west and south, however, it is much more sandy.

A belt of chaparral more or less dense, extends somewhat interruptedly from a distance south of town, northward along the bay shore to near Victoria, thence eastward to south of Inez and thence down along the east side of the northwest area of the bay. On the Port Lavaca side this chaparral belt is composed largely of *Acacia farnesiana*[P], *Condalia obovata*[P], *Zizyphus obtusifolius*[P], *Berberis trifoliata*[P], and *Opuntia engelmanni*[P]; but on the opposite shore there is a considerable infusion of *Prosopis juliflora*, due no doubt to the fact it has not been cut off, as I am assured has been the case about Port Lavaca. On this east side of the bay, on Mitchell's Point, there is a tract of elm woods, and here occur *Tillandsia usneoides*[P] and *Phoradendron flavescens*[P], which also were not found elsewhere near Port Lavaca. No timber grows near this locality except a narrow fringe along portions of the larger creek some miles west of town. The greater part of the prairie is practically without shrubby vegetation, except for such places as the Thomas Ranch, some seven or eight miles southwest of Port Lavaca, where hedges of *Rosa setigera* have been planted as windbreaks for cattle. These hedges have grown so rank that they threaten to appropriate much more of the prairie than was intended, and are from ten to twenty feet in height, and from ten to thirty feet in thickness. On this same ranch there is a considerable area of live oak brush, two to twelve feet in height and in places very dense. Some of the less ephemeral ponds support a considerable growth of *Sesbania cavanillesii*.

Oberholser visited O'Connorport (now known as Port O'Connor), near Port Lavaca, in 1900 and wrote the following report describing the countryside:

O'Connorport is situated at Alligator Head, the point of land forming the southern extremity of the western shore of Matagorda Bay, and is almost due north of the northeastern end of Matagorda Island. The land about here is a sandy gently rolling prairie, only a few feet above sea level. Properly speaking, there are no streams, though the entire country is more or less intersected by shallow washes, through which water runs only after rains, and which lead usually into bayous. These bayous in some cases extend a considerable distance into the land from the bay, the largest of them being Powderhorn Bayou, about 9 miles north of Alligator Head and near the site of the former town of Indianola. There are a few small and more or less ephemeral ponds scattered over the prairie. The shore of the bay is bordered by a low bluff, at places only a foot or two in height, but increasing northward. The south shore, facing the bayous that separate the mainland from Matagorda Island, is bordered by extensive tide flats, the favorite feeding place of thousands of shore birds. The bay proper is itself very shallow, the slope from shore being for a long distance exceedingly slight, a condition similar to that existing at Port Lavaca.

With the exception of portions of the land lying along the shore, and frequent areas, more or less extensive, scattered throughout the region, the entire face of the country is covered with a dense growth of live oak brush. This grows to the height of twelve or fifteen feet on many of the small low mounds which are a curious feature of portions of this region, forming thus conspicuous clumps rising above the level of their surroundings. Other than this live oak growth there is little shrubby vegetation. Some *Opuntia* grows in the open places along the shore; and *Sesbania cavanillesii* is found in some of the ponds. At one place some twelve miles west of the settlement there is considerable *Persea borbonia* and *Balodendron arboreum*[P].

The federal agents worked extensively in the Semiarid Lower Sonoran region of South Texas, along with the interface of the South Texas Plains and Gulf Prairies and Marshes vegetation regions, which they generally characterized as a mesquite plain. William Lloyd provided the following description of the Nueces County area in 1891:

This county lies due north of Cameron and has for its boundary the great sandy area that runs up from the Laguna Madre touching the Rio Grande between Carrizo and Laredo. It is well watered in the maps but with exception of a swift, deep little treeless creek called the Santa Gertrudis, it has no other running water than the Nueces River that divides it from San Patricio County on the north. This river, noted as being a paradise for hunters for the last twenty years, is well-wooded and deep and runs into Nueces Bay, a deep long inlet of Corpus Christi Bay. The country from Santa Rosa to Alice on the north and to Banquete is all prairie, broken by a few sandy elevations, and sparsely covered with mesquite. A peninsula jutting into Nueces Bay and another from Portland are flats full of swamps and saltwater lakes. A large saltwater reach also forms the mouth of the Oso on the Laguna Madre.

A large portion of the southern part of the county, especially near the mouth of the Nueces and along the Laguna Madre, is still covered with a dense scrubby jungle almost impassable that if not continually kept cleared will encroach on all sides and gives considerable shelter to the smaller mammals.

The soil is rich black loam mixed with layers of sand around Corpus and seems to offer the most inviting field for the operations of thousands of *Geomys* and it is hard to say how many *Scalops*[M]. When the country was unsettled they worked under disadvantages but now they have increased largely and every ploughed field or additional clearing only adds an extra foothold for them which they are not slow to avail themselves of.

South winds prevail and though snowstorms do occur they are so rare as not to militate against the general mildness of the atmosphere which is now in the luxuriant growth of vegetable life.

In extreme South Texas, Bailey and the agents encountered tall stands of native sabal palm forests (*Sabal mexicana*) and dense forests of cedar elm and Texas ebony stretched along the Rio Grande from its delta in the Gulf of Mexico inland to Brownsville and beyond for several kilometers. This area was characterized by thick woods of mesquite, willow, huisache, and retama in the resacas, and the resacas still flooded once or twice a year after heavy rains. William Lloyd wrote the following description of the resaca habitat when he visited Brownsville in 1891:

The Rio Grande from about 10 miles north of Brownsville to its mouth is characterized by its numerous outlets, "resacas," that act as drainage channels for the river whenever it has an overflow which is always in the Fall, September and October. This overplus forms numerous lagoons which generally last all the year though of lessening extent; and, as the waters evaporate they become brackish and even in a few instances decidedly or entirely salty. They are the breeding homes of many of the waders and even of some of the sea-birds.

After passing Brownsville the river is so winding that a straight line drawn to its mouth would cross about equal parts of Texas and Mexican territory and the channel changes every season in half a dozen or so places keeping the border owners of land in continuous tribulation.

All the country was a jungle along the river, but it is now so thickly settled, that most of it has long since been cleared, and the timber is now principally in occasional strips along the river, with exception of the Palmetto and smaller brush-covered hills on the Mexican side.

Reaching to the edge of the river, or separated from it by the strips of timber already referred to, is a wide plain sometimes covered with small brush, or with coarse grama grass and still further south is a soft sandy expanse covered with a rank bunch grass. Alkali flats occur devoid of the slightest vegetation or life, except crabs, beds of mollusks, and grasshoppers and the birds that prey on them or a traveling jackrabbit or coyote. The bare level is occasionally relieved by sand dunes 30 to 50 feet high, which however are always shifting and changing.

The ground, generally speaking, away from the red clayey soil of the river is a rich black loam free from rocks or pebbles of any kind, in fact there is not a rock bluff or cliff in the whole locality. The temperature for the summer months is 95-98 but outside towns is always modified by the sea breeze. On the Mexican side of the river the sand does not commence until one reaches Bagdad, that once was a considerable settlement but now, like Clarksville which was on Texas side at the mouth, has been totally swept away by the sea and river combined overflow. The flats also in Mexico are studded with a series of detached low elevations covered with brush and are veritable oases in the desert around. They are usually covered with mesquite, chaparral, yuccas and cacti.

Padre Island is about 100 miles long and most of the lower half about 2 miles broad. It is a simple collection of sand dunes without any timber whatever on the

Pt. Isabel end though there is said to be huisache etc. near the Corpus Christi end where the island is 4 miles broad. On the sea side there is no life except sea birds and crab as the strong wind blowing on shore keeps the sand particles in perpetual agitation preventing any vegetation from taking root. On the lagoon side (Laguna Madre) is a strip of soft beach about a mile wide in its usual distance, and like the sea-beach destitute of everything. All animal life is in the middle division where the north side of the hills are all covered with grass and low vegetation.

Of the immediate area around Brownsville, in Cameron and Starr counties, Lloyd wrote the following description in 1891:

The principal trees of the Lower Rio Grande Valley to a point some few leagues north of Brownsville are the Hackberry, Elm, Ash, Mesquite, Huisache, and Mulberry. The Ebony have increased in size and, though not reaching the same dimensions as on the Rio de San Juan, still are a most important tree with its hard wood and deep shade. The only Cypress in the region under consideration are some in a pasture a little south of Rio Grande City and a few Anacua near Brownsville. Retama, Lignum, Vital, the Colima and Palo Brazil are common along the lagunas, and Coma (a well known fruit) are scattered amongst the chaparral. Two *Koeberlinias* occur on the edge of the region subject to flood and Willow generally along the river and are used (one species) for basket making. The Palmetto, referred to by Major Emory, has only a few specimens standing south of Santa Maria, then none occur until the tropical fauna of Brownsville is reached; they have been used for thatching jacals and now for making sombreros.

A few Yucca of large size—the Maquez—both planted and in a wild state, and the prickly pear which line all the belt beyond the valley of the river with a few *Echinocactus*[P] and one species of *Mammillaria*[P] are the sole representatives of the cactus family.

Sunflowers grow up in rank patches filling all the dry lagunas until they are leveled by another rise, and wild ipecac is abundant in the southern part of the district.

Corn fields are continuous for nearly the whole length of the river, with a few patches of cotton. Sugar has been tried but now abandoned.

A wild blackberry ripening in May and June is also abundant in the lower part of the region.

Texas has some of the longest barrier islands in the world, beginning with Galveston Island in the north and continuing southward to Matagorda, San Jose, Mustang, and Padre Islands. During the biological survey of the state, one or more of the federal agents visited all of these islands except for Mustang and San Jose.

Harry Oberholser visited Matagorda Island in 1900. His survey report describes the physiography and vegetation of the island, and he made a list of the birds, reptiles, and mammals he observed. According to his field trip report,

Matagorda, or Saluria, Island is a low strip of land, some 40 miles long and from half a mile to about four miles wide, extending southwestward along the coast of Texas from the entrance of Matagorda Bay to beyond San Antonio Bay. It encloses San Antonio Bay and the various lagoons which connect this with Matagorda Bay. The northwestern end was the only portion visited, but the island is practically the same throughout its entire length. Viewed from the top of the lighthouse it is seen to consist of a few parallel ridges running length-wise, with broad though slight depressions between, the highest parts of the land being only a few feet above the sea. Along the Gulf shore is a series of sand dunes, some of them from ten to fifteen feet in height. The inner shore of the island is bordered by broad, bare sand flats, some of them covered at ordinary high tide, others only occasionally under water. Between these tide flats, which are, it may be noted, favorite resorts for water birds and waders of all kinds, between these and the sand dunes lies the main portion of the island, covered for the most part with grass, and suitable for grazing. The soil here is a sandy loam so much like that of the neighboring mainland as to suggest the idea that this island was once a part of the mainland, rather than of marine origin, or at least is not of very recent formation.

There is little chaparral on Matagorda Island, a few small groves of *Prosopis juli-flora* being about the only approach to such. *Tamarix gallica*[P] is about the only other shrubby vegetation that is at all common, this growing principally on the sand dunes or along their inner border. Many of the sand dunes are to a greater or less extent covered with a growth of wiry grass. Matagorda Peninsula, which separates the bay from the Gulf, and of which Matagorda Island is an interrupted continuation, has according to information obtained, considerable more chaparral than the island, besides being more sandy.

William Lloyd traversed all of Padre Island by foot in November of 1891. His account of that trip was one of the few intact physiographic reports that Bailey included in the 1905 publication.

South Texas Plains

The agents worked at several sites throughout the plains region of southern Texas (Figs. 66-74) and along the Rio Grande throughout the Texas-Mexico borderland region. William Lloyd wrote the following account of the area around Eagle Pass (Maverick County) in 1890:

The country round Eagle Pass is a mesquite flat, sometimes rising into small eleva-tions and everywhere cut into deep ravines, which though often only 10 feet across are several hundred yards long. The soil near the Rio Grande is composed of fine

Figure 66. Texas bluebonnets near Beeville, Bee County, 1900. Courtesy National Archives, 22-WB-41-B1321.

Figure 67. Edge of live oak and post oak forest near Beeville, Bee County, 1900. Courtesy National Archives, 22-WB-41-B1315.

Figure 68. Grassy opening in chaparral, Cotulla, La Salle County, 1900. Courtesy National Archives, 22-WB-41-1424.

Figure 69. Hedgehog cactus and chaparral, San Diego, Duval County, 1900. Courtesy National Archives, 22-WB-40-B1354.

Figure 70. San Diego Creek, near San Diego, Duval County, 1900. Courtesy National Archives, 22-WB-40-B1365.

Figure 71. Yucca in chapparal, fifteen miles below Laredo, Webb County, 1900. Courtesy National Archives, 22-WB-40-B1419.

Figure 72. Looking north up Rio Grande from fifteen miles below Laredo, Webb County, 1900. Courtesy National Archives, 22-WB-41-B1381.

Figure 73. Chaparral composed entirely of creosote bush, below Laredo, Webb County, 1900. Courtesy National Archives, 22-WB-41-B1408.

Figure 74. Palm grove near Brownsville, Cameron County, 1911. Courtesy National Archives, 22-WB-41-12922.

sand and principally overgrown with thistles, burrs, sunflowers, and crabgrass. It has a depth of over 20 feet, before one reaches the limestone strata below, and on the river changes into soft red clay, which gives the water its muddy color. There are also a few mesquite growing along the banks, with here and there a large hackberry, a mulberry, a few pecans, with an occasional wild china. Elm Creek, S of Eagle Pass about 4 miles, is, as its name denotes, bordered with some fine elms for the last mile or two of its course where it is one continuous pool of spring water. Fossil wood is abundant here in large pieces with the bark well defined and, in Eagle Pass, I saw 6 fossilized nuts that appeared pinon from this region. Mineral Paint, red, is also common and there is a large coal mine in the immediate neighborhood.

South of Eagle Pass opposite to the Rio Chiquito which empties a volume of clear blue-green water into the Rio Grande, of same extent as Sycamore Creek, were once large plantations of corn, etc. The river has been exceptionally high this year and has washed away all vestiges of crops and covered everything with an extra layer of sand one foot high. Here, all the country south to Laredo, there is no water whatever, except that of the River and the hills are covered with small brush instead of mesquite and often are almost entirely bare, but by the river the Bermuda grass has already shown itself above the top of the sand and will in time probably cover the whole region.

In that same year, in November and December, Lloyd traveled from Eagle Pass to Laredo with Clark Streator and wrote the following account of the countryside:

The country all lies at a low altitude, the Rio Grande falling about 200 ft in the 120 or 130 miles from Eagle Pass south to Laredo. The country, on the Texas side, is as a rule with few hills or elevations over 40 or 50 feet, though the Rio Grande valley proper, is bordered by a chain of hills running continuously parallel with the river and from 1/4 to 1 miles from it of perhaps 100 ft. high. The only elevation in the country is "Las dos Hermanos," two circular hills of red sandstone and gravel of 600 feet high, and to the W and S one can see the blue ridges toward Monterey. Water is the great want beyond the Rio Grande which is practically worthless for collecting near it, owing to the great September floods.

Sous Creek, which meanders through the country and has a few willow on it, flows through a shrubless, treeless plain, and did not seem to have any sign of animal life whatever, owing perhaps to the countless bands of sheep and goats that go to the few water holes for water and have trampled all herbage and soil into a dusty mass.

San Lorenzo Creek has also a few pools of stagnant water and no doubt at times owing to the thick belt of timber, is tenanted by mammals that can feed on acorns, but this year the crop together with the pecans and hackberries have failed.

Santa Isabel has a few pools of saline water near its mouth, and is undrinkable, but mammals seem to like it, and it was by this creek, with its peculiar salty grasses, that we were most successful.

Most of the hills and a large proportion of the low lands are so gravelly as to render it hopeless to expect any mammals, as they avoid such places keeping to the fine sand which are found in places in the river bottom, and along the NE border of Webb County, and also in the cactus flats that, as yet, are not choked up with rank grass and small mesquite.

For the last 60 miles the road and country, from the Rio Grande E and S are broken by barrancas of little depth, in which the rain floods have twisted into all sorts of shapes. They are of soft earth and are utterly devoid of animal life.

Oberholser worked around the city of Laredo in April and May of 1900 and wrote the following description of the area:

Laredo is situated on the Rio Grande, about 200 miles from its mouth; and is in the neighborhood of 400 feet above the level of the sea. Behind the city, and some distance back from the river rise broken hills, between which and the stream the country is comparatively level. Farther down the hills approach in places much nearer the river. In fact, the whole region is a succession of hills and valleys, very little of which, except that close to ranch houses and a considerable area just north of Laredo, is under any kind of cultivation. The Rio Grande here runs for the most part between nearly vertical banks of varying height and where not too steep covered at least to some extent with trees or bushes. At Laredo the river bed is possibly a third or a half mile wide, and much of it is bare except at high water; but lower down it narrows in places to 100 or 150 yards, and is usually there all under water. The river is a swift, muddy, treacherous stream, and where narrow, rises of fifty feet in vertical measurement are not unusual. Its banks are sometimes pure sand, sometimes cliffs of solid rock. The hills in this region are usually rocky, almost always gravelly or stony, but the depressions between them are commonly sandy, and frequently cut into deep arroyos by the heavy and violent rains. These arroyos are of course most numerous in the vicinity of the river, and are there a conspicuous feature of the landscape.

This entire region is covered with chaparral, somewhat scattering and stunted on many of the stony hills, but rank and dense on most of the lower areas. The most abundant shrubs composing this growth are *Prosopis juliflora*, *Lippia ligustrina*[P], *Larrea mexicana*[P], *Leucaena pulverulenta*, *Diospyros texana*, *Condalia obovata*[P], *Zizyphus obtusifolius*[P], *Koeberlinia spinosa*, *Leucophyllum texanum*[P], and *Opuntia engelmanni*[P]; the first two of which occur much less commonly on the hills. *Acacia farnesiana* is not of frequent occurrence but is found in some of the arroyos. *Parkinsonia aculeata*[P] (now in bloom), *Celtis occidentalis* and *Ulmus crassifolia* grow commonly along the river banks; where also the mesquite most flourishes, attaining sometimes great size, 45 feet in height, with a diameter at base of over three feet, the largest one measured being 3 feet, 1-1/2 inches through.

In May and June of 1891, Lloyd traveled along the Rio Grande Valley from Laredo to Rio Grande City, in Webb and Starr counties, and wrote the following description of the countryside:

The country from Laredo to Rio Grande City is a country differing widely from that to the north of it—principally in the wide river valleys which are entirely wanting above that place.

Above, the Rio Grande runs in a deep channel carved out of the reddish clay and alluvial soil which line its banks, and which add to the dirty look of the water and which always carries a large amount of soil in solution (a bucketful yielding about half an inch). It is nearly an average width all the way from Laredo 80 to 100 yards and is rarely approachable owing to its high banks, except at the mouth of its numerous arroyos (dry) which, like the river are for the most part entirely treeless. The road in many places passes through belts of fine sand, one north of Carrizo being about 20 miles in extent from north to south. This soil is the only part of the region under consideration that is rich in mammals such as armadillo, badger and gopher, but the great dearth of water in all the region renders it practically inaccessible at least with a wagon. There is no water whatever from Laredo south with exception of the wells at the Mexican ranches that lie away from the road, until after leaving

Roma and approaching Rio Grande City, where after a continued rainfall are large depressions in the land that become veritable lakes and last quite a while, until they change into marshes. These lakes may perhaps be a continuation of the great chain that extend from the Nueces in Zavalla County and are supposed to touch the Rio Grande below Laredo. Some are of large extent, one said to be 40 miles around, and as I mentioned some time before, are almost entirely unknown.

Edwards Plateau

The Edwards Plateau had already undergone extensive landscape and habitat change by the time Bailey and agents arrived in the area (Goetze, 1998). Before European settlement, the vegetation of the region was about half forest and half grassland. By the time Bailey and the federal agents worked there, most of the area was heavily overgrazed by cattle, goats, and sheep, and most of the grasses had been depleted and replaced by less desirable woody shrubs. Bailey and the agents visited 21 different places in this region including Kerrville, Llano, Austin, Fredericksburg, and Rock Springs, as well as several locations along the Mexican border (Figs. 75-86). While visiting Howard Lacey's Ranch, 12 miles (19 km) east of Kerrville, Bailey wrote the following report in May of 1899:

> The ranch of Howard Lacey, where we have worked for a week, is on a little branch of Turtle Creek, 12 miles west of Kerrville. The altitude of Kerrville is 1750 feet, of Lacey's Ranch approximately 2000 feet. The country is exceedingly rough and deeply eroded through horizontal layers of limestone. The surface lies in flat topped mesas extending back and becoming wider toward the source of drainage; in steep, terraced or cliffy slopes; and in narrow-bottomed stream valleys. The highest mesas are 300 or 400 feet above the stream valleys. There are some rocky canyons and numerous sheer walls and cliffs. As usual in limestone formations the cliffs are much broken and full of cracks and caverns. Most of the land is too rough and stony for cultivation, though the flat parts of the valleys are very mellow and fertile. The soil is mostly clay and very sticky when wet. Springs and streams are abundant and the water is generally good.
>
> The country is mostly covered with scrubby timber which becomes dense and of considerable size in the canyon and gulches and scattered and dwarfed on the mesas. There are many open strips and bare or grassy slopes, and the flat valley land is now mostly cultivated. The characteristic vegetation of the mesa tops is juniper, shin oak, bastard oak, blackjack, mimosa, acacia, a few species of *Opuntia*, and *Yucca*. The canyons and gulches are characterized by the turkey oak (*Q. texana*), bastard oak (*Q. durandii*[P]), walnut, pecan, elms, *Celtis*, basswood, benzoin bushes, *Cornus*, viburnum, black cherry, *Rhamnus*, and grape and *Smilax* vines. The open valleys are especially characterized by live oaks (*Q. virginiana*), a few scattered junipers, some yuccas and cactus, mimosa bushes, and along the streams sycamores and willows. The physical features are unusually varied and offer conditions suitable to an extensive flora and fauna. The mesa tops are almost arid, the canyons and gulches are damp

Figure 75. Hills along Pecos River near Ft. Lancaster, Crockett County, 1901. Courtesy National Archives, 22-WB-30-B2726.

Figure 76. Looking up valley of Pecos River near Ft. Lancaster, Crockett County, 1901. Courtesy National Archives, 22-WB-30-2722.

Figure 77. Oak-juniper countryside near Kerrville, Kerr County, 1906. Courtesy National Archives, 22-WB-30-9045.

Figure 78. Oak-covered hills near Kerrville, Kerr County, 1906. Courtesy National Archives, 22-WB-30-9044.

Figure 79. Field of flowers near Rock Springs, Edwards County, 1900. Courtesy National Archives, 22-WB-41-B1482.

Figure 80. Chaparral near Rock Springs, Edwards County, 1900. Courtesy National Archives, 22-WB-41-B1475.

Figure 81. Looking up Nueces River near Barksdale, Edwards County, 1900. Courtesy National Archives, 22-WB-30-1496.

Figure 82. Hills just north of Barksdale, Edwards County, 1900. Courtesy National Archives, 22-WB-30-1497.

Figure 83. Pond in side canyon along Rio Grande, near Langtry, Val Verde County, 1901. Courtesy National Archives, 22-WB-30-B2748.

Figure 84. Pecos High Bridge from east side, Pecos River, Val Verde County, 1901. Courtesy National Archives, 22-WB-30-2779.

Figure 85. Looking down Pecos Canyon from Pecos High Bridge, Val Verde County, 1901. Courtesy National Archives, 22-WB-30-2778.

Figure 86. Large spring near Del Rio, Val Verde County, 1901. Courtesy National Archives, 22-WB-30-2795.

and cold or hot according to slope and form, while the open valleys are warm and fertile. Birds and mammals are unusually numerous in both species and individuals. Most of the good land is occupied by small farms. Corn, wheat, sorghum cane, some cotton, garden vegetables, and fruit are raised. Hogs, sheep, goats, and cattle are pastured on the rough land. Stock raising is evidently the main industry of the region.

In June of 1900, Oberholser penned the following description of the landscapes around Rock Springs in Edwards County:

Rock Springs lies on the divide between the Nueces and Llano Rivers, about 50 miles east of Devil's River, and is about 2300 feet above sea level. The entire surrounding country is more or less uneven, though not hilly, this being due to the many shallow valleys that have been cut down from the general level. The name Rock Springs does not convey an accurate impression, for there are no springs about the town, and although it lies at the ultimate sources of the Nueces River, these streams are merely channels for surface water after rains, the real canyons not beginning for ten or twelve miles to the southward. There are in this vicinity a few vegetation-choked ponds, usually dry in the summer, but while they contain water are resorts for various birds. The whole region is exceedingly rocky, and the soil, except for occasional somewhat sandy flats, is black and heavy. Very little of the land is under any kind of cultivation.

   With the exception of frequent, though not often extensive, open grassy areas, the entire face of the country is covered with a growth of live oak brush, this being the predominant vegetation. Other common components of this chaparral are two species of *Acacia*, *Xanthoxylum pterota*[P], *Berberis trifoliata*[P], and *Zizyphus obtusifolius*[P]. Occasional patches of *Prosopis juliflora* occur, but it does not appear to grow among the live oaks. Scattered throughout the live oak chaparral there are frequent live oak trees, as well as a few *Celtis occidentalis*, and some *Juniperus*. *Centaurea americana* and a species of lupine are very abundant.

Along the Rio Grande, J. H. Gaut visited Del Rio in May-June 1903, and wrote the following physiographic report:

Del Rio, Texas, a flourishing town, with a population about four thousand, is situated near the Rio Grande in the southwestern part of the state. To the east, west, and south are rich farming districts, intersected throughout by numerous irrigating ditches, which are supplied with water from a group of springs about a mile northeast. Valuable crops are raised of which Johnson grass hay and corn seem to predominate.

   Toward the southeast and west, the land is one gradual slope to the Rio Grande—a distance of about three miles. The soil is a sandy substance. Where the land is left in its natural state and the plant life is not disturbed by cultivation, *Prosopis juliflora*,

Opuntia engelmanni[P], *Eysenhardtia amorphoides*[P], and *Artemisia ludoviciana* are found on the higher ground while low down in the vegas or made land *Nicotiana glauca*, *Fraxinus velutina*, *Salix*, *Milla azedarach*[P] appear to flourish. The latter together with sunflowers are extremely abundant close to the water. North of Del Rio the surrounding country is of a gravelly mesa nature, connecting to the sandy river land by means of stony slopes and shallow arroyos. On these slopes live *Karwinskia humboltiana*[P], *Guaiacum angustifolia*[P], *Dalea formosa*, *Acacia amentacea*[P], *Macrorhamnus ericoides*[P], and *Lycium berlandieri*, while in the arroyos are *Brayodendron*[P] and *Prosopis juliflora*. On the vast stony mesa country and flats, *Leucophyllum texanum*[P] and *Karwinskia humboltiana*[P] are the most abundant species of plants, while the following are numerous: *Koeberlinia spinosa*, *Larrea mexicana*[P], *Macrorhamnus ericoides*[P], *Berberis trifoliolata*[P], *Opuntia engelmanni*[P], and *Dalea formosa*.

About two miles west of Del Rio is a small stream called Cienegas Creek, running south, which empties into the Rio Grande. Another stream, San Felipe Creek, quite a large affair, runs through a section of the town and eventually empties into the river about three miles south, while about twelve miles east of the town is another stream called Sycamore Creek, which empties into the river. Along the two latter streams are many pecan and hackberry trees, some of which grow to be enormous and afford fine shade from the heat of the sun.

This season has been very wet about Del Rio, hence therefore the weeds are very dense in the low places.

Two months earlier, Gaut had visited Langtry (Val Verde County), about a hundred miles upriver from Del Rio, and wrote the following description:

Langtry, a small place situated close to the Rio Grande in southwestern Texas, is reached by the G.H.S.A. Railway. The surrounding country on all sides, for many miles, is very rough, a rolling aspect with deep rocky canyons. Some of these canyons are so deep and rocky in places, that for miles on either side they are inaccessible except by means of ropes. The most characteristic of this is Pump Canyon, a short walk west of the town. In this gorge are several large springs from which the railway company secures water to supply its engines, and the inhabitants of the town. Due south the Rio Grande has formed another deep gorge which is accessible at intervals by means of benches.

Low down, just above the level of the river, are small sandy "vegas" or flats on which *Baccharis glutinosa*[P] predominates over all other plant life. However, *Nicotiana glauca* and *Fraxinus velutina* and *Prosopis juliflora* are quite numerous. On the benches high up *Opuntia engelmanni*[P], *Selaginella lepidophylla*, *Karwinskia humboltiana*, *Euphorbia antisyphilitica*, *Dalea formosa*, *Guaiacum angustifolia* are plentiful, while *Leucophyllum texanum*, *Agave lechuguilla* and *Fouquieria splendens* are rather numerous. In the beds of the canyons, *Sophora secundiflora*, *Karwinskia humboltiana*[P], and *Fraxinus velutina* and *Prosopis juliflora* are found among the rocks and gravel. Along the slopes are found *Opuntia engelmanni*[P], *Echinocactus longihamatus*[P], *Cereus enneacanthus*[P], *Fouquieria splendens*, *Brayodendron*[P] and *Guaiacum angustifolia*[P]. On the gravelly hillsides in places *Agave lechuguilla* form a regular network and hinder the traveling of animals considerably. *Opuntia engelmanni*[P] seems to flourish in such places, as well as *Dalea formosa*, *Acacia amentacea*[P], *Dasylirion texanum*, *Macrorhamnus ericoides*[P], *Lycium berlandieri*, *Leucophyllum texanum*[P], *Larrea mexicana*[P], *Mortonia scabrella*, *Echinocereus pectinatus*, *Echinocactus texensis*, *Echinocactus horizonthalonius*, *Cereus paucispinus*[P].

About ten miles west of Langtry, a vast flat country is found. Here may be seen *Larrea mexicana*[P], *Prosopis juliflora*, *Lycium berlandieri*, *Koeberlinia spinosa*, *Flourensia cernua*, and *Mammillaria heyderi*.

Along the western edge of the Edwards Plateau and bordering the Rolling Plains, in the region of the Concho Valley and Stockton Plateau, it is evident that by the end of the nineteenth century, woody species had increased in density and had begun to appear on the once near-treeless grasslands. Small mesquite began to appear widespread on the plains, although some areas remained largely shrubless. They also saw scrub oaks and junipers on the hillsides, where observers 40-50 years earlier had not noted their presence (Maxwell, 1979b).

Bailey traveled through the Concho Valley in May 1899. He described the vegetation from San Angelo (see photos of San Angelo on pp. 406-408) to Big Spring, through the North Concho Valley: "San Angelo is an open, mesquite plain in the genuinely arid region. There are great stretches of smooth surfaces with only short grass and little desert plants, but much of the country is covered with a scattered growth of small mesquites." Between San Angelo and Sterling City he noted: "Mostly mesquite plains with short grass and scattered shrubs of *Lyssium*[P], *Condylia*[P], *Zizyphus*, *Acacia*[P], *Berberis*, *Mimosa*, *Cactus*[P], and *Yuccas*. The first buttes are near Water Valley where we had dinner and the next set near Sterling. These buttes seem to be covered with shin oak and some have juniper on them." William Bray accompanied Bailey on this trip and elaborated on the butte slopes around Sterling City:

> On three sides of Sterling City, plus or minus three miles (4.8 km) distant are buttes covered by oak scrub and *Juniperus*. The slope leading up these buttes is . . . very thin soiled with the flora listed at San Angelo . . . , including *Microrhamnus ericoides*[P] (javelina bush). North of Sterling six miles (9.6 km) the road to Colorado City crosses a part of one of the buttes where the characteristic *Q. undulatus* (wavyleaf oak) and *Juniperus pinchotii* (red-berry juniper) occur abundantly. Beyond this lies a flat mesquite plain extending some 20 miles (32 km), having occasional *Opuntia arborescens* (tree cholla) and wide stretches of needle grass.

Harry Oberholser traveled from San Angelo to Ozona, following Spring Creek, in April 1901. He described this region as follows:

San Angelo lies in west central Texas, close to the junction of the North and South Concho Creeks which together form the Concho River. These streams have in many places cut down considerably below the level of the country, and their banks are in many places rocky and precipitous. The country is generally rolling immediately about the town, though some miles away in various directions rise rocky hills. Close to the town much of the land is under cultivation, but farther away it is covered with a growth of low chaparral. This is composed chiefly of *Prosopis juliflora*, *Opuntia engelmanni*[P], *Acacia*, *Berberis trifoliata*[P], *Diospyros texana*, *Zizyphus obtusifolius*[P], and *Ephedra nevadensis*. There is a considerable fringe of timber along the streams, this composed chiefly of *Hicoria pecan*[P], *Quercus*, and *Celtis occidentalis*.

To the southwest of San Angelo the country continues as a rolling chaparral-covered plain for several miles, then being interrupted by rocky hills covered with a growth of oak brush and low junipers. Beyond this the hills increase in number toward Sherwood, the valleys and mesas between them being covered with a growth of chaparral similar to that about San Angelo, the mesquite predominating in much of the level lower ground. Three streams drain this region, all of them similar in character; rocky, with low banks, and in many places with a fringe of timber. Spring Creek, which rises southwest of Sherwood, is fed by springs some six or eight miles from this town, and though the valley of the creek reaches for many miles farther southwest it is at even this season merely a dry bed. Near and beyond Sherwood the hills are less high and finally merge into the high plains to the southwest, the general level of this portion (Sherwood to 18 miles southwest) rising more rapidly than that farther to the northeast.

These high, almost level, grassy plains, evidently a southern extension of the Staked Plains, extend from about 20 to 45 miles southwest of Sherwood toward Ozona, broken only occasionally by shallow valleys. They are scarcely anywhere entirely free from shrubby growth, but in many places, particularly along the eastern side, are practically open, being only very sparingly dotted with *Prosopis juliflora* and a low species of *Acacia*[P], with an occasional bunch of *Opuntia*. Toward the western side, however, the brush increases in quantity, and in places forms a fairly dense chaparral. The only shrubby plants noted on the top of the plains are as follows, all but the first five being of uncommon occurrence: *Prosopis juliflora*, *Opuntia*, *Zizyphus obtusifolius*[P], *Acacia*[P], *Nolina texana*, *Ephedra nevadensis*, *Yucca g. stricta*, and *Berberis trifoliata*[P]. On the stony slopes of the shallow valleys which cut the plains along their southwestern side there is much *Juniperus*, *Celtis occidentalis*, and *Diospyros texana*, with a few other shrubs unidentifiable at this season on account of lack of foliage.

West of these high plains toward the Pecos River the whole country is much cut up by deep canyons, with hills and sometimes high mesas between them. There are practically no running streams, the beds of all the canyons being entirely dry excepting after hard rains. These canyons have almost always steep rocky sides, having been scored out through almost always horizontal beds of limestone; and the appearance of most of the country is thus rocky and broken. Westward from Ozona, which lies a few miles beyond the high plains and not far from the head of one of the canyons, the vegetation is much the same as that on the northeastern side of the high plains. Many of the rocky slopes have an abundant growth of *Juniperus*, and the bottoms of the canyons are nearly always covered with chaparral consisting of the customary species. Between Ozona and Fort Lancaster the influence of the Rio Grande–Pecos region is seen in the appearance of *Sophora secundiflora*[P], *Koeberlinia spinosa*, *Dasylirion texanum* and others.

Trans-Pecos and Chihuahuan Desert

Bailey and the federal agents visited twenty-four sites in this vegetative region (Figs. 87-108). From their descriptions, the impact of overgrazing on the rangelands was evident, but the riparian habitats were relatively unaltered compared to today. Already, there was some evidence that stands of desert scrub were expanding at the expense of grassland habitat; lumbering of the "islets" of coniferous and hardwood timber at the summits of desert mountains was underway; and freestanding natural water was rare throughout the region.

Bailey and his field party were the first to survey the Big Bend of Texas in what is known today as Big Bend National Park, having been established in 1944. In this region of diverse habitats and vegetation, varying from desert valleys and grassy plateaus to wooded mountain slopes, they discovered extensive plant and animal diversity. During the time of their visits, most of the land was native range in large holdings, typically used for livestock grazing (cattle, sheep, and some Angora goats). Cultivated areas were confined largely to the irrigated valleys.

Bailey and Fuertes were particularly impressed with the magnificent Santa Elena Canyon where Terlingua Creek flowed into the Rio Grande. In his 27 June 1901, field trip itinerary, Bailey wrote:

> Fuertes and I walked down to mouth of canyon—2 miles from camp. Took photographs and explored it up 4 or 5 miles from the end. The river has cut its way down through about 1700 feet of stratified limestone as the rock rose slowly with a clean escarpment facing NE and a flat topped plateau dipping back to the SW. . . . There is not much talus at the bottom of the cliff and the walls are nearly vertical as also are the walls of the canyon. There is apparently no place where either cliff or canyon wall can be climbed to the top without going 5 miles back from the river to get on the plateau or 5 or 6 miles up the canyon to climb the walls. From the escarpment the big open valley, mainly hot and barren desert, stretches away down the river and into the rough badland country toward the Chisos Mountains. Its bareness is relieved by the Mexican farms of corn, melons, etc. along the valley near the mouth of Terlingua Creek and by mesquites and cottonwoods along the river flats. The canyon is impressive on account of its sheer walls and narrowness—often apparently higher than wide and with walls leaning in as commonly as out. At present the river is low and we waded, swam, and climbed along shore up 4 or 5 miles into the canyon—to where

Figure 87. Franklin Mountains, near Ft. Bliss, El Paso County, 1901. Courtesy National Archives, 22-WB-30-3776.

Figure 88. Looking over El Paso, El Paso County, and Juarez, Mexico, 1901. Courtesy National Archives, 22-WB-30-3787.

Figure 89. Looking up one of the main gulches at the head of Dog Canyon, Guadalupe Mountains, Culberson County, 1901. Courtesy National Archives, 22-WB-30-3759.

Figure 90. Gulch in head of Dog Canyon, Guadalupe Mountains, Culberson County, 1901. Ponderosa pine the typical tree. Courtesy National Archives, 22-WB-30-3764.

Figure 91. Sierra Blanca, Hudspeth County, 1901. Courtesy National Archives, 22-WB-30-3147.

Figure 92. Davis Mountains from east side of Toyahvale, Reeves County, 1901. Courtesy National Archives, 22-WB-30-3079.

Figure 93. Looking up Limpia Canyon from mouth of cave in cliff, Davis Mountains, Jeff Davis County, 1901. Courtesy National Archives, 22-WB-30-3136.

Figure 94. Fort Davis, Jeff Davis County, 1901. Courtesy National Archives, 22-WB-30-3144.

Figure 95. Limpia Canyon, 20 miles below Fort Davis, Jeff Davis County, 1901. Trees in bottom *Juglans rupestris* (walnut). Courtesy National Archives, 22-WB-30-3702.

Figure 96. Near lower end of Limpia Canyon, Jeff Davis County, 1901. Courtesy National Archives, 22-WB-30-3704.

Figure 97. Mouth of Tornillo Creek (now Big Bend National Park), Brewster County, 1901. Courtesy National Archives, 22-WB-30-B3617.

Figure 98. Looking down Rio Grande from above Boquillas (now Big Bend National Park), Brewster County, 1901. Courtesy National Archives, 22-WB-30-B3615.

Figure 99. Badlands, east base Chisos Mountains (now Big Bend National Park), Brewster County, 1901. Courtesy National Archives, 22-WB-30-3663.

Figure 100. South end of Santa Elena Canyon (now Big Bend National Park), Brewster County, 1901. Courtesy National Archives, 22-WB-30-3015.

Figure 101. Lower entrance to Santa Elena Canyon (now Big Bend National Park), Brewster County, 1901. Courtesy National Archives, 22-WB-30-3016.

Figure 102. Looking up Santa Elena Canyon from terrace on south side (now Big Bend National Park), just above the mouth, Brewster County, 1901. Courtesy National Archives, 22-WB-30-3669.

Figure 103. Primitive irrigation dam and ditch near mouth of Terlingua Creek (now Big Bend National Park), Brewster County, 1901. Courtesy National Archives, 22-WB-30-3011.

Figure 104. Chisos Mountains (now Big Bend National Park), Brewster County, 1901. Courtesy National Archives, 22-WB-30-B35548.

Figure 105. Chisos Mountains (now Big Bend National Park), Brewster County, 1901. Camp in foreground. Courtesy National Archives, 22-WB-30-3651.

Figure 106. Looking up valley below Pine Canyon, Chisos Mountains (now Big Bend National Park), Brewster County, 1901. Courtesy National Archives, 22-WB-30-2913.

Figure 107. Peña Colorado Creek, near Marathon, Brewster County, 1901. Courtesy National Archives, 22-WB-30-2828.

Figure 108. Cottonwood grove on Peña Colorado Creek, near Marathon, Brewster County, 1901. Courtesy National Archives, 22-WB-41-B2824.

the walls were lower—less sheer—not over 1000 feet. Next to the Grand Canyon of the Colorado it is the best I have seen—with its square walls it is far more impressive than the Yellowstone Canyon. One side is cold and shady, the other scorching hot. Springs come out of the banks. There are grand places to camp on sandy bars, in cool, deep caverns or under shadowy cliffs on little mesas.

The grandeur of this incredible place remains today, although some of the wildlife species recorded by Bailey and Fuertes (e.g., bighorn sheep) no longer occur there. Scattered throughout the rugged topography of the Chihuahuan Desert in the Trans-Pecos are island-like, forest-covered highlands. Some of these restricted projections rise only 1,600 feet (488 m) or so above the arid lowlands. Other mountain masses are higher, extending upward from the desert floor a thousand or so meters. These support montane habitats with luxuriant stands of oaks, pines, sometimes firs, and other boreal growth. Bailey and the agents visited the most massive mountains along the Front Range, beginning with the Chisos Mountains near the Rio Grande and continuing northward to the Davis Mountains, in Jeff Davis County, and the Guadalupe Mountains on the Texas-New Mexico border.

Of the Chisos Mountains, which today constitute the heart of Big Bend National Park, Bailey wrote the following description in June of 1901:

The Chisos Mountains are a small group of very old volcanoes pushed up through broken and tilted strata, then worn down until the hard, basaltic caves and old lava rims stand up as towers, pinnacles, cliffs and crests above the crumbling scattered debris of the lower slopes. Many of the old craters are almost obliterated by succeeding ones and by erosion but half a dozen large old rims may be traced for part of their circumference while many little craters, fairly well preserved, stand out from the main mass. As a result of their origin and age the mountains are extremely rough and steep and full of deep gulches, high cliffs and precipitous walls. A few old lava streams extend from them, one of the most conspicuous forming a plateau ridge toward the Rio Grande, but the lava flow seems not to have been extensive. The long axis of the group is east and west and a gap only 7000 feet high partly separates them into east and west groups. Our main camp is on the east slope of the eastern group, or rather in the central crater of the eastern part of the group at an altitude of 6000 feet, about 5 miles west of Rock Spring and more than 1000 feet higher. Half a mile above our camp a spring of good water comes out in the bed of the stream and a quarter of a mile farther up a pretty little water fall about 300 feet high trickles over the precipice. Below us the rugged amphitheater almost closes in, having only a narrow gap opening out over the valleys to the SE. Crossing the ridge just back of our camp one goes down into the long north and south canyon that partly separates the two groups and then either up the steep slope or around up the winding canyon that brings you into the east side of Mt. Emory or on top of the high, timber covered ridges south of Mt. Emory.

Mt. Emory (9000 feet) is a rather barren peak at the west end of the group, over-looking Terlingua and the Rio Grande Valley. From its sides the ridges extending east and south range from 8000 to 8500 feet high and form an extensive timber covered plateau, cut through by the central gulch. As this gulch slopes down to the north and then to the east it gives the coldest possible conditions and carries import-ant species of trees and birds. The water in it is not permanent, or more important species, including mammals would be found in it.

In general the water of the mountains is scarce and ephemeral. Snow is said never to last long on the mountains. Sudden, violent rain and hail storms of short duration occur filling the pools and rock basins with water that sinks quickly into the stony slopes, not to reappear until some hard stratum brings it to the surface far out on the valley. The impervious rock bottom of the long gulch above keeps a little stream trick-ling over the falls back of our camp, only to be lost in the gravel below. There seems to be no permanent stream of any size. Little springs (aptly called "seeps") are found here and there over the lower slopes of the mountains. More of the accessible water is held in rock basins, where it stands from one shower to another unless they are far apart.

The soil of the mountains, where held on tops of the ridges or on the lower slopes, is a firm, red, volcanic soil of good fertility. Extensive slopes are bare masses of slide rock above the angle of stability.

From the foothills at 5500 to 6000 feet the mountains are generally covered with more or less open and scrubby timber to the tops. In the cool canyons it becomes more dense and on dryer slopes more scarce. Oaks, pines, and junipers are the dom-inant trees with cypress and Douglas spruce in one gulch. The giant century plant blooming over the lower slopes and common to the top of the mountains is a striking feature of the vegetation. The abundance of acorns, pine nuts, sweet juniper berries, and various beans attract many birds and mammals.

The main part of the mountain lies in Upper Sonoran zone. Lower Sonoran species reach to 5500 on the east and to 6000 on hot south slopes. Upper Sonoran species reach to the top of the mountains except on the cool north slopes and in cold gulches which carry mainly Transition forms. I find no trace of Canadian species of plants or animals.

Luxuriant grass covers almost the whole of the mountains and would make them important grazing country if water was more abundant. By digging wells, improving springs, and some piping all of the grass can easily be utilized and if not over stocked and killed out it will support much stock. There are one or two little ranches near the base of the mountains and other ranches for a distance of 50 miles draw their timber from them. In course of time every bit of timber on the mountains will be needed in the valleys and become valuable. The yellow pines, cypress (*Cupressus*), and junipers are all valuable timber to the ranchmen.

Of the Davis Mountains, situated along the Front Range about halfway between the Guadalupe Mountains to the north and the Chisos Mountains to the south, Bailey wrote the following description in July of 1901:

The Davis Mountains are a very old volcanic uplift. They stand as a group, not a range, with Mt. Livermore and a few other half basaltic cores rising as sharp peaks in the middle, but only a few hundred feet higher than the tops of the main ridges. The mountains owe their present form to ages of erosion that have cut deep gulches through the brown lava rock and left high ridges radiating from the center. As much of the rock is a roughly formed basalt, many of the gulches are canyons with cliff or pillared walls that become deeper toward the eastern edge of the mountains on account of the low base level of the Pecos Valley. The steeper ridges are bare, brown lava rock but in general they are covered with soil and a good growth of vegetation, grass, shrubs, and scrubby forest. From the east base the low, desert valley of the Pecos stretches out of sight, to the west the high, short-grass plains come up to the edge of timber on the foothills. To the north and south low ridges reach away in the line of the axis of the Guadalupe and Ord and Santiago Mountains; which, with the Corrozone [Corazon] and Chisos Mountains make a somewhat continuous chain with the ranges of eastern Mexico (Coahuila).

Although a much eroded group of mountains, permanent water is scarce, especially in the central part. A few springs break out in the gulches and good wells are found at most of the ranches. Lower down in the canyons the streams are usually permanent for considerable distance, as we found in Limpia and Musquez canyons, but not with enough water to extend beyond the canyons except in time of rains. The rain comes in irregular showers that are quickly swallowed up in the stony soil and do not warrant a heavy growth of vegetation. Scrubby timber of oaks, junipers, and nut pine covers a great part of the mountains, with yellow pine and *monticola* (?) on the cold upper slopes. Generally the timber is open and scattered with grassy spaces intervening, but in many places it is filled up with dense scrub of brush and shrubby oaks. A shrubby form of *Quercus hypoleucoides* forms extensive thickets. Maples, madrones, walnuts, and grapevines help to fill up the gulches. The big *Agave* grows on barren ridges along the top of Limpia Canyon and well up into the mountains.

With the exception of a few little farms in the canyons where small crops are raised the whole region is devoted to stock raising. The ranches are large with much rough country between for range and extensive areas where little stock ranges on account of scarcity of water. Much good grass has been utilized by means of reservoirs and wells and it is safe to say that all will be in course of time. The rains came late this year but at the present time grass is coming on fresh and green, though still short and with a fresh, spring appearance. The mountain plateau catches more rain than the lower country and is therefore a more productive stock region. It is also cooler and a desirable ranch country for its delightful climate.

Bailey was particularly impressed with the grandeur and magnificence of the Guadalupe Mountains, where he described the Transition Life Zone in Texas and lamented the poor land use in the region. Guadalupe Mountains National Park was established by Congress in 1967 to preserve the fragile biological equilibrium between the fauna and flora of the Chihuahuan Desert in the lowlands and the Rocky Mountains at the higher elevations (Genoways and Baker, 1979).

Of the Guadalupe Mountains, Bailey wrote the following description in August of 1901:

The Guadalupe Mountains are approximately 100 miles long by 20 to 30 wide, the main ridges ranging from 7000 to 8500 feet and the peak reaching 9500. They lie half in New Mexico and half in Texas, the peak and highest part of the range lying south of the line. They are practically a southern continuation of the Sacramentos with only a low gap to cut them off. To the south they run toward the Davis Mountains but are separated from them by a wider gap. Each range forms a link in the chain, heading toward the mountains of eastern Mexico. The Davis Mountains are volcanic. The Guadalupes, like the Sacramentos, are of stratified limestone, faulted and abrupt on the west and dipping gradually toward the Pecos Valley on the east. The crest of the range is along the western edge. The north half is wider and plateau like while the part lying in Texas is narrower and deeply cut by canyons. South of Guadalupe Peak the range breaks down abruptly at the Point and straggles southward only as low ridges. The peak is given as 9500 feet high, which appears to be approximately correct. The ridge on the north side of McKitterick Canyon is 8600 feet (by the aneroid), that on the south side about 9000, and the peak, 10 miles farther south, still higher. North of the Texas-New Mexico line the crest of the range runs from 7000 to 8000 feet, apparently getting down to about 6000 near its northern end.

Water is scarce throughout the mountains and especially so on the higher parts and along the west slope. Rain falling on the open edges of tilted strata runs back toward the Pecos Valley. There is almost no permanent water except below 6000 feet in the canyons on the east slope. On top of the range the few ranches depend on artificial or natural tanks that are filled irregularly by the rains. There are no ranches on the west slope, owing to the difficulty of making tanks on the open edges of tilted strata, although the abundant grass and beautiful ranch valleys have tempted many such efforts. One little "seep spring" at the head of Dog Canyon furnishes a few buckets of water daily to a small mining camp and small natural tanks in the rocks hold water for some days after each rainstorm. A few springs occur in the big salt basin valley at the west base of the mountains.

An abundant growth of the best of forage grasses covers the mountains and valleys at their sides. From about 5800 feet on the east slope and 6500 feet on the west the mountains are covered with an almost continuous, but scattered growth of timber, mainly junipers, nut pines, and oaks. In cold canyons above 6000 and over all but the hot slopes above 7000 yellow pines and Douglas spruce are the common trees, with maples, hornbeam, and chestnut oak in the gulches. *Pinus monticola* is in the gulches from 7000 feet up and over open north slopes from 7800 feet up. Several species of live and deciduous oaks grow as low, round topped trees up to about 7000 feet, and several species of shin oaks grow in great abundance from 7000 feet to the top, forming extensive chaparral. The big *Agave applanata* is a conspicuous tree all over the mountains above 6000 feet, or at least up to 8600. Yuccas, *Dasylirion*, and cactuses grow over rocky slopes where they can crowd out other plants.

A conspicuous feature of the plateau top of the mountains is beautiful, grassy, park like, or orchard like, forests of scattered junipers, nut pines, and oaks. The trees being widely spaced take beautiful forms and the clean green sward below gives the effect of landscape gardens.

The only important industry of the mountains is stock raising, to which the country is admirably adapted in all but water supply. When proper reservoirs are built the number of cattle on the range can be doubled without over crowding. Snow lies but a few days at a time on most of the range and cattle get green grass all winter. In time of storms there is abundance of shelter in canyons and on warm slopes.

Some mines about the heads of Dog and McKitterick Canyons yield gold, silver, and copper in paying quantities but as the ore has to be hauled to Carlsbad (75 miles) and then shipped to El Paso there is little profit above cost of transportation and milling.

Bailey visited El Paso, a town of 10,000 inhabitants in far West Texas, in December of 1889 and wrote the following physiographic description of the area:

El Paso is on the east side of the Rio Grande just below where it has cut through a cross range of low mountains, leaving high rocky banks and a very rough country near the river. A mile or two back from the river the mountains are probably 1000 feet above the river on each side. They are rough and bare, of coarse granite and limestone, without timber or much other vegetation. Where there is a little soil over the rocks some yuccas, *Agaves*, cacti, *Fouquieria*, and a few scrubby shrubs grow. The foothills are pretty well covered with *Larrea mexicana*[P], *Fouquieria splendens*, cacti, and small bushes. Numerous deep cuts and washes have been worn from the high land down to the river, making a very uneven surface and leaving in places broad tables (mesas) between. On the east side of the river the mountains rise abruptly from the plain, which stretches from their base away east to the horizon. This is level, sandy and pebbly, and covered with the evergreen *Larrea mexicana*[P].

The flat river bottoms form a third well marked region. From one to several miles wide, they are hardly above high-water-mark, in places sandy and again of the hardest clay. Here and there are bare patches where nothing grows and the ground is crusted with alkaline, but about half of the flats is covered with cottonwood timber and most of it with small brush, some mesquite and a number of saline plants, as *Atriplex*, *Salicornia*, saltgrass, and others. Several species of *Chenopodium* and wild sunflowers furnish food for the rodents as well as birds. Part of this flat land is irrigated and a variety of good crops raised. This season has been unusually dry and lack of water has been a disadvantage. A large canal for irrigation is being built along here now.

High Plains

Probably no region of Texas has changed more since the time of the biological survey than has the Staked Plains or Llano Estacado region of the Texas High Plains. Bailey and the agents visited the area just prior to the time of major human settlement and agricultural development, and they were struck by the flatness of the land, its extensive grasslands, and the lack of water (Figs. 109-116). While traveling from Colorado City to Amarillo (via Gail, Lubbock, and Canyon) in 1899, Bailey wrote the following description of the region in one of his survey reports:

> The escarpment of the real Llano Estacado begins just west of Gail but our road does not rise onto it for about 15 miles farther north. To the north the country rises slowly but steadily to Lubbock, Hail Center, Canyon, and Amarillo but appears as a flat plain, covered with short grass, free of bushes or trees, and mainly with no surface drainage. At Lake Tahoka, Lubbock, and Hail Center slight valleys open out to the east but most of the country drains into local depressions that are ponds after each rain and some permanent lakes. Most of the mesquites, thorn bushes, and cactus are left below the escarpment and soon disappear from the higher levels. Short-grass and low prairie, or plains, plants cover the top of the plains and give the characteristic smooth carpet appearance to the region. The country is mostly fenced up in pastures of enormous extent, often containing several hundred square miles, and cattle are scattered over the whole region. Except on areas where there has been no rain for a long time the range is not overstocked and the grass is abundant.

Farther to the north around Hereford (Deaf Smith County), at the northern extent of the Staked Plains, Oberholser wrote the following description of the countryside in 1901:

> Hereford is located on the Staked Plains at an altitude of about 3800 feet, in the southeastern portion of Deaf Smith County, on the line of the Pecos Valley Railroad, about 45 miles southwest of Amarillo, Texas.
>
> For long distances in any direction from the town the country is a level or slightly rolling grassy plain, at intervals interrupted by wide, scarcely abrupt, and usually not very deep grassy valleys, or "draws," as they are known in local parlance, which lead to the head waters of the Red River, but contain water only after heavy rains. Close to the south side of Hereford there is, however, a spring-fed creek, the Tierra Blanca, which for some fifteen miles is a running though sluggish stream. Along the sides of its broad grassy valley there are frequent outcrops of friable white limestone appearing in low isolated cliffs usually close to the present stream bed. The margins

Figure 109. Sand hills fifteen miles east of Texline, Dallam County, 1903. Courtesy National Archives, 22-WB-60-B6184.

Figure 110. Prairie at Texline, Dallam County, 1903. Courtesy National Archives, 22-WB-54-B6186.

Figure 111. River bottom and buttes near Miami, Roberts County, 1903. Courtesy National Archives, 22-WB-30-B6172.

Figure 112. Town of Miami, Roberts County, 1903. Courtesy National Archives, 22-WB-30-B6173.

Figure 113. Sand hills south of Dimmitt, Castro County, 1901. Courtesy National Archives, 22-WB-30-3157.

Figure 114. Staked plains south of Dimmitt, Castro County, 1901. Courtesy National Archives, 22-WB-30-3156.

Figure 115. Tierra Blanca Creek at Hereford, Deaf Smith County, 1901. Courtesy National Archives, 22-WB-30-3186.

Figure 116. Stock pond and windmill on Staked Plains south of Dimmitt, Castro County, 1901. Courtesy National Archives, 22-WB-30-3154.

of this creek are overgrown with a dense mass of cattails and rushes, while on the cliffs occurs the nearest approach to a natural growth tree anywhere to be found—a few scrubby bushes of *Celtis occidentalis*. Scattered over the plain are numberless surface lakes, more or less ephemeral, and ranging from a hundred feet to half or three quarters of a mile across. Most of these have no growth in the water, but a few are filled with grass and rushes.

Some 45 miles to the south of Hereford the general evenness of the country is broken by an extensive area of low sand hills. Here the soil changes from the dark colored loam of the environs of Hereford to a pale sand which supports a much less dense growth of grass. The highest of these hills probably is not much over 30 feet above the plain; and all are usually covered with a growth of rank weeds and bushes of *Rhus aromatica*, with sometimes a few low trees of *Celtis occidentalis*. The trend of this range of sand hills seems to be about northeast to southwest, and beyond in the latter direction they become higher and less distinctive in character as the general face of the region becomes more rolling. After crossing them to the southeast the country more resembles that immediately about Hereford.

Among the low hills to the southwest and about 75 miles from Hereford are a number of salt, alkali lakes, some of them several miles in extent. Their beds are almost perfectly level and where exposed are encrusted with a heavy saline deposit. They are dry in seasons of drought, but at ordinary times contain some water. Immediately about the margins scarcely anything but salt grass flourishes but a few feet away the effect of the proximity of the salty ground is apparently lost. In the one visited, the water, even when the lake is full, can scarcely be anywhere more than a few feet in depth. Aside from herds of cattle, the only living animals seen about this lake were a few sandpipers, plovers, and swallows.

This entire region is apparently Upper Sonoran in its zonal affinities. The portion about Hereford is characterized by the absence of shrubby vegetation which begins to appear about 40 miles to the southward, in the shape of scattered bushes of *Mimosa biuncifera*, this increasing in amount to the southwest until it forms in many places a low chaparral. About 50 miles south of Hereford scattered scrubby bushes of *Prosopis juliflora* begin to be observable, and continue to occur along the road leading southwest.

To the north, Bailey and the agents documented landscapes at several places of the Texas Panhandle, including Lipscomb, Texline, Washburn, Tascosa, Canadian, Wheeler, and Amarillo. This was mostly cattle and ranching country, with few settlements, and the agents were impressed with the prairies in the region. A. H. Howell wrote the following description of the country around Lipscomb in 1903:

The country over this region is high rolling prairie, intersected by numerous streams. The soil on the higher portions is heavy and of a dark color, and supports a rich growth of grasses and an abundance of flowering plants. There are no shrubs or trees on the high prairie, with the exception of a few small patches of 'shin oak' about 3 miles north of Higgins. This is said to be the northern limit of this species. The soil in the creek bottoms and on some of the lower hillsides is sandy, and in such situations, sagebrush (*Artemisia filifolia*) flourishes in abundance, the wild plum (*Prunus angustifolia*) forms dense thickets, and 'skunk brush' (*Rhus trilobata*) covers many of the smaller dunes. The cottonwood grows sparingly along the streams, and to some extent in the 'draws' where the water is near the surface. It also joins with the elms (*Ulmus* sp.) and western soapberry (*Sapindus saponaria*) to form groves of timber covering several acres. A few willows occur, mostly shrubby, though occasionally reaching the size of a tree.

There are no very high hills or deep canyons in this region. Most of the creeks are sandy and of small size. Wolf Creek, which flows east just north of Lipscomb, is the largest, and drains the whole region. It has cut a broad bed which is nearly dry at this season. The smaller creeks flowing into Wolf Creek contain more water than Wolf Creek. They run for short distances as narrow brooks through meadow grasses, then widen to form deep holes several hundred yards in length, containing fish, turtles, and water snakes. The most extensive creek I saw is the one known as First Creek, entering Wolf Creek from the north about 15 miles west of Lipscomb. This is broader than the others, and consists at this season of a series of wide, deep holes, separated from one another by meadowy stretches where the water has dried entirely out. A water plant (*Batrachium divaricatum*) grows in abundance in most of these holes, and furnishes the staple article of food to the muskrats which here abound. In the region about this creek, numerous springs are found in the 'draws.' The sand bluffs along First Creek are higher and steeper than elsewhere in the region.

Similarly, Howell described the countryside around Canadian (Hemphill County) in the same year as follows:

Canadian is situated on the south bank of the Canadian River, which at this season is a very insignificant stream, but which in high water covers an area of half a mile in width.

The soil in the valley is very sandy, and supports only a scanty growth of sagebrush, wild plum bushes, 'skunk brush', various grasses and wild flowers, and a few cottonwood trees. The wind has piled the sand into curiously shaped hills on many of which there is no vegetation whatever.

In the bottoms, between the sandhills and the prairie which rises gradually from the valley, are numerous ponds, mud flats and meadows. Considerable hay is cut from the meadows, some of them being wet nearly all summer, others dry for most of the year. The sandhills described above are principally on the south shore of the river, at a point where the river makes a bend. On the north shore the conditions are somewhat different. The meadows are broader, and there are extensive marshy areas grown up to rushes. There are also good sized tracts of timber on this side, most of it on more or less swampy ground. The cottonwood is the prevailing tree, with a few elms, hackberries and willows. Shrubs noticed were *Cephalanthus occidentalis, Cornus asperifolia, Prunus angustifolia.* Just back of the meadows the sandhills begin, the soil being of a more or less yellow color and grown over sparsely with sagebrush (*Artemisia filifolia*) and a few small shrubs.

Clear Creek flows into the Canadian from the north, and several fine ranches are located along its lower course, the water being used for irrigating the fields and gardens. I spent 4 days at Mr. Studer's ranch close to the river, and trapped in the sand hills and about the borders of the fields. Mesquite grows sparingly on the western slopes of some of the hills west of Canadian. The largest brush seen was about 10 feet high.

Vernon Bailey visited the area around Washburn (Armstrong County) in 1892 and wrote the following description of the region, which included Palo Duro Canyon:

Washburn is on the Staked Plains at an altitude of 3538 feet. There is nothing to break the smooth, even monotony of almost level prairie, meeting the horizon on all sides. There is no drainage, save the basin like depressions here and there over the prairie, and these so shallow as to appear, when dry, like their surroundings. The swells are so long and gradual as not to appear like ridges. A low, thick growth of grass covers the country, giving an almost velvety smoothness of surface. Here and there a thistle, an *Asclepias*, or *Euphorbia* rises a foot or so, but so scarce as to be hardly noticed. An occasional house, with windmill and fences, or cattle and horses grazing are all that break the monotony. The grass is becoming yellow and dry. During the heat of the day there is such a dancing of air that the ground can be seen for but a short distance and every object appears distorted.

The soil is a good mixture of clay and sand, but the rainfall is too slight to make farming a success. However, many crops are raised with considerable success. Most of the grain for use on the ranches and some vegetables are raised. Cattle and horses are the principal products.

Thirteen miles south of Washburn the Prairie Dog Fork of Red River flows through an abrupt canyon of about 1000 feet in depth. It is cut abruptly through the prairies like the canyon of Snake River, Idaho, and much of the way it does not drain the land back a mile from its edge. The sides of the canyon are argillaceous rock, clay, and sandstone, broken and rough. The top of the canyon is about a mile wide, the south side several hundred feet the highest. The bottom is sandy and smooth. The river bed is small and now almost dry.

The whole canyon is lined with trees, brush, and plants not found outside in the surrounding country, and furnishes the only timber for a long distance. Junipers are the most important and useful tree, being used for fence posts and fuel, and are very common along the sides of the canyon. A few cottonwoods grow along the bottom, also *Celtis* and Chinaberry tree. *Prosopis* fills the canyon and runs over an edge of prairie as also a little oak. *Lycium* is common, several species of *Rhus*, *Ptelea*, *Cercocarpus*, *Atriplex canescens*, 5 species of cactus, *Yucca angustifolia*[P], *Artemisia* and others entirely new to me.

At Mobeetie (Wheeler County) along the eastern edge of the Panhandle, Oberholser in 1901 wrote the following description of the countryside:

Mobeetie, Wheeler County, lies in the extreme eastern portion of the region known as the "Panhandle." It is just below the top of the Staked Plains proper, and has an altitude of approximately 2400 feet. The general surface of the country is heavily rolling, with occasional areas of nearly level land. The two streams, Sweetwater Creek and its tributary Graham Creek, flow in broad shallow valleys, and have at some places low cut banks. The former is here a running stream, though shallow and of irregular volume; the latter is for the greater part dry, with water only in deep holes. The soil in many places is very sandy, and there are many brush-covered sand hills in this vicinity.

This entire region is a grass country, and there are few uncovered areas. On large tracts, however, there is a dense growth of low live-oak brush (*Quercus undulata*) with not infrequent "mottes" of the same, this vegetation mostly confined to the sandy areas. Along the streams there is considerable timber, composed chiefly of *Populus monilifera*[P], *Ulmus crassifolia*, *Salix nigra* and *Celtis occidentalis*. There are also in many places thickets containing much *Cephalanthus occidentalis*, *Cornus asperifolia*, *Vitis*, and *Prunus americana*. In other respects the flora of this locality is very similar to that about Hereford, Texas, and is essentially Upper Sonoran.

Rolling Plains

To the east of the High Plains and Panhandle, the Rolling Plains represent a transitional grassland that progresses from the true tallgrass prairies in the east to the short-grass prairies and desert grasslands to the west and southwest (Figs. 117-125). Originally the vegetation in this country was predominantly tall- and mid-grasses. Massive cottonwoods lined the streams in the north, and pecans and walnuts were common in the south. Bailey and the other agents spent some time in the Rolling Plains, especially the area that stretches today along Highway 287 from Childress to Vernon to Henrietta, as well as areas around the larger towns of Abilene and Wichita Falls.

Of the country around Abilene (Tebo, Clyde, and Baird), Merritt Cary wrote the following description in 1902:

The country is an immense prairie dog town extending north to the Red River, and nearly 100 miles to the south. There are rolling plains partly covered by mesquites. The mesquites grow quite high, and look more like trees with trunks than bushes. It is a farming country, where they grow hay, cotton, and a little corn.

Figure 117. Pond on prairie near Lipscomb, Lipscomb County, 1903. Courtesy National Archives, 22-WB-30-B6130.

Figure 118. First Creek, near Lipscomb, Lipscomb County, 1903. Home of *Ondatra zibethicus* (muskrat). Courtesy National Archives, 22-WB-30-B6143.

Figure 119. Studer's Ranch, near Canadian, Hemphill County, 1903. Irrigation ditch occupied by musk-rat. Courtesy National Archives, 22-WB-30-B6166.

Figure 120. Sweetwater Creek near Mobeetie, Wheeler County, 1903. Courtesy National Archives, 22-WB-30-B6177.

Figure 121. Broken country along Little Wichita Creek, near Archer, Archer County, 1900. Courtesy National Archives, 22-WB-30-1535.

Figure 122. One mile south of San Angelo, Tom Green County, ford on South Concho Creek, 1901. Courtesy National Archives, 22-WB-30-B2667.

Figure 123. Hills west of San Angelo, Tom Green County, 1901. Courtesy National Archives, 22-WB-41-B2688.

Figure 124. Scattered brush near San Angelo, Tom Green County, 1901. Courtesy National Archives, 22-WB-41-B2662.

Figure 125. Town of San Angelo, Tom Green County, 1901. Courtesy National Archives, 22-WB-41-B2659.

Clyde is situated at the extreme eastern edge of the plains. It is a level country for the most part, and covered to a large extent by a dense growth of post oak. In the immediate vicinity of the town there is an open plain upon which a colony of prairie dogs have their abode. This dense oak "shrub" extends west nearly ten miles where it merges into the open, mesquite plain. East of Clyde about three miles the level country comes to an end, and "breaks" down into a rocky, hilly country several hundred feet lower.

The principal shrubs in the vicinity of Clyde are *Forestiera*, *Zizyphus obtusifolia* (rare), *Prosopis juliflora* (rare), *Yucca stricta*[P], *Opuntia* sp., and *Celtis occidentalis*. *Eriogonum* and *Solanum rostratum* are found in sandy places.

Cotton is grown to a large extent at Clyde, and there are two gins. Cane is also grown. The country is thickly settled.

Baird is seven miles east of Clyde, in a rocky valley, surrounded by oak-covered hills. Shrubs noted here and not at Clyde are *Dalea formosa*, *Opuntia engelmanni*[P], *Opuntia leptocaulis*[P], *Acacia*[P] sp., *Ulmus* sp. (smooth leaved). Other plants were *Grindelia* sp. and *Euphorbia maculata*.

A fair amount of cotton is grown around Baird.

In 1892, Bailey described the physiography in the vicinity of Wichita Falls as follows:

From the western edge of the Upper Cross Timbers at Belcher west along the Ft. Worth and Denver Railroad to where it crosses the Prairie Dog Fork of Red River and from Wichita Falls to Seymour on the Brazos River the country is very similar. The characteristic features are gently rolling, dry plains with short grass and along the streams narrow strips of timber, and over much of the prairie shrubby mesquite—*Prosopis*.

Along the Wichita River at Wichita Falls there is an almost continuous belt of timber composed of cottonwoods, elms, *Celtis*, ash, and an occasional pecan tree. In places the belt is broken and no trees occur for some distance.

Southwest to the Brazos River at Seymour the prairie is mostly scattered over with mesquites and a *Lycium*, *Opuntia*, and *Yucca angustifolia*[P]. I found no timber along the Brazos, save the upland trees of mesquite, *Xanthoxylum*[P], and cedar elm, though B.H. Dutcher writes me from Brazos (Palo Pinto Co.) that the Upper Cross Timbers extend 20 miles above there on the river.

North of the Wichita River after leaving the river valley there is more smooth prairie covered only with grass though an abundance of mesquites in valleys along the streams as far as the crossing of the Prairie Dog Fork of Red River where mesquite and *Artemisia* meet. At Clarendon there is some mesquite but *Artemisia* predominates.

The soil is a mixture of red clay and sand, usually baked hard, or when wet, muddy, except where the sand predominates as it does along the Brazos River and the Pease River and the Prairie Dog Fork of the Red River. Short grass (*gramma* and others) covers the whole region and furnishes good grazing. Stock raising is the chief industry. Farming is beginning to be carried on quite extensively, but crops have suffered much from drouth this year and are poor. Corn is small.

J. A. Loring penned this description of the physiography in the area between Henrietta and Childress in 1894:

Henrietta, Texas, is situated near the Wichita River just beyond the line of timber called "Cross-timber." It is a beautiful prairie country and fine for wheat raising. The soil is a reddish clay. The first mesquite trees seen in going west, were about two miles southwest of Henrietta, and the first cottonwood trees occurred there. Close to the Wichita River is a large tract of timber. Among the common trees found are *Ulmus americanus*[P], *Celtis mississippiensis*[P], sumac and oak. *Smilax rotundifolia* was very common. At Vernon, which is on the Pease River, the country is somewhat similar although there is more sand on the riverbanks, more cottonwoods, large tracts of mesquite. Here too, sagebrush was first found, on some small sandhills near the river.

Childress is on the open prairie like the two preceding towns. Just east of the village is an immense grove of mesquite, reaching for miles and miles, while to the north is a row of sandhills covered more or less with sagebrush, mesquite, and a species of yucca.

The Red River was a bed of dry sand when I reached Newlin, while on its banks was a row of sandhills covered with sagebrush and mesquite, while the surrounding country was open prairie.

Cross Timbers and Prairies

Fingers of deciduous forest cross the prairie in northeastern Texas. This configuration, arranged on a north-south axis, creates alternating belts of grassland and forest known as the Cross Timbers (Figs. 126-131). The distinctive pattern of this vegetation became a milepost in the westward march across North America. Pioneers left the eastern deciduous forests and entered the prairie, but they encountered additional forested areas—the Cross Timbers—before reaching the grasslands that extended thereafter to the Rocky Mountains (Bolen, 1998). Bailey recognized the significance of the Cross Timbers as an area of faunal transition when he penned the following report:

Influence of the Cross Timbers on the Texas Fauna

Among the many natural barriers to distribution of animal life, large and continuous bodies of timber are of no small importance. Where timber joins the prairie or open country it usually marks the limit of range of many species of animals, both those species inhabiting the timber and depending on it for protection and on its products for sustenance and likewise the species living on the prairies and subsisting on foods not found in the timber. A multitude of other causes might be brought up as influencing the range of different species of animals: different soils; the range and abundance of food plants; moisture; distance from water; climatic conditions; with birds, the manner of building nests; and with mammals, the kind of homes selected; not to mention the more general effects of latitude, and the great barriers of bodies of water and mountain ranges.

The distribution of timber itself depends so directly upon soil, moisture and climate that its effect on faunal areas may well be considered secondary but it is none the less real and evident.

In mapping the distribution of birds and mammals the boundary of the range of a species is often traced along the margin of a forest. Innumerable examples might be cited, but among them one strip of timber of particular interest, namely the Upper Cross Timbers of Texas. This represents the lower Cross Timbers, this the Upper. The lower Cross Timbers covers a smaller area and is of less importance. Both are narrow strips of timber extending down from the Red River nearly to the center of the state. The adjoining and nearby surrounding country is prairie. East of the lower Cross Timbers and between the two is the black "waxland", a prairie of very rich and sticky soil. In the timbers the soil is sandy and the surface hilly. The timber does not occur along streams alone as I had supposed, but is continuous over hills and through valleys, apparently following the strips of sandy land. Oaks form the bulk of the timber, the commonest being *Quercus stellata*, *parvifolia*[P], *nigra*, *macrocarpa*[P], and *rubra*. Other common trees are *Ulmus alata* and *americana*, hickory, pecan, walnut,

Figure 126. Rolling prairie with artificial pond in distance, Henrietta, Clay County, 1900. Courtesy National Archives, 22-WB-30-1523.

Figure 127. Rocky hill eight miles east of Henrietta, Clay County, 1900. Courtesy National Archives, 22-WB-30-1527.

Figure 128. Little Wichita Creek, near Henrietta, Clay County, 1900. Courtesy National Archives, 22-WB-30-1538.

Figure 129. Brazos River, near Graham, Young County, 1900. Courtesy National Archives, 22-WB-30-1543.

Figure 130. Oak scrub near Granbury, Hood County. Courtesy National Archives, 22-WB-41-B2658.

Figure 131. Red River near Benvanue, Clay County, 1900. Courtesy National Archives, 22-WB-30-B1530.

cottonwood, sycamore, hackberry, mulberry, redbud, persimmon, and of course many other species.

The Upper Cross Timbers join the Great Plains on the west and cut off the eastward extension of plains species, especially affecting the distribution of certain mammals.

After collecting on the prairie east of the lower Cross Timbers, Mr. Dutcher and I made somewhat thorough collections at different points on the strip of prairie between the two and then crossing to the western edge of the Upper Cross Timbers, we were able to determine with considerable accuracy what species passed through and what were brought to a stand by the timbers.

Fox squirrels, gray squirrels, and flying squirrels were, of course, restricted to the timbers. Prairie dogs are exceedingly numerous from the western edge of the Upper Cross Timbers westward but are not found east of them at any point in Texas. The little pocket mouse (*Perognathus flavus*) and the kangaroo rat (*Perodipus ordii*[M]) were common west of the timbers but none were found on the east side. Two species of ground squirrels (*Spermophilus spilosoma*[M] and *S. mexicana*[M]) were found on the west side but not on the east. The cottontail rabbit (*Lepus sylvaticus*[M]) was common over all eastern Texas and throughout the timbers, but on passing out upon the prairie west of the Upper Cross Timbers another smaller species (*Lepus nuttalli*[M]) was met with great abundance.

Thus we found 6 species of mammals reaching their eastern limit along the western border of the upper Cross Timbers and 4 reaching their western limit at the edge of the plains bordering them on the west. Doubtless a more thorough search would add other species to this list. So far as the Upper Cross Timbers extend therefore they may be taken as the boundary of the Great Plains fauna.

Blackland Prairies

Bailey and his agents encountered the Grand Prairie and the Blackland Prairies, which lie in the transition area between the forests and grasslands of Texas (Figs. 132-133). They spent little time here probably because the landscape was already so altered that it held little promise of yielding wildlife. The main belt of the Blackland Prairie formed an essentially continuous north-south strip of grasslands from the Red River to Austin and south of San Antonio. The region was mainly a tallgrass prairie at the time of European settlement.

Of the area around Austin, William Bray wrote the following description in 1899:

The biological conditions in the vicinity of Austin are associated with three main structural features (1) the southeastern escarpment and deeply eroded edge of the plateau of the plains (2) the black prairie which joins the plains at the fault line extending from Austin northwestward, and southwest to San Antonio (3) the valley of the Colorado River which after emerging from its canyon in the hills, widens into

Figure 132. Bosque River near Waco, McLennan County, 1907. Courtesy National Archives, 22-WB-30-9886.

Figure 133. Bosque River near Waco, McLennan County, 1907. Courtesy National Archives, 22-WB-30-9883.

a rather broad alluvial bottom. The first region embraces the vegetation of Kerrville and Lacey's ranch. The second presents an alternation of exposed chalk slopes or ridges or knolls with rich areas of deep black waxy soil. The flora of the former is conspicuously xerophytic of species found west and southwest. The latter being much richer is largely in cultivation and is attacked mainly by weeds, though a considerable indigenous element related to that of north central Texas and Indian Territory or to east Texas occurs. The valley of the Colorado possesses a flora much like that at West Point and Richmond but very much less luxuriant than the latter because of less rainfall.

Lying fifty to one hundred feet above the river about Austin are a series of gravelly clay terraces which offer the open texture favorable to "post oak" formations, and accordingly these areas are thickly covered by *Q. minor, Q. marilandica, Q. virginiana*, and other trees and shrubs in less abundance.

Post Oak Savannah

The next vegetative region encountered by Bailey and the federal agents was the post-oak woodlands (Figs. 134-137). These woodlands, which occupy a narrow strip that is nowhere more than 96 km wide, are located in the central part of eastern Texas and extend in a southwesterly direction, forming a peninsula surrounded by prairies. The topography is level to gently rolling and slopes gently from the northwest to the southeast. Vegetatively, the post-oak region can best be described as an ecotone between the eastern deciduous forest and the tallgrass prairie. The area supports a stunted, open forest dotted with small tallgrass prairies. The dominant plants of the overstory are post oak and black-jack oak and to a lesser extent winged elm and black hickory. The land in the post-oak region is used primarily for farming and ranching.

In 1902, as he traveled from Houston to Austin, Harry Oberholser penned the following description of the countryside through the southern reaches of the post-oak country:

From Houston, which lies on the intermittently wooded coast plain, to within 5 miles of Hempstead, Texas, the country is principally open prairie, becoming more and more rolling. Hempstead is three or four miles east of the Brazos River, but not in its valley, and in a region of alternating oak woods and cultivated areas. There are here few running streams, though the dry watercourses are numerous, the surface of the land toward the river being much scored by them. The valley of the Brazos at this point is comparatively narrow, and nearly flat, and below its level the river channel has been cut with almost perpendicular banks twenty to forty feet high. The forest that undoubtedly once covered these bottom lands has been largely removed, and much of the land is in pasture or under cultivation.

West from Hempstead to Carmine the general face of the country is much the same, with exceptionally an area of rolling, unbroken, prairie such as exists south of Chappell Hill. On this prairie there are a few small springs, about which a tall rank

Figure 134. Live oaks and Spanish moss, Columbus, Colorado County, 1904. Courtesy National Archives, 22-WB-41-7228.

Figure 135. Live oaks on edge of cultivated field, Verhelle, DeWitt County, 1907. Courtesy National Archives, 22-WB-40-9892.

Figure 136. Guadalupe River, Verhelle, DeWitt County, 1907. Courtesy National Archives, 22-WB-30-9896.

Figure 137. Guadalupe River, near Cuero, DeWitt County, 1907. Courtesy National Archives, 22-WB-30-9894.

grass, similar to *Nolina texana*, grows in profusion. So much of the region about Brenham is now in farms that the larger part of the timber left standing is along the streams. The country in the immediate vicinity of this town is more rolling than that about Hempstead, and to the northward a short distance becomes even somewhat rocky and broken.

The woodland of *Quercus marilandica* and *Q. minor* increases noticeably beyond Carmine, these two species making up by far the larger part of the arboreal vegetation. Between Ledbetter and Giddings much live-oak occurs, principally in mottes, but beyond the latter place it disappears. Beyond Giddings as far as Paige the country is more open, the post-oak and blackjack woods being replaced largely by mesquite chaparral; thence to Elgin the region is one of almost solid post and blackjack oak woods, with scattering areas of open mesquite chaparral in which *Opuntia engelmanni*[P] occurs to some extent.

The country about Elgin is in general appearance much like that near Brenham—decidedly rolling, with frequent shallow narrow valleys cut by streams which run only during seasons of rain, and which are fringed by a growth of timber. The greater part of the land is either under direct cultivation, or in more or less grassy "pastures" grown up to *Prosopis juliflora*, *Zizyphus obtusifolius*[P], *Opuntia leptocaulis*[P], and *Opuntia engelmanni*[P]—a characteristic assemblage indicating the predominance of arid Lower Sonoran influence, and marking rather definitely the eastern limit of this predominance.

Beyond Elgin toward Austin the country becomes more level, prairies preponderating, though mostly under cultivation.

In the post oaks between Austin and Bryan, and continuing to Waco, Oberholser in the same year wrote the following description:

From Austin to Round Rock there is practically no change in the aspect of the country, but farther eastward the prairies are more cultivated, the junipers disappear, and the mesquite becomes less abundant. Between Taylor and Thorndale, however, mesquite is numerous, as well as *Opuntia engelmanni*[P] locally, but beyond the latter place both are not common. East of Rockdale there are many areas of post oak woods, which increase as the Brazos is approached.

Gause is situated near the Brazos River, in the southeastern part of Milam County, about ten miles southeast of Hearne. From here to Waco, a distance of some 60 miles, the rolling uplands are partly woodland, partly prairie, and in many places largely under cultivation. The Brazos River, here a large stream, flows in a tortuous course between almost perpendicular earth banks some twenty or thirty feet in height. Its level valley is considerably below the surrounding country, and varies from one to ten miles in width. The bottom lands, though probably once largely if not entirely wooded, are now devoted principally to the raising of cotton and corn.

Piney Woods

The easternmost part of Texas, adjacent to the Louisiana border, is called the Piney Woods, an area where Bailey and the agents did a considerable amount of work, particularly in the southeastern part of the region known as the Big Thicket (Figs. 138-143). The Big Thicket forest is a mixture of evergreens, both conifers and hardwoods. Bailey provided the following description of the Big Thicket, writing in his 1904 report:

> The Big Thicket of southeastern Texas is a well known and well defined area mainly within the counties of Montgomery, Liberty, Hardin, Tyler, Jasper, and Orange. From the Sabine River it extends 100 miles west, not as a solid body but following the flood lands, or "bottoms," of every river, creek, and bayou with endless branches that alternate with or surround strips or plots of open pine woods or grassy prairie. While the actual thicket does not cover more than about half of the area assigned to it, the thousand interlacing and connecting branches tie it together in a perfectly homogenous body so far as its influences effects the distribution of species of plants, mammals, birds, etc. Similar conditions prevail in a less marked degree along the river bottoms over most of the eastern half of Texas, so the boundaries of the Big Thicket are largely based on degree of density and abundance of species. Few if any species are restricted to it and its greatest significance is as a last resort for such vanishing forms as *Ursus luteolus*[M], *Canis ater*[M], *Campephilus principalis*, etc.
>
> The wide, flat river bottoms are frequently flooded and water left standing in shallow pools and basins. The general drainage is so slight that water often stands knee deep in the swamps during rainy seasons. As a result of abundance of moisture and rich alluvial soil a dense growth of palmettos, vines, and bushes fill in the base of the forest with an almost impenetrable jungle. Numerous vines several of them thorny help to make progress through the jungle slow and painful. The Big Thicket timber is mainly deciduous, save for the magnolia and holly. The pine timber is mainly in open woods on the adjoining higher and dryer ground, but in places the pines are scattered through dense portions of the thicket woods. The bald cypress and tupelo gum are restricted to the swampy ground, bayous or stream banks.

Any biologist who ever worked in the Big Thicket would appreciate this description. In 1974, Big Thicket National Preserve was established to protect the remaining biodiversity in this unique region.

Interestingly, prairies once were common where only trees now stand in eastern Texas. For presumably rational reasons, early settlers called many places in the southeastern Texas forests "prairies," for example Tarkington Prairie in Liberty County where Bailey and the federal agents visited. Today, though plowed fields and mowed pastures abound, hardly a vestige of land remains that would not have trees if left alone (Truett and Lay,

Figure 138. Old cypress gnawed by bear, Saratoga, Hardin County, 1904. Courtesy National Archives, 22-WB-51-7237.

Figure 139. Edge of clearing in Big Thicket at Bragg, Hardin County, 1904. Courtesy National Archives, 22-WB-41-7247.

Figure 140. Railroad cut, Big Thicket at Bragg, Hardin County, 1904. Courtesy National Archives, 22-WB-41-7248.

Figure 141. Big Thicket near Dan Griffin's place, Hardin County, 1904. Courtesy National Archives, 22-WB-41-7249.

Figure 142. Pine woods near Saratoga, Hardin County, 1904. Courtesy National Archives, 22-WB-41-7244.

Figure 143. Tarkington Prairie, Liberty County, 1904. Courtesy National Archives, 22-WB-41-7242.

1984). Vernon Bailey visited Sour Lake in Hardin County in 1902 and wrote the following survey report regarding his observations:

> Sour Lake station lies about twenty miles west of Beaumont on the open prairie at an elevation of about 45 feet. The prairie at this point is dry and covered with a rich pasture grass, excellent for cattle grazing. Except for this grass there is very little vegetation on the prairie, but here and there are wooded "islands" with good timber—mostly oak of various species, black and sweet gum. There is a small sprinkling of short-leafed pines and one large "island" southeast from town is covered with a fine growth of long-leafed pine. In the lower country north and west of town a short ways there is an abundance of the small scrub "palmetto" and a few yuccas. To the north of Sour Lake lies the "big brush," a heavy thicket and almost wilderness. The country is very thinly settled but the recent discoveries of oil nearby and the value of the lands as rice fields as soon as properly irrigated are rapidly bringing numbers of people here. At points both east and west from the station are lower lands, in places marshy.

CHAPTER 4

Twentieth and Twenty-First Century Changes in Texas Landscapes and Land Uses

T HE ARRIVAL OF MODERN EUROPEANS INITIATED MAJOR LANDSCAPE changes upon the North American continent (Doughty, 1983; Weniger, 1984). Europeans influenced natural ecosystems by replacing native herbivores with domestic livestock and by confining them with fences; altering the frequency, intensity, location, and extent of fires; developing watering areas where such areas had not previously existed; and cultivating the most productive land, thus relegating herbivorous animals to less productive areas (Smeins, 1983). These and other impacts were superimposed upon existing changes fostered by indigenous Native Americans, changes that were already in progress at the time the Europeans arrived.

The aforementioned activities dramatically altered the kinds, numbers, and distributions of native animal populations. By the 1800s, faunal conditions in much of North America bore little resemblance to the communities of early times. What may be termed "wilderness wildlife" had been greatly reduced. The cougar (*Puma concolor*), wolf (*Canis lupus*), elk (*Cervus canadensis*), and wild turkey (*Meleagris gallopavo*) were largely gone from the eastern US, and many lesser species had become locally or regionally scarce or absent. This was in part due to lack of adequate protection, as well as to widespread changes in habitat.

When Bailey and his crew of field agents began their survey work in Texas in the 1880s, the habitats and wildlife resources of the state were beginning to change dramatically from the conditions that had existed during the early days of exploration and settlement. Although it is beyond the scope of this book, Del Weniger (1984, 1997) has written an excellent account about what Texas was like, including the lands, water, and animals, up to 1860, just prior to the period of the biological survey. Robin Doughty (1983) also has written about environmental change and land-use practices in Texas during the nineteenth and twentieth centuries, and Smeins et al. (1997) provided an overview of long-term changes emphasizing the central and northwestern parts of the state.

Comparison of old photographs taken by the survey field agents with modern landscapes from the same areas serves to document local landscape change, and from this evidence it is obvious that the landscapes of many sites have changed dramatically since the turn of the twentieth century, and perhaps irreparably. Some of the major changes are discussed briefly in this chapter, contrasting the conditions described and photographed by the federal agents with the opinions of land-users, biologists, and conservation professionals regarding current conditions.

People have a tendency to view the natural world as static because so many important natural changes are slow and therefore not obvious in their personal experiences over several decades or at the spatial scales they normally experience. The truth is that change is a normal part of the environment, a factor to which the components of natural communities are more or less adapted. Variation in daily and seasonal weather changes, short- and long-term shifts in species composition, and gradual climatic changes occur in nearly all communities, and natural communities rapidly adapt to such oscillations. However, changes caused by humans are different from those to which natural communities are adapted, often causing drastic alterations of land-use and landscape patterns, including their inherent abiotic and biotic elements.

Now climate change threatens to exacerbate the conditions mentioned above. Arid and semiarid regions, such as those that characterize most of Texas, have been among the most fragile and susceptible to human and climate alteration. Prior to European colonization, human disturbance in Texas was dominated by Native American occupants and was minimal (Doughty, 1983). Rather it was the rapid population increase and the spread of people across the state throughout the nineteenth and twentieth centuries (Fig. 144a-e) that greatly accelerated landscape changes. Due to increasing anthropogenic activities, wildlife populations suffered severe losses or even extinction. The most important early causal activities were habitat destruction and overexploitation, followed later in the twentieth century by pollution and the introduction of invasive species to the environment. Habitat destruction involving agricultural land conversion, deforestation, overgrazing, energy development, urbanization, and the gradual process of habitat fragmentation were the major causes of biodiversity loss. Human-induced rapid environmental change also altered animal behavior and species interactions in the wild, which spurred species declines, including extinctions and range shifts (Jackson and Sax, 2010).

Population growth presents a formidable challenge to conservationists because fish and wildlife resources and people share near identical needs for two critical commodities—

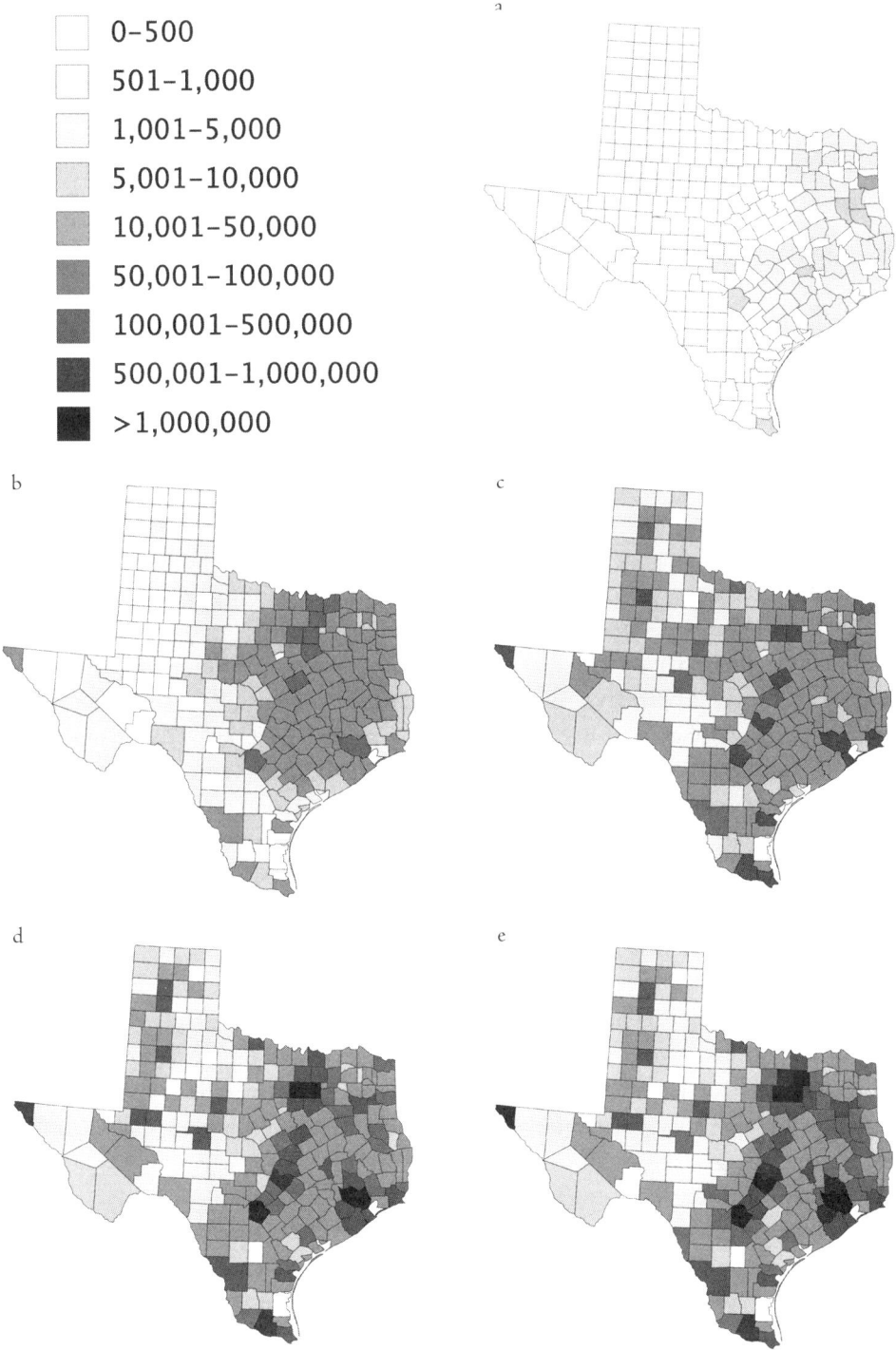

0–500

501–1,000

1,001–5,000

5,001–10,000

10,001–50,000

50,001–100,000

100,001–500,000

500,001–1,000,000

>1,000,000

Figure 144a-e. Population of Texas by county in 1850 (a), 1900 (b), 1950 (c), 2000 (d), and 2018 (e).

water and land. The following discussion considers these landscape factors along with other changes that have impacted Texas.

Overview of Factors Causing Major Changes to Texas Landscapes

1. Land Conversion and Development

Although humans contribute to landscape changes in many ways, herein we focus on what we consider to be the four greatest transformations to the Texas landscape in modern times—population growth and urban sprawl, oil and gas development, wind energy development, and high-fenced game ranching (Fig. 145a-d). Each of these anthropogenic factors impacts the Texas landscapes in a unique manner, as discussed below.

During the twentieth century and continuing into the current century, land cover in Texas has been altered principally by human activity, including farming and agriculture; ranching and raising of livestock; logging of forests and depletion of hardwood bottomlands; suppression of natural fire; rapid and continuous expansion of the oil and gas industry; wind farm development; and construction associated with expanding urbanization. In particular, urban development, including urban sprawl and the development of vacation homes and winter destinations, has had a huge impact on the landscape.

The Texas population expanded from 3 million people in 1900 to more than 20 million in 2000, almost a seven-hundred percent increase in one century (Fig. 146). In the first two decades of the twenty-first century, it increased another 48 percent. Most of this growth (86 percent) has occurred in the 25 highest growth counties (Texas Land Trends, 2019). Today, there are more than 29 million people living in the state, and it is projected that by 2050 the population could approach 50 million people (Texas Demographic Center, 2019). Among the locations in Texas where "urbanization" has greatly altered the landscape, the Dallas–Fort Worth complex, the Austin to San Antonio corridor, the greater Houston area, the Juarez–El Paso area, and the Lower Rio Grande Valley are probably the most intensive (Fig. 147a-c), but almost every region of the state has been severely impacted by urban development.

Texas lands are almost entirely privately owned, with more than 95 percent of the land in private ownership (Lopez, 2014). Within the state, 141 million acres (57 million ha)—83 percent of the total land area—are farms, ranches, and forests (the so-called "working lands"), which provide an economic impact of more than $100 billion annually (Texas Department of Agriculture, 2019). Of Texas's working lands, the current acreage available as wildlife habitat is unknown, but in 1997 it was estimated to be approximately 133 million acres (54 million ha) (Texas Audubon Society, 1997). These private lands, which support our vital wildlife resources and much of the native flora and fauna, are under increasing land conversion pressure driven by rapid population growth, suburbanization, and rural development. From 1982 to 2010, the USDA National Resources Inventory (NRI) reported the conversion of more than 4.1 million acres (1.6 million ha) of Texas working lands to urban uses (USDA, 2013). Conversion rates are even higher today, with 2.2 million acres (890,000 ha) lost from 1997 to 2017 and 1.1 million acres (445,000 ha)

Figure 145a-d. Urbanization, oil development, wind energy farms, and high fencing represent some anthropogenic impacts on landscapes and wildlife in Texas. Public domain photos.

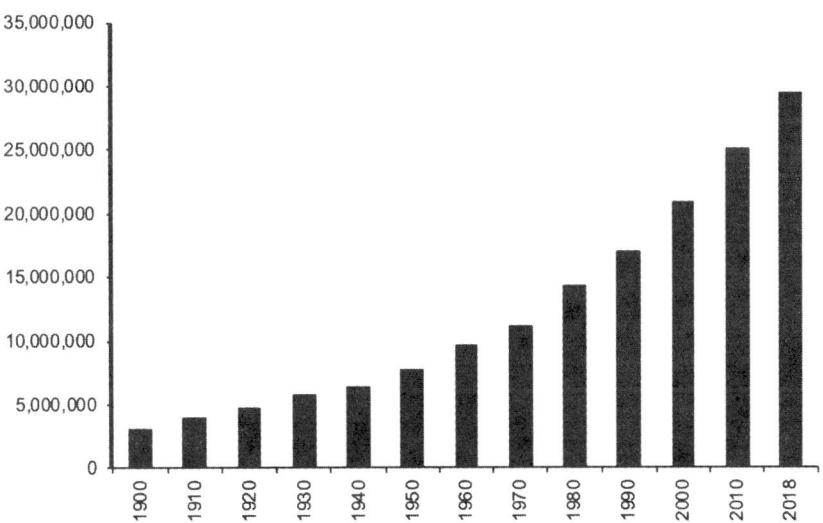

Figure 146. Total population growth of Texas, 1900–2018.

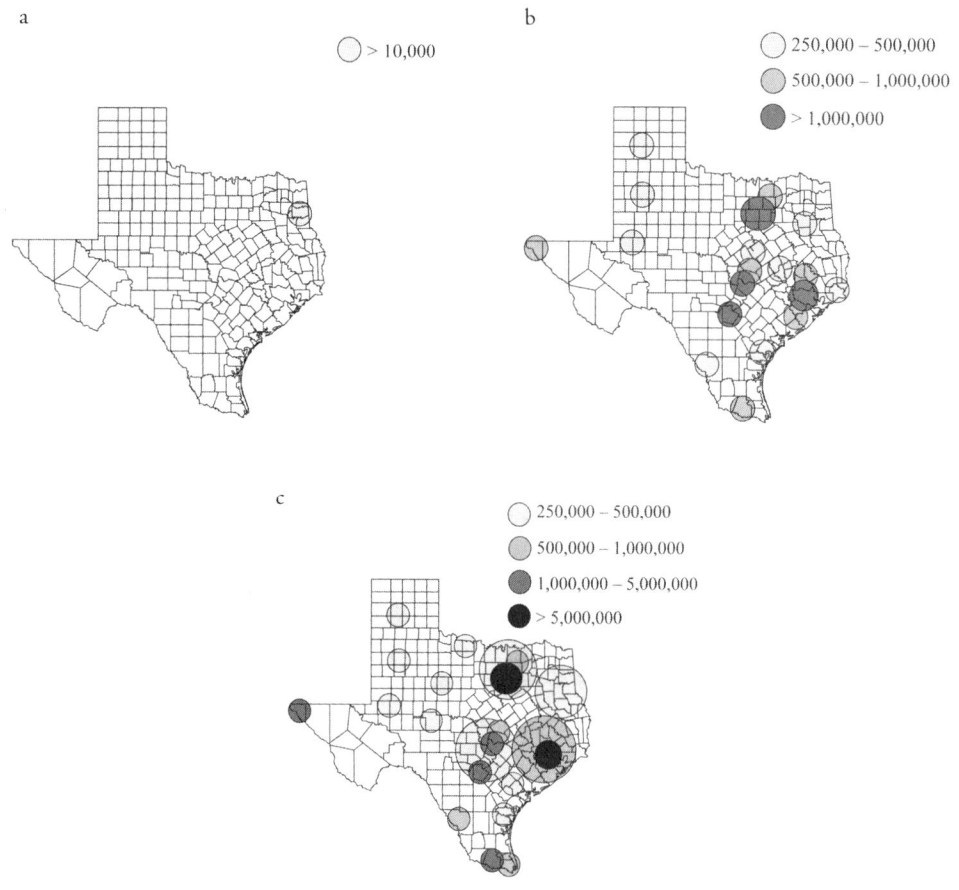

Figure 147a-c. Illustration of human population centers in Texas for 1850 (a), 2019 (b), and projected for 2050 (c).

of that lost in the last five years (Texas Land Trends, 2019). Texas now leads the nation in the loss of working lands. This dramatic shrinking of open-space land has serious implications for the conservation of natural resources, including clean air and water, fish and wildlife habitat, ecosystem services, and outdoor recreation.

The process of land conversion changes both land use and land cover. Land-cover changes represent differences in the area occupied by vegetative cover types or habitats through time. Land-use activities may alter the relative abundance of natural habitats and result in the establishment of new land-cover types. The introduction of new cover types can, in some cases, increase the variety of plant and animal species by providing a greater diversity of habitats. Natural habitats, however, are often reduced by land conversion, leaving less area available for native species. Plant and animal species that are not native, the so-called invasive species, may gain a foothold and out-compete the native species. Also, the spatial patterns of habitat may be altered, resulting in the fragmentation of once continuous habitat (see discussion below). Finally, land-use activities may change the natural pattern of environmental variation by causing changes in natural disturbance events. In general, the chances of losing native animal and plant species and disrupting ecological functions increase when the patterns of natural habitats and disturbance are altered. All of these processes came into play in the twentieth and early part of the twenty-first century.

Among the many human activities that cause habitat loss, urban development produces some of the greatest local extinction rates and frequently eliminates the large majority of native species (McKinney, M., 2002). Urban growth also replaces native species with widespread "weedy" nonnative species, a process known as biotic homogenization that reduces the biological uniqueness of local ecosystems. The conservation challenge of urban sprawl is its current and growing geographical extent. Habitat loss to urbanization is both drastic and increasingly widespread (McKinney, M., 2006). Texas is now losing more rural land to urbanization than any other state, in large part because of the exploding growth of metropolitan areas, especially around Austin, San Antonio, Dallas, and Houston. Between 1997 and 2012, Texas's population grew by 36 percent and the top 25 fastest growing counties grew by 87 percent. This population growth consumed approximately 1 million acres (405 thousand ha) of rural land that was absorbed by urban sprawl (Vanetta and Satija, 2014). This constitutes a major loss of habitat for wildlife and further exacerbates the problem of biotic homogenization.

Another significant land conversion issue in the twenty-first century involves oil and gas development: new technological advances in horizontal drilling and high-volume hydraulic fracturing (known as "fracking") have proliferated in the shale formations of northeastern, north-central, western, and southern parts of the state. The space and infrastructure required by these activities are transforming hundreds of thousands of acres of the state into industrialized landscapes, with drilling projected to continue. Texas now produces more crude oil than any other state and is responsible for more than one-third of the nation's total oil production.

Oil and gas drilling activity grew steadily throughout the twentieth century, even in "boom and bust" cycles; with new technology fueling recent expansion, these impacts upon the landscape have intensified. Anyone who has ever flown over the Permian Basin region of West Texas, near Midland and Odessa, would have noticed a landscape that looks like graph paper stretching for hundreds of square miles across the land, with each intersection marking an oil well or a gas well (see Fig. 145b). The large operations of conventional well sites have been replaced by hundreds of well pads dotting the landscape. Each requires the transportation of water, chemicals, and equipment to and from these pads as well as the removal of wastewater, which has greatly increased roadways to handle the massive truck traffic (TAMEST, 2017).

There is a broad range of possible effects of these activities—soil erosion; reductions in water quantity and quality; reduced flow and increased siltation in streams; habitat loss and fragmentation; changes in native vegetation; air, noise, and light pollution; and increases in heavy road traffic. The cumulative effects of these types of changes may represent threats to native plants and wildlife, especially to endangered or threatened species that live in places where technology is expanding and where animals face direct and indirect harm. The actual footprint of the well pads, pipelines, roads, and other infrastructure, although significant, is smaller than the ecological footprint. The loss of habitat connectivity affects species and the ecosystem beyond the direct loss of vegetation because it typically creates more "edge" habitat that benefits common generalist species at the expense of rare and vulnerable species. It also can create conduits for invasive species that displace native species and deleteriously affect critical ecological functions, such as biomass production.

The most comprehensive information on species-specific impacts of the oil and gas industry has been compiled for the dunes sagebrush lizard (*Sceloporus arenicolous*) and the lesser prairie chicken (*Tympanuchus pallidicinctus*), with extensive studies of changes to their habitats, life cycles, and requirements (Bookhout, 2012; Eckstein and Snyder, 2013). Both species have been under consideration for listing as threatened species, and both have been covered by voluntary conservation plans overseen by state agencies. Unfortunately, the plan for the lizard, which had been in effect for six years, was recently declared a failure and the state has gone back to the drawing board to develop new strategies for its protection. In May 2021, the USFWS proposed to list two distinct populations of the lesser prairie chicken under the ESA.

Unfortunately, for most areas of Texas, there is little information regarding impacts of oil and gas activities on vegetative resources, agriculture, and wildlife and their habitats. No comprehensive and integrated assessment of the large-scale impact of shale development on Texas land resources has been conducted. Since Texas lands are almost entirely privately owned and the fracking of shale formations takes place largely on private lands, these areas can be difficult to assess given that they generally are not sites of formal environmental impact studies. Nonetheless, the fracking process is controversial and has been linked to contaminated drinking water, methane gas escaping from wells into the atmosphere, and earthquakes, and we are only now just beginning to understand how fracking harms wildlife (Lohan, 2019).

Another recent development that is impacting the landscape in the twenty-first century, particularly in western Texas, is the rapid expansion of wind energy platforms. Texas produces the most wind power of any US state, and the state has many wind farms from over forty different projects that stretch from the Panhandle to the Lower Rio Grande Valley: these wind farms now produce more than 20 percent of the electricity generated in the state. The Roscoe Wind Farm, near the town of Roscoe in Nolan County, is the state's largest (and the world's third-largest) wind farm. It is made up of 627 turbines that sit on 250 square miles (648 sq km) of farmland. Wind turbines typically consist of three 116-foot (35 meter) blades mounted on a 212-foot (65 m) tower for a total height of 328 feet (100 m), although there are some that now reach 345 feet (105 m) high. The direct land use for wind turbines is about three-quarters of an acre (one-third hectare) per megawatt of rated capacity. That means a 2-megawatt wind turbine would require 1.5 acres (0.6

ha) of land, and there are nearly 15,000 wind turbines now operating in Texas for electricity. Although the green benefits are enormous in comparison to fossil fuel extraction and burning, the turbines are known to cause mortality among wildlife, particularly birds and bats (see Chapter 6), and more research is needed to determine a sound cost/benefit assessment.

High fencing of ranch properties is yet another change to the Texas landscape that has been implemented in the twentieth and twenty-first centuries. All over the state, ranchers have been putting up eight-foot (2.4 m) fences to keep the deer inhabiting the ranchers' properties from roaming, allowing the owners to charge more for hunting leases. The practice is not new, having started in the Hill Country and South Texas in the 1930s on ranches such as the Y.O. Ranch near Kerrville, where high fences were erected to contain exotic game imported from Africa for the purpose of hunting, thus providing a new revenue stream that could supplement ranching income. What changed toward the end of the twentieth century, and continues virtually unabated today, is using the same type of fence for white-tailed deer, a native species that occurs over the entire state. Deer ranching is now being replicated all across Texas and especially in South Texas and the Hill Country. The state now leads the world outside of South Africa in high fencing, with at least 4 million of the 16 million acres (1.6 million of 6.5 million ha) of deer habitat in South Texas now high-fenced. High fences and genetics, combined with the huge constituency of hunters who will pay handsomely to hunt deer, are effecting a massive change in land use. For large swaths of the state, what is happening amounts to the de facto privatization of deer, a wildlife resource that is defined by law as the property of the people, and the redefinition of hunting as a sporting amusement reserved for people who can afford hefty fees. It also has caused an increasingly bitter controversy among landowners, hunters, and government regulators (Patoski, 2002). The conservation implications for high fencing are discussed in Chapter 7.

2. Changes in Water and Wetlands

Texas possesses tremendous surface and groundwater resources. It is dominated by several river systems and more than 11,247 named streams and tributaries, very few of which are free-flowing today, that course along 80,000 miles (128,748 km) of streambed (McKinney, L. D., 2002). All but four rivers—the Canadian, Red, Sulphur, and Cypress—generally flow from west to east and eventually drain into one of seven major estuaries, or several minor associated ones, that line the margin of Texas's nearly 400 miles (644 km) of coastline.

Most changes in aquatic systems in Texas can be traced to the construction of dams, for either water storage or flood control, and other developments on or near waterways, such as diversion structures and drainage of wetlands. All of these changes occurred in the twentieth century, after Bailey and the federal agents had surveyed the state, and they continue as major challenges in the twenty-first century.

Of all the water resources in Texas, rivers are by far the most seriously threatened, followed closely by loss of surface waters and aquifer reduction. Rivers link our land and water ecosystems, and they are the great integrators that connect and meld the state's distinctive ecological regions. Adequate stream flows and good water quality are essential to their health and to the health of the ecosystems they pass through. Additionally, while water is essential for human health, economic growth, and quality of life, both surface and underground water are essential for our fish and wildlife resources. Yet today Texas faces the possibility that its rivers could be de-watered and its vibrant aquatic ecosystems irreparably harmed. There are growing concerns that existing water supplies cannot sustain population growth of the magnitude projected for this century.

Today, every major river basin in Texas has been impounded: nearly seven thousand small dams form a network of small reservoirs (storage capacity of 10 acre-feet [12,000 cubic meters] or more) for livestock watering and soil stabilization (*Texas Almanac*, 2018). In 1913 there were only eight reservoirs in Texas; today, more than 200 major dams have been constructed to provide flood control and municipal water supplies. These impoundments have altered and substantially degraded our hardwood bottomland and riverine/streamside landscapes as well as our coastal waterways.

Natural flow regimes, which are a key element in maintaining Texas's diverse aquatic ecosystems, exhibit tremendous variability across the state as a result of flash floods, seasonal periods of low flow, and extended periods of drought. Maintaining stable base flow is difficult because most water rights issued by the state do not contain provisions for instream flows and freshwater inflow maintenance. Today, most river basins in Texas are fully or over-appropriated, and currently permitted diversions have the capacity to reduce stream flows significantly below levels necessary to maintain in-stream use (Sansom, 2008). Ironically, although there has been a reduction in flow in most springs and streams from perennial to intermittent, the total amount of surface water in the state is greater today because of the construction of thousands of tanks, large reservoirs, and man-made dams for irrigation, flood control, and water storage.

Texas has seven major and three minor estuarine systems located along its Gulf coastline, and these are characterized by high biological productivity. Freshwater inflows are critical to the health of coastal estuaries. Most marine species are estuarine-dependent during at least some portion of their life cycle, particularly the larval or juvenile stages. Water quality of freshwater inflows is a matter of concern, as available freshwater inflows have become more nutrient rich and subject to influences of non-point source pollution.

Texas originally had 281 major and historically significant springs, other than saline springs (Brune, 1975, 2002). These have been very important to Texas from the time of its first inhabitants. Many springs afforded stops on stagecoach routes, power for mills, water for medicinal treatment, municipal water supplies, and recreational parks. They also provided vital habitat for many species of fish and wildlife. As many as 63 of the major springs have now completely failed and the water flow of many others has been severely reduced. A variety of factors caused the decline of spring flows. Clearing of forestland and heavy grazing of pastures in the late nineteenth and early twentieth centuries began to reduce recharge. The drilling of wells during this period greatly reduced the artesian pressure of springs. Heavy well pumping of underground water for agricultural, municipal, and industrial purposes during the twentieth century, along with the construction of surface reservoirs that inundated some springs, continued their decline and disappearance. In combination, these factors have made, and continue to make, many springs endangered (Brune, 1975, 2002).

The state also has significant groundwater resources, including major and minor aquifers, the most significant of which is the Ogallala Aquifer—one of the largest groundwater aquifers in the world—that underlies the Texas Panhandle as well as parts of seven other states. Groundwater depletion has become a serious problem in the Panhandle because of over-pumping of the Ogallala Aquifer for irrigating agricultural crops. Water law in the state dictates that landowners own the water below their property, unlike in many other states, where water is more regulated. For this reason, it is more difficult to manage the water resources of the state to meet the multiple needs of people and the environment.

Water is one of the main ingredients in "fracking" for oil (see above), and this is creating pressure on water resources. Drillers shatter layers of shale rock with high-pressure water, sand, and chemicals to start the flow of hydrocarbons from a well. Both drinkable and undrinkable sources of water are used to crack shale rock to release oil and gas, and concerns are emerging among residents who live where fracking is prevalent that the oil and gas boom is requiring more water than is available. Shale wells are swallowing twice as much water as they did a few years ago, with estimates of around 10 million gallons (38 million liters) required for each well. Much of this drilling is happening in the driest parts of the state that are most susceptible to droughts, such as the Permian and Delaware basins of West Texas and the Eagle Ford formation in South Texas. There is little surface water in these regions, so drillers often tap underground aquifers—some fresh, some brackish and undrinkable—for the water they need. It has been estimated that as much as 90 percent of total water use in some sparsely populated counties might be associated with fracking. Groundwater declines of as much as 100 to 200 feet (30 to 60 m) have been reported in parts of the Eagle Ford and Permian oil basins.

Oil and gas companies have to dispose of water that returns after hydraulic fracturing, called flowback water, and the water that comes out of the rock itself, called production water. Often that wastewater gets trucked to disposal wells, where it is pumped underground far below freshwater aquifers. Disposal wells are common across the state, and they have been linked to an upswing of earthquakes in some regions, especially in North Texas. Clearly, there is a need for more in-depth analysis of risk related to water management decisions associated with the oil and gas industry.

The state's variable weather, particularly its susceptibility to both short and long-term drought, is compounding these problems both for people and wildlife. Drought is a normal condition in Texas. A Texas Water Commission study found that a drought of six months to a year is more likely to occur somewhere in Texas than average precipitation during the same time period. A three-month drought is likely to occur in some part of the state every nine months. Droughts lasting six months or longer will likely occur once every sixteen months, and year-long droughts are likely somewhere in the state once every three years (*Texas Parks and Wildlife Magazine*, 2000). Our native wildlife can be severely affected by droughts, which can result in lower reproductive success, increased susceptibility to predators because of sparse grass and ground cover, and a serious decline in physical condition, particularly for younger animals. Fortunately, today many species of wildlife benefit from the construction of livestock watering troughs and farm ponds that can provide water in areas that historically had little water during droughts.

For all the above reasons, water availability and quality will be one of the defining natural resource issues of the state in the twenty-first century. Texas is at a crossroads: the choices made now will determine whether the state can meet future water needs to support its economy, sustain quality of life of its citizens, and protect its natural resources (Sansom, 2008). Water availability, or scarcity, will have profound impacts on continued growth as well as the ecological health of natural resource systems.

Of all the natural habitats in Texas, wetlands are the most dependent on water quantity and quality. In the not-too-distant past, wetlands were regarded as wastelands; it was common practice to drain them, fill them, or treat them as dumping grounds. Today, we know that wetlands provide many important services to the environment and to the public. They offer critical habitat for many species of amphibians, reptiles, birds, and mammals, including endangered and threatened species, that are adapted to aquatic environments; they purify polluted waters; and they help check the destructive power of floods and storms.

There are two major types of wetlands in Texas: (1) coastal wetlands, including marshes and estuaries; and (2) inland terrestrial wetlands, including bottomland hardwoods, forests, shrub swamps, marshes, and lakes in eastern Texas; springs and riparian vegetation in central Texas; playa lakes, saline lakes, and riparian habitats in western Texas; and ponds, potholes, and resacas in southern Texas. These interior wetlands account for 80 percent of the total wetland acreage in the state.

Wetlands, hardwood bottomlands, and riparian areas have been impacted by land conversion and land use throughout Texas. Although wetlands originally comprised less than five percent of the state's total area, the state suffered significant losses of wetlands in the twentieth century (Tiner, 1984). Estimates indicate the state lost one-half of its coastal wetlands and 60 percent of its terrestrial wetlands in the past 200 years (*Texas Environmental Almanac*, 2000). This trend continues today as much of the remaining acreage is being seriously degraded by saltwater intrusion due to construction of canals, channels, drainage ditches; by land subsidence and groundwater depletion; by inadequate freshwater inflows due to upstream water projects (dams) and the alteration of natural hydrology; and by pollution from industry, shipping, and urbanization. The wetlands associated with the playa lakes and prairie potholes of the High Plains and Panhandle have suffered from the dramatic lowering of the Ogallala Aquifer and increased sedimentation as irrigated agriculture expanded in this region (Luo et al., 1997).

According to a 2014 report released by the USDA Natural Resources Conservation Service, Texas has 5.5 million acres (2.2 million ha) of wetlands remaining on privately owned lands, tribal and trust lands, and state and local government-owned lands. Approximately 43 percent of these wetlands occur on forested land, 17 percent are on rangelands, and 18 percent are on cropland, pasture, or CRP land, which means that 78 percent, or almost 4.3 million acres (1.7 million ha) of all freshwater and tidal wetlands in Texas are on agricultural or silvicultural lands (Texas A&M AgriLife Extension, 2015).

The "Texas Coastal Wetlands: Status and Trends" report issued by the USFWS estimates that coastal Texas saw a net loss of 210,590 acres (85,223 ha) of wetlands from 1955 to 1992 (Moulton et al., 1997). The greatest threat for wetland loss is from development as the Houston–Galveston region expands to accommodate population growth. The eight-county greater Houston metropolitan area lost around 23,000 acres (9,308 ha) of natural freshwater wetlands between 1992 and 2010 (Jacob et al., 2014), with forested wetlands and freshwater marsh wetlands comprising the vast majority of non-tidal freshwater wetlands lost in the region. More than 70 percent of wetland loss was attributed to development projects. Unfortunately, these trends are continuing to accelerate today.

Forested wetlands in Texas are found primarily in the bottomland hardwoods of East Texas. Since the beginning of the twentieth century, Texas has lost up to 63 percent of its original bottomland forests due to reservoir development, timber clearing, and attendant land use changes (*Texas Environmental Almanac*, 2000). Large-scale loss and degradation of riparian landscapes throughout the state have resulted from the construction of impoundments; overgrazing by livestock, which has destabilized vegetation and resulted in arroyo cutting and gullying of the landscape; and the introduction of alien plants such as salt cedar and Russian olive. Old growth forests in the Big Thicket are believed to have declined 85-98 percent just since 1960 (White et al., 1998).

3. Changes in Rangeland Vegetation from Overgrazing, Brush Encroachment, and Fire Suppression

During the nineteenth and twentieth centuries, brush and cacti continued to cover many areas of the state that were formerly prairie, grassland, or savannah. The main causes of this change were natural factors in association with human activities that produced large-scale (macro) vegetative changes, primarily woody plant invasion (Ansley and Hart, 2012). The spread of species such as mesquite, juniper, scrub oak, and prickly pear can be attributed to the livestock industry with its tendency for overgrazing and the suppression of wildfires that had formerly so often swept the western plains (Harris, 1966; Lehman, 1969; Fuhlendorf and Smeins, 1997; Smeins et al., 1997). The initial suppression of natural surface fires by livestock grazing graded into a period of active fire suppression of all fires by land management agency personnel shortly after the beginning of the twentieth century. The increased dominance of woody plants was fairly rapid (50 to 100 years), exponential, and frequently associated with a threshold where return to grassland dominance requires major inputs.

Figure 148. Widespread grazing by domestic livestock has had a major impact on land-cover in Texas. Courtesy the Southwest Collection, Texas Tech University.

Historically, brush species were considered a nuisance because the primary focus was on maximizing livestock production. But, even with the most rigorous efforts to control these species, they continued to dominate rangeland landscapes. Recently, as the focus of natural resource managers shifted toward consideration of multiple factors, including wildlife, water, and biodiversity, the value of woody plants and cover has become more appreciated. Rangeland ecologists have written extensively about the "brush problem" in Texas, including beneficial and detrimental impacts as well as fire ecology and history (e.g., see Smeins, 1983; Archer, 1994, 1995; Fuhlendorf et al., 1996; Fuhlendorf, 1997; Smeins and Fuhlendorf, 1997; Smeins et al., 1997; Ansley and Hart, 2012). The following account has been adapted from this body of work.

Widespread grazing of domestic livestock in the late nineteenth century caused major cumulative effects on the ecology of Texas (Fig. 148). The extremely high historical stocking rates and concomitant overgrazing led to significant alterations in the species composition of vegetation. Brush encroached slowly as cattle, deer, and other wildlife spread seed around established clusters. As brush encroached, the production and diversity of grasses declined from tall to mid to short and finally annual grasses; the landscape eventually transitioned from grassland to shrubland, and it would not return to grassland unless there was a major disturbance such as a fire. Cool-season grasses and other preferred forage

Figure 149. Natural fires were once common in Texas. During the twentieth century, fire suppression negatively impacted many Texas ecosystems. Courtesy Carlton Britton, Texas Tech University.

species declined, while unpalatable weedy species, shrubs, and non-indigenous plants increased. The year round, high-intensity grazing of open ranges that occurred in the past led to marked reductions in herbaceous plant and litter cover. Overgrazing also was a major contributor to soil erosion, flooding, and arroyo cutting, and it directly reduced the ability of grasses to compete with woody plants. Other factors such as drought further increased the spread of woody plants, especially when coupled with fewer fires and cattle grazing. Overall, livestock grazing and fire suppression, both interacting with fluctuations in climate cycles, had a major impact on land cover in Texas.

When Texas entered the Union in 1845, it was the largest prairie state, and fire was a major factor in shaping its landscape and land cover (Fig. 149). It is well documented that Native Americans routinely and indiscriminantly used fire as a management tool in the state prior to European settlement. Although there is no persistent record of the extent and frequency of fire in the state during the time of the Native Americans, evidence suggests it was relatively high, and that this practice had significant and long-term impacts on the composition and structure of the native vegetation. Frequent, low-severity wildfires facilitated landscape and habitat diversity by providing opportunities for the establishment and maintenance of early successional species and communities. Under conditions of fire suppression, on the other hand, landscape complexity was made simpler, some early- and mid-successional plant communities were eliminated, and shade-tolerant tree and shrub populations expanded rapidly. Today, the vegetation and soils of many parts of the modern Texas landscape have been altered to an extent that even if we knew the original vegetation, it would be questionable if those communities could be restored under the existing soil conditions and with the altered fire regimes that now exist.

With the elimination of most natural fire across the state, prescribed burning has become a common management tool widely used by foresters, range and wildlife managers, ranchers, and other landowners to manage excessive natural fuels under very specific and safe conditions. The practice is used across the state, and with proper oversight and management, prescribed burning is useful for ecosystem management at the local level by reducing wildfires, rejuvenating wildlife habitat, and providing control of invasive brush. It is better than almost any other practice to improve habitat for deer and other wildlife species.

In the twenty-first century, wildfires have become more common as people have continued to spread over the landscape and drought conditions associated with warmer climates have prevailed. In just the last decade, wildfires have decimated different areas of the state. The 2011 wildfire season was unprecedented. Corresponding with a terrible drought, the state experienced some of the largest, most destructive blazes in its history with over 31,000 fires burning an estimated four million acres (1.62 million ha), 2,947 homes, and more than 2,700 other structures. The worst fire in Texas history occurred that year in Bastrop County, killing two people and destroying more than $325 million in property as well as burning much of Bastrop State Park and the adjacent Lost Pines forest, one of the most unique plant communities in the state. In 2017, grassland fires in the Texas Panhandle north of Amarillo burned almost 500,000 acres (202,000 ha), killing five people as well as thousands of head of livestock (Balaskovitz, 2017).

Another major change has been the encroachment of brush species. The most obvious culprit in terms of the increase in woody plants was the spread of mesquite and juniper (Owens, 1997). Though the presence of mesquite and juniper was documented by the earliest European settlers, most of the southern Great Plains at that time is thought to have been grassland or very open savanna (grassland with scattered woody plants).

Most early-twentieth-century historical accounts published on mesquite in the southwestern United States contend that mesquite had spread into South Texas from Mexico (*Life Magazine*, 1952) and then into the southern plains (Bray, 1906; Malin, 1953). However, early- and mid-nineteenth-century descriptions of the South Texas Plains and the Rolling Plains indicated that large mesquite trees formed open groves on many different range sites, especially in low areas, along intermittent streams, and river bottoms (Maxwell, 1979b; Rappole et al., 1986). In the late 1800s, cattle herds increased rapidly, and during drives from south to north they would rest and eat mesquite beans from under mesquite groves before moving to another location miles away the next day, thus spreading seed between established mesquite patches and expanding the overall range of the species (Smeins et al., 1997). In addition to cattle drives, human settlements and the establishment of new ranches in the northwestern portions of Texas led to the continual presence of cattle in new regions (Ansley and Hart, 2012). Railroads also were a mechanism for the expanded distribution of mesquite through rapid and extensive movement of livestock in the late nineteenth and early twentieth centuries (Frey and Frey, 2000).

The overall effect statewide was an increase in mesquite distribution and density in a northwesterly direction. During the time that Bailey and the federal agents worked in Texas, mesquite occurred in many places in southern and western Texas, but the density and geographic extent were markedly less than later in the twentieth century. The extent of mesquite increased by 1.3 million acres (526,000 ha) between 1948 and 1963, with a total of more than 56.7 million acres (230,000 sq km) occupied by 1963 (Smith and Rechenthin, 1964). The latest estimate is that there are 50-60 million acres (202,000-243,000 sq km) of mesquite in Texas (*Texas Environmental Almanac*, 2000). Mesquite also is a menace because of the amount of water it uses. Sparse stands do not deplete soil water drastically, but dense stands growing along streams and rivers, above shallow water tables, and in deep soil use water extravagantly (Sosebee, 2010). There are several approved herbicides that when applied in appropriate concentrations to either the stems or trunks of the trees, depending on their size and shape, result in effective control.

Juniper has been a component of Texas plant communities for thousands of years. The basic geographic range has probably not changed greatly since European settlement, although increases in density have occurred, and in some instances juniper has spread into habitats where it previously was absent or limited in abundance (Smeins et al., 1997). Historically, juniper is believed to have been restricted to rocky outcrops and north-facing slopes where it was protected from intense grass fires (Wolff, 1948; Ellis and Shuster, 1968). Following settlement, the amount of juniper in some areas was purposefully reduced to provide open areas for additional grazing, cultivation, for building materials (fence posts, railroad ties, charcoal production), and other reasons. The overall result was a period from the late nineteenth century until the 1950s when mature juniper stands were greatly depleted, particularly on the Edwards Plateau (Smeins et al., 1997).

Significant spread along lower topographic slopes did not begin until the late nineteenth century, with the greatest increase occurring after 1900 (McGinty, 1997). Juniper was reported on approximately 19 million acres (77,000 sq km) in Texas in 1948 (Wolff, 1948), and had increased to more than 21 million acres (85,000 sq km) by 1963 (Smith and Rechenthin, 1964). Juniper coverage in northwestern Texas increased by 61 percent between 1948 and 1982, and this expansion continues today (Ansley and Hart, 2012). Juniper secretes a toxin into the soil that prevents other plants from growing; this strategy, known as allelopathy, reduces competition with other plants for limited resources, particularly water. As juniper density increases, production of herbaceous vegetation decreases, which in turn reduces livestock production from rangeland. Juniper can be of value for wildlife as a food source, nest site, or by providing escape cover and thermal protection. However, large, dense stands of juniper are not very beneficial to either livestock or wildlife.

Two other notable trends regarding our rangelands emerged in the twentieth century and continue to impact landscapes and wildlife today. The most significant has been the conversion of native rangelands and croplands to improved pastures, composed almost

Figure 150. Agricultural and urban development have severely fragmented and reduced in size the remaining stands of natural habitat in Texas. Courtesy Michael Tewes, A&M University–Kingsville.

entirely of non-native grasses, which are not good for wildlife. These monocultures fail to produce the variety of plants that are needed by wildlife species for food and cover. Non-native pastures now account for more than 11 million acres (44,500 sq km) and represent the third largest land use category in the state (Riskind, 2015). Another more recent trend has been a shift to "wildlife management" following state legislation in 1996 that created this official land use category for tax appraisal purposes. Since then, lands classified as wildlife management have increased to 5.3 million acres (21,400 sq km), whereas grazing lands have steadily decreased since 1997, losing more than 4.5 million acres (18,200 sq km) to other land uses (Texas Land Trends, 2019). In contrast to fire suppression, brush encroachment, and overgrazing, this trend toward wildlife management of rangelands has the potential to have a positive impact on wildlife habitat.

Multiple factors impact rangeland vegetation and must be considered when designing economically and ecologically effective rangeland management programs. New and more refined methods of managing rangelands have been discovered by range scientists, and land managers have developed a better understanding of stocking rates and rotational grazing systems. Rangeland can be improved with good management and favorable climatic conditions (Box, 1996).

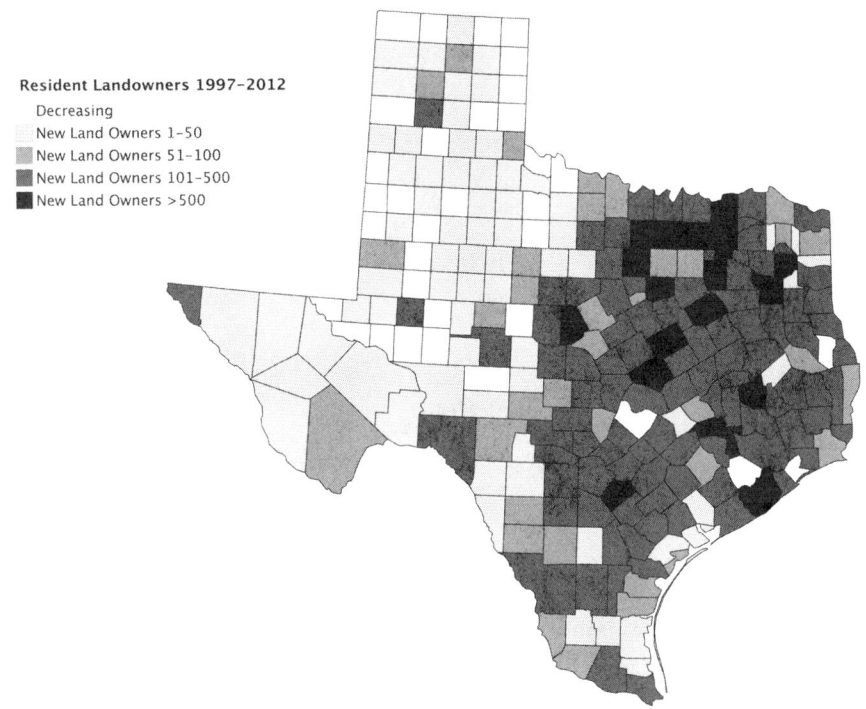

Fig 151. Total change in resident landowners by county, 1997-2012. Modified from Lund et al., 2017.

4. Habitat Degradation Caused by Land Fragmentation

As a result of land conversion and development (discussed above), Texas's working lands are undergoing a fundamental change. Natural landscapes are increasingly threatened by suburbanization, rural development, and land fragmentation (Fig. 150) driven by rapid population growth (Lopez, 2014). Fragmentation of family-owned farms and ranches has been identified as the greatest single threat to wildlife habitat, water supply, and the long-term viability of agriculture in Texas (Kjelland et al., 2007).

Habitat fragmentation is a process by which stands of native vegetation become smaller and discontinuous due to the clearing of land for various purposes, such as agricultural, residential, or commercial use (Hobbs et al., 1993). These types of anthropogenic land transformations cause habitat fragmentation by converting relatively undisturbed landscapes into a mosaic of remnant habitat patches surrounded by different land uses. Habitat fragmentation has been called the most serious threat to biological diversity and a primary cause of the present extinction crisis. Its impacts on animals, plants, and their habitats are numerous—it can disrupt dispersal and movements of animals, increase predation, disturb animal social structure, diminish habitat health by inhibiting migratory grazing and natural fires, and reduce biological diversity of native species (Kjelland et al., 2007). The effects will vary depending on the size and shape of the remnant and the different abilities of species to move across land between remnants (Saunders et al., 1991).

In general, the smaller the fragment, the higher the probability of extinction due to loss of genetic diversity, catastrophes, or an imbalance in relations of competing species, predators, and their prey (Soulé et al., 1988; Soulé and Mills, 1998).

As cities spread and urban dwellers sought land ownership outside the confines of city limits, land-tenure systems started to change during the latter half of the twentieth century (Sansom, 1995) and the trend continues virtually unabated today. Many rural areas, especially those near urban centers, are going through rapid transition in land ownership and use, with many parts of the state experiencing reductions in land ownership sizes (see Lund et al., 2017). The number of absentee landowners has been stable in recent years, and the number of resident landowners has been increasing (Fig. 151), but property sizes are decreasing as working lands are being sold and divided into smaller parcels. This trend is due largely to a reduction in owner-operated farms and ranches; the rural property demands of a rapidly increasing urban population; and transfer of estates to a new generation of landowners. From 1997 to 2017, Texas gained about 1,000 new working farms and ranches annually, but the average ownership size declined from 581 acres (235 ha) in 1997 to 509 acres (206 ha) in 2017. Discounting forestlands, Texas had approximately 248,000 farm ownerships in 2017, up from 228,000 in 1997 (Texas Land Trends, 2019). Under conditions like this, land holdings exhibit accelerating fragmentation, and the increase in decision-makers can result in less integrated land management decisions.

With respect to land use and land cover, ranches greater than 2,000 acres (809 ha) are more likely to remain as native rangeland, whereas midsize properties (500-2,000 acres; 202-809 ha) are more likely to contain a high proportion of cropland, and areas fragmented into properties less than 500 acres (202 ha) are most likely to be converted to non-native pasture (Wilkins et al., 2003). Today in Texas, small ownerships (<500 acres; <202 ha) represent 85 percent of ownerships but only 16 percent of all working lands (Texas Land Trends, 2019). Shifts in land cover from rangelands to croplands and non-native pastures affect ecosystem services and increase the potential for pollution externalities, such as elevated nutrient runoff leading to increased water contamination (Kjelland et al., 2007), and they can lead to degradation of wildlife habitats and biodiversity.

5. Introduction of Invasive Plants and Animals

Invasive species are plants, animals, or other organisms that are introduced to a given area outside their original range and subsequently cause harm in their new homes. Because they have no natural enemies to limit their reproduction, they usually spread rampantly. Invasive species are one of the leading threats to biodiversity and impose enormous costs to agriculture, forestry, fisheries, and other human enterprises, as well as to human health.

Texas has been invaded by a number of harmful exotic plants and animals. To date, more than 800 aquatic and terrestrial species have invaded Texas, and experts predict the trend will continue to increase. The state is at the geographic crossroads of the introduction of invasive species, primarily due to its major maritime and land-based ports of entry, corridors between states, major trucking routes, and its long border with Mexico.

It is beyond the scope of this book to address all of the invasive species in Texas, so we have focused on just some of the plants and animals that are having the greatest impact on Texas wildlife. Chapter 5 includes a discussion of the invasive mammals that have become established in Texas.

Already by the turn of the twentieth century, non-native plants had successfully invaded Texas landscapes. While working in the coastal prairies, Bailey and the agents photographed several landscapes in which McCartney's rose (*Rosa bracteata*) had been used to establish major hedgerows. But this was only the beginning, and by the end of the century, non-native plants would be common all over the state.

The salt cedar or tamarix (*Tamarix* spp.), in particular, has inflicted damage to our native landscapes and habitats. Salt cedar was brought over from Eurasia and planted across the western United States by government agencies in the early 1900s for erosion control. It proved to be a vigorous invader of moist pastures, rangelands, and riparian habitats, having spread to almost every river, stream, creek, and wash in the arid southwest, infesting more than a million acres (404,686 ha). It is of little forage value, provides no seed source for native species, and other than as nesting sites for some birds, it is of limited value to wildlife. Even worse, this tree displaces native hardwoods and consumes tremendous amounts of water by filling riverbeds and sucking springs dry. Salt cedar has an extremely high rate of evapotranspiration. One plant alone may use 200 gallons (757 liters) of water a day. Annual water losses often result in a substantial decline in water tables where salt cedar is abundant (Duncan et al., 1999), and the scaly-leafed tree also adds large amounts of salt to the soil and river.

One of the most striking examples of the rapid spread of salt cedar and its impact on local habitat change can be seen from photographs taken by the federal agents in the Big Bend region of the state. One of their photographs, taken at a working cattle ranch and farm near the mouth of Santa Elena Canyon, depicts open habitat and cottonwood trees along the river (see Fig. 100, Chapter 3, p. 385). As this photograph shows, no salt cedar was present along the river. Today this region is covered with a dense stand of river cane (*Phragmites* sp.) and introduced salt cedar trees, with few native cottonwoods. Interestingly, river cane is an invasive plant that was already well established in West Texas before the time of the biological survey. Other areas where salt cedar has gotten out of control include the Pecos and South Canadian rivers. In fact, it is so bad along the South Canadian that water flow into Lake Meredith has almost stopped. Where possible, the state has begun spraying salt cedar with herbicides, but landowner rights and access difficulties have deterred the program.

Since its introduction, salt cedar has proven to be very resilient and difficult to control, easily recovering from burning, efforts at mechanical control (root plowing, bulldozing, mowing, and shredding), and herbicide treatment. In 1997 a large-scale ecosystem restoration program (the Pecos River Ecosystem Project) was initiated on the Pecos River in western Texas where salt cedar had created a monoculture along the banks of the river by replacing most of the native vegetation (Hart et al., 2005). Local irrigation districts, private landowners, federal and state agencies, and private industry worked together to formulate and implement a restoration plan, with a goal of reducing the effects of salt cedar and restoring the native ecosystem of the river. An initial management phase utilizing state-of-the-art aerial application of the herbicide Arsenal® began in 1999 and continued through 2003. Initial mortality of salt cedar averaged about 85-90 percent. Monitoring efforts were initiated at the onset of the project to include evaluating the effects of salt cedar control on salinity of the river water, efficiency of water delivery down the river as an irrigation water source, and estimates of water salvage. Water salvage estimates showed a significant reduction in system water loss after salt cedar treatment.

Recently, a successful biological control program using subtropical salt cedar beetles (*Diorhabda elongata*) has been implemented to remove these trees from the Rio Grande and Pecos rivers, from the Canadian River in the Texas Panhandle, and in the Rolling Plains (Burns, 2013; HardingCounty.org, 2019; Texas A&M AgriLife Research and Extension Center at Amarillo, 2019). The beetles, imported from several parts of Europe and Asia, attack salt cedar foliage and cause the trees to defoliate. The early results have been very promising, with the beetles surviving the winter, spreading in geographic distribution, and producing area-wide impacts. Unfortunately, however, the beetles have spread so rapidly and successfully into Arizona, and consumed so many salt cedar, that they have become a threat to an endangered subspecies, the Southwestern Willow Flycatcher (*Empidonax traillii extimus*), which relies on some of the salt cedar for roosting and nesting sites (Fonseca/Associated Press, 2019). There are no studies yet to indicate whether the Asian beetles have impacted the salt cedar enough in Texas to impact the flycatcher, which occurs in the Trans-Pecos region of the state.

A situation similar to that of salt cedar in the western part of the state exists along the upper Texas coast where the Chinese tallow tree (*Triadica sebifera*, formerly known as *Sapium sebiferum*) from Asia has proliferated substantially during the twentieth and twenty-first centuries. Chinese tallow is a deciduous woody tree, introduced into the United States sometime between 1900 and 1910, which has now become established along the Atlantic and Gulf Coasts in floodplain, freshwater wetland basins, coastal prairie, abandoned rice fields, mixed bottomland hardwood forests, as well as disturbed sites (Conway et al., 1999). Its initial introduction into Texas, in and around Houston, was based upon the economic potential of its seed oil, but it quickly expanded its distribution and now covers much of the coastal prairie from the Louisiana border to the semi-arid regions of south Texas near Kingsville. The invasion of Chinese tallow into coastal Texas has changed much of the region from coastal prairie to monotypic tallow woodlands, displacing native plant and wildlife species (Conway et al., 1999). The rapid forestation of Chinese tallow has contributed significantly to the degradation of wetlands along the Gulf Coast (Pase, 2005). This rapid-growing tree creates a sterile environment of dense thickets that have limited value to wildlife. Seed dispersal is facilitated by water and by birds that feed on the fruit of the plant (Conway et al., 2002). Other than providing cover, it is virtually useless to mammals. Recent studies have shown that exposure to Chinese tallow leaf litter has negative effects on tadpole survival (Cotten et al., 2012), which could impact amphibian populations. Conservationists battle Chinese tallow with herbicides, selective burning, and digging up seedlings, but none of these remedies are totally suitable for comprehensive control (Conway et al., 1999; Pase, 2005; Pile et al., 2016).

Invasive plants cause an estimated 13 billion dollars of damage per year in Texas, and they can have deleterious consequences for both fish and wildlife. They impact nearly one-half of the species on the federal threatened and endangered lists, but a complete discussion of all of the problems is beyond the scope of this book.

The red imported fire ant, *Solenopsis invicta*, was introduced into the United States through the port of Mobile, Alabama, in the early 1930s. Despite intensive control efforts, the species has infested much of the United States, including most of Texas. Fire ants entered Texas in 1957 and spread across the state at about 30 miles (48 km) annually until the dry conditions in West Texas slowed them in the 1980s (Huffaker, 2000). They colonize most easily in ecologically disturbed habitat. Fire ant mounds dot the Texas landscape following spring rains. The ants have spread when they were transported with dirt or nursery plants, and they can float for days during a flood. Areas that have been disturbed by clearing trees and undergrowth attract huge numbers of the ants (Huffaker, 2000).

Fire ants are devastating to some wildlife, particularly creatures that breed, nest, or raise offspring on or near the ground. The ants kill newborn fawns, rabbits, and other small mammals, and they have seriously reduced populations of quail, ducks, and other birds, both by attacking hatchlings and by eating insects that would nourish them (Huffaker, 2000). There are numerous scientific reports of fire ants causing damage to reptiles, birds, and mammals in Texas (Wilson and Silvy, 1988; Allen et al., 1994; Drecs, 1994). They have been shown to alter habitat use and to reduce the carrying capacity of the habitat for some small mammal species (Smith et al., 1990). Many conservationists believe that in time, *S. invicta* will wreak havoc on Texas wildlife. Scientists are now exploring new ways of controlling fire ants without harming the environment. The Texas Legislature has appropriated money to support these efforts and progress has been made, although no foolproof procedure has been found that will work across the entire state.

A particularly scary incident of a biological invasion occurred in 1993 when a brown tree snake (*Bioga irregularis*) was found in crated household goods shipped from Guam to Ingleside Naval Station near Corpus Christi (McCoid et al., 1994). Fortunately, the snake was killed before it escaped into the wild, but this incident illustrates how easy it can be for animals to become introduced outside of their native range. The brown tree snake, a New Guinea native, has caused significant economic, biological, and human safety problems wherever it has been introduced in the South Pacific islands, resulting in the extirpation of nine of Guam's eleven native birds and threatening the entire native forest-dwelling avifauna on many islands: clearly this is not a species we would ever want to have permanently established in Texas. Comparably, in the Everglades of Florida, Burmese python snakes (*Python bivittatus*) were introduced in the early part of the twenty-first century, and they have spread throughout the region and other places in that state, wreaking havoc on the native wildlife.

A number of initiatives are underway in Texas to mitigate the impacts of invasive species. The Texas Invasive Species Institute (headquartered at Sam Houston State University), an initiative of the Texas State University System, is the first comprehensive research effort in the state focused on early detection and elimination of multiple invasive species (Texas Invasive Species Institute, 2019a, b). The Invaders of Texas Program, an initiative of the USDA, is an innovative campaign whereby volunteer "citizen scientists" are trained to detect the arrival and dispersal of invasive species in their own local areas and deliver that information into a state-wide mapping database (USDA, 2019). The Texas Parks and Wildlife Department (TPWD) enforces laws to protect the state against the introduction of exotic and invasive species and has been especially aggressive in addressing the problems of aquatic invaders.

The Compounding Impacts of Environmental Pollution, Diseases, and Climate Change

Three threat multipliers have compounded the rapid changes that have occurred in Texas landscapes. Environmental pollution, animal-borne diseases and zoonoses, and climate change, in concert with the other land-use changes discussed above, are impacting human health and wildlife and are likely to remain significant issues throughout the twenty-first century.

1. Environmental Pollution

Toxic pollutants, including synthetic chemicals, oil, toxic metals, and acid rain, have become one of the primary ways in which humans have caused drastic modifications of wildlife habitat. Unfortunately, Texas has become one of the most polluted states in the US. The state now ranks number one in water pollution, and its waterways are regarded as the nation's fourth most polluted. There is a tendency among some Texans to regard the air, water, and soil surrounding them as waste receptacles with little consideration for the ecological consequences of their actions; as a result, wildlife populations and their habitats have been and are today confronted with a bewildering array of pollutants that have been released into the environment either by intent or accident. In some instances wildlife populations have suffered severe losses or even faced extinction due to this pollution. For example, the bald eagle, peregrine falcon, and brown pelican all nearly became extinct before scientists discovered that the synthetic chemical DDT was the cause of devastating reproductive failure in these species (Davis, 2019).

Among mammals, bats are highly sensitive to some insecticides, especially the chlorinated hydrocarbons. Bats feeding on insects treated with these chemicals slowly accumulate pesticide residues in their fat during late summer and fall, which eventually results in their death as this fat is broken down and used during hibernation and migration (Clark, 1981). To date, only outright mortality on a local population level has been identified as a threat to bats from pesticides, but subtle—and equally devastating—effects are possible on such aspects as reproduction, acoustic behavior, and hibernation metabolism (Schmidly, 1991; Ammerman et al., 2012).

In the last two decades there has been a growing awareness of the possible effects in humans and wildlife from exposure to chemicals that can interfere with the endocrine system: the so-called endocrine disruptors (Center for Biological Diversity, 2019; EPA, 2019). Endocrine disruptors interfere with natural hormone functions, affecting reproduction, development, and growth. Wildlife may accumulate many of the long-lasting chemicals (e.g., PCBs, dioxins, and organochlorine pesticides) in their fat, often passing these chemicals along to offspring and predators. Endocrine disruptors are present in many common household products: plastics, shampoos, cosmetics, cushions, pesticides, and canned foods. These agents enter waterways via wastewater effluent and urban and agricultural runoff. Known sources of contamination include spray-drift and runoff of pesticides from agriculture, livestock waste runoff from confined animal feeding facilities, leaching from municipal landfills, and septic systems. Despite concerns, however, establishing cause and effect of endocrine disruptors in wildlife has been difficult and even controversial. In humans, endocrine disruptors are suspected of factoring in the decline of sperm counts in men over the last 50 years. Unfortunately, there is very little information available about the impact of endocrine disruptors on Texas wildlife, but this is likely to become a more significant issue for the remainder of the twenty-first century.

A class of pesticides called neonicotinoids—neonics, for short—are some of the most controversial in use. These pesticides are highly effective insect-killing chemicals widely applied to seeds before they go in the ground, and they work by disrupting the central nervous system of insects (Bale, 2019). Use of neonicotinoids has been linked to the decline of honeybees and wild bees around the world, as well as birds. Less is known about their impact on mammals and other kinds of native wildlife.

Another aspect of pollution in Texas involves methane, which is one of the greenhouse gases, along with carbon dioxide, that is a major contributor to climate change. Methane is particularly problematic because its atmospheric impact on global climate change is 34 times greater than carbon dioxide over a 100-year period (Yvon-Durocher et al., 2014). Texas is the leading emitter of methane gas in the US, with the primary sources coming from the extraction and processing of coal and natural gas; the digestive process of livestock and other agricultural practices; and from landfills, which emit methane as waste decomposes. Recently, controversy has developed around the impact of the cattle industry on methane emissions, leading some environmentalists to call for the elimination of red meat in human diets as a way to limit climate change. This issue, and its potential impact on biodiversity, is discussed in more detail in Chapter 7.

Concern has been expressed about the potential polluting impact of "gas flaring" and its possible effects on human and wildlife health in the oil-producing regions of the state. From 2017 to 2018, gas flaring nearly doubled in the Permian Basin, and satellite analysis has shown that natural gas flaring in West Texas is severely unreported. Gas flaring burns unwanted or excess gases and liquids released during normal or unplanned over-pressuring operations in the oil and gas industry. High levels of benzene, a known carcinogen, have been recorded in shale areas, creating speculation that it could impact the health and environment of landowners who live near flared wells. There is almost no information about the impact of toxic fumes from flaring on wildlife, but this is clearly an area that needs further exploration. Flaring also releases methane, a greenhouse gas that contributes directly to global warming. According to the May 2021 UN-backed report *Global Methane Assessment*, drastically cutting methane emissions is necessary to avoid the worst impacts of climate change; implemented immediately, such cuts could reduce the rate of global warming by 45 percent by 2030. The EPA of the Biden administration is expected to propose aggressive federal methane mandates for oil and gas industries, focused primarily on addressing methane leaks from nearly a million oil and gas wells and on blocking the venting and flaring of methane gas. The technology exists to prevent or capture most methane emissions from oil and gas sites, making this a "cheap and easy fix," according to the Natural Resources Defense Council, that would have a significant impact on slowing global warming (Dlouhy, 2021).

2. Animal-borne Diseases and Zoonoses

Diseases are devastating wildlife around the world, and although the species affected and the effects of each disease may be unique, developing successful management strategies and garnering adequate resources to control and treat these diseases is a universal challenge (Carmichael, 2012). For example, one particular wildlife disease that is proving devastating to many North American bats (and that may impact Texas bat diversity in the near future) is white-nose syndrome (see Chapter 6). Equally important is a specific category of wildlife diseases, collectively called zoonotic diseases, which can be transmitted to humans. Zoonotic diseases often are not manifested in a wildlife host, but they can cause illness and even fatality in humans; other zoonoses, such as rabies, can be detrimental to both wildlife and humans. Zoonotic viruses are becoming more prevalent as humans invade remote regions of the planet. The New World and Old World tropics are the major sources for these types of zoonoses, presumably due to the more stable environments surrounding tropical regions that allow them to serve as incubator sites where viruses can survive year-round. This human invasion of the tropics, coupled with a reverse scenario (tropical climates shifting toward more temperate areas as a result of climate change), has been predicted to produce an increase in the number of zoonotic diseases (especially tick and mosquito-borne) in the US (Weise, 2019).

Of recent concern is the number of tropical zoonoses that are being increasingly detected on a worldwide basis. With aggressive transmission beginning in 2020, the SARS-CoV-2 virus that causes coronavirus disease (COVID-19) created a worldwide pandemic, killing 4.55 million people, including more than 684,800 in the US and more than 62,000 in Texas (as of 24 September 2021). Current data indicate that this virus may have laid dormant in bats in China until December of 2019, when apparently the disease was transmitted to humans via bat meat being sold in a local meat market (at the time of publication, this was a working hypothesis for the mode of transmission). There is some concern among Texans that its own bat fauna might harbor this coronavirus or some other undescribed virus. However, it appears that nearly all severe coronaviruses have been restricted to the Old World bats; consequently, Texas citizens should not be fearful of contracting these bat-borne coronaviruses from native bat fauna. What is of primary concern, however, is human-to-human transmission of viruses and the potential for worldwide spread of disease with modern air travel and globalization, as evidenced in 2020 and 2021.

Texas is susceptible to a variety of other zoonotic diseases because of its location in the transition zone between tropical and temperate climates and the fact that its fauna are a mixture of tropical and temperate species. Therefore, in Texas the biggest concern should be focused on zoonoses (such as hantaviruses, arenaviruses, rabies, Chagas' disease, Lymes' disease, plague, avian influenza, etc.) that have been dispersing from the New World Tropics to the more temperate regions that are typical of much of our state. Most of these zoonoses have been in Texas for numerous years and are occasionally epizootic, but generally they are not of major concern. For example, there are occasional flare-ups of rabies and hantavirus, and during those times, a few people may get exposed; however, the infection rate never reaches a pandemic level in the state. However, with climate change and increasing human invasion into previously unoccupied tropical regions, it is possible that these "normal" zoonoses or perhaps new zoonoses will be of greater human concern.

Relative to wildlife diseases, exotics and introduced species such as feral hog and aoudad, among others, present challenges for humans and native wildlife. Feral hogs are known to harbor pseudorabies, swine brucellosis, swine flu, tuberculosis, bubonic plague, tularemia, hog cholera, foot and mouth disease, and anthrax, as well as internal parasites such as kidney worms, stomach worms, roundworms, whipworms, liver flukes, and trichinosis. It is of concern that diseases and parasites such as these may be transmitted from exotics to many native wildlife species and even to domestic livestock. Further, given that swine-to-human transmission is relatively easy, zoonotic experts are concerned that feral hogs may pose a threat to human health as well. For example, it is generally thought that most common influenza (flu) viruses get their start in a domestic pig-to-human transmission. Over the last few years, the influenza virus accounted for 60–80,000 fatalities per year in the US alone. Although most influenza viruses get their start in Southeast Asia, there is some concern that similar influenza strains could originate in the feral hog population in the United States.

Similarly, aoudad and domestic sheep pose a threat to native wildlife (such as bighorn sheep) via transmission of pneumonia, blue tongue, and other wildlife diseases that may cause sickness and death. We are only beginning to get a handle on the potential transmission of wildlife diseases and zoonoses between native wildlife, exotics, domestic livestock, and humans. This will most certainly be a growing area of research in the twenty-first century.

Recently, chronic wasting disease (CWD), a contagious neurological disease affecting deer and elk that is always fatal and has no known treatment, has been detected in Texas, and this is particularly worrisome. CWD was first documented in free-ranging mule deer in far West Texas in 2012, and it has been continually spreading, creating a serious challenge for wildlife managers (Lightfoot, 2012, 2015; TPWD News 2019c, 2020a, 2021a, c). Attempts to pseudo-domesticate big game species for augmentation of hunting and trait development (trophy antlers in white-tailed deer, for example) is adding a new chapter on wildlife diseases.

If CWD were to increase in frequency in Texas deer populations, it could have a major impact on active sporting industries as well as on several predator species that depend upon deer as prey items (Carmichael, 2012). Further, closely related prion diseases, such as Creutzfeldt-Jakob disease and Bovine Spongiform Encephalopathy (BSE, or Mad Cow Disease), are transmittable to humans; consequently, CWD would be of great concern if it were to jump from deer to cattle or other livestock species. With these recent discoveries, TPWD has stepped up its efforts to strategically monitor the situation, including sampling hunter-harvested deer at a greater level. As a result of the surveillance program, as of June 2021, CWD has been detected in 228 white-tailed deer, red deer, mule deer, and elk in thirteen counties (Dallam, El Paso, Hartley, Hudspeth, Hunt, Kimble, Lavaca, Lubbock, Mason, Matagorda, Medina, Uvalde and Val Verde) across the state, with many of the infected animals connected to deer breeding facilities and release sites in Uvalde and Hunt counties (TPWD News, 2021c).

The recent discovery in the US and Texas of rabbit hemorrhagic disease virus 2 (RHDV2), a highly infectious and lethal form of viral hepatitis that impacts lagomorphs, has the potential to devastate rabbit populations in the state. As of April 2021, the TPWD had confirmed this disease in the wild rabbit populations of Brewster, Cottle, Culberson, El Paso, Gaines, Hale, Hockley, Hudspeth, Jeff Davis, Lubbock, Pecos, Presidio, Randall, Terrell, and Ward counties (TPWD News, 2021b).

RHD is highly contagious and spreads between rabbits through contact with infected individuals or carcasses, and it can persist in the environment for a very long time, making disease control efforts extremely challenging in wild populations. The disease is 90 percent fatal, and entire food chains could be severely impacted if staple prey species, such as jackrabbits and cottontails, decline.

3. Climate Change

Compounding the challenge of environmental pollution and diseases, climate change has become an even bigger threat entering the twenty-first century. The United Nations International Panel on Climate Change (IPCC) issued a 2018 report indicating that the impact of human-induced warming is worse than previously feared, and that time is running out to address the problem. According to this report, the burning of fossil fuels (coal, oil, and gas) on an industrial scale has already raised global temperatures by about one degree Celsius (1.8 degrees Fahrenheit). The IPCC has calculated that a further temperature rise of one degree, which is considered inevitable, would be challenging but manageable. However, an increase of about two degrees, which is entirely possible, would be globally disastrous, resulting in increasing drought conditions beyond the tolerance of humans in certain regions and dramatic rises in sea level that would flood coastal areas across the globe.

Texas's climate is changing, and the state could be in the crosshairs of climate change in the future (see Schmandt et al., 2011 for a treatment of the subject). Most of the state has warmed between one-half and one degree (F) in the past century. In the eastern two-thirds of the state, average annual rainfall is increasing, yet the soil is becoming drier due to rising rates of evapotranspiration. Rainstorms are becoming more intense, and floods are becoming more severe. Along much of the coast, the sea is rising almost two inches per decade. Warmer sea temperatures have fueled several major deadly hurricanes that have devastated portions of the Texas coast during this century, including Allison in 2001, Rita in 2005, Ike in 2008, and Harvey in 2017. In the coming decades, storms are likely to become more severe and summers are likely to become increasingly hot and dry, creating problems for agriculture, wildlife, and possibly human health (EPA, 2016). Higher temperatures and drought are likely to increase the severity, frequency, and extent of wildfires. The combination of more fires and drier conditions may expand deserts and otherwise change parts of the Texas landscape. Given how the climate crisis produced the devastating wildfires seen in Australia during 2019–2020, a similar situation could conceivable occur someday in Texas.

Chapters 5 and 7 include more detailed discussions of how climate change is currently impacting the state and what might occur in the future, especially with regard to conservation and wildlife.

An Overview of Ecoregional Changes

The timing and sequence of land-use and climate changes in Texas have varied among ecoregions, but each of the regions has changed markedly since Bailey and the federal agents visited the state. Among the most altered places are the prairies and wetlands, the riparian and riverine ecosystems, and the rangelands of the Edwards Plateau, Rolling Plains, South Texas Plains, and the Texas Panhandle (Sansom, 1995), but no place in the state has remained immune from man's impact. In the accounts that follow, we highlight some of the major changes in each ecoregion since the biological survey of the late nineteenth and early twentieth centuries. With the landscape and environment of Texas continuing to change, there will no doubt be greater demands on every acre of land to continue to produce commodities even while maintaining environmental integrity.

An account of the current ecological conditions in the ecoregions can be found in the recently published *The Natural History of Texas* (Chapman and Bolen, 2018), and David Riskind (2015) has provided a historical summary of changes in vegetation and land use for each of the regions. Much of the material presented below has been adapted from these two excellent publications.

Gulf Coastal Prairies and Marshes and Gulf Barrier Islands

Of all the areas in Texas, the historical coastal prairie, which extends from Mexico to the Louisiana border, has seen the greatest industrial development during the twentieth century. The world's second largest petrochemical complex and some of the nation's busiest port facilities also are located along the coast. The population of Houston in 1900 was less than 50,000. Today, Houston is the nation's fourth largest city, and Harris County is the nation's second most populated county. More than six million people (plus a good number of "winter Texans") populate the coastal area of Texas, and this number continues to grow.

Few areas remain in the pristine state of a century ago, and conspicuous man-induced impacts are numerous. The original vegetation types of the upland prairies and woodlands were tallgrass prairies and post-oak savannahs, but today, primarily due to fire suppression, these regions have been invaded by trees and shrubs such as mesquite, acacia, various kinds of oaks, and the introduced Chinese tallow tree in many places along the upper Texas coast. Cattle and horses were introduced all across the region and the resulting overgrazing has caused significant ecosystem change. Many of the marshes and coastal areas have been converted to livestock grazing and to farms in the last century. Land subsidence poses a major concern along the Upper Texas coast with areas around Galveston Bay sinking by more than 1.5 feet (0.5 m)—and as much as 10 feet (3 m) in some places—just in the last half of the twentieth century. A related concern—sea-level rise—threatens many low-lying areas along the coast. Oil and chemical spills remain an ongoing hazard because of the extensive nature of this industry along the coast. Urbanization and industrial developments have eliminated habitats; water and air pollution are significant problems; natural fire has been all but eliminated; and agriculture has destroyed most of the native prairies, substantial acreages of marsh, and much of the forested and brush habitats. Reservoir construction has robbed the rivers of their sediments, and jetties and navigation channels have disrupted long-shore transport, exacerbating erosion and subsidence problems. Freshwater inflows have been reduced along the Gulf Intracoastal Waterway, and other dredging, irrigation, and flood control projects have had a huge influence on historical water circulation patterns. Taken together, all of these changes have had significant negative impacts on wildlife habitat in the region.

As if this weren't bad enough, disastrous hurricanes have battered the region with increased frequency in the twenty-first century, exacerbating the many problems described above. In 2017, Hurricane Harvey hit Houston and the upper Texas coast with the worst flooding in the history of the state, devastating people and wildlife. Although it is difficult to assess the damage to wildlife populations and habitats since this recent event, the damage that sewage, chemicals, and saltwater can have on the environment from this massive flooding event was extensive, and Harvey was Houston's third "500-year flood" in three years. If humans do not address the causes of a warming climate, these types of catastrophes are projected to become more frequent.

While highly impacted, the coastal system remains quite productive for a wide variety of fish and wildlife species. The near-coastal forests are critically important for the nation's songbird resource, as the vast majority of these species utilize this habitat during their trans-Gulf/circum-Gulf migrations. The forested habitats in the region are world famous for the spring "fallouts" when twenty or more species might be seen in a single tree at one time. A number of rare species of animals and plants occur across the variety of habitats along the Texas coast. All of the critically endangered Attwater's prairie chicken (*Tympanuchus cupido attwateri*) and almost all of the wintering whooping crane (*Grus americana*) are completely dependent upon coastal habitats in Texas. The native brush habitats of the lower coast make up the northeastern range of the endangered ocelot (*Leopardus pardalis*). Recovery of all of these species is highly (and in some cases entirely) dependent upon habitat conservation and restoration activities in the region.

The status of the Texas Gulf barrier islands has become particularly precarious because of rising sea levels, increasing intensities of hurricanes and tropical storms, and reduction in sand supply. For these reasons, most of the barrier islands are diminishing in size even as the human population living on them continues to grow: from 1990 to 2000, the average population density on Texas's barrier islands increased by 14 percent.

The islands vary considerably given the extent of human impact upon them. Galveston Island has the longest history of human impact, with the first permanent settlement founded in 1839. In contrast, Matagorda and San Jose islands have received very little recent human impact, with only a handful of permanent residents located on Matagorda Island. San Jose Island is under private ownership and is managed as pastureland. Mustang Island has a small port, Port Aransas, founded in the 1850s. The remainder of the island is used mostly as pasture, with intervening housing developments, although not to the extent of Galveston Island. Mustang Island State Park, south of Port Aransas, and an adjacent land purchase by The Nature Conservancy have preserved almost 5,000 acres (2,023 ha) of habitat on that island. Most of Padre Island is protected as Padre Island National Seashore, with a small amount of human habitation at the southern and northern ends of the island with numerous hotels, boat channels, and residential sites.

Data from recent field studies (Hice and Schmidly, 1999, 2002; Jones and Frey, 2013) have shown that the barrier islands of Texas have a depauperate fauna when compared to the fauna of the Texas coast. A number of factors have influenced the faunal composition of the barrier islands, including hurricane events, the mainland species pool upon which the islands draw, and the islands' degree of isolation from the mainland.

South Texas Plains

The South Texas Plains and brush country is a predominantly dry area encompassing almost 21 million acres (8.5 million ha) that stretches from the edges of the Hill Country into the subtropical regions of the Lower Rio Grande Valley. Grasslands apparently dominated the landscape in pre-settlement times, although woody plants (trees and shrubs) were often present in thickets, upland areas, major drainages and river bottoms (see Chapter 3). Mesquite was present throughout the region, but at a much lower density than today. Natural fires helped to maintain the region as a savannah and control woody plant densities on the prairie.

As early settlers occupied the area in the nineteenth century, changes to the landscape and vegetation communities were inevitable as grazing intensified during the transition from open range to fenced ranches. Brush densities and distribution increased because of overgrazing, lack of natural fires, soil compaction, and periodic droughts. This kind of habitat was commonly encountered when Vernon Bailey and the federal agents worked in the area during the late nineteenth century. Today, the region is commonly characterized by a mixture of thorny brush, including mesquite, acacias, hackberry, prickly pear, and other shrubs (Archer, 1989).

Because of overgrazing, elimination of fires, and other factors, the plant communities were altered to such a degree that ranches in the South Texas brush country during the first half of the twentieth century faced serious brush problems that often required control. Ranchers regarded increased brush density as detrimental to livestock operations, and attempts to control it were intensive and widespread, involving various techniques such as root-plowing, roller chopping, discing, chemical spraying, and range reseeding to increase grass production.

In the late 1960s, land managers discovered that extensive brush control was detrimental to wildlife. During the same period, white-tailed deer hunting began to increase in popularity, giving landowners an economic incentive to provide quality habitat for deer. Consequently, brush removal was reduced or applied in a less detrimental manner to improve the habitat for deer and other wildlife species. Because of the tremendous growth (and profits) in hunting trophy deer, many South Texas ranches are now under high fencing and practice intensive deer management, including genetic manipulation of populations (see Chapter 7 for a discussion).

South Texas brushlands form a sizeable part of some the largest privately owned ranches in the United States—notably, the 825,000-acre (333,866 ha) King Ranch and the 500,000-acre (202,343 ha) Kenedy Ranch—and this factor thereby excludes human intrusions into much of the region. This protection, along with the relatively low density of human populations elsewhere, clearly helps maintain the region's extraordinary diversity of animals, plants, and habitats. The legendary King Ranch has become a major force in wildlife conservation, and it probably constitutes the largest block of wildlife habitat in the state (see Hodge, 2003, for a history of this ranch and its contributions to wildlife conservation in Texas).

A subtropical bottomland forest, which included trees interspersed with scattered patches of palms and subtropical woodlands, at one time extended along the course of the Rio Grande from Del Rio southward to the Gulf of Mexico (today this is referred to as the Lower Rio Grande Valley, LRGV). Today, less than five percent of the native Tamaulipan thornscrub and less than three percent of riparian forest, scattered in narrow strips along the riverbank, remains (Schlyer, 2018). There are only 37 acres (15 ha) of protected palm forest compared to approximately 40,000 acres (16,200 ha) that once covered the delta region. Some deep lakes dot the region, as well as short-lived "resacas." A resaca is a former channel of the Rio Grande that has been cut off, like an oxbow. Resacas occasionally fill with silt and water, creating marshes and ponds. Bailey and the federal agents found resacas and ponds prevalent when they visited the area, but today they have been much reduced in size and scope.

This region is home to many tropical species whose ranges extend down into Mexico and Central America, numerous grassland species that range northward, and several desert species also commonly found in the Trans-Pecos region. Many wild and rare species of plants and animals, such as the ocelot, occupy the region. Urban development and agriculture use threaten the existing wildlife habitat as does the expansion of the border wall along the border with Mexico (see below).

Ranching is one of the primary industries in this ecoregion, although truck farming and citrus production are practiced in many places. Bird-watching and game hunting leases for white-tailed deer, mourning dove, and northern bobwhite are a major industry and source of income over much of the region.

In the twenty-first century, parts of South Texas have become important for oil and gas production, which has impacted the landscape. The Eagle Shale Formation in South Texas runs from the US–Mexico border north of Laredo in a narrow band extending northeast for several hundred miles. The most active part of the region for drilling is mainly in McMullen, Maverick, Dimmitt, LaSalle, Karnes, Live Oak, and Atascosa counties. Typically there are 900-1,000 active rigs in the region.

Edwards Plateau

The Edwards Plateau region consists of a relatively flat plateau in the west transitioning to rolling to very hilly terrain in the east (the Hill Country); the region is dissected by several river systems, including the Colorado, Pedernales, and the Guadalupe. The vegetation is characterized by grasslands, live oak–mesquite savannahs, and juniper-oak woodlands. However, open grasslands and savannas were much more common before settlers arrived. Today, mid-grass and tallgrass communities have been replaced by short-grass communities. Overgrazing and overbrowsing by domestic livestock and wild herbivores, as well as fire suppression, have contributed greatly to the observed changes. The expansion of Ashe juniper (*Juniperus ashei*) also has had a tremendous impact on the habitat, causing a decrease in plant species diversity and an increase in soil erosion. Ranching is the primary agricultural industry, with areas under cultivation largely confined to the valley bottoms and areas with deeper soils.

The region's major environmental threat—urbanization—has accelerated in recent years. Rapid expansion of the human population, subdivisions, and 1- to 10-acre "ranchettes" together have degraded native vegetation and threatened the integrity of both surface and underground water systems. There are continued efforts to preserve large tracts of undisturbed habitat—such as the Balcones Canyonlands National Wildlife Refuge— to protect rare and endangered species that occupy the region. Several state, federal, and private entities have purchased and set aside sites for managing endangered species.

The eastern edge of the Edwards Plateau—the "Hill Country" in the vicinity of Austin and San Antonio—is an area where major conservation problems have developed as a result of human impacts during the twentieth century due to the incredible population growth. Throughout the twentieth century, and continuing into the twenty-first century, Ashe juniper increased in the area because of a reduction in wildfires across the region. Increasing demands for water by metropolitan areas and agricultural and industrial enterprises caused water flow to be reduced in many springs and rivers, and it has severely depleted the Edwards Aquifer. Several species of animals and plants have suffered from a loss of forest and riparian habitats, degradation of karst (cavernous) habitats, and water quality impacts.

The central and western areas of the Edwards Plateau, while less disturbed than in the east, also have been subjected to severe land and habitat fragmentation. Only tiny remnants of the native landscape survive today, and the average tract size in many counties has dropped in this generation alone from thousands of acres to fewer than two hundred. These areas, which once provided large blocks of land for wildlife habitat and outdoor recreation, now consist of tiny plots of introduced vegetation that struggle to sustain the native wildlife.

Numerous karsts, sinkholes, and caves have been leached out of the limestone rock in the region and serve as a unique habitat for many kinds of animals. The Devil's Sinkhole, near Rocksprings (Edwards County), Texas, is the most prominent sinkhole in all of the state. It is home to an estimated three million free-tailed bats as well as a large population of cave swallows (Graves, 2008). Numerous caves occur throughout the region, especially along the southeastern border, and these are home to some of the largest bat colonies in the world. In addition to bats, the caves and sinkholes provide ideal habitat for skunks, raccoons, snakes, salamanders, and a host of invertebrate creatures, many of which are considered rare and endangered.

Trans-Pecos (Mountains, Basins, and Chihuahuan Desert)

The Trans-Pecos ecoregion, which encompasses all of Texas west of the Pecos River, is the only part of the state where mountain and desert habitats are found, making it among the most ecologically complex regions in the state. Its contrasting habitats include semidesert grasslands, intermixed with desert shrubs on the plateaus of the Chihuahuan Desert, and surrounding island mountains of pine-oak–juniper woodlands. Elevation of the region varies from 2,500 feet (762 m) to 8,749 feet (2,667 m) at Guadalupe Peak, the highest point in Texas. Human populations within the Trans-Pecos ecoregion are sparse, and ranching remains the primary industry just as it was during the time that Bailey and the federal agents visited the region.

All early accounts provide evidence that grasslands lightly interspersed with shrubs and desert succulents dominated the region before human settlement in the nineteenth century (Richardson, 2003). Over time, grasslands decreased and gradually gave way to increases in woody plant abundance. Causal factors included overgrazing of grasslands by livestock, suppression of grassland fires, and short and long duration drought periods in association with plant competition, changes in climate, and erosion of topsoil in areas where vegetation has been removed. Healthy grassland savannahs exist today on sites where wildfires have occurred or where prescribed burning is practiced, as well as on ranches that have been conservatively grazed and properly managed for decades. Most of these healthy grassland savannahs occur at moderate to high elevations (cooler temperatures and greater average rainfall) in Hudspeth, Jeff Davis, Presidio, and Brewster counties (Richardson, 2003).

Grazing pressure was severe in this region during the time Bailey and the agents worked there, and this initiated a desertification process that continues today. Even though livestock numbers in these areas have stabilized well below historical levels, these are marginal grazing habitats, and the combination of reduced fire frequency and continued topsoil erosion has sustained a land-cover decline in which much of the remaining grassland is being converted to desert shrubland (Bogan, 1998).

The following environmental changes have occurred over the Chihuahuan Desert region as a whole: reduction in surface and ground water; alteration of riparian habitats caused by the spread of invasive plants such as salt cedar; arid lands in ever-larger sectors irrigated for crop production; natural forage subjected to grazing by larger numbers of domestic livestock; islets of coniferous and hardwood timber at the summits of desert mountains subjected to lumbering; and introduction of noxious plants and animals (Baker, 1988). Semidesert grasslands in good condition are rare today in western Texas.

Sadly, air pollution also has begun to impact the deserts and mountains of the Trans-Pecos. A large coal-burning power plant in Mexico, known as Carbon II, is located 125 miles (200 km) southeast of Big Bend National Park. At peak periods of operation, and under certain climatic conditions, the smoke generated by this plant has degraded visibility by up to 60 percent in Big Bend National Park, and the effects also have been noticed at McDonald Observatory in the Davis Mountains (Bartlett, 1995). There is some concern that this air pollution could impact rainfall patterns in the region. Daniel Rosenfeld (2000), an atmospheric scientist at Hebrew University of Jerusalem, has published research that demonstrates that air pollution can spread hundreds of miles downwind of a large industry's plume of pollution and can alter clouds and natural precipitation. Even remote, protected national parks are no longer immune to the impacts of modern society.

Many of the original vegetative areas of the Trans-Pecos have been preserved in their near-natural state through the establishment of national and state parks. These include Big Bend National Park, Guadalupe Mountains National Park, Davis Mountains State Park, Balmorhea State Park, Big Bend Ranch State Park, and the Chinati Mountains State Natural Area. In addition, there are several TPWD wildlife management areas (Black Gap, Elephant Mountain, and Sierra Diablo) that protect natural habitat.

The US Fish and Wildlife Service, National Park Service, US Geological Survey, and Texas Parks and Wildlife Department formally created the Big Bend Conservation Cooperative in 2010. That group works with various agencies in Mexico (Secretariat of Environment and Natural Resources, National Water Commission, and National Institute of Ecology and Climate Change), nonprofit organizations (World Wildlife Fund and The Nature Conservancy), and business corporations (Coca-Cola and CEMEX) to address issues regarding the Rio Grande and wildlife in the Big Bend region that are threatened by habitat loss, climatic change, and exotic invaders (Gaskill, 2015).

High Plains

The High Plains is the largest level plain of its kind in the US and rises from 2,700 feet (823 m) on the east side to more than 4,000 feet (1,219 m) on the west side in some areas along the New Mexico border. The Caprock Escarpment divides the High Plains from the lower Rolling Plains to the east. The escarpment rises abruptly in places nearly 1,000 feet (305 m) above the lower plains (Riskind, 2015). Notable canyons occur along the northeastern boundary with the Rolling Plains, including Tule and Palo Duro along the Caprock. The region also is home to unique ephemeral lakes known as "playas." A playa is a relatively small wetland with a closed watershed (Luo et al., 1997).

Since the High Plains was settled late in the nineteenth century, around the time that Bailey and the federal agents worked there, virtually all lands have been disturbed by cultivation or overgrazing. Changes in the native vegetation have been substantial. Even the playa lakes have been impacted with the advent of extensive irrigation that has lowered the water table and removed most of the natural springs that once existed in the region. Bailey was absolutely correct in predicting that the High Plains would, with irrigation, become a major agricultural area. Today, more than one-third of the total crop production in Texas, and more than one-third of the cotton raised in the country, occurs in this region. Almost 80 percent of the region now consists of agricultural fields, a stark contrast from the days of the biological survey when there were very few farms or towns in this area.

There are approximately 30,000 playa wetlands in the southern High Plains region (Osterkamp and Wood, 1987) that collectively cover almost 340,000 acres (1,376 sq km) (Riskind, 2015). These wetlands provide critical ecological and societal benefits for the region, serving as recharge sites for the underlying Ogallala Aquifer, water storage during flood events, irrigation water for crops, and water for livestock as well as vital wildlife habitat. During the twentieth century, almost all of the land area surrounding the playas was either cultivated or subjected to grazing by livestock. Consequently, the playas have lost volume through sediment deposition and, unless the problems of soil erosion are reversed, it has been estimated that sediments could fill all cropland playas by the end of the twenty-first century (Luo et al., 1997).

In the latter two decades of the twentieth century, a portion of the High Plains region was converted from agriculture back to productive grassland habitat, primarily through the Conservation Reserve Program (CRP). This initiative deserves special mention as it represents an example of a government program that has been good for private landowners and, in turn, has produced positive environmental benefits.

The CRP was established by the US Department of Agriculture in 1985 as a private-lands reclamation program designed to conserve and improve soil and water resources by taking highly erodible land out of crop production and establishing suitable vegetative cover in its place. In exchange for a yearly rental payment, farmers who enrolled in the program agreed to remove environmentally sensitive land from agricultural production and utilize plant species that will improve environmental health and quality. Throughout the southern High Plains, under the supervision of the Soil Conservation Service, landowners have established grasslands, either of native or introduced grasses, on previous cropland. After establishment, no other maintenance is required, no grazing by domestic livestock or harvesting of vegetation is allowed, and the site must remain in the CRP for 10 to 15 years. The CRP has improved water quality, reduced soil erosion, and increased natural habitat that has resulted in an increase of species diversity in small mammals and other kinds of wildlife (Hall and Willig, 1994). As of 2017, Texas had 2,520,821 acres (10,201 sq km) enrolled in the program, the largest of any state in the country. Unfortunately, the CRP in Texas has allowed seeding with weeping lovegrass, a fast-growing, nonnative species that, although drought resistant, crowds out native plants. The CRP was extended and updated in the 2018 Farm Bill with a provision for climate change mitigation added to the other elements of the program.

Rolling Plains

The Rolling Plains ecoregion lies to the east of the High Plains ecoregion. The two ecoregions together constitute the southern extension of the Great Plains into the central US, and they have experienced many of the same impacts to wildlife habitat. Similar to the High Plains, the Rolling Plains has suffered from the over-pumping of the Ogallala Aquifer that has reduced flows in the Canadian River. Exotic tree species have impacted native riparian habitat, and there has been a loss of grassland habitat as a result of agricultural practices. The elimination of fire and the impact of overgrazing have resulted in the growth and spread of brush (cedar, mesquite, scrub oak, and prickly pear) across the entire region, and this has severely impacted native prairies and watersheds. Today, only scattered remnants of the original prairie are intact and these are dominated by sideoats grama, little bluestem, and blue grama; much of the region has been widely invaded by "increaser" grasses, as well as mesquite, shinnery oak, tree cholla, prickly pear, redberry juniper, and other brush species.

During the time that Bailey and the federal agents traversed this area more than a century ago, it remained under the control of a few large ranches, so urbanization and fragmentation developed slowly compared to many other regions in the state. Before settlement, lightning-caused hot fires scorched large areas every five to twenty years, often enough to check the spread of mesquite, cacti, and redberry juniper. As fire suppression accompanied settlement, the intensity and frequency of fires was reduced, allowing species like redberry juniper to expand into the grasslands. Further, a combination of overgrazing, droughts, and invasion of salt cedar has reduced or in some cases eliminated cottonwood bottomlands, thus negatively impacting many riparian areas in the region.

More than 75 percent of the land of the Rolling Plains is rangeland, with crop and livestock production constituting the other major land uses. Ample wildlife habitats have been created by the mixing of these varying croplands and rangelands. Thus, many wildlife species such as quail, mourning dove, white-tailed deer, and turkey live in this region, providing good recreational hunting opportunities (Riskind, 2015). Several caves occur along the northwestern escarpment of the region where it abuts with the Llano Estacado. These caves house large bat populations and provide habitat for other wildlife species as well.

Much of the Rolling Plains is held in private ownership, which limits access to many areas. However, four state parks offer public access to the landscape. These include Palo Duro Canyon State Park, the second largest park in the state; Caprock Canyons State Park, home to the Texas State Bison Herd; Copper Breaks State Park, which houses part of the state's official herd of longhorn cattle; and the newly established (but not yet open to the public) Palo Pinto Mountains State Park, about halfway between Abilene and Fort Worth. The ecoregion also has several wildlife management areas (Matador Wildlife Management Area, Taylor Playas Wildlife Management Area, and Gene Howe Wildlife Management Area) and national grasslands (Rita Blanca Grassland) that provide good wildlife habitat. With its ideal topography and wind patterns, the region has become important for producing wind energy, and wind turbines are prevalent throughout the area, especially from Snyder to Abilene.

Cross Timbers and Prairies

As its name depicts, and as Bailey's description implies, the Cross Timbers is a rolling to hilly region where prairies and timber intermingle. Straddling Texas and Oklahoma, it was a relatively narrow but thickly wooded area that was once dense enough to have been considered a natural barrier to settlement, remaining largely uninhabited even by Native Americans until the nineteenth century (Francaviglia, 2000). Bailey recognized this and wrote a lengthy report about its significance as an area of faunal transition; excerpts from that report are included in Chapter 3.

Most of the Cross Timbers has experienced some anthropogenic impacts, and few areas can be regarded as untouched. About 75 percent of the land area is used today as range and pasture, having been altered by plowing and overgrazing. Past mismanagement and cultivation have caused the uplands in this vegetative region to be invaded by shrubby vegetation along with weedy annual and perennial grasses. Grazing, farming, and other disturbances have altered the grassland vegetation so that little remains of the original Cross Timbers prairie (Bolen, 1998). There are many impoundments on the rivers that traverse this region, and this has resulted in a loss of riparian habitat.

Richard Francaviglia (2000) published an extensive account of the natural and cultural history of the Cross Timbers. Before modern settlement, early explorers characterized the region as a "forest of cast iron" in reference to its nearly impenetrable forest of stunted oak trees. Since then, large cities have grown up on the eastern boundary of the region in Dallas, Fort Worth, Arlington, and Waco, and the population pressure and associated development has increased steadily throughout the twentieth century. Variations in Cross Timbers vegetation are the result of both natural and cultural conditions, with humans being significant agents who have wielded fire, axes, plows, and bulldozers in transforming the vegetation at both local and regional levels. The region has survived into the twenty-first century as discontinuous patches of forest—from copses or mottes of but a few trees to sections of hundreds of acres or more. These areas exist in an extremely complex pattern as they are interspersed with farms, pastures, and urban developments.

Blackland Prairies

The original Blackland Prairies had elements of both the true and coastal tallgrass prairies along with some unique elements of its own, but because of precipitation and soil moisture retention characteristics, it took on a lowland grassland appearance even on well-drained uplands. About 98 percent of the region was cultivated to produce cotton, sorghum, corn, wheat, and forages beginning in the late nineteenth century and continuing into the twentieth century. In the 1950s, pasture and forage crops for livestock production increased in the region, but now only about 50 percent of the area is used as cropland; about 25 percent is tame pasture; and the remaining area is used as rangeland. As a result of cultivation, overgrazing, and other imprudent land-use practices, there are few if any remnants of climax vegetation remaining in the region. Most estimates indicate that less than one-half of one percent (about 5,000 acres; 2,023 ha) of the original Blackland Prairies remain intact. Except for a scattering of extremely small patches that escaped cultivation, locations historically managed as hay meadows represent the only remaining opportunity to protect sizeable areas of unplowed ground. This has limited conservation opportunities in the region to land easements or purchases. Both The Nature Conservancy and the Native Prairies Association of Texas represent nongovernmental organizations spearheading such efforts.

Of the small remnants of native vegetation that remain for grazing or for native hay production, the majority are in Lamar County or nearby in northeastern Texas (Sharpless and Yelderman, 1993). Another center of remnant prairies is located around Temple and Waco in Bell and McLennan counties. Virtually nothing remains in the southwestern region of the Blackland Prairies between the Colorado and San Antonio rivers in the vicinity of Austin and San Antonio.

The Blackland Prairies originally had numerous wooded belts along the streams that traversed the region. Settlements spread rapidly as farmers discovered the productive possibilities of the black prairie soils. Early immigrants from Germany and Czechoslovakia settled entire communities and made the Blackland Prairies the foremost cotton-producing region of the state. So valuable was the land for farming that row crops were usually planted to the immediate roadsides, and only the most necessary fences were built. Wooded and brush-covered bottomlands were cleared to the stream banks. The agricultural activity of this region was largely responsible for the growth of some of Texas's largest cities, including Dallas, Fort Worth, Waco, Temple, Austin, and San Antonio. Most of the Blackland Prairies region no longer provides suitable habitat for a diversity of wildlife. Today, crop production and cattle ranching are the two principal agricultural industries of the region.

Post Oak Savannah

This region, which lies between the Piney Woods to the east and the Blackland Prairies to the west, is best described as an ecotone between the eastern deciduous forest and the tallgrass prairie. The original landscape was shaped by naturally occurring wildfires, intentional fires set by Native Americans for hunting and grass production, climatic fluctuations, and grazing by nomadic bison herds of the southern plains. Settlement eventually brought an end to the open range, and although the economy of the region prospered, the natural vegetation characteristics were drastically altered. Today, the area supports a stunted, open forest dotted with small tallgrass prairies. A distinctive lowland forest habitat in the region is limited to the floodplains of the major streams.

Over the past century, heavy human use has severely impacted the entire region and forever altered the once-pristine prairies. Today, only small, scattered patches of true prairie remain in this region. Farming, livestock ranching, conversion of native grasslands to improved pastures, and suppression of range fires reduced and fragmented the native savannah grassland. In the absence of periodic fires, woody vegetation encroached on the remaining tracts, and the upland sandhills developed dense stands of oaks, shrubs such as yaupon holly, and the invasive eastern red cedar. Much of the region is still in native or improved pastures, although small farms are common. Improved pastures are commonly seeded to Bermuda grass, dallisgrass, vaseygrass, carpetgrass, and clovers. About 50 percent of the total area has been cleared, and about one-half of the cleared areas are planted with crops. Clearing of the wooded areas creates extensive edges that are further emphasized by the practice of clearing fairly small, irregular areas.

The phenomenon of land fragmentation is well demonstrated by changes in the post-oak belt. Fragmentation of lands north and west of the Houston metropolitan area has resulted in long-term declines in the wildlife resources of the region. The ecoregion has been severely altered by the introduction of large herds of grazing cattle and horses and by the introduction of nonnative grasses and forb species planted to provide forage for livestock (Riskind, 2015).

Piney Woods

The Piney Woods represent the westernmost extension of the Eastern Deciduous Forest, a vast woodland that once covered essentially all of the eastern United States. Far from just pinelands, the region encompasses a rich variety of natural woodlands including remnants of once extensive longleaf pines, galleries of loblolly pines and deciduous forests on mesic slopes, extensive stands of bottomland hardwoods bordering streams and rivers, and bald cypress–tupelo swamps. Moreover, shrubby bogs, natural prairies, desert-like dunes of white sand, and cleared agriculture lands punctuate the landscapes, all distinguishing the Piney Woods as one of the most diverse regions in Texas. Though rapidly diminishing, the bottomland hardwood forests of oak, hickory, elm, sweetgum, sugarberry, and ash—the most diverse and richest wildlife habitats left in Texas—are located in the Piney Woods.

The Piney Woods were severely impacted by humans during the twentieth century. Early in the century, the largest oilfields in the state were located there. But lumbering was the main form of land use. Overcutting was widespread over much of the region, as was the uncontrolled burning of pine woodlands. Practically all of the virgin pine timberland has been cut over and is now producing a second or third crop of timber. Farming enterprises are scattered throughout the region, although most of the farms are comparatively small. For much of the twentieth century, hog farming was very popular in this region. Overstocking, particularly with hogs, has been detrimental to range and forest management in the region. In many places, hogs became free ranging and severely damaged long-leaf pine seedlings. Many of the streams and rivers have been dammed, which has flooded much of the hardwood bottomland habitat. Joe Truett and Dan Lay (1984) have written a thorough and dramatic account of the natural history of the East Texas Piney Woods.

The status of longleaf pine forests in the southeastern United States, including East Texas, has become a matter of serious concern to ecologists and conservationists. Before the first settlers ventured into Texas, longleaf pine savannahs dominated an area of about 5,000 square miles (13,000 sq km), encompassing an estimated 60-79 million acres (24.3-32 million ha), in the south-central portion of the Piney Woods that extended eastward from the Louisiana border to the Trinity River and southward to the coastal prairies. Fires maintained the uncluttered landscape and stimulated the growth of grasses and forbs while inhibiting invasions of shrubs and hardwoods. Because of its high quality, longleaf pine quickly became the most valuable and readily marketed timber resource in Texas. Annual harvest commonly reached up to 750 million board feet (1.8 million cubic meters) during the early 1900s, and most stands of longleaf pine had disappeared by the 1920s, having been replaced by plantations of faster growing species, such as loblolly pine.

Longleaf pine communities today cover no more than 3 percent of their original range. Their demise can be attributed to over-harvesting for wood products, problems associated with thousands of free-ranging hogs that annually uproot tens of thousands of seedlings, and fire suppression that allows for the invasion of oak forests (Bolen, 1998).

The southeastern portion of the Piney Woods includes an area known as the Big Thicket, a legendary area of impenetrable vegetation that remained untamed until long after other parts of the region had given way to settlement. It was known as the "old bear hunters' thicket" because hunters there used dogs to chase Louisiana black bear and feral pigs (Chapman and Bolen, 2018). Early travelers in southeastern Texas described the Big Thicket, which originally exceeded two million acres (809,371 ha), as an extensive, impenetrable, mixed mesic woodland that stopped them as they traveled inland across the Gulf Prairie. Bailey and some of the other federal agents worked extensively in the Big Thicket during the biological survey at Sour Lake and Tarkington Prairie. In 1974, the US Congress established Big Thicket National Preserve, within a four-county region of southeastern Texas, to ensure the protection of the fragile and delicate fauna and flora of this region.

Nationwide, several species became extinct as a result of forest changes and human uses during the twentieth century, including the passenger pigeon (*Ectopistes migratorius*), heath hen (*Tympanuchus cupido cupido*), and Carolina parakeet (*Conuropsis carolinensis*). An even larger number of subspecies and wildlife populations were substantially diminished. At least two subspecies of Texas mammals became extinct in the Piney Woods of southeastern Texas (see Chapter 5).

Many species that were threatened with extinction in 1900, however, have come back in abundance. Due to actions that were set in motion in the early decades of that century, today most forest-wildlife species are both more abundant and more widespread than they were at the turn of the twentieth century. Among mammals, examples include beaver (*Castor canadensis*), black bear (*Ursus americanus luteolus*), otter (*Lontra canadensis*), and white-tailed deer (*Odocoileus virginianus*) (see Chapter 5). Since the 1930s, forest wildlife that can tolerate a relatively broad range of conditions (so-called "habitat generalists") has increased, and most forest-wildlife species fit this category due to the natural dynamics of forests that cause frequent disturbances in the natural regime (McCleery, 2011). Some species with specialized habitat requirements are of concern today though, including the red-cockaded woodpecker (*Leuconotopicus borealis*) and gopher tortoise (*Gopherus polyphemus*), both natives of fire-created southern pine savannahs and woodlands. The red-cockaded woodpecker needs both mature forest and other specific habitat conditions, such as open savannahs and woodlands created by frequent ground fires.

Although there have been some gains in the management of Texas's forests since 1900, a variety of environmental trends are not positive and problems remain, the most significant of which are: (1) the rising loss and fragmentation of forests due to residential subdivision and urban developments; (2) the loss and deterioration of the forest and grassland habitats that once were created by frequent, low intensity fire; (3) the increasing impact of climate change (increased wildfires, insects and disease, and stresses on habitats and species); (4) the effects of air pollution on forests in some areas (greater Houston metropolitan area); and (5) introduced exotic plants, animals, and diseases (McCleery, 2011). There is a long history of damage to forests from introduced biological agents such as white pine blister rust, the caterpillar of the moth *Lymantria dispar* (formerly known as the gypsy moth), and more recently the Asian long-horned beetle. Since 2010, the World Wide Fund for Nature has listed the Piney Woods region as one of the critically endangered ecoregions of the United States (World Wildlife Fund, 2019).

The Special Situation of the Borderlands

One region of the state warrants special attention insofar as wildlife habitats and wildlife diversity are concerned. This is the borderland region between Texas and Mexico, which stretches for 1,250 river miles (2,012 km) along the Rio Grande from the Gulf of Mexico to El Paso (Gehlbach, 1981). Bailey and the federal agents worked at several places along the Texas–Mexico border, from El Paso to Brownsville, and Edgar Mearns, the naturalist and surgeon on the second US–Mexico boundary survey, wrote a detailed account of the mammals of the Mexican border from Fort Clark and Fort Hancock, Texas, westward to the Pacific Ocean in 1907.

Gehlbach (1981), Steinhart (1994), and McKinney (2012) have published modern accounts of the natural history of the borderlands, including its many unique species and habitats, and the human impacts in the region. Earlier in the century, overgrazing, compounded by short-term climatic cycles, exacerbated man-made vegetative changes. More recently, irrigation, industrial development, and an increase in urban populations have put a different face on the human presence in the borderlands.

The nonprofit watchdog group American Rivers (American Rivers, 2019) named the Rio Grande among America's most endangered rivers in 1993, 1994, 2000, 2003, and 2018 because of overuse and outdated water management practices. Massive water diversions and proposed flood control projects have substantially changed the river's character. Diversions for municipal and agricultural use already claim nearly 95 percent of the Rio Grande's average annual flow to such an extent that long stretches of the river are now dry. In 2001, the river failed to reach the Gulf of Mexico for the first time, and in 2002, it happened again. Channelization and flood control measures have altered the river channel and its banks, and reservoir construction and inundation have flooded hundreds of acres of riparian and canyon habitats. Pollution problems along the entire length of the Rio Grande include hazardous waste dumps, municipal and industrial effluent, irrigation return flows, and municipal runoff. Overallocation, invasive species, and drought are other serious issues that impact the river.

The upper Rio Grande essentially stops at El Paso, where the river is entirely diverted from its channel for human use. Where the river is dry, sediments and vegetation have clogged the channel. The river does not begin to flow again until approximately 250 miles (402 km) downstream at the confluence with the Rio Conchos near Presidio. The Conchos supplies the Rio Grande with two-thirds of its flow below that point, although Granero Dam on the Rio Conchos in Chihuahua, Mexico, greatly restricts water flow into the Rio Grande at certain times of the year.

The Rio Grande corridor provides essential wildlife habitat to numerous species as it traverses the Chihuahuan Desert region. Many wildlife species depend on surface water to survive. In addition, flowing water and associated riparian habitats act as vital dispersal corridors for many species, and, conversely, as effective distributional barriers for other species. Species abundance and composition have been altered as some species have undergone local extinctions, other species have changed their distributional limits, and new species have invaded the region. Schmidly and Jones (2000) summarized the twentieth-century changes in mammals and mammalian habitats along the Rio Grande / Rio Bravo from Fort Quitman to Amistad reservoir.

The southern part of Texas along the Rio Grande, known as the Tamaulipan Brushland or Lower Rio Grande Valley, has been especially altered since the time of the biological survey. The natural habitats in the lower valley virtually disappeared in the twentieth century. This was once a densely vegetated, sometimes marshy, woodland—a wet oasis in the middle of a long stretch of dry scrub country. A large forest of Mexican palmetto (*Sabal mexicana*) extended from the coast to approximately 80 miles (130 km) inland as late as 1852. Spanish explorer Alonso Álvarez de Pineda, in fact, named the river Rio de las Palmas (Palm River) in 1519 because of the forest that surrounded it (Wright, 2019). Virtually nothing of this habitat type remains today. Dense brush in this unique ecosystem provided food, nest sites, and cover for many wildlife species, including a large number of neotropical origin. Good riparian and gallery woodlands extended upriver—in fact, where conditions allowed, all the way into the deep canyon country of the Big Bend.

Historically, much of the delta encompassed a diverse mix of grasslands, brushlands, marshy habitats, and riparian woodlands and gallery forests. By 1937, the extent of the sabal palm forest was confined to a limited area near the Rio Grande in and around Brownsville (Leonard et al., 1991). Since 1937, clearing for agriculture has destroyed most of the palms except for the grove on the Rabb Ranch, which was purchased by the Audubon Society in 1975, and for the Southmost Ranch Tract, purchased by The Nature Conservancy of Texas in 1999. These are the last remnants of the relatively extensive riparian woodland forest.

The most important human-caused components of environmental change in the lower valley have been water diversion and flood control, brushland clearing, human population increases, contaminants, and continued dredging of the Intracoastal Waterway in Laguna Madre (Rappole et al., 1986; Chapman et al., 1998). Particularly significant has been the conversion of native brushland and riparian woodlands for agricultural uses and the subsequent large-scale fragmentation of remaining native habitats.

Since the 1920s, more than 95 percent of the original native brushland has been converted to agricultural or urban use, and more than 90 percent of the riparian habitat along the Rio Grande has been cleared. The resaca environment also has been declining, largely because of drainage for irrigated agriculture. Brush clearing, pesticide use, and irrigation practices associated with agriculture have been intensive (Jahrsdoerfer and Leslie, 1988). Water development for flood control and municipal use resulted in extensive clearing of brush, alteration of riparian habitats, and changes in water flow in the Rio Grande, and industrialization degraded water quality. Brushland habitats were converted to rangeland using herbicides, mechanical clearing, and fire. Arturio Longoria (1997) has written a riveting account of his experiences in the "brushlands" of South Texas and the extensive changes that occurred in the region during the twentieth century.

Conserving and managing wildlife in the borderlands region requires close collaboration and coordination with Mexico. Although the picture is often grim, conservation efforts on both sides of the border currently are gaining momentum as preserves have been developed and damaged habitat has been reclaimed. In the United States, private and federal agencies such as The Nature Conservancy and the US Fish and Wildlife Service work to acquire land and nurture the recovery of endangered species.

Of particular concern about the future of the borderlands are proposed plans to build a continuous wall along the entire US border with Mexico, significantly expanding the limited sections of border wall that already exist. Texas perhaps has more at stake in this debate than does any other US state due, in part, to the physical-geographic fact that Texas comprises over two-thirds (1,254 mi; 2,018 km) of the total US–Mexico border length (1,954 mi; 3,145 km) and encompasses the largest remaining area of the borderlands, the Trans-Pecos portion of the Chihuahuan Desert (LaDuc et al., 2019).

Although most of the public discussion about "the wall" has focused on its cost and human impacts, expanding the physical barriers along the border between Texas and Mexico will have substantial negative effects on wild species and ecosystems (see Fowler et al., 2018). Substantial amounts of habitat would be degraded or destroyed by the construction of barriers and the roads running alongside them. Natural areas are particularly at risk of having border barriers built across them. One of the impacted ecosystems of most concern is Tamaulipan thorn-scrub, remnants of which occur in South Texas on higher ground along the river. This diverse and formerly widespread ecosystem is now rare, having been replaced by agricultural and urban land uses (Leslie, 2016). Many of the plant and animal species dependent on this ecosystem would lose some of their last remaining US habitat. One of the at-risk species in South Texas would be the endangered ocelot (*Leopardus pardalis*), which would lose habitat as a result of barrier construction. The two remaining ocelot populations are already isolated from Mexico and from each other by other types of habitat fragmentation and are experiencing a loss of genetic variability as a result (Janecka et al., 2011; Tewes, 2017).

Similar habitat loss and fragmentation would occur in western Texas. For example, a border barrier would separate the black bears (*Ursus americanus*) in Big Bend National Park from the population in Mexico, making the park population too small to persist (Hellgren et al., 2005). The same would be true for the reintroduced bighorn sheep (*Ovis canadensis*) populations that are now known to move back and forth between Mexico and Texas.

Larger mammal species would likely be the most vulnerable, but smaller mammal, reptile, and amphibian species may be blocked even if gaps are provided for animal passage (McCorkle, 2011). Species cut off from the Mexican portion of their populations would have smaller effective population sizes, which in turn further increases the probability of extirpation or extinction (Lasky et al., 2011). Animal movement among habitat fragments within Texas would also be inhibited (Jahrsdoerfer and Leslie, 1988).

The construction of a wall could have negative economic implications as well. Riparian vegetation in South Texas is a hotspot for bird diversity and ecotourism. If ecotourism declines substantially because access to preserves has been impeded, there may be negative economic impacts on the region. On the other hand, if the barriers are not far enough from the river, they may trap wildlife escaping from floods and may even act as levees, which tend to increase downstream flooding. Negative impacts could be lessened by limiting the extent of physical barriers and associated roads, designing barriers to permit animal passage, and substituting less biologically harmful methods, such as electronic sensors, for physical barriers (Fowler et al., 2018).

So much concern has been raised about the proposed wall that more than 2,500 scientists co-signed a paper describing the significant harm to wildlife posed by infrastructure on the US–Mexico border (Peters et al., 2018). The border wall project was also unusual in being exempt from environmental reviews. Its construction became a major political issue, pitting environmentalists against the administration of former President Donald Trump, with both opponents and proponents invoking the environment—one side to stop the barrier, and the other to build it (Puko, 2020). The subsequent Biden administration immediately halted construction on the wall, but they have not yet made any announcement about removing any of the sections previously constructed. Most recently, Texas Governor Greg Abbott announced plans to continue building the Texas–Mexico border wall using a combination of state funds and donations from the public. The debate over the wall, and the environmental issues associated with it, will likely play out in court and may take years to resolve.

Twentieth and Twenty-First Century Changes to the Texas Mammal Fauna

The wildlife diversity and composition of Texas have undergone significant changes in the twentieth and twenty-first centuries. The jaguar and gray wolf are two of the more charismatic species of the many taxa eliminated from Texas. Jaguar and gray wolf, public domain photos.

IN EXAMINING HOW LANDSCAPE AND LAND USE CHANGES HAVE IMPACTED the biota of Texas, we have focused on the mammalian fauna because it has been well documented and serves as a good surrogate for other groups of animals. Bailey's 1905 publication focused almost entirely on mammals because that was his, as well as Merriam's, primary interest. Therefore, we have a good baseline of the mammal fauna before the state's landscapes and habitats became significantly altered. A 7th edition of *The Mammals of Texas* was recently published (Schmidly and Bradley, 2016), summarizing the most recent natural history information that is available for the 202 species of mammals that occur in the state today. So, there is a good "before" and "after" standard on which to document the changes in the Texas fauna.

The mammal fauna of Texas changed substantially during the twentieth and early twenty-first centuries, including a proliferation of extinctions, numerous species added to rare, endangered, or threatened species lists, geographic range reductions and expansions (with population declines or increases) of some species, and a few new elements added to the fauna, especially nonindigenous species. Chapter 5 addresses significant trends in Texas's mammal populations since the time of the Biological Survey, and Chapter 6 summarizes the current population and conservation status of mammals and selected reptiles in the state.

Significant Trends in Texas Mammal Populations

D URING THE TWENTIETH CENTURY, AS THE NATURAL HABITATS OF Texas were being obliterated or altered by farming, ranching, timbering, the development of the energy industry, and the urbanization impacts of its ever-growing population, the wildlife resources of the state suffered greatly. This chapter chronicles changes to the terrestrial Texas fauna, emphasizing mammals, following the period of the biological survey up to the first two decades of the twenty-first century.

On a macroscale, the diversity of terrestrial mammals changed substantially during this time and these trends likely will continue throughout the current century. There has been a substantial turnover in species composition, involving both a loss and gain in species, since 1900. A significant number of mammals are now extinct and a growing number of species are regarded as endangered, threatened, or having critical status (imperiled or vulnerable). Almost one-third (31.6 percent) of the extant mammal species (44 out of 139 native species) either have had subspecies or metapopulations become extinct, or they appear to be rare and face some sort of problem that causes conservationists to consider them potentially imperiled or vulnerable. In actuality the number of species in this situation may be much higher than this estimate would suggest. This is because we have not been able to do enough monitoring of populations to obtain usable information on population trends for most species of mammals.

Among the most significant wildlife population trends during the century, seven are particularly important and worthy of discussion:

- Proliferation of extinctions;

- Declines in geographic distribution and population abundance;

- Range expansions and regional faunal changes;

- Documentation of additional faunal elements;

- Increase in the number of threatened, endangered, and rare species;

- Introductions of nonindigenous, invasive species; and

- Growth in the commercialization of wildlife.

In addition to the discussion about population trends, we have included in this chapter a review of some of the earliest work on the possible impacts of climate change on Texas mammals.

Proliferation of Extinctions

Species extinctions increased dramatically during the twentieth century. When Bailey published his work in 1905, the only terrestrial extirpated mammals were the American bison (*Bos bison*) and elk (*Cervus canadensis*). During the course of that century, the grizzly bear (*Ursus arctos*), gray wolf (*Canis lupus*), red wolf (*Canis rufus*), black-footed ferret (*Mustela nigripes*), jaguar (*Panthera onca*), margay (*Leopardus wiedii*), and bighorn sheep (*Ovis canadensis mexicana*) all joined the list of extirpated species, and now it appears that the jaguarundi (*Puma yagouaroundi*) must be added as well. One other species, the American black bear (*Ursus americanus*), was thought to have gone extinct but now has established breeding populations in the state.

The ten terrestrial species that became extinct were all large herbivorous or predatory species. This same pattern has been observed time and time again throughout the globe—large-bodied, wide-ranging species with complex social systems are highly vulnerable to extinction. Populations of at least two of these extinct species (elk and bighorn sheep) have been reintroduced, although the individuals introduced are not from the original stock or subspecies that were once native to the state. A third species, the bison, is reared in captivity by the State of Texas and by private individuals.

A variety of factors can cause extinction, but in the case of these species, exploitation and habitat alteration by humans probably had more to do with their disappearance than any other single factor. Market-style overhunting definitely seems to have caused the disappearance of the elk and American bison. Predator control activities probably had much to do with the extirpation of the gray wolf and the jaguar. The red wolf disappeared as a result of predator control efforts and the genetic effect of interbreeding with coyotes (*Canis latrans*). The black-footed ferret disappeared primarily as a result of destruction of black-tailed prairie dog (*Cynomys ludovicianus*) towns, which removed most of the ferret's natural food supply. The big factors in the decline of the bighorn sheep were unregulated hunting, competition with domestic sheep, and the use of net-wire fences that prevented

Figure 152. Texas red wolf (*Canis rufus*), at the National Zoo, 1939. Courtesy National Archives, 22-WB-73-B57021.

the sheep from wandering about from one mountain range to another. The margay and grizzly bear were probably only marginal in Texas and never represented by established breeding populations, and the jaguarundi was overwhelmed by habitat loss and fragmentation. The following accounts summarize the status of the native populations of mammals that became extinct in the state at the end of the nineteenth and early part of the twentieth centuries.

Red and Gray Wolves, *Canis rufus* and *Canis lupus*

The situation with wolves and coyotes (genus *Canis*) is especially interesting. At the end of the nineteenth century, gray wolves (*C. lupus*) occupied the western part and red wolves (*C. rufus*) (Fig. 152) the central and eastern parts of the state. Coyotes (*C. latrans*) occurred statewide but they were less abundant than today (McCarley, 1962). With the onset of massive land clearing and predator control, coyotes increased in numbers and began to hybridize with red wolves. Wolves began to decline statewide. By the 1940s, gray wolves had been all but eliminated in Texas, and when red wolf numbers began to decline as a result of predator control activities, coyotes moved in from the west to fill the vacant predatory niche. The two canines, being closely related and genetically similar, began to interbreed, and the remnant red wolf populations gradually declined. In 1967, the USFWS listed red wolves as endangered under the US Endangered Species

Protection Act due to their rapid population decline, and subsequently, red wolves were among the first species listed on the 1973 Endangered Species Act (ESA), the country's landmark environmental law (McCarley and Carley, 1979).

The last pure group of the Texas subspecies of red wolf (*C. r. rufus*), found along the Gulf Coast south of Houston, seems to have been genetically swamped by coyote genes by 1970 (Paradiso, 1965, 1968; Paradiso and Nowak, 1972). The eastern Texas subspecies (*C. r. gregoryi*) survived for a few more years in extreme southeast Texas and southern Louisiana but it subsequently disappeared in the wild (McCarley and Carley, 1979). With red wolves on the brink of extinction, recovery was initiated though trapping what were believed to be the last wild red wolves along the Gulf Coast of Louisiana and Texas in the 1970s. More than 240 canids were trapped from Louisiana and Texas between 1973 and 1977. Forty individuals were selected for captive breeding, of which 17 were deemed 100 percent red wolf. However, only 14 wolves successfully reproduced and became the founders from which all red wolves in the recovery program descended (Heppenheimer et al., 2018). Due to the successful captive breeding population, red wolves were restored to the landscape in North Carolina less than a decade after being declared extinct in the wild. This historic event represented the first attempt to reintroduce a wild-extinct species in the US and set a precedent for returning wild-extinct wildlife to the landscape (Heppenheimer et al., 2018). Texas was not included among the list of states scheduled to receive reintroduced red wolves. Because of the large density of people on the upper Texas coast, it was assumed that reintroductions of red wolves could never be successful in that region.

After the early introductions, the North Carolina experimental population of red wolves was reduced by the USFWS in response to negative political pressure from the North Carolina Wildlife Resource Commission and a minority of private landowners. Further, gunshot-related mortalities increased the probability that wolf packs deteriorated before the breeding season, thereby facilitating the establishment of coyote-wolf breeding pairs. Consequently, the North Carolina population has fewer than 40 surviving members and red wolves are once again on the brink of extinction in the wild. Recently, a federal judge in North Carolina issued a permanent injunction against the US Fish and Wildlife Service's shoot-to-kill authorization, noting that the agency violated a rule passed by Congress to resurrect, protect, and conserve the species (Fears, 2018). The judge's ruling also noted that wildlife are not the property of landowners but belong to the public and are managed by state and federal governments for the public good.

There are still inquiries from Texas citizens who claim to be in possession of or to have seen a red wolf, but to date none of the reports or specimens has proven to be a red wolf. However, the possibility remains that canid individuals with substantial red wolf ancestry have naturally persisted in isolated areas of the Gulf Coast. Recent images of Galveston Island canids piqued the interest of local naturalists, and two genetic samples were taken from road-killed individuals. After researchers extracted and processed the DNA, in the form of so-called ghost alleles, they compared the samples to each of the legally recognized wild species of the genus *Canis* that occur in North America. Their results revealed that the Galveston Island animals were more similar to captive red wolves than typical southeastern coyotes (Heppenheimer et al., 2018).

The most likely scenario to explain the presence of red wolf genes in contemporary coyotes is through past hybridization between the two species. It is likely that as red wolf populations declined, the more numerous coyotes became an alternative choice for mates. As pure red wolves declined in abundance and eventually went extinct in the wild, a proportion of their alleles remained in the local coyote population. Although the frequency of red wolf alleles continues to decline over time, their imprint may be seen for many generations. This consequence is similar to that proposed for hybridization between modern humans and Neanderthals, where it has been suggested that remnant Neanderthal alleles comprise one to two percent of the human population (Prüfer et al., 2014). In addition to sharing genes unique to the captive red wolf populations, the Galveston animals carried a unique genetic variation not found in any of the known canines of North America. The findings from this study are not totally surprising given Galveston Island's location and isolation from the mainland, and it could mean that Texas may, after all, be an appropriate location for future red wolf reintroduction efforts (Fuller-Wright, 2018).

Associated with the vast herds of bison that once roamed the plains of western Texas were considerable numbers of gray or lobo wolves (*Canis lupus*) that preyed upon them. This predator-prey relationship between wolves and bison had existed more or less in balance for eons until the arrival of the white man who destroyed the bison and populated the ranges with cattle. Deprived of their natural prey, the lobos turned their attention to livestock and thereby incurred the wrath of the ranchers. The long-drawn-out battle between ranchers and lobo wolves was more or less a draw until the advent of World War I, when the demand for increased cattle production to supply meat for our armed forces led the federal government to back the ranchers with "government trappers." Ironically, following the resignation of C. Hart Merriam as director, it was the US Bureau of Biological Survey that employed and supported the federal trappers (Fig. 153). These actions quickly led to the extermination of gray wolves in Texas and surrounding states, although they still occurred in northern Mexico in small numbers.

The last gray wolves known to have been taken in Texas were two killed in December of 1970 in Brewster County (Scudday, 1972; Schmidly and Bradley, 2016). It is possible that a few individuals still cross over into the Trans-Pecos region from Mexico, but the species is endangered there, too. Gray wolves have been successfully reintroduced into Yellowstone National Park, southern Arizona, and southwestern New Mexico, as well as Minnesota, Idaho, and Washington in recent years. There was some discussion of reintroducing them to Big Bend National Park, as well, but any action on that proposal is years away and is likely to be strongly questioned by representatives of the livestock industry. In late 2018, House Republicans passed legislation to remove gray wolves in the lower 48 states from the list of species shielded by the Endangered Species Act, but the measure did not make it through the Senate. The removal of the act's federal protection would

Figure 153. Wolves (*Canis lupus*) and coyotes (*Canis latrans*) killed by government agent on C. T. Mitchell Ranch, near Marfa, Presidio County, 1929. Courtesy National Archives, 22-WB-50-B4102M.

leave laws regulating the killing of wolves up to the states. It would lift restrictions on logging, grazing, and construction activities in wolf habitats that were previously prohibited by the act or required consultation with the USFWS.

Grizzly Bear, *Ursus arctos*

The grizzly or brown bear (*Ursus arctos*) probably occurred sparingly in western Texas from the Panhandle to the Trans-Pecos up until the days of early exploration of the region (Jones, 1993). However, as described in Chapter 1, only one specimen from the Davis Mountains has ever been recorded in Texas, an interpretation that recently has been reaffirmed by Stangl et al. (2014). The grizzly bear is regarded as threatened throughout its former range in the continental United States and in Mexico.

Black-footed ferret, *Mustela nigripes*

The last confirmed Texas records of black-footed ferrets (*Mustela nigripes*) (Fig. 154) were from Dallam County (1953) and Bailey County (1963; Campbell, 1995). Its historic range presumably included roughly the northwestern third of the state in the Panhandle, much of the Trans-Pecos, and a considerable part of the Rolling Plains. This distribution corresponded with that of the ferret's principal prey, the black-tailed prairie dog. The decline of the black-footed ferret has been attributed to conversion of prairie habitat

Figure 154. Black-footed ferret (*Mustela nigripes*) at Crow Indian Reservation, Montana, 1927. Courtesy National Archives, 22-WB-59-B3480M.

to agriculture and systematic efforts over the past one hundred years to eradicate prairie dogs. Thus, ferrets were unintentional victims of prairie dog eradication efforts, and their numbers dwindled as prairie dog towns were plowed under and prairie dogs were poisoned (Thorne and Williams, 1988).

In 1979 the species was thought to be extinct in the wild throughout its range. In 1984, however, one surviving colony, numbering about 130 individuals, was discovered at Meeteetsee, Wyoming. That colony subsequently suffered an epidemic of canine distemper. In 1986 the remaining 18 ferrets known to have survived were captured and put into a captive breeding program. In the fall of 1991, the first group of captive-born ferrets was released in the Shirley Basin area of Wyoming. An additional 83 captive-born ferrets were released in Shirley Basin during the fall of 1992. These initial attempts at reintroduction met with limited success, but in the last few decades much has been learned about the successful breeding of ferrets as well as their ecology, and reintroduction efforts are now meeting with more success. Approximately 150-220 ferrets are released to the wild each year.

Today, there are about 350 ferrets living in the wild (although numbers have been as high as 1,000) at twenty-nine sites across eight US states, Canada, and Mexico. About 3,000 black-footed ferrets are necessary to fully recover the species. Although the captive breeding and reintroduction capabilities continue to improve, habitat availability and the lack of stable prairie dog towns remain a limiting factor in the effort to restore this species to the North American landscape (Schmidly and Bradley, 2016). It is now believed that most existing prairie dog colonies are either too small or isolated from one another to support black-footed ferrets. However, larger prairie dog colonies in the northern Panhandle of Texas could potentially provide suitable habitat to support reintroduced populations of these endangered mammals (Campbell, 1995), and discussions are underway to attempt a reintroduction in Texas, most likely on the Rita Blanca National Grasslands. Similarly, a few years ago, a private ranch in Hockley County was known to support a colony of about 10,000 prairie dogs—thus, some prairie dog colonies of sufficient size and abundance might still exist in the Panhandle as potential reintroduction sites for the black-footed ferret in Texas.

Jaguar, *Panthera onca*

The distribution of the jaguar once extended well into central Texas, including much of the Edwards Plateau and Trans-Pecos, as well as along the southern and southeastern parts of the state. There are many records and sightings that date from the late 1800s and early 1900s (Figs. 155, 156), and this large cat actually was regarded as common in some areas; however, the last documented record from the state was in the early 1950s (Tewes, 1990). This species is now regarded by the IUCN as "near threatened," with populations decreasing due primarily to habitat fragmentation throughout its relatively broad range in Central and South America. An assessment of the jaguar's distribution in the 1980s placed the northernmost limit of established populations in central Mexico (Swank and Teer, 1989), but due to their protected status, their current range is known to extend northward to the state of Sonora, Mexico, south of Arizona.

Although a few animals stray into the US, as evidenced by several confirmed sightings in Arizona since 2000, there is no evidence that habitat in the southwestern US is crucial for survival of the species (Rabinowitz, 1999). Nevertheless, a group of conservation scientists are advocating for the reintroduction of jaguar to their native habitats in the southwestern US (Sanderson et al., 2021). They argue that the species once occupied this territory and was extirpated by human actions that should no longer pose a threat and that the proposed recovery area provides suitable ecological conditions. Such a proposal is not likely to receive a warm reception in Texas, however, where most of the land is privately held.

Margay, *Leopardus wiedii glauculus*

It is interesting that Bailey made no mention of the possible occurrence of the margay in Texas. This species is known on the basis of a single specimen taken near Eagle Pass in the 1850s. However, that specimen originally was recorded in Spencer Fullerton Baird's 1859 account of *North American Mammals* as an ocelot. It was not until 1914 that Ned Hollister, who was one of the federal agents who assisted Bailey, correctly identified the specimen as a margay (Hollister, 1914). Some people have speculated the Eagle Pass margay was actually a pet brought to the border and sold. Whether this is true or not is difficult to

Figure 155. A jaguar (*Panthera onca*) taken near Goldthwaite, Mills County, 1903. (See Chapter 2, pages 165-166 of *Biological Survey of Texas* for an account of this animal's collection.) Public domain photo.

Figure 156. A jaguar (*Panthera onca*) killed in South Texas near San Benito, Cameron County, 1946. Courtesy Laguna Atascosa Refuge, Rio Hondo, Texas.

Figure 157. Jaguarundi, found dead on a highway near Harlingen, Cameron County. Courtesy Laguna Atascosa National Wildlife Refuge, Los Fresnos, Texas.

determine, but it is certain that margays no longer occur in Texas. This small spotted cat is regarded by IUCN as "near threatened" throughout the remainder of its range from Mexico to central South America.

Jaguarundi, *Puma yagouaroundi*

This small, unspotted cat once occurred in small numbers in the dense, thorny thickets of extreme southern Texas in Cameron, Hidalgo, Starr, and Willacy counties, but it is now considered to be extinct (Fig. 157). The last documented record in the United States was in 1986 when a road-killed individual was salvaged two miles (3.2 km) south of Brownsville in Cameron County. Jaguarundis were thought to be represented in the lower Rio Grande Valley by no more than 15 individuals in 1990 (Tewes, 1990). With numbers that low, it was doubtful to survive, and most experts think it is now extinct (Michael Tewes, personal communication). Occasionally, claims of sightings are still reported in Texas, but no professional biologist has encountered one and no specimens are available since the 1986 record. Additionally, although many people have reported jaguarundis in central Texas (and other states), these felines have never been confirmed north of extreme South Texas. As with the jaguar and ocelot, predator control and habitat destruction took their toll on this species (Schmidly and Bradley, 2016).

Figure 158. A pile of more than 40,000 bison (*Bos bison*) hides awaiting shipment from Dodge City, Kansas, 1874. Bison, once numerous over the plains and prairies of Texas, were eliminated by 1889. Courtesy National Archives, 22-WB-50-B4102M.

Bison, *Bos bison*

Millions of bison once thundered across the prairies and plains of the American West. Just consider what one observer saw on an 1871 cattle drive through the Texas Panhandle:

> On a plain about halfway between the Red Fork and the Salt Fork [Brazos River] we had to stop our herds until the buffalo passed. Buffalo, horses, deer, antelope, wolves and some cattle were all mixed together and it took several hours for them to pass ... so that we could proceed with our journey. I think there were more buffalo in that herd than I ever saw of any living thing (Robbins, 2005).

For all practical purposes, bison were gone before Bailey and the federal agents roamed the state. When protection of the bison was under consideration by the Texas Legislature after the Civil War, General Phil Sheridan opposed it. He won his point by convincing the legislators that the sooner the bison were eliminated, the sooner the Native Americans who depended upon them for food and clothing would be starved into submission. Sure enough, before 1900, these intentions became reality for both the bison and the Native Americans, as the bison were killed and nearly all the Native Americans were killed or forcibly removed from their tribal lands. The largest slaughters of bison took place in the 1870s (Fig. 158). Fueled by the European demand for robes and industrial leather, the slaughter of the Texas bison herd began in earnest in the spring of 1874. By late 1878, buffalo hunters had killed more than 3.5 million of these animals across the Texas plains (Robbins, 2005). From Fort Griffin (Shakelford County) in the winter of 1877-1878 more than 1,500 outfits killed in excess of 100,000 animals in the months of December and January alone (Davis, 1961). Hundreds of thousands of carcasses were strewn over the landscape. There is even a record of one legendary hunter, J. Wright Mooar, who claimed to have killed more than 22,000 head of buffalo, killing a white buffalo (albino) near Snyder in Scurry County (Snyder was known as "Hide Town" in those days) in 1876 (Anderson, 2000). The odds of a true albino calf being born in the 1800s was about 1 in 10 million. The last verified report of wild bison was from the northwestern part of the Panhandle (Dallam County) in 1889 (Jones, J. K., Jr., et al., 1988b).

Though no longer wide-ranging, bison herds prevail on many national and state sanctuaries in the West. In Texas, bison are kept in captivity on many farms and ranches, and in 1996, a visionary plan was set in motion to rebuild a herd of southern plains bison and reestablish them in the wild. This plan was made possible more than 100 years earlier by the foresight of legendary rancher Charles Goodnight, owner of the JA Ranch in the Palo Duro Canyon area. Mary Ann Goodnight, the wife of Charles, realized that the slaughter was sending the buffalo to the brink of extinction. She urged her husband to save a few of the animals, and he captured five wild bison calves to start a protected herd in 1876. There, for more than a century, the animals and their offspring ranged freely over the JA Ranch and surrounding ranches of West Texas. These animals were the last genetically pure examples of the original stock of the southern plains bison and are unique from all other bison in the world today (Robbins, 2005; Roe, 2011).

In 1994, TPWD became concerned about the future of the Goodnight herd, as its size had declined to fewer than 30 individuals. In 1996 it was determined by genetic testing that the animals were indeed unique, and efforts were initiated to preserve the herd, even though there was a relatively small amount of cattle genetic introgression in a few individuals of the herd. The herd was donated to the TPWD in 1997 by the JA Ranch (Robbins, 2005). The animals were transferred to breeding facilities at Caprock Canyons State Park in the winter of 1997–1998, and a Texas State Bison Herd was established. Genetic testing had revealed that the Goodnight herd had dangerously low genetic variation and heterozygosity levels compared to the federal herds and most of the state and private herds. This was a result of the limited number of original founders, multiple population bottlenecks over the past 120 years, and chronically small population size, coupled with genetic drift and inbreeding. Population viability analysis revealed there was a 99 percent chance of extinction of the herd within the next 41 years (Halbert et al., 2004). The continued existence of this historically important bison population appeared doubtful without the introduction of new genetic variation from another plains bison herd. So, in 2003 media mogul and conservationist Ted Turner donated three young bulls from one of his New Mexico herds, and the bulls were bred to several females from the Goodnight herd to inject some genetic variability without altering the unique DNA of the herd (Robbins, 2005; Roe, 2011). The short-term goal of TPWD was to preserve these invaluable animals and allow the population to grow. The long-range goal is to restore the southern plains bison on a 100,000-acre refuge within its historic Panhandle range as part of a native prairie ecosystem. The size of the herd was 70-80 animals in 2011 (Roe, 2011) and today it is estimated to contain 156 animals (Beard, 2018).

Bighorn sheep, *Ovis canadensis mexicana*

Prior to the mid-1800s, bighorn sheep were abundant and widely distributed in the western United States, including Texas (O'Brien et al., 2014). During this time frame, desert bighorns, subspecies *O. v. mexicana*, occupied the rugged, mountainous terrain of West Texas (Cook, 1994). Approximately 1,500 bighorn were estimated to inhabit the Trans-Pecos ecoregion in the mid-1880s, but they experienced a consistent decline thereafter caused by unregulated hunting, disease, interspecific competition with domestic sheep, and predation.

Bighorn sheep were still wide-ranging over many parts of the Trans-Pecos during the time of the biological survey. However, despite a ban against hunting them in 1903, they steadily declined in numbers and distribution. Bailey (1905) estimated the number of sheep in Texas had declined to 500 individuals by the beginning of the twentieth century. He attributed the primary cause to the introductions of large numbers of domestic sheep and goats. The mountain sheep competed directly with the domestic sheep and goats for food, were subjected to heavy losses from domestic sheep diseases, and were limited in their movements by net wire fences. By 1945, the total estimated population was less than one hundred and all of them were concentrated in the Beach, Baylor, and Sierra mountains north of Van Horn (Davis, 1961). The last native bighorn sheep were seen in the Sierra Diablo Mountains in 1959, when the total population was estimated at fourteen.

Restoration efforts by the Texas Parks and Wildlife Department began in the mid-1950s at the Black Gap Wildlife Management Area (Cook, 1994), and although early efforts were of limited success due to disease and predation, recent efforts have been more successful and bighorn sheep now occur in several mountainous areas of the Trans-Pecos region (Locke et al., 2005). The reintroduced animals, however, were not of the native desert bighorn but came primarily from populations of two other subspecies (*O. c. canadensis* and *O. c. nelsoni*). Since the 1980s, wild sheep of these two subspecies have been released in the Sierra Diablo–Baylor–Beach Mountains complex, the Sierra Vieja Mountains, the Van Horn Mountains, Elephant Mountain Wildlife Management Area, Big Bend Ranch State Park, and Big Bend National Park where small, wild populations have been established. Currently, there are an estimated 1,500 individuals in ten mountain ranges in the Trans-Pecos, and it appears that the number of individuals and their distribution are expanding (Schmidly and Bradley, 2016).

Success in maintaining large, self-sustaining desert bighorn herds will depend upon large blocks of available habitat with sufficient escape terrain. Bighorn sheep select habitat based on factors such as proximity of steep-sloped escape terrain, forage availability, and horizontal visibility. They typically avoid areas with dense vegetation, which hampers horizontal visibility and leaves them vulnerable to ambush predators (O'Brien et al., 2014). Limiting contact with domestic livestock and exotic sheep species (e.g., *Ammotragus lervia*), to prevent disease transmission and interspecific competition, is absolutely essential, as is predator management to allow populations time to become established, and availability of freestanding water to sustain the animals (Locke et al., 2005; Dolan, 2006; Drew et al., 2014; O'Brien et al., 2014). Time and patience also are required. Whereas individual bighorns from historical populations were largely migratory, it has been documented that translocated individuals initially are not (Jesmer et al., 2018). After multiple decades, however, translocated populations seem to gain knowledge about local foraging conditions and increase their propensity to migrate.

Wapiti or elk, *Cervus canadensis*

The wapiti or elk once ranged over parts of northern and western Texas, although there seem to be no actual specimens to document its modern occurrence there (Jones, 1993), other than prehistorical archeological evidence (Pfau, 1994; Shaffer et al., 1995). These elk originally were assigned to the species *Cervus merriami*, Merriam's elk, which was later reduced to a subspecies of the more wide-ranging American elk, *Cervus canadensis* (Fig. 159). Merriam's elk resembled the American elk, but it was paler and more reddish in color, with a more massive skull and more erect antlers. It once roamed in the Guadalupe Mountains, and perhaps other forested areas in West Texas, but it appears to have been extirpated there before 1900. The elk currently occupying the Guadalupe Mountains are descendants of 44 animals of another subspecies, *C. e. nelsoni*, brought in from the Black Hills of South Dakota and released in McKittrick Canyon in 1928 by Judge J. C. Hunter. These elk increased at a maximum rate of 10 percent a year to a peak population size of approximately 350 in the mid-1960s. A severe reduction in population size commenced thereafter with only 108 individuals reported in a 1976-1978 census and a further decrease to 58 in 1983. The reason for this population decline is unclear (McAlpine, 1990) but a few of the elk do remain in the park today (John Karges, personal communication).

Presently, free-ranging elk exist in West Texas in five small herds in the Guadalupe Mountains (Culberson County), Glass Mountains (Brewster County), Wylie Mountains (Culberson County), Davis Mountains (Jeff Davis County), and Eagle Mountains (Hudspeth County), and recently they have been observed in the Texas Panhandle (Dallam County). The total population of elk in the Trans-Pecos was estimated in 1995 at 330, but today the herds are thought to contain as many as 3,500 individuals, with many others kept in deer-proof pastures on scattered ranches across the state.

Figure 159. Old antlers of Merriam's elk (*Cervus canadensis*), photographed at a ranch in the Sacramento Mountains, New Mexico, 1903. Courtesy National Archives, 22-WB-57-B5519.

There is a move to have elk listed as a game species so that the populations and harvest limits can be managed more effectively. At this time it is unclear whether the Trans-Pecos elk are a result of an expansion of the Guadalupe herd (subspecies *canadensis*), if they represent natural immigrants from the Rocky Mountain herd in New Mexico (subspecies *nelsoni*), or if they are progeny of escaped pen-raised individuals (Dunn et al., 2017). The elk herd from the Dalhart area in Dallam County could be either *canadensis* or *nelsoni*.

Subspecies now extinct in Texas

In addition to the above ten extinct species, at least four subspecies probably became extinct during the twentieth century. These included the Presidio mole (*Scalopus aquaticus texanus*), the Big Thicket hog-nosed skunk (*Conepatus leuconotus telmalestes*), the Louisiana vole (*Microtus ochrogaster ludovicianus*), and Bailey's pocket gopher (*Thomomys bottae baileyi*).

William Lloyd, one of the federal agents, obtained the only specimen of the Presidio mole, *Scalopus aquaticus texanus*, in 1887 from Presidio County, which at that time consisted of the present West Texas counties of Presidio, Jeff Davis, and Brewster. No other specimen of mole has been observed, reported, or documented anywhere in West Texas. Rollin Baker (1951) trapped a new species of mole, *S. montanus* (now considered to be a subspecies of *S. aquaticus*; Yates and Schmidly, 1977b) in the Sierra del Carmen, Coahuila,

Mexico, just south of the Big Bend region. He hypothesized that *S. a. texanus* could have been trapped at high elevations in old Presidio County as well as near the river. However, given the extensive collecting in this region during the twentieth century without uncovering the Presidio mole, it seems likely that this taxon is now extinct.

In his 1905 publication, Vernon Bailey described the Big Thicket population of the hog-nosed skunk as a distinct subspecies, *Conepatus mesoleucus telmalestes*, noting it was the most common skunk that he and the federal agents trapped in that area. No other specimens of this skunk were obtained or reported until 5 March 1960, when Gerald Raun and B. J. Wilks (1961) picked up a specimen dead on the road in Waller County. William B. Davis (1945) had assumed that the Big Thicket population had been wiped out, but Howard McCarley (1959) believed that *Conepatus* was extant although rare in some sections of the area. From 1977 to 1982, five years of continuous fieldwork in Big Thicket National Preserve yielded no evidence of this skunk anywhere in the region (Schmidly, 1983). Since it has been at least six decades since anyone has made a confirmed sighting of the Big Thicket hog-nosed skunk, we presume that it is now extinct. The reason for its disappearance remains a complete mystery.

The Louisiana vole (*Microtus ochrogaster ludovicianus*) is known on the basis of a single specimen secured by Ned Hollister in the coastal prairie region of Sour Lake on 16 July 1902. Numerous attempts during the twentieth century to collect this species by professional mammalogists have been unsuccessful, and it is now thought to be extinct throughout its range in both Texas and Louisiana.

C. Hart Merriam described *Thomomys baileyi* in 1901 on the basis of specimens collected by Bailey at Sierra Blanca in Hudspeth County (Merriam, 1901). Subsequently, E. A. Goldman (1938), another biological survey scientist, described a second race of *baileyi*, *T. b. spatiosus*, from Alpine and Paisano in Brewster County. In 1966, Syd Anderson synonymized *T. baileyi* with the more wide-ranging species, *T. bottae* (Anderson, 1966). Subsequent attempts by mammalogists, including both DJS and RDB, to collect gophers that fit the description of *baileyi* have failed. It now appears that *baileyi* is extinct, having been replaced ecologically by another pocket gopher, *Cratogeomys castanops* (see discussion in Chapter 6).

Declines in Geographic Distribution and Population Abundance

A notable number of Texas mammals, including species of all sizes and life history traits, have undergone drastic range reductions and today occupy a mere scant portion of their former range. Examples of species that are now severely reduced in distribution or abundance include the pronghorn (*Antilocapra americana*), black-tailed prairie dog (*Cynomys ludovicianus*), common muskrat (*Ondatra zibethicus*) and, to some extent, the mountain lion (*Puma concolor*)—although recently, the distribution of the latter species has been expanding in the state.

Pronghorn, *Antilocapra americana*

Pronghorn once occurred over the western two-thirds of Texas, but during the late nineteenth and early twentieth centuries their range decreased significantly because of increasing human populations and land development, specifically agriculture (Leftwich, 1977). The great herds that once roamed the Trans-Pecos and Panhandle regions were reduced to a mere handful by the time Bailey and the federal agents completed their work in Texas. Bailey counted only 32 pronghorns in a 95-mile (153-km) railroad journey from Canyon, Texas, to Portales, New Mexico. In South Texas, he encountered a few pronghorn west of Alice, and small bands roamed near Cotulla and Rock Springs. But pronghorn had disappeared from the vicinities of Alpine and Marfa and from the outwash plains of the Davis Mountains. A 1924 survey of pronghorns showed a statewide population of only 2,407 (Nelson, 1925).

The long-term decline in pronghorn populations appears to be associated with overgrazing of grasslands by domestic livestock, unregulated hunting, extensive cultivation of prairie habitat, and drought. Studies have shown that pronghorn populations and fawn production are affected by both long-term and short-term drought (Simpson et al., 2007). Recently, worm infestations (Barber's pole worms and blood worms) have been documented to cause mortality in malnourished animals (Lightfoot, 2010).

Pronghorn habitat appears to be in limited supply because of anthropogenic and environmental influences. Optimum habitat is open-country grassland or shrubland communities, with varying successional stages, mainly caused by fire; they do not do well in woodlands or agricultural areas (Lightfoot, 2010; Duncan et al., 2016). However, the Panhandle population appears to be benefitting from row crop production and this supplemental food source may be the key to long-term survivability in the region. The pronghorn is now restricted to isolated patches of suitable habitat from the Panhandle to the Trans-Pecos (see Fig. 39a–b, Chapter 2, p. 151). It is a desirable game species, but despite extensive management efforts, including restocking programs begun in the 1940s and continuing even today, pronghorn numbers have suffered multiple periods of decline over the years. This has been especially true of the Trans-Pecos population, which declined from 17,226 animals in 1987 to 2,751 in 2010. However, recent data suggest this population has expanded to approximately 6,500 individuals, largely thanks to improved success in translocating animals to the Trans-Pecos from the larger and more stable populations in the Panhandle since 2011 (Woodward, 2020). As pronghorn numbers declined, TPWD also reduced the number of hunting permits. Hopefully, these efforts will be successful and populations will continue to recover and stabilize.

Black-tailed prairie dog, *Cynomys ludovicianus*

No Texas mammal suffered more from population decline during the twentieth century than did the black-tailed prairie dog. This highly gregarious rodent creates colonies or "towns" that can range anywhere from one to 1,000 acres (0.4-400 ha). Once widespread throughout the Great Plains states, prairie dog colonies are now in decline. It has

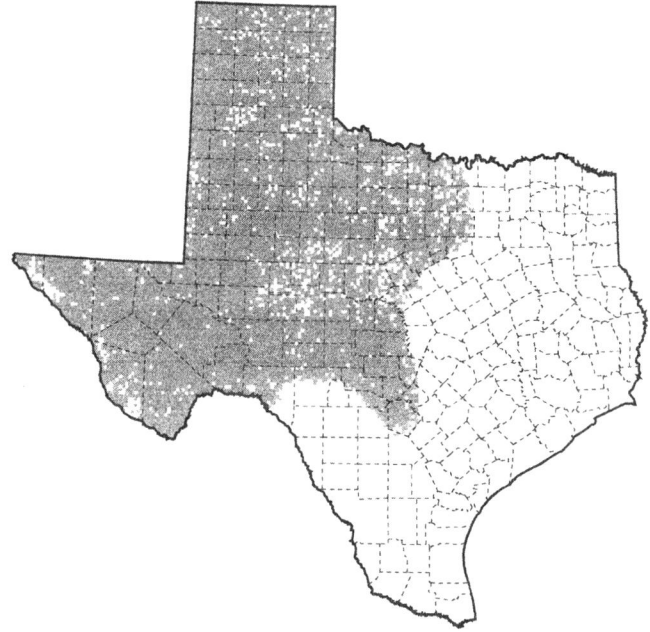

Figure 160. The predicted distribution of the black-tailed prairie dog (*Cynomys ludovicianus*) in Texas. This distribution map is based on the known historic range and the available appropriate habitat for the species.

Figure 161. Poisoning crew and dead prairie dogs (*Cynomys ludovicianus*), Means Brothers Ranch, Jeff Davis County, 1936. Courtesy National Archives, 22-WB-65-B43487.

been estimated that in the early 1900s, prairie dog colonies covered 100 to 250 million acres (40-101 million ha) throughout North America. The largest expanse of black-tailed prairie dog colonies occurred in Texas where their distribution covered one-third of the entire state, or approximately 90,000 square miles (233,100 km²) (Fig. 160). In the 1890s, Vernon Bailey estimated that more than 800 million prairie dogs inhabited the western part of the state and that they were consuming as much range vegetation as would three million cattle. One 25,000 square mile (65 km²) area just east of the Staked Plains from San Angelo to Clarendon was described as one continuous "dog" town with approximately 400 million inhabitants! Such concentrations were a heavy drain on range vegetation, and ranchers enlisted the federal government to combat them. Using mainly strychnine-treated grain, the ranchers, along with government rodent control specialists (employed by the US Biological Survey), poisoned millions of prairie dogs (Fig. 161). By 1960, the once overwhelming populations had been reduced to scattered, small colonies (Davis, 1961).

Sylvatic plague devastated many of the remaining small colonies and contributed significantly to the decline of black-tailed prairie dog populations across much of their range. By the end of the twentieth century, it was estimated that 98 percent of the population had been lost, and that only 300,000 prairie dogs remained in Texas (Long, 1998). These gloomy predictions may have overestimated the size of the population because most surviving colonies were fragmented and covered less than 50 acres (20 ha). Records indicate that prairie dog habitat declined 61 percent just in the last two decades of the twentieth century. For that reason, the National Wildlife Federation petitioned the USFWS in 1998 to list the black-tailed prairie dog as a threatened species with all of the rights and privileges thereof, but the petition was denied.

In response to the petition to list the black-tailed prairie dog as threatened under the ESA, an inventory of the species was undertaken in Texas, including historical and current distributions (Singhurst et al., 2010). The results produced historical records from 114 Texas counties; remote sensing and roadside ground-truthing confirmed current colonies in 73 counties. An estimate of 3,180 colonies of prairie dogs in Texas occupying 146,500 acres (59,300 ha) was developed. Current populations were concentrated on the Great Plains Shortgrass Prairies ecosystem, but colonies also were found on four other ecosystems and three anthropogenic systems. The inventory revealed that the black-tailed prairie dog population had receded from the southern and eastern boundaries of the historical range in Texas (Singhurst et al., 2010).

A listing of this species as threatened or endangered at the beginning of the twentieth century would have been considered ludicrous. The *Austin Daily Statesman*, on 8 August 1899, published the following quote about prairie dogs:

> During the recent session of the legislature, the western members tried to secure an appropriations bill to kill the millions of prairie dogs that are infesting and laying waste all over the Panhandle. They failed, so now W. L. Grogan of Sweetwater was

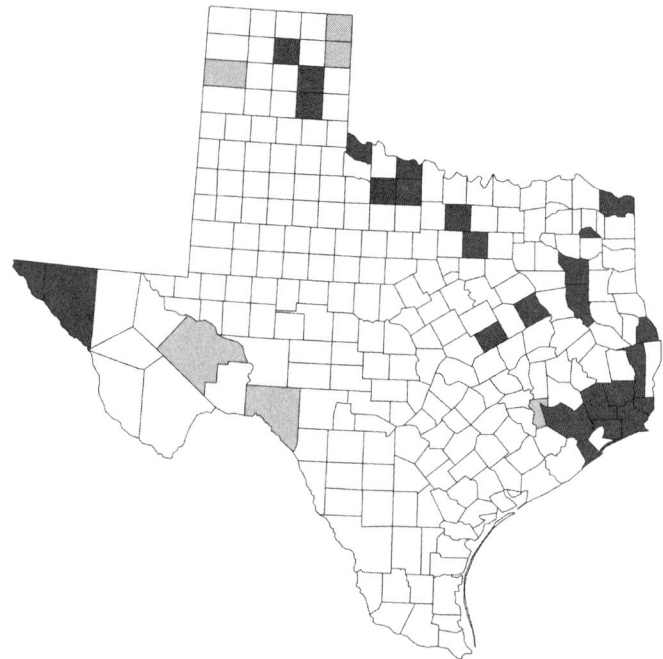

Figure 162. The historic distribution of the muskrat (*Ondatra zibethicus*) in Texas. The light shading represents counties where Bailey (1905) recorded the species, and the dark shading indicates current records for the species.

in Austin yesterday and stated that the people out on the plains have rigged up big mouse traps with long projecting noses which they sink in the holes of the prairie dogs leaving the trap above ground. The trap holds up to 10 prairie dogs. Thousands are being caught. The war waged on prairie dogs is proving effective.

Common Muskrat, *Ondatra zibethicus*

Another interesting case is that of the common muskrat. Muskrats are principally marsh inhabitants and live near creeks, or along the edges of rivers, lakes, drainage ditches, and canals that support small populations in places where requisite food and shelter are available (Fig. 162). In some regions of the state they appear to have declined or even disappeared, whereas in other regions they have invaded and increased in abundance. Remarkably, neither Bailey nor the other field agents collected any muskrat along the upper Texas coast that is the region of the species' greatest abundance today. The only mention of muskrat is an account from Waller County by J. H. Gaut, who reported them from the Brazos River and its tributaries. Previously, John James Audubon had reported them on Galveston Island in 1824 (Geiser, 1930). Del Weniger (1997) noted that none of the early explorers reported muskrats from the Texas coast, and they were certainly not common there during the biological survey. However, by 1936 muskrats produced 54 percent of the fur-trapping income for Jefferson, Chambers, and Orange counties (Lay and O'Neil, 1942). Bailey and his co-workers collected extensively in Jefferson County but never reported trapping or seeing muskrats. The most likely explanation for these observations is that the 1900 and 1915 hurricanes, which covered much of the prairies and marshes of the upper Texas coast, decimated muskrat populations and it took them almost three decades to fully recover.

Today, muskrats are reasonably common along the upper Texas coast but are rare along the tributaries of the Canadian River in the Panhandle and along the Rio Grande near El Paso. Bailey and the other field agents found them to be abundant at the beginning of the century in the Canadian River drainage. Frank Blair (1954b) reported a dense population in the tule marshes of Moore and Bugby creeks in Hutchinson County. However, J. Knox Jones Jr., Clyde Jones, and associates, while conducting an extensive survey of the mammals of the northern Texas Panhandle during the 1980s, did not find any evidence of muskrats at the sites where they had been previously reported (Jones, J. K., Jr., et al., 1988b). According to Clyde Jones (personal communication), most of the creeks were dry and the tule marshes were greatly reduced in scope. Fred Stangl and his associates reported finding the bones of a muskrat in the regurgitated pellets of a great-horned owl near a stream in the vicinity of Clarendon in Donley County (Stangl et al., 1989), suggesting a few muskrats still remained in the Panhandle. Interestingly, the late T. Boone Pickens, who owned the Mesa Verde Ranch along the Canadian River in Roberts County, began in the 1990s to restore natural surface water all over his property and a large population of muskrats soon occupied the habitat.

The subspecies *O. z. ripensis*, the Pecos River muskrat, occurred in numerous areas of the Trans-Pecos, along both the Pecos River and the Rio Grande and at springs such as those at Ft. Stockton and Balmorhea during the time of the biological survey. As human population and irrigation wells increased, many of the springs of the Pecos River drainage were destroyed. As the springs dried up, the populations of muskrats that they supported died out. The last records of muskrats from the Pecos drainage were from 1980 when six specimens were obtained from Balmorhea in Reeves County and one specimen from Orla in the same county but closer to the New Mexico–Texas border (Evans, 2015; Falcone et al., 2019). There are no contemporary records from the Big Bend region of the Rio Grande, and it has been suggested that modern reduction in the flow of the Rio Grande between El Paso and Presidio drastically modified suitable muskrat habitat and reduced muskrat populations along the Lower Rio Grande (Hafner et al., 1998). Competitive exclusion resulting from the presence of the highly invasive nutria (*Myocastor coypus*) apparently forced the muskrat from Lake Amistad in Val Verde County (Schmidly and Ditton, 1978).

Today the drainage ditches near El Paso represent the last stronghold of this subspecies in the Trans-Pecos. However, there appears to be minimal genetic differentiation between these populations (subspecies *O. z. ripensis*) and those farther up river near Albuquerque, New Mexico (subspecies *O. z. osoyooensis*). For this reason, there is less urgency than previously thought to manage the former taxon as a unique subspecies, especially given the genetic data aligning muskrats from the Pecos River with populations along the Rio Grande (Falcone et al., 2019).

The IUCN lists the common muskrat as a species of least concern. It does not appear on the federal or state lists of concerned species, although at one time *O. z. ripensis* was considered by the USFWS as a category 2 species for listing. The decline of permanent natural surface water, especially the drying up of freshwater springs as a result of irrigation, followed by the reduction of true marsh habitat, has caused their demise. The muskrat could be vulnerable to the rapid expansion and spread of the nutria. Further, the muskrat scored "vulnerable" to climate change on the Middle Rio Grande due to inherent riparian habitat degradation caused by drought (Friggens et al., 2013). This is a species that should be carefully monitored in the future.

Mountain lion, *Puma concolor*

Mountain lions, or "cougars" as they often are called, have the most widespread distribution of any terrestrial mammal in the Western Hemisphere, ranging from the southernmost tip of South America through Central and North America. However, their distribution in North America has been reduced over the last 100 years (Anderson et al., 2010; Logan and Sweanor, 2001), and in the US they are now limited to the western states, including Texas, with an isolated population in Florida (Harveson et al., 1999).

In Texas, populations were historically distributed throughout the state, but population size and geographic distribution declined over time, largely due to predator control and loss of habitat (Russ, 1997; Logan and Sweanor, 2001; Holbrook et al., 2012b). For example, livestock ranching was an important industry in Texas during the 1800s and mid-1900s and predator removal was widely practiced (Lehmann, 1969; Wade et al., 1984). Additionally, mountain lion habitat in central and southern Texas became increasingly fragmented during the past century due to agricultural practices, urbanization, and energy development (Holbrook et al., 2012b).

Mountain lion populations are not formally managed in Texas, receiving the designated status of a nongame or varmint species since 1970 (Harveson et al., 1997; Russ, 1997). Given that mountain lions are not important fur-bearing mammals (trapping efforts are predominantly for depredation control; Young et al., 2010) and that the reporting of harvest data is not required by law, it is difficult to estimate population levels. The most recent harvest data (Fig. 163) we could obtain was from 1998 (Texas Wildlife Services, US Department of Agriculture, Animal Plant Health Inspection Service) and indicated that levels of harvest have fluctuated over time. Ongoing research suggests that populations in western and southern Texas have been restricted by low survival (Harveson et al., 1997; Young et al., 2010) and low reproductive rates (Harveson et al., 1997; Pittman et al., 2003). Further, a preliminary genetic analysis found that mountain lions in southern Texas have low levels of genetic diversity and appear to be isolated from those in western Texas (Walker et al., 2000).

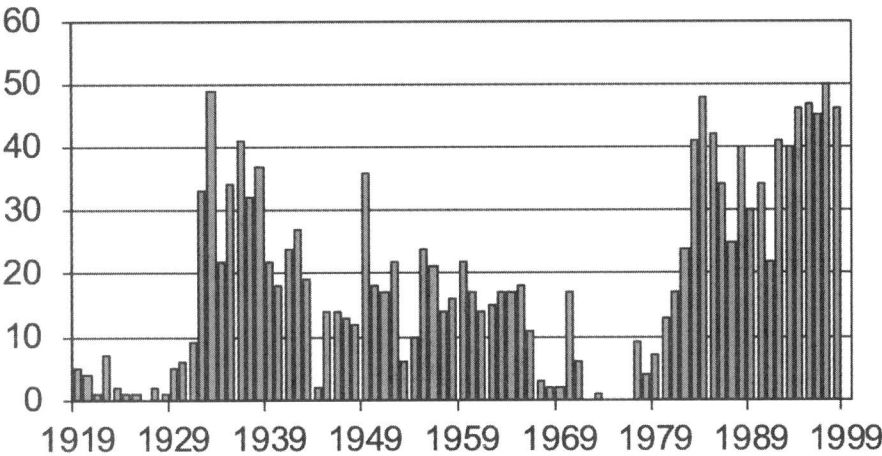

Figure 163. Mountain lion (*Puma concolor*) taken in Texas, 1919-1998, by the Texas Wildlife Damage Management Service.

With the slowing down of predator control efforts since about 1970, it appears that mountain lions are repopulating portions of their former range, including the Edwards Plateau, Panhandle, and Gulf Coast areas as well as the Big Thicket. Sightings of individuals have been reported from all of these regions in recent years (Schmidly and Bradley, 2016). Today, breeding populations are known to persist primarily in the Trans-Pecos and South Texas Plains ecoregions of western and southern Texas (Schmidly and Bradley, 2016). An increasing number of mountain lions have been reported in Big Bend National Park since the 1990s (Holtcamp, 2008), which likely supports the largest population of cougars in the state. A genetic study of mountain lions in Guadalupe Mountains National Park indicated there was a high number of transient animals and, perhaps, an unstable population in the park as a result of intense hunting pressure outside of the protected area (Gilad et al., 2011). Despite these recent gains, a radio-telemetry study of captured cougars in Big Bend National Park, Big Bend State Ranch, and Guadalupe Mountains National Park revealed that human-caused sources of mortality (e.g., shooting and trapping) were greater than natural sources of mortality, and that rates of survival were among the lowest in the US for all three areas (Young et al., 2010).

The *Dallas Morning News*, in a series of articles in late 2020 (2, 7, 10, 15, and 16 December), covered the fate of a mountain lion sighted and later killed in the greater Dallas Metroplex. The animal was first documented on a camera trail video near Rowlett, a town that straddles Dallas and Rockwall counties, on November 22. Supposedly, the same lion was sighted a few days later near Lipan in Hood County, and the sheriff there claimed that it had possibly maimed a man, although this allegation was later refuted by

TPWD experts (Scudder, 2021). The lion was then sighted again near Princeton, Collin County, which is about twenty miles from the first report near Rowlett. The male mountain lion was killed a few days later by a hunter in Hunt County. This example documents the fate of many mountain lions that wander outside of protected or wilderness areas and are subsequently shot by hunters or landowners.

Mountain lions have begun in recent years to receive some recognition for their ecological, aesthetic, and sporting value. However, current management practices are controversial and relate to whether or not the species should be listed as nongame with no regulation of take or listed as a game species with regulation of harvest within the state (Klepper, 2005). Maintaining connectivity among mountain lions in New Mexico, Texas, and perhaps Mexico will likely have a positive influence on regional persistence by sustaining large effective population sizes. But the change that would most benefit the conservation of mountain lions in Texas would be to classify the species as a game animal, as it is in New Mexico, with a limited harvesting season.

Range Expansions and Regional Faunal Changes

Since the beginning of the twentieth century, a number of mammals have expanded their ranges in Texas. Notable examples include the armadillo (*Dasypus novemcinctus*), the northern pygmy mouse (*Baiomys taylori*), the prairie vole (*Microtus ochrogaster*), and the porcupine (*Erethizon dorsatum*). At the time that Bailey and the federal agents roamed the state, all of these species had restricted distributions, whereas today they are much more wide-ranging and common. In addition, several species of bats appear to have increased their geographic range in the latter part of the twentieth and early part of the twenty-first centuries.

Armadillo, *Dasypus novemcinctus*

Bailey mapped the distributional limits of the armadillo as between the Colorado and Guadalupe rivers, with extralimital records from Colorado, Grimes, and Houston counties. By 1914 the armadillo had crossed the Brazos River and moved to the Trinity River, and along the coast the species had already reached the Louisiana line in Orange County. The northward and eastward range expansions continued over the next forty years, and by 1954 the armadillo was known from everywhere except Red River and Lamar counties in extreme northeastern Texas. By 1958, it was known from these latter two counties and today is abundant everywhere in the region (Fig. 164a-b). The range to the west appears to be established about as far west as the 20 inch (50 cm) limit of annual precipitation in Texas and Oklahoma, where it has remained stationary since 1994 (Taulman and Robbins, 2014). The zone covering the range of 16-20 inches (40-50 cm) of annual precipitation may represent a sink region where individual armadillos can survive temporarily, but a permanent population will probably not be able to become established (Taulman and Robbins, 2014).

Apparently, pioneering resulting in range expansion of the armadillo was most successful in riparian habitat, and invasion was especially rapid parallel to rivers, which served

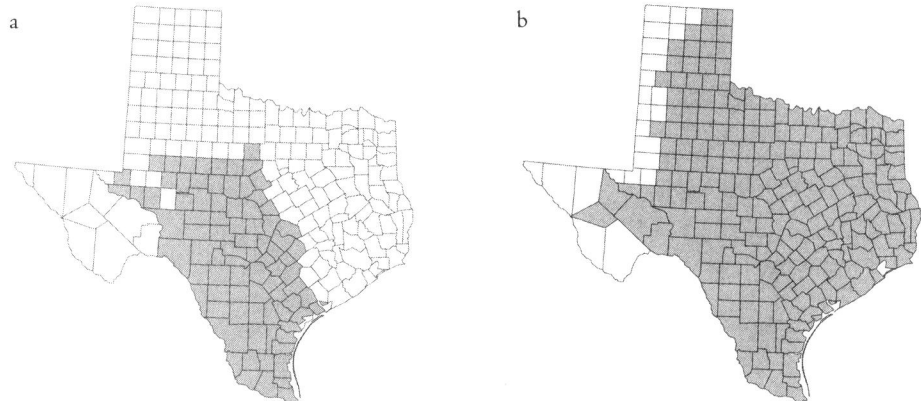

Figure 164a-b. The historic (a) and current (b) distribution of the nine-banded armadillo (*Dasypus novemcinctus*) in Texas.

as dispersal conduits. Average invasion rates have been calculated as from 2.5 to 6 miles (4-10 km) per year in the absence of obvious physical or climatic barriers. Possible reasons for the armadillo's northward expansion since the nineteenth century include progressive climatic change, encroaching human civilization, overgrazing, and decimation of large carnivores (Cleveland, 1970; Humphrey, 1974).

Bats

Documentation of range expansions has been very prevalent for bats in recent decades. Three species, the American perimyotis (*Perimyotis subflavus*), the evening bat (*Nycticeius humeralis*), and the Seminole bat (*Lasiurus seminolus*), previously thought to be confined to the eastern half of the state (Schmidly, 1983), have now been collected in far western Texas (Dowler et al., 1992; Yancey et al., 1995b; Dowler et al., 1999; Brant and Dowler, 2000; Ammerman et al., 2012) (Figs. 165–167). These examples suggest the Rio Grande may be serving as a dispersal corridor for eastern species of mammals to gradually expand their ranges westward during the last decade (Schmidly and Jones, 2000).

The western red bat, *Lasiurus blossevillii*, previously known only from Presidio County (a July 1988 record), was recently documented as a fatality at a wind farm along the Rio Grande in Starr County in southern Texas (Weaver et al., 2020). This bat could have dispersed eastward along the Rio Grande or, more likely, northward out of Tamaulipas, Mexico, where it is common (Schmidly and Hendricks, 1984).

The western yellow bat, *Dasypterus xanthinus*, is an example of a bat that only recently invaded Texas toward the end of the twentieth century (Fig. 168). Previously known from the southwestern US in Mexico and Arizona, as well as from the Mexico Plateau and Baja California, this species was first documented in the Big Bend of Brewster County in October 1996 (Higginbotham et al., 1999), where apparently a resident population resides (Ammerman et al., 2012). It has since been reported from other sites in the

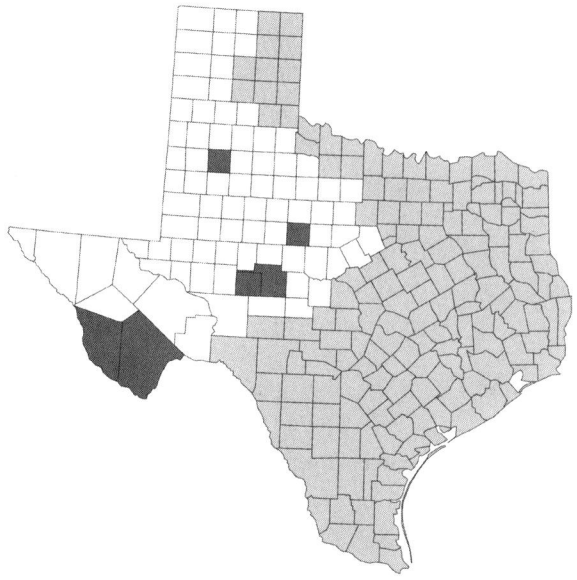

Figure 165. Current distribution of American perimyotis (*Perimyotis subflavus*) in Texas. Light shading represents historic distributions, whereas darker shading represents recent records.

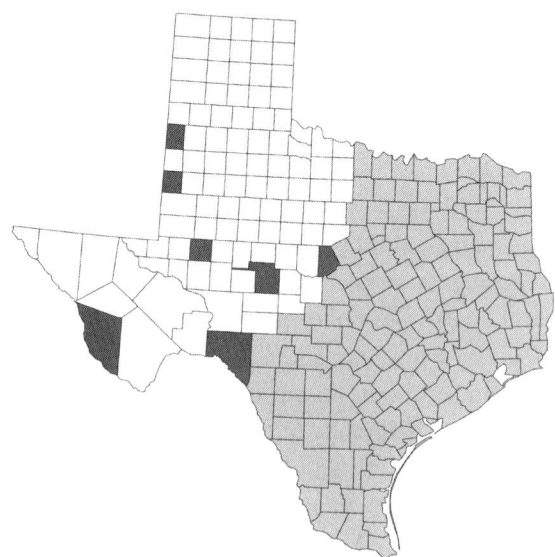

Figure 166. Current distribution of the evening bat (*Nycticeius humeralis*) in Texas. Light shading represents historic distributions, whereas darker shading represents recent records.

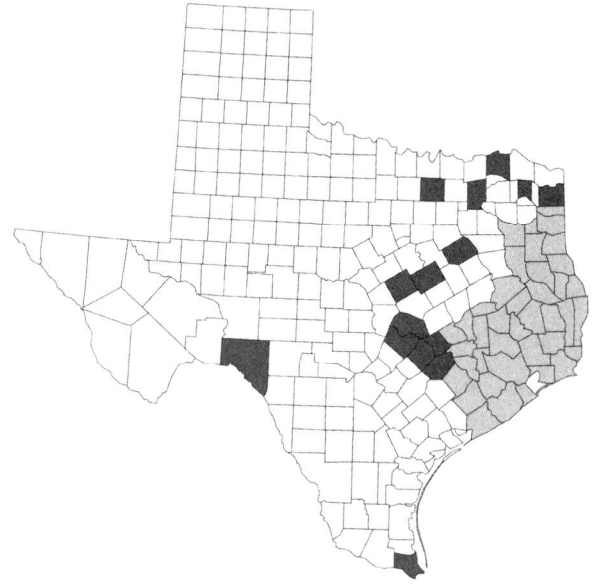

Figure 167. Current distribution of the Seminole bat (*Lasiurus seminolus*) in Texas. Light shading represents historic distributions, whereas darker shading represents recent records.

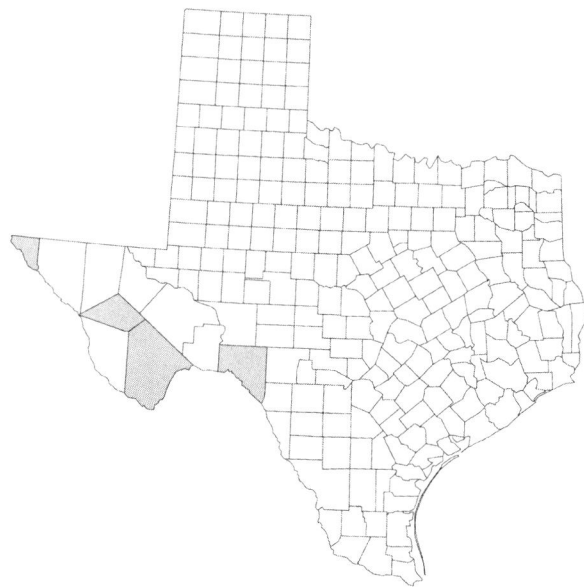

Figure 168. Current distribution of the western yellow bat (*Dasypterus xanthinus*) in Texas.

Chihuahuan Desert region from Black Gap Wildlife Management Area (Bradley et al., 1999a), east of Big Bend National Park, to as far north as the Davis Mountains (Jones et al., 1999) in Jeff Davis County, and to the east as far as Val Verde County (Weyandt et al., 2001) along the Rio Grande.

The first Texas record of the lesser long-nosed bat (*Leptonycteris yerbabuenae*) was only recently reported from El Paso County (Krejsa et al., 2020), although the voucher specimen that is the basis for the record was obtained in 2010. This specimen represents the eastern-most record of the species in the US.

Northern pygmy mouse, *Baiomys taylori*

The northern pygmy mouse represents another example of a southern species, characteristic of the tropical lowlands of Mexico, expanding its range northward and westward. The federal agents indicated it was restricted to the coastal and southern mesquite-chaparral regions. Numerous studies have documented recent (post-1950) expansion of that Gulf Coast population north and west across the Balcones Escarpment and the Red River into Oklahoma (Hart, 1972; Choate et al., 1990; Tumlinson et al., 1993; Schmidly, 2004; Green and Wilkins, 2010) from its initially limited distribution (Light et al., 2016). Since the early twentieth century, the species has successfully invaded the oak-hickory association, the Blackland Prairies, the Cross-Timbers, the Rolling Plains, and most recently the High Plains (see Fig. 32a–b, Chapter 2, p. 84). These mice have a preference for grassy areas, and they are commonly found in old fields, pastures, and along railroad and highway rights-of-way (Diersing and Diersing, 1979; Cleveland, 1986; Hollander et al., 1987; Jones and Manning, 1989; Choate et al., 1990; Brant and Dowler, 2002). Their range expansion has been attributed to rapid population size increases (Abuzeineh et al., 2011), expansion of preferred blackland prairie ecotype following degradation of the oak-hickory association and piney woods, and dispersal along highway corridors (Green and Wilkins, 2010). Apparently, this small mouse continues to expand its range, as it has recently been documented in Hunt County in northeastern Texas just south of the Red River and the boundary with eastern Oklahoma (Green and Wilkins, 2010).

Prairie vole, *Microtus ochrogaster taylori*

A microtine rodent, the prairie vole appears to have invaded the Texas Panhandle toward the end of the twentieth century from nearby Oklahoma, and subsequently spread southward on the High Plains all the way to Lubbock County (Schmidly and Bradley, 2016). The details of this discovery are explained more fully later in this chapter.

Porcupine, *Erethizon dorsatum*

The porcupine (*Erethizon dorsatum*) represents another example of a mammal that has expanded its range in Texas during the twentieth century. It is largely an inhabitant of forested areas in the west and prefers rocky areas, ridges, and slopes. Porcupines wander about a great deal and may be found irregularly in areas that appear wholly unsuited to them. Bailey and the federal agents recorded them only in one county in the Panhandle and in Jeff Davis and Brewster counties in the western part of the state (see Fig. 47, Chapter 2, p. 234). Today, they occur east to Bosque, Travis, and Van Zandt counties, and individuals recently were recorded from Webb and Hidalgo counties in South Texas. None of the early explorers or naturalists, until the report of Bailey, documented their occurrence. Weniger (1997) apparently was not aware of Bailey's observations when he wrote: "The earliest record of porcupines here seems to be those reported from the Davis Mountains of the Big Bend in 1940." Weniger (1997) speculated that the spread of porcupines may have been facilitated by the practice of deicing highways with salt. The "salt drive" of these animals has been well documented, and salt availability is a limiting factor in determining where they live and a clue to much of their behavior.

Other Rodents

Unfortunately, not much is known about micro-scale changes in rodent diversity such as reductions or expansions of geographic ranges, changes in species abundance and community structure, and extinctions of local populations and subspecies. But from the evidence at hand, it seems obvious that the faunal composition in several areas of the state changed during the twentieth century.

This is especially true on the High Plains, or Llano Estacado, and the northern Texas Panhandle. Bailey and the federal agents worked in these regions at the end of the nineteenth century when most of this country was comprised of large ranches with little human settlement. Today, this region, and especially the Llano Estacado, has been converted to farmland for crop production. Larry Choate (1997) documented the mammal fauna of the Llano Estacado, and J. Knox Jones Jr. and associates (1988b) did likewise for the northern Panhandle. Comparing their results with those of Bailey and the federal agents reveals some interesting differences. Several species, including the western harvest mouse (*Reithrodontomys megalotis*) and the Plains harvest mouse (*Reithrodontomys montanus*), which are common in this region today, were never trapped or sighted by Bailey or his co-workers (see Fig. 43, Chapter 2, p. 189). It is highly unlikely, given the prowess of the federal agents as field collectors, that they would have failed to collect these small mammals had they occurred in the region. Therefore, it is tempting to speculate that the occurrence of these species in the High Plains and Panhandle is a later twentieth-century phenomenon, perhaps arising as an artifact of human intervention in the form of hauling hay from the central regions of the state to feed livestock in these areas during periods of drought.

There are other interesting examples of species being transplanted by man from one region of the state to another, resulting in viable, breeding populations outside of a species' normal geographic range. One of the best cases of this involves tree squirrels of the genus *Sciurus*. Bailey and the federal agents mapped the distribution of the gray squirrel, *Sciurus carolinensis*, as occurring no further west than the eastern edge of the Balcones Escarpment. Gray squirrels have been introduced in many places outside of their natural range. A thriving population in the city of Lubbock was established in the 1970s, more than 400 miles (644 km) from the species' normal range.

Interestingly, the biological survey field agents did not document the occurrence of fox squirrels (*Sciurus niger*) in the Panhandle. Mammalogists from Texas Tech University found them to be common in the deciduous riparian vegetation, mostly cottonwoods, along the Canadian River and its major tributaries (Jones, J. K., Jr., et al., 1988b). Given the extensive amount of fieldwork the federal agents conducted in this region, it is not likely they would have overlooked the occurrence of fox squirrels. This would suggest that fox squirrels occupied this region sometime during the twentieth century, dispersing along the Red River drainage system. The best fox squirrel habitat is mature oak-hickory woodland broken into small, irregularly shaped tracts and connected by strips of woodland. In the eastern part of their range, they occur primarily in the upland regions, whereas in the western part of the state they are restricted more or less to river valleys that support various fruit and nut trees. Also, there is good evidence that they have been introduced in several places on the Llano Estacado (Frey and Campbell, 1997).

Documentation of Additional Faunal Elements

Even though Bailey and the federal agents did a thorough job of documenting the mammal fauna, mammalogists working in the state throughout the twentieth century have continued to document additional species and subspecies. Twenty-six other species of mammals, listed below, have been documented since the biological survey. (Note: This list does not include species or subspecies that have undergone taxonomic reassignment and are now referred to by different scientific names.)

Order Eulipotyphla.—*Blarina hylophaga*, Elliot's short-tailed shrew.

Order Chiroptera.—*Nyctinomops femorosaccus*, pocketed free-tailed bat; *Nyctinomops macrotis*, big free-tailed bat; *Leptonycteris nivalis*, Mexican long-nosed bat; *Leptonycteris yerbabuenae*, lesser long-nosed bat; *Choeronycteris mexicana*, Mexican long-tongued bat; *Diphylla ecaudata*, hairy-legged vampire bat; *Corynorhinus rafinesquii*, Rafinesque's big-eared bat; *Dasypterus ega*, southern yellow bat; *Dasypterus xanthinus*, western yellow bat; *Euderma maculatum*, spotted bat; *Lasionycteris noctivagans*, silver-haired bat; *Lasiurus blossevillii*, western red bat; *Myotis occultus*, southwestern little brown myotis; *Myotis austroriparius*, southeastern myotis; *Myotis ciliolabrum*, western small-footed myotis; *Myotis septentrionalis*, northern long-eared myotis; *Myotis thysanodes*, fringed myotis; and *Myotis volans*, long-legged myotis.

Order Rodentia.—*Geomys knoxjonesi*, Jones' pocket gopher; *Geomys jugossicularis*, Hall's pocket gopher; *Ochrotomys nuttalli*, golden mouse; *Peromyscus nasutus*, northern rock deermouse; *Peromyscus truei*, piñon deermouse; and *Sigmodon fulviventer*, tawny-bellied cotton rat.

Order Carnivora.—*Mephitis macroura*, hooded skunk.

Of these 26 species, more than two-thirds are bats. Bats were undoubtedly the least understood group of mammals at the time of the biological survey. Only 17 taxa were included in the survey publication. Today, we know of 34 species that occur in the state. The primary reason bats were so little understood was the difficulty in obtaining specimens. Historically, the only methods available for capturing bats were shooting them with shotguns in the early evening and hand-capturing them in their daytime hiding places. These time-consuming and uncertain methods obtained only a fraction of the total number of species of bats occurring in an area. In 1937, A. E. Borell, while studying mammals in the Big Bend region, developed a method of capturing bats by stringing fine wire across a water tank frequented by bats coming to drink after dark (Borell, 1937). The bats would hit the wires and drop into the water, and they could then be collected when they swam to the side of the tank. This method was more effective and economical than shooting, and allowed bats to be banded and released unharmed if desired. However, slow-flying species, such as *Antrozous* and *Parastrellus*, often avoided the wires, and the technique could be used only on relatively small, man-made water tanks.

In 1954, Walter Dalquest, a mammalogist at Midwestern State University in Wichita Falls, Texas, published a technique for capturing bats using Japanese silk "mist nets" (Dalquest, 1954). The use of this technique would soon revolutionize the study of bats and today it is still the most effective method available for capturing bats. The mist nets are most commonly strung across ponds after dark but also can be set among bushes or trees, or across the openings of caves or other roost sites to capture bats as they leave their roosts in the evening. Mist nets offer the following advantages to the collector: bats can be taken in large numbers in them, species otherwise difficult to collect are easily obtained, and individuals can be released alive.

Mammalogists documented several new taxa of bats and rodents during the latter half of the twentieth century. The first occurrence of the western yellow bat (*Dasypterus xanthinus*) in Texas was documented in the 1990s. Specimens were obtained in the Chisos Mountains of Big Bend National Park (Higginbotham et al., 1999), Black Gap Wildlife Management Area (Bradley et al., 1999a), the Davis Mountains near Fort Davis (Jones et al., 1999), and more recently from Val Verde and El Paso counties (Schmidly and Bradley, 2016), places where many mammalogists had previously collected bats without finding this species (see Fig. 168, p. 501).

Another bat, the lesser long-nosed bat (*Leptonycteris yerbabuenae*), was added to the Texas fauna with the discovery of a voucher specimen from El Paso County (Krejsa et al., 2020). Previously, the species was known in the US from California, Arizona, and New Mexico. The individual from Texas, obtained ten years ago, was thought to be a vagrant and not indicative of an established population.

An example of the discovery of a subspecies new to the Texas fauna involves the prairie vole, *Microtus ochrogaster taylori*, which was first discovered in two counties (Hansford and Lipscomb) in the northern Panhandle along the Oklahoma border in 1988 (Jones, J. K., Jr., et al., 1988b). The only other record of this species was of another subspecies, *M. o. ludovicianus*, recorded from the Big Thicket region by Bailey and his co-workers. A. H. Howell, one of the leading federal agents with the biological survey, worked extensively in Lipscomb County in 1903, collecting 20 species of small mammals, none of which were voles. Recently, however, Ray Matlock and his students at West Texas A&M University examined rodent remains in owl pellets from throughout the Texas Panhandle and discovered that the prairie vole was ubiquitous throughout the region. Although the occurrence of *M. o. taylori* in Texas probably represents another invasion during the twentieth century, the subspecies has now spread all the way into Lubbock County (Schmidly and Bradley, 2016).

Probably the most remarkable discovery of a new species of mammal during the twentieth century was the documentation of a third species of cotton rat (*Sigmodon fulviventer*, the tawny-bellied cotton rat) from near Fort Davis in Jeff Davis County in the spring of 1991 (Stangl, 1992a, b). Previously known only from southeastern Arizona and southwestern New Mexico, not only did this isolated population represent a new species of mammal for the state, but Stangl described it as a new subspecies, *S. f. dalquesti*, named after his mentor, Walter Dalquest.

The extent of the tawny-bellied cotton rat's range and relative abundance remain unknown. The population was documented from an area where many mammalogists and their students previously had conducted extensive fieldwork, documenting on numerous occasions two other species of cotton rats, the hispid cotton rat (*Sigmodon hispidus*) and the yellow-nosed cotton rat (*S. ochrognathus*). Recent attempts to collect *S. f. dalquesti* at its type locality have proven unsuccessful, and the status of this taxon remains an enigma.

Increase in the Number of Threatened, Endangered, and Rare Species

Several land mammals are viewed as having some sort of biological issue that threatens or potentially threatens their existence. These are species that, in the opinion of biologists and conservation groups, currently face or likely will face serious conservation problems in the future.

Many of the species in jeopardy share life history attributes that make them especially vulnerable to local extinction events. Some species, such as bats, which have low reproductive rates, are slow to recover from population declines caused either by catastrophic events or by habitat destruction. Others, such as many of the carnivores and larger herbivores, are large in body size and have an extensive home range that, coupled with low population densities and their trophic level, make them highly vulnerable to human disturbance. Many have a confined geographic range, limited to a handful of locations, which makes them highly vulnerable to local extinction events.

Another category of vulnerability includes those species dependent on some highly specialized but scarce resource. The organism may be a masterpiece of adaptation, but it is vulnerable because of one requirement within its habitat that must be met. An example would be the Mexican free-tailed bat, *Tadarida brasiliensis*. This species is highly mobile and can range widely in its feeding, but it is absolutely dependent on caves or other suitable roosts with the appropriate temperature and humidity for passing the day and rearing its young. Natural caves were the original source of refugia. If, for some reason, caves become unavailable to these bats, such as through commercial exploitation and mining, then the species could become extinct.

History of Endangered and Threatened Mammal Listings

Scientific documentation about the conservation status of mammals started early in the twentieth century and increased dramatically during the final decades of the millennium. Vernon Bailey was the first to draw serious attention to the plight of certain species and the need for immediate action. This was followed in 1917 by a paper written by H. P. Attwater, from Houston, a contemporary naturalist of Bailey (see Chapter 1), who argued for the need to take action to prevent the further decimation of the fauna. In the closing paragraph of his paper, Attwater wrote:

> I have no personal interests to serve, no 'axe to grind,' and I concede the right of any one to disagree with me, but I am absolutely certain that Texas people are making a great mistake in permitting the wanton and reckless destruction of valuable and useful wild life and that the future interests of the State demand that it be put a stop to, before it is too late (Attwater, 1917).

Glover Allen (1942) drew continent-wide attention to the problem of extinct and vanishing mammals in the western hemisphere. His list of species and subspecies of concern included 18 taxa from Texas. In the late 1940s and throughout the 1950s, a series of articles appeared in the *Texas Game and Fish* magazine about the declining status of many mammals (Burr, 1949a, b, c; Baughman, 1951; Slaughter, 1960), and Rollin Baker (1956) called attention to some of the most serious problems in East Texas.

Following the passage of the Endangered Species Act (ESA), written accounts about the status of mammals began to proliferate. Culbertson (1974) performed a status evaluation of Texas mammals, focusing on those species that required immediate attention. Baker (1977) and Findley and Caire (1977) prepared lists of rare, endangered, and poorly known species from the Chihuahuan Desert region, including Texas, and Rappole and Tipton (1987) prepared a summary of information on the small, terrestrial mammals that had been identified as potentially declining, threatened, or endangered. The latter authors surveyed mammalogists to determine their list. J. Knox Jones Jr. (1993) provided a synopsis of threatened and endangered species law and published a list of taxa warranting conservation concern. Linda Campbell (1995) discussed the life history and management of all of the endangered and threatened vertebrates in Texas, including mammals on the federal and state lists. Finally, the last two editions of *The Mammals of Texas* (Schmidly, 2004; Schmidly and Bradley, 2016) contained explanations of the conservation status of all Texas mammals. Much of the content for this chapter has been taken from the latest edition of *The Mammals of Texas*.

The legal protection of plants and animals considered to be endangered or threatened was not an issue during the time of Bailey and the federal agents. The concept of federal and state laws to protect wildlife was just taking root at the beginning of the twentieth century (see Jones, 1993, for a thorough discussion). The first significant step in federal wildlife law was the Lacey Act of 1900 that prohibited interstate transportation of "any wild animals or birds" killed in violation of state law. At that time most of the laws in effect had to do with the protection of game animals.

One of the earliest efforts to give legal protection to non-game animals in Texas occurred in 1917. In that year, the Texas legislature passed a general law (H.B. No. 40) making it a misdemeanor to kill or injure bats because of their perceived value in controlling malarial mosquitoes (Schmidly, 1991). A provision of this law (Section I, Article 887a) stated, "If any person shall willfully kill or in any manner injure any winged quadruped known as the common bat, he shall be deemed guilty of a misdemeanor and upon conviction shall be fined a sum of not less than five ($5.00) dollars nor more than fifteen ($15.00) dollars."

The first federal legislation in this area was the Endangered Species Preservation Act of 1966, replaced soon thereafter by the ESA of 1969, and culminating in congressional passage of the ESA of 1973, which was subsequently reauthorized in 1988 and amended in 1992, 1996, 2004, and 2009. Now, state and federal agencies as well as private organizations have developed lists of rare and endangered mammals. The USFWS publishes a list of endangered and threatened species in the *Federal Register*, and the TPWD has a list of protected nongame wildlife. These are the official lists governed by federal and state law, statutes, and regulations.

The ESA is the centerpiece of federal efforts to conserve biological diversity (NRC, 1995). Its aim is to prevent the extinction of plant and animal species by regulating a wide range of activities affecting plants and animals designated as endangered or threatened. By definition, an *endangered species* is an animal or plant listed by regulation as being in danger of extinction. A *threatened species* is any animal or plant that is likely to become endangered within the foreseeable future. A species must be listed in the *Federal Register* as endangered or threatened for the provisions of the act to apply. Any species or subspecies of plant or animal may be eligible for protection, including species found outside the United States. Distinct populations of vertebrate animals also may be protected under the act. Enforcement of the ESA falls under the authority of the USFWS.

The ESA prohibits the following activities involving endangered species: importing into or exporting from the United States; taking (includes harassing, harming, pursuing, hunting, shooting, wounding, trapping, killing, capturing, or collecting) within the United States and its territorial seas; taking on the high seas; possessing, selling, delivering, carrying, transporting, or shipping any such species unlawfully taken within the United States or on the high seas; delivering, receiving, carrying, transporting, or shipping in interstate or foreign commerce in the course of a commercial activity; and selling or offering for sale in interstate or foreign commerce. The ESA also provides for protection of *critical habitat* (habitat required for the survival and recovery of the species) and the creation of a *recovery plan* for each listed species.

One of the more controversial aspects of the ESA developed in the early 1990s when the Solicitor's Office of the DOI issued the so-called Hybrid Policy, which ruled that hybrids between endangered species, subspecies, or populations could not be protected. Further, the Code of Zoological Nomenclature states that hybrids cannot be recognized in the animal kingdom. See Jones et al. (1995) for a discussion. Therefore, a strict interpretation and enforcement of the Hybrid Policy would mean that the red wolf could not be protected under the Endangered Species Act, because recent genomic studies have suggested that red wolves might be the result of hybridization between gray wolf subspecies and coyotes (however, see Chapter 2 and the section on red wolves earlier in this chapter for an alternative view). Despite this ruling, the USFWS has continued to tentatively list red wolves as endangered. This would appear not to be a problem in Texas because red wolves have been considered to be extinct in the state for more than half a century. However, the recent discovery of canid populations on Galveston Island containing individuals with apparent red wolf genes implies that red wolf genes remain in some coyote populations (see earlier discussion of red wolves). The fact that red wolf genes still occur in Texas canids raises the question: should they be protected, or are they the product of past hybridization? Situations like this have caused legal scholars to argue for modification and a more modern interpretation of the Hybrid Policy (Erwin, 2017).

The ESA provides for listing plant and animal species into the following categories: listed endangered species, listed threatened species, proposed endangered species, proposed threatened species, candidate species (awaiting listing), and delisted species (species removed from endangered or threatened list due to extinction, taxonomic change, or abundance). Proposed species are those species for which a proposed rule to list as endangered or threatened has been published in the *Federal Register*. Candidate species are those species for which the USFWS has on file sufficient information on biological vulnerability and threats to support issuance of a proposed rule to list but issuance of the proposed rule is precluded. In earlier versions of the ESA, candidate species were classified into two categories: Category 1 species qualified for listing based on available data, and Category 2 species required additional information as to their status. The 1996 act, as amended, dissolved Category 2, categorizing all of those taxa as Candidate Species.

In 2019, the administration of President Donald Trump announced some of the broadest, and most drastic, changes to enforcement of the ESA (Department of Interior, 2019). The proposed changes were strongly opposed by environmental groups and most wildlife scientists on the grounds they would strip vital protections from threatened species, allow major destruction of critical habitat, remove agency accountability to protect wildlife or address threats to wildlife from climate change, and encourage policy makers to calculate the perceived economic costs (but not the benefits) of ESA protection of plants, fish, and wildlife (note: previously, economic factors were intentionally not considered in listing decisions). On 4 June 2021, the Biden administration announced that it would reverse most of these policies, including the requirement to take the economic cost of conserving species into account when deciding on whether to put a plant or animal on the endangered species list—a requirement that many environmentalists had claimed violated both the letter and the spirit of the law (Grandoni and Fears, 2021).

In 1973 the Texas legislature authorized the Texas Parks and Wildlife Department to establish a list of endangered animals in the state. Endangered species are those species that TPWD has named as being "threatened with statewide extinction." Threatened species are those species that TPWD has determined are likely to become endangered in the future. State laws and regulations pertaining to endangered or threatened species are contained in Chapters 67 and 68 of the Texas Parks and Wildlife Code and Sections 65.171–65.184 of Title 31 of the Texas Administrative Code (T.A.C.). In 1988 the Texas legislature authorized TPWD to establish a list of threatened and endangered plant species for the state. Laws and regulations pertaining to endangered or threatened plant species are contained in Chapter 88 of the TPWD Code and Sections 69.01–69.14 of the T.A.C. TPWD regulations prohibit the taking, possession, transportation, or sale of any of the animal or plant species designated by state law as endangered or threatened without the issuance of a permit. Listing and recovery of endangered species in Texas are coordinated by the Wildlife Diversity Branch in the Wildlife Division of TPWD. The most recent update of the TPWD list was made in 2020 (https://tpwd.texas.gov/huntwild/wild/wildlife_diversity/nongame/listed-species/media/fedState-ListedSpeciesComplete-3302020.pdf).

Endangered and Threatened Mammals on USFWS and TPWD Lists

According to the USFWS, Texas ranks sixth in the nation in terms of the number of endangered species living within its borders. Seventy-two plants and animals protected under the ESA have been recorded from Texas, a number that reflects the state's large size, diverse array of habitats, and growing development pressures. The USFWS is required by law to assess periodically the status of America's endangered species and to report its findings to Congress. The most recent report, compiled in 2013-2014, provided data on 37 of Texas's 45 listed animals and all 27 of its listed plants (USFWS, 2016). According to this report, 39 percent of the endangered species found in Texas are still declining, only 24 percent are judged to be improving or stable, and for the remaining 37 percent the Service lacks the resources to determine how they are faring.

Currently, TPWD lists 16 native terrestrial Texas mammals as endangered or threatened, but at least four of the mammals on that list are already extinct in the state (Table 8). Mammals included in the federal and state lists are distributed throughout the state. There is no obvious geographic pattern or concentration of occurrence of these species,

Table 8. Native terrestrial Texas mammals with critical status as defined by Texas Parks and Wildlife Department (TPWD), US Fish and Wildlife Service (USFWS), and International Union for Conservation of Nature (IUCN). These lists were compiled in 2020.

Scientific Name, Common Name	TPWD (state status)	USFWS (federal status)	IUCN[4] (global status)
Sylvilagus robustus, Davis Mountains Cottontail	Not Listed	Not Listed	Vulnerable
Choeronycteris mexicana, Mexican Long-tongued Bat	Not Listed	Not listed	Near Threatened
Leptonycteris nivalis, Mexican long-nosed Bat	Endangered	Endangered	Endangered
Corynorhinus rafinesquii, Rafinesque's Big-eared Bat	Threatened	Not Listed	Least Concern
Dasypterus ega, Southern Yellow Bat	Threatened	Not Listed	Least Concern
Euderma maculatum, Spotted Bat	Threatened	Not Listed	Least Concern
Canis lupus[1], Gray Wolf	Endangered	Endangered	Least Concern
Canis rufus[1], Red Wolf	Endangered	Endangered	Critically Endangered
Leopardus pardalis, Ocelot	Endangered	Endangered	Least Concern
Leopardus wiedii[1], Margay	Not Listed	Endangered[2]	Near Threatened
Panthera onca[1], Jaguar	Endangered	Endangered	Near Threatened
Puma yagouaroundi, Jaguarundi	Endangered	Endangered	Least Concern
Spilogale putorius interrupta, Plains Spotted Skunk	Not Listed	Under Review	Vulnerable (species)
Mustela nigripes[1], Black-footed Ferret	Not Listed	Endangered	Endangered
Nasua narica, White-nosed Coati	Threatened	Not Listed	Least Concern
Ursus americanus, American Black Bear	Threatened	T/SA[3] (eastern); Not Listed (western)	Least Concern

Scientific Name, Common Name	TPWD (state status)	USFWS (federal status)	IUCN[4] (global status)
Ursus americanus luteolus, Louisiana Black Bear	Threatened	Not Listed	Not assessed
Ursus arctos[1], Grizzly or Brown Bear	Not Listed	Threatened	Least Concern
Bos bison, American Bison	Not Listed	Not listed	Near Threatened
Oryzomys couesi, Coues's Rice Rat	Threatened	Not Listed	Least Concern
Peromyscus truei comanche, Palo Duro Mouse	Threatened	Not Listed	Least Concern
Geomys arenarius, Desert Pocket Gopher	Not Listed	Not Listed	Near Threatened
Dipodomys elator, Texas Kangaroo Rat	Threatened	Under Review	Vulnerable
Dipodomys spectabilis, Banner-tailed Kangaroo Rat	Not Listed	Not Listed	Near Threatened
Sigmodon fulviventer, Tawny-bellied Cotton Rat	Threatened	Not Listed	Least Concern

[1] Species considered by the authors to be extinct in Texas.
[2] The USFWS status of endangered applies to populations in Mexico; former US population (subspecies *cooperi*) is not listed.
[3] Threatened, similarity of appearance.
[4] The IUCN lists five non-native mammals (scimitar-horned oryx, *Oryx dammah*; lesser kudu, *Ammelaphus imberbis*; blackbuck, *Antilope cervicapra*; eastern Thomson's gazelle, *Eudorcas thomsoni*; and aoudad, *Ammotragus lervia*) that have been introduced into Texas as either extinct, near threatened, or vulnerable in their native range.

suggesting that the conservation pressures on Texas's rare and endangered resources are statewide and not just regional or local in nature. Rare mammals occur in 54 of the 91 plant communities in Texas (Diamond et al., 1997; Cameron et al., 1997). Thirteen plant communities contain state or federal endangered mammals. These results demonstrate that no single habitat can be targeted for conservation of rare mammals in Texas.

Six land mammals currently are listed as endangered in Texas by both the USFWS and TPWD. Two of them (the Mexican long-nosed bat, *Leptonycteris nivalis*, and the ocelot, *Leopardus pardalis*) still have extant populations in the state. The other four species (red wolf, *Canis rufus*; gray wolf, *Canis lupus*; jaguarundi, *Puma yagouaroundi*; and jaguar, *Panthera onca*) listed as endangered by both agencies are now extinct in Texas. The black-footed ferret, *Mustela nigripes*, is also extinct in the state and is no longer listed as threatened or endangered by TPWD, although it remains on the USFWS list.

Ten taxa of land mammals are regarded as threatened by the TPWD: southern yellow bat (*Dasypterus ega*), spotted bat (*Euderma maculatum*), Rafinesque's big-eared bat (*Corynorhinus rafinesquii*), Texas kangaroo rat (*Dipodomys elator*), Coues's rice rat (*Oryzomys couesi*), tawny-bellied cotton rat (*Sigmodon fulviventer*), Palo Duro deermouse (*Peromyscus truei comanche*), American black bear (*Ursus americanus*), Louisiana black bear (*U. americanus luteolus*), and white-nosed coati (*Nasua narica*). The margay (*Leopardus wiedii*) and the grizzly bear (*Ursus arctos*) are now extinct in Texas and are no longer listed by TPWD as threatened, but they continue to be listed by the USFWS as threatened.

The plains spotted skunk, *Spilogale putorius interrupta*, and Texas kangaroo rat, *Dipodomys elator*, currently are under review for listing by the USFWS. Three other species previously have been considered candidates for listing as endangered or threatened on the federal list. They were the Davis Mountains cottontail (*Sylvilagus robustus*), the eastern hog-nosed skunk (*Conepatus leuconotus*; now considered to be conspecific with the western hog-nosed skunk), and the swift fox (*Vulpes velox*) and possibly its former conspecific, the kit fox (*Vulpes macrotis*). Six mammals (Mexican long-tongued bat, *Choeronycteris mexicana*; southeastern myotis, *Myotis austroriparius*; southwestern little brown myotis, *Myotis occultus*; northern long-eared myotis, *Myotis septentrionalis*; western bonneted bat, *Eumops perotis*; and yellow-nosed cotton rat, *Sigmodon ochrognathus*) were previously listed as Category 2 species in the USFWS list, meaning that the information at hand indicated it was possibly appropriate to list the species as endangered or threatened but that substantial biological data were not available to support a proposed ruling. The 1996 ESA, as amended, dissolved Category 2 and moved these taxa to the Candidate Species list.

Rare and Threatened Texas Mammals on IUCN "Red List"

The International Union for Conservation of Nature (IUCN), founded in 1964, publishes the *IUCN Red List of Threatened Species* (*Red List*), which is the most comprehensive inventory of the global conservation status of biological species. In 2012 the IUCN completed an update of the status for all 5,488 mammal species then recognized in the world (however, a total of 6,399 extant mammal species were reported by Burgin et al., 2018). The IUCN list includes information about each species' geographic range, ecology and habitat, extinction risk, and overall conservation status according to eight categories: extinct, extinct in wild, critically endangered, endangered, vulnerable, near threatened, least concern, and data deficient. In the case of mammals, this list was put together by the IUCN staff and a group of partners that included Arizona State University, Texas A&M University, University of Rome, University of Virginia, and the Zoological Society of London.

The most recent edition of *The Mammals of Texas* (Schmidly and Bradley, 2016) included for the first time the IUCN's *Red List* categorization for each Texas mammal. Species considered to be critically endangered, endangered, vulnerable, and near threatened were included along with the species listed by USFWS and TPWD. The IUCN list included three terrestrial species (Mexican long-nosed bat, *Leptonycteris nivalis*; red wolf, *Canis rufus*; and black-footed ferret, *Mustela nigripes*) as endangered. In the near-threatened category, the IUCN lists six native Texas species (Mexican long-tongued bat, *Choeronycteris mexicana*; margay, *Leopardus wiedii*; jaguar, *Panthera onca*; American bison, *Bos bison*; desert pocket gopher, *Geomys arenarius*; and banner-tailed kangaroo rat, *Dipodomys spectabilis*). The IUCN list includes the Texas kangaroo rat, *Dipodomys elator*, Davis Mountains cottontail, *Sylvilagus robustus*, and spotted skunk, *Spilogale putorius*, as vulnerable.

The status of the mammals included on the USFWS, TPWD, and IUCN lists is discussed in Chapter 6.

Introductions of Nonindigenous, Invasive Species

Throughout the twentieth century a number of devastating species have been introduced into the state, including plants (salt cedar and Chinese tallow), insects (Africanized honeybees and red imported fire ant), fish (grass carp), and birds (English sparrows and starlings), just to mention a few examples in non-mammal cases. The character and composition of our landscapes have changed substantially as a result of these biological invaders (see Chapter 4 for a discussion of this problem). In this section we focus on the large number of non-native mammals that have been introduced in the state and the impact they are having on our native species. Nonindigenous mammals, which were hardly ever encountered by Bailey and the federal agents, now openly range over much of the state.

Old World Rats and Mice

Early explorers and settlers brought Old World rats (genus *Rattus*) and mice (genus *Mus*) to the United States. By the time Bailey worked in Texas, Old World rats and mice already were established in every city and town in the state as well as occurring in the feral state. But this was just about the extent of introduced mammals. Since 1900, the number of mammals introduced has increased at a staggering rate.

Red fox, *Vulpes vulpes*

As Bailey noted, red foxes were not native to Texas nor anywhere else in the United States. It is generally agreed that the species was introduced from England by a group of Maryland planters in 1730 for the purpose of providing the sport of fox hunting in that colony. Whether any of the descendants of those first individuals ever reached Texas is not known (Weniger, 1997). Texas got its own infusion of the species later, when at least 145 individuals were released in the central and eastern parts of the state in the 1890s (Doughty, 1983). As Bailey commented, they quickly became locally common. Today, red foxes occur throughout eastern and north-central Texas, as far west as the Panhandle and central Trans-Pecos, and they seem to be getting increasingly abundant. Highest densities occur in the Piney Woods region.

Nutria, *Myocastor coypus*

The next major mammalian introduction involved the nutria, a native rat of South America, which was introduced into the southeastern United States in 1938 in the vicinity of Avery Island, Louisiana. They were brought to the US for a fur industry that never took off. Since the early 1940s, the nutria has spread throughout the southeastern United States and eastern Texas, often to the detriment of the native muskrat, *Ondatra zibethicus*, which occupies much the same habitat. Nutrias also were thought to represent a "cure-all" for ponds choked with vegetation. They do reduce many kinds of aquatic plants, but they will not eat "moss" (algae) and many of the submerged plants. The trouble is that once established in a lake, their high reproductive capacity soon results in overpopulation. They become so numerous that the available food supply will not satisfy them, and then trouble begins. The animals move into places where they are not wanted or where they destroy vegetation that is valuable for such wildlife as waterfowl and muskrats. Thus, they became a threat to the agriculture industry and to the water supply. Currently, nutria populations are moderately high and on the increase in Texas. They have continued to expand their range and probably will spread throughout the state. Specimens have now been reported from Val Verde and Terrell counties (Hollander et al., 1992), and from Big Bend National Park in Brewster County (Milholland et al., 2010).

Feral hogs, *Sus scrofa*

Invasive feral hogs, also termed wild pigs, feral pigs, feral swine, or wild boars, are widely distributed and destructive throughout at least 35 states in the US (Corn and Jordan, 2017), and they are expanding their range rapidly (Katz, 2019). Feral swine include free-living descendants of domestic swine, introduced Eurasian wild boar, and their hybrids. Their expansion has been attributed to intentional and accidental introductions by humans, high reproductive potential (more than four times that of native ungulates), lack of predators, and human alterations to the landscape that increased their habitat availability (Snow et al., 2017). Current population estimates for the US are 6.9 million feral swine, with more than 2.6 million occurring in Texas alone (Kinsey, 2020). Today, they are regarded as the most abundant introduced ungulate in the US, and the IUCN considers them as one of the Earth's 100 worst invasive species. In 2010, it was estimated that 753,646 feral hogs were harvested in Texas (Timmons et al., 2012), which represented about 29 percent of the estimated population. However, it is estimated that annual population control efforts would need to continuously achieve 66–70 percent population reduction just to hold the wild pig population at its current level (Timmons et al., 2012; Dzieciolowski et al., 1992).

Feral swine are responsible for significant damage including negative effects on native plant and animal communities (including sensitive species), agriculture, and livestock, and they pose a significant disease risk for humans, livestock, and native wildlife. For example, they have been implicated in destroying habitat of the endangered Houston toad, *Bufo houstonensis*, in the Lost Pines region of central Texas (Brown et al., 2015), and in 2019 a woman was attacked and killed by multiple feral hogs in Anahuac, Texas, about 40 miles (64 km) east of Houston (Bogel-Burroughs, 2019).

Bailey and federal agents encountered feral hogs in only a few regions of the state, but they have now spread to 253 of the 254 counties in the state, with the one exception (as of 2019) being El Paso County (Kinsey, 2020). Feral hogs now constitute one of the most serious conservation threats in Texas. Because of their prolific reproductive potential—three or four piglets in every litter and about 1-1.5 litters a year—and adaptability, feral hogs pose a danger to other wildlife and especially other mammals. For example, where the range of the feral pig overlaps that of the collared peccary (*Pecari tajacu*), there is some indication that the two species may compete to the detriment of the native species. Likewise, along the Texas coast they have been implicated in the decline of the armadillo, and they could be responsible for the demise of the hog-nosed skunk. Feral hogs alter their diet during drought or seasonal challenges, and in many cases, their food choices overlap with those of native species. Feral hogs disturb (root up) the ground and vegetation when feeding on plant roots and soil invertebrates, and pig rooting reduces the number of plant species in an area.

Besides their growing population, feral hogs pose another serious problem: disease spread. They play a major role in the transmission of such diseases as brucellosis, pseudorabies, and swine fever. Currently, slowing the population explosion of these animals can be achieved only through hunting, death through natural causes, or trapping and removal of the hogs. TPWD allows an open season on hunting these animals without any limits to the number of individuals harvested, and as of 1 September 2019, a new law (SB 317) allows any person (resident or nonresident) to hunt feral hogs on private lands (with landowner consent) without a state-issued hunting license. It is hoped that this new law will assist in the control of feral hog populations. Further, steps are underway by TPWD and other agencies and universities to discover other methods of population control, including toxic baits (Snow et al., 2017), pharmaceuticals for fertility control (Campbell et al., 2006), hog-proof fencing (Lavelle et al., 2011), and aerial gunning (Campbell et al., 2010). There is evidence that some ranchers believe they could be an underutilized resource and the amount of damage could be reduced if landowners treated them more as an economic resource by taking advantage of hunt-leasing opportunities (Adams et al., 2005).

Domestic cats and dogs

Domesticated species of mammals (cats and dogs), while not technically "alien," often become feral and free ranging and thereby become a serious threat to native wildlife. Free-ranging domestic cats, usually called feral cats, prey on ground-nesting birds, songbirds in trees, and small mammals, and feral dogs are known to run down deer and livestock. Feral cats have been introduced globally and have contributed to multiple wildlife extinctions on islands. A recent scientific survey estimated that feral cats kill 1.3-4.0 billion birds and 6.3-22.3 billion mammals annually, with un-owned cats, as opposed to owned pets, causing the majority of this mortality (Loss et al., 2013). This study concluded that free-ranging cats cause substantially greater wildlife mortality than previously thought and are likely the single greatest source of anthropogenic mortality for US birds and mammals.

Japanese snow monkey, *Macaca fuscata*

Amazingly, the most recent example, and perhaps the most dangerous, of an exotic mammal introduced in Texas involves a group of Old World monkeys called Japanese snow monkeys or macaques that were originally established in captivity but later established a free-living population in South Texas. The monkeys were descendants of a small troop brought to Texas in 1972 to save them from destruction in Kyoto, Japan, where they had become a pest. The initial troop of about 150 animals was brought to a ranch near Encinal, La Salle County, where they remained until 1980; they were then moved to a ranch near Dilley, Frio County, where they were allowed to roam with minimum human interference or control (Racine, 2000). The number in the troop slowly increased to between 500 and 600 monkeys, and several eventually escaped and began living in the wild, which caused considerable consternation among ranchers and wildlife officials. They adapted well to the climate in South Texas and established a free-ranging, breeding population.

Eventually, several monkeys were shot by hunters, causing a public outcry with TPWD caught in the middle. Subsequently, funds were raised to properly house the monkeys, and TPWD clarified their legal status as refugee monkeys that could not be hunted. However, this did not completely prevent others from escaping; for example, one was sighted in Kerr County, near the Kerr Wildlife Management Area, in 1998. For additional details about their status in Texas, see Schmidly and Bradley (2016). The introduction of free-ranging primates onto rangeland in Texas represents one of the most irresponsible acts of species introductions in the history of the state because wild monkeys pose a potential health hazard to people and to native wildlife.

Exotic ungulates or the so-called "Texotics"

Texas has the most widespread and abundant populations of nonindigenous ungulates (often referred to as "Texotics") within the United States (Teer et al., 1993) (Fig. 169a–d); because of its subtropical-like terrain, lax regulations, and abundance of private land, the state has become the epicenter of the exotic game industry in the country (Ferguson, 2021). The number of species and their populations have proliferated since their first known introduction on the King Ranch in southern Texas in the late 1920s (Jackson, 1964). Landowner and hunter interest in stocking exotic game has grown rapidly in recent years due to the potential economic return to landowners, the aesthetic value of the animals, and the demand for recreational harvest opportunities for these species (Demarais et al., 1998). The interest in the subject has become so popular that *Texas Monthly* magazine published an article that was featured on the front cover of the February 2021 issue (Ferguson, 2021).

Figure 169a-d. Exotic ungulates that are free-ranging and common in Texas: (a) Axis deer, (b) aoudad, (c) nilgai, and (d) blackbuck. Photos a-c courtesy Texas Parks and Wildlife Department; photo d courtesy John and Gloria Tveten.

TPWD developed a statewide survey instrument beginning in 1963 to document the status of exotic populations; results of the 1963 through 1988 surveys were summarized by Mungall and Sheffield (1994). During this time, the exotic population in Texas grew from about 14,000 animals of 13 species to more than 164,000 animals and 67 species. A 1994 TPWD survey (Traweek, 1995), the last time one was conducted, produced an estimate of about 195,000 exotic animals and 71 species, which amounted to a 19 percent increase during the six-year interval after 1988. The estimated population of free-ranging exotics in 1994 was about 77,000 (39 percent of total), with 50 percent in the Edwards Plateau and 42 percent in the South Texas Plains, respectively (Traweek, 1995).

As a result of tremendous demand from private landowners throughout the state, the exotic wildlife industry has continued to experience tremendous growth. Today, the Kerrville-based Exotic Wildlife Association estimates that 5,000 ranches across nearly all of rural Texas are home to 2 million non-native mammals from 130 different species, pumping 14,000 jobs and $2 billion into rural economies (Ferguson, 2021). This dollar value is up from an estimated $1.3 billion that was reported in a 2007 Texas A&M University report (Anderson et al., 2007).

Among the numerous exotic ungulates, 11 species have large enough populations in a free-living condition so that they must now be considered permanent additions to our mammal fauna. These are: axis deer (*Cervus axis*), fallow deer (*Cervus dama*), sika deer (*Cervus nippon*), nilgai (*Boselaphus tragocamelus*), aoudad sheep (*Ammotragus lervia*), black-buck (*Antilope cervicapra*), eastern Thomson's gazelle (*Eudorcas thomsonii*), sable antelope (*Hippotragus niger*), scimitar-horned oryx (*Oryx dammah*), common eland (*Taurotragus oryx*), and greater kudu (*Tragelaphus strepsiceros*). Aside from feral hogs, axis deer are thought to be the most common non-native game species in Texas.

Although exotics have become a significant source of revenue for landowners, there is considerable concern among game biologists and mammalogists about the long-term impacts of these introduced animals on native wildlife (Middleton, 2007). This controversy pits the private sector, eager to diversify its agricultural base, against traditional sportsmen and government agencies worried about impacts of such activities on indigenous free-ranging wildlife, particularly ungulates (white-tailed deer) and their habitat (Demarais et al., 1990).

Some of the issues that concern conservationists and wildlife professionals about wildlife ranching and farming include disease-related interactions (e.g., chronic wasting disease) between commercial livestock and native pen-raised cervids, competitive interactions between native and exotic big game, and potential consequences of interbreeding between native and exotic big game, which could alter the genetic makeup of affected populations (Samuel and Demarais, 1993). But overpopulation is probably the single greatest "negative" concerning exotics in Texas (Middleton, 2007).

There are indications that high densities of exotic ungulates, combined with overgrazing by native deer and livestock, may severely impact rare native plants. Bob O'Kennon, a rancher in Gillespie County in the Hill Country, established a 15-acre native-plant sanctuary on his ranch by installing fencing to keep out both exotic and native ungulates. In an eighteen-month period following establishment of the enclosure, he recorded 30 species of plants that previously had not been recorded in the county (Schmidly, 2002).

There is fairly conclusive evidence of displacement of native ungulates by nonindigenous species (Demarais et al., 1998). Research at the Kerr Wildlife Management Area conducted by TPWD biologists has shown that sika deer and axis deer displaced white-tailed deer within experimental 96-acre enclosures after eight or nine years (Harmel, 1980), and wildlife ecologists also found white-tailed deer absent from several larger (4,000-acre plus) high-fenced ranches that contained exotics after about the same length of time (Middleton, 2007). The most often cited mechanisms underlying potential competition are dietary overlap and competition for limited forage resources (Demarais et al., 1998). The theory is that sika and axis deer outcompete white-tailed deer (Fig. 170a-b) because they compete directly for the forages most needed by whitetails, and they also do well on lower quality forage that cannot sustain whitetails (Feldhamer and Armstrong, 1993).

Variation in reproductive characteristics of the three species of deer also may explain some of the competitive differences (Demarais et al., 1998). Whitetail females typically produce two fetuses with peak fawning in June of each year; sika and axis deer produce only one fawn and fawning season peaks during May. The effect of earlier fawning dates

a

b

Figure 170a-b. Two photos illustrating the differences in body condition in (a) white-tailed deer (*Odocoileus virginianus*) and (b) axis deer (*Axis axis*) following a severe drought near Junction, Texas, that occurred in 2011. In the two photographs, taken in September, it is clear that the native whitetails were approaching starvation due to the lack of forbs and other preferred forage, whereas the introduced axis deer were able to maintain good body condition by switching their diet to grasses and browse. Courtesy Derrick Ard, Texas Tech University Center at Junction.

and producing only one fawn improves the recruitment rate for sika and axis deer over white-tailed deer.

The potential exists for a more widespread displacement in areas where deer-proof fences do not confine exotics (Butts, 1979). Some wildlife ecologists think there may be as many exotics in central Texas as there are white-tailed deer, and that the exotics may now constitute a majority of the free-ranging animals in the Hill Country. It is these loose herds of common exotics that pose the greatest potential threat to native wildlife and plants (Middleton, 2007).

In addition to these biological concerns, the hunting of exotic animals, confined under conditions of husbandry by fencing and with many of them rare or endangered in their native ranges, has pitted preservationist animal-rights groups against more conservationist and utilitarian rancher-landowners (Fernandez, 2017; Condy, 2018). To many animal-protection groups, the management of rare and endangered species for purposes of breeding and hunting is repulsive and creates a legal and ethical gray area for hunting ranches. Hunting advocates disagree and say the breeding and hunting of exotic animals help ensure the survival of species. They see themselves not as an enemy of wildlife conservation but as an ally. For example, they maintain there are more blackbuck antelope in Texas than there are in their native India because of the hunting ranches. In addition, Texas ranchers have in the past sent exotic animals, such as scimitar-horned oryx, back to their home countries to build up wildlife populations there.

Exotics are here to stay, and their presence challenges Texas wildlife managers to minimize habitat degradation, disease, competition, and possibly hybridization with native wildlife while at the same time retaining the economic and aesthetic attraction of such introductions. These are issues that have emerged in the latter half of the twentieth century, but they loom even larger for the twenty-first century. Because exotics are classified for purposes of management in the same way that domestic livestock are, TPWD does not regulate them, although it will provide guidance to landowners who want to manage both exotics and wildlife, especially white-tailed deer.

Conservationists and the public have begun to argue for more accountability and transparency over this industry. The *Dallas Morning News* published an editorial in 2019 (*Dallas Morning News*, 2019) calling for an official census, much more state oversight, and more research and general awareness about the impact of these animals. This is long overdue, as the last survey to determine the number of species and animals in the state was conducted more than twenty-five years ago in 1994. But, for any such study to be useful, it must address biological as well as economic considerations.

Growth in Commercialization of Wildlife

Hunting on both private and public lands this century has trended toward commercial or fee-hunting, in which those who own or manage the land receive compensation for their services (Teer et al., 1993). Commercial lease hunting has produced many positive benefits for wildlife conservation in Texas. Leasing of private land for hunting of white-tailed deer and numerous other game species has become a widespread and profitable practice for many landowners throughout the state. In 1991, private landowners received an estimated income of $163 million from hunting leases in Texas; the total today is likely to be several hundred million dollars. The 2011 survey of hunting, fishing, and wildlife-associated recreation reported that almost 5.9 million Texas residents engaged in fishing, hunting, or wildlife-watching activities, and that state residents and non-residents spent $6.2 billion on wildlife-associated recreation (USDI/USFWS/USDC/USCB, 2011). Of that total, $1.8 billion was spent for hunting. By profiting from the practice, landowners also have instituted habitat management programs that benefit a wide variety of species. Also, the sale of hunting licenses by TPWD has resulted in the availability of public funds to support management programs for game animals.

In spite of these positive benefits, a few wildlife conservationists, notably Valerius Geist from the University of Calgary in Canada, have published a series of articles in professional wildlife journals severely criticizing the Texas system of wildlife commercialization (e.g., Geist, 1985, 1988, 1989, 1991, 1994), specifically involving game ranching and hunt leasing, on the basis of philosophical grounds. Among the drawbacks of game ranching and market hunting is its potential impact on predators, which are not compatible with this practice, and the reliance of landowners upon exotics that potentially could threaten native ungulates via disease transmission and genetic pollution.

There are also bio-political and social equity issues at play when considering how to provide access to public resources in a system that is privately controlled. Paid hunting excludes many with low incomes, particularly ethnic minorities. Many wildlife professionals are concerned about the long-term erosion of public support for wildlife conservation and management in such a scenario. Game ranching was not even on the radar at the beginning of the twentieth century. It grew as an enterprise in the latter half of the century as demand for hunting opportunities increased and private landowners developed a need for income diversification. While the practice is certainly here to stay and seems to be working at the present time, it is an issue that bears watching as we seek to manage wildlife resources in the twenty-first century.

Another aspect of wildlife commercialization is the growing international traffic in illegal wildlife trade that has become one of the most serious conservation issues in the world. A recently published study has documented more than 8,000 wild species in the global wildlife trade market, encompassing almost 30 percent of all mammal, bird, amphibian, and reptile species (Scheffers et al., 2019). This increase in demand, driven by exotic pets, furs, jewelry, and body parts for traditional medicine, is a major factor in the decline of many species globally. In recent years, the market for illegal wildlife trade has moved online, and in 2013 TPWD joined with the USFWS, other states, and three Asian countries to crack down on internet wildlife crimes, resulting in 61 state and federal cases in Texas. The illegal sale and exploitation of wildlife resources is a global problem that has a direct negative effect on Texas and could lead to the loss of Texas native species, either through the harvest of native species or introduction of nonindigenous invasive species (Roe, 2019).

Unfortunately, Texas lacks a good, up-to-date record on the extent of international trade in wildlife in the state. A survey by Jester et al. (1990) evaluated the commercial use of nongame species, which prior to that time had been overlooked and virtually unknown.

This report, although inconclusive in terms of the information available, documented several instances of unregulated commercial use of many terrestrial wildlife species. The major demand exists in the rattlesnake market, pet industry, food industry, apparel and decoration industry, the Asian folk medicine market, and various opportunistic markets. The pet industry utilizes more species of nongame species than all of the other industries combined. Reptiles, amphibians, mammals, and birds are all utilized by this trade, including songbirds, game birds, and game animals. The apparel, decoration, and food markets utilize select species, primarily reptiles and mammals. TheAsian folk medicine market utilizes venomous snakes and their parts. The opportunistic markets specialize in the use of live coyotes, cougars, jackrabbits, and feral pigeons.

The survey by Jester et al. (1990) suggests that the trade in nongame wildlife may be substantial. Markets for nongame wildlife and wildlife products are diverse, and the economic magnitude of this trade cannot be dismissed as inconsequential. Such commercial trade results in a few people realizing profits without returning anything back to the resource that is the property of all citizens. Until 1999, the only requirement to collect and commercialize nongame wildlife was a hunting license. However, the 76th Legislature adopted a new section of the Parks and Wildlife Code, Chapter 67, which provides the commission with authority to establish any limits on the taking, possession, propagation, transportation, importation, exportation, sale, or offering for sale of nongame fish or wildlife and establishes the requirement of a permit for the collection and sale of nongame wildlife.

In 2007, TPWD split "nongame" animals into two groups: those that can be collected and sold by dealers, and those that cannot (Roe, 2019). The "white list" names 84 species of lizards, amphibians, snakes, and mammals that can be collected from the wild. There are only eight mammals on this list: Texas antelope squirrel (*Ammospermophilus interpres*), black-tailed prairie dog (*Cynomys ludovicianus*), Merriam's kangaroo rat (*Dipodomys merriami*), eastern flying squirrel (*Glaucomys volans*), black-tailed jackrabbit (*Lepus californicus*), spotted ground squirrel (*Xerospermophilus spilosoma*), thirteen-lined ground squirrel (*Ictidomys tridecemlineatus*), and rock squirrel (*Otospermophilus variegatus*). The "black list" spells out the species that cannot be collected or sold. This list includes 180 species, of which 85 are nongame mammals. Of the mammals, all five shrews and moles (Order Eulipotyphla) are included as are 27 bats (Order Chiroptera), 52 rodents (Order Rodentia), and 1 carnivore (Order Carnivora).

The Possible Impact of Climate Change on Texas Mammals

In the last decade of the twentieth century, Guy Cameron, formerly of the University of Houston, and his students made an early attempt to use models of climate-vegetation association to predict mammalian distributions in the state under current conditions, as well as two future climate scenarios that would occur if carbon dioxide levels doubled as then predicted (Cameron and Scheel, 1993; Scheel et al., 1996). Both future climates were warmer, but one was wetter and the other drier than current conditions. Under both future climatic conditions, the results suggested that all temperate vegetation in Texas would eventually be lost. Under drier conditions, steppe and scrub habitats would appear in the state; under wetter conditions, subtropical habitats would appear.

Cameron's analysis revealed that, in general, rodents would be the most adaptable group of mammals to global climate change. Half of rodent species were predicted to decrease under drier climates and 63 percent were predicted to decrease under wetter climates. For both insectivores and lagomorphs, no species were predicted to go extinct in the state, but species ranges were predicted to contract under both drier and wetter conditions and species diversity was predicted to decrease in the southern part of the state. Ranges of bats dependent on fixed roosts such as caves and rock crevices were predicted to decrease substantially under future climatic conditions, although a few were predicted to increase under both climatic scenarios. Under both future climates, species richness was predicted to decrease in western Texas because of the loss of cavity-roosting bats but to increase in southern and central Texas with the expansion of tree-roosting bats. Again, no species were predicted to go extinct under new climatic conditions.

However, there is a caveat to all of this work: it was conducted more than twenty-five years ago and climate models have changed and improved considerably since that time. For this reason, there needs to be a reevaluation of how mammal diversity and distributions might react to changing climate conditions in the state over the remainder of the twenty-first century using today's more enhanced and accurate prediction models. Recently, scientists have begun to use ecological niche models to project availability of suitable bioclimatic conditions for wildlife species under various climate scenarios derived from recently generated general circulation models (e.g., Salas et al., 2017a, b). Studies of select bird, amphibian, and reptile species using this approach have predicted different distributions of suitable climatic conditions and a significant reduction of present-day distributions for many of the species studied. Using these new approaches with more modern climate change models could produce different predictions about the impact of climate change on mammal species in Texas.

Interestingly, the severe week of arctic winter weather that hit Texas in February 2021, which arguably could be associated with weather extremes caused by climate change, produced a dramatic toll on wildlife in the state. Thousands, if not tens of thousands, of native deer, exotic ungulates, and free-tailed bats died during this extreme cold spell. This represents another example of how extreme weather events, which quickly cause animals to exceed their thermoneutral zone of survivabililty, can negatively impact wildlife populations.

CHAPTER 6

Summary of the Current Status of Mammals and Selected Reptiles in Texas

T HERE ARE 139 EXTANT NATIVE TERRESTRIAL SPECIES IN TEXAS (Schmidly and Bradley, 2016). Adding extinct, introduced, exotic, and free-ranging (feral) domestic species brings the total to 175 terrestrial species of mammals in the state. Fifty-seven of the 139 extant native terrestrial species (41 percent) are represented by more than one subspecies. Seven species and approximately 30 subspecies of terrestrial Texas mammals are thought to be endemic and confined in distribution to the state. This diversity suggests that Texas harbors tremendous genetic and morphological variability in its mammal fauna. Texas also harbors 61 species of snakes and lizards, but a holistic treatment of these species is beyond the scope of this book. Therefore, our comments and discussion are limited to those reptile species that Bailey reported on in his 1905 work; that section is presented at the end of this chapter, following the discussion of the mammalian fauna.

The great diversity of mammalian fauna in Texas, and the fact that for most species there are no ongoing, standardized inventories, makes it difficult to predict trends in mammal populations and species in the state. Using the best data available, as well as the opinions of many professional mammalogists and conservationists, the following summary assesses the current status of the major groups of mammals in the twenty-first century.

As we look back at the twentieth century and prepare for the remainder of the twenty-first, many issues will continue to influence the status of the mammal fauna. The propensity of over-exploitation, largely through over-harvesting and unregulated hunting, severely impacted wildlife at the beginning of the twentieth century. As we forge ahead into the current millennium, loss of habitat, landscape fragmentation, and the commercialization of wildlife undoubtedly will affect our landscapes, habitats, fauna, and flora.

There are several species or populations of terrestrial mammals in Texas that, although not currently on any official list of at-risk species, nevertheless are considered by professional mammalogists to possibly be declining and, thus, may warrant protection in the future. Certainly their situation bears watching; in some cases, considerable additional data are needed to establish the facts necessary to arrive at a meaningful and biologically defensible position as to their status. A synopsis of their status, organized by classification order, is provided below (adapted primarily from Schmidly and Bradley, 2016). Species that already are extinct in Texas have been addressed in Chapter 5.

Since this manuscript was submitted for publication, the Texas Parks and Wildlife Department (TPWD) has published a list of "Species of Greatest Conservation Need in Texas" (SGCN), based on the state's revised Texas Conservation Plan. In that listing, developed through expert consultation and public feedback, species are ranked according to a conservation status system established by NatureServe (https://www.natureserve.org/), using the following categories: critically imperiled, imperiled, vulnerable, apparently secure, and secure (TPWD News, 2021d). The Texas SGCN list includes 66 taxa of mammals (51 species, 15 subspecies), from the following orders: Artiodactyla (2), Carnivora (14), Insectivora (3), Chiroptera (23), Rodentia (22), and Lagomorpha (2) (see https://tpwd.texas.gov/huntwild/wild/wildlife_diversity/nongame/tcap/sgcn.phtml). Of these, almost three-quarters (73 percent) are considered likely to be already extirpated, critically imperiled, imperiled, or vulnerable. Only 24 percent of the taxa listed are considered to be secure or apparently secure. Although the following accounts do not provide the specific SGCN status of each species, most of the mammals on the SGCN list are discussed herein. However, in our opinion, several of the listed taxa, even though they may be locally declining, are quite common throughout some regions of the state (e.g., *Dasypterus ega*), whereas several species that are extremely rare and possibly critically imperiled or endangered do not appear on the list (e.g., *Lasiurus blossevillii*, *Myotis occultus*, and *Mephitis macroura*). In the future, we strongly recommend that the TPWD group that developed the SGCN list work in close concert with mammalogists who have expertise in systematic, population genetic, and conservation research regarding the species and subspecies that appear on this list.

Order Didelphimorphia (Opossum)

The opossum (*Didelphis virginiana*) occurs statewide except in the xeric counties of the western High Plains and Trans-Pecos regions. The species is extremely common throughout the riparian and wooded regions of East Texas. It is listed by the IUCN as least concern and appears to be increasing in numbers throughout the United States, including Texas. By dispersing along streams and rivers, it expanded its range throughout the twentieth century, and it adapts well to most human conditions. It does not appear on the federal or state lists of concerned species and does not need any conservation action in Texas.

Order Cingulata (Armadillos)

Anyone who has driven across central and eastern Texas and noticed the numerous carcasses of road-killed nine-banded armadillos (*Dasypus novemcinctus*) would likely not be concerned about the status of this species. However, concerns recently have been expressed about possible alarming declines in armadillo populations in many places in Texas. In 1999, Dr. Rollin Baker, a retired mammalogist living in Eagle Lake, undertook a survey of professional mammalogists in Texas and found that almost all agreed that armadillos were rare at best when compared with populations decades ago. Dr. Baker (personal communication), who is now deceased, concluded that armadillo numbers along the Texas coast were truly down, and he suggested this may correlate with the dramatic upsurge of feral hogs across the state, as they are known to feed on newborn armadillos. The trouble with population declines is that rarely does anyone—even trained observers—notice them until the creatures involved are practically gone. Therefore, there are reasons to be vigilant in beginning to monitor the status of our official state mammal, the armadillo.

Nine-banded armadillos are the only free-ranging vertebrates other than humans known to exhibit naturally occurring infections of *Mycobacterium leprae*, the causative agent of leprosy. Although little is known about the ecological consequences in wild populations, there is at least some reason to believe that it could be detrimental to the long-term viability of populations. A recent study of an armadillo population in Mississippi, with both leprosy-positive and leprosy-negative individuals, seems to suggest that leprosy has minimal impacts on individuals, which is a surprising and unexpected result given the substantial impact of infection documented in the laboratory (Morgan and Loughry, 2009).

Order Lagomorpha (Rabbits and Hares)

Five species of rabbits and hares occur in Texas. The lone hare, the black-tailed jackrabbit (*Lepus californicus*), occurs statewide and appears to have few conservation challenges. Of the four rabbits, however, two species bear watching in the future because of their restricted distribution in combination with major habitat pressures.

Sylvilagus robustus, the Davis Mountains cottontail, has a restricted distribution in the central core of mountains (Guadalupe, Davis, and Chisos) in Trans-Pecos Texas. It also is known from the Guadalupe Mountains in New Mexico and from the Sierra de la Madera of Coahuila, Mexico. This species was elevated to the status of a distinct species by Ruedas (1998), previously having been considered as a subspecies of the wide-ranging species, *S. floridanus*. This taxonomic interpretation has been reaffirmed by recent DNA sequence analysis (Nalls et al., 2012). The IUCN listed this species as endangered in 2008, but it does not appear on the USFWS or TPWD lists. Given its restricted distribution, this rabbit almost certainly will require conservation efforts in the future.

In 1998, Luis Ruedas (Portland State University) expressed concern that populations of *S. robustus* had declined dramatically in recent decades. Ruedas recommended state listing for this species, as well as listing as endangered by the IUCN, pending the completion of more detailed studies about its natural history, including assessment of population status and life history parameters, long-term ecological studies, and population genetic analyses. The population in the Guadalupe Mountains declined for unknown reasons to about fifty individuals in the late 1940s. Few individuals have been verified there since the 1960s, although one specimen was collected in May 2000 (Ruedas and Dowler, 2018). In the Davis Mountains, specimens collected in 1996 and 1997 constituted the first records in 20 years from that area. However, Robert Bradley and his students collected eleven specimens from Mt. Livermore and one from Elephant Mountain in 1998 and 1999, and they hypothesized that this species might be more common than initially thought at high elevations in the Davis Mountains and perhaps the Del Norte Mountains. Several specimens from the Chisos Mountains also have been collected in the past fifteen years. The Davis Mountains cottontail probably occurred historically in low densities and population numbers, fluctuating as a function of precipitation, resulting in greater sensitivity to threats and leading to local extirpations and extinctions. The narrow elevation range where the species occurs (4,500-7,700 ft; 1,400–2,347 m) suggests it likely will be sensitive to global warming (Ruedas and Dowler, 2018). Careful monitoring of this species is needed throughout the isolated mountain ranges of the Trans-Pecos and northern Mexico to continually ascertain its overall conservation status.

The recent discovery of rabbit hemorrhagic disease in cottontail and jackrabbit populations from several counties in the Trans-Pecos (see Chapter 4), including within close proximity to the largest remaining population of the Davis Mountains cottontail in Jeff Davis County, would be disastrous if it were to infect these rare rabbits.

The swamp rabbit (*Sylvilagus aquaticus*), as the name suggests, inhabits poorly drained river bottoms and coastal marshes. Even though it is widely distributed throughout the eastern one-third of the state, it seldom occurs in large numbers. Although its IUCN status is listed as least concern, and it does not appear on the federal lists of concerned species, this species appears to be declining in numbers. Optimum habitat in much of the rabbit's range is shrinking with the drainage of wetlands and clearing of bottomland hardwood forests; consequently, there is much concern about the future conservation status of this rabbit in Texas. In the Hill Country, the species is threatened by habitat fragmentation. There is an imperative need to monitor populations in the future.

Order Eulipotyphla (Shrews and Moles)

There are four species of shrews in Texas, and although all of them are relatively uncommon, one especially bears watching in the future. Elliot's short-tailed shrew, *Blarina hylophaga*, has a disjunct distribution in Texas, having been recorded in Montague County on the Red River near Oklahoma, in Bastrop County in the central part of the state, and in Aransas County along the lower Texas coast. A recently published taxonomic revision referred the Aransas County and Bastrop County specimens to the subspecies *B. h. plumbea*; the population from Montague County was assigned to *B. h. hylophaga* (Reilly et al., 2005). The only place where the species has been taken in any numbers is in Aransas County (Schmidly and Brown, 1979) where it was commonly found in the oak mottes on Aransas County Wildlife Refuge. A number of management proposals have been implemented for this habitat, including burning, clearing, and grazing, but the potential effects of these activities on the shrew are unknown. Although this shrew does not appear on any official list of threatened or endangered species, given its limited distribution and population status this is a species that should be carefully monitored in the future. The subspecies *B. h. plumbea*, which is endemic to the state, is listed as critically imperiled in the Texas Conservation Plan by TPWD.

The other three shrews in Texas—southern short-tailed shrew (*Blarina carolinensis*), least shrew (*Cryptotis parva*), and Crawford's desert shrew (*Notiosorex crawfordi*)—are relatively common and more broadly distributed. Likewise, the only species of mole in the state (eastern mole, *Scalopus aquaticus*) is almost statewide in distribution and shows no signs of population decline, although one subspecies, *S. a. texanus*, is now extinct in the state (see Chapter 5).

Order Chiroptera (Bats)

Thirty-four species of bats occur in Texas, including all but twelve of the species and more than 50 percent of the subspecies in the United States. As many as thirteen of these—almost one-third the total number—appear on some sort of endangered, threatened, or rare species watch-list and many more bear careful watching. Drastic reductions in bat populations have been reported in recent years not only in the United States but worldwide as well. Throughout the Southwest and Mexico, there is evidence that many bat populations are declining. The life history strategy of bats, which features low fecundity (they typically rear only one young per year and hence are slow to recover from major population declines), specialized roosting requirements, and (in most cases) insectivorous habits, make them especially vulnerable to environmental threats resulting from pesticide use, habitat destruction, and roost disturbance. In addition, many species form large aggregations, which are vulnerable to mass destruction.

Bats occur in all the major ecological regions of Texas, but the Big Bend region of the Trans-Pecos, with its topographic pattern of high mountains and desert lowlands, supports more kinds of bats (20 species) than any other part of the state. Several extremely rare and unusual bats occur in this region, and there is concern about their conservation status. Four species of bats in Texas are included on at-risk, protected lists either by TPWD or the USFWS, but several others require considerable additional data to arrive at a meaningful and biologically defensible position as to their status.

Family Molossidae, free-tailed bats

Two large free-tailed bats of the family Molossidae, the western bonneted (*Eumops perotis*) and the big free-tailed bat (*Nyctinomops macrotis*), have restricted distributions in Texas and have been rarely observed (Schmidly, 1991; Ammerman et al., 2012; Jones and Weaver, 2018). Although neither species is on federal or state lists of concerned species, additional data, particularly about population size, are needed to establish a meaningful and biologically defensible position as to their status. A smaller free-tailed bat, the pocketed free-tailed bat (*Nyctinomops femorosaccus*), occurs in the Big Bend region where it is a common year-round resident.

The Brazilian free-tailed bat (*Tadarida brasiliensis*) is the most common bat in Texas, occurring statewide in caves, buildings, and bridges in urban areas. Its migratory life history strategy normally has the bats migrating south to Mexico, where breeding takes place, in late October and November with the onset of colder temperatures in Texas. With the onset of warmer weather in March, the bats return to roosts in Texas where females give birth in large nursery colonies. One of the most popular places to observe these bats is the Ann W. Richards Congress Avenue Bridge in Austin where the bats have become a major tourist attraction. In the last three to five years, thousands of these bats have been expanding their stay at the bridge because Austin's average winter temperature has been getting warmer (Bradshaw, 2019). Instead of leaving the bridge in November, many bats now stay through December and January, and some may remain throughout the year. Typically, bat season in central Texas starts in March, when the bats return from Mexico, but now some bats are coming back as early as January and February in response to the warmer weather. This is a good example of how climate change can influence life history patterns of mammals.

Family Phyllostomidae, New World leaf-nosed bats

Four New World leaf-nosed bats occur in Texas and each of them bears watching from a conservation perspective. *Leptonycteris nivalis*, the Mexican long-nosed bat, is a colonial, cave-dwelling bat that usually inhabits deep caverns. It was first discovered in the United States in 1937 in a cave in the Chisos Mountains of Big Bend National Park (Borell and Bryant, 1942). Even today, aside from a few records of foraging individuals, the only known colony of these bats in the United States is from the large cave on Mt. Emory in the park and from the Chinati Mountains in Presidio County (Schmidly, 1991; Ammerman et al., 2012). Yearly estimates of population size at the Big Bend colony ranged from zero to as many as 10,650 individuals (Easterla, 1972). However, recent observations in the cave cast doubt on the reliability of data in years when the population estimate was zero. Not only do the bats use the "main" room of Emory Peak Cave but they also roost in deep passageways where they are known to go undetected (Ammerman et al., 2012). It has been suggested that the colony size of these bats in Texas likely fluctuates with food availability in northern Mexico and Texas. A documented decrease in colony size at Emory Peak Cave in 2008 was probably due in large part to a decline in the number of flowering agave plants that year (Ammerman and Tabor, 2008). Some scientists believe that the colony forms in years when overpopulation or low food supply in Mexico forces the bats to move northward. However, even considering natural fluctuations and different methods of estimating numbers, there still appears to be a downward trend in the numbers of bats at the Big Bend colony, and population declines also have been documented in Mexico (Campbell, 1995).

For these reasons, the US Fish and Wildlife Service added this bat, along with its closely related congener, the lesser long-nosed bat, *Leptonycteris yerbabuenae*, to the federal endangered species list (Shull, 1988). Long-nosed bats are nectar-feeding species that utilize the pollen and insects from night-blooming century plants (*Agave* sp.). Destruction of caves and loss of food sources, particularly agave plants, through agriculture, ranching, and human development, are detrimental to populations. There are concerns about the loss of their habitat in Mexico and the destruction of caves throughout the range of the species in that country (Wilson, 1985).

The lesser long-nosed bat, only recently recorded from Texas (see Chapter 5), made history in 2018 when it became the first bat to be removed from the US endangered species list (Greshko, 2018). Thirty years ago, when it was listed, only a few thousand individuals remained, scattered among fourteen roosts in the US and Mexico. The population declines were due primarily to the loss of habitat, including the destruction of roost sites, land conversion for agriculture, and the harvesting of agave to make mescal and tequila. To obtain the sugars from the heart of the agaves that produce these liquors, the plants are harvested and destroyed before they flower, thus eliminating the opportunity for the bats to feed on them and, ironically, limiting the reproduction of the very plants on which the tequila producers rely for their product (Bradley and Bradley, 2018).

In 1994, a cooperative, bi-national program of recovery, specifically aimed at the lesser-nosed bat, was established between Mexico and the United States that included research, active environmental education, and direct conservation actions. Rodrigo Medellin, a mammalogist at the National Autonomous University of Mexico (UNAM) and a leading researcher of these bats, began working with tequila producers to join the conservation efforts to protect nectar-feeding bats by allowing five percent of the agave crop to flower. Cooperating tequila producers are certified as "bat-friendly" and use a bat-friendly label on their bottles as a marketing tool. This practice became very popular with consumers, thus benefitting both the tequila producers and the bats (Bradley and Bradley, 2018). These efforts, combined with protection of roost sites, have been so successful in aiding the recovery of the lesser long-nosed bat that the species was delisted as an endangered species in Mexico in 2015 and the United States in 2018. There are now an estimated 200,000 bats in at least 75 roosts between the two countries. Unfortunately, the Mexican long-nosed bat (*Leptonycteris nivalis*), from the Big Bend region of Texas, remains listed as endangered in Mexico and the United States. Only time will tell whether conservation efforts will help this species survive (Bradley and Bradley, 2018).

Another nectar-feeding bat, the Mexican long-tongued bat (*Choeronycteris mexicana*), for a long time was known in Texas on the basis of a single individual photographed in Hidalgo County, leading to speculation that it was only of accidental occurrence in the state or a southern species that was gradually making its way into the state. Now, it has been recorded in seven counties, suggesting that a tenuous, seasonal population occupies the southernmost portions of the state (Schmidly and Bradley, 2016). The IUCN status of this bat is listed as near threatened. Although widely distributed, it is dependent on a highly fragile habitat (agave) and is thought to be in significant decline due to increased human populations and habitat conversion. It is not included on the federal or state lists of concerned species, but it is listed as endangered by the Mexican government, and one could argue that it should be afforded the same status in Texas (Ammerman et al., 2012).

Of certain accidental occurrence is the one specimen of hairy-legged vampire bat (*Diphylla ecaudata*) found in a partially inundated railway tunnel in Val Verde County in 1967. The nearest record to Texas is from 450 miles (724 km) to the southwest in Tamaulipas, Mexico, where the species is more frequently encountered.

Family Vespertilionidae, Vesper bats

There are 25 species of vesper bats in Texas, making it by far the most common assemblage of bat species in the state. The various species occur throughout the state, and a few of them appear to have major conservation concerns.

Corynorhinus rafinesquii, Rafinesque's big-eared bat, occurs throughout the southeastern United States, reaching the westernmost portion of its range in the pine-oak and longleaf pine vegetation regions of East Texas where it has been found only in small numbers at scattered localities (Ammerman et al., 2012). It was first recorded in 1965 (Michael and Birch, 1967), and it remains one of the rarest species of bats today, having been recorded from only fifteen counties. Its favored roosting sites include partially lighted, unoccupied buildings and other man-made structures such as wells and cisterns. This species currently is not listed by USFWS, although previously it had been considered as a candidate species. It is currently listed as threatened by TPWD. This bat bears special watching because of its scarcity, the lack of knowledge about its population levels, and the considerable potential that exists for degradation of roosting and feeding sites by commercial logging practices in its preferred habitat. There is a real need to determine the effects of modern timber management practices on this species as these bats appear to be extremely sensitive to human disturbance at their roost sites, which could negatively impact their population (Schmidly and Bradley, 2016). A closely related species, *Corynorhinus townsendii*, which is known from the northern High Plains and adjacent Rolling Plains and from the Trans-Pecos and western Edwards Plateau vegetation regions, appears to have declined in Texas as a result of the blasting of old mine tunnels to permanently seal them off. This practice destroys not only significant numbers of the bats but also permanently removes many of their roosting sites.

There are three species of yellow bats (genus *Dasypterus*) in Texas and all bear watching in the future. Interestingly, individuals of all three species recently were documented as fatalities at a wind farm in Starr County in southern Texas, which is the first report of sympatry among all three *Dasypterus* species in the state (Weaver et al., 2020).

Northern yellow bats, *D. intermedius*, are tree-roosting bats comprised of two subspecies in Texas, *D. i. intermedius* from Victoria County southward and *D. i. floridanus* from Bexar and Travis counties northeastward through southeastern Texas. The two subspecies appear to be geographically separated; the former prefers to roost in Spanish Moss and the latter subspecies in palm trees (Schmidly and Bradley, 2016). A recent molecular genetics study confirmed genetic differences between the two subspecies, although gene flow is evident (Decker and Ammerman, 2020). The study of bat mortality at the Starr County wind farm produced an estimated annual mortality of more than 1,100 northern yellow bats, which is of concern because the species does not appear to be common anywhere in the state (Weaver et al., 2020).

The western yellow bat, *D. xanthinus*, which was known from Mexico and extreme southwestern New Mexico, southern Arizona, and southern California, has been reported recently (1990s) from western Texas, including Brewster County (Bradley et al., 1999a; Higginbotham et al., 1999; Dixon, 2000), the Davis Mountains (Jones et al., 1999), and from Val Verde and El Paso counties (Schmidly and Bradley, 2016). There is also a recent record from the Starr County wind farm (Weaver et al., 2020; Chipps et al., 2020a). The status of this species is listed by the IUCN as least concern because of its wide distribution, presumed large population, and occurrence in a number of protected areas. At this time the species is not listed on the federal or state lists of concerned species. However, its complete distribution and population abundance must be studied before its conservation status can be accurately determined. There appears to be an established population in the Big Bend region of Brewster County, where it is encountered most often in the fall.

Dasypterus ega, the southern yellow bat, is a neotropical species that reaches the United States in southern Texas where it has been recorded from seven counties in the Southern Texas Plains and Gulf Coastal Plains ecoregions (Schmidly and Bradley, 2016). It has been suggested that this bat is expanding its range in the United States because of the increased usage of ornamental palm trees in landscaping (Ammerman et al., 2012). In the vicinity of Brownsville, numbers of them inhabit a natural grove of palm trees (*Sabal texana*), and they appear to be a permanent resident of that area (Baker et al., 1971). Although concerns have been raised about the conservation status of this bat, primarily because of low population size and concerns about its habitat (Schmidly and Bradley, 2016), this bat does not appear on the USFWS, IUCN, or TPWD lists. A recent molecular genetics study has confirmed evidence of some hybridization between *D. ega* and *D. intermedius* in southern Texas (Chipps et al., 2020b).

Euderma maculatum, the spotted bat, although unmistakable in appearance, is one of the least understood of American bats, primarily because of its relative scarcity, at least in mammal collections. There have been scattered records of this bat throughout the western United States dating back to 1891, but it has been taken with regularity only in California, Arizona, New Mexico, southern Utah, and southern Colorado. It was first found in Texas in 1967 in Big Bend National Park (Easterla, 1970). Additional specimens have since been obtained from several localities within the park, but no specimens have been captured outside the park area (Ammerman et al., 2012). This infrequency of capture has caused much confusion and speculation regarding its status. It is listed as threatened by TPWD because of its restricted range and apparent low population abundance. The IUCN lists its status as "least concern."

There are three species of tree-bats of the genus *Lasiurus* in Texas, and one of them bears watching in the future. *Lasiurus blossevillii*, the western red bat, was known only from the Sierra Vieja in Presidio County (Genoways and Baker, 1988) until a second record was documented in Starr County (Weaver et al., 2020). In the southwestern US (Arizona and New Mexico) and northeastern Mexico (Tamaulipas), this is a relatively common species in riparian forest canopies associated with streams in arid mountain ranges. Because it is rare and has such a patchy distribution, more work is needed to determine its status and whether a resident population occurs in the state. The eastern red bat, *L. borealis*, is much more common, having been recorded in the Davis Mountains as well as the Sierra Vieja (Jones and Bradley, 1999) and across the rest of the state. The third species of tree bat, the Seminole bat (*L. seminolus*), remains locally abundant throughout eastern Texas and may be expanding its range westward based on a recent record from Val Verde County (Schmidly and Bradley, 2016).

Nine species of myotis bats (genus *Myotis*) occur in Texas, making this the most specious genus in the state. Most appear to be in good shape, with the possible exception of a few. *Myotis austroriparius*, the southeastern myotis, occurs in the southeastern United States, extending westward into eastern Texas where it was known previously in low population numbers from a few sites in the Piney Woods region. In 1996, three county records extended the range of this species westward to Leon, Freestone, and Walker counties, and a specimen collected in Comanche County in 1995 extended its known range approximately 150 miles (241 km) west in the state (Schmidly and Bradley, 2016). Although it is primarily a cave bat in other parts of its range, in Texas these bats roost primarily in live, hollow, bottomland hardwood trees close to slow-moving rivers and in fabricated structures such as abandoned houses and culverts. It was first recorded in Texas in Bowie County in 1962 (Packard, 1966). In 1991, Schmidly noted nine county records and it is now known from 24 counties. Although it is not included on the TPWD list, previously it had been considered a species of concern by the USFWS. Also, at one time the Texas Organization of Endangered Species (which no longer exists) listed it as a watch-list species because of perceived low population density. Although these bats now appear to be more abundant in Texas and Arkansas than was once believed, major declines in populations have been documented in other states over the past several decades, particularly in Florida, where most of the large maternity colonies are located. This species is considered endangered in Illinois, Indiana, and Kentucky, threatened in South Carolina, and a species of special concern in North Carolina. There has been speculation that this species could be threatened by clearing of bottomland hardwood habitats and by destruction of major cave roosting sites, and for those reasons it requires special monitoring in the future (Schmidly and Bradley, 2016).

Other species of *Myotis*—*M. occultus* and *M. septentrionalis*—are known from Texas on the basis of just one or two records. *Myotis occultus* had been known only from Ft. Hancock in Hudspeth County in the early 1900s, but a second specimen was recently recorded in El Paso County in 2011 (Krejsa et al., 2020). *Myotis septentrionalis* is known only by a record from Winter Haven, Dimmit County, in 1942 (Schmidly, 1991; Ammerman et al., 2012). Probably these represent wanderers, and neither species now is considered a part of the permanent Texas fauna.

Two Special Conservation Concerns of Texas Bats

In the twenty-first century, bats are being subjected to a double whammy of potential threats from the erection of wind turbines on rapidly expanding wind farms for electricity production and from the expansion into Texas caves of a fungus that has rapidly spread across the United States, producing a deadly condition in bats known as white-nose syndrome. Collectively, these two developments have the potential to be devastating to many bat populations across the state.

Impact of Wind Energy

Over the past 10 years, wind energy production has increased dramatically in Texas, especially in the western portions of the state, where wind farms may contain anywhere from 10 to 1,000 wind turbines. As a result of this growth, Texas is the top wind power state in the United States. What is not understood very well is the impact of these facilities on wildlife, particularly birds and bats, although the story that is emerging suggests that they are serious threats to the populations of both.

The development and expansion of wind energy facilities is now regarded as one of the key current threats to bats in North America, and in some regions of the United States wind turbines have been implicated in the deaths of up to 600,000 bats per year (Hayes, 2013). Most often, bats are killed when they approach too closely to the vortex caused by the rotating blades. The changes in air pressure basically cause the bat's lungs to explode, a condition known as barotrauma. In addition to the direct effect of wind turbines killing bats, indirect effects may occur as well. Bats have low reproductive rates, and adult females generally give birth to a single individual once a year. This results in bat populations growing slowly and an inability to quickly rebound after rapid declines in population size. Bat populations therefore rely on high adult survival rates to compensate for low reproductive rates and prevent declines. Therefore, substantial cumulative impacts of wind-energy development on certain species, especially tree-roosting species, are expected and these populations would be slow to recover from any population declines.

Several review articles have been published about the impact of wind-energy development on the mortality of bats (e.g., Kunz et al., 2007; Kuvlesky et al., 2007; Arnett et al., 2008; Baerwald et al., 2009; Smallwood, 2013). Bat-fatality rates appear to be variable across sites and regions. Unfortunately, there have been few studies of bat mortality in Texas, which is surprising given that the state leads the United States in installed wind energy capacity (Ellison, 2012). Miller (2008), in an unpublished master's thesis at Texas Tech University, studied wildlife fatalities at a facility in Scurry County in the southern portion of the Texas Panhandle from September 2006 to September 2007. During this study, which involved both birds and bats, 56 bat fatalities were recorded, with free-tailed bats representing the majority of all fatalities (94 percent) followed by hoary bats (4 percent) and eastern red bats (2 percent). The reported mortality rate at this site was above the average bat mortality rate reported for other wind generation projects in the United States.

The *San Antonio Express-News* reported on a study conducted at a wind farm site near Sarita, Kenedy County, where 2,309 bats were killed in 2009 (McDonald, 2011). Although these mortality rates were higher than expected, the study did not identify the species involved, but it did report that no endangered or threatened species had been killed. This study was paid for by the wind energy company and it was not peer reviewed, causing some skepticism about the validity of the conclusions.

Recently, a group of scientists associated with a private consulting company in San Antonio, Texas State University (TSU), and two national energy laboratories (Sandia National Laboratory and the National Renewable Energy Laboratory) conducted mortality studies at the Las Vientas Wind Farm along the Texas–Mexico border in Starr County (see Weaver et al., 2020). The site, which encompasses 54,000 acres (22,000 ha) and 255 wind turbines, is operated by Duke Energy Renewables (DER). The data from their study, which was conducted over a year-long period in 2017–2018, revealed that eight species of bats suffered fatalities at the wind farm, with Brazilian free-tailed bats (*Tadarida brasiliensis*) experiencing the highest annual fatalities (almost 6,000 deaths) followed by northern yellow bats (*Dasypterus intermedius*) with 1,100 deaths. The other six species had fewer than 500 deaths each.

In an attempt to reduce bat mortalities, DER has been working with NRG Systems (a subsididary of ESCO Technologies, Inc.) and TSU to implement an innovative Bat Deterrent System that during field testing has reduced bat fatalities by 50 percent (Amanda Jones, personal communication). The deterrent equipment is mounted on the nacelle (the covering that houses the generating components of the turbine). Once installed, it emits continuous ultrasonic energy in the same frequency range as the bat's sonar. When bats enter the airspace where the deterrent units are operating, the ultrasonic energy essentially disrupts their bio-sonar, making it difficult for them to find food sources and navigate their surroundings, thus effectively minimizing their interactions with the wind turbines (Duke Energy, 2019).

Studies to date concerning the overall impact to bat populations in the US suggest that tree bats (genera *Lasiurus*, *Aeorestes*, and *Lasionycteris*) are the most common victims at wind-generating facilities. It appears that tree-roosting bats are attracted to the wind towers, either mistaking them for roosting sites or perhaps lured by the sound of the motor or rotating blades. However, at wind farms in West Texas, where tree bats are scarce, fatalities appear to be minimal (Parlos et al., 2019), although migratory free-tailed bats (genus *Tadarida*) are regularly killed. Clearly there is a need for more detailed studies of wind farms in Texas to determine a more complete understanding of their impact on bat populations in the state.

Impact of a Rapidly Spreading Fungal Disease

Possibly an even more alarming situation with regard to Texas bats involves the recent discovery of the fungus that causes white-nose syndrome (WNS) among some bats living in Texas caves (Oko, 2019). WNS, named for the white fungus that grows on the nose and wings of bats, largely affects bats that hibernate in caves because the fungus that causes the disease (*Pseudogymnoascus destructans*, or Pd) thrives in the cold. Scientists largely believe that Pd is transmitted from bat to bat, though there is evidence suggesting that humans inadvertently carry it to clean sites from infected areas. The disease causes bats to lose their fat reserves long before winter ends—they wake up too frequently during hibernation and starve to death. WNS was first documented in the northeastern United States in 2007 and has now been recorded in 33 states and seven Canadian provinces. From western Oklahoma, where it was confirmed in 2015 (Brennan et al., 2015; Creecy et al., 2015), the fungus has jumped the state line, and it was first documented in the eastern Texas Panhandle in 2017 (see Graves, 2019a). Cave and bat samples collected from Childress, Collingsworth, Cottle, Hardeman, King, and Scurry counties all tested positive for Pd in three species—Townsend's big-eared bat (*Corynorhinus townsendii*), cave myotis (*Myotis velifer*), and the American perimyotis (*Perimyotis subflavus*). In the spring of 2018, the fungus made its way to Central Texas (Blanco and Kendall counties) where it was detected on a Mexican free-tailed bat (*Tadarida brasiliensis*). By 2019, a detection in Liberty County was the first in East Texas, and recent detections in Frio and Victoria counties represent the most southerly detections of the fungus (TPWD News, 2019a).

Although the fungus responsible for WNS has been detected in at least 21 Texas counties, until recently no sign of the white-nose syndrome disease had been observed (TPWD News, 2019a). There was some thought that Texas bats might be less susceptible to the cold-loving fungus because, with its warmer winters, many Texas bats either do not hibernate or do not hibernate as long (Graves, 2019a). However, on 5 March 2020, a *Myotis velifer* from Gillespie County tested positive for WNS (TPWD News, 2020b), and subsequently, infected bats have been confirmed from 18 central Texas counties, including Uvalde County, now the southernmost case of WNS in North America (*Texas Parks and Wildlife Magazine*, 2020). This instance of WNS is quite concerning, because there are at least 13 species of bats in the state that are known to occupy and use torpor in subterranean habitats (caves, mines, bunkers, culverts, etc.) in the winter months, making them potentially susceptible to WNS. These include: the pocketed free-tailed bat

(*Nyctinomops femorosaccus*), ghost-faced bat (*Mormoops megalophylla*), pallid bat (*Antrozous pallidus*), Townsend's big-eared bat (*Corynorhinus townsendii*), big brown bat (*Eptesicus fuscus*), southeastern myotis (*Myotis austroriparius*), western small-footed myotis (*Myotis cilliolabrum*), cave myotis (*Myotis velifer*), southwestern little brown myotis (*Myotis occultus*), fringed myotis (*Myotis thysanodes*), long-legged myotis (*Myotis volans*), Yuma myotis (*Myotis yumanensis*), and American perimyotis (*Perimyotis subflavus*). Cumulatively, these species constitute more than a third of the total species of bats in the state.

There certainly should be concern that all of these bats could be seriously impacted by WNS because of the important role that bats play in the ecosystem by consuming large numbers of insects. Bats are major predators of night-flying insects, especially moths that produce the larvae of many agricultural pests. Recent estimates suggest the value of bats to farmers in the United States to be about $3 billion annually, and the agricultural value of insect control by bats in Texas has been estimated at $1.4 billion annually. This value includes reduced crop loss to insect pests, reduced spread of crop diseases, and reduced need for pesticide application.

The value of bats to Texas agriculture becomes even more apparent if you consider the potential impact of just a single species, the Mexican free-tailed bat, *Tadarida brasiliensis*. It has been estimated that Texas is home to more than 100 million of these bats that disperse nightly from caves and highway structures such as bridges and culverts. This species consumes important crop pests such as the adult state (moths) of the fall army worm, cabbage looper, tobacco budworm, corn earworm, and cotton bollworm. A single Brazilian free-tailed bat can consume twenty bollworm moths in a single night. Such a foraging bout translates into two cents per bat per night of ecosystem services that are provided by the bats because additional pesticides would not need to be applied to achieve the same yield of cotton. This ecosystem service, when extrapolated across all of the Brazilian free-tailed bats potentially foraging over cotton-producing areas, translates to an annual agro-economic value on cotton of up to $1.725 million. This is substantial when compared to the value of the crop in this region of $4.6 million to $6.4 million per year (Anonymous, 2018).

Texas has a large number of extensive caves in the central part of the state that may harbor millions of Brazilian free-tailed bats. Bracken Cave, located near New Braunfels along the southeastern edge of the Hill Country, is thought to be the world's largest bat colony, serving as a nursery colony for adult female Brazilian free-tailed bats and their young, and housing perhaps as many as 20 million to 40 million bats during the summer. Likewise, the largest urban colony of bats in North America is located at the Ann W. Richards Congress Avenue Bridge in Austin. This colony, which numbers about 2.5 million bats in the summer, has become one of the major tourist attractions in that city. Fortunately, Brazilian free-tailed bats do not hibernate for long durations during the winter and are therefore unlikely to be severely affected by WNS, but because of their migratory patterns they could become a vector for spreading the disease through a large region of the state and into Mexico.

Texas is extremely fortunate that the major organization devoted to the conservation of bats, Bat Conservation International (BCI), is located in Austin. In cooperation with the USFWS, TPWD, and a number of research universities in the state, efforts are now underway to understand the impact of wind farms and WNS on Texas bats and to determine the best ways to mitigate their impacts.

Order Carnivora (Carnivores)

Texas has a relatively rich terrestrial carnivore fauna, with 20 native and one introduced species (red fox) distributed across six families and 17 genera. Because of the interplay of trophic level, life history strategy, and size, terrestrial carnivores are extremely vulnerable to extinction. Within the last century, Texas has lost the red wolf (*Canis rufus*), gray wolf (*Canis lupus*), jaguar (*Panthera onca*), jaguarundi (*Puma yagouaroundi*), margay (*Leopardus wiedii*), black-footed ferret (*Mustela nigripes*), and grizzly bear (*Ursus arctos*) as well as many black bear (*Ursus americanus*) populations (see Chapter 5). The ocelot (*Leopardus pardalis*) is now included on the endangered species lists, and several other species (e.g., coati, *Nasua narica*; hooded skunk, *Mephitis macroura*; and hog-nosed skunk, *Conepatus leuconotus*) appear to be in serious jeopardy. Some species, such as the raccoon (*Procyon lotor*), gray fox (*Urocyon cinereoargenteus*), and striped skunk (*Mephitis mephitis*), have adapted well to urbanization and seem to be increasing in numbers and distribution in the state. The following discussion assesses the status of many of the species in each of the six recognized carnivore families, with an emphasis on those under some sort of conservation threat.

Family Canidae, Dogs, Foxes, and Wolves

Two of the five species of canids that occur in Texas warrant special consideration about their conservation status. These two species, the swift fox (*Vulpes velox*) and kit fox (*Vulpes macrotis*), were considered to be conspecific in the recent past (see Annotations 201 and 202 in Chapter 2). These two species once were abundant throughout the short-grass and mid-grass prairies of North America but declined with expansion of human settlement (Egoscue, 1979). Populations were reduced by habitat destruction and the indiscriminate use of traps and poison baits to control large carnivores, principally wolves (*Canis lupus*) and coyotes (*Canis latrans*). Swift and kit fox populations may have begun to increase by the mid-twentieth century due to the elimination of poisoning campaigns, but they still remain below historic levels (Samuel and Nelson, 1982). The swift fox was a candidate for endangered species listing by the USFWS from 1992 to 2001; it is unclear what the current conservation status of the foxes is, although IUCN lists them both as least concern.

The late Warren Ballard and his students at Texas Tech University conducted extensive studies of natural history and conservation status of the swift fox, and most of this account is taken from their work (see Darden et al., 2003; Kamler et al., 2003a, b, c; McGee et al., 2006; Nicholson et al., 2006; Schwalm et al., 2012). Current distributional records indicate that populations have been significantly reduced from their historical distribution. The swift fox once occupied approximately 79 counties on the High Plains and Panhandle, but recent surveys indicate they now occur only in Dallam and Sherman counties, where their numbers have been declining. These two factors indicate that the species may be at risk of extirpation in Texas. The two principal causes of limited distribution and density include mortality related to coyotes and loss of habitat. Unlike most canids, swift foxes are habitat specialists, selecting for short-grass prairie habitats, rarely using dry-land agricultural fields, completely avoiding irrigated agricultural fields, and nearly completely avoiding CRP grasslands.

Because of their specialized habitat selection, protection of native short-grass prairies might be necessary for their long-term existence in Texas. Also, it has been demonstrated that placement of artificial escape dens where these foxes occur contributes to increasing swift fox distributions and population sizes by reducing coyote mortality (McGee et al., 2006). Swift foxes currently are classified as furbearers in Texas and thus receive no protection other than regulations pertaining to trapping seasons. Given their precarious position, it would seem prudent for TPWD to at least place them on the state list of rare and threatened species.

The two other foxes that occur in Texas, the gray fox (*Urocyon cinereoargenteus*) and the introduced red fox (*Vulpes vulpes*), as well as the coyote (*Canis latrans*), are broadly distributed and common throughout their range in the state, with the gray fox having been recently recorded from the barrier islands off the coast (Jones and Frey, 2008). None of these species currently warrants conservation concern. Coyote populations have expanded in suburban, urban, and even downtown areas to the point that in places like the Dallas–Fort Worth metroplex, conflicts between coyotes and people are becoming increasingly common.

Family Felidae, Cats

With the exception of the bobcat (*Lynx rufus*) and the mountain lion (*Puma concolor*), the wild felines in Texas did not fare well during the twentieth century. *Leopardus pardalis*, the ocelot, was still common during the time of the biological survey, but predator control, habitat destruction, and urbanization combined to greatly reduce its range and numbers. Historically, its distribution included Texas, Arkansas, Louisiana, and Arizona, but it was extirpated from the vast majority of its US range during the twentieth century, and by the 1960s its distribution was restricted to South Texas. The remaining distribution in the US has been reduced to two isolated populations, with fewer than 80 individuals, in three counties (Willacy, Kenedy, and Cameron) of southern Texas (Tewes, 2019). The species is listed as endangered on both the state and federal lists and by the IUCN.

Habitat loss and fragmentation have led to population reductions and losses in genetic diversity across ocelot populations (Janecka et al., 2011, 2014). Genetic variation in the Texas populations is significantly less than that seen in Mexico, and the distribution of genetic markers suggests no recent gene flow among the two US populations and with those in northeastern Mexico (Janecka et al., 2014). Studies have shown that the effective population size of Texas ocelots is below the critical value recommended for short-term

Figure 171. A road-killed ocelot, Kenedy County, 2016. Public domain photo.

viability (Janecka et al., 2008). This is because of the disappearance of dense thornshrub communities, human-caused mortality, and demographic uncertainty (Janecka et al., 2011).

Vehicle-caused mortality seems to be the primary anthropogenic factor causing ocelot deaths (Fig. 171). Applications of remedial tactics within transportation corridors to promote safer felid movements have been proposed to minimize ocelot mortality, including cat underpasses (e.g. culverts), which have been constructed for ocelots in southern Texas (Tewes and Blanton, 1998; Tewes and Hughes, 2001; Haines et al., 2005).

The Lower Rio Grande Valley of southern Texas has the most impoverished and rapidly growing border population of humans in the US (Fulbright and Bryant, 2002). This growth is increasing the rate of habitat fragmentation and threatening the preservation of ocelot habitat (Haines et al., 2006b, c). Telemetry studies (Haines et al., 2006a) have revealed that ocelots prefer closed habitat and large continuous patches of native thorn scrub, although they also will make use of small patches and corridors of this habitat type (Jackson et al., 2005). Thus, they are more specialized in their habitat requirements than other felids, such as the bobcat (Tewes, 1986; Emmons, 1988). In contrast to the ocelot, bobcats have maintained a wide distribution, high abundance, and population connectivity despite continued legal harvesting and frequent road-related mortality (Janecka et al., 2016).

Michael Tewes and his students and collaborators at the Caesar Kleberg Wildlife Research Institute in Kingsville, Texas, have been studying the ecology and conservation of ocelots for four decades. They have documented aspects of its biology and natural history, and through their research they have elucidated the elements necessary for a conservation plan to preserve this species (Tewes, 2019). Future management and conservation strategies to ensure its survival will have to include plans for habitat restoration and offsetting the erosion of genetic variation in isolated populations (Harveson et al., 2004). The two Texas populations share a close genetic and phylogenetic affinity with populations of the subspecies *L. p. albescens* in northern Mexico, and that seems to be the best option for translocation of animals to improve the genetic diversity of US populations (Janecka et al., 2007).

The USFWS, TPWD, and The Nature Conservancy of Texas have been working for several years in a cooperative effort to restore habitat by linking the few remaining blocks of contiguous thorn scrub and riparian forest in the Lower Rio Grande Valley. Combinations of different recovery strategies will be needed to effectively reduce the probability of extinction in the US. Short-term recovery strategies should include reducing ocelot road mortality (the most important goal) and translocation of ocelots in the US from northeastern Mexico to replenish genetic variation. Long-term recovery strategies should include the restoration of habitat between and around existing ocelot habitat patches and the establishment of a dispersal corridor between ocelot breeding populations (Haines et al., 2006c). The involvement of private landowners in the ocelot recovery process will be vital to accomplish these goals (Haines et al., 2006b).

The bobcat, *Lynx rufus*, is a medium-sized carnivore that lives in both remote and urban habitats. There is no indication that bobcat populations have declined in the twentieth and twenty-first centuries. In fact, there is growing evidence of a high density of bobcats in urban landscapes despite earlier assumptions that they require large areas of habitat and are sensitive to fragmentation. Two recent studies (Young et al., 2019a, b) of bobcats in the Dallas–Fort Worth Metroplex have revealed a sizeable population (at least one bobcat per km^2) that makes use of natural habitat areas within urban areas, such as agricultural fields and creeks, and avoids highly anthropogenic features, such as roads. While there is strong evidence that urbanization and loss of habitat is a leading threat to many large mammalian species, that does not appear to be the case for bobcats, much like their canid counterpart, the coyote (*Canis latrans*).

The mountain lion, *Puma concolor*, the largest felid in the state, also appears to be faring better than many other large carnivores. The latest information regarding the status of mountains lions is discussed in Chapter 5.

Family Mephitidae, Skunks

Something is definitely impacting the diversity and ranges of skunks in Texas, and nearly all of the species are in need of serious study to determine the factors that may be causing their decline. Although the striped skunk (*Mephitis mephitis*) has continued to increase in numbers and geographic range, the other species of skunks have declined for reasons that are not completely understood, and several of these species bear special monitoring to determine the causes of their apparent decline. The hog-nosed skunk (*Conepatus leuconotus*) has totally disappeared from the Big Thicket, where a distinct subspecies (*C. l. telemelastes*) had been recognized, and from much of the South Texas Plains region. Although they are still considered common in the central part of the state, especially in the Hill Country, there is a growing consensus among professional mammalogists that the overall population level of hog-nosed skunks in Texas has declined drastically during the past few decades. The IUCN lists the hog-nosed skunk as a species of least concern, although it reports this species as declining throughout its range. It does not appear on the federal or state lists of concerned species. Relatively little is known about the ecology and behavior of these animals. Several possible explanations for their decline have been proposed, including the conversion of brushy habitats to row-crop agriculture, competition with feral pigs, and the use of pesticides that may limit the insect food source for the skunks. Recently, eight road-killed *Conepatus leuconotus*, a live-trapped one, and two recorded on camera traps, as well as one road-killed animal in Tamaulipas, Mexico, have been reported (Holbrook et al., 2012a). All of the Texas specimens came from Brooks and Hidalgo counties in South Texas. These records suggest that hog-nosed skunks are persisting along the Gulf Coast region of southern Texas and northern Mexico but this species should be carefully monitored in the future.

The status of the hooded skunk (*Mephitis macroura*) in Texas at the present time is uncertain. No hooded skunks were documented by Bailey (1905) in his biological survey of the state, and the first specimens reported in Texas were from Pecos County in 1925 (Patton, 1974) and Jeff Davis County in 1940 (Blair, 1940). The hooded skunk also was reported from the vicinity of Balmorhea in Jeff Davis County in the early 1970s (Patton, 1974). In total, the hooded skunk has been reported from only six counties (Brewster, Jeff Davis, Pecos, Presidio, Reeves, and Ward) in the Trans-Pecos region of West Texas. The last record of the species in Texas was of a dead individual obtained in 1999 at the Davis Mountains State Park in Jeff Davis County (Yancey et al., 2017). Some mammalogists have suggested that hooded skunks may be extinct in the state (Schmidly and Bradley, 2016), but others are of the opinion that, at least in some cases, the hooded skunk is sometimes mistaken for the striped skunk in areas where these species are sympatric (Pacheco, 2014).

The hooded skunk currently is not listed by state or federal agencies as threatened or endangered but probably warrants some form of protection, at least within the part of its range that occurs in the US (Schmidly and Bradley, 2016). The IUCN lists this skunk's status as least concern and increasing in numbers, although this does not seem to be true in Texas. The Texas Conservation Action Plan for the Chihuahuan Desert and Arizona-New Mexico Mountains Ecoregions Handbook (Connally, 2012) lists Limpia Creek in the Davis Mountains as a priority habitat for conservation. It also suggests that hooded skunks should be a "species of greatest conservation need" within the Trans-Pecos region. Documentation of additional records of this skunk is critical for making future listing and conservation decisions (Yancey et al., 2017).

Another example of troubled skunks involves the two species of spotted skunks in the state, the western spotted skunk (*Spilogale gracilis*) in the Trans-Pecos and adjacent regions and the eastern spotted skunk (*S. putorius*) in the east and north (Jones, 1993). Once relatively common, these two species are now rare in some areas, apparently in response to the degradation of prairie habitat in the state. A recent review of their range in Texas indicates that the entire state is occupied by at least one species and areas of historical range overlap still remain (Dowler et al., 2008). However, their current status is unknown, and the genus is badly in need of detailed study. A recent study of habitat use by western spotted skunks has shown they have a decided preference for areas characterized by large mesquite (*Prosopis glandulosa*), and that these types of areas are now often brush-controlled for management of livestock, thereby threatening their habitat (Neiswenter and Dowler, 2005).

The fluctuating nature and overall decline of eastern spotted skunk (*S. putorius*) populations over the past century have prompted concern over their conservation status. Population declines have been documented throughout the species range and Texas. Although there is no strict consensus on the cause of their decline, possible contributing factors include overharvesting in the fur trade, large-scale changes in agricultural practices that occurred throughout the twentieth century, disease, pesticide use, and altered predator guilds (Shaffer et al., 2018). For example, the modernization of farming methods served to reduce habitat and prey availability, as dilapidated farm buildings, fence rows, creek bottoms, and wood piles—habitats historically abundant with spotted skunks—were cleared for industrial farming purposes. Coupled with other anthropogenic activities, such as oil and gas extraction and urban sprawl, it is easy to understand why this skunk has been the subject of recent conservation concern. In response to the documented population declines and lack of sightings, the IUCN now regards the eastern spotted skunk as vulnerable, and the USFWS is currently considering it for listing as an endangered species (Shaffer et al., 2018).

Robert Dowler and his students at Angelo State University currently are conducting a review of the current and historical records of both of these skunks in the state, a habitat and fragmentation assessment to identify and quantify available habitat, field surveys to verify existing populations, and genetic analyses of populations across the range of the species. Their genetic studies reveal that genetic variability in the eastern spotted skunk is lower than that seen in common carnivores (e.g., striped skunks and raccoons) but slightly higher than in the endangered black-footed ferret.

Family Mustelidae, Weasels, Otters, Mink, and Badgers

As our natural surface waters have declined throughout the twentieth and early twenty-first centuries, so have populations of mink (*Vison vison*) and long-tailed weasels (*Mustela frenata*) that depend on this type of habitat (Schmidly and Bradley, 2016). Both appear to have declined in abundance since the twentieth century and do not appear to be common anywhere in the state, which makes it difficult to determine their population status. The IUCN lists both as species of least concern, and neither appears on the federal or state list of concerned species. However, based on their uncommon and rare status across the state, both species should be monitored in the future.

Lontra canadensis, the northern river otter, is presently known from the major watersheds in the eastern one-fourth of the state. In the last few decades of the twentieth century, concern was raised about the disappearance of northern river otters in many portions of their range as a result of habitat loss, heavy trapping pressure, and drowning in fish traps. In response to this concern, TPWD prepared reports on the status of the otter in Texas. These reports suggest that otters are now increasing in abundance in much of their remaining suitable habitat. In fact, TPWD trapping reports (2002-2013) indicate that river otters have expanded their range into 20 additional counties in East Texas and now occur in two counties in the Rio Grande Valley. The highest density of inland otter was documented in the Sabine and Angelina-Neches River drainage of the Piney Woods region. The IUCN lists the northern river otter as a species of least concern, and it does not appear on the federal or state lists of concerned species. Apparently, the reestablishment and abundance of beaver and the improved habitat diversity and productivity associated with beaver activity have benefited the otter. Human-induced changes in habitat, such as impoundments, canals, and levees, also are providing improved conditions, suggesting that this species may no longer need monitoring (Schmidly and Bradley, 2016).

The American badger, *Taxidea taxus*, is distributed throughout much of the state except for the extreme eastern part, and it appears to be extending its range eastward as a result of land-clearing operations and increased artificial grasslands. The fur of the badger ordinarily does not command a high price and, because of this, relatively few are trapped. Badgers locally are abundant at many places in the state, and populations appear to be stable. They have been expanding their range and seem to be reasonably adaptable to human conditions, although land clearing and conversion and habitat fragmentation represent potential threats where development is accelerating. The IUCN places the species in the least concern category, and it does not appear on the federal or state lists of concerned species. Reduction of its primary food sources (prairie dogs and ground squirrels), due to increased agricultural land use, is a primary concern, which means that it should be carefully monitored in the future.

Family Procyonidae, Raccoons, Ringtails, and Coatis

Three species of this family occur in Texas, and historically the raccoon and ringtail have been important furbearers. Ringtails, *Bassariscus astutus*, are common throughout the rocky habitats of the western and central parts of the state, but they are less abundant in eastern and southern Texas and the Panhandle. Trapping and camera trap data often document that they are more common than realized. There does not appear to be any serious threats to them, although continued habitat fragmentation in the Hill Country could certainly be a long-term issue. The same appears to be true for the raccoon (*Procyon lotor*), which occurs statewide and is one of the most common carnivores in Texas, having

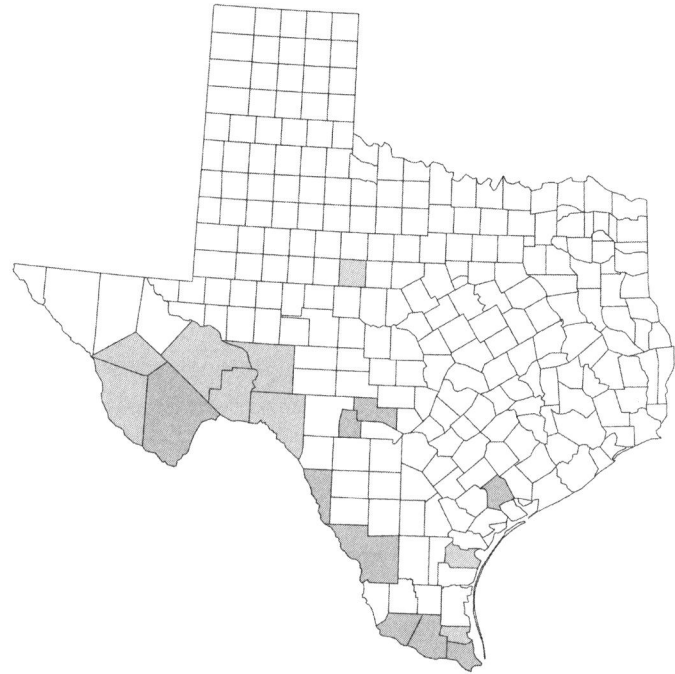

Figure 172. Possible distribution of the white-nosed coati (*Nasua narica*) in Texas, based largely on unconfirmed records, sightings, and other anecdotal evidence.

adapted well to human conditions. Corn from deer feeders has provided them with a supplemental food source, which enables animals to better survive the winter months.

The status of the other procyonid in the state, the white-nosed coati (*Nasua narica*), has been an enigma for decades. Schmidly et al. (2016a) reviewed documentation of the historical records as well as numerous reports of recent sightings, very few of which are "confirmed" with specimens or photographs; most of the sightings have come from amateur naturalists and must be considered "anecdotal." Long-term mammal surveys by professional mammalogists and camera trap studies from the expected range of the coati in Texas (Fig. 172) have not produced any specimens or other documentation of the species. Almost all of the sightings have been of solitary individuals from within 100 miles (161 km) of the borderlands and/or of released pets. The most numerous sightings are from Big Bend National Park just across the river from the Maderas del Carmen/Sierra del Carmen Mountains of northern Coahuila, Mexico, where a small breeding population has now become established. There is no evidence that a breeding population of wild coatis now lives in Texas, although a recent conservation corridor established along the borderlands could eventually result in the species expanding its range into the state. The IUCN lists the white-nosed coati as a species of least concern, but the species is listed as threatened by the TPWD. It is not listed by the USFWS because of the relatively stable populations in southwestern Arizona. Although it is widespread throughout Mexico and

Middle America, coati populations have been seriously impacted by the degradation and loss of much of the riparian woodland habitat over much of their range. These animals require a sizeable area of habitat to maintain a viable population. This is a species that needs continuous monitoring in Texas.

Family Ursidae, Bears

The American black bear (*Ursus americanus*) experienced a significant range contraction from its widespread North American distribution during the nineteenth and twentieth centuries due to anthropogenic factors such as habitat fragmentation, unrestricted harvesting, and predator control (Laliberte and Ripple, 2004). This contraction resulted in many isolated populations, particularly in the southern and southwestern United States and northern Mexico, and may have facilitated loss of genetic variation within populations and increased genetic differentiation among populations (Van Den Bussche et al., 2009). Despite this recent fragmentation, many populations have been increasing in size since the 1980s, especially within the southern United States (Pelton et al., 1999).

The black bear was thought to be extinct in Texas (and reported that way in the first edition of this book), but a plethora of recent sightings from 1988 to the present indicate that a natural restoration of populations may be underway as a result of a coalescence of biogeographic, ecological, and sociological factors (Skiles, 1995; Taylor, 1999; Onorato and Hellgren, 2001). Using genetic evidence from mitochondrial DNA haplotypes, it has been demonstrated that the presence of bears in adjacent but geographically isolated mountain ranges in northern Coahuila, Mexico, facilitated the colonization of populations in the Big Bend region (Onorato et al., 2004). Evidence is clear that black bears are reproducing in a small number of counties. Breeding populations now exist in the Chisos, Guadalupe, Davis, and Glass mountains as well as the limestone canyon areas around Comstock in Val Verde County. During a recent three-year camera-trap study at Big Bend Ranch State Park, Frank Yancey and Stephen Kasper documented a persistent breeding population of black bear in the Solitario region of the park in typical low elevation Chihuahuan Desert scrub habitat. This is considered the northernmost reproductively viable population of American black bears in the Trans-Pecos region, and it is not in an area spatially separated from other black bear populations (Yancey and Kaspar, 2019). The source for these populations is the Sierra Carmen Mountains in Mexico, where a major conservation program to protect black bear populations has been underway for several decades involving landowner initiatives and encouragement from the Mexican government (Doan-Crider and Hellgren, 1996).

The continued recolonization of island populations of black bears in western Texas will depend upon dispersing individuals from the larger mountain ranges in northern Coahuila. To maintain the smaller populations in western Texas, dispersal must not be impeded by anthropogenic factors (e.g., hunting, poaching, development, high-speed highway construction, a border wall). Black bears are classified as threatened in Texas and are protected from hunting. In Mexico, a hunting moratorium for black bears was declared in 1986, and they subsequently were listed as an endangered species (Onorato et al., 2004).

In addition to the sightings of black bear near the border of Mexico, there have been several recent sightings throughout the Hill Country, and in 2012, a black bear was sighted north of Dalhart in the far northern Panhandle region. This individual presumably was a wanderer from the area of Raton, New Mexico. Jessica Light and her students at Texas A&M University have used records from TPWD, iNaturalist, and VertNet to document a total of 242 black bear sightings from 42 Texas counties, 27 of which were not previously included in the distribution map for the species in the 2016 edition of *The Mammals of Texas* (Schmidly and Bradley, 2016). Combining these records with those of Schmidly and Bradley (2016) indicates that black bears may be distributed in as many as 70 counties across Texas (Keane et al., 2019).

American black bears have been sighted in 31 counties in East Texas in both the northern part of the region and in the Big Thicket area of southeastern Texas, primarily as the result of individuals that have wandered into the state from release sites in Louisiana and Oklahoma (Willingham, 2001; Patoski, 2006; TPWD News, 2020c). Historically, the Big Thicket population (subspecies *U. a. luteolus*) was one of 16 subspecies of the black bear in the US, with a distribution that included the coastal plain of southern Mississippi, Louisiana, and East Texas (Hall, 1981; Kennedy et al., 2002). The southern portion of the Big Thicket area was apparently one of their last strongholds (Truett and Lay, 1984). Bailey already regarded them as rare at the beginning of the twentieth century. They were eliminated over most of the region during the period from 1850 to 1910 by hog-raisers who felt that the bears were a threat to their free-ranging razorbacks. The big slaughter began about 1883 when two men in Liberty County alone killed 182 bears in a two-year period. Bailey became acquainted with one of them, Ab Carter, who told him his story (see Chapter 1).

Almost all of the bears in East Texas were gone by 1940, except for a few confirmed sightings in the Big Thicket of Hardin County and the dense woodlands of Matagorda County (Anonymous, 1945). A confirmed kill also was reported in the late 1950s near Livingston in Polk County (Fleming, 1980). Since then, TPWD has been recording and investigating sightings and mortalities. From 1993 through 1997, fifteen reliable sightings were recorded in counties bordering Louisiana, Arkansas, and Oklahoma (Cox, 1996; Garner and Willis, 1998). At that time, there was no evidence of a resident, breeding population, and the Texas sightings were considered transient, mainly consisting of young males seeking home territories (Cox, 1996). However, given the number of recent sightings of sows with cubs being reported, it now appears that many of these sightings include resident individuals and not just transient bears (Patoski, 2006; Schmidly and Bradley, 2016).

Because the subspecies *U. a. luteolus* had declined so precipitously during the twentieth century, the subspecies was declared endangered throughout its range in 1992. During the early twentieth century, much of its habitat was lost to agricultural development.

Since the subspecies was added to the endangered species list, government agencies and private landowners have restored more than 700,000 acres (283,280 ha) of critical black bear habitat. These efforts have been so successful that scientists estimate the number of Louisiana black bears has doubled since the subspecies was listed. Delisted in 2016, an estimated 500 to 750 Louisiana black bears now live in the wild, and it appears that this recovery has allowed breeding populations to become reestablished in the Big Thicket of southeastern Texas (Willingham, 2001; Patoski, 2006).

Together, these data seem to signal the return of permanent black bear populations in Texas (see Schmidly and Bradley, 2016). This case is noteworthy because natural recolonization of historical range by large carnivores is uncommon in today's world of habitat fragmentation, disturbance, and destruction (Mladenoff et al., 1995; Forbes and Boyd, 1996).

The grizzly bear, *Ursus arctos*, was never common in Texas, being documented on the basis of a single specimen obtained on a bear hunt in the Davis Mountains at the end of the nineteenth century. The hunt is described in Chapter 1, and the status of the species is briefly discussed in Chapter 5.

Three Special Aspects of Carnivore Conservation

Carnivores come in a wide variety of shapes and sizes, ranging from small weasels to large bears. They are the major predators of other mammals, including domestic livestock, which has put them in conflict with man and has led to massive campaigns to eradicate populations. They also are the major group of Texas animals harvested for direct commercial use and sale of hides and pelts. Carnivores are highly important members of ecological communities. They control pests, such as rodents, and even consume large quantities of insects.

Animal Damage Control

Nuisance animal control covers a wide range of issues, including predation on livestock, damage to property, and health and safety concerns in urban and suburban areas. Several social issues compound these problems. One of these is opposition to trapping, based on reservations about steel leg-hold traps or animal-rights objections to killing furbearers to produce clothing.

A healthy balance between predator and prey is considered a simple and indisputable concept among ecologists, but for most humans, predation is considered a good thing only when the circumstances place the interests of people on the winning end of the ordeal (Klepper, 2005). In Texas, where the domestic livestock industry and property ownership have dominated the dialogue about the state's native predators, control of predators has been a long established practice (see Wade et al., 1984). Historically, widespread poisoning and trapping of predators resulted in the destruction of many non-target animals and ultimately contributed to the extirpation of the gray wolf and red wolf. Today, control efforts result in the removal of thousands of predators in Texas each year, but the techniques used are safer, fewer non-target animals are lost, and often the offending individuals can be selectively taken.

Damages caused by predatory animals are governed by Chapter 825 in Title 10 of the Texas Health and Safety Code, which states, "the state [of Texas] shall cooperate through the Texas A&M University System with the appropriate federal . . . agencies in controlling coyotes, mountain lions, bobcats . . . and other predatory animals . . . to protect livestock, food and feed supplies, crops, and ranges." The US Department of Agriculture (USDA), Animal and Plant Health Inspection Service (APHIS), Wildlife Services (WS) program is the federal agency responsible for managing conflicts with animals. Pursuant to the Texas Health and Safety Code, the Texas A&M University System, through the Texas A&M AgriLife Extension Service and the WS program, have signed a Memorandum of Understanding (MOU) to conduct a cooperative program to alleviate damage caused by predators. In addition, the Texas Wildlife Damage Management Association (TWDMA), which consists of local cooperative groups, including county governments, private associations, and/or individuals, also signed the MOU. The cooperative program created by the MOU is known as the Texas Wildlife Services Program (TWSP).

To provide efficient program support and assistance, the TWSP has divided Texas into districts for the purposes of implementing a program to manage predatory animals. Predatory mammals as defined by the statute include the following carnivores: coyote (*Canis latrans*), feral/free-roaming dogs (*Canis familiaris*), gray fox (*Urocyon cinereoargenteus*), red fox (*Vulpes vulpes*), feral/free-ranging cats (*Felis domesticus*), bobcats (*Lynx rufus*), mountain lions (*Puma concolor*), hog-nosed skunks (*Conepatus leuconotus*), hooded skunks (*Mephitis macroura*), striped skunks (*Mephitis mephitis*), western spotted skunks (*Spilogale gracilis*), eastern spotted skunks (*Spilogale putorius*), and raccoons (*Procyon lotor*). Although not a carnivore, the Virginia opossum (*Didelphis virginianus*) also is listed as a "predatory" mammal.

At the beginning of the twentieth century, most predator control efforts in Texas involved farmers and ranchers shooting, trapping, or poisoning predators (primarily coyotes and wolves) to protect their livestock (primarily sheep and goats). Many landowners paid bounties to "professional coyote killers" for each coyote, wolf, or other large predator killed. Organized predator control began in the Edwards Plateau region, the largest area of sheep concentration, in the early 1900s. Beginning in 1915, the federal government became directly involved with the establishment of the Predator and Rodent Control branch of the Bureau of the Biological Survey. This department later became known as the Animal Damage Control Program of the US Department of Agriculture.

By the 1920s, many of the inner Edwards Plateau counties were almost free of coyotes and wolves, and by the 1950s these predators were thought to be extirpated from most of the region. After a coyote population irruption in the early 1960s, coyotes began to reestablish themselves on the periphery of the Plateau. This encroachment process accelerated through the 1980s and 1990s, and coyotes now have become common across the entire state (Nunley, 1995). Coyote reestablishment resulted from changes in land use—away from sheep and goat production—and a reduction in the use of toxicant controls, such as strychnine, Compound 1080 (sodium monofluoroacetate), and the M-44 cyanide device by the EPA in 1972. Compound 1080 and the M-44 device are still in use in Texas, although the EPA requires specialist training for certification of applicators.

A major controversy has developed in the twenty-first century about carnivore conservation and the impact of widespread lethal control of native mammals, particularly by the federal government targeting carnivores in the western states (see Bergstrom, 2017). The concern is that the high level of human-caused mortality may have negative unintended consequences for native ecosystems and biodiversity. For these reasons, a consensus is emerging among ecologists that nonlethal methods of preventing livestock depredation by large carnivores (such as the use of guardian animals and livestock protection collars) may be more effective, more defensible on ecological, legal, and wildlife-policy grounds, and more tolerated by society than lethal methods (Bergstrom, 2017). Nevertheless, despite these arguments and the scientific evidence and public support of them, the Trump administration in 2019 reauthorized government officials to continue using the controversial M-44 device (Tobias, 2019), and it is not yet determined if the current Biden administration will reverse this decision.

Avocational Fur Trapping

The harvest of furbearing mammals fluctuated dramatically throughout the twentieth century. Laws pertaining to furbearing animals are enacted by the Texas legislature to ensure the conservation of the fur resource or to protect human health or property. Enforcement and management of rules and regulations is delegated to TPWD. Texas law requires that a "trapper" (one who takes a fur-bearing mammal or the pelt of such an animal) possess a trapping license. The trapping season for most species opens on December 1 and closes on January 31, lasting only two months. Eighteen species of Texas mammals are officially considered to be furbearers (https://tpwd.texas.gov/regulations/outdoor-annual/hunting/fur-bearing-animal-regulations/definitions): Virginia opossum (*Didelphis virginiana*), red fox (*Vulpes vulpes*), swift fox (*Vulpes velox*), kit fox (*Vulpes macrotis*), gray fox (*Urocyon cinereoargenteus*), raccoon (*Procyon lotor*), ringtail (*Bassariscus astutus*), mink (*Mustela vison*), badger (*Taxidea taxus*), spotted skunks (*Spilogale putorius* and *S. gracilis*), striped and hooded skunks (*Mephitis mephitis* and *M. macroura*), hog-nosed skunk (*Conepatus leuconotus*), northern river otter (*Lontra canadensis*), North American beaver (*Castor canadensis*), common muskrat (*Ondatra zibethicus*), and nutria (*Myocastor coypus*). Formerly, coyotes (*Canis latrans*) and bobcats (*Lynx rufus*) were considered to be furbearers (Schmidly, 1984b), but they are no longer on the list. Coyotes can now be taken at any time by any means; bobcats can as well, but they must have a pelt tag (required by CITES).

Fur prices are not static from season to season. They fluctuate according to supply and demand, and during periods of high demand a substantial crop of fur-bearing mammals has been harvested in Texas. For example, the increase in prices paid for raw furs during the 1970s produced a 665 percent increase in licenses sold to trappers. Tabulations from a trapper survey revealed that more than 2.5 million furbearers were harvested, producing an income in excess of $30 million. During this period the average price paid for furs of most species increased substantially. For some species, the value of their fur tripled over this nine-year period. The bobcat possessed the most valuable pelt, followed by the otter and red fox. Other valuable furbearers were the gray fox, coyote, and raccoon. The mushrooming of license purchases caused concern over the possible adverse effects of increased harvest pressure on furbearers.

Special concern developed about the harvest of bobcat during the 1970s because of federal export restrictions from the 1973 Convention on International Trade in Endangered Species of Wild Fauna and Flora (CITES). After a 1977 ban on international trade of various species of endangered spotted cats, the value and demand for the fur of native cats, such as the bobcat, greatly increased. From 1976 to 1982, bobcat pelts in Texas sold for an average of $67.50, and approximately 97,000 bobcats were harvested. A decade later, bobcat pelts were worth an average of $25.75 and the annual harvest declined to 4,657.

TPWD began collecting data on annual bobcat harvests in 1976, and Bluett et al. (1989) analyzed the geographic distribution of the harvests from 1978 to 1986. They found that 54 percent of the bobcats harvested in Texas during this period were taken in the Trans-Pecos, on the South Texas Plains, and on the Edwards Plateau. Regional harvest densities (number of bobcats harvested per 100 square kilometers) during this period were greatest in the Piney Woods, Cross Timbers and Prairies, and South Texas Plains. It was concluded that populations may have been overharvested in some local areas and that the illegal transport of pelts from Mexico to Texas may have elevated harvest estimates in some border counties.

Today, participation in trapping, a historically and biologically significant wildlife management practice, has been declining, and it does not enjoy strong public support (Armstrong and Rossi, 2000). Several factors appear to be affecting participation, including anti-trapping sentiment, low recruitment of new trappers, low prices being paid for pelts, the negative image of trappers and trapping, and the issue of land access. The factors affecting the future of avocational trapping are not mutually exclusive. For example, low fur prices affect recruitment because the economic enticement to participate is not present and competing activities fill the void in time and interest (Armstrong and Rossi, 2000).

Wildlife Killing Contests

One aspect of predator control that should be seriously questioned is wildlife killing contests. These are competitions where the goal is to kill as many individuals of the target species as possible within a set timeframe, in order to win money and prizes. The usual targets in these hunts are predators, especially coyotes, foxes, and bobcats. There is no comprehensive federal law prohibiting these contests and only a few states have placed restrictions on them. Texas is thought to have more of these contests than any other state, with more than 600 documented. For example, in 2021 the West Texas Big Bobcat Contest, based out of San Angelo in Concho County, had 378 teams entered and paid out $86,940 in prize money.

There are no ecological or wildlife management justifications for these contests. The typical rationale used to justify them is to protect livestock from predators, but there is no evidence to support this. Most of the animals obtained are discarded and no scientific data are collected that could be used to monitor populations. Predators have important ecological roles in ecosystems, and unless predation on livestock is documented as a serious problem in an area, it is difficult to defend this mass extermination of predators. Producing piles of dead predators to win a monetary prize is not hunting, and among many in the public, these events give hunting and hunters a bad reputation. Texas should join the growing list of states, such as New Mexico, that have banned these contests.

Order Artiodactyla (Even-toed Ungulates)

Mammals in this order include big game species—white-tailed deer, mule deer, pronghorn, elk, peccary, aoudad sheep (introduced in Texas), and desert bighorn sheep. Big game mammals came under intense anthropogenic pressure throughout the latter half of the nineteenth and early part of the twentieth centuries. By the early 1900s, overexploitation resulting from unregulated market hunting had become a serious threat to them. Trapping and unrestricted killing decimated many game species. Today, hunting of game species is an important management tool regulated by state law, and the revenue from hunting has become an effective market incentive for landowners to manage for wildlife habitat.

Seven species of artiodactyls are native to Texas, and native populations of three of them (the bighorn sheep, *Ovis canadensis*; bison, *Bos bison*; and elk, *Cervus canadensis*) were extirpated (note: bison descended from the Goodnight herd could be considered native; see Chapter 5). Recent introductions of these big game animals by private individuals and captive breeding programs sponsored by TPWD have successfully reestablished populations of all three species. With the possible exception of the bison, none of the replacement populations are of the native subspecies that originally occupied the state (see Chapter 5). In addition, at least 123 species of ungulates not native to Texas have been imported into the state since 1930. For the most part, these exotic animals have been confined on private ranches; however, 11 species have escaped, reproduced, and now exist in parts of Texas as free-ranging, feral populations that constitute a part of the local fauna (see Chapter 5). At least four exotics, eastern Thomson's gazelle (*Eudorcas thomsoni*), scimitar-horned oryx (*Oryx dammah*), aoudad (*Ammotragus lervia*), and blackbuck (*Antilope cervicapra*), are considered already extinct or vulnerable in their native range; the first three are native to North Africa and the blackbuck is from India and Pakistan. The four are listed by the IUCN, respectively, as near threatened, extinct in the wild, vulnerable, and near threatened. None of these exotics are listed as vulnerable species by USFWS or TPWD, with the exception of *Oryx dammah*, which is listed as Endangered by USFWS; however, the status only applies to the native populations in northern Africa.

The restoration of populations of large game mammals and other species of special public interest has been a major component of conservation strategy in Texas. There also are some excellent examples of native species and subspecies that were substantially reduced in numbers by over-hunting and trapping but which have subsequently recovered over most of their range as a result of hunting regulations and restocking or other restoration efforts. Probably the most notable examples in Texas of successful restocking programs involve the white-tailed deer (*Odocoileus virginianus*), pronghorn (*Antilocapra americana*), and bighorn (*Ovis canadensis*). The situation of the latter two species is discussed in Chapter 5.

Native white-tailed deer (*Odocoileus virginianus*) were virtually eliminated in Texas by the end of the nineteenth century because of indiscriminate slaughter by commercial meat and hide hunters (see Doughty, 1983, for a detailed discussion). The TPWD developed a program to successfully restock deer from central or southern Texas into other regions of the state during the 1930s, '40s, and '50s. By the late 1950s and early 1960s, deer populations had peaked in many areas and habitat decline due to excessive deer numbers began to occur. During the 1960s, cattle numbers increased drastically and much native pasture was converted to coastal Bermuda grass and other forages that are poor deer food. In addition, the brush on many abandoned farmlands had grown beyond the reach of deer. These factors led to a decline in deer numbers in many areas by the 1970s. However, the white-tailed deer is now being successfully managed with the participation of private landowners, and it is the most numerous and economically important big game animal in Texas. Although declining in the Trans-Pecos, the population seems to be healthy and stable or even increasing in other regions of the state. The 1997 statewide population estimate was 3,359,000 deer, and the number currently hovers around 4.5 million animals (Graves, 2019b).

The status of the three other big game mammals can be summarized as follows. Mule deer (*Odocoileus hemionus*) populations are considerably lower and of more concern than those of white-tailed deer. There are some indications that hybridization with or replacement by white-tailed deer, or both, are occurring in some regions. The aoudad or Barbary sheep (*Ammotragus lervia*) is an introduced species that has increased in number in Palo Duro Canyon and several West Texas mountain ranges. The overall range of the peccary (*Pecari tajacu*) in Texas has declined substantially, although populations of the species still thrive along the Rio Grande and in adjacent areas of brush country and in the Big Bend region.

Population restoration involves more than returning extirpated fauna to their former range. Biological and law enforcement concerns must be part of a holistic approach that enhances the success of these increasingly popular efforts, and in Texas it is necessary to achieve the cooperation of private landowners. From a population standpoint, it is better to emphasize the recovery of endemic stocks, as a result of protective legislation and habitat management, rather than supplementing populations with nonnative individuals. Conservation management initiatives seeking to restore threatened, endangered, or depleted populations should consider the degree to which reintroductions can contribute to the longevity of the population and potentially alter the population's gene pool. For example, Ellsworth et al. (1994) found that while restocking of white-tailed deer in the southeastern US may have affected local genetic stocks in some instances, by and large it contributed insignificantly to the genetic composition of extant populations. These authors suggest that despite massive (and expensive) restocking efforts, the abundance of white-tailed deer today is primarily attributable to the recovery of native herds.

Order Perissodactyla (Odd-toed Ungulates)

Horse-like mammals of this order are not native to Texas, and for the most part are not sufficiently numerous across the entire state to exercise any major effects on the vegetation, or the wildlife, except in limited areas where they have been released in large numbers and run wild or where intensive pasturing may be practiced. The feral ass or burro (*Equus asinus*) apparently only recently invaded the Big Bend region of West Texas from Mexico and from ranches where they were housed and escaped. A large feral herd (>50 individuals) is known to inhabit the southern reaches of Big Bend Ranch State Park. The establishment of these animals is potentially a major problem because they can become a destructive menace. Their droppings foul springs and creeks in a landscape where water is limited and precious, and they can compete with native wildlife for limited forage, thus decreasing available food resources for some animals, disrupting the food chain for others, and threatening native plants. Big Bend State Ranch Park is one of the major sites selected for the reintroduction of bighorn sheep, and burros are known to adversely affect desert bighorns via competition and disease transmission. In 2008, TPWD began to remove the burros by shooting them, and a public outcry ensued. The removal program was stopped, and TPWD is now seeking nonlethal options for removing the burros. The same problem has been in place for several decades in nearby Big Bend National Park. The government is prevented from shooting the burros in the park and must capture and place them in adoption programs.

The mustang, or wild horse (*Equus caballus*), was first brought to Texas in 1542 by early Spanish explorers. At one time they were very common in the state, but wild mustangs are rare today. In Big Bend National Park, herds from Mexico periodically will cross the river and graze in the riparian areas of the park, where they trample vegetation and foul the water. The problem there, however, is nothing like that seen on public lands in Arizona and Nevada, where these animals run wild in large numbers and have exceeded the carrying capacity of public lands.

Order Rodentia (Rodents)

There are 66 native species of rodents in Texas, making this the most diverse group of mammals in the state, and it appears that most of them do not have conservation issues. However, a few species and subspecies with known restricted distributions and narrow habitat requirements require monitoring and study to determine their conservation status.

Long-term monitoring of most rodent populations has not been conducted, primarily because these mammals are inconspicuous and mostly nocturnal. The effort needed to obtain reliable estimates of population trends would be extraordinarily costly. Thus, quantitative assessments of population changes have been made for only a few species, and these typically are based on occasional and geographically restricted studies. Nonetheless, the patterns that do emerge suggest that habitat alterations can have negative impact on populations of many species, as well as positive impact in some cases, such as the initiation of the government programs (e.g., Conservation Reserve Program) and habitat improvement efforts currently practiced by private landowners in the state.

For these reasons, it is difficult to predict to what extent conservation problems exist for most species of small rodents. For certain, we should be concerned about the status of any unique species. Six species of mammals (one kangaroo rat and five pocket gophers) have most, or all, of their known geographic range confined to the mainland part of the state. For these species, Texas is the key to their survival. They are: the Texas kangaroo rat (*Dipodomys elator*), Attwater's pocket gopher (*Geomys attwateri*), the Texas pocket gopher (*Geomys personatus*), the Llano pocket gopher (*Geomys texensis*), and Strecker's pocket gopher (*Geomys streckeri*). Each of these species lives in extremely small and easily altered habitats, and their status should be monitored carefully.

In addition, the entire US distribution of another two rodents is confined to Texas, these being the Mexican spiny pocket mouse (*Liomys irroratus*) and the Coues' rice rat (*Oryzomys couesi*). Both of these are tropically distributed types that are common over much of Mexico and reach their northern distributional limits in Texas, where they are typically restricted to a few counties along the Texas-Mexico border.

Family Castoridae, Beavers

Because of the high commercial value of their pelts, beaver (*Castor canadensis*) figured prominently in the early exploration and settlement of western North America, including Texas. Thousands of their pelts were harvested annually, and by the 1850s, beavers were reduced to very low population numbers over a considerable part of Texas. However, these animals were able to survive on the remote streams of the upper and western Hill Country, along the Devils River, along the edge of the Panhandle, and around El Paso (Weniger, 1997). By the early 1900s, the federal agents found them to be increasing in numbers, and shortly thereafter strict harvest regulations were imposed and restocking of depleted populations became common practice. Today, beaver are found over much of the state where suitable aquatic habitat prevails, and their populations appear to be increasing.

Family Cricetidae, New World Mice, Rats, and Voles

There are 30 species of cricetid rodents in Texas, making this the largest family of mammals in the state. Very few of them appear to have serious conservation issues, except for the few mentioned below. Woodrats (genus *Neotoma*), grasshopper mice (genus

Onychomys), and harvest mice (genus *Reithrodontomys*), of which there are collectively 10 species, all appear to be in good shape and not to have serious conservation concerns.

Very few species of voles (microtine rodents) live in Texas, and all of them require watching from a conservation perspective. Already, one of the subspecies of the prairie vole (*Microtus ochrogaster ludovicianus*) has become extinct in the Big Thicket region of the southeastern part of the state, and the other subspecies (*M. o. taylori*) apparently is a recent invader in the northern Panhandle where it now has been recorded from the Texas-Oklahoma border all the way south to Lubbock County (Manning and Jones, 1988; Choate and Killebrew, 1991; Poole and Matlack, 2007; Schmidly and Bradley, 2016). The Mogollon vole (*Microtus mogollonensis*) is restricted to the higher elevations of the Guadalupe Mountains in Culberson County. The IUCN lists this vole as a species of least concern, and it does not appear on the federal or state lists of concerned species. The species appears to be in good shape in Texas, but it could be threatened by a catastrophic local event such as a massive forest fire within the park (Schmidly and Bradley, 2016). The woodland vole (*Microtus pinetorum*) is the most widely distributed vole in Texas with a highly scattered and localized distribution in the eastern and central parts of the state. Nowhere does it appear to be common, and continued degradation of grassland habitat could greatly impair its status in the state. There are no recent records from the Hill Country or other places where it was once taken, and where grasslands have been converted to shrubland. Although it does not appear on the federal or state lists of concerned species, the woodland vole bears careful monitoring in the future.

Oryzomys couesi, Coues' rice rat, occurs in the United States only in four counties (Cameron, Hidalgo, Kenedy, and Willacy) in extreme South Texas. This is a common small mammal from Mexico southward to Panama, where it is in no danger. The IUCN lists it as a species of least concern, and it does not appear on the federal lists of concerned species; however, it is regarded as threatened in Texas by TPWD because of threats to its habitat. Resacas bordered by cattail-bulrush marsh and subtropical woodlands, the preferred habitat of this rat, are essentially confined to the Lower Rio Grande Valley region, and this habitat type is declining largely because of drainage for irrigated agriculture.

Nine species of deer mice (genus *Peromyscus*) occur in Texas and all of them appear to be in good shape except for a couple of species with restricted distributions in the western part of the state. *Peromyscus truei comanche*, the Palo Duro mouse, originally was described as a separate species by Frank Blair (Blair, 1943). The species has had a varied taxonomic history, shifting between species and subspecies status. The latest taxonomic arrangement, based on a composite of morphological and genetic characters, arranges it as a distinct subspecies of the wide-ranging species, *Peromyscus truei* (Schmidly, 1973b). Its geographic range, which is completely isolated from the other populations of *P. truei*, encompasses three counties—Armstrong, Briscoe, and Randall—where it has been trapped in Palo Duro Canyon State Park and Caprock Canyons State Park (Yancey et al., 1996). There it occupies high rocky ledges clothed with juniper along the breaks of the Llano Estacado. Taking into account the numerous side canyons, indentations, and contours along the edge of the Llano, this mouse must have at least 100 miles (161 km) of more-or-less continuous distribution. Because of its limited distribution, the Palo Duro mouse presently is regarded as threatened by TPWD. It was a candidate for Category 2 listing by the USFWS but currently it is not being considered for listing. At this time, there appear to be no significant threats to its specific habitat. Moreover, there are two state parks within the geographic range to serve as refugia (Yancey et al., 1996).

Another subspecies of the piñon mouse, *Peromyscus truei truei*, also occurs in Texas, and it is much rarer than *P. t. comanche*. It is known by only four specimens from Guadalupe Mountains National Park and five from along the northern edge of the Llano Estacado, just inside Texas from New Mexico in Deaf Smith County. To be consistent, all known populations of *P. truei* in Texas should be regarded as threatened until further information is available that clarifies their status (Jones, 1993).

The rock mouse, *Peromyscus nasutus*, which occupies a few mountainous habitats in the southern and western Trans-Pecos (Franklin, Guadalupe, Davis, Chinati, and Chisos mountains) is another poorly known species that requires mentioning. These mice have been documented in small numbers from the mesic canyons associated with the forested woodlands of the largest mountain ranges in the Trans-Pecos. Its status bears watching in the future.

There are three species of cotton rats (genus *Sigmodon*) in Texas and two of them bear watching in the future. The hispid cotton rat (*Sigmodon hispidus*) occurs state-wide and is one of the most common rodents in the state. Yellow-nosed cotton rats (*Sigmodon ochrognathus*) live primarily in high-elevation grasslands as well as upland slopes with scattered bunches of grasses in mountain riparian habitat in the main mountain range of the Trans-Pecos region. At one time, this species was thought to be rare and possibly in need of listing and special protection. Recent trapping evidence, however, suggests it is becoming more abundant and widespread throughout its range in the Trans-Pecos. Collecting records show that it occupies a number of nonmontane habitats (Yancey and Jones, 1996; Yancey, 1997; Heaney et al., 1998; Schmidly and Bradley, 2016). This could be another species that may be expanding in association with the increasing aridity in western Texas. Recent DNA studies revealed very little genetic differentiation among localized populations of the yellow-nosed cotton rat, suggesting that in the recent past it may have been more abundant and widespread than at the present time (Carroll et al., 2002). The IUCN lists *Sigmodon ochrognathus* as a species of least concern, and it does not appear on the federal or state lists of concerned species. However, given its restricted distribution, this is a species that warrants monitoring in the future.

The status of another species of cotton rat, the tawny-bellied cotton rat (*Sigmodon fulviventer*), is enigmatic in Texas because the extent of its range and relative abundance remain unknown (see Chapter 5). Given that it is known from only a single location in the Davis Mountains—collected on only one occasion—and subsequent attempts to document it at the same place (and surrounding areas) have failed suggests that it is extremely rare or perhaps already extirpated (Schmidly and Bradley, 2016). The Texas Parks and Wildlife Department recently listed the species as threatened. Caleb Phillips and his students at Texas Tech University have conducted extensive trapping efforts to document other locations in the Trans-Pecos where this rare rodent might occur. To date, they have not identified any additional populations.

Family Erethizontidae, New World Porcupine

During the first half of the 1900s, the porcupine (*Erethizon dorsatum*) in Texas was restricted to the northernmost Panhandle and parts of the Trans-Pecos (Bailey, 1905; Taylor and Davis, 1947; Hall and Kelson, 1959). By the mid-1900s, observations suggested that the range was expanding into the southern Panhandle and onto the western Edwards Plateau (Milstead and Tinkle, 1958). Subsequent accumulation of voucher specimens over the past several decades have documented the expansion of the porcupine eastward across the southern Rolling Plains (Dalquest and Horner, 1984), the Edwards Plateau (Ilse and Hellgren, 2001; Baird et al., 2009), and Webb County in South Texas (Goetze and Miller, 2015; see Chapter 5). The species has presently come to occupy most of the western two-thirds of the state all the way to the eastern edge of the Edwards Plateau (Baird et al., 2009).

Family Geomyidae, Pocket Gophers

Texas is home to 11 species and at least 25 subspecies of pocket gophers belonging to the genera *Thomomys*, *Geomys*, and *Cratogeomys*. Several of these taxa have highly localized ranges and could be vulnerable to localized extinction events. An obvious example is a subspecies of Botta's pocket gopher, *Thomomys bottae limpiae*, which has a limited distribution at lower elevations in the Davis Mountains (Blair, 1939; Davis, 1940). This subspecies has not been found at its type locality since 1968, having been replaced by another gopher species, the yellow-faced pocket gopher, *Cratogeomys castanops* (Reichman and Baker, 1972). The replacement of *Thomomys bottae* populations by *Cratogeomys castanops* in the southwest has been noted by several authors during the twentieth century (Nelson and Goldman, 1934; Davis, 1940; Baker, 1953). This phenomenon could be considered as strictly "natural," one that normally occurs in the evolution and replacement of species over geological time. The problem, however, has been complicated by extensive overgrazing during the past century in the Davis Mountains, which has changed the plant community, increased runoff, and caused more xeric conditions (Reichman and Baker, 1972). These xeric conditions, in turn, favored *Cratogeomys* over *Thomomys*. This situation appears to have driven one of the subspecies of *Thomomys bottae*, *T. b. baileyi*, to complete extinction as discussed in Chapter 5.

Recent DNA studies (Jolley et al., 2000; McAliley and Sudman, 2005; Sudman et al., 2006; Chambers et al., 2009) have confirmed at least nine species of plains pocket

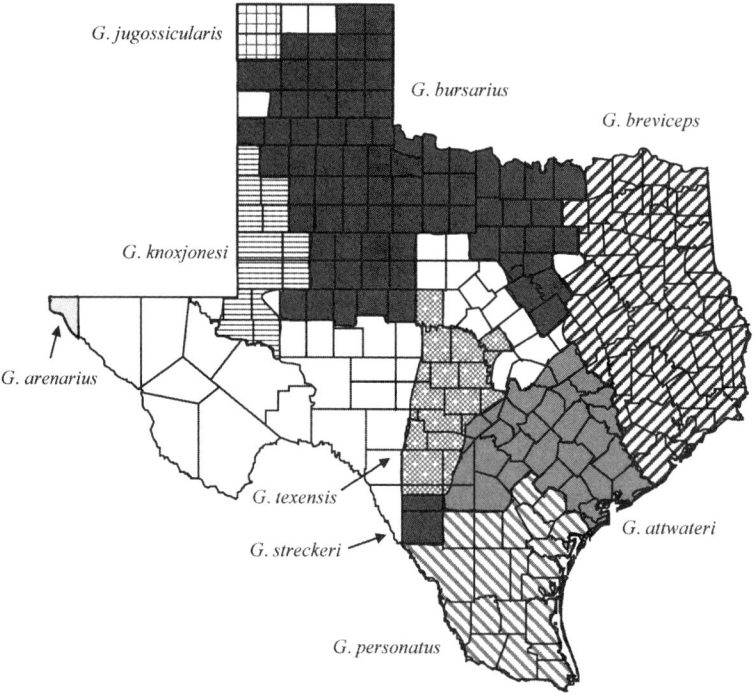

Figure 173. Distribution of nine species of pocket gophers of the genus *Geomys*. At the time of the Biological Survey of Texas, most of these species were recognized by Bailey as subspecies.

gophers (genus *Geomys*) in Texas (Fig. 173) and further work may reveal even more than that. Conservation or convergence of morphological characters, presumably resulting from adaptations to a fossorial lifestyle, has led to difficulties in differentiating these species based on morphological characteristics (Mauk et al., 1999). Furthermore, many of them have restricted distributions and their conservation status is uncertain. Two South Texas species, the Texas pocket gopher (*G. personatus*) and Strecker's pocket gopher (*G. streckeri*), especially bear watching. Although neither of these species currently appears on any official rare or threatened listing of mammals, a subspecies of the Texas pocket gopher, *G. p. maritimus*, is known only from its type locality at Flour Bluff, near Corpus Christi (Williams and Genoways, 1981). Since the only locality is within the greater Corpus Christi metropolitan area, it obviously could be threatened by urbanization. *Geomys streckeri* is a recently described species that is known from only two localities in South Texas. There is a need to document its full range and ascertain its population status. It should be carefully monitored because of its rarity and limited geographic range (Schmidly and Bradley, 2016).

Geomys attwateri, Attwater's pocket gopher, occurs in the south-central part of eastern Texas, from Milan County southward to Matagorda and San Patricio counties, and southwestward to Atascosa County. This species still appears to be locally common, although it is absent from some places where it was previously present. The same appears to be true for the Llano pocket gopher, *Geomys texensis*, which has a highly localized distribution in the Hill Country, and for Hall's pocket gopher, *G. jugossicularis*, which occurs in only a few counties in the extreme northern Panhandle. All three of these gopher species should be monitored for long-term changes in distribution and population abundance.

Family Heteromyidae, Pocket Mice and Kangaroo Rats

Five species of kangaroo rats (genus *Dipodomys*) occur in Texas, and all appear to be in good shape with the possible exception of two. The large Texas kangaroo rat, *Dipodomys elator*, was discovered by the federal agents in Clay County at the turn of the twentieth century. Now, the species has been documented in a band just south of the Red River from Motley County in the west to Montague County in the east, but with substantial populations apparently only in Hardeman, Wichita, and Wilbarger counties (Martin and Matocha, 1972; Dalquest and Horner 1984; Jones and Bogan, 1986; Jones, C., et al., 1988). Its historic range may have included southwestern Oklahoma but the species probably no longer occurs there (Moss and Mehlhop-Cifelli, 1990). The IUCN lists the species as vulnerable. The TPWD lists it as threatened because of its restricted geographic range and because of habitat alteration for agricultural purposes. The USFWS has taken a somewhat more cautious view of the status, all the while seeking additional information on *D. elator*. Several state and federally sponsored studies have been completed in recent years (e.g., Goetze et al., 2007, 2008, 2017; Nelson et al., 2009, 2016; Stasey et al., 2010; Pfau et al., 2019), and a follow-up study currently is being sponsored by the Texas Comptroller and TPWD. Heavily grazed rangeland and the eroded sides of well-worn rangeland roadways may provide optimum habitat for this species, in the same manner that overgrazing and trampling by bison may have done in the past (Stangl et al., 1992). Where it occurs the rodent is often common, even abundant, and certainly occupies its habitats to the limits imposed by food and other factors (Dalquest and Horner, 1984). Its limited geographic range, land-use practices within its range, and its selection of rather specific habitat—short grasses associated with mesquite on clay or sandy loam soils—are thought to contribute to its vulnerability (Stangl et al., 1992).

Another large kangaroo rat, the banner-tailed kangaroo rat (*Dipodomys spectabilis*) appears to still be common throughout its range in western Texas, and in some places it is locally abundant. The IUCN lists it as near threatened, however. It does not appear on the federal or state lists of concerned species. Degradation or loss of grassland habitat could severely affect its status, and this is a species that should be carefully monitored in the future (Schmidly and Bradley, 2016).

Family Sciuridae, Squirrels and allies

With the possible exception of the prairie dog (*Cynomys ludovicianus*; see Chapter 5), flying squirrel (*Glaucomys volans*), eastern gray squirrel (*Sciurus carolinensis*), and the gray-footed chipmunk (*Tamias canipes*) discussed below, all of the species of this family in the state appear to be in good shape from a conservation perspective. The concern about the flying squirrel is that forestry and timber harvesting practices that eliminate old-growth habitats could be harmful to its long-term status. Unfortunately, we know very little about its population abundance, making it difficult to accurately predict its conservation status.

The most serious conservation problem with squirrels in Texas is associated with the eastern gray squirrel. A drastic reduction in suitable habitat occurred throughout the twentieth century as a result of detrimental land-use practices, such as logging hardwood forests, employing certain practices of timber stand improvement, establishing pine plantations, overgrazing domestic livestock, flooding bottomland habitats through reservoir impoundments on major streams and rivers, and draining lowland bottomlands. The future of this species in Texas will depend on the acreage remaining in hardwood stands and the abundance of mast supplies and dens. In a few areas these squirrels have adapted to urban areas and appear to be doing quite well (Schmidly and Bradley, 2016).

Texas, unlike its neighboring state to the west, New Mexico, has only a single species of chipmunk, and it has a very restricted range in the Trans-Pecos region. *Tamias canipes*, the gray-footed chipmunk, is known from two areas in West Texas (Guadalupe Mountains and Sierra Diablo; Davis and Schmidly, 1994) along with three areas in New Mexico (Findley et al., 1975). It is restricted to the forests of pine, oak, and fir found at the highest elevations in Texas (5,900-8,200 ft; 1,800-2,500 m). Though the animal is common and presumably well protected in Guadalupe Mountains National Park, there are questions concerning its well-being elsewhere within its restricted range.

Comments on the Status of Select Reptiles in Texas

Bailey briefly discussed the occurrence of Texas reptiles, limiting his comments to 61 species of snakes and lizards. His treatise ignored turtles (one of the major extant reptilian groups). Further, Bailey did not provide information for amphibians, a group that generally is treated hand-in-hand with reptiles (collectively termed "herps," the general name given to amphibians and reptiles) by biologists, those of Bailey's time as well as modern times. It is not clear why Bailey provided only anecdotal information for these taxonomic groups. Given that Bailey employed the expertise of Leonhard Stejneger, head of the Division of Reptiles and Batrachians at the National Museum, perhaps Bailey believed he personally lacked the professional expertise to sufficiently report on these taxonomic groups. Or perhaps, as with birds (Oberholser and *The Bird Life of Texas*), there were plans for a more detailed coverage of reptiles and amphibians at a later date. Or, it may be that Bailey simply lacked the personal interest in pursuing these taxonomic groups.

Currently, there are at least 230 species of native and exotic species of amphibians (27 salamanders and 43 frogs/toads) and reptiles (1 crocodilian, 31 turtles, 52 lizards, and 76 snakes) in Texas (Dixon, 2013). The herpetofauna of the state ranks among the greatest in North America (north of Mexico) and contains 36 percent of the 632 species of amphibians (194 salamanders and 102 frogs/toads) and reptiles (2 crocodilians, 59 turtles, 120 lizards, and 155 snakes) occurring in this region (Crother et al., 2012).

Given the incomplete coverage of amphibians and reptiles presented in Bailey's work, it is difficult to undertake a direct comparison of modern-day conservation concerns to those that would have been appropriate for the late 1890s and early 1900s. However, there currently are 17 species of amphibians (10 salamanders, 1 newt, 1 siren, and 5 frogs/toads) and 25 species of reptiles (8 turtles, 1 tortoise, 12 snakes, 1 gecko, and 3 lizards) known to occur in Texas that currently are listed as either threatened or endangered by the TPWD or the USFWS. Of these, Bailey reported a single specimen for four species (reticulate collared lizard, *Crotaphytus reticulatus*; speckled racer, *Drymobius margaritiferus*; Texas indigo snake, *Drymarchon melanurus*; and timber rattlesnake, *Crotalus horridus*), indicating that these species may not have been abundant during the time of the Biological Survey. Regarding the Texas horned lizard, *Phrynosoma cornutum*, Bailey reported this species as being common across much of western and southern Texas; however, in more recent times it appears to have undergone major reductions in overall distribution and population numbers (see below).

If one broadly considers the 42 threatened or endangered species of amphibians and reptiles in Texas, it appears that four general factors—habitat loss, loss of water, human activities, and commercial collecting—are shared among species whose distribution, population size, or both, have declined in recent years. Below, we briefly discuss each of these four categories and give an example of an amphibian or reptilian species or group that has been negatively impacted by these events. We also discuss three recent concerns—climate change, introduction of exotic species, and disease—that may negatively impact amphibian and reptile populations in the future.

Habitat Loss and Fragmentation

Habitat loss is one of the most dire situations facing many of the wildlife species in Texas and on the worldwide scale. The Texas horned lizard, *Phrynosoma cornutum*, is one of the best examples of a reptile species that historically was abundant in numbers and distributed widely across much of western and southern Texas but is today considered to be threatened within the state. Several hypotheses have been proposed for the decline of the Texas horned lizard, but habitat loss, habitat fragmentation, and fire suppression due to agricultural and urbanization practices are among the most frequently cited reasons (Horned Lizard Conservation Society, 2019; Wild Lens Inc., 2019). It is noteworthy that Dixon (2013) suggests that long-term pesticide usage (primarily for controlling fire ants) has reduced the abundance of harvester ants, a major prey item of the Texas horned lizard.

Loss of Surface Water and Springs

One of the most highly threatened vertebrate groups in Texas is the genus *Eurycea*, encompassing cave and spring salamanders. Members of this group are generally specific to a particular spring or wet cave system and consequently have a limited geographic distribution (see Chippindale et al., 2000; Dixon, 2013; TPWD, 2019b). Unwise water use and other human activities have jeopardized the stability of these fragile habitats, especially in the rapidly growing region of central Texas (Austin Monitor, 2019).

Human Activities

For many species, human activities present the primary conservation concern. These activities, such as urbanization, agricultural expansion, pollution, and habitat destruction or conversion, generally produce a detrimental impact on wildlife populations. Some activities may seem relatively minor (establishment of roads, transmission lines, windmills, etc.); however, for a species operating within a narrow habitat requirement or whose life history traits are closely tied to a specific set of biological parameters, a minor deviation from the "norm" may produce a catastrophic outcome. For example, the dunes sagebrush lizard (*Sceloporus arenicolus*) has garnered recent attention due to the increase in oil and gas-related activities. This species occupies a rather narrow habitat range (four counties in Texas) and concern has mounted that the recent boom in drilling and pipeline construction in that region may prove detrimental to this species (Price, 2018; Comptroller.texas.gov, 2019; Skolos et al., 2019; USFWS, 2019). It appears that the dunes sagebrush lizard is dependent on the specifics of the shinnery oak/sand dunes, and removal of the shinnery during sand mining for fracking activities has reduced the natural habitat for this species.

Commercial Collecting

Native populations of kingsnakes (*Lampropeltis getula*), rat snakes (*Callopeltis obsoletus*), and several turtles have been reduced significantly during the last 25 years or so as a result of commercial collecting (primarily for the pet trade). The TPWD has passed legislation regulating some commercial collecting activities, such as a ban on collection of native turtles from public waters. Further, TPWD has initiated studies on the movement of red-eared sliders (*Trachemys scripta elegans*), common snapping turtles (*Chelydra serpentine*), and two species of soft-shelled turtles (*Apalone mutica* and *Apalone spinifera*) to determine the impacts of such collecting in public waters.

In another example of "commercial collecting," many communities sponsor an annual "rattlesnake roundup" that blends an old-time tradition (rattlesnake removals) with modern-day fund-raising opportunities (at least 25 such roundups occur or have occurred on an annual basis in Texas). These communities sponsor an event or "festival" in which snake hunters compete in various categories (most snakes collected, largest, etc.) and attendees view a variety of events (milking, handling, eating, etc.) and products (belts, wallets, etc.). A long-time justification for the need for such roundups has been the milking of snakes for venom to supply the pharmaceutical anti-venom market; however, most anti-venoms are now produced synthetically, thereby reducing the necessity for these activities. Nonetheless, such events may generate a substantial amount of income for a community. For example, an economic impact report (prepared by Sara T. Page Consulting, LLC) indicated that in 2015 the Sweetwater, Texas, rattlesnake roundup sponsored by the Jaycees attracted 21,314 visitors and generated $8,392,864 in revenue for the community (sweetwatertexas.org, 2015).

Despite the monetary potential of these roundups, the potential harm to ecosystems by removal of these top-level predators should be considered. For example, although it is unclear how many snakes are captured each year during these events, in 2016, the Sweetwater Festival reported 24,262 pounds of rattlesnakes; if one assumes that an average rattlesnake weighs between 1 and 3 pounds, then approximately 8,000 to 24,000 snakes were harvested. It is unknown whether these "roundups" have a long-term negative impact on rattlesnake populations compared to other unintentional deaths such as highway deaths. It is noteworthy that in 2017, organizers of the Sweetwater Festival capped the number of allowable captured pounds of snakes at 6,500 (although the pounds reported after the event exceeded the cap, at 7,958 pounds). From 2017 to 2021, pounds captured have trended downward (to 3,682 pounds in 2021). It is unclear whether this trend is a result of decreased participation (fewer snake hunters) or an indication of declining rattlesnake populations in the area.

Other detrimental environmental factors associated with these roundups include the use of gases to flush snakes from their dens. This practice often leads to contamination of the hibernaculum and the surrounding soil and ground water. Further, the harm to "non-target" species of snakes that use these hibernacula has not been adequately considered. TPWD has been involved in actively monitoring these roundup events, and in 2009 TPWD instituted a study (Snake Harvest Working Group; see Davis, 2016) to evaluate the impact of snake roundups, gassing, and the cultural and economic impacts to communities, and to provide solutions and recommendations about this activity.

Climate Change and Sex Determination

For many species of amphibians and reptiles (and fishes), the sex of an individual is determined by the temperature under which the embryo develops, rather than by a particular set of sex chromosomes as in birds and mammals (Bull, 1983). Biologists now worry that the sex ratio for some species may become skewed as more males or more females are produced due to an increase in mean annual temperatures as a result of climate change. This phenomenon already has been documented in a few species of turtles (Hawkes et al., 2007; Fuentes et al., 2011).

Introduced and Exotic Species

Dixon (2013) lists nine species of amphibians and reptiles that have been introduced into Texas that have become successful to the point of having established breeding populations. With the exception of the Mediterranean gecko (*Hemidactylus turcicus*), most of these introduced species have not successfully expanded their range beyond the initial area of introduction. To date, it does not appear that these introduced species compete with native species, although Dixon (2013) mentions that the rough-tailed gecko (*Crytopodion scabrum*) may outcompete another introduced species, the common house gecko (*Hemidactylus frenatus*). Although not a topic of immediate concern for Texas, the status of these introduced species should be monitored carefully to ensure that they do not begin to replace native species of amphibians and reptiles.

Diseases

The chytrid fungi (*Batrachochytrium dendrobatidis* and *Batrachochytrium salamandrivorans*), which were discovered during the late 1980s and early 1990s, have dramatically impacted some amphibian populations throughout much of the tropics. These fungi die at soil temperatures above 30 degrees Celsius; however, increased cloud cover in tropical areas (resulting from evaporation due to increased temperatures associated with climate change) prevents the sun from warming the soil surface (Pounds et al., 2006). If temperatures continue to increase due to climate change, it is possible that chytrid fungi may invade more temperate zones and begin to impact amphibian species in Texas.

Conservation Issues, Challenges, and Strategies

Conservation issues and challenges facing Texas will be solved, in part, by a new generation of scientists that incorporate the most recent research technologies, such as GPS tracking and next-generation DNA methods. Current research on two of the most charismatic Texas mammals, the endangered ocelot (top) and the desert bighorn (bottom), demonstrates some of these strategies. Ocelot, photo courtesy Michael Tewes, Texas A&M University–Kingsville. Desert bighorn, photo courtesy Emily Wright, Texas Tech University.

R APIDLY INCREASING HUMAN PRESSURES HAVE CHANGED THE natural landscapes and biota of Texas in both positive and negative ways, although concerns about the loss of biodiversity in the state have emerged as a central issue. The persistence of biodiversity and ecosystem services will depend on the adoption of effective conservation initiatives at a pace and scale that match or exceed environmental threats. This fourth and final section of the book includes two chapters that examine some of the projects, programs, initiatives, and policies that have been implemented, or will need to be implemented, in order to address the serious challenges for managing biodiversity in Texas in the twenty-first century. In Chapter 7 we describe eleven action steps that are important to protecting our faunal and floral resources from the continued onslaught of development and economic growth. A concluding chapter—Chapter 8—reflects the new political realities both in Texas and the nation along with some bold suggestions that could, if implemented, help address growing environmental and conservation concerns. The suggestions include consideration of effective initiatives developed outside of Texas that, in the future, could be adaptable to the unique political and socioeconomic realities of the state.

CHAPTER 7

Twenty-First Century Successes and Challenges for Wildlife Conservation in Texas

I T IS NOW OBVIOUS THAT THE TWENTY-FIRST CENTURY WILL BE AS different from the twentieth century as the latter was from the nineteenth century, perhaps even more so given the accelerating pace of population growth, changes in lifestyle, and advances in technology. During the twenty-first century, the future of wildlife in Texas and other states will be decided—including, directly or indirectly, how much and what kind of wildlife diversity survives. Furthermore, the management of wildlife and other types of natural resources will change substantially and become increasingly more complex in terms of both the clientele served and the socioeconomic and political environment in which management will occur.

Conservation pressures in the twenty-first century will come from a variety of sources. Habitat loss and degradation / fragmentation, introduction of exotic species, pollution and toxins, and other factors will continue to take a significant toll on wildlife resources. Climate change will likely exacerbate the loss and degradation of wildlife diversity by increasing the rate of species extinction, changing population sizes and species distributions, modifying the composition of habitats and ecosystems, and altering their geographic extent.

Essentially the problem involves proliferating human land uses and consumption that are powerfully changing the form and shape of the landscape. People now constitute a pressure on the global environment that is evident everywhere. There are no longer any unoccupied frontiers; every square centimeter of the earth's surface is affected by the activities of humans. This results in insufficient habitat for many species or situations in which habitats are isolated in separate pieces too small or too unstable to sustain viable populations of species and thus biological diversity. The study of biogeography reveals that species richness is a function of land area. All environmental variables being equal, the greater the area, the more species it supports. Thus, as habitats are fragmented and isolated into small islands, they lose the capacity to support wildlife diversity.

Challenges for Managing Wildlife Diversity

In Texas, the more traditional view of natural resource management, which advocates the practice/science of making land produce valuable products (wildlife, rangeland, timber, or recreation), has been the preferred model for conservation. This perspective advocates direct or indirect management of resources through manipulation to produce or favor certain targeted goals (Bailey, 1984). Management, in this sense, is on a sustainable-yield basis so that the resource may be harvested periodically without reducing the base stock. However, it is questionable how sustainable the current Texas model of wildlife conservation will be in the twenty-first century. With the large population increases predicted for the state (the population is projected to increase by 41-52 percent from 2018 to 2050; Texas Demographic Center, 2019) and with most of that increase projected to be in large, expanding urban areas and cities where there is limited contact with nature and the outdoors, will voters support a model for conservation that is very different from much of the remainder of the United States? The latter is a critical factor, especially given the public's declining support for hunting or other consumptive taking of wildlife. And, with the Texas wildlife economic system focused primarily on one species, white-tailed deer, there is a legitimate concern about whether the current Texas model will work to protect biodiversity and its important ecosystem resource component.

A recent fifty-state study of wildlife management in the US, entitled *America's Wildlife Values* (Manfredo et al., 2018), which was based on a massive public opinion poll, reveals what the concerns are. The study compared "traditionalists," who believe animals should be used to benefit humans through things like medical research and hunting, with "mutualists," who put animals on a more equal footing with humans. According to the results, traditionalists now make up about 28 percent of the US population compared to 35 percent who identify as mutualists. In Texas, these percentages were 30 and 35 percent, respectively. Western states, including Texas, experienced a 5.7 percent decline in traditionalists and a 4.7 percent increase in mutualists between 2004 and 2018. These changes likely reflect the increasing urbanization and different way people in cities learn about wildlife. Interestingly, there were differences among different segments of the population, with Hispanics having twice as many mutualists as traditionalists.

The results of the study suggest that beliefs held by Americans in general, and Texans in particular, about wildlife management are changing, possibly at a rapid pace (Colorado State University, 2019). Only 33 percent of Texans who participated in the survey indicated an interest in hunting, compared to 77 percent of respondents who expressed an interest in wildlife viewing. This has caused concern in some circles that hunters (and by extension, hunting) could become an endangered "species," and it suggests those involved in the management and conservation of Texas's wildlife diversity should pay more attention to public attitudes and interests that clearly seem to be shifting from a consumptive, hunting focus toward a nonconsumptive, biodiversity perspective.

Future progress will require major shifts in attitudes and values about biodiversity and an ability to overcome substantial economic, social, and political challenges. A new model will be required that benefits all people and all species. In order to fulfill this broader mandate, all citizens really must contribute. It is critical that we begin to view biodiversity as a public good that provides such benefits as clean air and fresh water, and that this view is integrated not just into policies but also into society and individuals' day-to-day decisions. It is easy to take nature for granted because we have received its benefits for free, but the costs of conserving biodiversity are massively outweighed by the benefits (see Rands et al., 2010).

Eleven Steps for Strengthening Conservation in Texas

In this chapter, we outline eleven steps for Texas to successfully transition toward addressing the issue of biodiversity loss. These steps are articulated around four primary strategies: identify and assess areas of biodiversity concern for conservation; convene multiple interest groups and institutions to achieve consensus on methods to conserve biodiversity; review state policy and legal mechanisms that may affect biodiversity; and educate the public about what the loss of biodiversity will mean to the state.

1. Develop a Statewide Biodiversity Initiative

The states are far better positioned than the federal government to protect and restore the nation's plants, animals, and ecosystems. The major federal law aimed at protecting threatened and endangered species, the Endangered Species Act (ESA), has struggled to stem the tide of species endangerment, despite some well-publicized successes (see Chapter 5). The federal government's scope is limited because it owns only 30 percent of the nation's land, and only about 50 percent of endangered species occur on federal lands; thus, most of the country's biologically important lands are on private property.

For these reasons, some of the best tools for biodiversity conservation are in the hands of the states where key land-use decisions that contribute to biodiversity loss are made at the state and local levels. As pointed out by Wilkerson (1999), statewide initiatives to protect biodiversity offer a variety of advantages. States usually are large enough to define planning units that encompass significant portions of ecological regions and watersheds. Many national environmental laws are implemented through state programs and regulations (or lack thereof). Laws addressing utility siting and regulation, agricultural land preservation, real property taxation and investment, and private forestry management also are developed and administered at the state level. State agencies, universities, and museums have collected large quantities of biological data, which often are organized and accessible at the state level. These programs collect and store data on types of land ownership, land use and management, distribution of protected areas, population trends, and habitat requirements. These computer-based resources, along with species data collected and managed by state natural resource agencies, nonprofit conservation organizations, and research institutions, comprise a large proportion of the available knowledge on the status and trends of the nation's plants, animals, and ecosystems. Finally, people identify with their home states and take pride in the states they are from. This sense of place provides a basis for energizing political constituencies to make policy decisions, such as voting for bond issues that fund open-space acquisition and taking private voluntary actions.

To develop scientifically sound and socially acceptable wildlife conservation, stakeholders must understand the necessity of blending appropriate models of conservation with their respective values to achieve overall resource management. In the twenty-first century, proactive approaches are helping to diminish earlier conflicts, and now there are many good examples of private landowners, government agencies and biologists, and nongovernmental organizations working positively together to find proactive solutions to complex conservation issues without compromising private property rights. Partnerships between agencies and conservation organizations and businesses, as well as cooperative efforts among landowners, have become en vogue to accomplish more effective conservation with limited conservation funding (Lightfoot, 2013).

The end of the twentieth century, specifically in the 1990s, was a particularly contentious time in Texas conservation circles. The perceived threat of the application of the Endangered Species Act convinced many private landowners that their property rights were under threat and that the government, both at the state and federal level, could not be trusted on conservation matters. Consequently, a conflict developed among private landowners, private wildlife managers, and professional wildlife scientists associated with government agencies (state and federal), nongovernmental organizations, and university scientists working in conservation. At the heart of the conflict were the government and its representatives, charged with protecting the people's resources, and the private sector who felt those efforts were threatening their property rights. The professional wildlife community was being challenged by "interest-group myopia" in which strong positions were being staked out on the extremes of the spectrum of approaches to conservation, and it was becoming difficult to find a common ground where the majority of people could seek compromise for the common good. Advocacy groups had taken over and polarized the debate because no one had offered society an acceptable alternative to their winner-take-all strategies. Wildlife professionals, who have devoted their careers to the management and conservation of wildlife resources, often were caught in the middle of these contentious battles.

Most conservation efforts in Texas have occurred behind the scenes on private property. Groups like the Texas Wildlife Association, The Nature Conservancy of Texas, Texas Audubon Society, Ducks Unlimited, the National Wild Turkey Federation, the Texas Chapter of the Wildlife Society, the Texas Quail Coalition, the Texas Bighorn Society, Texas Master Naturalists, and many others have helped make wildlife conservation happen in Texas (Holtcamp, 2013). As their names imply, many of them are focused on providing conservation for a targeted wildlife species. Others have broader or holistic interests. The commonality among these organizations is their dedication to conservation through grassroots collective effort.

These conservation entities do not operate in a vacuum but rather under an umbrella known as the Texas Conservation Action Plan (TPWD, 2012), which is under the stewardship of the Texas Parks and Wildlife Department. Each state in the US has now completed a Wildlife Action Plan or Comprehensive Wildlife Conservation Strategy to improve the stability and recovery of species that are in decline, already listed as threatened or endangered, and/or are representative of the diversity and health of the state's wildlife.

Texas, under the leadership of TPWD, completed its plan in 2005 and submitted it for approval by the US Fish and Wildlife Service in order to be eligible for $3 million in annual funding for habitat protection and assessment of nongame species. The plan consists of a series of 11 regionally-specific ecoregion handbooks and a statewide/multi-region handbook (TWPD, 2012). The various handbooks contain information on species of greatest conservation need, regionally important habitats, local conservation goals and projects, regional and statewide activities, contact information for conservation partners, and maps to assist with interpretation. Recently, TPWD published a list of "Species of Greatest Conservation Need" in Texas based on updates from the conservation action plan handbooks (see Chapter 6). The activities in each handbook are starting points from which to engage landowners, land-use planners, natural resource professionals, and the public in regional and local community-based conservation (see Holtcamp, 2006, for more details about the background and organization for the Texas plan).

As part of the state action plan, neighboring landowners are joining forces to create wildlife partnerships by forming landowner-driven wildlife management associations (WMAs) focusing on collaborative efforts that benefit wildlife. These organized efforts share resources and knowledge to achieve wildlife conservation across property lines. In places like East Texas, where land fragmentation has created a patchwork of habitats, these cooperatives help piece the landscape back together for the benefit of wildlife. With more than 2.5 million acres (1.01 million ha) enrolled in Texas, the goals of WMAs include improving cover and food for wildlife, inventorying species, controlling grazing and stocking rates, planting trees and shrubs, and allowing diverse weeds and forbs to grow (Todd and Ogren, 2016). Today, there are more than 185 WMAs in Texas, as well as a statewide network, the Texas Organization of Wildlife Management Associations (Todd and Ogren, 2016) that promotes their work and shares success stories. These associations are also capable of tackling some of the bigger habitat management jobs, like prescribed fire projects, native grassland restoration, and watershed protection.

Approaches to land management involving local stakeholders have been referred to as "relationship-scale conservation" (see Brooks et al., 2015). The objective of this approach to conservation is to allow stakeholders and managers to achieve outcomes that are mutually beneficial for people and resource protection. Potential outcomes may involve collective action that results in consensus regarding future conditions that can help alleviate environmental, social, economic, and political forces that threaten conservation and human relationships. Another aspect of relationship-scale conservation is that it is best practiced at the regional level where it can be integrated with local customs, values, and land uses, and where people can have a vested interest and become direct participants in the decision-making process. Implementing relationship-scale conservation should become a higher priority for wildlife managers and other conservation professionals, and it needs to be applied more often and at all levels of organization. It has been a major component of all of the Nature Conservancy projects for decades (see below).

2. Recognize the Changing Nature of Our Clientele and the Need for Philosophical Adjustments in Attitudes and Values

There is an absolute need to make conservation a higher priority in state and local government actions. To achieve this will require the cultivation of a strong public conservation ethic. Such an ethic would include an understanding of how each of us is connected to the natural world, a commitment to placing conservation higher on the political agenda, and a deep appreciation for the value of fish, wildlife, and other species (i.e., for biodiversity). Unless people recognize the link between their consumptive choices and biodiversity losses, the diversity of life on the planet will continue to decline.

Thus, there is a need for a discourse about needs, relevance, and future directions and priorities. This requires a meaningful strategic plan that has broad public appeal and involves holistic conservation (and not just game management). The Texas Conservation Action Plan outlines the species, subspecies, populations, and habitats in the major ecological regions that are threatened or endangered, and the TPWD strategic plan contains a blueprint for conservation action in this century. But both of these efforts fall far short in terms of identifying new strategies and actions that will be required to address the many conservation challenges in the twenty-first century.

The concept of biodiversity must be emphasized and mainstreamed across local and state government agencies. There, too, must be political engagement at the highest levels of state government to embrace these needs. The state must take more concrete steps to fortify the laws, regulations, and policies that affect biodiversity. Until biodiversity protection is integrated into the fabric of the state's laws and institutions, habitat for plant and animal populations will continue to be lost, fragmented, and degraded.

The last governor in the state to seriously consider the importance of conservation as a priority was George W. Bush, who appointed a task force of citizens to examine the needed changes to address conservation of Texas's wildlife resources. In 2000, Governor Bush established the Governor's Task Force on Conservation with a charge to: (1) examine the impact of fragmentation on lands in Texas and the wildlife habitat located therein; (2) make recommendations as to appropriate incentives and tools available to assist landowners in more effectively conserving and managing lands in their stewardship; (3) make recommendations as to how all Texans may benefit from the many forms of economic activity associated with natural resources, including hunting, fishing, other forms of outdoor recreation, and nature tourism; and (4) provide specific recommendations as to how the state, in partnership with other government entities, private landowners, and community-based groups, can better meet the conservation and outdoor recreation needs of the state in the future.

One of us (DJS) had the pleasure of serving as one of the 13 task force members, including participating on the report-drafting subcommittee. The task force issued its final report, "Taking Care of Texas: A Report from the Governor's Task Force in Conservation," (Dinkins, 2000) in November of 2000 with three categories of primary recommendations: (1) private lands: incentives, partnerships, and stewardship; (2) public lands: planning, repairing, developing, and meeting future needs; and (3) water: assuring, protecting, and managing for conservation. The recommendations in the report were intended to serve as practical objectives and strategies for the state's leaders to ensure the future of our cultural, historical, and natural resources by using sound science, good planning, responsible management, assessment measures, respect for landowners, local participation, and economic incentives to bolster the protection of our unique landscapes, wildlife, and water.

Following publication of the report, Governor Bush initiated action, particularly on the recommendations related to private lands, and considerable progress has been made in the first two decades of this century. However, once Bush left office in 2001 to assume the presidency of the country, little was achieved during the tenure of subsequent governors, particularly as it related to public lands and water conservation, and in both of these areas the state is lagging in significant accomplishments.

Part of a Texas Tech University study of conservation needs in the twenty-first century (see below) included extensive assessment of public attitudes regarding conservation and outdoor recreation. The resulting surveys demonstrated the strong support of the Texas public for both concepts. In fact, the results of the 2000 comprehensive public opinion survey were startling—97 percent of those surveyed stated it was either very important (79 percent) or somewhat important (18 percent) that natural areas exist in Texas for enjoying and experiencing nature; 94 percent felt it was either very important (73 percent) or somewhat important (21 percent) that fish and wildlife populations are being properly managed and conserved in Texas; 94 percent felt it was either very important (69 percent) or somewhat important (25 percent) that ecologically important habitats and lands in Texas are being protected and preserved; and a staggering 100 percent felt it was either very important (93 percent) or somewhat important (7 percent) that Texas's water resources are safe and well protected (Duda, 2000). The public's concern for the plight of wildlife diversity is reflected in the growing number of private conservation groups, environmental groups, and scientific societies devoting attention to wildlife and other natural resources.

Societal values regarding wildlife resources have adjusted in the twenty-first century to reflect concern for both consumptive (hunting and fishing) and non-consumptive (wildlife watching) uses, for threatened and endangered species protection, and to stem the tide of shrinking global biodiversity. The 2011 National Survey of Fishing, Hunting, and Wildlife-Associated Recreation in Texas revealed that 6.3 million Texas residents and nonresidents 16 years old and older (not quite a third of the population at that time) fished, hunted, or wildlife-watched in the state (USDI/USFWS/USDC/USCB, 2011). Of the total number of participants, 2.2 million fished, 1.1 million hunted, and 4.4 million participated in non-consumptive activities (wildlife-watching including observing, feeding, and photographing wildlife). Interest in nature tourism has grown in Texas as rural communities look for ways to diversify local economies and landowners look for ways to diversify agricultural income. One of the largest components of nature tourism is wildlife-watching and associated activities. Wildlife tourism engenders strong public support for conservation, it educates hundreds of thousands of people about the value and importance of biological diversity, and it generates funds for local conservation efforts and state government. An important challenge, therefore, is to solicit the monetary and political support of the growing number of non-consumptive users toward the management and conservation of wildlife resources on public and private lands.

The economic value of hunting and fishing, nature tourism, and outdoor recreation in Texas is significant. These activities increased substantially in the latter part of the twentieth century and have continued to expand in the current century so that they are now one of the fastest growing sectors of the travel industry. In 2018, hunting, fishing, canoeing, hiking, bird and mammal watching, and other outdoor recreational activities generated $28.7 billion in consumer spending and directly supported 277,000 Texas jobs; hunting and fishing alone generated purchases of 3.1 million licenses, permits, stamps, and tags worth about $103 million in revenue. License fees are used to help fund the restocking of fish, wildlife management, habitat restoration, land conservation, game wardens, and other TPWD operating expenses. As many as 14 million Texans annually enjoy some form of outdoor recreation. It is a critical sector of the state's economy with far-reaching benefits that stretch beyond the simple financial impact (Outdoor Industry Association, 2019).

3. Place More Emphasis on Conservation Education

It is crucial to educate future generations of Texans about the importance of the state's enormous biological diversity and what it will take to protect and sustain it, particularly in this time when so many people are becoming disconnected from nature. Biodiversity should be managed as a public good, but it is narrow-minded to dwell exclusively on its material benefits to people. Discussions about human development and ecosystem services need to delve deeper and be communicated more effectively. Natural history and conservation education must be expanded at all levels, from preschool children to political leaders. Educators must explicitly recognize the importance of teaching people of all ages about basic ecological and evolutionary concepts—and getting them outdoors. Education must focus on whole organisms and ecosystems so that we don't risk losing interest in the living world by generations of students of all ages (Noss et al., 2012).

It is especially important to get children, of all ethnic groups and socioeconomic strata, to experience the outdoors. It is no secret that kids now spend more time inside playing on computer and phone screens and less time outside playing in the woods (Gonzalez, 2019). Fewer youth today get to experience the real adventure of discovering the natural world around them. Those kinds of experiences are getting harder and harder to find outdoors, especially in our cities and suburbs. The average large US city devotes just 9.3 percent of lands to parks, and only 65 percent of the average city's population lives within walking distance of a park (Gonzalez, 2019).

Conservationists need to take advantage of nature's novelty to get more kids hooked on exploring outdoors. Electronic screens will always be a temptation, but we need to balance those indoor activities by putting nature back into young people's daily routines. Doing so improves their health, creativity, and self-confidence. And it helps nature because, lest we forget, the children of today will be the next stewards of the natural world.

One way of achieving these goals would be to make conservation education mandatory at all levels of public education. A Conservation Education Strategy Tool Kit, which contains resources developed by the Association of Fish and Wildlife Agencies (AFWA), is now available to support conservation educators who offer fish and wildlife-based programs. This guide is designed for classroom teachers and non-formal educators to support high-quality conservation education efforts that align with academic standards as well as the AFWA Conservation Education Core Concepts and Framework. The guide was developed to help K–12 teachers introduce their students to the methodologies used for scientific field research and guide them through the process of conducting field investigations using these scientific practices (Association of Fish and Wildlife Agencies, 2019).

Other states in the US are beginning to see the importance of this and are initiating adjustments in their educational curricula. For example, New Jersey in 2021 became the first state in the nation to require all public schools to adopt climate change education into its curriculum. The New Jersey program will address issues such as how climate change is impacting wildlife on Earth based on a proactive approach that seeks to empower students to understand the problems and seek positive solutions.

With 86 percent of the population now concentrated in cities and towns, Texas has become an urbanized state (Figs. 174, 175). As this trend continues, it is imperative that

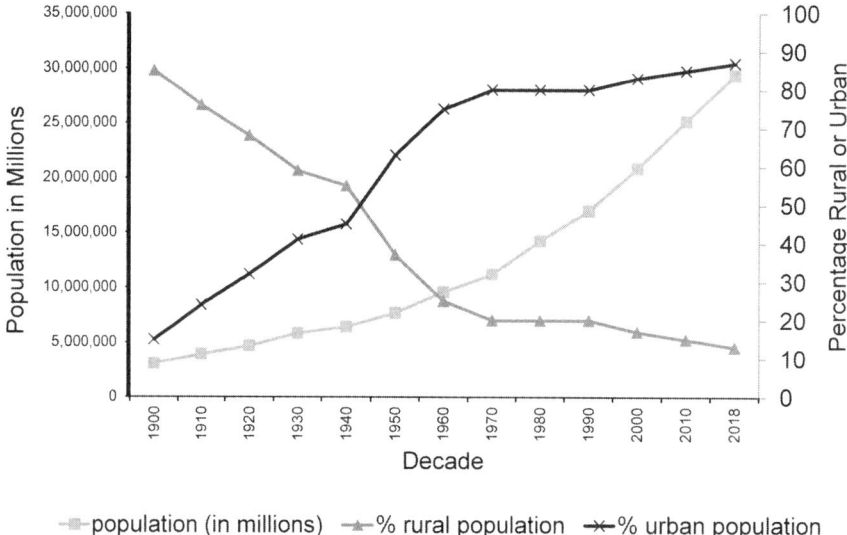

Figure 174. The relative changes in the Texas population as rural populations have declined and urban populations have increased over the last 100+ years.

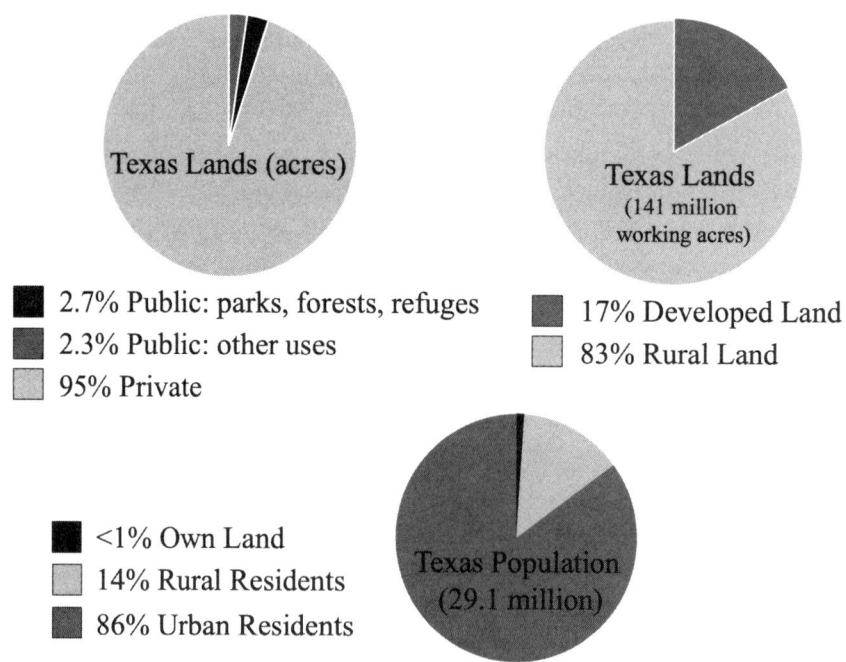

Figure 175. Top left: public vs private land in Texas. Top right: developed vs rural land in Texas. Bottom: rural vs urban population in Texas. Data for these figures were based on information obtained from the 2010 US Census.

resource managers do a better job of educating urban audiences relative to proper wildlife conservation. Most of our citizens are not involved in day-to-day management of natural resources and are unaware of the technology being used, or available, for managing those resources on a sustainable basis. As the urban public assumes a more active role in public policy decision-making, it is critical to the natural resource base and the state's economy that educational programs be designed and supported that address scientific management for continued environmental quality and the necessary flow of goods that must come from our lands and waters. People must learn what it takes to produce the things they desire and at what costs and consequences. There is a need to understand more about how people get their information and form their opinions. Unfortunately, many people today receive their information either from television and online in the news media, which in recent years has become increasingly politicized, or, in a trend that is especially alarming, from social media, rather than from the primary literature or scientists and other experts in the various fields.

People can be educated to understand what the continuation—or destruction—of wildlife means to their future and that of their descendants, and they may be persuaded to act on their resulting concern in ways respectful to the diversity of life and to their own cultural values. However, to be effective, we must reach beyond our traditional clientele—the hunters and landowners—to include all segments of society. We must also reach beyond the traditional commodity-oriented values to address the entire spectrum of values that society places on wildlife resources. Unless we broaden our clientele base, the public confidence to gain the political power and resources to conserve wildlife diversity will not be forthcoming. One of the most critical educational challenges in Texas will involve engaging all demographic groups with regard to conservation issues. Projections for Texas suggest that the population of the state in the year 2030 will be about 46 percent Hispanic, 36 percent Anglo, 10 percent African American, and 8 percent from other ethnic groups. In 1990, the breakdown was 60 percent Anglo, 26 percent Hispanic, 12 percent African American, and 2 percent from other groups (US Census Bureau, 2001). These changing demographics must be considered when educating Texans about wildlife conservation issues.

One of the most successful efforts in public education about natural resource conservation in Texas has been the Texas Master Naturalist (TMN) program that was established in 1998 and has flourished through the first two decades of the twenty-first century. The TMN program is a joint effort between TPWD and the Texas A&M AgriLife Extension Service. The program provides an opportunity for concerned adult citizens to learn about the natural environment and seek ways to better their communities. To gain the title of "Texas Master Naturalist," participants must complete a minimum of 40 hours of natural resource training, 40 hours of service, and eight hours of advanced training offered through the program within their first year (for a complete explanation of the program along with its expansive curriculum see Haggerty and Meuth, 2015). On its twentieth anniversary in 2018, the program reached 4.4 million volunteer service hours, valued at more than $100 million, delivered by 9,329 volunteers in 48 recognized local chapters located throughout Texas and impacting more than 226,200 acres (91,540 ha) of land across the state (Haggerty and Meuth, 2015).

4. Continue to Emphasize and Increase the Participation of Private Landowners

With about 95 percent of Texas in private lands, it will be difficult to conserve wildlife diversity in the state without the continued support and participation of landowners. The future of most wildlife habitat is in private hands. Because owning land is expensive, this creates an inseparable linkage between conservation and economics (Guynn and Steinbach, 1987).

During the first two decades of the twenty-first century, rapid population growth, driven in part by urban migrants seeking a more relaxed lifestyle, has begun to transform property in many rural areas (see Golden et al., 2013). Affluent young professionals, young families, and urban retirees are more representative of these new landowners. It has been estimated that as many as 75,000 urban Texans own at least 100 acres (40 ha) of land in a rural county (Feldpausch and Higginbotham, 2006). These absentee landowners are often innovative in their approach to land management, and they represent a critical audience for wildlife managers. They bring new values to rural areas because they often have a more protective view of wildlife than do long-term rural residents. In the twenty-first century, it will be important to encourage and not discourage the participation of private landowners (both old-timers and newcomers) in wildlife management.

Numerous factors including loss of open space, urban growth, fragmentation as a result of estate taxes, and loss of land productivity are threatening the state's most prime habitat. With traditional agriculture declining, landowners have diversified and increased economic returns by converting land from agriculture to wildlife production. Hunting leases generate income for landowners and, in turn, leave the land in "wildlife" condition as opposed to monocultures of crop or grazing lands. Ecotourism also encourages income development from wildlife. These trends can be expected to continue with increased emphasis on the commercialization of wildlife. Exotic game ranching, deer farming, lease hunting, and various forms of ecotourism, which create markets for wildlife and outdoor recreation, have become popular among landowners. When managed successfully on an acceptable ecological basis, wildlife ranching fulfills a vital role in the conservation of natural resources for the direct benefit of future generations and at the same time yields a sustained crop for the landowner (White, 1986).

In 1995, Texas voters approved an amendment to Article 8, Section 1-d-1 of the Texas Constitution that made wildlife management a qualified use for open space property tax appraisal. The wildlife management use property tax appraisal (WMU) designation allows qualified land under active wildlife management to be appraised at its prior agricultural value. In 1997, shortly after implementation, about 92,000 acres (37,231 ha) qualified for wildlife valuation, but by 2012 more than 3.3 million acres (1.3 million ha) were enrolled, demonstrating the popularity of the program (Lopez, 2016). The WMU has become one of the most successful conservation tools available in Texas.

Several state and federal agencies, as well as private organizations, offer incentive programs that provide technical or cost-share assistance to private landowners for voluntarily enrolling environmentally sensitive land or wildlife habitat in conservation programs to protect or enhance natural resources. For example, TPWD has implemented several initiatives and incentives for private land conservation in Texas over the last several decades. One of the most successful has been the Texas Parks and Wildlife Department-assisted Wildlife Management Plans (WMPs). Through this program, TPWD offers free technical assistance regarding wildlife management activities to private landowners to help improve habitat for native Texas species. TPWD staff help landowners prepare a wildlife management plan that states wildlife management goals, identifies key habitat resources, and prescribes habitat management practices to achieve the identified goals. The program has grown substantially from 2000, when there were 2,727 WMPs approved that covered approximately 11.5 million acres (4.6 million ha), to 2018 with 7,158 WMPs covering approximately 31.2 million acres (12.6 million ha). Another successful TPWD program has been the Managed Lands Deer Program (MLDP) that provides landowners and land managers with additional flexibility to manage deer populations, improve habitats, and provide greater hunting opportunities under the guidance of department biologists. In the past twenty years, the MLDP has grown from 800 tracts of land on 3 million acres (1.2 million ha) to more than 12,000 tracts of land on 28 million acres (11.3 million ha). Other programs sponsored by TPWD include the Landowner Incentive Program (LIP), Pastures for Upland Birds (PUB), and the Texas Prairie Wetlands Project (see the TPWD website for an explanation of these programs as well as others designed to assist landowners).

In addition to the TPWD programs, the Partners for Fish and Wildlife program is a voluntary technical and financial assistance program administered by the US Fish and Wildlife Service (USFWS). It provides technical assistance to restore important fish and wildlife habitat for federal trust species, including migratory birds, threatened and endangered species, and various forms of marine life. As of 2017, more than 1,000 partnerships in Texas have helped restore or enhance over 71,000 acres (28,733 ha) of wetlands, 7,000 acres (2,833 ha) of riparian habitat, and 380,000 acres (153,780 ha) of uplands.

Another successful program is the Conservation Reserve Program (CRP), administered by the US Department of Agriculture–Natural Resources Conservation Service (USDA–NRCS). This is a voluntary program established in the 1985 Farm Bill to provide financial assistance to farmers to install approved cover on highly erodible cropland or pastureland (see Chapter 4 for a discussion). The 2018 Farm Bill continued this program with a slight increase in eligible acreage. Other federal assistance programs available from the USDA–NRCS include the Wildlife Habitat Incentive Program (WHIP), Wetlands Reserve Program (WRP), and the Grassland Restoration Incentive Program (GRIP). An explanation of eligibility and requirements for these programs can be found on the NRCS website.

State and federal agencies also cooperate with nonprofit land trust organizations in Texas to promote private land conservation tools. In this new century, land trusts and conservation easements have gained popularity among landowners. A land trust is a local, regional, or national nonprofit organization that protects land for its natural, recreational, scenic, historic, or productive value. Land trusts work with willing landowners to help them meet their long-term land conservation objectives by accepting donated properties and easements or by purchasing land and easements for conservation purposes; thus, they work to conserve by permanently protecting special lands and waters from development for the benefit of the people, economy, and wildlife of our state. One of the most successful land trusts has been the Texas Ag Land Trust founded by the Texas and Southwestern Cattle Raisers Association, the Texas Farm Bureau, and the Texas Wildlife Association. The mission is to help Texans conserve their agricultural lands, their wildlife habitat and natural resources, and sustain long-term stewardship of private lands for the benefit of all of Texas (see website at txaglandtrust.org).

The Texas Land Trust council was formed in 1999, in partnership with TPWD, as a support organization for all land trust organizations in Texas (see website at http://texaslandtrustcouncil.org). The council is made up of more than thirty land and water conservation organizations working across the state to promote and sustain the conservation efforts of land trusts. Today in Texas, land trusts have collectively protected 1.6 million acres (647,497 ha).

One of the tools that land trusts employ to help accomplish their mission is conservation easements. A conservation easement is a legal agreement a property owner makes to restrict the type and amount of development that may take place on his or her property (Francell, 1998). Each easement's restrictions are tailored to the particular property and to the interests of the individual owner. People grant conservation easements to protect their land from inappropriate development while retaining private ownership. By granting an easement in perpetuity, the owner may be assured that the resource values of his or her property will be protected indefinitely, no matter who the future owners are. Easements include covenants or restrictions that limit subdivision, define building, prohibit destructive practices, and foster appropriate resource management. Easements also can include covenants and restrictions to limit development and protect unique habitat, rare species, open space, and scenic landscapes (The Nature Conservancy, 1999). Granting an easement also can yield tax savings.

Collectively, the various programs described above have assisted more than 7,000 landowners in implementing wildlife management plans on more than 29 million acres (11.7 million ha), representing about 17 percent of the state's total land area. This calculates to about 20 percent of the state's available working lands (144 million acres; 58.3 million ha) as being under some form of wildlife management.

The Farm Bill is the primary federal funding source for private land conservation, which it achieves by funding critical conservation initiatives that assist farmers and landowners to voluntarily and proactively conserve land (Clemens, 2018). Congress added the first conservation title to the Farm Bill in 1985, providing assistance for education, training, and financial support for conservation efforts. The Farm Bill must be reauthorized every five years, and the 2018 version retained support for two broad categories of programs (Lichtenberg, 2018). The first category includes paid land diversion programs that shift acres from agricultural production to conservation uses. The second category consists of working lands programs that pay farmers to adopt or continue to use conservation practices. The 2018 bill places the wide range of paid land diversion programs into two major initiatives: the previously mentioned Conservation Reserve Program (CRP) and the Agricultural Conservation Easement Program (ACEP), which provides matching funds that land trusts can use to purchase conservation easements on agricultural lands, grasslands, and wetlands. Three programs account for the bulk of spending on working lands initiatives: the Environmental Quality Incentives Program (EQIP), the Conservation Stewardship Program (CSP), and the Regional Conservation Partnership Program (RCPP). EQIP provides financial and technical assistance for activities that benefit air quality, water quality, soil and water conservation, and wildlife habitat. The CSP pays farmers an annual fee to maintain a suite of approved conservation management practices on working farmland. The RCPP funds projects at a regional or watershed scale that are undertaken by governmental or nonprofit entities to support innovative solutions to natural resource challenges such as water quality, flood prevention, resilience during drought, and soil conservation (Land Trust Alliance, 2018). TPWD has a Farm Bill Coordinator who works with landowners and NRCS conservationists to assist in the implementation of programs that promote and enhance wildlife and wildlife habitat on private lands.

5. Halt the Current Trend Toward the Privatization of Wildlife and Follow the Principles of the North American Model of Conservation and the Public Trust Doctrine

In the twenty-first century, a conflict is emerging that is poised to threaten the integrity of wildlife and fisheries as a public resource (Yardley, 2002). At the center of this controversy is the ownership of wildlife. In particular is the issue of intensive deer management driven by the promotion of privatization of wildlife and creation of markets that sell public wildlife resources. Intensive deer management involves high fencing of property followed by supplemental feeding of deer and augmented by selective breeding of deer to produce unnaturally large antlers (Knox, 2011). In other words, it is about trophy deer management to produce big antlers and big money. Although TPWD efforts have focused on habitat preservation that would support all wildlife, many landowners have focused increasingly on deer management because of the disproportionate value of this species to hunters and therefore landowners (Adams et al., 2000, 2004).

Many Texas landowners now sell trespass privileges to their properties through various forms of hunting leases and agreements, and this practice has become big business. During the twentieth century, game ranching and hunt leasing became the primary model for allocation of wildlife resources in Texas. The model arose after a stiff trespass law was passed in 1925 that promoted the allocation of wildlife by the pocketbook. Based on the right of access to wildlife, the "Texas-style" system resulted in the commercialization of wildlife as a commodity sold, in many cases, for substantial amounts of money.

Through these leases, the landowner grants the hunter (or in some cases a fisherman) permission to: access their property; live or camp on the property; erect hunting blinds, feeders, or other structures; and harvest game. These arrangements vary and may range from the harvesting of certain species (e.g., deer or quail) to all species that may be legally taken under the laws of the TPWD and USFWS; also, the arrangement may be for a special hunting season or year-long. In some extravagant cases, the landowner may provide landing strips, lodging, meals, entertainment, and guiding services to accommodate the lessee.

The arguments for paid hunting tend to run along three lines. First, that private initiative provides better hunting for financial reward and, at the same time, better preserves habitat by preventing landowners from having to sell to developers, who would destroy natural habitat that is forever lost to other land uses. Second, that raising wildlife for sport hunting is an attractive income source and productive alternative, or supplemental option, to raising cattle in periods of decline in the livestock industry, thereby helping keep family land intact and providing an incentive to restore the land to its original state with native trees, brush, and grasses. And, third, that the transfer of wildlife into de facto private control saved many wildlife species from extinction in the state, an argument along the lines that "what is good for a deer is also good for a deermouse and many species of songbirds."

Others have argued that leasing of hunting rights to benefit private landowners in the long term will become deleterious for conservation by reducing participation rates and causing low-income participants (such as the young, early married, minorities) to forgo hunting and concern for conservation. For example, it has been noted that Texas has but half the deer hunters of Wisconsin, yet almost five times the number of deer and three times the human population. Valerius Geist, a Canadian wildlife biologist and conservationist, has been one of the more vocal critics of the Texas system and has published extensively on the subject (e.g. Geist, 1988, 1989, 1991).

A controversy about this system has developed around the rapidly growing practice of "high fencing" (deer-proof, game-proof enclosures, etc.) to privately isolate many species of wildlife, especially large native ungulates that, according to Texas law, are a public resource. Perhaps as early as the 1930s, but certainly by the 1950s, Texas landowners were experimenting with high fences to better manage white-tailed deer populations. Interestingly, in most of these instances, high fences served as a means to fence deer away from a property rather than the initial assumption that such fencing would be used to enclose deer inside private property and away from other landowners. This counterintuitive scenario is a product of vast portions of Texas being overpopulated with white-tailed deer. If landowners or biologists desired to improve deer herds (through management practices), they discovered that they first must reduce the local deer population and then

586

find ways to prevent deer from neighboring properties from moving in to fill the vacuum. Consequently, high fencing became the go-to solution for most deer management practices and then expanded in popularity as the exotic market took off in the 1980s. Today, high fencing is a quite common occurrence in the Hill Country and South Texas as a means not only to protect the landowner's financial investment through implementation of habitat improvements under a wildlife management plan but also to protect expensive individual animals in the cases of trophy deer and exotic species.

Currently, Texas is second only to South Africa in total number of acres under high fences. In South Texas, approximately three million of the 16 million acres (1.2 of 6.47 million ha) of deer habitat are under high fence (Patoski, 2002). Similar numbers of high-fenced acres are probably representative for much of the Hill Country and perhaps other parts of the state. Many of these high-fenced properties have provided a valuable service to Texans and to numerous species of wildlife. First, high fencing offers the conscientious landowner a tool to reduce overpopulated deer herds in an effort to improve range conditions and the overall health of the deer herd. If done correctly, and with modern wildlife management techniques, previously overgrazed habitats can be returned to more productive conditions. These restored ranges benefit not only white-tailed deer but also cattle and untold numbers of wildlife, ranging from quail to mice and butterflies. Second, through high fencing and the resulting "healthier" deer herd (larger body size and antlers) the landowner can lease his/her property for hunting as described above, and ultimately be rewarded for land stewardship principles.

It is true that high fences restrict the movements of large native ungulates (white-tailed deer, mule deer, elk, and pronghorn), introduced or exotic species (feral hogs, axis deer, nilgai, blackbuck, etc.), peccary, and perhaps some carnivores (mountain lions and coyotes) depending on size of wire openings and height of net-wire. However, in totality, few species of Texas wildlife are probably impacted by high fencing; and therefore, movement or containment of most species is not necessarily a primary concern, especially given the large number of white-tailed deer in Texas—the species at the center of most high fencing debates. However, given the economic importance of many of these species and the debate surrounding landowner rights, high fencing has become controversial considering that it prevents wildlife from being able to freely roam from one property to another.

With both of these practices (hunting leases and high fencing) there are obvious benefits to landowners and, typically, to wildlife, although several ethical questions have arisen. First, given that wildlife in Texas belongs to the people of Texas (TPWD Code Section 1.011; see statutes.capitol.texas.gov, 2019), does a landowner have the right to "fence in" wildlife? By extension, does a landowner have the right to prevent the movement of wildlife from one property to another? According to TPWD Code Section 1.013 (statutes. capitol.texas.gov, 2019), the answer to both questions is yes. Second, does a landowner have a right to profit from a natural resource that belongs to the people of Texas? The answer to this question also is yes. Third, do you have a right to hunt, photograph, or enjoy the "people's wildlife" on private land? The answer to this question is no—unless you own the land or you obtain trespass permission from the landowner. The answers to these and other questions pose an interesting dilemma concerning wildlife from both a human access and financial standpoint.

As discussed above, some Texas landowners practice management techniques to enhance the production of desired traits, specifically antler size. These landowners may implement the culling of less desirable bucks, establish harvest quotas based on size or age, or import "superior" breeding stock (big-antlered bucks). In return, they may charge a higher trespass fee to recoup their investment. Generally, these operations are small in scope and do not necessarily draw much attention or ire from the general public. However, in recent years, some Texas landowners have turned "big antlers" into big business. These "deer breeder" operations often employ biologists, geneticists, animal nutritionists, veterinarians, and other skilled professionals; they may construct elaborate pens and facilities for breeding (often using artificial insemination); they may organize fawning events; and they may spend exorbitant amounts for "breeder bucks." In return, these operations can command top dollar for hunts (some exceed $10,000 for trophy-class deer), semen ($4,000-5,000 per straw in some cases), and sires (tens to hundreds of thousands of dollars per trophy breeding buck). In fact, an atypical white-tailed buck known as "Stickers" is estimated to have been purchased for approximately $1 million, although the actual price is not publicly available (*Woods 'N Water Magazine*, 2008).

Deer breeding operations are legal under Texas law and, similar to livestock, "pen-raised" deer can be bought and sold. Generally, most Texans do not begrudge an individual from making a profit on these types of operations as long as they involve pen-raised progeny. Where it gets problematic is when a deer breeder releases a "breeder buck" into the wild, generally onto a high-fenced ranch, to genetically improve antler and body size in the native herd. This scenario then elevates further when the breeder buck mates with a native white-tailed deer (owned by the people of Texas), and the landowner then sells the right to harvest the offspring of that mating for several hundred or even several thousands of dollars. Under that scenario, the landowner makes a larger than normal profit using native deer owned by the people of the state, who receive no money and are prevented from harvesting the resource themselves because the offspring in question is isolated by the high fence and on private property.

Under the current Wildlife Codes of Texas this is perfectly legal, but is it ethical? Should a landowner be allowed to profit, significantly in some cases, from the people of Texas or should the landowner be required to share his/her profits with the people, since the people own the wildlife? An analogy is that the people of Texas own a number of school buses, police cars, ambulances, and fire trucks: can an individual sell any of those vehicles, or other types of state property, and then pocket the proceeds? If a "breeder buck" mates with a native white-tailed doe (owned by the people of Texas) and the landowner charges $5,000 to harvest the offspring, shouldn't the people of Texas receive an appropriate amount of the proceeds? Should landowners be required to pay a tax, purchase a permit, or completely run their operation with pen-raised deer and zero influx of native individuals? Where do the rights of the people and landowner intersect? Currently, there are as many as 9 million acres (3.6 million ha) of Texas land under high fence (Todd and Ogren, 2016). Many of these properties have been eligible for the Wildlife Management Property Tax Appraisal program and for TPWD Assisted Wildlife Management Plans at no cost. Some are beginning to question the fairness of offering public tax discounts to these parcels, since the tracts offer little public benefit to state wildlife outside of those fences (Todd and Ogren, 2016).

The closet efforts to privatize native fish and wildlife are now being criticized by wildlife professionals as potentially eroding public support for conservation initiatives, leading to the slippery slope for more species to be privatized, degrading the outdoor hunting experience, and favoring short-term economic gain over long-term conservation (Bryant, 2016). Unfortunately, the Texas legislature has reacted to industry lobbying efforts to promote intensive deer management by adopting regulations that support the activity. They have allowed special harvest permits that enable landowners to cull bucks and does to alter the age and sex composition of herds and reduce herd size (a practice historically reserved for hunters rather than landowners), and they have endorsed captive breeding programs to correct perceived "weaknesses" in the genetic makeup of herds to more quickly provide high-quality bucks that are more expensive to hunt. As a result, hunter access to land is becoming limited in much of Texas to the number of hunters who can afford and be accommodated for "high quality" and expensive buck hunts. This, in turn, is leading to many deer hunters, and especially young hunters, being excluded from hunting on private lands. Many wildlife professionals have predicted that too much emphasis on intensive deer management under high fencing (this has been termed "antler religion"; see Knox, 2011) could ultimately cause the demise of traditional hunting. The hunting tradition is already in jeopardy through retirement of older hunters, low recruitment of youth, and lack of public interest and support for trophy hunting (Adams et al., 2004).

Industry groups and wealthy landowners and individuals should not be allowed to privatize the wildlife that belongs to the citizens of Texas. Government, namely the Texas legislature, should not hand over ownership of a public resource to private entities. The Texas Foundation for Conservation has been organized by wildlife professionals to resist any efforts to privatize native fish and wildlife for the benefit of a few (Bryant, 2016).

The North American Model of Wildlife Conservation, and its cornerstone Public Trust Doctrine, decrees that wildlife belongs to all citizens and its management is entrusted to the government to benefit present and future generations (The Wildlife Society, 2007). The model also is centered on the principle that fish and wildlife management must be based on sound science, and that science, rather than emotions or short-term economics, should drive regulations. In the twenty-first century, there must be a recommitment to the North American Model of Wildlife Management, which demands that wildlife remain a public resource. Otherwise, activities such as high fencing by landowners wanting to raise trophy deer for expensive hunts, which is nothing more than an attempt at de facto privatization of a public resource, will eventually undermine public credibility in much-needed conservation programs.

In 2016, two individual captive deer breeding permit holders filed a lawsuit against TPWD in connection with Chronic Wasting Disease (CWD) management rules adopted by the state agency to protect all deer in the state. The lawsuit sought to establish that deer held by breeders in captivity are private property and therefore exempt from TPWD regulation. Fortunately, on 2 October 2020, the Texas Supreme Court upheld a lower court ruling that deer in Texas are not private property, upholding the Public Trust Doctrine that wildlife belongs to everyone. Furthermore, in the age of CWD and possibly other wildlife diseases, the ruling means the state agency has regulatory powers to manage wildlife resources to ensure that the best interest of wildlife and the public are protected (Dorsett, 2020).

6. Emphasize Sustainable Resource Systems and Ecosystem Management within Integrated Landscapes

For most of the twentieth century, wildlife management focused almost exclusively on one element of the wildlife fauna: game species (deer, quail, dove, turkey, and so forth). Most monetary resources and time were spent on attempting to manage lands for these species almost to the total exclusion of nongame species. Then, toward the latter part of the last century, attention began to focus on nongame species deemed threatened or endangered. As part of this shift, conservation biologists gradually moved to a focus on "surrogate species" to serve as shortcuts to guide decision-making (see Caro, 2010). Attention was paid to three types of species: *umbrella species* (those that occupy such large tracks of habitat that saving them will automatically save many other species); *flagship species* (normally a charismatic large vertebrate that can be used to anchor a conservation campaign because these species arouse public interest and sympathy); and *keystone species* (those whose presence and fluctuations are believed to reflect those of other species in the community). In Texas, mammalian examples of these three types of species, respectively, would be the prairie dog (*Cynomys ludovicianus*), ocelot (*Leopardus pardalis*), and beaver (*Castor canadensis*).

Although environmental monitoring and single-species approaches can help us divert problems before species become extinct, this is a labor-intensive and extremely expensive solution. In this new century, a new approach that focuses on collaborative solutions at the landscape and ecosystem level is being advocated as the best solution to expanding the scope of conservation management to achieve economies of scale and efficiency. Managing populations of particular species of interest will cause us to fall farther and farther behind in meeting the challenges of preserving biodiversity, as more and more species fall below a threshold of endangerment and as funding in no way keeps pace with their individual needs. The best way to deal with this challenge is to manage entire ecosystems, if not whole landscapes, by unified methods designed to save all their inhabitants at one time (Simberloff, 1998). A recent study (Blanco et al., 2021) has indicated that ecological assemblages, associated with persistent ecosystems, are more resistant to environmental change than are species compositions, which are transitory in comparison to their ecological roles.

This calls for a broader management approach that considers ecosystems and land-scapes and all biological resources instead of a limited subset of species and their habitats. In a state like Texas, with its enormous biological diversity, this is really the only long-term approach that is feasible because it provides a conceptual basis for mixing economic, social, and ecological theory with the application of human, biological, and financial resources to manage resources at the local, land-enterprise scale. There are not enough financial or human resources, or the time, to take a single-species approach toward managing all of our wildlife resources. However, managing habitat on this scale is challenging in Texas because of the privately owned nature of the land, the dwindling size of rural land parcels, and the economic value of the wildlife involved (Todd and Ogren, 2016). To be successful it demands that competition for resources must give way to cooperative management strategies, where conservation and resource management are linked in sustainable resource systems.

Texas is a land that has been split into many privately owned pieces, estimated at 248,000 farms and countless urban and suburban tracts (Lund et al., 2017). Although state, federal, or other public entities own some 8.55 million acres (3.46 million ha) of Texas (5 percent), hundreds of thousands of farmers, ranchers, and other private individuals own the balance, which comprises more than 162.45 million acres (65.7 million ha; 95 percent); of this, 17 percent is developed and 83 percent is rural land (see Fig. 175, p. 580). Of the publicly owned lands, 4.63 million acres (1.87 million ha; 54 percent) represent parks, forests, and refuges that are available for conservation and outdoor recreation (see Table 9). Most of the rural land is held in large cattle ranches where grasslands and the soils that sustain them are the key ingredients. To manage habitat among so many landowners is daunting. It cannot be done with a single one-size-fits-all remedy; there are simply too many different ecosystems, markets, family circumstances, and financial situations involved (Todd and Ogren, 2016). And it cannot be done by edict as Texas has few regulatory tools for land management. These converging trends argue for more coordinated and intensive habitat management. One approach that has emerged to address this need has been the formation of landowner-driven wildlife management associations (WMAs) as described above in the section on the Statewide Biodiversity Initiative.

With the right information and motivation, private landowners have demonstrated their willingness to develop wildlife habitat and protect species. An excellent example in Texas can be seen in the efforts of a private landowner, a government agency, and nongovernmental entities to work together to protect critical habitat for the endangered ocelot in South Texas (see Chappell, 2010). After years of implementing a brush-clearing program to increase grazing range for cattle, a well-known South Texas rancher, the late Frank Yturria, became interested in ocelots and decided to begin reestablishing native brush to support conservation efforts. Subsequently, in 1989, he sold a 500 acre (202 ha) easement on his ranch, home to one of only two documented ocelot breeding grounds in the country, to the USFWS. That fragment of native thornscrub has proven to be a

Table 9. Parks, forests, refuges, and other public lands in Texas available for public recreational use. Figures are current for 2019 unless denoted by an asterisk (*), which indicates values are from 2000.

	Area (acres)
Federal Lands[1]	
National Park Service	1,201,670
US Forest Service	755,365
US Fish and Wildlife Service	527,418
US Army Corps of Engineers *	469,754
Bureau of Reclamation *	10,510
Subtotal Federal Lands	**2,964,717**
State Lands[2]	
Texas Parks and Wildlife	
Wildlife Management Areas	714,094
State Parks	631,256
Natural Areas	63,155
Historic Sites	5,971
Subtotal Texas Parks and Wildlife	**1,414,476**
Texas Forest Service	7,306
Subtotal State Lands	**1,421,782**
Local Parks	
Cities and Counties	230,920
River Authorities	15,172
Subtotal Local Parks*	**246,092**
Total parks, forests, and refuges	4,632,591
Total land area of Texas	171,902,000
Percent of Texas land in parks, forests, and refuges	2.7

[1] This table does not account for all federal public lands, such as Department of Defense, BLM, NASA, and Department of Energy, which are not considered parks, forests, refuges, or other lands available for public recreational use.

[2] This does not include other state-owned lands, such as those held by Texas state universities and colleges.

critical "source habitat" for ocelots in Texas (Michael Tewes, personal communication). In 2007 and 2009, Yturria granted additional easements of 698 acres (282 ha) and 1,300 acres (526 ha), respectively, to The Nature Conservancy. Those easements, although comprised of open rangeland at the time, are adjacent to the original 500 acres of dense brush and, with protection, are now returning to native brushland and expanding critical habitat for ocelots. Most recently, the USFWS acquired 7,400 acres (2,995 ha) of open rangeland from the Yturria family. That protected rangeland is expected to provide additional ocelot habitat as it gradually returns to native brushland over the next 20 to 40 years, with the potential to significantly expand the core population of ocelots in South Texas (Michel Tewes, personal communication). These conservation easements, totaling 9,898 acres (4,006 ha), provide a perfect example of cooperation between private landowners and wildlife conservationists. In a further demonstration of the power of individuals who are committed to wildlife conservation, Mr. Yturria endowed the Frank Daniel Yturria Chair for Wild Cat Studies at Texas A&M–Kingsville to fund continued ocelot research.

The difficulties with the single-species approach can be seen in the application of the ESA in Texas (Gilliland and Mays, 2003). Although the ESA clearly has helped to preserve the natural diversity of wildlife in the United States (e.g., see Greenwald et al., 2019), it has had a checkered history in Texas, coming under strong criticism by many landowner associations and citizens concerned with the protection of private property rights. To address these concerns, in 1995 the Environmental Defense Fund worked with the USFWS to develop a program, known as "safe harbor," to aid private landowners who want to participate in species protection programs. The safe harbor policy assures landowners that if they undertake actions that are beneficial to endangered species, they will not incur added regulatory burdens as a result of their good deeds. In other words, participants who improve their property for the benefit of endangered species retain the right to undo those voluntary improvements should they wish to make some other use of their land in the future (Earley, 1996).

Texas is among the nation's leaders in the creative use of safe harbor agreements (USFWS, 2011). For example, more than a million acres (404,686 ha) of privately owned ranch land has been enrolled in a safe harbor program aimed to restore the endangered northern aplomado falcon (*Falco femoralis septentrionalis*) in South Texas. About 16,500 acres (6,677 ha) of grassland have been enrolled in a safe harbor program for the Attwater's prairie chicken (*Tympanuchus cupido attwateri*) along the Texas coast. A similar program has now been approved for the red-cockaded woodpecker (*Leuconotopicus borealis*) in East Texas. In short, some of America's rarest birds are gaining a new lease on life as a result of safe harbor programs. Importantly, this form of habitat protection also benefits other forms of wildlife and promotes more of an ecosystem approach.

7. Expand Protected Area Acquisition and Outdoor Opportunities for Urbanites

To effectively conserve wildlife diversity in the future, it is clear that greater attention must be devoted to increasing protected areas, or lands set aside exclusively or primarily to conserve wildlife. During the time of the biological survey, there were no parks or preserves in Texas. Today, there are hundreds of resource-based parks and preserves in the state managed by state and federal agencies, as well as numerous local parks managed by cities, counties, and river authorities for purposes of outdoor recreation. Texas has

125 state parks, forests, historic sites, and natural areas that total approximately 700,000 acres (283,280 ha). In addition, there are 51 wildlife management areas with approximately 715,000 acres (289,350 ha) owned or leased. National forests, national parks, and USFWS wildlife refuges make up another 2.5 million acres (1 million ha). Additionally, United States Army Corps of Engineers' lands, Bureau of Reclamation lands, and municipal and county-owned parkland total nearly another half a million acres. Thus, the total land area in public domain for conservation and outdoor recreation in Texas is about 4.6 million acres (1.86 million ha), or about 2.7 percent of the total land area of the state (Table 9). This percentage is less than one-quarter the percentage of the total land area on the globe set aside in protected areas. In other words, Texas has a much smaller percentage of its land devoted to protected areas than is found across the entire globe. Realistically, Texas should seek to have at least 15 percent of the state set aside as protected land for conserving biodiversity by the end of the twenty-first century, which will bring the state more in line with global goals.

Parks and preserves serve as living laboratories for education and scientific investigation and as places for outdoor recreation activity. They also serve to protect natural diversity, although their capacity for preservation is limited by a number of factors, including the size of the protected area, their geographic distribution, the system and area configuration, the amount and kind of site development, management objectives and practices, and outside environmental influences (Carls, 1984). A study of the 14 largest national parks in western North America concluded that none of them were large enough to retain an intact mammalian fauna (Newmark, 1987), and no protected area in Texas is as large as the smallest of the 14 parks used in that study.

Our parks and preserves are scattered throughout the state, although the geographical distribution is far from proportional (Fig. 176). The heaviest concentration of areas is found in the Trans-Pecos, Piney Woods, Gulf Coast, and South Texas Plains regions. The many sites in the Trans-Pecos clearly indicate the bias for siting parks and preserves in areas of scenic and more "spectacular" natural history. Concentrations in the eastern and southern part of the state reflect the great public demand for outdoor recreation in centers of population. The Rolling Plains and High Plains natural regions tend to be underrepresented by parks and preserves. These areas are generally more remote from centers of population, and they lack the scenic qualities usually favored for park development. Most conspicuous by absence from the system are protected areas of native grassland and prairie.

Although they represent only a small part of the Texas landscape, these parks and preserves, and the federal, state, local, and private entities that oversee them, play increasingly vital roles in the conservation of biodiversity in Texas. For some regions, such as the Rio Grande Delta, they represent what is left of the native habitats. Clearly, there is a need for more parks and preserves and additions to existing areas to protect rare or unique natural resources and to better represent the natural regions of the state. Given their concerns about conservation and scenic beauty, Bailey and the

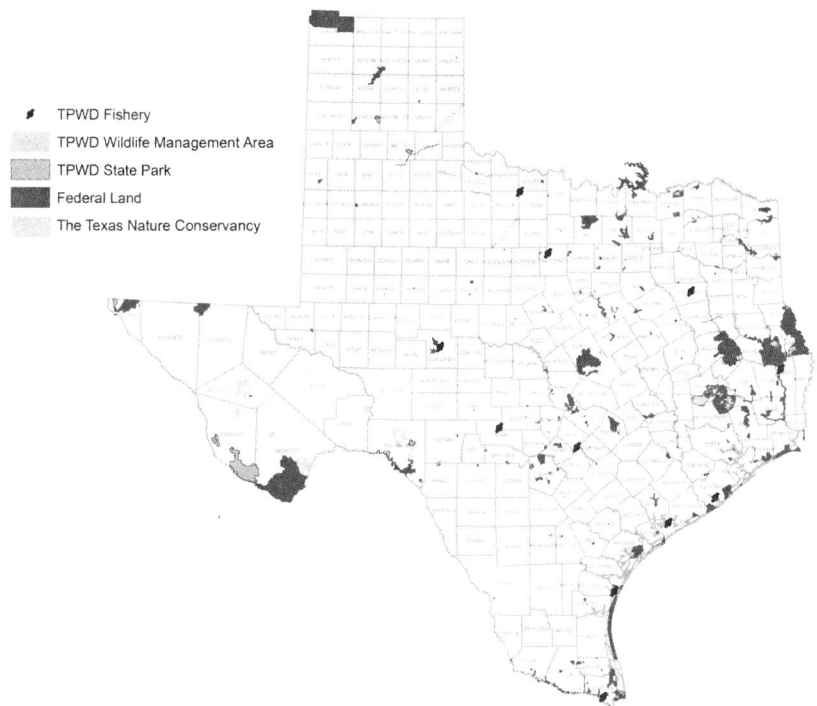

Figure 176. Distribution of public lands in Texas. These lands are a combination of TPWD properties, land managed by the US government, and lands owned by the Texas Nature Conservancy. Figure modified from a PDF available from tpwd.texas.gov (TPWD, 2019a).

federal agents, if they were alive today, surely would endorse such a strategy for the twenty-first century.

The Nature Conservancy (TNC) has been one of the most effective conservation organizations in Texas working to protect ecologically important lands, waters, and rare species. It first became active in Texas in the 1950s and formerly established the Texas Chapter in 1964 but really expanded its work in the state in the latter part of the twentieth century and the early part of the twenty-first century. Using a science-based approach, TNC works with governmental and nongovernmental partners, as well as with farmers and ranchers, to protect rare species and characteristic habitats under threat, and it promotes agricultural practices that support healthy lands and waters throughout the state. To date, the TNC in Texas has established 38 Texas nature preserves and conservation properties and more than 130 easements, thereby protecting more than 900,000 acres (364,217 ha) in the state (The Nature Conservancy, 2019). The interested reader can go to their website to get a list of their projects in Texas. Among their most recent successful projects are the Davis Mountains Preserve Project in West Texas (Jeff Davis County), the Barton Creek Habitat Preserve and the Balcones Canyonlands Habitat Conservation Plan in the Hill Country near Austin, and the Government Canyon State Natural Area near San Antonio.

One of the biggest challenges to the management of protected lands is their small size and isolation, which makes them vulnerable to island effects and climate disruption over the long term. An important component in identifying areas for conservation priority is to consider the corridors between conservation areas. Attention to planning and management of the areas between conservation reserves must become a priority for future reserves. The goal should be to expand protected areas to include a cross-section of all major ecosystems in the state and to link them via conservation corridors so they are more effective (Adams and Dove, 1989). Habitat connectivity allows wildlife to move across their range seasonally or disperse to unoccupied habitat. Protected migratory pathways also are essential to maintain genetic diversity in populations and avoid harmful genetic isolation. The more habitat strongholds are connected, the better the chances are for wildlife to survive the stresses of habitat loss, fragmentation, and climate disruption.

Three scientifically based strategies are essential in making decisions about protected area acquisitions. First, it is necessary to identify and locate the species and communities most in peril. Second, efforts must be made to carefully define a "portfolio" of the most important places to protect. And third, we must learn how these ecosystems function and this knowledge must be used to develop and carry out long-term conservation strategies. Added to this must be efforts to build long-lasting and positive relationships with both public and private partners. Conservation projects are much more effective when they are locally based and involve private landowners.

In addition to their conservation value, Texas's state parks and protected areas serve as a major source of outdoor recreation in the state, and for that reason they are incredibly popular with Texas citizens, particularly urbanites. For example, in the 2017 fiscal year, there were 10 million park visitors, a 20 percent increase over 2012 (Patoski, 2018). State parks have become one of the biggest tourist attractions in Texas and the economic engine of many rural counties. The population of the state has been booming (increasing by more than seven million since 2000), and as stated many times in several chapters of this book, most of the growth has been in urban areas. Many newcomers are seeking outdoor experiences, and the system needs new parks to accommodate them.

For whatever reason, the Texas Legislature has not provided sufficient funding to accommodate the need for more parks and to address the backlog of maintenance in the current system. TPWD has had to approach projects piecemeal to make any significant additions. In the last decade, it added 80,000 acres (32,375 ha) of prime land, including the 17,000-acre (6,880-ha) Powderhorn Ranch near Port O'Connor on the Texas coast, and a stunning 17,000-acre (6,880-ha) tract on the Devils River. And, in 2019 it earmarked $12.5 million to establish Palo Pinto Mountains State Park in the Rolling Plains and Cross Timbers region west of Fort Worth. Although these property acquisitions represent a step in the right direction, they are a "drop in the bucket" of what is really needed.

In November 2001, TPWD issued a sweeping study of Texas's outdoor resources. It was intended to inform the agency's planning over the next 30 to 50 years. The study, overseen by one of us (DJS) in his role as president of Texas Tech University, was prompted by a legislative review criticizing TPWD for lacking a long-term vision. The Texas Tech study noted the great diversity of animals and plants in the state and the need to accommodate the state's rapidly growing human population while conserving Texas's natural resources.

The Texas Tech study used science and data-based arguments to emphasize the importance of gradually acquiring additional public land around large urban areas to strengthen participation in outdoor recreation, hunting, and conservation, particularly for youth and people with incomes insufficient to access large hunting leases and for minorities who for the most part had not been involved in conservation issues but were quickly making up a larger segment of the Texas population. It recommended that an additional 1.4 million acres (567,000) of parkland were needed inside the Houston–DFW–San Antonio population triangle—more than twice the amount statewide at the time—to accommodate the state's long-term needs. The response from the governor-appointed TPWD commissioners and the ascendant GOP leadership was tepid. They simply lacked the will to commit the funds to make ambitious land purchases near the big cities. Instead, they set a much more modest goal of opening "four parks of at least 5,000 acres (2,023 ha) each inside the urban triangle by 2030" (Patoski, 2018). This unwillingness to add state parks and protected areas is short-sighted and disenfranchises millions of Texans, particularly the rapidly growing Hispanic population, from appreciating wildlife and the outdoors, and it makes them less likely to support important conservation initiatives in the future.

8. Address the Crucial Issues of Water, Land Fragmentation, and Climate Change

Land fragmentation and the misuse of water already had become the dominant conservation issues by the end of the twentieth century, and their importance continues on into the twenty-first century. Texas is no longer a state where economy and culture are defined principally by the land. With the continuing influx of new residents and a population increasingly shifting to cities and their suburbs, Texas has become a primarily urban society. Many landowners whose families have lived on the land for generations have come under tremendous pressures to sell their farms and ranches for development. As a result, Texas, like other states across the country, is in jeopardy of losing its legacy of families who live and work on the land—the traditional stewards of our natural heritage. Approximately 41 to 48 percent of rural landowners are now "absentee landowners," meaning they own property in rural areas but mainly reside in cities (Stanley, 2015).

The fragmentation of large family-owned farms and ranches poses perhaps the greatest single threat to our wildlife because it places once plentiful habitat for native plants and animals increasingly at risk. For this reason, it is crucial to find ways to keep large continuous tracts of land together and to find ways for all landowners, including absentee landowners, to participate in conservation. Tremendous progress has made through the creation of incentives, such as purchase of development rights programs (see above), that allow landowners to sell the development rights to their land by granting conservation easements to a government entity or nongovernmental conservation organization, yet retain all other rights of ownership, including the right to continue ranching, farming, hunting, and fishing.

Management of water will be a critical conservation issue in Texas in the twenty-first century. Water is the limiting factor for all aquatic life, plants, and wildlife. Rivers link our land and water ecosystems, but we have pumped so much water out of the ground that it has caused a "slow desiccation" of river ecosystems. With Texas's population expected to double in the first thirty years of the twenty-first century, there is an urgent need to maintain sufficient water for adequate flows to rivers, lakes, and estuaries to maintain the fish and wildlife that depend on them. Historically, the allocation of water rights in Texas has not taken into account the needs of the state's ecosystems. The state's current water statutes and regulations require that environmental needs be considered in the overall picture, but they do not assure minimal instream flows to sustain the health of rivers and estuaries (for a thorough discussion of Texas water rules and regulations, see Texas A&M AgriLife Extension and Texas Waters Resources Institute: twri.tamu.edu). As Texas attempts to meet increased water needs, it must be careful not to impair the ecological health of these ecosystems, which form the natural infrastructure of our state. To help accomplish this, a Land and Water Strategic Plan was developed by TPWD in 2002 that is periodically updated (last update in 2013). Texans must learn to treat every drop of water as precious, and with the rapid onset of climate change, and predictions of a significantly warmer and drier climate, the need for a comprehensive strategy has become magnified.

Undoubtedly, the single greatest challenge to Texas in the twenty-first century will be the impact that climate change will have on the Texas landscape and ultimately upon its wildlife species. There are hundreds if not thousands of scientific articles being written about the potential for significant changes in the climatic patterns—in fact there are so many that instead of citing the primary literature (as we have tried to do throughout this book) it is far simpler to suggest that the reader search online scholarly databases for the most recent information.

Climate change is occurring as a direct result of rising levels of greenhouse gases in the atmosphere and reduction of the ozone layer. In the last 100 years, the earth has warmed 1.3-1.6 degrees Fahrenheit (0.7-0.9 degrees Celsius). The National Climate Assessment, a stunning report released by 13 federal agencies and the White House in 2018, showed that climate change has already had devastating impacts on our environment and economy, and that costs could amount to hundreds of billions of dollars by the end of the twenty-first century (US Global Change Research Program, 2018).

According to the United States Environmental Protection Agency, Texas could be among a string of "Deep South" states that will experience the worst effects of climate change. Regardless of the climate model used, Texas is predicted to experience a challenging assortment of climate changes that scientists have concluded are already spinning off from a man-made atmospheric warming trend.

Research by the Union of Concerned Scientists (US 2019) suggests there will be dangerous heat increases across the US, with Texas bearing the brunt of that increase, and a recent study of Texas climate trends seems to confirm that conclusion (Dawson, 2020). Looking ahead to 2036, the year that Texas will be 200 years old, the study projects that the state will experience substantial increases in extreme heat and extreme rainfall, with average temperatures about 1.6 degrees Fahrenheit (0.9 degrees C) warmer than the 2000–2018 average and 3 degrees Fahrenheit (1.7 degrees C) warmer than the 1950–1999 average. The number of 100 degrees Fahrenheit (38 degrees C) days has more than doubled over the past 40 years and could nearly double again by 2036. At the same time, extreme cold could become more severe. In February of 2021, Texas experienced a week of record-breaking cold that impacted both people and wildlife, including vast die-offs of many bats and exotic ungulates, as discussed earlier. Extreme rainfall is expected to become more frequent and severe, resulting in increased urban flooding. And, with the forecasted sea level rise, the risk of hurricane storm surge in coastal areas could double. Finally, despite increased precipitation, other factors are expected to increase drought severity, resulting in severe and more extended wildfire seasons. As temperatures rise and the human population continues to expand, more of the state will risk becoming vulnerable to desertification, the permanent degradation of land that was once arable. Climate change is a threat multiplier because it takes many of the risks we already face today and makes them worse.

To illustrate the predicted climate for Texas, we selected six cities (Lubbock, El Paso, Austin, Texarkana, Houston, and Brownsville) that more or less are strategically positioned to represent the geographic regions of the state. We then used the website and data assembled by the University of Maryland Center for Environmental Science (2019) to provide the reader with an idea of what the geographic climates might look like in 2080 if we continue to emit carbon at the current levels (Table 10).

From these climate models, it appears that the western half of the state will become an extension of the Chihuahuan Desert as warmer/drier habitats extend into the Southern High Plains of the Texas Panhandle and portions of the Hill Country. Predictions suggest that this region will receive perhaps 50 percent of its current precipitation amounts and will see a 5-14 degrees Fahrenheit increase in mean annual temperatures (see data for Lubbock, El Paso, and Austin). The South Texas Plains (see data for Brownsville) will be warmer and wetter and will more closely resemble the Tamaulipan Thorn Scrub habitats currently seen in the Mexican state of Tamaulipas. Models predict the Big Thicket region of northeastern Texas and coastal regions of the state (see data for Houston and Texarkana) to be warmer but the least impacted in terms of precipitation (although precipitation may be greater in some places). Houston and the coastal areas will be much warmer and in some cases the heat index is predicted to be dangerously high (120 degrees Fahrenheit [49 degrees Celsius] by some models—see Leahy, 2019). There is a likelihood of stronger and more frequent coastal storms and flooding, such as tropical storm Imelda—that storm dropped more than 40 inches (102 cm) of rain in some areas of southeastern Texas (Iracheta, 2019) in one of the most significant rainfall events ever recorded that was not the result of a hurricane.

Table 10. Current and projected (year 2080) climate data for six Texas cities.[1]

City	Current Annual Mean Temperature	Annual Mean Temperature Predicted 2080	Current Annual Mean Precipitation	Annual Mean Precipitation Predicted 2080	Current Average Annual High	Average Annual High Predicted 2080
Lubbock	60.7	75.1	19.2	8.0	74.3	86.7
El Paso	64.7	71.3	9.7	7.9	77.5	88.5
Austin	69.4	74.1	34.3	16.9	79.8	86.8
Texarkana	63.9	69.6	52.0	62.5	74.5	78.0
Houston	69.1	77.0	45.3	41.1	78.3	87.5
Brownsville	74.6	82.9	27.4	46.3	83.7	85.8

[1] Cities were chosen to geographically represent various regions of the state (Maryland Center for Environmental Science, 2019). For each Texas city, the projected climate data are based on another city that currently reflects what the climate of these Texas cities is predicted to be like in the future. Temperatures are in degrees Fahrenheit; precipitation is in inches. City comparisons are as follows: Lubbock to Phoenix, AZ; El Paso to Maricopa, AZ; Austin to Nuevo Laredo, Mexico; Texarkana to New Orleans, LA; Houston to Ciudad Mante, Mexico; Brownsville to Acapulco, Mexico.

Changes in temperature and rainfall patterns will cause changes in the distribution of plant communities, which in turn would affect the distribution of wildlife diversity across the landscape. A northwest to southeast gradient of freeze days and a west to east gradient of increasing soil moisture, both of which influence species distribution, represent two major influences of climate on regional vegetation in Texas (Owen and Schmidly, 1986). With the higher temperatures expected from climate change, evapotranspiration rates will increase and critical levels of soil moisture will shift to the east along this gradient.

Such drastic climatic changes would impact current wildlife species inhabiting the High Plains, Hill Country, and South Texas Plains. Based on the climate predictions, we would expect the flora and fauna currently occupying the Trans-Pecos region to extend northward to perhaps the Texas Panhandle and Red River regions. The grassland flora and fauna currently representative of the Plains region could possibly disappear from the state. Species adapted to warm and humid conditions could increase their distribution into the South Texas Plains and Hill Country, and the East Texas flora and fauna is expected to experience the least impact.

Even if these predictions are only partly correct, Texas will be in for a major restructuring of its current land-use practices. For example, agriculture may cease to exist throughout the Southern High Plains and Red River areas due to increased temperatures, reduced rainfall, and continued decline of the Ogallala Aquifer. Desertification of the Texas Panhandle and portions of the Hill Country may actually benefit wildlife (although it would be a different species composition—more like the current fauna of the

Trans-Pecos) as farmland is abandoned and Chihuahuan Desert habitat becomes more prominent.

Climatic variability and change affects biodiversity by modifying species habitats and interactions. The changes could be damaging, leading to a reduction in species populations as habitats disappear. Climate change also can lead to shifts and contractions in species distributions. In addition, shifts toward warmer temperatures could influence disease dynamics and trigger outbreaks. While some species could adapt to climate changes because of their genetic variability and ability to disperse, other species may not be mobile enough to adapt to local climate stress through dispersal. A recent study suggests that, although species may show adaptive responses, these responses may not occur fast enough to ensure population viability in the long term (Radchuk et al., 2019). In other words, many species are not tracking climate at a pace that is sufficient to guarantee survival.

Climate change can impact mammals in a number of ways. Temperature increases and changes in precipitation could directly affect some species, depending on their physiology and tolerance of environmental change. Climate change also could alter a species' food supply or its reproductive timing, indirectly affecting its fitness. Because of their body size, morphology, and ecology, mammals would not be expected to adapt as a group to environmental shifts from changing climates. Most mammals would not be able to avoid the effects of climate changes, with both positive and negative effects possible. Mammals generally utilize a variety of disjunct resources. They need places to hide, eat, drink, and breed, and in many cases these places are distinct and may change seasonally. Thus, there are many opportunities for climate change to disrupt mammalian life histories. Most mammals are also highly mobile and, compared with perennial plants, have relatively short life spans (generally <20 years). Thus, if climates become unsuitable, mammalian responses could be rapid and likely detrimental.

McCain (2019) published an assessment of the predicted impact of climate change on mammals in the United States and Canada using a "trait-mediated model." Her results predicted species responses to climate change that range from negative responses, such as elevational range contractions, upward shifts, and decreases in abundance; to positive responses, such as range expansions and increases in abundance; and finally no detectable responses. She also concluded that responses will vary among and within species and that many of the changes will be correlated with natural history traits. The traits predicted to show the strongest links to climate change are body size (large mammals will respond more often and most negatively to climate change), activity times (few mammals with flexible activity times will respond to climate change), and spatial distribution (high-latitude and high-elevation mammals will respond most often to climate change). McCain's model predicts that 15 percent of mammal species, including species from most orders, will be at high and very high risk of experiencing negative impact. Finally, she predicts that large, high-altitude, and high-elevation carnivores and artiodactyls, many of the larger squirrels (e.g., ground squirrels and prairie dogs), and montane hares, pikas, and cottontails will be the most vulnerable and subject to conservation pressures. Although many of the species targeted in McCain's (2019) study do not occur in Texas (pikas, for example), the pattern of vulnerability can be extended to the flora and fauna of the state. For example, montane forms such as the Davis Mountains cottontail (*Sylvilagus robustus*) and yellow-nosed cotton rat (*Sigmodon ochrognathus*) would be expected to decline or even become extinct.

Another recent study of 538 plant and animal species distributed worldwide predicts that one-third of them could face extinction by 2070 due to climate changes (Román-Palacios and Wiens, 2020). The same study predicts that the hottest daily high temperatures in summer will be the key variable that explains whether a population will go extinct, and that most species will not be able to "escape" to a cooler climate in order to avoid extinction. Finally, the study concludes that niche shifts will be far more important for avoiding extinction than dispersal.

It is important to remember that climate change represents just one of a set of stressors. When combined with other changes that are challenging fauna and flora, such as land development, habitat fragmentation, invasive species, chemical stressors, and direct exploitation, it becomes easy to visualize dire consequences for biological diversity. Comprehensive assessments in each of the state's ecological regions will be needed to develop science-based management practices for wildlife and plant communities to compensate for these possible scenarios (see Packard et al., 2011).

9. Strengthen Scientific Research and Science-Based Decision-Making in Wildlife Management and Conservation Biology

One thing is for certain: in the twenty-first century, Texas will need science and research to help solve wildlife-related issues. Outside of human ingenuity and determination, scientific research may be the most important tool in the wildlife biologist's toolbox. Science has an invaluable role in resource management: to provide basic knowledge, to yield objective and unambiguous information on what is possible, to help develop sound strategies to meet goals, and to show the costs and consequences of alternative strategies. Doing all this requires a judicious mix of science and technology development that ranges from the most basic to the most applied. But we must remember that public policy and direction in a democratic society ultimately come from a political process that weighs scientific information with other considerations. For most of the twentieth century, science had too little to do with decision-making on conservation issues—most of that was played out through the media and in the courtroom. In the twenty-first century, the demand is increasing for expert opinion on how to manage wildlife resources that will be increasingly used by people. Therefore, not only is more research necessary, but also more effective ways to transfer this information to people and policymakers will be required.

Without delving too deep into "what is science," a few points need to be clarified and emphasized. First, science is the search for the truth—it is based on facts and has no room for personal opinions, subjectivity, and especially political conformity. Although biologists may have their own views or preferences, those opinions should be set aside and scientifically gathered data should be used in an objective manner relative to decision-making processes. Second, science, and by extension research, should be viewed as positive endeavors that help people make good decisions. Often research gets viewed as a negative enterprise because the conclusions may necessitate some sort of change on our part. Humans by nature resist change and therefore oftentimes we fear research because we anticipate that the results will require us to do something different—even if it benefits us in the long run. Our goal with this strategy is to strongly encourage more scientific research to provide the citizens of the state of Texas with the most complete data (information) possible. Also, we want individuals, communities, interest groups, state agencies, and politicians to rely on the best science available so that the best possible decisions can be made for the future of managing and conserving our state's natural resources.

Wildlife managers today are often forced to act on the basis of incomplete information, or in some cases no information at all. In setting priorities for conservation it is not uncommon to rely on "expert opinion" or models built around assumptions. This gives the pretense, and the false security, of an objective system when in fact much of the process is subjective. If we are to conserve wildlife diversity in Texas, it will be necessary to have an adequate information base upon which to make management decisions.

For example, of the roughly 6,400 recognized mammal species on planet Earth, it has been determined that there are only 106 published scientific studies (of 5,287 articles reviewed) that have adequately looked at the explicit link between the survival and reproduction rates of mammals and climate change factors, and those studies covered only 87 mammal species, none of which are known from Texas (Paniw et al., 2021). That number is shockingly low and means we only have studies exploring the imminent effects of climate change on just over one percent of the total mammalian species.

Unfortunately, the only groups of Texas wildlife species for which sufficient information or resources exist to effectively manage them at the present time are the commercially important species, primarily the game species and some of the large charismatic species. Game species, which constitute a small overall percentage of the Texas fauna, historically have received the primary attention of hunters, landowners, and professional wildlife managers. For most nongame species, the information base and resources have been lacking to acquire the information needed for effective long-term management, although public concern for these species is increasing substantially in the twenty-first century. To have effective conservation of species and manage protected habitats, we must learn more about all the species in Texas and their interactions within ecosystems. This will require accelerating the effort to discover, describe, and conduct natural history studies for more species in the state in order to provide effective guidance. Information and monitoring systems are needed to provide an indication of when a species appears to be in some danger (Kim and Knutson, 1986; NRC, 1993). Only if we know there is a problem can we attempt to develop a solution. The solutions effected will be dependent on how much is known about natural history.

It has been more than a hundred years since the last biological survey of Texas (published in 1905), and this did not include fishes, amphibians, or the state's vast and important invertebrate fauna. A well-planned and carefully executed follow-up survey would provide the kind of baseline data necessary to assess wildlife population trends and habitat conditions. It also would contribute much life history knowledge about the lesser-known components of our fauna. Texas has the resources within its university faculty, state biologists, and scientists who work for private organizations to develop a database network and coordinated approach to conduct biological surveys that would assess the status of wildlife populations.

At the same time, wildlife research must improve in quality and broaden its scope if societal issues are to be addressed adequately. Essentially all of the major scientific challenges require changes in the way researchers organize themselves as well as improvements in technology. It will be necessary to enhance and promote interdisciplinary research to provide new technology and different research approaches. As wildlife issues become more complex, the need for interdisciplinary research will become even greater. Unfortunately, funding opportunities for this kind of research, particularly at the federal level, have dwindled recently as agency budgets have been stagnant or even reduced. If this problem persists, basic research, education, and ultimately the management of resources will be impacted.

A major opportunity to increase interdisciplinary research will be in the interaction of traditional wildlife biology with biotechnology and molecular biology. As species become more endangered and ecosystems more impoverished by habitat fragmentation, pollution, and other human impacts, methods to keep species artificially, to maintain and boost their reproduction, and eventually to return them to the wild will play a larger role in conservation. The field of ex situ care has grown with the involvement of zoos and zoo-parks (Oldfield, 1989). Techniques include capture, translocation, selective breeding, artificial rearing, genetic management, and disease treatment.

Texas is home to a major program, the Fossil Rim Wildlife Center, located on a 1,700 acre ranch near Glen Rose in Somerville County, that is involved with this type of conservation work. Founded in 1973 and opened to the public in 1984, Fossil Rim has more than 1,000 animals from 50 species, mostly large ungulates from Africa, that are endangered or threatened in their native range. The Center, which is accredited by the Association of Zoos and Aquariums (AZA), brings the expertise of many large-scale zoological and environmental institutions to address issues related to the conservation of endangered species (e.g., the scimitar-horned oryx) through study, management, and recovery plans. The Center also is involved with some conservation and breeding programs that benefit native Texas species, such as the red wolf, Mexican gray wolf, and Attwater's prairie chicken. Most of the work is financed through tourism, with the major attraction being a scenic wildlife drive that includes a 9.5 mile guided or self-guided tour of the Center (Kimble, 2014).

The late twentieth and early twenty-first centuries have witnessed a proliferation of advances in research methods for studying wildlife as a result of new technology (see Bradley and Dowler, 2019, for a discussion of research advances in mammalogy). Many of the scientific improvements have involved simple advancements in field equipment, and others have come from major breakthroughs in biochemistry (DNA sequencing) or electronics (computers). Some of the major equipment breakthroughs include improvements in traps for capturing animals, methods for marking captured animals, the use of radio-telemetry to follow animals in their environment, the development of Global Positioning Systems (GPS) to improve the accuracy of field studies, and the use of cameras as "traps" to document animals in nature.

Biotechnology and genetic engineering, especially, have much to offer conservation efforts, with examples becoming evident on a regular basis. Scientists recently announced the cloning of the first US endangered species, a black-footed ferret (*Mustela nigripes*) duplicated from the genes of an animal whose remains were frozen more than 30 years ago in the early days of DNA technology. These slinky predators once roamed the grasslands of the Texas Panhandle where they thrived among the extensive prairie dog colonies in the region. Perhaps they will flourish again one day, as cloned individuals!

Advances in technology also have changed how science is conducted. Through the internet and smartphones, we now have access to a great deal of information that, in the past, would have been inaccessible or would have taken a very long time and great effort to share worldwide. This highly accessible communication has opened new avenues for collaboration among scientists across the globe, and research teams have expanded as a result. In addition, this unprecedented access to information has broadened opportunities for the public to participate in data acquisition (see discussion of citizen science, below).

An important area of research in need of strengthening is that of human-wildlife interactions. The needs of people drive the use and the misuse of wildlife resources. Previous efforts to understand how people think about and act on wildlife resources have been minimal, and yet most controversies and shortages ultimately arise from human activity. Cooperative efforts between natural and social scientists in wildlife are increasing in the twenty-first century and leading to increased knowledge and problem solving in this area.

Scientific research is essential to evaluating various trade-offs for coping with environmental problems, such as climate change. For example, there is an ongoing movement (see Fairlie, 2019) encouraging humans to lower their consumption of red meat because livestock (cattle, sheep, goats, and pigs) are thought to contribute approximately 14 percent of the worldwide greenhouse emissions, primarily through methane release (note: the oil and gas industry contributes about 30 percent of methane emissions). The take-home message has been that the cattle industry, in particular, is contributing significantly to global warming and climate change (Gerber et al., 2013). Consequently, the Intergovernmental Panel on Climate Change of the UN called for changes to the human diet, and several politicians jumped on board, with some calling for drastic reductions in beef cattle production (e.g., Representative Alexandria Ocasio-Cortez, D-NY) followed by a worldwide diet switch to other sources of protein such as synthetically produced meats, as well as fruits, vegetables, and other human-grown crops that require monoculture agriculture (Schiermeier, 2019).

The politicization of this topic has resulted in a high level of sensationalism, but despite the hype, the scientific assumptions behind this controversy are far from certain. As mentioned in Chapter 4, methane is thought by many scientists to be a major contributor to greenhouse gases heating up the planet, equal to, if not greater than, the impact of carbon dioxide (Höglund-Isaksson et al., 2020). However, several other recent studies suggest that the impact of methane on the environment should not be equated with carbon dioxide because the two molecules degrade at different rates and appear to have different impacts on climate (Allen et al., 2018). It is thought that each year approximately 558 metric tons of methane is produced globally, with 188 metric tons coming from agriculture; however, nearly that entire quantity (548 metric tons) is broken down through oxidation and absorbed by plants and soils as part of the sink effect (Stocks, 2019). The implication here is that the net effect of methane, particularly the amount produced by agriculture, could be virtually inconsequential relative to global warming.

As mentioned, much of the finger-pointing for methane production has been directed at the cattle industry. The US cattle herd is estimated to be approximately 95 million individuals and undoubtedly, this group of animals contributes the largest proportion of the US livestock-produced methane. However, this may not necessarily be a new or even increasing source of methane. In the mid-1800s, approximately 30 million bison occupied the central regions of the US, and they produced methane in a similar fashion as the modern-day cow (at least non-feedlot cattle). Over the last 150 years, humans essentially have replaced the bison population (and other native grazers such as elk, pronghorn, etc.) with cattle. Although 95 million and 30 million may not seem to be comparable on the surface, bison are much larger than cattle (presumably producing more methane by weight). In addition, cattle breeds have been selected for faster weight gain and other nutritional advantages (presumably their digestive systems are more efficient and would produce less methane than bison), which suggests that methane production today could be approximately the same level as it was in the mid-1800s. There are even some animal nutritionists who think that the feedlot diet (higher protein) may produce less methane than cattle raised solely on native vegetation (Lozano, 2019; but see below). In other words, cows of today may be no worse than the bison of 150 years ago relative to methane production. What is clear, however, is that increased methane production in the US is largely attributable to the growth in the oil and gas industry, industrialization, and the increase in the human population from approximately 3.1 million in 1850 to 333.4 million today.

Such considerations raise a few questions. For example, is it a sound biological strategy to encourage humans to consume more grains, fruits, and vegetables while discouraging a "red meat" diet? Some argue that eliminating cattle from our diet would lower methane production and help alleviate global warming. But, what would be the carbon cost of substituting a native grassland with cropland suitable for growing fruits, grains, or vegetables? Native grasslands or even improved pastures (native species replaced with higher producing varieties) have very low carbon footprints because they require little or no maintenance. Occasionally, improved pastures are fertilized, shredded/mowed, or sprayed for weeds, so it not entirely a carbon-zero system. But in contrast, croplands are fairly expensive to maintain relative to carbon footprints (as well as other negative impacts on the environment). These crops (such as corn) typically require irrigation, plowing, planting, tillage, harvesting, fertilizer application, and other trips across the field with farm equipment (powered by fossil fuels). No-till agriculture helps reduce the carbon footprint but growing crops is still carbon- and, in some cases, water-expensive.

Crops such as fruits, grains, and vegetables are short-lived compared to native grassland species. Not only are they typically annual species (require replanting each year), their growing season has been shortened substantially (through artificial selection) compared to most native species. For example, corn and soybeans are planted in May and harvested in August or early September (time may vary by latitude). So, crop species function as a carbon sink (reduction) for only 4-5 months, whereas native grasses may be productive from late March to late October (again, time varies by latitude). Consequently, it would seem that native grasslands generate a smaller carbon footprint and may function for a longer period of time as a carbon sink.

Grazing cattle on native grasslands would produce a very low carbon footprint, especially if the methane production does not contribute significantly to climate change due to the rapid degradation rate of methane. About the only carbon expenditure from raising cattle on native pasture would be roundups, transporting to market, and processing. If we ate grass-fed beef, as opposed to grain-fed beef, the total carbon footprint would be very low, certainly compared to a fruit, grain, or vegetable diet as described above. Consequently, it may be that the cost (carbon footprint) of producing beef is much lower than the cost of growing cereal grains. However, most cattle are "finished" in a feedlot where they are fed a high-protein diet to increase weight and improve flavor. As described above, that process would generate an increase in the carbon footprint as the protein source is typically corn and other grains. Consequently, the net effect may vary substantially depending on time spent eating grass and grains in a feedlot.

Without getting too bogged down in the details, a simple comparison of the biodiversity present in a native grassland (the source of our red meat diet) compared to a cornfield (the source of our cereal grain diet) is instructive. In making this comparison, we will use a standard Texas measure of land size, a section (640 acres; 259 ha), and then estimate the biodiversity observed in each habitat. For this "quick and dirty" comparison we used Lubbock County as the study site because of our personal in-depth biological knowledge of this local area; other areas of the state would display varying degrees of local biodiversity. Table 11 illustrates the significant loss of biodiversity when a native grassland is converted to a monoculture crop such as corn. This comparison suggests that perhaps habitat for as many as 2,000 native species may be lost by converting a grassland to a monoculture cornfield. Further, in 2017 we lost 1.7 million acres (688 thousand ha) of grasslands in the Great Plains, alone, to crop production. It is unclear how detrimental the conversion of grasslands to agriculture is to climate change.

Table 11. Comparison of estimated species diversity in a cornfield versus a native grassland in Lubbock County, Texas.

Taxonomic category	Species in a cornfield	Species in a native grassland
Plants	1–3 (corn plants and a few weed species that survive the herbicide treatments)	50–100
Insects	1–3 (a few pest insect species that survive the insecticide treatments)	>500
Nematodes and other invertebrates excluding insects	1–10 (few survive insecticide and nematode treatments)	<1,000
Birds	~10 (a few bird species will feed in fields on waste corn following the harvest)	354 (actual number of species in Lubbock County)
Mammals	~10 (a few mammal species will feed in fields on waste corn following the harvest)	54 (actual number of species in Lubbock County)
Reptiles and amphibians	0	41 (actual number of species in Lubbock County)
	Total ~1–50 (with most species occupying the cornfield only following the harvest)	Total >2,000 (most species are year-round residents)

From these scenarios, it seems that having a grassland community, while also eating a few cows, deer, quail, etc., may provide for a lower carbon footprint and certainly provides for a much greater diversity in wildlife than planting a cornfield and eating the corn. Of course, the comparison is not exactly fair, because we can feed many more people from one acre of corn than we can from the beef and other animals harvested from one acre of grassland habitat. However, the point is that by attempting to "do good" we may inadvertently be doing more harm. In examining these scenarios, it becomes clear that additional research is needed in order to better understand which practices best achieve a balance between human needs and wildlife conservation. For example, although raising cattle may produce methane that could contribute to climate change, growing native grasses (for cattle production) helps lower atmospheric carbon; perhaps this is an acceptable carbon balance. This scenario provides an excellent opportunity for the beef industry, agriculture, and research groups to generate models that provide for an environmentally friendly food supply.

Implementing more scientific research to address all of the efforts described above will require more resources, both in terms of funding and manpower, in order to get the job done. Current efforts, in both of these areas, are inadequate, although progress is being made on both fronts. Funding is now being made available by some Texas state agencies to allow for university naturalists and conservation professionals to begin to monitor the status of various species in the state, especially those that are deemed endangered, threatened, or potentially at risk. For example, TPWD's Nongame and Rare Species Program offers grants to support implementation of the Texas Conservation Action Plan for the study of rare species and communities in greatest need (TPWD, 2019c). Likewise, the Office of the Texas Comptroller has established an Economic Growth and Endangered Species Management grant fund to study species under consideration for ESA protection (Comptroller.texas.gov, 2017). The purpose is to obtain the most accurate information about their populations, range, habitat, and natural history to provide a better basis for decisions that can have major economic consequences (Benton and Wright, 2017).

An exciting new proposal, the Recovering America's Wildlife Act (RAWA), is now before the US Congress. If passed, this bipartisan legislation would dedicate $1.3 billion annually to the states for at-risk species research and protection (The Wildlife Society, 2019). Based on its population and geographic size, Texas would be eligible for about $50 million per year. The funding, which would come from royalties currently collected on oil, gas, and mineral production on federal lands and waters and thus requires no new taxes or diversion from other federal programs, would be used for the highest-risk species with the goal of keeping them off the endangered species list (Stein et al., 2018). Many at-risk species have not been adequately studied due to the vagaries of research funding, and part of this proposed money would fund scientific research at universities and colleges to provide the information and knowledge to improve decision-making about the status of species. Texas is home to more than 1,300 of the 12,000 species identified nationwide as Species of Conservation Need, and H.R. 3742 is being called a once-in-a-generation opportunity to save these wildlife species and to provide more regulatory certainty for businesses, land developers, the oil and gas industry, and governmental agencies (TPWD News, 2019b).

Two other pieces of legislation passed and signed into law in 2020 have important impacts for wildlife conservation. America's Conservation Enhancement Act (ACE) supports proven efforts to preserve important ecosystems, such as wetlands, and promotes innovative ways of addressing the growing threats of invasive species and wildlife diseases such as CWD. The Great American Outdoors Act fully funds the land and water conservation on an annual basis of $900 million and provides $9.5 billion over the next five years to address deferred maintenance on federal land. However, because Texas has so little federal land, it will not be as impacted as other states.

All of these efforts are going to require more "boots on the ground" in the form of field biologists trained in natural history, ecology, and wildlife management with the ability to conduct field studies. And this is where a problem is beginning to develop. The first two decades of the twenty-first century witnessed a decline in the perception of the importance and practice of research and education in natural history at many universities, which has impacted the talent level required in the future for conservation to succeed (see Schmidly, 2005). In particular, there has been a precipitous decline in the level of education that involves natural history field studies, with more of an emphasis shifting to molecular biology laboratory work.

David Wilcove and Thomas Eisner (2000), in an article published in the *Chronicle of Higher Education Review*, made the following statement of concern about this important matter: "This deinstitutionalization of natural history looms as one of the biggest scientific mistakes of our time, perpetrated by the very scientists and institutions that depend upon natural history for their well-being. What's at stake is the continued vibrancy of ecology, of animal behavior and botany, of much of molecular biology, and even of medicine and biotechnology."

Citizen science has emerged in recent years as a major way to approach the manpower problem of natural history. Citizen science has been defined as public participation in scientific research projects, usually by volunteers, who collaborate with scientists and researchers to increase scientific knowledge (Vliet and Moore, 2016). The volunteers who participate in citizen science have varying levels of expertise. They may be children who are exposed to science through school projects, high school students who participate in science clubs, amateur scientists with little formal training, community groups organized around a science interest, educators, or naturalists. There are now a number of citizen-science projects that include wildlife-monitoring programs, where individuals share photos and observations of species in a particular area online. There are a number of beneficial aspects to citizen-science projects: they showcase the scientific process and highlight the fact that science is all around us; they allow for large amounts of data collection around the world that would be very expensive to generate without the use of volunteers; and they can advance projects more rapidly due to the large number of people that are involved (Vliet and Moore, 2016).

Natural history is one area where citizen science has been particularly popular. For example, iNaturalist is one of the five top citizen-science communities in the US with more than one million users (see https://www.inaturalist.org/). Bioblitzes represent a good example of how citizen science can help to advance our understanding of wildlife diversity. Also known as a biological inventory or biological census, a bioblitz is an event that focuses on finding and identifying as many species as possible in a specific area over a short period of time (*National Geographic Kids*, 2019). Another successful example of citizen science in natural history in Texas involves the TPWD program "Nature Trackers" that is designed to help document where species are being found in the state. While the highest priority is to document target species, especially those that are rare or endangered, they seek data on all species. The program encourages the use of game cameras and observations of tracks and roadkill to help detect hard-to-find species. Using the iNaturalist application, user volunteers can enter data through the website where it is added to other data on species that the department tracks and uses to supplement the Texas Natural Diversity Database (TPWD Nature Trackers, 2019).

10. Enhance International Cooperation with Mexico

Because Texas shares a large border with Mexico, international cooperation also is important, as wildlife do not respect politically drawn boundaries. A relatively large expanse of public lands exist along the front range of mountains of the Trans-Pecos, beginning with Big Bend National Park, Big Bend Ranch State Park, and Black Gap Wildlife Management Area along the border, extending northward through the Chinati Mountains, Elephant Mountain Wildlife Management Area and the Davis Mountains Preserve, and ending with Guadalupe Mountains National Park along the Texas–New Mexico border. This region constitutes a virtual continuous corridor along the front-range from Mexico to New Mexico. The conservation potential of this vast corridor, given the large number of rare plants and animals in this region, is immense.

An effort is now underway to link the northern end of this Texas Trans-Pecos corridor with a major conservation program in Mexico, known as the El Carmen–Maderas del Carmen Project in Coahuila, Mexico, just across the river from Big Bend National Park. The USFWS, NPS, US Geological Survey, and TPWD signed an agreement to create the Big Bend Conservation Cooperative in 2010 (Gaskill, 2015). That group works with various conservation and environmental agencies in Mexico, as well as private landowners, universities from both countries, the World Wildlife Fund, The Nature Conservancy, and corporations such as Coca-Cola and CEMEX (a worldwide building materials company based in Mexico) to provide and protect a corridor on both sides of the Rio Grande that will allow wildlife to move freely within an intact ecosystem (McKinney, B. R., 2006, 2012; McKinney and Villalobos, 2004). Currently, the corridor includes over 400,000 acres (161,874 ha) of land in the Sierra del Carmen and Madera del Carmen, and 47,000 acres (19,020 ha) of "wilderness" with no development at all along the Rio Grande Gorge. Progress is being made as bighorn sheep, black bear, mule deer, and beaver are thriving in the corridor, and a number of other rare species, such as the coati, are beginning to make a comeback (Schmidly et al., 2016a).

This US–Mexico corridor eventually could be an excellent location for a major "rewilding" effort. Rewilding, a conservation strategy that has caught on in recent years, particularly in Europe and Africa, involves large-scale conservation aimed at restoring and protecting natural processes and core wilderness areas by providing connectivity between such areas and protecting or reintroducing apex predators and keystone species, such as wolves, mountain lions, and jaguars (Soulé and Noss, 1998). The concept incorporates what has been termed the three C's of conservation: cores, corridors, and carnivores. Large predators are important for their role in driving ecosystems by their interactions with other species and the manner in which these effects cascade through the whole ecosystem. Predators require large core areas of wilderness and they need safe corridors so they can migrate without coming into conflict with humans. Rewilding is seen by many as a way to stem the loss of biodiversity and the functions and services that biodiversity provides to humanity (Foreman, 2004; Nogués-Bravo et al., 2016). Recently, the concept has been reenvisioned to emphasize all biodiversity and not just carnivores, with a focus on restoring ecosystems and reversing biodiversity loss by allowing wildlife and natural processes to reclaim areas no longer under human management (IUCN Rewilding Thematic Group, https://www.iucn.org/commissions/commission-ecosystem-management/our-work/cems-thematic-groups/rewilding). However, to implement this approach in Texas would involve solving long-standing conflicts and issues with private landowners that operate in the buffer areas around these protected places.

A significant problem with the development of future cooperative wildlife projects between the US and Mexico would be the implementation of a border wall separating the two countries along the Rio Grande. A fully completed steel wall would eliminate the migration of most species of wildlife, as discussed in Chapter 4, effectively ending the current cooperative efforts that are underway. The construction of the border wall was halted when the Biden administration replaced the Trump administration, but the governor of Texas has threatened to continue with the building of the barrier using state and private donor funds.

11. Increase Funding for Conservation Projects and Natural Resource Protection

It is absolutely essential that the state identify sustainable funding sources to acquire more protected areas and open spaces for outdoor public recreation. In 2001, the Texas Parks and Wildlife Department contracted with Texas Tech University to conduct a study "Texas Parks and Wildlife for the 21st Century." The objectives of the study were to: 1) conduct a major public opinion survey of Texans to gain a better understanding of their values and attitudes toward natural and cultural resources and outdoor recreation in Texas; 2) inventory current holdings of public lands for conservation and outdoor recreation purposes; and 3) conduct a needs assessment of future requirements for public land to ensure that historical, cultural, and natural resources will be adequately provided, maintained, and conserved for future generations of Texans.

The strategic plan study was commissioned in part because a Sunset Commission review had criticized the TPWD for lacking a vision. The overall population of Texas had increased by seven million since 1990, and many of the newcomers were seeking outdoors experiences. But land acquisition had virtually stopped by 2000. Also, there was concern about how disconnected the projected demographic changes taking place in Texas were from the mission of TPWD. The public survey part of the study revealed, for example, that most Hispanics did not know the state parks system existed, yet they were all for more access to parks and the mission, once they found out about it.

The public opinion part of the study revealed overwhelming support on the part of Texans (in excess of 90 percent of the sampled population) to conserve water and wildlife as well as to protect and preserve ecologically important habitats and lands in Texas (Duda, 2000). The needs and supply analysis revealed that Texas had fallen far short of its 1963 goal of 45 acres (18 ha) of publicly owned state parks per 1,000 residents. At that time, the state had about 31.5 acres (12.7 ha) of state parks per 1,000 people, and 20.4 acres (8.3 ha) of wildlife management area per 1,000 residents for a combined total of 51.9 acres (21 ha) of combined parks and wildlife land per 1,000 people (Loomis Austin, Inc.,

2000a, b, c). A professional needs conference held in 2000 recommended that Texas have 55 acres (22.3 ha) of parks and wildlife preserve land per 1,000 people, which would have placed the state in the 75th percentile among all states at that time (Loomis Austin, Inc., 2000c).

Unfortunately, the report (Schmidly et al., 2001) with all of the recommendations was for all practical purposes completely ignored. The commissioners of TPWD, the governor, and the legislature did not buy in. Under the leadership of the governor (Rick Perry at that time), the idea of the state buying more parkland was just not acceptable and there was concern about the government taking actions that might remove landowners' rights. Less than one percent of the people of Texas are major landowners, but that one percent is powerful and highly connected politically. A lot of them opposed government solutions to problems, so there really was no appetite for government to get involved with anything in Texas. For the complete story about the controversy surrounding the 2001 report, see Patoski (2018).

As a consequence, the idea of acquiring additional parks and protected areas for the public to enjoy and to better protect biodiversity was put on hold for about a decade until new state leadership began to take actions along the lines recommended in the 2001 report. Those actions have focused mainly on the backlog of maintenance at the state's parks and to a lesser extent on the acquisition of protected land (only three new parks have been opened since 2000). More land is becoming accessible, but not the 1.4 million additional acres (567 thousand ha) recommended by the Texas Tech study, although tens of thousands of new acres have been acquired for the public to explore. The population of the state continues to grow, which means that we are continually falling behind in meeting these targeted ratios.

Land acquisition is a powerful tool for conserving biodiversity, and beginning to accomplish this in Texas will be absolutely essential to biodiversity conservation. But the cost to acquire land for the purposes mentioned above has increased substantially since the 2001 Texas Tech study, and it seems clear that additional funding resources will be required to make this happen on the scale that is necessary to protect our biological resources. A mechanism will be needed that can generate enough money over the long term, is sustainable, has growth capability, and is not administratively difficult or costly to manage (Cerulli, 2013). These funds must be in addition to traditional sources of funds for fish and wildlife resources (sales of hunting, trapping, and angling licenses, and federal excise taxes on sporting arms, ammunition, and fishing equipment). Healthy fish and wildlife populations are important to the quality of life for all of the state's citizens, and the broader public should help with funding.

Many conservationists believe that the extinction crisis can be stopped, but in order to do so we will have to provide more funding for conservation and tackle the crises of climate change and species loss together, taking measures that fix both and not just one. A campaign, known as "Closing the Gap," has been proposed to secure support from government, business, and the finance sector to provide funding for the conservation of biodiversity and the mitigation of climate change (Solberg, 2021). In 2019, the Bush School of Government & Public Service at Texas A&M University provided a report to the Boone & Crockett Club of Texas about potential methods to fund wildlife conservation in Texas (Bush School of Government & Public Service, 2019). The state estimated a need of $20 million annually to address the Texas Conservation Action Plan. The Texas A&M report identified three categories of possible solutions—consumer-funded, industry-funded, and voluntary-funded—with each of those further divided into subcategories of user fees, sales taxes, green taxes, personalization fees, and sumptuary taxes. The report concluded the most viable options were the consumer and industry-funded approaches. Increased user fees at state parks and wildlife management areas, water surcharges, sales tax increases, and online sales tax, as well as taxes on outdoor recreation equipment and cameras, taxes on carbon emissions, real estate transfer taxes, and transportation project taxes were recommended as the most viable for long-term sustainability.

In 1993, Texas lawmakers passed legislation allowing up to 94 percent of state sales tax proceeds on sporting goods to go to the Texas Parks and Wildlife Department for state parks, with the remaining 6 percent designated for the Texas Historical Commission (THC) that maintains Texas's 22 historic sites. In the following decades, however, the state allocated an average of just 40 percent of the sporting goods sales tax revenue to the parks system and used the rest to help balance the state budget; from 1993 to 2017, Texas collected nearly $2.5 billion in revenue from the sporting goods sales tax, but lawmakers allocated only about $1 billion of that to state parks (Anchondo, 2019). Then, in 2019, Senate Bill 26 was approved by voters—this bill amended the Texas Constitution to automatically appropriate these funds for the budgets of the Texas Parks and Wildlife Department and the Texas Historical Commission, thus preventing other uses of these funds by the state. In the future there will be more funding and greater predictability in annual budgets for these agencies, allowing for much-needed repairs and maintenance to existing parks and sites as well as the establishment of new parks.

There also are opportunities to engage funding solutions that involve private corporations and companies as well as NGOs to expand outdoor recreation opportunities. For example, tax credits could be given to companies that sponsor youth hunting and fishing days, as well as for companies that buy hunting leases and make them available at lower costs to the public. Actions such as these could help turn around the current trend of few young people participating in hunting and fishing, activities that traditionally have been a great way to introduce them to the importance of conservation and natural history. A high priority should be to continue working with The Nature Conservancy and other NGOs to support their efforts to acquire more protected land. Property tax credits could be issued for gifts of land that serve these purposes.

In summary, Texas habitats and landscapes have undergone thousands of years of man-made changes generated both by Native American occupants and generations of western settlement. The associated changes have caused deep, broad, and long-term habitat and landscape alterations in the state. Although some fish and wildlife species are thriving thanks to careful, science-based management, many face increasing challenges and are in steep decline. Invasive species, emerging diseases, habitat loss and fragmentation, and extreme weather threaten many fish and wildlife populations at a scale inconceivable just a few decades ago. With more than one million species at risk of extinction globally, there has never been a more urgent time to fundamentally rethink how we engage with nature, conserve at-risk species, and restore degraded wildlife habitat. Implementation of the 11 action steps outlined above will greatly contribute to reversing these trends and providing for a more sustainable future for Texas's native biota.

CHAPTER 8

Concluding Thoughts

EARLY SETTLERS IN TEXAS DURING THE LATE SEVENTEENTH AND early eighteenth centuries found the countryside well populated with wildlife. Interestingly enough, deer were less abundant than now, but the plains supported huge herds of bison and pronghorn. Tallgrass prairies teemed with prairie chickens. The wooded areas supported large populations of wild turkeys, black bears, mink, and beaver. These animals served those early settlers mainly as a seemingly inexhaustible source of food and clothing. As the human population grew, in the latter part of the eighteenth and throughout the nineteenth centuries, lands in the eastern section of the state were cleared, plowed, and planted to crops. Thousands of livestock, mainly cattle, were brought in to stock the open ranges of the west, and later barbed wire fences were installed. All of these events altered the original balance between wildlife and its habitat.

This adjustment between wildlife and natural vegetation on the one hand and humans and their use of the land on the other progressed steadily throughout the twentieth century, and it has continued into the first two decades of the twenty-first century. Only the highly adaptable species of wildlife are likely to survive moving forward; those that cannot adjust to changing conditions will be doomed, which already has happened to a number of species, unless we strengthen our commitment to conservation and improve the way in which we manage our landscapes.

That wildlife is in trouble, not only in Texas but around the world, can be illustrated by these recent predictions regarding the conservation status of mammals:

+ Many mammal species are going to become extinct during the next five decades, with the possibility that the largest surviving mammal could be the cow; the few remaining giants, such as rhinos and elephants, are in danger of being wiped out rapidly (Davis et al., 2018). Of 177 mammals studied in detail, all have lost 30 percent or more of their geographic range and more than 40 percent of the species have experienced severe population decline (Ceballos et al., 2017).

- In a recently published study, 75 mammals have been identified as being on the brink of extinction, with populations of fewer than a thousand individuals (Ceballos et al., 2020).

- It has been estimated that mammal habitats have declined globally by 5 to 16 percent and regionally up to 25 percent (Baisero et al., 2020).

- The average total biomass of mammals is predicted to drop by 25 percent over the next 100 years as there is a shift towards smaller mammals, as the most adaptable species (small, fast-lived, highly fertile, insect- and seed-eating creatures, such as rodents) that thrive in a variety of habitats will survive, whereas other mammals that live longer and require specific environmental conditions will be at high risk of extinction (Cooke et al., 2019).

We are in the midst of a biodiversity crisis of historic proportion. Since 1970 there has been a nearly 70 percent decline in populations of mammals, birds, reptiles, amphibians, and fish; and by the year 2070, research suggests the Earth could lose a third or more of its species if steps are not taken now to stop the depletion (Solberg, 2021). One-third of America's plant and animal species are thought to be vulnerable, with one in five imperiled and at high risk of extinction. Emblematic of these declines, more than 1,600 US species are now receiving protection under the Endangered Species Act, and 422 of these are vertebrate animals (Stein et al., 2018).

Over its multibillion-year history, the planet has suffered five mass extinctions, the last of which occurred about 66 million years ago, when a giant asteroid (believed to have landed near the Yucatan Peninsula) set off a chain reaction that wiped out the dinosaurs and roughly three-quarters of the other species on Earth. In a 2014 book, *The Sixth Extinction* (Kolbert, 2014), the writer Elizabeth Kolbert warned of a devastating sequel, with plant and animal species on land and sea already disappearing at a ferocious clip, their habitats destroyed or diminished by human activities. This time, she made clear, the asteroid is humans—and we will pay heavily for our folly. This loss of biodiversity—the other living things with which we share Earth—has become one of the most serious aspects of the environmental crisis, and it affects the well-being of humans by interfering with crucial ecosystem services such as crop production and water purification and by destroying humanity's beautiful, fascinating, and culturally important living companions (Ceballos et al., 2015a, b). The problem has become so serious that scientists have coined a term, the Anthropocene, for the current geological epoch during which human activity has been the dominant influence on climate and the environment (Ruddiman, 2013; Malhi, 2017). And, very alarmingly, there is good evidence that these mass extinction events are accelerating (Ceballos et al., 2020).

Unsustainable human activities and population growth have pushed nature to the brink, with more plants and animals threatened with extinction now than at any other period in human history. This unprecedented and accelerating deterioration of nature, most of which has occurred over the past 50 years, has been driven by changes in land and sea use, exploitation of living beings, climate change, pollution, and invasive species. These five drivers are, in turn, underpinned by societal behaviors ranging from consumption to governance. Because people have behaved as though Earth was designed solely for human habitation, crop development and energy exploitation have raised extinction rates and put the natural world on the brink of collapse.

The scale of the threats to the biosphere and all of its lifeforms can seem so great that it is difficult to grasp even for well-informed experts. Nevertheless, it is important to avoid a fatalistic perspective because there are so many successful interventions to prevent extinctions, restore ecosystems, and encourage more sustainable activities at both local and regional scales. The path forward should start with stopping species extinctions. Addressing climate change is obviously a cornerstone to success, as is fundamental changes to global capitalism, education, and equality.

The solution to this problem has led many conservationists to call for a "Global Deal for Nature," which proposes that we protect 30 percent of the remaining habitat on earth in reserves to stem the loss of biodiversity (Dinerstein et al., 2019). Currently, about 16 percent of the planet is set aside as protected areas (in the US it is closer to 12 percent, and in Texas it is only about 3 percent), so the new recommendation would require a considerable expansion of protected areas. Other prominent scientists, including the conservation biologist E. O. Wilson, have argued for a more ambitious effort, termed the "Half Earth" concept (see Wilson, 2016), which would set aside 50 percent of the earth as protected areas to save biodiversity. The Biden administration has adopted the goal of conserving at least 30 percent of US lands and ocean by 2030 (referred to as the "30 x 30" plan) in hopes of curbing global warming while preserving some of the nation's most scenic places for future generations to enjoy as well as protecting more of our biodiversity. The Biden 30 x 30 plan calls for expanding collaboration with private landowners and state and local governments, with less emphasis on putting more land under federal protection (Biden-Harris Administration Report, 2021).

On a global scale, these numbers may not be as far-fetched as first imagined. A recent study of spatial landscape maps has revealed that nearly half of Earth still remains in areas without large-scale intensive care (Riggio et al., 2020). In Texas, however, where more than 95 percent of the state is under private ownership, both the 30 and 50 percent calculations for land set aside exclusively as protected areas seem completely unachievable, although reaching the global average of 16 percent possibly could be achieved this century. However, under the Biden plan, which includes collaboration with private landowners, Texas fares much better because there are currently 29 million acres of private land under some sort of management to enhance wildlife habitat (see Chapter 7). If these lands are factored in, then Texas has nearly 20 percent of its land under some sort of conservation collaboration.

Texas, and its biodiversity, will not be immune from the conservation crisis. State fish and wildlife agencies have identified 15,000 species nationwide in need of proactive conservation action, and Texas is home to more than 1,300, slightly less than 10 percent, of these species (Bryant, 2018). A significant number of Texas mammals are now extinct and a growing number of species are regarded as endangered and threatened (see Chapters 5 and 6). Forty-four species of extant native mammals are considered to be endangered, threatened, near threatened, imperiled, or vulnerable by the Texas Parks and Wildlife Department (TPWD), the US Fish and Wildlife Service (USFWS), or the International Union for the Conservation of Nature (IUCN). This means that they either have become extinct or have had subspecies or metapopulations disappear, or they are rare and appear to face some sort of problem that potentially threatens their existence. This number constitutes almost one-third of the total diversity of native terrestrial mammals in the state (see Chapter 5). In actuality the number of species in this situation may be much higher than this estimate would suggest. This is because we have not been able to do enough monitoring of populations to obtain usable information on population trends for most species of mammals.

During the late nineteenth and twentieth centuries, the state lost most of its native, large mammal megafauna (carnivores and herbivores, see Chapter 5). This reduction is crucial because many megafauna function as keystone species and ecological engineers, generating strong cascading effects in the ecosystems in which they occur (see Ripple et al., 2016). And each year we see more and more species either threatened with extinction or declining in numbers and geographic range. On a macro scale, the diversity of terrestrial mammals changed substantially during the twentieth century, and these trends are continuing into the current century. There has been a substantial turnover in species composition, involving both a loss and gain in species, since 1900.

The ecology and natural history of Texas changed profoundly during the twentieth century. As we have documented throughout this book, substantial changes can be demonstrated in land use and landscapes (including fauna and flora), and anthropogenic pressures changed from the beginning to the end of the century. In the beginning most of the pressures came from over-harvesting and unregulated, market hunting. By the end of the century the problem came from the rapidly expanding human population. Population growth and demographic shifts progressively increased as a form of conservation pressure and there is no let-up in sight. As Carter Smith, Executive Director of TPWD, eloquently put it: "The State is growing by leaps and bounds, and we must contend not only with burgeoning pressures on our fish and wildlife populations and their habitats, but also with a citizenry that is more urban, more diverse, and more disconnected from the outdoors than any previous generation" (personal communication).

Pressures on critical fish and wildlife habitats and populations are greater than ever before. Energy development, large-scale agriculture production, pollution, increasing threats from invasive species, and deteriorating water quality all present significant changes to the maintenance of sustainable populations of fish and wildlife. The compounding impact of climate change threatens to dramatically impact scores of species and has introduced considerable uncertainty to the management of fish and wildlife. In 2018, citing 6,000 scientific studies, the United Nations released a report asserting that the world has 12 years to prevent a "climate change catastrophe." That report also noted the close connection between the loss of biodiversity and climate change because intact ecosystems (with their associated biodiversity) provide the best defense against climate change (Martin and Watson, 2016).

Unfortunately, not much progress has been made on a global scale to reduce the emission of greenhouse gases and address other factors associated with climate change. This has caused more than 11,000 scientists to issue a warning that we are rapidly approaching a "global tipping point" where it will become more difficult, if not impossible, to cope with the problem and its various ramifications (Ripple et al., 2019). Further contributing to the situation is the fact that many Americans, among them numerous Texans, discount human-caused climate changes despite the overwhelming scientific evidence. Consequently, the national government and the Texas state government have been slow to address the problem. Texas is way behind the curve in this regard. For example, state lawmakers slashed funding to pollution control programs at the Texas Commission on Environmental Quality by 35 percent between 2008 and 2018, even as the state budget grew by 41 percent over that period (Roldan, 2019).

Political psychologists have argued that "solution aversion" has set in in many sectors of society. The basic idea behind this concept is that people are less likely to believe that something is a problem if they have an aversion to the solutions associated with the problem (Singal, 2014). Many Americans tend to favor free markets and limited government, and because generally speaking the best-known potential solutions to the problems posed by climate change involve increased regulation, they are hesitant to support needed actions unless they are associated with free-enterprise principles. At the same time, many people are willing to go only so far in engaging in habits that conserve energy and reduce emissions that are warming the planet (Flesher and Swanson, 2019). Actions that would involve significant lifestyle choices in areas such as diet or transportation are a tougher sell. Fortunately, there is evidence that the American public is beginning to wake up to what is at stake. A recent public opinion poll revealed that in the last five years, the percentage of people calling climate change a "crisis" has jumped from 23 to 38 percent (*Washington Post*, 2019). A 2019 poll of Texans undertaken by the *Texas Tribune* and The University of Texas revealed that two-thirds of registered voters believe that climate change is happening, but Texans are deeply divided along party lines on the question of what the government should do about it (The Texas Politics Project, 2019).

Governmental and nongovernmental organizations, colleges and universities, private landowners, and citizens in the state have worked hard to make progress in efforts to preserve the state's wildlife heritage, but if we fail to address the climate crisis there will be limits to what can be accomplished in the long term. TPWD has made great strides in implementing a more holistic approach that emphasizes ecosystems and monitoring biodiversity trends, and other state agencies (e.g., the General Land Office) are beginning to monitor and provide scientific information on rare and threatened species. Likewise, many nonprofit organizations (such as The Nature Conservancy, National Audubon Society, and Bat Conservation International, to mention a few) have set up operations in the state and have established and promoted programs to protect the state's unique places and species. The Texas Conservation Alliance and its 48 member organizations (e.g., Texas Wildlife Association, Environmental Defense Fund, Texas Commission on Environmental Quality, Texas Land Conservancy) have worked to build grassroots coalitions that can influence decision makers. Cities and municipalities across the state have made plans for coping with climate change and providing more green space for people and wildlife. Private landowners have been provided incentives and recognized for their efforts to conserve natural resources. Colleges and universities have established teaching and research programs that emphasize natural history and are providing the scientific evidence needed to support conservation efforts. And, the public has responded with substantial volunteer programs (such as the Master Naturalist Program) to assist with all of these activities.

Partnerships involving the aforementioned entities, as well as international countries, organizations, and businesses, have been formed to more effectively utilize precious resources. However, the evidence shows that we are not winning the struggle and more will be required as we move through the twenty-first century.

Ensuring a more sustainable future will necessitate a suite of actions, which we identified in Chapter 7 as 11 action steps. Conserving biodiversity, which recognizes neither ownership nor boundaries, calls for good science, first-rate technology, excellent management, and a broad constituency willing to make some concessions to save it. To maintain biological diversity in this century and beyond will require financing, public support, and, above all, the integrated management skills that even the most sophisticated high-tech farming systems lack. And farm production systems pale in complexity compared to the natural systems we must preserve, repair, and reconstruct. We face a monumental task, far beyond our existing capabilities. But now is the time to look ahead, coordinate, and plan, before our options are further narrowed.

The nature that society saves in the new millennium will reflect our values. The primary value judgment of conservation is that naturally evolved basic elements—genomes, species, communities, landscapes—are fundamentally more valuable than artificial ones (Angermeier, 2000). Humans value ecosystems for various reasons including extractable goods, ecological services, and beauty. Nature fills certain physical, intellectual, and spiritual needs that may be essential to human survival (Kellert, 1995). Yet many people do not appreciate the importance of native biodiversity in providing these features, and they tacitly allow loss of biodiversity.

Conservationists must help reestablish the connection of society to natural ecosystems by promoting recognition of a broad array of ecological values and their relation to natural biotas. Value systems give rise to ethics, which guide human behavior. The ethics guiding our use of ecosystems reflect our respect for nature, and ethics founded on respect for nature is a vital social mechanism for affirming the intrinsic value of ecosystems and limiting their alteration (Angermeier, 2000).

As Texas approaches the daunting challenge of conserving its biodiversity, there is an urgent need to make conservation a higher priority in state and local government actions. To avoid repeating past mistakes, we must understand what has happened to our fauna and flora and what it will take to sustain it in the future. We must ensure there is a knowledgeable citizenry that understands the unique complexities of the state and has the political interest and the will to support a sound system of wildlife management and conservation.

Thinking in terms of long-term conservation of wildlife diversity in Texas, there are a number of crucial challenges that must be addressed in the twenty-first century. To prevent and reduce threats to biodiversity, more substantial conservation efforts will be needed and proactive policies will be essential. In this century, we must begin to effectively plan for various land uses within the dynamics of landscapes. Most regional landscapes include a matrix of semi-natural habitats that surround natural reserves and support some sort of land use (Packard and Schmidly, 1991). These buffer areas are ideal for conservation-based rural development where products and services can be produced from the land without destroying the long-term sustainability of the resource base. Most landscapes also contain highly developed and managed lands that have long since been converted from their original condition to agricultural or industrial purposes. However, by practicing restoration ecology and wise land management, these areas have much to contribute toward conserving our natural resources. Natural areas, such as those in the Big Bend Corridor of West Texas, are perfect places to implement rewilding concepts that allow wildlife the freedom to flourish and habitats to regenerate naturally (see Chapter 7). Finally, it must be remembered that large urban centers are now a major part of our landscapes. Most of the people are concentrated in these areas, and this is where much of the political clout resides relative to decision-making on conservation issues.

Michael Rosenzweig, in his 2003 book *Win-Win Ecology*, has argued for the importance of practicing reconciliation ecology, a branch of ecology that studies ways to encourage biodiversity in human-dominated ecosystems, based on the theory that there is not enough area for all biodiversity to be saved with designated nature preserves (Rosenzweig, 2003a). By managing for diversity in ways that do not decrease human utility of the system, it becomes a "win-win" situation for both human use and native biodiversity (Rosenzweig, 2003b). Aspects of reconciliation ecology can be found in many of the management programs currently practiced on private land in Texas as well as a growing number of urban environments. More conservationists are beginning to call for a focus on "net positive outcomes for nature," an approach that proposes policy shifts away from conservation targets that focus on avoiding losses toward processes that consider net outcomes for biodiversity (Bull et al., 2019).

The eleven detailed actions we presented in Chapter 7 offer several long-term solutions necessary for conserving the natural history of Texas, including actions focused on both private and public lands. Our hope is that these "working points" can serve as the rallying cry for a new century of change for Texas natural history.

Of the eleven steps, Texas is furthest behind in the most critical one: the acquisition of protected areas. Without progress in this area, over the long term there is little hope of retaining the biological integrity of the state. Protected areas are the cornerstone of biodiversity conservation (Dinerstein et al., 2017). Where networks of protected areas are large, connected, well managed, and distributed across diverse habitats, they sustain large populations of threatened and functionally important species and ecosystems more effectively than other land uses (Gray et al., 2016). Protected areas also play an important role in climate-change mitigation (Melillo et al., 2015). Currently, Texas only has 2.7 percent of its land protected, and those areas are not distributed adequately across our ecoregions. The Biden administration 30 x 30 plan, with its emphasis on voluntary conservation efforts by farmers and ranchers, could be an effective way to expand protected areas in Texas.

Protected areas often are too small, too isolated, or both to buffer species from human influence outside their borders. Therefore, wildlife movement corridors that connect protected areas also are vital to long-term success. Corridors can be established using a variety of strategies, such as land easements or through the efforts of land trusts and NGOs to purchase development rights. Texas, with its numerous rivers and streams that traverse the state, could use these various mechanisms to provide for both corridors and protection of critical water resources along major waterways, thus providing multiple benefits. In planning for protected areas, however, it will be important to take into account the potential long-term effects of climate change to these areas (Carrasco et al., 2021). A wide variety of strategies to locate and manage protected areas will be required.

For conservation outside of protected areas to succeed, the utilization of wildlife and their habitat should result in some economic benefit to local people (Bowyer et al., 2019). Currently, we see this in Texas in the form of ecotourism and trophy hunting that have been established successfully in the rural sectors of the state. This is one area where the state is way ahead of the curve, primarily because of the growing importance of wildlife resources to the economic success of the rural ranching and farming sector.

Questions have been raised about the long-term viability of wildlife conservation and economic development—as to whether or not they can be compatible (The Wildlife Society, 2003). For example, it has been documented that nearly all species listed by the USFWS as threatened or endangered have declined because of human economic activity (Czech and Krausman, 1997). Natural resource extraction tends to remove, destroy, or deplete wildlife habitat components (i.e., food, water, cover, and space; NRC, 1970). Agricultural, extractive, and industrial infrastructure reduces the space available to wildlife, and pollution degrades the other components of species' habitats. For all these reasons, some wildlife professionals (Czech, 2000) have argued that an alternative to neoclassical economic growth, with a gradual transition toward a steady-state economy characterized by stable human population and per capita consumption, may be necessary to ensure wildlife conservation over the long term. This will be a bitter pill to swallow in Texas because most Texans have come to cherish economic improvement, as reflected by income levels, job and wage growth, and short-term and long-term GDP, as the *sine qua non* of progress, with the environment and ecological integrity a secondary, minor concern. There is a strong sense of individualism and risk-taking among Texans, with an attitude of "we'll do what we want today and worry about the future later." But that kind of thinking will not be productive in terms of facing the challenges of climate change and the loss of biodiversity.

Another of the eleven critical action steps stressed the importance of strengthening scientific research in support of natural history and conservation. But what good will it do if the research never becomes part of the policy debates and decisions? For too long we have seen politics and ideology trump science in making key decisions about natural resources and the environment. There is a misconception among many scientists that if enough evidence is generated and put in the hands of policy-makers, the problem will be solved, although we know from behavioral science that translating research into practice is not quite that simple. Conservation researchers must learn to better navigate the spaces between research and implementation. The key to this is timely and improved cooperation among conservation-relevant science, politics, and practice.

Sound, well-executed science will be needed to inform effective public policy if Texas is to respond in sufficient time to avoid a major setback in terms of biodiversity and natural resources. Recently a group of highly recognized climate-change scientists in the state offered to brief the governor about the scientific evidence for climate change and the long-term implications of inaction, but the governor ignored them. Future leaders in the state would do well to appoint a science board that would provide advice to the legislature and state leaders about issues relative to conservation, natural resources, and biology (organisms, water, climate, threatened and endangered species, etc.). Membership should include leading faculty from the state's universities and colleges, state agency scientists, and those who work for NGOs. Such a board could also advise state leaders about the funds needed for research and protection of the state's natural resources.

One of the easiest of the eleven action steps to implement, but certainly one of the most important, is to strengthen conservation education across all sectors in the state. This issue should receive mandatory treatment at the elementary, middle, and high school levels such that learning about biodiversity, natural resources and ecosystems, and other conservation issues is incorporated into the curriculum. Other countries are beginning to take the lead in this area. For example, Italy has become the first country in the world to require climate change and sustainable development to be included in the public education curriculum for every grade level (Diskin, 2019). And, as mentioned in Chapter 7, New Jersey has become the first state in the US to require all its public schools to adopt climate change education, including its many ramifications, into its curriculum. We would suggest in Texas that at least one week per school-year be devoted to such topics. The governor could designate a "conservation week" to help generate enthusiasm for such activities. Collegiate courses in conservation and natural resources could be offered at a reduced tuition rate. Education opportunities in conservation and natural resources could be expanded to underrepresented groups (females and minorities) through programs offered by the Texas Master Naturalists. The state through TPWD could expand its education programs and increase the number of workshops, seminars, and demonstrations. Summer work programs could enlist high school and college students to help supply the workforce needed to teach such courses. Organizations such as the Boys and Girls Clubs, Lions Clubs, and other philanthropic groups could be enlisted to provide access to the broadest clientele.

We cannot emphasize enough the importance of continuing to support landowners in their conservation efforts. As surprising as it seems, many landowners remain leery that the "Feds," as they are referred to, will seize their lands if a threatened or endangered species is suspected of being nearby. This fear halts cooperation and research. Landowners should be made to feel that they are already doing something good if they are home to a declining species. The fact that it remains on their property means their land practices benefit the species. State and federal personnel need to articulate the message that we want to work with the landowner—not to be at odds. State and federal agencies have made progress in this area in the form of providing advice, financial incentives, and tax breaks to landowners, but more can be done to ensure their full cooperation in conservation efforts. For example, assisting with simple property improvements, such as constructing ponds, drilling water wells, installing fencing, and maintaining roads, would be very appealing to landowners, thereby encouraging a commitment to improving biodiversity. Similarly, continuing the "safe harbor" provisions of the ESA and the wildlife tax exemption will encourage landowners to conserve wildlife on farm and ranchlands.

But even while measures are being taken to assist landowners, a halt must be brought to the privatization of wildlife. No one should begrudge a landowner from leasing hunting rights. In fact, attaching dollar signs to wildlife may be one of the better ways to ensure that biodiversity survives because it incentivizes good habitat management. However, if profits are to be made from privatization of wildlife, then those profits should be subject to a tax, especially if the privatization is based on genetically modifying (selective breeding under high fence conditions) a naturally occurring species that is a public resource. With a continued emphasis on deer farming and the importation of deer for breeding under the restrictions and crowded conditions of high fencing, the spread of chronic wasting disease (CWD) in Texas is a potentially devastating situation, both ecologically and economically. Indeed, CWD has now been detected in several Texas counties, including deer farming operations (see Chapter 4), potentially threatening the nearly $2 billion annual deer hunting industry in the state. Fortunately, TPWD prevailed in a lawsuit challenging its authority to implement management regulations for CWD on the grounds that deer held in captivity are private property and not subject to TPWD regulation (see Chapter 7).

Finally, it must be noted that there is "no free lunch" when it comes to providing for conservation programs. In this century, more funds are going to be needed to purchase more land for protected areas, parks, and wildlife management areas, as well as for funding research, education, and other initiatives. The highest levels of local and state government, the private sector, and NGOs and other conservation organizations must be brought together to close the biodiversity funding gap. A recent report (Deutz et al., 2020) issued by The Nature Conservancy, the Paulson Institute, and Cornell University's Atkinson Center for Sustainability outlines a roadmap for private sector financing and government actions (legislation, policies, and regulations) to encourage investment in nature protection and conservation. Many of the ideas, plans, and strategies contained in this report incorporate free enterprise investment approaches that could succeed in Texas.

Fortunately, public opinion polls show that Americans of all walks of life overwhelmingly support efforts to conserve natural resources, including wildlife diversity and improved access to outdoor recreation. Public opinion surveys have shown that Texas citizens strongly support government funding to protect wildlife species and the lands and waters they inhabit. In a general population survey of Texans for the report, "Texas Parks and Wildlife for the 21st Century," researchers found that more than 90 percent of those surveyed thought that it was important that natural areas exist in Texas for enjoying and experiencing nature, that fish and wildlife populations are being properly managed and conserved in Texas, and that ecologically important habitats are being protected and preserved; 100 percent felt it was important that Texas's water resources are safe and well protected (see Chapter 7). And, a recent national poll conducted by the Defenders of Wildlife and Earthjustice (a nonprofit environmental law organization) found that 90 percent of Americans support the ESA.

As it approaches its bicentennial in 2036, Texas needs a vision for the future, and the protection and sustainability of its natural resources should be a part of that vision. Tom Luce, a prominent Texas citizen from the Dallas area, has started Texas 2036, a nonprofit that aims to use data and research to help Texas solve some long-term structural issues important to its future (see website at https://texas2036.org/). With Margaret Spellings, the US Secretary of Education under George W. Bush, serving as its chief executive, the early focus of the nonprofit has been on education and workforce issues, which are certainly important, but the agenda should be expanded to include the integrity and sustainability of natural resources if the state's quality of life is to be sustained. As Texas approaches its 200th birthday, it needs to find the right solution to protect its biodiversity and cope with climate change in order to ensure the best environment for future generations while allowing for managed growth and development.

Texas and Texans would benefit from a big goal to promote protection of its biodiversity and natural ecosystems (land, water, wildlife). James Collins and Jerry Porras in their 2004 book, *Built to Last: Successful Habits of Visionary Companies*, coined the phrase "Big Hairy Audacious Goal" (BHAG, pronounced "Bee Hag"), which they described as a highly ambitious 10-to-30-year goal to progress toward an envisioned future (Collins and Porras, 2004). The thesis of their book was that highly successful organizations often have bold missions, or BHAGs, that are strategic and emotionally compelling. Think of the moon mission in the 1960s!

Texas needs a BHAG to address the growing concerns of environmental degradation and the loss of biodiversity and natural ecosystems. A BHAG should be clear and compelling and serve as a unifying focal point of effort. It needs a clear finish line, so the organization can know when it has achieved the goal. It must engage people—reach out and grab them in the gut. It should be tangible, energizing, and highly focused. People should "get it" right away with little or no explanation.

It has been twenty years since the highest levels of Texas government have called for action to conserve the state's natural and biological resources and seriously addressed its ever-increasing environmental problems. Consequently, we have continued to fall further and further behind to the point that Texas is now the number one emitter of CO_2 in the US (twice the rate of California). In fact, a recent study (Symons, 2019) has demonstrated that increased emissions from Texas alone have erased reductions from all states in the west and northeast combined; in other words, one state's emissions have now cancelled out climate progress across the entire country! Texas's carbon emissions surged 20 percent higher from 2009 through 2016, an increase of 108 million tons. Texas also ranks first in terms of violating water pollution rules (Sadasivam, 2018), and comes in fortieth with regard to natural environment rankings (*US News and World Report*, 2019), and it sits right in the crosshairs of climate change events. These trends, if they continue, will eventually have devastating consequences for the state's ecosystems, biodiversity, and quality of life.

We outlined many possibilities to secure additional funding for such an initiative in Chapter 7. The 2019 86th Texas Legislature authorized, and Texas voters approved, a constitutional amendment (SJR 24) to ensure that parks and historic sites get the maximum funding available from an existing sales tax on sporting goods. While this will greatly aid the maintenance backlog and hopefully future development of state parks and outdoor recreation, it is unclear whether these funds could be used for the purchase of protected areas. Of the many alternatives available for that purpose, consideration should be given to initiatives that have proven successful in other states. For example, the state of Missouri voted to implement a one-eighth of a cent sales tax to help support conservation issues. In 2014, 75 percent of Florida voters approved the largest conservation measure ever adopted by a single state by passing Amendment 1, which earmarked $1 billion a year to conservation efforts for the next 20 years with funding coming from the state's property taxes. Likewise, Texas lawmakers could dedicate a portion of the state's record oil and gas revenues to a permanent trust fund for habitat and water restoration projects in the state. Potential projects could include preserving open space, purchasing conservation easements, managing invasive species, creating healthy soil and other sustainable agriculture projects on private lands, and enhancing wildlife habitat in areas impacted by residential, energy, mineral, or industrial development. Such a fund could also leverage state money as a match for federal conservation grants.

An even more ambitious approach would be for the state to create a conservation initiative similar to what it did in the health care field in 2007 when voters approved a $3 billion state bond program to set up the Cancer Prevention and Research Institute (CPRIT). This positioned Texas as the nation's second-largest public funder of cancer research, trailing only the federal government's National Cancer Institute. In 2019, the 86th legislature authorized another constitutional amendment to provide an additional $3 billion over ten years to continue the program, which Texas voters approved in November 2019. Just imagine what could be accomplished if Texas established a conservation fund of $3 billion to provide the needed funding for the acquisition of land for biodiversity conservation and to further expand outdoor recreation opportunities for its growing population as well as enhance research and educational programs in this critical area. Such a BHAG (we propose calling it "Saving the Nature of Texas") would set the state apart in terms of protecting its natural resources.

A well-thought-out and ambitious BHAG such as we have described could succeed and unite Texans, although there would be challenges with implementation. To have any chance of success would require: demonstrating the need for increased funding; active gubernatorial and legislative support that reaches out to voters, particularly in urban centers where a majority reside; support within the agency charged with leadership implementation; and support from the business community, legislature, and active nongovernmental organizations. Before launching any funding effort, a massive public education and media program would be required to educate Texans about the critical importance of such an initiative. The program could be modeled after the successful "Don't Mess with Texas®" trash cleanup program. A similar approach under the umbrella of "Saving the Nature of Texas" would stimulate universities, government agencies, NGOs, and the private sector to seek progress towards an envisioned future where biodiversity and our environment are protected for future generations.

Texas is blessed with an extraordinary diversity of wildlife, and it is one of the most biodiverse states in the country. However, these resources face a multitude of threats. Texas stands on the front line of climate change, which means its wildlife and ecosystems are potentially susceptible to the mass extinction crisis that is threatening species across the globe. The system of wildlife management that was developed in the twentieth century has served Texans well in the past, but it requires updating if the state is to overcome these new challenges. This modified system should be based on the management and conservation of the public's wildlife in accordance with the North American Model of Wildlife Management and the Public Trust Doctrine—as a trust resource with intrinsic and ecological value for the equitable benefit, use, and enjoyment for all Texans, including future generations.

Texans are not going to want to live in an impoverished world devoid of biodiversity and its considerable wildlife resources. The late Barry Lopez, an American author whose work was known for its humanitarian and environmental concerns, put it very succinctly when he made this statement in a 2001 article in *Orion Magazine*: "Recognize that politics without field biology, or a political platform in which human biological requirements form but one plank, is a vision of the gates of hell" (Lopez, 2001). In our view, Texans, if given the choice, will reject the vision of an ecologically impoverished state and opt for a healthy alternative in which resources remain a vital part of the living environment. It is toward that end that we prepared this book.

Afterword

I T IS HARD TO IMAGINE THE TEXAS THAT VERNON BAILEY AND HIS fellow naturalists of the Bureau of Biological Survey explored at the end of the nineteenth century and the beginning of a new one. A Texas essentially without cars and trucks, with few metropolitan areas, and a population that most often was directly tied to the land and its wildlife. Bailey's objective was to describe the mammals, birds, reptiles, and plants of the largest state in the union over the period from 1889 to 1905 (although the description of Texas birds was left to his colleague, Harry Oberholser). In that time, sometimes working individually and at other times with a field crew, the team of field biologists covered all the major ecological regions of one of the most biologically diverse states. The product of this endeavor became *Biological Survey of Texas* published by Bailey in 1905, reprinted in its entirety in this book. The authors take us into the lives of these early naturalists and the roles they played in that first major description of the state's fauna. The quotations from their field notes and reports along with the reproduced archival photographs allow us to step back in time. They give us a glimpse of Texas from the naturalists' perspective—from sites in the eastern Piney Woods and coastal prairie to the Panhandle High Plains and the deserts and mountains of the west.

The authors again update the taxonomy used in the original biological survey, given that much of it changed from the last edition of *Texas Natural History*. Perhaps more important, they detail the changes that have occurred in Texas landscapes and in the mammals and reptiles of Texas over more than 120 years. These changes include the extinction of some species and the decline of many others, most of which coincided with the shrinking natural areas of the state as agriculture, oil and gas exploration, and urban development increased through the twentieth century. In 1900, during the fieldwork for the *Biological Survey of Texas*, seventy cities in the United States were larger than San Antonio, the largest Texas city at that time with just over 53,000 people. In 2021, Texas now has claim to three of the ten largest cities in the US (Houston, San Antonio, and Dallas). This growth has come with a cost in terms of a population now largely urbanized and without a direct connection to nature. The growth also has converted increasingly more natural areas to cleared land to meet the needs of the burgeoning population of Texas.

In just short of two decades since the publication of David Schmidly's *Texas Natural History: A Century of Change*, many of the wildlife conservation concerns presented therein continue unabated today. There is no doubt that the biological diversity documented by Bailey is seeing stresses that he could not have imagined. This new edition of the book describes in detail those threats to the state's wildlife and plant communities. One such conservation concern that few in the public consider is the introduction of exotic wildlife and the trend towards the effective "privatization" of wildlife resources through high-fenced ranches across the state. These prevent free interbreeding of native species and may have serious and long-term genetic repercussions on wildlife populations. Another concern is the low number of protected natural areas, relative to much of the rest of the United States. With about 95 percent of Texas privately owned, the protected areas are severely limited. Global issues such as climate change will also take a toll on Texas biodiversity.

The predictions made for the future of biodiversity in Texas are grim but familiar to those of us who are wildlife biologists. The good news is that there is hope for the future. As the authors have detailed in their eleven steps for strengthening conservation in Texas, there are tangible solutions to some of these conservation concerns. Despite the fact that state and federally protected areas are quite limited in the state, many ranchers and other landowners are concerned about the loss of biodiversity and the danger of encroaching development. Their efforts to preserve large tracts of our original Texas habitats are a bright hope for the future. An additional hope for maintaining the wild nature of Texas lies in educating the populace about the wealth of biodiversity in their state. The vast majority of the population is now living in urban or suburban areas and rarely gets to see the striking beauty of the landscape or appreciate the flora and fauna unique to the state. One tenet of conservation biology is that it is unlikely that people will care for things with which they are not familiar. In contrast, it is easy to care about the decline in biodiversity if you experience wildlife and plant communities firsthand on a regular basis. Our continuing efforts to educate the public on this issue are essential. This book is one part of that ongoing educational process.

Naturalists have always existed in Texas, and they will continue to study the natural history of organisms into the future using advanced technology of which Bailey could never have dreamed. Those studies will continue to be necessary as we track changes to the state's flora and fauna. Still, there has been a broad decline in both hiring of professional field biologists and in funding for natural history research over the last century. There are fewer field biologists in academic biology programs, and that translates to fewer new students now being trained to carry forward natural history research. For many of us in this field, this decline is especially disturbing as the need for a better understanding of our biodiversity and its conservation has never been greater. Nevertheless, it is clear that the authors of this book hope that biologists across Texas will continue the work begun by Vernon Bailey, seeking to expand our understanding of the shifts and changes to the biodiversity of Texas and, crucially, to introduce new generations of students to the biological bounty of this state.

ROBERT C. DOWLER, PHD
MR. & MRS. VICTOR P. TIPPETT PROFESSOR OF BIOLOGY
ANGELO STATE UNIVERSITY

Glossary

animal damage control. The act of control or removal of mammalian predators, rodents, and birds that are a nuisance to agriculture and people.

anthropocene. The name proposed for the current geological age and the period during which human activity has been the dominant influence on climate and the environment.

archival natural history. The practice of using information (photographs, field notes, physiographic reports, specimen catalogs, etc.) archived in natural history institutions to reconstruct historic landscapes and biotic conditions.

archive. A collection of historical documents or records providing information about a place, institution, or group of people.

at-risk species. A wildlife species that may become a threatened or endangered species because of a combination of biological characteristics and identified threats.

barrier island. A long, broad, sandy island lying parallel to a shore that is built up by the action of waves, currents, and winds, and that protects the shore from the effects of the ocean.

biodiversity. A term for the variety of life found on Earth (plants, animals, fungi, and microorganisms) as well as the communities they form and the habitats in which they live.

biogeography. The study of the distribution of different species around the planet and the factors that influence their distribution.

biological resources. Genetic resources, organisms or parts thereof, populations, or any other biotic component of ecosystems having actual or potential value or use to humanity.

biological species concept. Groups of actually (or potentially) interbreeding natural populations that are reproductively isolated from other such groups. Reproductive isolation implies that interbreeding between individuals of two species normally is prevented by intrinsic factors.

biome. A complex biotic community covering a large geographic area and characterized by the distinctive life forms of important climax species.

biota. A grouping of animals, plants, fungi, and other organisms that all share the same geographic region on Earth.

biotic province. A community occupying an area where similarity of climate, physiography, and soils results in the recurrence of similar combinations of organisms.

bottomland. Lowlands along streams and rivers, usually on alluvial floodplains that are periodically flooded.

brush. Various species of shrubs or small trees usually considered undesirable on rangelands.

brush management. The process and act of removing, reducing, or manipulating nonherbaceous plants from rangeland.

chemical stressor. A hazardous substance which, when released into the environment, damages living organisms or ecosystems and reduces their ability to cope with environmental changes.

chronic wasting disease (CWD). A contagious neurological, prion disease affecting deer, elk, and moose; produces a characteristic spongy degeneration of the brain causing emaciation, abnormal behavior, loss of bodily function, and death.

chytrid fungus. A primitive type of fungus, living exclusively in water or moist environments, that feeds on dead and rotting organic matter and is thought to be responsible for a worldwide decline in amphibian populations.

cladistics. A method of developing phylogenies based upon the branching sequences of evolution.

classification. The assignment of groups of organisms to taxa.

conservation biology. The management of nature and the Earth's biodiversity with the aim of protecting species, their habitats, and ecosystems from excessive rates of extinction and the erosion of biotic interactions.

conservation easement. A legally binding restriction placed on a piece of property to protect its associated resources, most commonly through an agreement between a landowner and land trust or unit of government, designed to limit certain types of uses or development from taking place on the land in perpetuity while the land remains in private hands.

conservationist. A person who advocates the wise use of a resource or the productive potential of a resource-generating system with the goal of maintaining its future availability or productivity. Conservationists typically adopt a utilitarian, land stewardship ethic that natural resources exist for the use of humans, who have an obligation to improve the natural condition and sustainably exploit resources for human economic benefit.

Conservation Reserve Program (CRP). A voluntary government program established in 1985 that offers financial assistance to farmers to establish approved cover on highly erodible cropland or pastureland.

consumptive uses of wildlife. The killing of wildlife, via hunting, fishing, and trapping, for sport, recreation, commercial use and sale, a means to control damage to private land and crops, as a food source or products for personal use, or as a population management tool.

Convention on Biological Diversity (CBD). A multilateral international treaty, informally known as the Biodiversity Convention, with three main goals: the conservation of biodiversity, the sustainable use of its components, and the fair and equitable sharing of benefits arising from genetic resources. The Convention was opened for signature at the Earth Summit in Rio de Janeiro on 5 June 1992 and entered into force on 29 December 1993.

cross-disciplinary. Relating to or representing more than one branch of knowledge; interdisciplinary.

cryptic species. One of two or more morphologically indistinguishable biological groups that are incapable of interbreeding.

deer farm. A high-fenced piece of land suitable for grazing that is populated with ungulates, and especially white-tailed deer, that are raised as livestock.

disjunct distribution. A discontinuous distribution or a geographic arrangement in which a species occurs in two or more isolated areas having no connection.

diversified farming. The practice of producing a variety of crops or animals, or both, on one farm, as distinguished from specializing in a single commodity.

ecoregion (or ecological region). Large unit of land containing a geographically distinct assemblage of species, natural communities, and environmental conditions.

ecosystem services. The direct and indirect contributions of ecosystems to human well-being, such as production of food and water, climate regulation, nutrient cycling and oxygen production, and recreational benefits.

ecotourism. Tourism directed toward exotic, often threatened, natural environments, intended to support conservation efforts and observe wildlife.

Endangered Species Act. US law, passed in 1973 and subsequently amended, that regulates the capture, possession, and sale of threatened and endangered species of animals and plants.

environmental degradation. Processes and activities, commonly accelerated or caused by human activities, that compromise the natural environment by reducing biological diversity and the general health of the environment.

exotic. An animal introduced from a foreign county; nonnative.

ex situ conservation. "Off-site" conservation, occurring within or outside a species' natural geographic range; individuals maintained ex situ exist outside an ecological niche.

extirpate. To destroy, make extinct, or exterminate.

Farm Bill. An omnibus, multi-year US law that governs an array of agricultural, natural resource, and food programs; the primary agricultural and food policy tool of the federal government that is eligible for renewal every five years.

fecundity. Similar to fertility, the natural ability to produce offspring. It is the major measure of fitness, under both genetic and environmental control.

feral. Pertaining to formerly domesticated animals now living in a wild state.

fieldwork (field research, field studies). The collection of raw data outside a laboratory, library, or workplace setting, involving direct observation of conditions, writing of field notes, and/or collection and preparation of specimens; typically involves studying free-living wild animals in their natural habitat.

fracking. The injection of fluid into shale beds at high pressure in order to release petroleum resources such as oil and gas for extraction.

fragmentation. The process whereby a large piece of habitat is divided into a number of smaller, isolated patches.

furbearer. A category of mammals harvested for direct commercial use and sale of hides and pelts; includes most small and large carnivores.

Gap Analysis Program (GAP). A proactive approach to analyzing biodiversity by seeking to identify gaps between land areas rich in biodiversity and areas that are managed for conservation.

gas flaring. The act of incinerating excess methane from an oil or gas well that cannot be transmitted by pipelines.

gene editing. The insertion, deletion, or replacement of DNA at a specific site in the genome of an organism or cell; usually achieved in the lab using engineered enzymes known as molecular scissors.

gene swamping. The specific situation where one genetic pool of a species is gradually replaced by that of another one, which can lead to the extinction of the original species.

genetic species concept (GSC). Groups of genetically compatible interbreeding natural populations that are genetically isolated from other such groups. Genetic isolation implies that each species is genetically distinct from other species and that the integrity of the respective gene pools is maintained.

Global Positioning System (GPS). A satellite navigation system used to determine the ground position of an object or place.

global warming (or climate change). An increase in the average temperature of the Earth's atmosphere, especially a sustained increase sufficient to cause climate change.

habitat. An ecological or environmental area that is inhabited by a particular species of animal, plant, or other type of organism.

habitat selection. Rules and processes used by organisms to choose among patches of habitat that differ in one or more variables that influence fitness, such as food availability or predation.

high fence. A tall fence (usually 8 feet) used to enclose a deer population that is managed for antler and body size.

improved pasture. A sown pasture with introduced species, usually grasses in combination with legumes; often fertilized and intensively managed for livestock.

iNaturalist. A citizen science project and online social network of naturalists, citizen scientists, and biologists built on the concept of mapping and sharing observations of biodiversity across the globe that is accessible via a website or from a mobile application.

increaser. A grass species that will increase under any type of mismanagement or disturbance.

integrated landscape management. Land management that encompasses multiple activities, such as agricultural production, provision of ecosystem functions and services (e.g., waterflow regulation and quality, pollination, climate change mitigation, etc.), protection of biodiversity, landscape beauty, and recreation value.

invasive species. A non-native species introduced into an ecosystem that causes or is likely to cause economic or environmental harm, including to human health.

IUCN. International Union for the Conservation of Nature, founded in 1964, which is the world's main authority on the conservation status of species. The IUCN Red List of Threatened Species (commonly known as the Red List) is the world's most comprehensive inventory of the global conservation status of biological species.

karst. A landscape formed by limestone, dolomite, or gypsum—soluble rocks deposited by ancient seas—and characterized by cracks, fissures, or sinkholes leading to underground caves or aquifers.

landscape. A social-ecological system consisting of a mosaic of natural and/or human-modified ecosystems with a characteristic configuration of topography, vegetation land use, and settlements influenced by the ecological, historical, economic, and cultural activities of the area.

landscape amnesia. A term coined by Jared Diamond to explain the phenomenon of forgetting how different the surrounding landscape looked 50 years ago because the change from year to year has been so gradual.

land stewardship. Responsible use and protection of the natural environment through conservation and sustainable practices.

land tenure. The relationship that individuals and groups hold with respect to land and land-based resources, such as minerals, pastures, and water. Land tenure rules define the ways in which property rights to land are allocated, transferred, used, or managed in society.

land use. The modification of natural environments or wilderness into built environments, such as settlements and semi-natural habitats (e.g., agricultural fields, pastures, and managed woods).

lease hunting. An agreement between a lessor (landowner) and leesee (e.g., hunter or angler) that allows the leesee to visit and hunt/fish on the lessor's land for a specified time period and price; lessees typically pay a per-acre or per-lease fee for the access rights.

leprosy. A chronic infectious disease of the skin, nervous system, and mucous membranes that is caused by the bacteria *Mycobacterium leprae*; transmitted via person-to-person contact.

life history. The traits of an organism that directly affect its survival and reproduction, such as age at first reproduction, number and size of offspring, and reproductive lifespan and aging.

life zone concept. A belt of vegetation and animal life that is similarly expressed with increases in altitude and latitude; similar to altitudinal zonation in mountainous regions of the western US; proposed by C. Hart Merriam in 1889 as a measure of describing areas with similar plant and animal communities.

lumper. A taxonomist who emphasizes the demonstration of relationships in the delimitation of taxa and who tends to recognize large taxa.

metapopulation. A regional population consisting of semi-isolated local populations; literally, a "population of populations."

mitrochondrial DNA. The DNA contained in the mitochondria of a cell; because offspring typically inherit only their mother's mitochondria, mitochondrial DNA is useful in tracing maternal lineages.

molecular genetics. The field of biology that studies the structure and function of genes at a molecular level and employs methods of both molecular biology and genetics.

multidisciplinary. Combining or involving several academic disciplines or professional specializations in an approach to a topic or problem.

National Museum of Natural History. A natural history museum administered by the Smithsonian Institution and located on the National Mall in Washington, DC; home to the largest group of professional scientists dedicated to the study of natural and cultural history in the world.

natural flow regime (of streams / rivers). The natural instream flow of water in a stream channel influenced by competing uses for water, such as irrigation, public supply, recreation, hydropower, and aquatic habitat.

natural history. The scientific study of plants and animals in their natural environments; concerned with levels of organization from the organism to the ecosystem and stresses identification, life history, distribution, abundance, and interrelationships.

natural resources. Naturally occurring raw materials obtained from the Earth than cannot be made by humans. **Biotic resources** come from living things, or organic material, and include plants, animals, and fossil fuels (coal, oil, and natural gas). **Abiotic resources** (air, sunlight, water, and minerals) originate from nonliving and inorganic materials. **Renewable natural resources** are those that can be replenished in our lifetime. A **nonrenewable resource** is a resource of economic value that cannot be readily replaced by natural means on a level equal to its consumption (e.g., most fossil fuels).

natural resource management. Management of natural resources focused on how management affects the quality of life for both present and future generations of humans.

neoclassical economic growth theory. Model of economic growth that has become the primary focus of modern-day economics, in which the emphasis is placed on the ease of substitution between capital and labor to ensure steady-state growth; assumes that competition leads to an efficient allocation of resources within an economy.

nonconsumptive uses of wildlife. Those activities in which wildlife is watched, studied, or recorded without being killed, such as bird-watching, sketching, and photography; can be harmful if observing wildlife at too close range during breeding seasons and high human uses of areas where endangered species may be negatively impacted.

nongame wildlife. Terrestrial and semi-aquatic vertebrates not formally hunted for sport; majority of wild vertebrates belong to this group.

nonindigenous. Not produced, growing, living, or occurring naturally in a particular region or environment.

overgrazing. The result of plants being exposed to intensive grazing for extended periods of time, or without sufficient recovery periods, to the point of damaging vegetation cover.

phylogenetic analysis. An analysis of the evolutionary history of an organism or groups of related organisms.

phylogenetic species concept. The concept of a species as an irreducible group whose members are descended from a common ancestor and who possess a combination of certain defining or derived traits; in other words, species at the "tip" of a phylogeny, representing the smallest set of organisms that share an ancestor and can be distinguished from other such groups.

phylogenetic systematics. The type of systematics practiced in the field of cladistics.

phylogeny. The evolutionary history of an organism or group of related organisms.

plant association. A grouping of plant species, or a plant community, that recurs across the landscape.

playa lake. Round, typically shallow, depressions in the ground in the southern High Plains that are ephemeral, meaning they only contain water at certain times of the year.

predator-prey relationship. An interaction between two organisms of unlike species in which one of them acts as predator that captures and feeds on the other organism that serves as the prey.

preservationist. A person who advocates the protection of an ecosystem or a species, to the extent possible, from disruptions by human use. They advocate a "biophilia" perspective that all living things have intrinsic value with a scientific basis for maintaining biodiversity with the government in a major role.

rangeland. Any extensive area of land that is occupied by native herbaceous or shrubby vegetation, which is grazed by domestic or wild herbivores.

range management. A professional field whose aim is to ensure a sustained yield of rangeland products while protecting and improving the basic range resources of soil, water, and plant and animal life.

reconciliation ecology. A branch of ecology that studies ways to encourage biodiversity in human-dominated ecosystems.

refugia. An area where special environmental circumstances have enabled a species or a community of species to survive after extinction in surrounding areas.

restricted geographic range. Species with a geographically limited area of distribution; as opposed to wide-ranging.

rewilding. Large-scale conservation aimed at restoring and protecting natural processes and core wilderness areas, providing connectivity between such areas, and protecting or reintroducing apex predators and keystone species.

riparian. Associated with the bank of a natural watercourse, such as a river or stream.

roost. A place where birds or bats rest or sleep.

safe harbor agreements. A voluntary agreement involving private or other non-federal property owners whose actions contribute to the recovery of species listed as endangered or threatened under the Endangered Species Act.

short-grass prairie. A semiarid climatic ecosystem located in the Great Plains of the US that is characterized by an abundance of short grasses, rivers, streams, and rocky soils.

splitter. A taxonomist who divides taxa very finely, to express every shade of difference and relationship, through the formal recognition of separate taxa and their elaborate categorical ranking.

subspecies. A geographically defined aggregate of local populations that differ taxonomically from other such subdivisions of the species.

sustainable resource use. The use of biological resources in a way and at a rate that does not lead to the long-term decline of biological diversity, thereby maintaining its potential to meet the needs and aspirations of present and future generations.

sylvatic plague. An infectious bacterial disease caused by the bacterium *Yersinia pestis* that primarily affects rodents such as prairie dogs and ground squirrels; the same bacterium that causes bubonic and pneumonic plague in humans.

tallgrass prairie. A fire-dependent ecosystem distinguished by tall grasses (up to 10 feet tall) and deep, rich soils; native to central North America although very little remains today due to expansive agricultural land use.

taxonomic revision. A novel analysis of the variation pattern in a particular taxon for the purpose of confirming or providing new insights about the identification and relationships among the entities within the taxon; analysis may involve any combination of variables, such as morphological, anatomical, genetic, and molecular.

taxonomy. The science of classifying organisms.

Texas Conservation Action Plan. A plan completed by the Texas Parks and Wildlife Department to provide a statewide "road map" for research, restoration, management, and recovery projects addressing Species of Greatest Conservation Need (SGCN) and important habitats.

Texotic. A general term for exotic ungulates that have been introduced onto Texas rangelands, often under conditions of high fencing.

trophy hunting. Hunting of wild game for human recreation. The trophy is the animal or part of the animal kept, and usually displayed, to represent the success of the hunt.

US Bureau of Biological Survey. A unit of the federal government established in 1885 as the Office of Economic Ornithology within the Department of Agriculture to study the food habits and migratory patterns of birds, especially those that had an impact on agriculture. This office gradually grew in responsibility and went through several name changes until finally renamed the Bureau of Biological Survey in 1905. The Bureaus of Fisheries and Biological Survey were transferred to the Department of Interior in 1939, and in 1940 they were combined and named the United States Fish and Wildlife Service.

VertNet. A National Science Foundation (NSF) funded effort to make biodiversity data available on the web.

wetlands. Areas, such as marshes or swamps, where water covers the soil, or is present either at or near the surface of the soil all year or for varying periods of time during the year, including the growing season.

wilderness wildlife. Native species (usually large herbivores or carnivores) characteristic of wilderness, in contrast to species that occupy areas where man and his activities dominate the landscape; examples include jaguars, grizzly bears, mountain lions, elk, and moose.

wildfire. A large, destructive fire that spreads quickly over woodland, grassland, or brush and can be difficult to extinguish.

wildlife commercialization. Exploitation of public wildlife resources by individuals for financial gain. Examples include fee hunting, guiding and outfitting for compensation, hunting derbies or contests, and sale of wildlife parts and products.

wildlife conservation. The practice of protecting wild species and their habitats in order to prevent species from going extinct.

wildlife corridor. A link of wildlife habitat, generally native vegetation, which joins two or more larger areas of similar habitat. Corridors are critical for the maintenance of ecological processes including allowing for the movement of animals and the continuation of viable populations.

wildlife diversity. A subset of general biodiversity, referring to living, non-domesticated animals, including feral animals, captive wild animals, and wild animals; often used as a general term to refer to the local indigenous vertebrate fauna.

wildlife management associations. Groups formed by landowners to improve wildlife habitats and associated wildlife populations.

wildlife (big game) ranching. The intentional raising of wildlife, especially ungulates, for any purpose, including hunting; domestic livestock may be raised simultaneously with game on a big game ranch.

zoological specimen. An animal or part of an animal preserved for scientific use to verify the identity of a species and to allow study; may include bird and mammal study skins, mounted specimens, skeletal material, animals preserved in liquid preservatives, and genetic tissues.

Literature Cited and References

Abramov, A. V. 2000. A taxonomic review of the genus *Mustela* (Mammalia, Carnivora). *Zoosystematica Rossica* 8: 357–364.

Abuzeineh, A. A., N. E. McIntyre, T. S. Holsomback, et al. 2011. Extreme population fluctuation in the northern pygmy mouse (*Baiomys taylori*) in southeastern Texas. *Therya* 2: 37–45.

Adams, C. E., R. D. Brown, and B. J. Higginbotham. 2004. Developing a strategic plan for future hunting participation in Texas. *Wildlife Society Bulletin* 32: 1156–1165.

Adams, C. E., B. J. Higginbotham, D. Rollins, et al. 2005. Regional perspectives and opportunities for feral hog management in Texas. *Wildlife Society Bulletin* 33: 1312–1320.

Adams, L. E., N. Wilkins, and J. L. Cooke. 2000. A place to hunt: Organizational changes in recreational hunting, using Texas as a case study. *Wildlife Society Bulletin* 28: 788–796.

Adams, L. W., and L. E. Dove. 1989. *Wildlife reserves and corridors in the urban environment: A guide to ecological landscape planning and resource conservation*. National Institute for Urban Wildlife, Columbia, Maryland.

Allen, C. R., S. Demarais, and R. S. Lutz. 1994. Red imported fire ant impact on wildlife: An overview. *Texas Journal of Science* 46: 52–59.

Allen, G. M. 1942. *Extinct and vanishing mammals of the Western Hemisphere*. Intelligencer Printing Co., Lancaster, PA.

Allen, J. A. 1891. Notes on new or little known North American mammals, based on recent additions to the collection of mammals in the American Museum of Natural History. *Bulletin of the American Museum of Natural History* 3: 263–310.

Allen, J. A. 1892. Description of a new species of *Perognathus* from southeastern Texas. *Bulletin of the American Museum of Natural History* 4: 45–50.

Allen, J. A. 1894. On the mammals of Aransas County, Texas, with descriptions of new forms of *Lepus* and *Oryzomys*. *Bulletin of the American Museum of Natural History* 6: 165–198.

Allen, J. A. 1896. On mammals collected in Bexar County and vicinity, Texas, by Mr. H. P. Attwater, with field notes by the collector. *Bulletin of the American Museum of Natural History* 8: 47–80.

Allen, J. A. 1919. Notes on the synonymy and nomenclature of the small spotted cats of tropical America. *Bulletin of the American Museum of Natural History* 35: 559–610.

Allen, M. R., K. P. Shine, J. S. Fuglestvedt, et al. 2018. A solution to the misrepresentations of CO2-equivalent emissions of short-lived climate pollutants under ambitious mitigation. *Climate and Atmospheric Science* (a *Nature* partner journal) 1, article 16. https://www.nature.com/articles/s41612-018-0026-8.

American Rivers. 2019. About America's Most Endangered Rivers® campaign. https://www.americanrivers.org/about-mer/.

Ammerman, L. K., C. L. Hice, and D. J. Schmidly. 2012. *Bats of Texas*. Texas A&M University Press, College Station.

Ammerman, L. K., and R. Tabor. 2008. *Monitoring the colony size and population fluctuations of the endangered Mexican long-nosed bat in Big Bend National Park using thermal imaging*. Division of Science and Resource Management, Big Bend National Park.

Anchondo, C. 2019. Lawmakers pass bills to better fund Texas parks, historic sites—if voters approve. *Texas Tribune*, May 28, 2019. https://www.texastribune.org/2019/05/28/texas-parks-historic-sites-funding-voter-approval/.

Anderson, C. R., F. G. Lindzey, K. H. Knopff, et al. 2010. Cougar management in North America. Pp. 41–54 in *Cougar ecology and conservation* (M. Hornocker and S. Negri, eds.). University of Chicago Press, Chicago, Illinois.

Anderson, D. P., B. J. Frosch, and J. L. Outlaw. 2007. Economic impact of the exotic wildlife industry. *Agricultural and Food Policy Center Report* 07-2, Texas A&M University, College Station. https://www.afpc.tamu.edu/research/publications/496/rr-2007-02.pdf.

Anderson, J. 2000. The white buffalo. *Texas Parks and Wildlife Magazine*, June 2000. Texas Parks and Wildlife Department, Austin, TX.

Anderson, S. 1966. Taxonomy of gophers, especially *Thomomys*, in Chihuahua, Mexico. *Systematic Zoology* 15: 187–198.

Angermeier, P. L. 2000. The natural imperative for biological conservation. *Conservation Biology* 14: 373–381.

Anonymous. 1945. *Principal game birds and mammals of Texas, their distribution and management*. Press of Von Boeckmann-Jones Co., Austin, TX.

Anonymous. 2018. NSRL combines forces with TxDOT to conserve Texas wildlife. Pp. 43–44 in *M Magazine*, Fall-Winter 2018. Museum of Texas Tech University, Lubbock, Texas. https://www.depts.ttu.edu/museumttu/about/M-fall-winter-2018-Final.pdf.

Ansley, J., and C. Hart. 2012. *Drivers of vegetation change on Texas rangelands*. AgriLife Extension, Texas A&M System L-5534. https://agrilife.org/vernon/files/2012/11/ANS2012-02-Ansley-Hart-TALES-L5534.pdf.

Archer, S. 1989. Have southern Texas savannahs been converted to woodlands in recent history? *American Naturalist* 134: 545–561.

Archer, S. 1994. Woody plant encroachment into southwestern grasslands: Rates, patterns, and proximate causes. Pp. 13–68 in *Ecological implications of livestock herbivory in the West* (M. Vavra, W. A. Laycock, and R. D. Pieper, eds.). Society for Range Management, Denver, CO.

Archer, S. 1995. Harry Stobbs memorial lecture, 1993: Herbivore mediation of grass-woody plant interaction. *Tropical Grasslands* 29: 218–235.

Armstrong, J. B., and A. N. Rossi. 2000. Status of avocational trapping based on the perspective of state furbearer biologists. *Wildlife Society Bulletin* 28: 825–832.

Arnett, E. B., W. K. Brown, W. P. Erickson, et al. 2008. Patterns of bat fatalities at wind energy facilities in North America. *The Journal of Wildlife Management* 72: 61–78.

Association of Fish and Wildlife Agencies. 2019. Conservation education strategy. https://www.fishwildlife.org/afwa-informs/ce-strategy/north-american-conservation-education-strategy.

Attwater, H. P. 1917. The disappearance of wildlife. *Bulletin of the Scientific Society of San Antonio* 1(3):

47–60.

Austin Monitor. 2019. Notice of intent to sue: Texas cave salamanders. http://www.austinmonitor.com/ wp-content/uploads/2019/02/19-02-27_NOI-Reinitiation-TxDOT-Cave-Salamanders.pdf.

Avibase. 2019. https://avibase.bsc-eoc.org/avibase.jsp?lang=EN.

Axtell, R. W. 1956. A solution to the long neglected *Holbrookia lacerata* problem, and the description of two new subspecies of *Holbrookia. Bulletin of the Chicago Academy of Sciences* 10: 163–179.

Baerwald, E. F., J. Edworthy, M. Holder, et al. 2009. A large-scale mitigation experiment to reduce bat fatalities at wind energy facilities. *The Journal of Wildlife Management* 73: 1077–1081.

Bailey, J. A. 1984. *Principles of wildlife management.* John Wiley and Sons, New York.

Bailey, V. 1902. Seven new mammals from western Texas. *Proceedings of the Biological Society of Washington* 15: 117–120.

Bailey, V. 1905. Biological survey of Texas. *North American Fauna* 25: 1–222.

Bailey, V. 1906. A new white-footed mouse from Texas. *Proceedings of the Biological Society of Washington* 19: 57–58.

Bailey, V. 1913. Two new subspecies of North American beavers. *Proceedings of the Biological Society of Washington* 26: 191–194.

Bailey, V. 1932. Mammals of New Mexico. *North American Fauna* 53: 1–412.

Baird, A. B., J. K. Braun, M. A. Mares, et al. 2015. Molecular systematic revision of tree bats (Lasiurini): doubling the native mammals of the Hawaiian Islands. *Journal of Mammalogy* 96: 1255–1274.

Baird, A. B., G. B. Pauly, D. W. Hall, et al. 2009. Records of the porcupine (*Erethizon dorsatum*) from the eastern margins of the Edwards Plateau of Texas. *Texas Journal of Science* 61: 65–66.

Baird, S. F. 1855. Characteristics of some new species of North American Mammalia, collected by the U.S. and Mexican Boundary Survey, Major W. H. Emory, U.S.A., Commissioner. *Proceedings of the Academy of Natural Sciences of Philadelphia* 7: 331–333.

Baird, S. F. 1859. *The Mammals of North America; the descriptions of species based chiefly on the collections in the museum of the Smithsonian Institution.* J. B. Lippincott and Co., Philadelphia, PA.

Baird, S. F., and C. Girard. 1852. Characteristics of some new reptiles in the Museum of the Smithsonian Institution, part 2. *Proceedings of the Academy of Natural Sciences Philadelphia* 6: 125–129.

Baird, S. F., and C. Girard. 1853. *Catalogue of North American reptiles in the Museum of the Smithsonian Institution. Part 1. Serpents.* Smithsonian Institution, Washington, DC.

Baisero, D., P. Visconti, M. Pacifica, et al. 2020. Projected global loss of mammal habitat due to land-use and climate change. *One Earth* 2: 578–585.

Baker, R. H. 1951. Two new moles (genus *Scalopus*) from Mexico and Texas. *Publications of the Kansas Museum of Natural History* 5: 17–24.

Baker, R. H. 1953. The pocket gophers (genus *Thomomys*) of Coahuila, Mexico. University of Kansas *Publications of the Museum of Natural History* 5: 499–514.

Baker, R. H. 1956. Remarks on the former distribution of animals in eastern Texas. *Texas Journal of Science* 3: 356–359.

Baker, R. H. 1977. Mammals of the Chihuahuan Desert region—future prospects. Pp. 221–225 in *Symposium on the biological resources of the Chihuahuan Desert region* (R. H. Waner and D. H. Riskind, eds.). National Park Service, Transactions and Proceedings Series 3, Washington, DC.

Baker, R. H. 1988. Future prospects for the depletion of mammalian populations in the Chihuahuan Desert region. Pp. 71–79 in *Third symposium on resources of the Chihuahuan Desert Region* (A. M Powell, R. R. Hollander, J. C. Barlow, W. B. McGillivray, and D. J. Schmidly, eds.). Chihuahuan Desert Research Institute, Alpine, TX.

Baker, R. H. 1995. Texas wildlife conservation—historical notes. *East Texas Historical Journal* 33: 59–72.

Baker, R. J., and R. D. Bradley. 2006. Speciation in mammals and the genetic species concept. *Journal of Mammalogy* 87(4): 643–662.

Baker, R. J., S. K. Davis, R. D. Bradley, et al. 1989. Ribosomal DNA, mitochondrial DNA, chromosomal and allozymic studies on a contact zone in the pocket gopher, *Geomys*. *Evolution* 43: 63–75.

Baker, R. J., and H. H. Genoways. 1975. A new subspecies of *Geomys bursarius* (Mammalia: Geomyidae) from Texas and New Mexico. *Occasional Papers of the Museum*, Texas Tech University 29: 1–18.

Baker, R. H., and B. P. Glass. 1951. The taxonomic status of the pocket gophers, *Geomys bursarius* and *Geomys breviceps*. *Proceedings of the Biological Society of Washington* 64: 55–58.

Baker, R. J., C. Jones, R. E. Martin, et al. 2007. History of the Texas Society of Mammalogists. *Special Publications of the Museum*, Texas Tech University 52: 1–60.

Baker, R. J., T. Mollhagen, and G. Lopez. 1971. Notes on *Lasiurus ega*. *Journal of Mammalogy* 52: 849–852.

Baker, R. J., J. C. Patton, H. H. Genoways, et al. 1988. Genic studies of *Lasiurus* (Chiroptera: Vespertilionidae). *Occasional Papers of the Museum*, Texas Tech University 117: 1–15.

Baker, R. J., L. W. Robbins, F. B. Stangl, Jr., et al. 1983. Chromosomal evidence for a major subdivision in *Peromyscus leucopus*. *Journal of Mammalogy* 64: 356–359.

Balaskovitz, R. 2017. Texas Panhandle wildfires take lives, burn nearly 500,000 acres. *Amarillo Globe News*, 7 March 2017. https://www.amarillo.com/news/local-news/2017-03-07/texas-panhandle-wildfires-take-lives-burn-nearly-500000-acres.

Bale, R. 2019. What role do pesticides have in killing nature? *National Geographic*, 12 December 2019. https://www.nationalgeographic.com/newsletters/animals/2019/12/what-role-pesticides-killing-nature-december-12.

Ballinger, S. W., L. H. Blankenship, J. W. Bickham, et al. 1992. Allozyme and mitochondrial DNA analysis of a hybrid zone between white-tailed deer and mule deer (*Odocoileus*) in west Texas. *Biochemical Genetics* 30: 1–11.

Bartlett, R. C. 1995. *Saving the best of Texas: A partnership approach to conservation*. University of Texas Press, Austin.

Baughman, J. L. 1951. Texas natural history—one hundred years ago. *Texas Game and Fish* 9(9): 14–16; 9(10): 18–21; 9(11): 6–9.

Baumgardner, G. D., N. O. Dronen, and D. J. Schmidly. 1992. Distributional status of short-tailed shrews (genus *Blarina*) in Texas. *Southwestern Naturalist* 37: 326–330.

Baumgardner, G. D., and D. J. Schmidly. 1981. Systematics of the southern races of two species of kangaroo rats (*Dipodomys compactus* and *D. ordii*). *Occasional Papers of the Museum*, Texas Tech University 73: 1–27.

Beard, D. 2018. 2018 Texas State Bison Herd annual roundup. *Caprock Courier*, 8 February 2018. https://www.caprockcourier.com/2018/02/08/2018-texas-state-bison-herd-annual-roundup/.

Beauchamp-Martin, S. L., F. B. Stangl, Jr., D. J. Schmidly, et al. 2019. Systematic review of Botta's pocket gopher (*Thomomys bottae*) from Texas and southeastern New Mexico, with description of a new taxon. Pp. 515–542 in *From field to laboratory: A memorial volume in honor of Robert J. Baker* (R. D. Bradley, H. H. Genoways, D. J. Schmidly, and L. C. Bradley, eds.). *Special Publications of the Museum*, Texas Tech University 71: xi+1–911.

Beckoff, M. 1977. *Canis latrans. Mammalian Species* 79: 1–9.

Benson, D. L., and F. R. Gehlbach. 1979. Ecological and taxonomic notes on the rice rat (*Oryzomys couesi*) in Texas. *Journal of Mammalogy* 60: 225–228.

Benton, J., and B. Wright. 2017. Economic growth and endangered species management. *Fiscal Notes*, Texas Comptroller of Public Accounts, June/July issue. (PDF downloaded from https://comptroller.texas.gov/economy/fiscal-notes/archive.php).

Bergstrom, B. 2017. Carnivore conservation: Shifting the paradigm from control to coexistence. *Journal of Mammalogy* 98: 1–6.

Biden-Harris Administration Report. 2021. Conserving and restoring America the beautiful. https://

www.doi.gov/sites/doi.gov/files/report-conserving-and-restoring-america-the-beautiful-2021.pdf.

Birney, E. C. 1973. Systematics of three species of woodrats (genus *Neotoma*) in central North America. *Miscellaneous Publications*, University of Kansas Museum of Natural History 58: 1–173.

Birney, E. C., and J. R. Choate, eds. 1994. Seventy-five years of mammalogy (1919–1994). *Special Publications of the American Society of Mammalogists* 11: 1–433.

Blair, W. F. 1939. New mammals from Texas and Oklahoma, with remarks on the status of *Thomomys texensis* Bailey. *Occasional Papers*, University of Michigan Museum of Zoology 403: 1–7.

Blair, W. F. 1940. A contribution to the ecology and faunal relationships of the mammals of the Davis Mountain region, southwestern Texas. *Miscellaneous Publications*, University of Michigan Museum of Zoology 46: 1–39.

Blair, W. F. 1943. Biological and morphological distinctness of a previously undescribed species of the *Peromyscus truei* group from Texas. *Contributions from the Laboratory of Vertebrate Biology*, University of Michigan 24: 1–8.

Blair, W. F. 1950. The biotic provinces of Texas. *Texas Journal of Science* 2: 93–117.

Blair, W. F. 1954a. A melanistic race of the white-throated packrat (*Neotoma albigula*) in Texas. *Journal of Mammalogy* 35: 239–242.

Blair, W. F. 1954b. Mammals of the Mesquite Plains Biotic District in Texas and Oklahoma, and speciation in the central grasslands. *Texas Journal of Science* 6: 235–264.

Blanco, F., J. Calatayud, D. M. Martín-Perea, et al. 2021. Punctuated ecological equilibrium in mammal communities over evolutionary time scales. *Science* 372(6539): 300–303.

Block, S. G., and E. G. Zimmerman. 1991. Allozymic variation and systematics of plains pocket gophers (*Geomys*) in south-central Texas. *The Southwestern Naturalist* 36: 29–36.

Bluett, R. D., M. E. Tewes, and B. C. Thompson. 1989. Geographic distribution of commercial bobcat harvests in Texas, 1978–1986. *Texas Journal of Science* 41: 379–386.

Bogan, M. A. 1998. Changing landscapes of the middle Rio Grande. Pp. 562–563 in *Status and trends of the nation's biological resources*, Vol. 2. (M. J. Mac, P. A. Opler, C. E. Puckett Haecker, and P. D. Doran, eds.). US Department of Interior, US Geological Survey, Reston, VA.

Bogel-Burroughs, N. 2019. Feral hogs attack and kill a woman in Texas. *New York Times*, 26 November 2019. https://www.nytimes.com/2019/11/26/us/texas-woman-killed-feral-hogs.html.

Bohlin, R. G., and E. G. Zimmerman. 1982. Genic differentiation of the chromosomal races of the *Geomys bursarius* complex. *Journal of Mammalogy* 63: 218–228.

Bolen, E. 1998. *Ecology of North America*. John Wiley and Sons, Inc., New York.

Bookhout, T., ed. 2012. Impacts of crude oil and natural gas developments on wildlife and wildlife habitat in the Rocky Mountain region. *Technical Review* 12–02, The Wildlife Society. https://wildlife.org/wp-content/uploads/2014/05/Oil-and-Gas-Technical-Review_2012.pdf.

Borell, A. E. 1937. A new method of collecting bats. *Journal of Mammalogy* 18: 478–480.

Borell, A. E., and M. D. Bryant. 1942. Mammals of the Big Bend area of Texas. *University of California Publications in Zoology* 48: 1–62.

Bowyer, R. T., M. S. Boyce, J. R. Goheen, et al. 2019. Conservation of the world's mammals: Status, protected areas, community efforts, and hunting. *Journal of Mammalogy* 100: 923–941.

Box, T. W. 1996. What is a healthy rangeland, and how would we know one? Pp. 3–10 in *Proceedings of a symposium on sustaining rangeland ecosystems* (W. D. Edge, ed.). Oregon State University Extension Service Special Report 953.

Boyd, R. A., R. C. Dowler, and T. C. Maxwell. 1997. The mammals of Tom Green County, Texas. *Occasional Papers of the Museum*, Texas Tech University 169: 1–27.

Bradley, L. C., B. R. Amman, J. G. Brant, et al. 2005. Mammalogy at Texas Tech University: A historical perspective. *Occasional Papers of the Museum*, Texas Tech University 243: 1–30.

Bradley, L. C., and R. D. Bradley. 2018. Century plants at the Museum put on a once in a lifetime display.

M Magazine, Museum of Texas Tech University, Fall–Winter 2018: 45–47.

Bradley, R. D., L. K. Ammerman, R. J. Baker, et al. 2014. Revised checklist of North American Mammals north of Mexico, 2014. *Occasional Papers of the Museum*, Texas Tech University 327:1–27.

Bradley, R. D., and R. J. Baker. 2001. A test of the genetic species concept: Cytochrome-*b* sequences and mammals. *Journal of Mammalogy* 82: 960–973.

Bradley, R. D., F. C. Bryant, L. C. Bradley, et al. 2003. Characteristics of hybridization between white-tailed deer and mule deer in northwestern Texas. *The Southwestern Naturalist* 48: 654–660.

Bradley, R. D., D. S. Carroll, M. L. Clary, et al. 1999a. Comments on some small mammals from the Big Bend and Trans-Pecos regions of Texas. *Occasional Papers of the Museum*, Texas Tech University 193: 1–6.

Bradley, R. D., S. K. Davis, and R. J. Baker. 1991. Genetic control of premating-isolating behavior: Kaneshiro's hypothesis and asymmetrical sexual selection in pocket gophers. *Journal of Heredity* 82: 192–196.

Bradley, R. D., and R. C. Dowler. 2019. A century of mammal research: changes in research paradigms and emphases. *Journal of Mammalogy* 100: 719–732.

Bradley, R. D., J. Q. Francis, R. N. Platt II, et al. 2019. MtDNA sequence data indicates evidence for multiple species within *Peromyscus maniculatus*. *Special Publications of the Museum*, Texas Tech University 70: 1–59.

Bradley, R. D., D. J. Schmidly, B. R. Amman, et al. 2015. Molecular and morphometric data reveal multiple species in *Peromyscus pectoralis*. *Journal of Mammalogy* 96: 446–459.

Bradley, R. D., D. J. Schmidly, and C. Jones. 1999b. The northern rock mouse, *Peromyscus nasutus* (Mammalia: Rodentia), from the Davis Mountains, Texas. *Occasional Papers of the Museum*, Texas Tech University 190: 1–3.

Bradshaw, K. 2019. Austin's bats are hanging around town longer. *Dallas Morning News*, 19 November 2019.

Brandley, M. C., A. Schmitz, and T. D. Reeder. 2005. Partitioned Bayesian analyses, partition choice, and the phylogenetic relationships of scincid lizards. *Systematic Biology* 54: 373–390.

Brant, J. G., and R. C. Dowler. 2000. A survey of the mammalian fauna of Devils River State Natural Area, Val Verde County, Texas. 18th Annual Meeting, Texas Society of Mammalogists. Abstract.

Brant, J. G., and R. C. Dowler. 2002. Reexamination of the range for the northern pygmy mouse, *Baiomys taylori* (Rodentia: Muridae), in northeastern Texas. *Texas Journal of Science* 54: 189–192.

Bray, W. L. 1906. Distribution and adaptation of the vegetation of Texas. University of Texas Press, Bulletin Number 42, Science Series Number 10: 1–108.

Brennan, R. E., Jr., W. Caire, N. Pugh, et al. 2015. Examination of bats in western Oklahoma for antibodies against *Pseudogymnoascus destructans*, the causitive agent of white-nose syndrome. *The Southwestern Naturalist* 60: 145–150.

Brooks, J. J., R. G. Dvorak, M. Spindler, et al. 2015. Relationship-scale conservation. *Wildlife Society Bulletin* 39: 147–158.

Brown, D. J., M. C. Jones, J. Bell, et al. 2015. Feral hog damage to endangered Houston toad (*Bufo houstonensis*) habitat in the Lost Pines of Texas. *Texas Journal of Science* 64: 73–88.

Brune, G. M. 1975. Major and historical springs of Texas. Texas Water Development Board, *Report* 189: 1–94. https://www.twdb.texas.gov/publications/reports/numbered_reports/doc/R189/R189.pdf.

Brune, G. M. 2002. *Springs of Texas*. Texas A&M University Press, College Station.

Bryant, F. C. 2016. Native fish and wildlife belong to all Texans. *San Antonio Express-News*. https://www.mysanantonio.com/opinion/commentary/article/Native-fish-and-wildlife-belong-to-all-Texans-10793989.php.

Bryant, F. C. 2018. How Texas can protect native species without restricting business growth. *Dallas Morning News*, 15 February 2018. https://www.dallasnews.com/opinion/commentary/2018/02/15/

how-texas-can-protect-native-species-without-restricting-business-growth/.

Bryant, M. D. 1945. Phylogeny of the Nearctic Sciuridae. *American Midland Naturalist* 33: 257–390.

Bull, J. J. 1983. *Evolution of sex determining mechanisms*. Benjamin/Cummings Publishing Company, Menlo Park, CA.

Bull, J. W., E. J. Milner-Gulland, P. F. E. Addison, et al. 2019. Net positive outcomes for nature. *Nature Ecology and Evolution* 4: 4–7.

Burgin, C. J., J. P. Colella, P. L. Kahn, et al. 2018. How many species of mammals are there? *Journal of Mammalogy* 99: 1–14 (with correction published 2019, *Journal of Mammalogy* 100: 615).

Burns, S. 2013. Saltcedar leaf beetles arrive in El Paso Valley. *Southwest Farm Press*, 29 July 2013. https://www.farmprogress.com/management/saltcedar-leaf-beetles-arrive-el-paso-valley.

Burr, J. G. 1949a. Conservation of Texas wildlife began almost a century ago. *Texas Game and Fish* 7(10): 5, 24–25.

Burr, J. G. 1949b. Game abundant when Mearns made survey. *Texas Game and Fish* 7(12): 9, 26.

Burr, J. G. 1949c. Texas teemed with all kinds of wildlife a century ago. *Texas Game and Fish* 7(11): 11, 29–30.

Burt, C. E. 1935. Further records of the ecology and distribution of amphibians and reptiles in the middle west. *American Midland Naturalist* 16: 311–336.

Bush School of Government & Public Service. 2019. Funding Wildlife Conservation in Texas. A report presented to the Boone & Crockett Club. https://bush.tamu.edu/psaa/capstones/2019/Graham%20Capstone%20Report_May%202019-2.pdf.

Butts, G. L. 1979. The status of exotic big game in Texas. *Rangelands* 1: 152–153.

Cameron, G. N., and D. Scheel. 1993. A GIS model of the effects of global climate change on mammals. *Geocartography International* 4: 19–32.

Cameron, G. N., J. M. Williams, and J. A. Robinson. 1997. Analysis of rare resources in Texas: Implications for management. Final report submitted to Endangered Resources Branch, Resource Protection Division, Texas Parks and Wildlife Department, Austin, 15 June 1997.

Campbell, L. 1995. *Endangered and threatened animals of Texas, their life history and management*. Texas Parks and Wildlife Press, Austin.

Campbell, T. A., D. B. Lang, and B. R. Leland. 2010. Feral swine behavior relative to aerial gunning in southern Texas. *Journal of Wildlife Management* 74: 337–341.

Campbell, T. A., S. J. Lapidge, and D. B. Lang. 2006. Using baits to deliver pharmaceuticals to feral swine in southern Texas. *Wildlife Society Bulletin* 34(4): 1184–1189.

Carleton, M. D. 1989. Systematics and evolution. Pp. 7–141 in *Advances in the study of* Peromyscus (Rodentia) (G. L. Kirkland, Jr., and J. N. Layne, eds.), Texas Tech University Press, Lubbock.

Carls, E. G. 1984. Texas natural diversity: the role of parks and reserves. Pp. 51–60 in *Protection of Texas natural diversity: An introduction for natural resource planners and managers*. Texas Agricultural Experiment Station, MP-1557, College Station.

Carmichael, C. 2012. Coordinating an effective response to wildlife diseases. *Wildlife Society Bulletin* 36: 204–206.

Caro, T. 2010. *Conservation by proxy: Indicator, umbrella, keystone, and other surrogate species*. Island Press, Washington, DC.

Carr, S. M., S. W. Ballenger, J. N. Derr, et al. 1986. Mitochondrial DNA analysis of hybridization between sympatric white-tailed deer and mule deer in West Texas. *Proceedings of the National Academy of Sciences* 83: 9576–9580.

Carrasco, L., M. Papeş, K. S. Sheldon, et al. 2021. Global progress in incorporating climate adaptation into land protection for biodiversity since Aichi targets. *Global Change Biology*. DOI:10.1111/gcb.15511.

Carroll, D. S., L. L. Peppers, C. Jones, et al. 2002. *Sigmodon ochrognathus* is a monotypic species: Evidence from DNA sequences. *The Southwestern Naturalist* 47: 494–497.

Cary, M. 1911. A biological survey of Colorado. *North American Fauna* 33: 1–256.

Cary, M. 1917. Life zone investigations in Wyoming. *North American Fauna* 42: 1–95.

Casto, S. D. 1992. Texan contributors to the Mississippi Valley Migration Study of 1884–1885. *Bulletin of the Texas Ornithological Society* 25(2): 51–63.

Cathey, J. C., J. W. Bickham, and J. C. Patton. 1998. Introgressive hybridization and nonconcordant evolutionary history of maternal and paternal lineages in North American deer. *Evolution* 52: 1224–1229.

Ceballos, G., P. R. Ehrlich, A. D. Barnosky, et al. 2015b. Accelerated modern human-induced species losses: entering the sixth mass extinction. *Science Advances* 1(5), e1400253.

Ceballos, G., P. R. Ehrlich, and R. Dirzo. 2017. Biological annihilation via the ongoing sixth mass extinction signaled by vertebrate population losses and declines. *PNAS* 114, e6089–e6096.

Ceballos, G., A. H. Ehrlich, and P. R. Ehrlich. 2015a. *The annihilation of nature*. Johns Hopkins University Press, Baltimore.

Ceballos, G., P. R. Ehrlich, and P. H. Raven. 2020. Vertebrates on the brink of extinction as indicators of biological annihilation and the sixth mass extinction. *PNAS* doi:10.1073/pnas.1922686117.

Center for Biological Diversity. 2019. Endocrine disrupters. https://www.biologicaldiversity.org/campaigns/pesticides_reduction/endocrine_disruptors/index.html.

Cerulli, T. 2013. Paying for state wildlife conservation. *Northern Woodlands Magazine*, 2 October 2013.

Chambers, R. R., P. D. Sudman, and R. D. Bradley. 2009. A phylogenetic assessment of pocket gophers (*Geomys*): Evidence from nuclear and mitochondrial genes. *Journal of Mammalogy* 90: 537–547.

Chapman, B. R., and E. G. Bolen. 2018. *The natural history of Texas*. Texas A&M University Press, College Station.

Chapman, D. C., D. M. Papoulias, and C. P. Onuf. 1998. Environmental change in south Texas. Pp. 268–272, 314 in *Status and trends of the nation's biological resources*, Vol. 1. (M. J. Mac, P. A. Opler, C. E. Puckett Haecker, and P. D. Doran, eds.). US Department of Interior, US Geological Survey, Reston, VA.

Chappell, H. 2010. Valley cat: Ocelot conservation and recovery in South Texas. *Texas Wildlife Magazine*, July 2010. Texas Parks and Wildlife Department, Austin.

Chippindale, P. T., A. H. Price, J. J. Weins, et al. 2000. Phylogenetic relationships and systematic revision of central Texas hemidactyline plethodontid salamanders. *Herpetological Monographs* 14: 1–80.

Chipps, A. S., A. M. Hale, S. P. Weaver, et al. 2020a. Genetic approaches are necessary to accurately understand bat-wind turbine impacts. *Diversity* 2020: 12, 236. https://doi.org/10.3390/d12060236.

Chipps, A. S., A. M. Hale, S. P. Weaver, et al. 2020b. Genetic diversity, population structure, and effective population size in two yellow bat species in south Texas. *PeerJ* 8, e10348. http://doi.org/10.7717/peerj.10348.

Choate, L. L. 1997. The mammals of the Llano Estacado. *Special Publications of the Museum*, Texas Tech University 40: 1–240.

Choate, L. L., J. K. Jones, Jr., R. W. Manning, et al. 1990. Westward ho: Continued dispersal of the pygmy mouse, *Baiomys taylori*, on the Llano Estacado and in adjacent areas of Texas. *Occasional Papers of the Museum*, Texas Tech University 134: 1–8.

Choate, L. L., and F. C. Killebrew. 1991. Distributional records of the California myotis and the prairie vole in the Texas Panhandle. *Texas Journal of Science* 43: 214–215.

Clark, D. R., Jr. 1981. Bats and environmental contamination: a review. *US Fish and Wildlife Service Resource Report* 235: 1–27.

Clemens, L. 2018. Seeds of change. *Texas Nature Conservancy Magazine*, Spring 2018: 55–57.

Cleveland, A. G. 1970. The current geographic distribution of the armadillo in the United States. *Texas Journal of Science* 22: 90–92.

Cleveland, A. G. 1977. First South Texas records of *Pappogeomys castanops*. *Texas Journal of Science* 29:

299.

Cleveland, A. G. 1986. First record of *Baiomys taylori* north of the Red River. *The Southwestern Naturalist* 31: 547.

Collins, J. T. 1991. Viewpoint: A new taxonomic arrangement for some North American amphibians and reptiles. *Herpetological Review* 22: 42–43.

Collins, J., and J. I. Porras. 2004. *Built to last: Successful habits of visionary companies.* 3rd ed. Harper Business, Harper Collins Publisher, New York.

Colorado State University. 2019. Americans' beliefs about wildlife management are changing. *Science Daily*, 25 April 2019. https://www.sciencedaily.com/releases/2019/04/190425143640.htm.

Comptroller.texas.gov. 2017. Economic growth and endangered species management. https://comptroller.texas.gov/economy/fiscal-notes/2017/june-july/endangered.php.

Comptroller.texas.gov. 2019. Candidate conservation agreement with assurances for the dunes sagebrush lizard. https://comptroller.texas.gov/programs/natural-resources/dslccaa/

Conant, R., and J. T. Collins. 1991. *A field guide to reptiles and amphibians of eastern/central North America*, 3rd ed. Houghton Mifflin, Boston/New York.

Condy, P. 2018. Some background on "conservation" in the exotic wildlife ranching industry. Exotic Wildlife Association blog. https://www.myewa.org/blog/background-on-conservation-in-the-exotic-wildlife-ranching-industry/.

Connally, W., ed. 2012. *Texas Conservation Action Plan 2012–2016: Chihuahuan Desert and Arizona–New Mexico Mountains handbook.* Texas Parks and Wildlife Department, Austin.

Conway, W. C., L. M. Smith, and J. F. Bergan. 2002. Avian use of Chinese tallow seeds in coastal Texas. *The Southwestern Naturalist* 47: 550–556.

Conway, W. C., L. M. Smith, R. E. Sosebee, et al. 1999. Total nonstructural carbohydrate trends in Chinese tallow roots. *Journal of Range Management* 52: 539–542.

Cook, R. L. 1994. A historical review of reports, field notes, and correspondence: The desert bighorn sheep in Texas. *Special Report to the Desert Bighorn Sheep Advisory Committee*. Contribution to Federal Aid Project Number W-127-R and W-123-D. Texas Parks and Wildlife Department, Austin.

Cooke, R. S. C., F. Eigenbrod, and E. A. Bates. 2019. Projected losses of global mammal and bird ecological strategies. *Nature Communications* 10, Article 2279.

Corn, J. L., and T. R. Jordan. 2017. Development of the National Feral Swine Map, 1982–2016. *Wildlife Society Bulletin* 41: 758–763.

Cotten, T. B., M. A. Kwiatkowski, D. Saenz, et al. 2012. Effects of an invasive plant, Chinese tallow, *Triadica sebifera*, on development and survival of anuran larvae. *Journal of Herpetology* 46: 186–193.

Cox, J. 1996. Bears at our borders. *Texas Parks and Wildlife Magazine* 54(6): 48–52.

Coyner, B. S., T. E. Lee, Jr., D. S. Rogers, et al. 2010. Taxonomic status and species limits of *Perognathus* (Rodentia: Heteromyidae) in the Southern Great Plains. *The Southwestern Naturalist* 55: 1–10.

Cracraft, J. 1983. Species concepts and speciation analysis. Pp. 159–187 in *Current Ornithology*, Vol. 1 (R. F. Johnston, ed.). Springer, Boston, MA.

Creecy, J. P., W. Caire, and K. A. Gilcrest. 2015. Examination of several Oklahoma bat hibernacula cave soils for *Pseudogymnoascus destructans*, the causative agent of white-nose syndrome. *The Southwestern Naturalist* 60: 213–217.

Crother, B. I., ed. 2012. Scientific and standard English names of amphibians and reptiles of North America north of Mexico, with comments regarding confidence in our understanding. Society for the Study of Amphibians and Reptiles, *Herpetological Circular* 39: 1–92.

Culbertson, K. F. 1974. *Rare, endangered, and peripheral mammals of Texas.* Unpublished MS thesis, Texas A&M University, College Station.

Culver, M., W. E. Johnson, J. Pecon-Slattery, et al. 2000. Genomic ancestry of the American puma (*Puma concolor*). *Journal of Heredity* 91: 186–197.

Cutright, P. R. 1956. *Theodore Roosevelt the naturalist*. Harper and Brothers, New York.

Czech, B. 2000. Economic growth as the limiting factor for wildlife conservation. *Wildlife Society Bulletin* 28: 4–15.

Czech, B., and P. R. Krausman. 1997. Distribution and causation of species endangerment in the United States. *Science* 277: 1116–1117.

Dallas Morning News. 2019. Editorial: African nations are struggling to save their wildlife. Here's how Texas can help. 23 July 2019. https://www.dallasnews.com/opinion/editorials/2019/07/23/african-nations-are-struggling-to-save-their-wildlife-heres-how-texas-can-help/.

Dalquest, W. W. 1954. Netting bats in tropical Mexico. *Transactions of the Kansas Academy of Science* 57: 1–10.

Dalquest, W. W., and N. V. Horner. 1984. *Mammals of north-central Texas*. Midwestern University Press, Wichita Falls, TX.

Darden, S. K., T. Dabelsteen, and S. B. Pedersen. 2003. A potential tool for swift fox (*Vulpes velox*) conservation: Individuality of long-range barking sequences. *Journal of Mammalogy* 84: 1417–1427.

Davis, J. 2019. Chapter 11: Wildlife and pollution (P. Moyle and D. Kelt, eds.). Marinebio.org, marine life, essays on wildlife conservation. https://marinebio.org/creatures/essays-on-wildlife-conservation/12/.

Davis, J. M. 2016. Snake harvest working group final report (January 4, 2016). Texas Parks and Wildlife Department, Austin. https://tpwd.texas.gov/huntwild/wild/wildlife_diversity/nongame/media/TPWD-SHWG-Report.pdf.

Davis, M., S. Faurby, and J. C. Svenning. 2018. Mammal diversity will take millions of years to recover from the current biodiversity crisis. *PNAS* 115: 11262–11267. doi:10.1073/pnas.1804906115.

Davis, W. B. 1940. Mammals of the Guadalupe Mountains of western Texas. *Occasional Papers of the Museum of Zoology*, Louisiana State University 7: 69–84.

Davis, W. B. 1942. The systematic status of four kangaroo rats. *Journal of Mammalogy* 23: 328–333.

Davis, W. B. 1945. Texas skunks. *Texas Game and Fish* 9: 18–21, 31.

Davis, W. B. 1960. The mammals of Texas. *Bulletin 41*: Texas Parks and Wildlife Department, Austin.

Davis, W. B. 1961. Vanished: A commentary on the extinct and threatened mammals of Texas. *Texas Game and Fish*, December: 15–22.

Davis, W. B. 1966. The mammals of Texas. *Bulletin 41*, Texas Parks and Wildlife Department, Austin.

Davis, W. B. 1974. The mammals of Texas. *Bulletin 41*, Texas Parks and Wildlife Department, Austin.

Davis, W. B., and H. K. Buecher. 1946. Pocket gophers (*Thomomys*) of the Davis Mountains, Texas. *Journal of Mammalogy* 27: 265–270.

Davis, W. B., and J. L. Robertson, Jr. 1944. The mammals of Culberson County, Texas. *Journal of Mammalogy* 25: 254–273.

Davis, W. B., and D. J. Schmidly. 1994. *The mammals of Texas*. Texas Parks and Wildlife Press, Austin.

Dawson, B. 2020. More extreme heat and extreme rainfall projected for Texas's 200th birthday. *Texas Climate News Magazine*, 30 March 2020. http://texasclimatenews.org/?p=17577.

Decker, S. K., and L. K. Ammerman. 2020. Phylogeographic analysis reveals mito-nuclear discordance in *Dasypterus intermedius*. *Journal of Mammalogy* 101: 1400–1409.

Demarais, S., J. T. Baccus, and M. S. Traweek, Jr. 1998. Nonindigenous ungulates in Texas: Long-term population trends and possible competitive mechanisms. *Transactions of the North American Wildlife and Natural Resources Conference* 63: 49–55.

Demarais, S., D. A. Osborn, and J. J. Jackley. 1990. Exotic big game: A controversial resource. *Rangelands* 12: 121–125.

Demboski, J. R., and J. A. Cook. 2003. Phylogenetic diversification within the *Sorex cinereus* group (Soricidae). *Journal of Mammalogy* 84: 144–158.

Department of Interior. 2019. Trump Administration improves the implementing regulations of the Endangered Species Act. Press release, 12 August 2019. https://www.doi.gov/pressreleases/

endangered-species-act.

Deutz, A., G. M. Heal, R. Niu, et al. 2020. Financing nature: Closing the global biodiversity financing gap. Published report by The Nature Conservancy, Paulson Institute (Chicago, IL) and the Cornell Atkinson Center for Sustainability (Ithaca, NY).

Diamond, D. D., C. D. True, and K. He. 1997. Regional priorities for conservation of rare species in Texas. *The Southwestern Naturalist* 42: 400–408.

Diamond, J. 2005. *Collapse: How societies choose to fail or succeed*. Viking Penguin Press, New York.

Dice, L. R. 1943. *The biotic provinces of North America*. University of Michigan Press, Ann Arbor.

Diersing, V. E. 1976. An analysis of *Peromyscus difficilis* from the Mexican-United States boundary area. *Proceedings of the Biological Society of Washington* 89: 451–466.

Diersing, V. E., and J. E. Diersing. 1979. Additional records of *Baiomys taylori* (Thomas) in Texas. *The Southwestern Naturalist* 24: 707–708.

Diersing, V. E., and D. F. Hoffmeister. 1974. The rock mouse, *Peromyscus difficilis*, in western Texas. *The Southwestern Naturalist* 19: 213–223.

Dinerstein, E., D. Olson, A. Joshi, et al. 2017. An ecoregion-based approach to protecting half the terrestrial realm. *Bioscience* 67: 534–535.

Dinerstein, E., C. Vynne, E. Sala, et al. 2019. A global deal for nature: Guiding principles, milestones, and targets. *Science Advances* 5: 1–17.

Dinkins, C. (chair). 2000. *Taking care of Texas*. A report from the Governor's Task Force on Conservation. Austin, TX.

Diskin, E. 2019. Italy becomes the first country to make climate change study mandatory in schools. Matador Network. https://matadornetwork.com/read/italy-climate-change-studies/

Dixon, J. R. 2013. *Amphibians and reptiles of Texas*. Texas A&M University Press, College Station.

Dixon, M. 2000. A new bat in Texas. *Texas Parks and Wildlife Magazine*, June 2000. Texas Parks and Wildlife Department, Austin.

Dlouhy, J. 2021. Biden nears methane crackdown dreaded by oil and gas industry. *Albuquerque Journal*, 25 September 2021. https://abqjournal-nm-app.newsmemory.com/?publink=49fe97928_1345f11.

Doan-Crider, D. L., and E. C. Hellgren. 1996. Population characteristics and winter ecology of black bears in Coahuila, Mexico. *Journal of Wildlife Management* 60: 398–407.

Dolan, B. F. 2006. Water development and desert bighorn sheep: Implications for conservation. *Wildlife Society Bulletin* 34: 642–646.

Dorsett, J. 2020. Texas Supreme Court reaffirms public owns state's white-tailed deer. *Texas Agriculture Daily*, Texas Farm Bureau, 10 November 2020. https://texasfarmbureau.org/texas-supreme-court-reaffirms-public-own-states-white-tailed-deer/.

Doughty, R. W. 1983. *Wildlife and man in Texas*. Texas A&M University Press, College Station.

Dowler, R. C. 1989. Cytogenetic studies of three chromosomal races of pocket gophers (*Geomys bursarius* complex) at hybrid zones. *Journal of Mammalogy* 70: 253–266.

Dowler, R. C., R. C. Dawkins, and T. C. Maxwell. 1999. Range extensions for the evening bat (*Nycticeius humeralis*) in West Texas. *Texas Journal of Science* 51: 193–195.

Dowler, R. C., C. E. Ebeling, G. I. Guerra, et al. 2008. The distribution of spotted skunks, genus *Spilogale*, in Texas. *Texas Journal of Science* 60: 321–326.

Dowler, R. C., T. C. Maxwell, and D. S. Marsh. 1992. Noteworthy records of bats from Texas. *Texas Journal of Science* 44: 121–123.

Dragoo, J. W., R. D. Bradley, R. L. Honeycutt, et al. 1993. Phylogenetic relationships among the skunks: a molecular perspective. *Journal of Mammalian Evolution* 1: 255–267.

Dragoo, J. W., J. R. Choate, T. L Yates, et al. 1990. Evolutionary and taxonomic relationships among North American arid-land foxes. *Journal of Mammalogy* 71: 318–332.

Dragoo, J. W., R. L. Honeycutt, and D. J. Schmidly. 2003. Taxonomic status of white-backed hog-nosed skunks, genus *Conepatus* (Carnivora: Mephitidae). *Journal of Mammalogy* 84: 159–176.

Drecs, B. M. 1994. Red imported fire ant predation on nestlings of colonial waterbirds. *Southwestern Entomologist* 19: 355–359.

Drew, M. L., K. M. Rudolph, A. C. S. Ward, et al. 2014. Health status and microbial (Pasturellaceae) flora of free-ranging bighorn sheep following contact with domestic ruminants. *Wildlife Society Bulletin* 38: 332–340.

Duda, M. 2000. *Texans' attitudes toward natural and cultural resources and outdoor recreation in Texas,* completed as part of the Texas Parks and Wildlife for the 21st Century. Prepared for Texas Tech University, Texas Cooperative Fish and Wildlife Research Unit, Lubbock, TX. Prepared by Responsive Management Inc., Virginia.

Duke Energy. 2019. Duke Energy Renewables to use new technology to help protect bats at its wind sites. News Release, 26 June 2019. https://news.duke-energy.com/releases/duke-energy-renewables-to-use-new-technology-to-help-protect-bats-at-its-wind-sites.

Duncan, K. W., S. D. Schemnitz, Z. N. Homesley, et al. 1999. Evolution of salt cedar management, Pecos River, New Mexico. Pp. 63–68 in Fourth symposium on resources of the Chihuahuan Desert region, United States and Mexico, 30 September–1 October 1993 (J. C. Barlow and D. J. Miller, eds.). Chihuahuan Desert Research Institute, Fort Davis, TX.

Duncan, N. P., S. S. Kahl, S. S. Gray, et al. 2016. Pronghorn habitat suitability in the Texas Panhandle. *Journal of Wildlife Management* 80: 1471–1478.

Dunn, C. D., M. R. Mauldin, M. E. Wagley, et al. 2017. Genetic diversity and the possible origin of contemporary elk (*Cervus canadensis*) populations in the Trans-Pecos Region of Texas. *Occasional Papers of the Museum,* Texas Tech University 350: 1–15.

Dzieciolowski, R. M., C. M. H. Clarke, and C. M. Frampton. 1992. Reproductive characteristics of feral pigs in New Zealand. *Acta Theriologica* 37: 259–270.

Earley, L. S. 1996. Safe harbor in the sandhills. *Wildlife in North Carolina,* October, 11–15.

Easterla, D. A. 1970. First records of the spotted bat in Texas and notes on its natural history. *American Midland Naturalist* 83: 306–308.

Easterla, D. A. 1972. Status of *Leptonycteris nivalis* (Phyllostomatidae) in Big Bend National Park, Texas. *The Southwestern Naturalist* 17: 287–292.

Eckstein, G., and J. Snyder. 2013. Endangered species in the oil patch: Challenges and opportunities for the oil and gas industry. *Texas A&M Law Review* 1: 379–409.

Edwards, C.W., C. F. Fulhorst, and R. D. Bradley. 2001. Molecular phylogenetics of the *Neotoma albigula* species group: Further evidence of a paraphyletic assemblage. *Journal of Mammalogy* 82: 267–279.

Egoscue, H. J. 1979. *Vulpes velox. Mammalian Species* 122: 1–5.

Ellis, D., and J. L. Schuster. 1968. Juniper age and distribution on an isolated butte in Garza County, Texas. *The Southwestern Naturalist* 13: 343–348.

Ellison, L. E. 2012. Bats and wind energy: A literature synthesis and annotated bibliography. US Geological Survey, Open-file Report no. 2012–1110. https://pubs.usgs.gov/of/2012/1110/OF12-1110.pdf.

Ellsworth, D. L., R. L. Honeycutt, N. J. Silvy, et al. 1994. White-tailed deer restoration to the southeastern United States: Evaluating genetic variation. *Journal of Wildlife Management* 58: 686–697.

Elrod, D. A., G. A. Heidt, D. M. A. Elrod, et al. 1996. A second species of pocket gopher in Arkansas. *Southwestern Naturalist* 41: 395–398.

Emmons, L. 1988. A field study of ocelots (*Felis pardalis*) in Peru. *Revue d' Écologie* (La Terre et la Vie) 43: 133–157.

EPA [Environmental Protection Agency]. 2016. What climate changes means for Texas. August 2016, EPA 430-F-16-045. https://www.epa.gov/sites/production/files/2016-09/documents/climate-change-tx.pdf.

EPA [Environmental Protection Agency]. 2019. What is Endocrine Disruption? https://www.epa.gov/endocrine-disruption/what-endocrine-disruption.

Erwin, A. 2017. Hybridizing law: A policy for hybridization under the Endangered Species Act. Discussion Paper No. 17-04, *Arizona Legal Studies*, James E. Rogers College of Law, University of Arizona.

Evans, J. 2015. Wild thing: Missing muskrat of the Pecos. *Texas Parks and Wildlife Magazine*, April 2015. Texas Parks and Wildlife Department, Austin.

Fairlie, S. 2019. A convenient untruth. *Resilience*. 10 May 2019. https://www.resilience.org/stories/2019-05-10/a-convenient-untruth/.

Falcone, J. H., P. M. Harveson, M. R. Mauldin, et al. 2019. Taxonomic and conservation status of the Pecos River muskrat. *Occasional Papers of the Museum*, Texas Tech University 359: 1–16.

Fears, D. 2018. Federal judge blasts Fish and Wildlife Service, says endangered wolves cannot be shot. *Washington Post*, 5 November 2018. https://www.washingtonpost.com/science/2018/11/05/federal-judge-blasts-fish-wildlife-service-says-endangered-wolves-cannot-be-shot/?noredirect=on.

Feldhamer, G. A., and W. E. Armstrong. 1993. Interspecific competition between four exotic species and native artiodactyls in the United States. Transactions of the North American Wildlife and Natural Resources Conference 58: 468–478.

Feldpausch, A., and B. Higginbotham. 2006. Texas absentee landowners managing for wildlife: their goals, interests, and education needs. Pp. 92–101 in *Proceedings*, 11th Triennial National Wildlife and Fisheries Extension Conference (R. M. Timm, C. A. Harper, B. J. Higginbotham, and J. A. Parkhurst, eds.). United States Department of Agriculture.

Ferguson, W. 2021. How Texas hunting went exotic. *Texas Monthly*, 21 February 2021. https://www.texasmonthly.com/travel/how-texas-hunting-went-exotic/.

Fernandez, M. 2017. Blood and beauty on a Texas exotic-game ranch. *The New York Times*, 19 October 2017. https://www.nytimes.com/2017/10/19/us/exotic-hunting-texas-ranch.html.

Findley, J. S. 1987. *The natural history of New Mexico mammals*. University of New Mexico Press, Albuquerque.

Findley, J. S., and W. Caire. 1977. The status of mammals in the northern region of the Chihuahuan Desert. Pp. 127–139 in Symposium on the biological resources of the Chihuahuan Desert region (R. H. Waner and D. H. Riskind, eds.). National Park Service, *Transactions and Proceedings Series* 3, Washington, DC.

Findley, J. S., A. H. Harris, D. E. Wilson, et al. 1975. *Mammals of New Mexico*. University of New Mexico Press, Albuquerque.

Fisher, A. K. 1893a. The Death Valley Expedition: A biological survey of parts of California, Nevada, Arizona, and Utah. Part II. Birds. *North American Fauna* 7: 1–158.

Fisher, A. K. 1893b. The hawks and owls of the United States in their relation to agriculture. US Department of Agriculture, *Biological Survey Bulletin* 3.

Fitch, J. H., K.A. Shrump, and A. U. Shump. 1981. *Myotis velifer. Mammalian Species* 149: 1–5.

Fleming, K. M. 1980. Texas bear hunting a thing of the past. *Texas Parks and Wildlife Magazine* 38(5): 12–15.

Flesher, J., and Emily Swanson. 2019. AP-NORC poll: Energy-saving habits vary in popularity. Associated Press, 27 September 2019. http://www.apnorc.org/news-media/Pages/AP-NORC-poll-Energy-saving-habits-vary-in-popularity.aspx

Fonseca, F./Associated Press. 2019. Tiny beetles munch through endangered songbird habitat. *Albuquerque Journal*, 26 July 2019. https://www.abqjournal.com/1345520/tree-eating-beetle-gains-ground-in-us-west-raising-concerns.html

Forbes, S. H., and D. K. Boyd. 1996. Genetic variation of naturally colonizing wolves in the central Rocky Mountains. *Conservation Biology* 10: 1082–1090.

Foreman, D. 2004. *Rewilding North America: A vision for conservation in the 21st century*. Island Press, Washington, DC.

Fowler, N., T. Keitt, O. Schmidt, et al. 2018. Border wall: Bad for diversity. *Frontiers in Ecology and the Environment* 16(3): 137–138.

Francaviglia, R. V. 2000. *The cast iron forest: A natural and cultural history of the North American Cross Timbers*. University of Texas Press, Austin.

Francell, J. 1998. *Conservation easements: A guide for Texas landowners*. Booklet, Texas Parks and Wildlife Department, Austin.

Frey, J. K., and M. L. Campbell. 1997. Introduced populations of fox squirrel (*Sciurus niger*) in the Trans-Pecos and Llano Estacado regions of New Mexico and Texas. *The Southwestern Naturalist* 42: 356–358.

Frey, T. C., and J. K. Frey. 2000. Railroads: An alternative mechanism for mesquite invasion in the southwest. Southwestern Association of Naturalists, 47th Annual Meeting, 20–22 April 2000, Abstract.

Friggens, M. F., D. M. Finch, K. E. Bagne, et al. 2013. Vulnerability of species to climate change in the Southwest: Terrestrial species of the Middle Rio Grande. *General Technical Report* 306. US Department of Agriculture, Rocky Mountain Research Station, Fort Collins, CO. https://www.fs.fed.us/rm/pubs/rmrs_gtr306.pdf.

Fuentes, M. M. B. P., C. J. Limpus, and M. Hamann. 2011. Vulnerability of sea turtle nesting grounds to climate change. *Global Change Biology* 17: 140–153.

Fuhlendorf, S. D. 1997. Why does brush dominate our rangelands? Paper presented at the symposia Brush Sculptors: Innovations for Tailoring Brushy Rangelands to Enhance Wildlife Habitat and Recreational Value. Texas Natural Resources Server, Texas A&M Agrilife Extension. https://texnat.tamu.edu/library/symposia/brush-sculptors-innovations-for-tailoring-brushy-rangelands-to-enhance-wildlife-habitat-and-recreational-value/why-does-brush-dominate-our-rangelands/.

Fuhlendorf, S. D., and F. E. Smeins. 1997. Long-term importance of grazing, fire, and weather patterns on Edwards Plateau vegetative change. In *Proceedings of the Juniper Symposium* (C. A. Taylor, ed.). Texas Agricultural Experiment Station Technical Report 97-1, San Angelo. Texas Natural Resources Server, Texas A&M Agrilife Extension. https://texnat.tamu.edu/library/symposia/juniper-ecology-and-management/long-term-importance-of-grazing-fire-and-weather-patterns-on-edwards-plateau-vegetation-change/.

Fuhlendorf, S. D., F. E. Smeins, and W. E. Grant. 1996. Simulation of a fire-sensitive ecological threshold in a case study of Ashe juniper on the Edwards Plateau of Texas, USA. *Ecological Modelling* 90: 245–255.

Fulbright, T. E., and F. C. Bryant. 2002. *The last great habitat*. Special Publication No. 1. Caesar Kleberg Wildlife Research Institute, Kingsville, TX.

Fuller-Wright, L. 2018. Red wolf DNA found in mysterious Texas canines. *Princeton University News*, 18 December 2018. https://www.princeton.edu/news/2018/12/18/red-wolf-dna-found-mysterious-texas-canines.

Gardner, A. L. 1973. The systematics of the genus *Didelphis* (Marsupialia: Didelphidae) in North and Middle America. *Special Publications of the Museum*, Texas Tech University 4: 1–81.

Gardner, A. L. 2016. The United States Biological Survey: A brief history, 1885–1940. Pp. 1–14 in United States Biological Survey: A compendium of its history, personalities, impacts, and conflicts (D. J. Schmidly, W. E. Tydeman, and A. L. Gardner, eds.). *Special Publications of the Museum*, Texas Tech University 64: 1–123.

Garner, N. P., and S. E. Willis. 1998. Suitability of habitats in East Texas for black bears. Eleventh International Conference on Bear Research and Management, Gatlinburg, TN.

Gaskill, M. 2015. Nature without borders. *Texas Parks and Wildlife Magazine*, April 2015. Texas Parks and Wildlife Department, Austin.

Gehlbach, F. R. 1981. *Mountain Islands and desert seas: A natural history of the U.S. Mexican borderlands.* Texas A&M University Press, College Station.

Geiser, S. W. 1930. Naturalists of the frontier. VIII. Audubon in Texas. *Southwest Review*, 16(1): 108–135.

Geiser, S. W. 1956. William Lloyd, British-American natural-history collector in Texas. *Field and Laboratory* 24(4): 116–122.

Geist, V. 1985. Game ranching: Threat to wildlife conservation in North America. *Wildlife Society Bulletin* 13: 594–598.

Geist, V. 1988. How markets in wildlife meat and parts, and the sale of hunting privileges, jeopardize wildlife conservation. *Conservation Biology* 2: 1–12.

Geist, V. 1989. Legal trafficking and paid hunting threaten conservation. *Transactions of the North American Wildlife and Natural Resources Conference* 54: 171–177.

Geist, V. 1991. Deer ranching for products and paid hunting: Threat to conservation and biodiversity by luxury markets. Pp. 554–561 in *The Biology of Deer* (R. D. Brown, ed.). Springer Verlag, New York.

Geist, V. 1994. Wildlife conservation as wealth. *Nature* 368: 491–492.

Genoways, H. H., and R. J. Baker, eds. 1979. Biological investigations in the Guadalupe Mountains National Park, Texas. *Proceedings and Transactions Series*, National Park Service 4: 1–442.

Genoways, H. H., and R. J. Baker. 1988. *Lasiurus blossevillii* (Chiroptera: Vespertilionidae) in Texas. *Texas Journal of Science* 40: 111–113.

Genoways, H. H., R. J. Baker, and J. E. Cornely. 1979. Mammals of the Guadalupe Mountains National Park, Texas. Pp. 271–332 in Biological Investigations in the Guadalupe Mountains National Park, Texas (H. H. Genoways and R. J. Baker, eds.). National Park Service, *Proceedings and Transactions Series* 4: xviii+1–442.

Genoways, H. H., M. J. Hamilton, D. M. Bell, et al. 2008. Hybrid zones, genetic isolation, and systematics of pocket gophers (genus *Geomys*) in Nebraska. *Journal of Mammalogy* 89: 826–836.

George, S. B., J. R. Choate, and H. H. Genoways. 1981. Distribution and taxonomic status of *Blarina hylophaga* Elliot (Insectivora: Soricidae). *Annals of the Carnegie Museum* 50: 493–513.

Gerber, P. J., H. Steinfeld, B. Henderson, et al. 2013. Tackling climate change through livestock—a global assessment of emissions and mitigation opportunities. Food and Agriculture Organization of the United Nations (FAO), Rome. http://www.fao.org/3/a-i3437e.pdf

Gilad, O., J. E. Janeka, F. Armstrong, et al. 2011. Cougars in the Guadalupe Mountains National Park, Texas: Estimates of occurrence and distribution using analysis of DNA. *The Southwestern Naturalist* 56: 297–304.

Gilliland, C. E., and M. Mays. 2003. Endangered Species Act: A landowner's guide. *Technical Report 1648*, Texas Real Estate Center, Texas A&M University.

Goetze, J. R. 1998. The mammals of the Edwards Plateau, Texas. *Special Publications of the Museum*, Texas Tech University 41: 1–263.

Goetze, J. R., and T. D. Miller. 2015. Two significant records of mammals from the Tamaulipan Biotic Province of Texas. *Texas Journal of Science* 64: 195–201.

Goetze, J. R., A. D. Nelson, D. Breed, et al. 2017. Texas kangaroo rat (*Dipodomys elator*) surveys in Copper Breaks State Park and surrounding areas in Hardeman County, Texas. *Texas Journal of Science* 67: 39–48.

Goetze, J. R., A. D. Nelson, and C. Stasey. 2008. Notes on behavior of the Texas kangaroo rat (*Dipodomys elator*). *Texas Journal of Science* 60: 309–316.

Goetze, J. R., W. C. Stasey, A. D. Nelson, et al. 2007. Mapping burrows of Texas kangaroo rats to examine soil types, vegetation, and estimate population size on an intensely grazed pasture in Wichita County, Texas. *Texas Journal of Science* 59: 11–22.

Golden, K. E., M. N. Peterson, C. S. DePerno, et al. 2013. Factors shaping private landowner engagement in wildlife management. *Wildlife Society Bulletin* 37: 94–100.

Goldman, E. A. 1923. Three new kangaroo rats of the genus *Dipodomys*. *Proceedings of the Biological Society of Washington* 36: 139–142.

Goldman, E. A. 1936. New pocket gophers of the genus *Thomomys*. *Journal of Washington Academy of Science* 26: 111–120.

Goldman, E. A. 1938. Six new rodents from Coahuila and Texas and notes on the status of several described forms. *Proceedings of the Biological Society of Washington* 51: 55–61.

Gonzalez, J. 2019. Nature every day: How to hook kids on the great outdoors. *Nature Conservancy Magazine*, Spring 2019.

Good, D. A. 1994. Species limits in the genus *Gerrhonotus* (Squamata: Anguidae). *Herpetological Monographs* 8: 180–202.

Grandoni, D., and D. Fears. 2021. Biden administration moves to bring back endangered species protections undone under Trump. *Washington Post*, 4 June 2021. https://www.washingtonpost.com/climate-environment/2021/06/04/biden-endangered-species/

Graves, R. A. 2008. When the Earth opens. *Texas Parks and Wildlife Magazine*, January 2008. Texas Parks and Wildlife Department, Austin.

Graves, R. A. 2019a. Bat killer white-nose syndrome may cause the most precipitous wildlife collapse of the past century. *Texas Parks and Wildlife Magazine*, January/February 2019, 31–35. Texas Parks and Wildlife Department, Austin.

Graves, R. A. 2019b. The state of whitetails. *Texas Parks and Wildlife Magazine*, November 2019.

Gray, C. L., S. L. L. Hill, T. Newbold, et al. 2016. Local biodiversity is higher inside than outside terrestrial protected areas worldwide. *Nature Communications* 7, Article 12306.

Green, N. S., and K. T. Wilkins. 2010. Continuing range expansion of the northern pygmy mouse (*Baiomys taylori*) in northeastern Texas. *The Southwestern Naturalist* 55: 288–291.

Greenwald, N., K. F. Suckling, B. Hartl, et al. 2019. Extinction and the U.S. Endangered Species Act. *PeerJ* 7, e6803. doi:10.7717/peerj.6803.

Greshko, M. 2018. First bat removed from U.S. endangered species list. *National Geographic*, 17 April 2018. https://www.nationalgeographic.com/news/2018/04/lesser-long-nosed-bats-conservation-delisted-endangered-animals-spd/.

Grinnell, J. 1919. Four new kangaroo rats from west-central California. *Proceedings of the Biological Society of Washington* 32: 203–206.

Groves, C. P. 2003. Taxonomy of the ungulates of the Indian Subcontinent. *Journal of the Bombay Natural History Society* 100: 341–362.

Guynn, D. E., and D. W. Steinbach. 1987. Wildlife values in Texas. Pp. 117–124 in *Valuing wildlife: Economic and social perspectives* (D. D. Decker and G. Goff, eds.). Westview Press, Colorado.

Hafner, J. C., and M. S. Hafner. 1983. Evolutionary relationships of heteromyid rodents. *Great Basin Naturalist Memoirs* 7: 3–29.

Hafner, D. J., M. S. Hafner, G. L. Hasty, et al. 2008. Evolutionary relationships of pocket gophers (*Cratogeomys castanops* species group) of the Mexican Altiplano. *Journal of Mammalogy* 89: 190–208.

Hafner, D. J., E. Yensen, and G. L. Kirkland Jr. 1998. North American rodents. Status survey and conservation action plan. International Union for Conservation of Nature (IUCN), Gland, Switzerland.

Haggerty, M. M., and M. P. Meuth (eds.). 2015. *Texas Master Naturalist Statewide Curriculum*. Texas A&M University Press, College Station.

Haines, A. M., L. I. Grassman, Jr., M. E. Tewes, et al. 2006a. First ocelot (*Leopardus pardalis*) monitored by GPS telemetry. *European Journal of Wildlife Research* 52: 216–218.

Haines, A. M., J. E. Janecka, M. E. Tewes, et al. 2006b. The importance of private lands for ocelot *Leopardus pardalis* conservation in the United States. *Oryx* 40: 90–94.

Haines, A. M., M. E. Tewes, and L. L. Laack. 2005. Survival and sources of mortality in ocelots. *Journal of Wildlife Management* 69: 255–263.

Haines, A. M., M. E. Tewes, L. L. Laack, et al. 2006c. A habitat-based population viability analysis for ocelots (*Leopardus pardalis*) in the United States. *Biological Conservation* 132: 424–436.

Haines, H. 1963. Geographical extent and duration of the cotton rat, *Sigmodon hispidus*, 1958–1960 fluctuation in Texas. *Ecology* 44: 771–772.

Haines, H. 1971. Characteristics of a cotton rat (*Sigmodon hispidus*) population cycle. *Texas Journal of Science* 23: 3–27.

Halbert, N. D., T. Raudsepp, B. P. Chowdhary, and J. N. Derr. 2004. Conservation genetic analysis of the Texas State Bison Herd. *Journal of Mammalogy* 85: 924–931.

Hall, D. L., and M. R. Willig. 1994. Mammalian species composition, diversity, and succession in Conservation Reserve Program grasslands. *The Southwestern Naturalist* 39: l-10.

Hall, E. R. 1951. American weasels. *University of Kansas Publications, Museum of Natural History* 4: 1–466.

Hall, E. R. 1981. *The mammals of North America*. 2nd ed. John Wiley and Sons, New York.

Hall, E. R., and K. R. Kelson. 1959. *The mammals of North America*. 2 vols. The Ronald Press Company, New York.

Hanson, J. D., J. L. Indorf, V. J. Swier, et al. 2010. Molecular divergence in the *Oryzomys palustris* complex: Evidence for multiple species. *Journal of Mammalogy* 91: 336–347.

Harding, L. E., and F. A. Smith. 2009. *Mustela* or *Vison*? Evidence for the taxonomic status of the American mink and a distinct biogeographic radiation of American weasels. *Molecular Phylogenetics and Evolution* 52: 632–642.

HardingCounty.org. 2019. Canadian River riparian restoration project. http://hardingcounty.org/canadian-river-riparian-restoration-project.html.

Harmel, D. E. 1980. The influence of exotic artiodactyls on white-tailed deer production and survival. Performance Report Job Number 20, Federal Aid Project Number W-109-R-3, Texas Parks and Wildlife Department, Austin.

Harris, D. R. 1966. Recent plant invasions in the arid and semi-arid southwest of the United States. *Annals of Association of American Geographers* 56: 408–422.

Hart, B. J. 1972. Distribution of the pygmy mouse, *Baiomys taylori*, in north central Texas. *The Southwestern Naturalist* 17: 197–216.

Hart, C. R., L. D. White, A. McDonald, et al. 2005. Saltcedar control and water salvage on the Pecos River, Texas, 1999–2003. *Journal of Environmental Management* 75: 399–409.

Harveson, L. A., B. Route, F. Armstrong, et al. 1999. Trends in populations of mountain lion in Carlsbad Caverns and Guadalupe Mountains National Parks. *The Southwestern Naturalist* 44: 490–494.

Harveson, L. A., M. E. Tewes, N. J. Silvy, et al. 1997. Mountain lion research in Texas: Past, present, and future. Pp. 40–43 in *Proceedings of the fifth mountain lion workshop* (W. D. Padley, ed.). California Department of Fish and Game, San Jose.

Harveson, P. M., M. E. Tewes, G. L. Anderson, et al. 2004. Habitat use by ocelots in south Texas: Implications for restoration. *Wildlife Society Bulletin* 32: 948–954.

Hawkes, L. A., A. C. Broderick, M. H. Godfrey, et al. 2007. Investigating the potential impacts of climate change on a marine turtle population. *Global Change Biology* 13: 923–932.

Hayes, M. A. 2013. Bats killed in large numbers at United States wind energy facilities. *BioScience* 63: 975–979.

Heaney, M. R., E. J. Cook, and R. L. Manning. 1998. Noteworthy record of the yellow-nosed cotton rat (*Sigmodon ochrognathus*) from Trans-Pecos Texas. *Texas Journal of Science* 50: 347–349.

Heffelfinger, J. R. 2000. Status of the name *Odocoileus hemionus crooki* (Mammalia: Cervidae). *Proceedings of the Biological Society of Washington* 113: 319–333.

Helgen, K. M., F. R. Cole, L. E. Helgen, et al. 2009. Generic revision in the Holarctic ground squirrel genus *Spermophilus*. *Journal of Mammalogy* 90: 270–305.

Hellgren, E. C., D. P. Onorato, and J. R. Skiles. 2005. Dynamics of black bear populations within a desert metapopulation. *Biological Conservation* 122: 131–140.

Henderson, W. C., and E. A. Preble. 1935. Fiftieth anniversary notes: Work and workers of the first twenty-five years. *The Survey* 16(4–6): 59–65. Bureau of Biological Survey, Washington, DC.

Heppenheimer, E., K. E. Brzeski, R. Wooten, et al. 2018. Rediscovery of red wolf ghost alleles in canid populations along the American Gulf Coast. *Genes* 9(12): 618. doi:10.3390/genes9120618.

Hice, C. L., and D. J. Schmidly. 1999. The non-volant mammals of the Galveston Bay region, Texas. *Occasional Papers of the Museum*, Texas Tech University 194: 1–23.

Hice, C. L., and D. J. Schmidly. 2002. The mammals of coastal Texas: A comparison between mainland and barrier island faunas. *The Southwestern Naturalist* 47: 244–256.

Higginbotham, J. L., L. K. Ammerman, and M. T. Dixon. 1999. First record of *Lasiurus xanthinus* (Chiroptera: Vespertilionidae) in Texas. *The Southwestern Naturalist* 44: 343–347.

Hinesley, L. 1979. Systematics and distribution of two chromosome forms in the southern grasshopper mouse, genus *Onychomys*. *Journal of Mammalogy* 60: 117–128.

Hobbs, R. J., D. A. Saunders, and A. R. Main. 1993. Conservation management in fragmented systems. Pp. 279–296 in *Reintegrating fragmented landscapes: Towards sustainable production and nature conservation* (R. J. Hobbs and D. A. Saunders, eds.). Springer Verlag, New York.

Hodge, L. D. 2003. The King's birthday. *Texas Parks and Wildlife Magazine*, October 2003. Texas Parks and Wildlife Department, Austin.

Höglund-Isaksson, L., A. Gómez-Sanabria, Z. Klimont, et al. 2020. Technical potentials and costs for reducing global anthropogenic methane emissions in the 2050 timeframe—results from the GAINS model. *Environmental Research Communications* 2 (2020): 025044.

Holbrook, J. D., R. W. DeYoung, A. Caso, et al. 2012a. Hog-nosed skunks (*Conepatus leuconotus*) along the Gulf of Mexico: Population status and genetic diversity. *The Southwestern Naturalist* 57: 223–225.

Holbrook, J. D., R. W. DeYoung, J. E. Janecka, et al. 2012b. Genetic diversity, population structure, and movements of mountain lions (*Puma concolor*) in Texas. *Journal of Mammalogy* 93: 989–1000.

Hollander, R. R. 1990. Biosystematics of the yellow-faced pocket gopher, *Cratogeomys castanops* (Rodentia: Geomyidae) in the United States. *Special Publications of the Museum*, Texas Tech University 33: 1–62.

Hollander, R. R., and K. M. Hogan. 1992. Occurrence of the opossum, *Didelphis virginiana* Kerr, in the Trans-Pecos of Texas. *Texas Journal of Science* 44: 127–128.

Hollander, R. R., J. K. Jones, Jr., R. W. Manning, et al. 1987. Noteworthy records of mammals from the Texas panhandle. *Texas Journal of Science* 39: 97–102.

Hollander, R. R., R. N. Robertson, and R. J. Kinncan. 1992. First records of the nutria, *Myocastor coypus*, in the Trans-Pecos region of Texas. *Texas Journal of Science* 44: 119.

Hollister, N. 1914. The spotted tiger cat in Texas. *Proceedings of the Biological Society of Washington* 27: 219.

Holtcamp, W. 2006. Texas wildlife action plan. *Texas Parks and Wildlife Magazine*, March 2006. Texas Parks and Wildlife Department, Austin.

Holtcamp, W. 2008. In search of America's lion. *Texas Parks and Wildlife Magazine*, April 2008. Texas Parks and Wildlife Department, Austin.

Holtcamp, W. 2013. Wildlife warriors. *Texas Parks and Wildlife Magazine*, January/February 2013. Texas Parks and Wildlife Department, Austin.

Honeycutt, R. L., and D. J. Schmidly. 1979. Chromosomal and morphological variation in the plains pocket gopher, *Geomys bursarius*, in Texas and adjacent states. *Occasional Papers of the Museum*, Texas Tech University 58: 1–54.

Hoofer, S. R., and R. A. Van Den Bussche. 2001. Molecular phylogenetics of the plecotine bats and allies based on mitochondrial ribosomal sequences. *Journal of Mammalogy* 82: 131–137.

Hoofer, S. R., and R. A. Van Den Bussche. 2003. Molecular phylogenetics of the chiropteran family Vespertilionidae. *Acta Chiropterologica* 5: 1–63.

Hoofer, S. R., R. A. Van Den Bussche, and I. Horacek. 2006. Generic status of the American pipistrelles

(Vespertilionidae) with a description of a new genus. *Journal of Mammalogy* 87: 981–992.

Horned Lizard Conservation Society. 2019. Horned lizards. www.hornedlizards.org/horned-lizards.html.

Horner, P., and R. Maxey. 1998. East Texas rare bat survey: 1997. Unpublished report, Texas Parks and Wildlife Department, Austin.

Howell, A. H. 1910. Notes on mammals of the Middle Mississippi Valley, with description of a new woodrat. *Proceedings of the Biological Society of Washington* 23: 23–33.

Howell, A. H. 1914. Revision of the American harvest mice (genus *Reithrodontomys*). *North American Fauna* 36: 1–97.

Howell, A. H. 1915. Description of a new genus and seven new races of flying squirrels. *Proceedings of the Biological Society of Washington* 28: 109–114.

Howell, A. H. 1918. Revision of the American flying squirrels. *North American Fauna* 44: 1–64.

Huffaker, B. 2000. The unconquered. *Texas Parks and Wildlife Magazine*, March 2000, 34–39. Texas Parks and Wildlife Department, Austin.

Hume, E. E. 1942. Basil Hicks Dutcher (1871–1922). Pp. 105–129 in *Ornithologists of the United States Army Medical Corps*. John Hopkins Press, Baltimore, MD.

Humphrey, S. R. 1974. Zoogeography of the nine-banded armadillo (*Dasypus novemcinctus*) in the United States. *BioScience* 24: 457–462.

Ilse, L. M., and E. C. Hellgren. 2001. Demographic and behavioral characteristics of North American porcupines (*Erethizon dorsatum*) in pinyon-juniper woodlands of Texas. *American Midland Naturalist* 146: 329–338.

Ingram, W., and W. W. Tanner. 1971. A taxonomic study of *Crotaphytus collaris* between the Rio Grande and Colorado Rivers. *Brigham Young University Science Bulletin* 13: 1–29.

Iracheta, M. 2019. The highest rainfall totals from Tropical Storm Imelda across southeast Texas. *Houston Chronicle*, 23 September 2019. https://www.chron.com/news/houston-texas/houston/article/The-highest-rainfall-totals-Tropical-Storm-Imelda-14461306.php.

Jackson, A. W. 1964. Texotics. *Texas Game and Fish* 23: 7–11.

Jackson, S. J., and D. Sax. 2010. Balancing biodiversity in a changing environment: Extinction debt, immigration credit and species turnover. *Trends in Ecology and Evolution* 25: 153–160.

Jackson, V. L., L. L. Laack, and E. G. Zimmerman. 2005. Landscape metrics associated with habitat use by ocelots in South Texas. *Journal of Wildlife Management* 69: 733–738.

Jacob, J. S., K. Pandian, R. Lopez, et al. 2014. Houston-area freshwater wetland loss, 1992–2010. Texas A&M AgriLife Extension and Sea Grant Texas, ERPT-002, TAMU-SG-14-303.

Jahrsdoerfer, S. E., and D. M. Leslie Jr. 1988. Tamaulipan brushland of the Lower Rio Grande Valley of South Texas: Description, human impacts, and management options. *Biological Report* No. 88(36), US Fish and Wildlife Service, Washington, DC.

Janecka, J. E., M. E. Tewes, I. A. Davis, et al. 2016. Genetic differences in the response to landscape fragmentation by a habitat generalist, the bobcat, and a habitat specialist, the ocelot. *Conservation Genetics* 17: 1093–1108.

Janecka, J. E., M. E. Tewes, L. L. Laack, et al. 2011. Reduced genetic diversity and isolation of remnant populations occupying a severely fragmented landscape in southern Texas. *Animal Conservation* 14: 608–619.

Janecka, J. E., M. E. Tewes, L. Laack, et al. 2014. Loss of genetic diversity among ocelots in the United States during the 20th century linked to human induced population reductions. *PLOS ONE* 9(2): e89384.

Janecka, J. E., M. E. Tewes, L. L. Laack, et al. 2008. Small effective population sizes of two remnant ocelot populations (*Leopardus pardalis albescens*) in the United States. *Conservation Genetics* 9: 869–878.

Janecka, J. E., C. W. Walker, M. E. Tewes, et al. 2007. Phylogenetic relationships of ocelot (*Leopardus pardalis albescens*) populations from the Tamaulipan biotic province and implications for recovery. *The*

Southwestern Naturalist 52: 89–96.

Jesmer, B. R., J. A. Merkle, J. R. Goheen, et al. 2018. Is ungulate migration culturally transmitted? Evidence of social learning from translocated animals. *Science* 361: 1023–1025.

Jester, S. L., C. E. Adams, and J. K. Thomas. 1990. *Commercial trade in Texas nongame wildlife*. Texas Agricultural Experiment Station, Texas A&M University System, College Station.

Jolley, T. W., R. L. Honeycutt, and R. D. Bradley. 2000. Phylogenetic relationships of pocket gophers (genus *Geomys*) based on the mitochondrial 12S rRNA gene. *Journal of Mammalogy* 81: 1025–1034.

Jones, A. K., and S. P. Weaver. 2018. Big free-tailed bat (*Nyctinomops macrotis*) discovered at a wind energy facility in the Lower Rio Grande Valley, Texas. *The Southwestern Naturalist* 63: 75–76.

Jones, C., and M. A. Bogan. 1986. Status report: *Dipodomys elator* Merriam, 1894. Office Endangered Species, US Fish and Wildlife Service, Albuquerque, NM.

Jones, C., M. A. Bogan, and L. M. Mount. 1988. Status of the Texas kangaroo rat (*Dipodomys elator*). *Texas Journal of Science* 40: 249–258.

Jones, C., and R. D. Bradley. 1999. Notes on red bats, *Lasiurus* (Chiroptera: Vespertilionidae), of the Davis Mountains and vicinity, Texas. *Texas Journal of Science* 51: 1–3.

Jones, C., L. Hedges, and K. Bryan. 1999. The western yellow bat, *Lasiurus xanthinus* (Chiroptera: Vespertilionidae) from the Davis Mountains, Texas. *Texas Journal of Science* 51: 267–269.

Jones, G. D., and J. K. Frey. 2008. First record of the gray fox (*Urocyon cinereoargenteus*) on Texas barrier islands. *Texas Journal of Science* 60: 225–227.

Jones, G. D., and J. K. Frey. 2013. Mammals of Padre Island National Seashore, Texas. *Special Publications of the Museum*, Texas Tech University 61: 1–63.

Jones, J. K., Jr. 1993. The concept of threatened and endangered species as applied to Texas mammals. *Texas Journal of Science* 45: 115–128.

Jones, J. K., Jr., R. D. Bradley, and R. J. Baker. 1995. Hybrid pocket gophers and some thoughts on the relationship of natural hybrids to the rules of nomenclature and the Endangered Species Act. *Journal of Mammalogy* 76: 43–49.

Jones, J. K, Jr., and C. Jones. 1992. Revised checklist of Recent land mammals of Texas, with annotations. *Texas Journal of Science* 44: 53–74.

Jones, J. K. Jr, C. J. Jones, and D. J. Schmidly. 1988a. Annotated checklist of Recent land mammals of Texas. *Occasional Papers of the Museum*, Texas Tech University 119: 1–26.

Jones, J. K., Jr., and M. R. Lee. 1962. Three species of mammals from western Texas. *The Southwestern Naturalist* 7: 77–78.

Jones, J. K, Jr., and R. W. Manning. 1989. The northern pygmy mouse, *Baiomys taylori*, on the Texas Llano Estacado. *Texas Journal of Science* 41: 110.

Jones, J. K., Jr., R. W. Manning, C. Jones, et al. 1988b. Mammals of the northern Texas Panhandle. *Occasional Papers of the Museum*, Texas Tech University 126: 1–54.

Judd, F. D. 1970. Geographic variation in the deer mouse, *Peromyscus maniculatus*, on the Llano Estacado. *The Southwestern Naturalist* 14: 261–282.

Kahn, P. H. 2002. Children's affiliation with nature: structure, developments, and the problem of environmental generational amnesia. https://depts.washington.edu/hints/publications/Childrens_Affiliation_Nature.pdf.

Kamler, J. F., W. B. Ballard, E. B. Fish, et al. 2003a. Habitat use, home ranges, and survival of swift foxes in a fragmented landscape: conservation implications. *Journal of Mammalogy* 84: 989–995.

Kamler, J. F., W. B. Ballard, R. L. Gilliland, et al. 2003b. Impacts of coyotes on swift foxes in northwestern Texas. *The Journal of Wildlife Management* 67: 317–323.

Kamler, J. F., W. B. Ballard, R. L. Gilliland, et al. 2003c. Spacial relationships between swift foxes and coyotes in northwestern Texas. *Canadian Journal of Zoology* 81: 168–172.

Katz, B. 2019. Feral pigs are invasive, voracious and resilient. They're also spreading. *Smithsonian*

Magazine, 18 December 2019. https://www.smithsonianmag.com/smart-news/feral-pigs-are-invasiv
e-voracious-and-resilient-theyre-also-spreading-180973824/

Keane, A., J. Light, and J. Evans. 2019. Collaborations with citizen science to update species distributions: Case study with *Ursus americanus* in Texas. Abstract 301, Abstracts of the 99th Annual Meeting of the American Society of Mammalogists.

Kellert, S. R. 1995. Concepts of nature east and west. Pp. 103–121 in *Reinventing nature? Responses to postmodern deconstruction* (M. E. Soulé and G. Lease, eds.). Island Press, Washington, DC.

Kellogg, R. 1946. A century of progress in Smithsonian biology. *Science* 104: 132–141.

Kendeigh, S. C. 1932. A study of Merriam's temperature laws. *Wilson Bulletin* 44: 129–143.

Kennedy, M. L., P. K. Kennedy, M. A. Bogan, et al. 2002. Taxonomic assessment of the black bear (*Ursus americanus*) in the eastern United States. *The Southwestern Naturalist* 47: 335–347.

Kim, K. C., and L. Knutson. 1986. Foundations for a national biological survey. Association of Systematics Collections, Museum of Natural History, University of Kansas, Lawrence.

Kimble, A. 2014. Fossil Rim Wildlife Center celebrates 30 years. News release, 17 May 2014. The Glen Rose Reporter, Glen Rose, Texas.

Kinsey, J. C. 2020. Ecology and management of wild pigs. Texas Parks and Wildlife Department, Austin. https://tpwd.texas.gov/publications/pwdpubs/media/pwd_bk_w7000_1943.pdf.

Kitchener, A. C., C. Breitenmoser-Würsten, E. Eizirik, et al. 2017. A revised taxonomy of the Felidae: The final report of the Cat Classification Task Force of the IUCN/SSC Cat Specialist Group. Cat News, Special Issue 11: 1–80. https://repository.si.edu/handle/10088/32616.

Kjelland, M. E., U. P. Kreuter, G. A. Clendenin, et al. 2007. Factors related to spatial patterns of rural land fragmentation in Texas. *Environmental Management* 40: 231–244.

Klauber, K. M. 1936. A key to the rattlesnakes with summary of characteristics. *Transactions of the San Diego Society of Natural History* 8: 185–276.

Klepper, E. D. 2005. King of the mountain. *Texas Parks and Wildlife Magazine*, June 2005. Texas Parks and Wildlife Department, Austin.

Knox, W. M. 2011. The antler religion. *Wildlife Society Bulletin* 35: 45–48.

Kofalk, H. 1989. *No woman tenderfoot: Florence Merriam Bailey, pioneer naturalist*. Texas A&M University Press, College Station.

Kolbert, E. 2014. *The sixth extinction: An unnatural history*. Picador Henry Holt and Company, New York.

Krausman, P. R., D. J. Schmidly, and E. D. Ables. 1978. Comments on the taxonomic status, distribution, and habitat of the Carmen Mountains white-tailed deer (*Odocoileus virginianus caminis*) in Trans-Pecos Texas. *The Southwestern Naturalist* 23: 577–590.

Krejsa, D. M., S. K. Decker, and L. K. Ammerman. 2020. Noteworthy records of 14 bat species in Texas including the first record of *Leptonycteris yerbabuenae* and the second record of *Myotis occultus*. *Occasional Papers of the Museum*, Texas Tech University 368: 1–10.

Kubatko, L. S., H. L. Gibbs, and E. W. Bloomquist. 2011. Inferring species-level phylogenies and taxonomic distinctiveness using multilocus data in *Sistrurus* rattlesnakes. *Systematic Biology* 60: 393–409.

Kunz, T. H., E. B. Arnett, B. M. Cooper, et al. 2007. Assessing impacts of wind-energy development on nocturnally active birds and bats: a guidance document. *The Journal of Wildlife Management* 71: 2449–2486.

Kuvlesky, W. P., Jr., L. A. Brennan, M. L. Morrison, et al. 2007. Wind energy development and wildlife conservation: Challenges and opportunities. *The Journal of Wildlife Management* 71: 2487–2498.

LaDuc, T. J., M. Terry, M. May, et al. 2019. Editorial: The proposed U.S.-Mexico border wall barrier: The writing is on the wall. *Texas Journal of Science* 7(1), Editorial 2.

Laliberte, A. S., and W. J. Ripple. 2004. Range contractions of North American carnivores and ungulates. *BioScience* 54: 123–128.

Land Trust Alliance. 2018. Farm Bill conservation programs: 2018 Farm Bill. www.landtrustalliance.org/

topics/federal-programs/farm-bill-conservation-programs/.

Lantz, D. E. 1908. The rabbit as a farm and orchard pest. Yearbook US Department of Agriculture for 1907, pp. 329–342, published 27 July 1908. A general account of the relations of rabbits to agriculture in the United States based on data gathered by the Biological Survey. Reprinted and issued as a separate.

Larivière, S., and L. R. Walton. 1998. *Lontra candensis. Mammalian Species* 587: 1–8.

Lasky, J. R., W. Jetz, and T. H. Keitt. 2011. Conservation biogeography of the US-Mexico border: A transcontinental risk assessment of barriers to animal diversity. *Diversity and Distributions* 17: 673–687.

LaVelle, M. J., K. C. Vercauteran, T. J. Hefley, et al. 2011. Evaluation of fences for containing feral swine under simulated depopulation conditions. *The Journal of Wildlife Management* 75: 1200–1208.

Lay, D. W., and T. O'Neil. 1942. Muskrats on the Texas coast. *Journal of Wildlife Management* 6: 301–311.

Leahy, S. 2019. "Off-the-chart" heat to affect millions in U.S. in coming decades. *National Geographic*, 16 July 2019. https://www.nationalgeographic.com/environment/2019/07/extreme-heat-to-affec t-millions-of-americans/.

Lee, D. N., R. S. Pfau, L. K. Ammerman. 2010. Taxonomic status of the Davis Mountains cottontail, *Sylvilagus robustus*, revealed by amplified fragment analysis. *Journal of Mammalogy* 91: 1473–1483.

Lee, T. E, Jr., and M. D. Engstrom. 1991. Genetic variation in the silky pocket mouse (*Perognathus flavus*) in Texas and New Mexico. *Journal of Mammalogy* 72: 273–285.

Lee, T. E., Jr., B. R. Riddle, and P. L. Lee. 1996. Speciation in the desert pocket mouse (*Chaetodipus penicillatus* Woodhouse). *Journal of Mammalogy* 77: 58–68.

Leftwich, T. J. 1977. *Man's past and present impact on the status and distribution of the Texas pronghorn*. Master's thesis, Texas Tech University, Lubbock.

Lehman, V. W. 1969. *Forgotten legions*. Texas Western Press, El Paso.

Leonard, R. I, J. H. Everitt, and F. W. Judd. 1991. Woody plants of the Lower Rio Grande Valley, Texas. *Miscellaneous Publication* 7, Texas Memorial Museum, University of Texas at Austin.

Leslie, D. M. 2016. An international borderland of concern: Conservation of biodiversity in the Lower Rio Grande Valley. *Scientific Investigation Report* 2016-5078, US Geological Survey, Reston, VA.

Levenson, H., R. S. Hoffmann, C. F. Nadler, et al. 1985. Systematics of the Holarctic chipmunks (*Tamias*). *Journal of Mammalogy* 66: 219–242.

Lichtenberg, E. 2018. Conservation programs in the 2018 Farm Bill. American Enterprise Institute, 29 October 2018.

Lidicker, W. Z., Jr. 1960. An analysis of intraspecific variation in the kangaroo rat *Dipodomys merriami. University of California Publications in Zoology* 67: 125–218.

Life Magazine. 1952. Man vs. mesquite. 18 August 1952 edition, *Life Magazine* 33(7): 69–70, 72.

Light, J. E., M. O. Ostroff, and D. J. Hafner. 2016. Phylogeographic assessment of the northern pygmy mouse, *Baiomys taylori. Journal of Mammalogy* 97: 1081–1094.

Lightfoot, S. 2010. The pronghorn prognosis. *Texas Parks and Wildlife Magazine*, September 2010. Texas Parks and Wildlife Department, Austin.

Lightfoot, S. 2012. New challenges for mule deer. *Texas Parks and Wildlife Magazine*, November 2012. Texas Parks and Wildlife Department, Austin.

Lightfoot, S. 2013. Wildlife warriors. *Texas Parks and Wildlife Magazine*, July 2013. Texas Parks and Wildlife Department, Austin.

Lightfoot, S. 2015. Deer sought for disease sampling. *Texas Parks and Wildlife Magazine*, October 2015. Texas Parks and Wildlife Department, Austin.

Locke, S. L., C. E. Brewer, and L. A. Harveson. 2005. Identifying landscapes for desert bighorn sheep translocations in Texas. *Texas Journal of Science* 57: 25–34.

Logan, K. A., and L. L. Sweanor. 2001. *Desert puma: Evolutionary ecology and conservation of an enduring carnivore.* Island Press, Washington, DC.

Lohan, T. 2019. We're just starting to learn how fracking harms wildlife. Ecowatch, 5 October 2019. https://www.ecowatch.com/how-fracking-harms-wildlife-2640821015.html.

Long, M. E. 1998. The vanishing prairie dog. *National Geographic* 193: 116–131.

Longoria, A. 1997. *Adios to the brushlands.* Texas A&M University Press, College Station.

Loomis Austin, Inc. 2000a. Inventory of conservation and recreation land in Texas: A report on the supply of natural resource areas, recreation areas, and historic and cultural resources, completed as part of the Texas Parks and Wildlife for the 21st Century. Prepared for Texas Tech University, Texas Cooperative Fish and Wildlife Research Unit, Lubbock, Texas. Prepared by Loomis Austin, Inc., Austin, TX, with the assistance of Glenrose Engineering, Capitol Environmental Services and the Texas Land Trust Council. LAI project no. 000304.

Loomis Austin, Inc. 2000b. Directions in land conservation and historic preservation: A report on the nationwide strategies and trends in natural and cultural resource protection, completed as part of the Texas Parks and Wildlife for the 21st Century. Prepared for Texas Tech University, Texas Cooperative Fish and Wildlife Research Unit, Lubbock, Texas. Prepared by Loomis Austin, Inc. and Capitol Environmental Services, Austin, TX. LAI project no. 000304.

Loomis Austin, Inc. 2000c. Proceedings of the professional needs analysis conference, completed as part of the Texas Parks and Wildlife for the 21st Century. Prepared for Texas Tech University, Texas Cooperative Fish and Wildlife Research Unit, Lubbock, Texas. LAI project no. 000304.

Loomis, J. A. 1982. *Texas ranchman: The memoirs of John A. Loomis.* Fur Press, Chadron, NE.

Lopez, B. 2001. The naturalist. *Orion,* Autumn 2001: 438–442.

Lopez, R. R. 2014. Status update and trends of Texas rural working lands. Texas A&M Institute of Renewable Natural Resources, *Texas Land Trends* 1(1): 1–14. http://txlandtrends.org/files/lt-2014-report.pdf

Lopez, R. R. 2016. How and why Texas is changing. Texas A&M AgriLife Research Extension, Texas A&M Natural Resources Institute. http://agrilife.org/sanangelo/files/2017/08/Current-Trends-for-Agricultural-Lands-in-Texas-Roel-Lopez.pdf

Loss, S. R., T. Will, and P. P. Marra. 2013. The impact of free-ranging domestic cats on wildlife of the United States. *Nature Communications* 4, Article number 1396.

Lozano, J. 2019. Texas Tech research aims to reduce cattle emissions, save water. lubbockonline.com. 31 March 2019; updated 3 April 2019. https://www.lubbockonline.com/news/20190331/texas-tech-research-aims-to-reduce-cattle-emissions-save-water.

Lund, A. A., L. A. Smith, A. Lopez, et al. 2017. Texas landowner changes and trends. Texas A&M Natural Resources Institute, *Texas Land Trends,* September 2017. http://txlandtrends.org/media/1018/ltchanginglandownerfinal2.pdf.

Luo, H. R., L. M. Smith, B. L. Allen, et al. 1997. Effects of sedimentation on playa wetland volume. *Ecological Applications* 7: 247–252.

Lyon, M. W., Jr. 1904. Classification of the hares and their allies. *Smithsonian Miscellaneous Collections* 45: 416–420.

Malhi, Y. 2017. The concept of the Anthropocene. *Annual Review of Environment and Resources* 42: 25.1–25.28.

Malin, J. C. 1953. Soil, animal, and plant relations of the grassland, historically reconsidered. *Science Monthly* 76: 207–220.

Manfredo, M. J., L. Sullivan, D. Carlos, et al. 2018. America's wildlife values: The social context of wildlife management in the U.S. national report from the research project entitled "America's Wildlife Values." Colorado State University, Department of Human Dimensions of Natural Resources, Fort Collins, CO. https://www.fishwildlife.org/application/files/9915/4049/1625/AWV_-_National_Final_Report.pdf.

Manning, R. W., and C. Jones. 1998. Annotated checklist of Recent land mammals of Texas, 1998. *Occasional Papers of the Museum*, Texas Tech University 182: 1–19.

Manning, R. W., C. Jones, J. K. Jones Jr., et al. 1988. Subspecific status of the pallid bat, *Antrozous pallidus*, in the Texas Panhandle and adjacent areas. *Occasional Papers of the Museum*, Texas Tech University 118: 1–5.

Manning, R. W., C. Jones, and F. D. Yancey. 2008. Annotated checklist of recent land mammals of Texas, 2008. *Occasional Papers of the Museum*, Texas Tech University 278: 1–18.

Manning, R. W., and J. K. Jones Jr. 1988. A specimen of the prairie vole, *Microtus ochrogaster*, from the northern Texas Panhandle. *Texas Journal of Science* 40: 463–464.

Manning, R. W., F. D. Yancey, II, and C. Jones. 1996. Nongeographic variation and natural history of two sympatric species of pocket mice, *Chaetodipus nelsoni* and *C. eremicus*, from Brewster County, Texas. Pp. 191–195 in *Contributions in mammalogy: A memorial volume honoring Dr. J. Knox Jones, Jr.* (H. H. Genoways and R. J. Baker, eds.). Museum of Texas Tech University, Lubbock.

Martin, C. O., and D. J. Schmidly. 1982. Taxonomic review of the pallid bat, *Antrozous pallidus* (Le Conte). *Special Publications of the Museum*, Texas Tech University 18: 1–48.

Martin, R. E., and K. G. Matocha. 1972. Distributional status of the kangaroo rat, *Dipodomys elator*. *Journal of Mammalogy* 53: 873–877.

Martin, T. G., and J. E. M. Watson. 2016. Intact ecosystems provide best defense against climate change. *Nature Climate Change* 6: 122–124.

Mauk, C. L., M. A. Houck, and R. D. Bradley. 1999. Morphometric analyses of seven species of pocket gophers (*Geomys*). *Journal of Mammalogy* 80: 499–511.

Maxwell, T. C. 1979a. *Avifauna of the Concho Valley of west-central Texas with special reference to historical change*. Unpublished PhD diss., Texas A&M University, College Station.

Maxwell, T. C. 1979b. Three men in Texas ornithology. *Bulletin of the Texas Ornithological Society* 12: 2–7.

Mayr, E. 1942. *Systematics and the origin of species, from the viewpoint of a zoologist*. Columbia University Press, New York.

Mayr, E. 1963. *Animal species and evolution*. The Belknap Press, Cambridge, MA.

McAliley, L. R., and P. D. Sudman. 2005. Genetic diversity within the Llano pocket gopher, *Geomys texensis*. *The Southwestern Naturalist* 50: 334–341.

McAlpine, S. 1990. Continued decline of elk populations within Guadalupe Mountains National Park, Texas. *The Southwestern Naturalist* 35: 362–363.

McBee, K., and R. J. Baker. 1982. *Dasypus novemcinctus*. *Mammalian Species* 162: 1–9.

McCain, C. M. 2019. Assess the risks to United States and Canadian mammals caused by climate change using a trait-mediated model. *Journal of Mammalogy* 100(6): 1808–1817.

McCarley, H. 1959. The mammals of eastern Texas. *Texas Journal of Science* 11: 385–426.

McCarley, H. 1962. The taxonomic status of wild *Canis* (Canidae) in the south central United States. *The Southwestern Naturalist* 7: 227–235.

McCarley, H. 1986. Ecology. Pp. 227–242 in *One-hundred years of science and technology in Texas* (L. J. Klosterman, L. S. Swenson, Jr., and S. Rose, eds.). Rice University Press, Houston, TX.

McCarley, H., and C. J. Carley. 1979. Recent changes in the distribution and status of wild red wolves *Canis rufus*. *Endangered Species Report No. 4*, US Fish and Wildlife Service, Albuquerque, NM.

McCleery, D. W. 2011. American forests: A history of resiliency and recovery. *Forest History Society Issue Series*, Durham, NC.

McCoid, M. J., T. H. Fritts, and E. W. Campbell III. 1994. A brown tree snake (Colubridae: *Boiga irregularis*) sighting in Texas. *Texas Journal of Science* 46: 365–368.

McConkey, E. H. 1954. A systematic study of the North American lizards of the genus *Ophisaurus*. *American Midland Naturalists* 51: 133–171.

McCorkle, R. 2011. Wildlife and the wall: What is the impact of the border fence on Texas animals?

Texas Parks and Wildlife Magazine, August 2011. Texas Parks and Wildlife Department, Austin.

McDonald, C. 2011. Wind farms and deadly skies: Turbines on Texas coast killing thousands of birds, bats each year. *San Antonio Express-News*, 27 February 2011. https://www.mysanantonio.com/living_green_sa/article/Wind-farmsand-deadly-skies-1032765.php.

McGee, B. K., W. B. Ballard, K. C. Nicholson, et al. 2006. Effects of artificial escape dens on swift fox populations in Northwest Texas. *Wildlife Society Bulletin* 34: 821–827.

McGinty, A. 1997. Juniper ecology. Accessed via Texas Natural Resources Server, Texas A&M Agrilife Extension. https://texnat.tamu.edu/library/symposia/brush-sculptors-innovations-for-tailoring-brushy-rangelands-to-enhance-wildlife-habitat-and-recreational-value/juniper-ecology/.

McKinney, B. R. 2006. Room to roam. *Texas Parks and Wildlife Magazine*, November 2006. Texas Parks and Wildlife Department, Austin.

McKinney, B. R. 2012. *In the shadow of the Carmens: Afield with a naturalist in the northern Mexican mountains*. Texas Tech University Press, Lubbock.

McKinney, B. R., and J. D. Villalobos. 2004. Overview of the El Carmen Project, Maderas del Carmen, Coahuila, Mexico. Pp. 37–45 in *Proceedings of the Sixth Symposium on the Natural Resources of the Chihuahuan Desert Region*, 14–17 October (C. A. Hoyt and J. Karges, eds.). The Chihuahuan Desert Research Institute, Fort Davis, TX.

McKinney, L. D. 2002. Water for the future. *Texas Parks and Wildlife Magazine*, July 2002. Texas Parks and Wildlife Department, Austin.

McKinney, M. 2002. Urbanization, biodiversity, and conservation. *Bioscience* 52(10): 883–890.

McKinney, M. 2006. Urbanization as a major cause of biotic homogenization. *Biological Conservation* 127: 247–268.

McLain, R. B. 1899. *Critical notes on a collection of reptiles from the western coast of the United States*. Privately published, Wheeling, WV.

Mead, R. A. 1968a. Reproduction in eastern forms of the spotted skunk (genus *Spilogale*). *Journal of Zoology* (London) 156: 119–136.

Mead, R. A. 1968b. Reproduction in western forms of the spotted skunk (genus *Spilogale*). *Journal of Mammalogy* 49: 373–390.

Mearns, E. A. 1896. Preliminary diagnosis of new mammals from the Mexican border of the United States. *Proceedings of the US National Museum* 19: 137–140.

Mearns, E. A. 1907. Mammals of the Mexican Boundary of the United States. *US National Museum Bulletin* 56.

Mecham, J. S. 1956. The relationship between the ringneck snakes *Diadophis regalis* and *D. punctatus*. *Copeia* 1956: 51–52.

Melillo, J., X. Lu, D. W. Kicklighter, et al. 2015. Protected areas' role in climate change mitigation. *Ambio* 45: 133–145. doi:10.1007/s13280-015-0693-1.

Mercure, A., K. Ralls, K.P. Koepfli, et al. 1993. Genetic subdivisions among small canids: Mitochondrial DNA differentiation of swift, kit, and arctic foxes. *Evolution* 47: 1313–1328.

Merriam, C. H. 1877. A review of the birds of Connecticut, with remarks on their habits. *Transactions of the Connecticut Academy* 4: 1–165.

Merriam, C. H. 1889. Descriptions of two new species and one new subspecies of grasshopper mouse. *North American Fauna* 2: 1–6.

Merriam, C. H. 1890a. Descriptions of twenty-six new species of North American mammals. *North American Fauna* 4: 1–34.

Merriam, C. H. 1890b. Results of a biological survey of the San Francisco Mountain region and desert of the Little Colorado, Arizona. *North American Fauna* 3: 1–126.

Merriam, C. H. 1893. Descriptions of eight new ground squirrels of the genera *Spermophilus* and *Tamias* from California, Texas, and Mexico. *Proceedings of the Biological Society of Washington* 8: 129–138.

Merriam, C. H. 1894. Laws of temperature control of the geographic distribution of terrestrial animals and plants. *National Geographic Magazine* 6: 229–238.

Merriam, C. H. 1895. Monographic revision of the pocket gophers, family Geomyidae (exclusive of the species of *Thomomys*). *North American Fauna* 8: 1–258.

Merriam, C. H. 1897. Suggestions for a new method of discriminating between species and subspecies. *Science* 5: 753–758.

Merriam, C. H. 1898a. Descriptions of six new ground squirrels from the western United States. *Proceedings of the Biological Society of Washington* 12: 69–71.

Merriam, C. H. 1898b. Life zones and crop zones of the United States. *Bulletin 10*, Biological Survey, US Department of Agriculture.

Merriam, C. H. 1901. Description of twenty-three new pocket gophers of the genus *Thomomys*. *Proceedings of the Biological Society of Washington* 14: 107–117.

Merriam, C. H. 1919. Criteria for the recognition of species and genera. *Journal of Mammalogy* 1: 6–9.

Merriam, C. H., V. Bailey, E. W. Nelson, et al. 1910. Fourth provisional zone map of North America. Frontispiece in American Ornithologists' Union Check-list, 3rd ed. American Ornithologists' Union, New York.

Michael, E. D., and J. B. Birch. 1967. First Texas record of *Plecotus rafinesquii*. *Journal of Mammalogy* 48: 672.

Middleton, R. 2007. Texotics. *Texas Parks and Wildlife Magazine*, April 2007. Texas Parks and Wildlife Department, Austin.

Miller, A. 2008. Patterns of avian and bat mortality at a utility-sealed wind farm on the southern High Plains. Master's thesis, Texas Tech University, Lubbock.

Miller, G. S. 1912. List of North American land mammals in the United States National Museum, 1911. *Bulletin US National Museum* 79: xiv+1–455.

Miller, G. S., Jr. 1929. Mammalogy and the Smithsonian Institution. *Smithsonian Report* 2995: 391–411.

Milstead, W. W., and D. W. Tinkle. 1958. Notes on the porcupine (*Erethizon dorsatum*) in Texas. *The Southwestern Naturalist* 3: 236–237.

Mittleman, M. B. 1942. A summary of the iguanid genus *Urosaurus*. *Bulletin of the Museum of Comparative Zoology*, Harvard 91: 105–181.

Mittleman, M. B. 1950. The generic status of *Scincus lateralis* Say, 1823. *Herpetological* 6: 17–20.

Mladenoff, D. J., T. A. Sickley, R. G. Haight, et al. 1995. A regional landscape analysis and prediction of favorable gray wolf habitat in the northern Great Lakes region. *Conservation Biology* 9: 279–294.

Morgan, R. E., and W. J. Loughry. 2009. Consequences of exposure to leprosy in a population of wild nine-banded armadillos. *Journal of Mammalogy* 90: 1363–1369.

Moss, S. P., and P. Mehlhop-Cifelli. 1990. Status of the kangaroo rat, *Dipodomys elator* (Heteromyidae), in Oklahoma. *The Southwestern Naturalist* 35: 356–358.

Moulton, D. W., T. E. Dahl, and D. M. Dall. 1997. Texas coastal wetlands: Status and trends, mid-1950s to early 1990s. US Fish and Wildlife Service, Southwestern Region, Albuquerque, NM. https://www.fws.gov/wetlands/documents/Texas-Coastal-Wetlands-Status-and-Trends-mid-1950s-to-early-1990s.pdf.

Mungall, E. C., and W. J. Sheffield. 1994. *Exotics on the open range*. Texas A&M University Press, College Station.

Nadler, C. F., R. S. Hoffman, J. H. Honacki, et al. 1977. Chromosomal evolution in chipmunks, with special emphasis on A and B karyotypes of the subgenus *Neotamias*. *American Midland Naturalist* 98: 343–353.

Nagy, Z., R. Lawson, U. Joger, et al. 2004. Molecular systematics or racers, whipsnakes, and relatives (Reptilia: Colubridae) using mitochondrial and nuclear markers. *Journal of Zoological Systematics and Evolutionary Research* 42: 223–233.

Nalls, A. V., L. K. Ammerman, and R. C. Dowler. 2012. Genetic and morphological variation in the Davis Mountain cottontail (*Sylvilagus robustus*). *The Southwestern Naturalist* 57: 1–7.

National Geographic Kids. 2019. BioBlitz. https://kids.nationalgeographic.com/explore/bioblitz/.

National Research Council. 1970. *Land use and wildlife resources.* National Academy, Washington, DC.

Neiswenter, S. A., and R. C. Dowler. 2005. Habitat use of western spotted skunks and striped skunks in Texas. *Journal of Wildlife Management* 71: 583–586.

Neiswenter, S. A., D. Hafner, J. E. Light, et al. 2019. Phylogeography and taxonomic revision of Nelson's pocket mouse (*Chaetodipus nelsoni*). *Journal of Mammalogy* 100: 1847–1864.

Nelson, A. D., J. R. Goetze, and J. S. Henderson. 2016. Vegetation associated with Texas kangaroo rat (*Dipodomys elator*) burrows in Wichita County, Texas. *Texas Journal of Science* 66: 3–20.

Nelson, A. D., J. R. Goetze, E. Watson, et al. 2009. Changes in vegetation patterns and its effects on Texas kangaroo rats (*Dipodomys elator*). *Texas Journal of Science* 61: 119–130.

Nelson, E. W. 1909. The rabbits of North America. *North American Fauna* 29: 1–314.

Nelson, E. W. 1925. Status of the pronghorned antelope, 1922–1924. *USDA Department Bulletin* 1346. Government Printing Office, Washington, DC.

Nelson, E. W., and E. A. Goldman. 1934. Pocket gophers of the genus *Thomomys* of the Mexican mainland and bordering territory. *Journal of Mammalogy* 15: 105–124.

Nelson, K., R. J. Baker, and R. L. Honeycutt. 1987. Mitochondrial DNA and protein differentiation between hybridizing cytotypes of the white-footed mouse, *Peromyscus leucopus. Evolution* 41: 864–872.

Newmark, W. D. 1987. A land-bridge island perspective on mammalian extinctions in western North American parks. *Nature* 325: 430–432.

Nicholson, K. L., W. B. Ballard, B. K. McGee, et al. 2006. Swift fox use of black-tailed prairie dog towns in northwest Texas. *Journal of Wildlife Management* 70: 1659–1666.

Nogués-Bravo, D., D. Simberloff, C. Rahbek, et al. 2016. Rewilding is the new Pandora's box in conservation. *Current Biology* 26: R87-R91. doi:10.1016/j.cub.2015.12.044.

Noss, R. F., A. P. Dobson, R. Baldwin, et al. 2012. Editorial: Bolder thinking for conservation. *Conservation Biology* 26: 1–4.

NRC [National Research Council]. 1993. *A biological survey for the nation.* National Academy Press, Washington, DC.

NRC [National Research Council]. 1995. *Science and the Endangered Species Act.* National Academy Press, Washington, DC.

Nunley, G. L. 1995. The re-establishment of the coyote in the Edwards Plateau of Texas. Pp. 55–64 in *Symposium Proceedings, Coyotes in the southwest: A compendium of our knowledge* (D. Rollins, C. Richardson, T. Blankenship, K. Canon, and S. Henke, eds.). Texas Parks and Wildlife Department, Austin.

O'Brien, J. M., C. S. O'Brien, C. McCarthy, et al. 2014. Incorporating foray behavior into models estimating contact risk between bighorn sheep and areas occupied by domestic sheep. *Wildlife Society Bulletin* 38: 321–331.

O'Brien, S. J., and E. Mayr. 1991. Bureaucratic mischief: Recognizing endangered species and subspecies. *Science* 251(4998): 1187–1188.

Oberholser, H. C. 1974. *The bird life of Texas* (E. B. Kincaid, Jr., ed.). 2 vols. University of Texas Press, Austin.

Oko, D. 2019. Can Texas bats be saved? *Texas Monthly*, August 2019. https://www.texasmonthly.com/news/texas-lead-way-saving-bats-nationwide/.

Oldfield, M. L. 1989. *The value of conserving genetic resources.* Sinauer Associates, Inc., Sunderland, MA.

Oldham, J. C., and H. M. Smith. 1991. The generic status of the smooth green snake, *Opheodrys vernalis. Bulletin of the Maryland Herpetological Society* 27: 201–215.

Onorato, D. P., and E. C. Hellgren. 2001. Black bear at the border: The recolonization of the Trans-Pecos.

Pp. 245–259 in *Large mammal restoration: Ecological and sociological challenges in the 21st century* (D. S. Maehr, R. F. Noss, and J. L. Larkin, eds.). Island Press, Washington, DC.

Onorato, D. P., E. C. Hellgren, R. A. Van Den Bussche, et al. 2004. Phylogenetic patterns within a metapopulation of black bears (*Ursus americanus*) in the American southwest. *Journal of Mammalogy* 85: 140–147.

Osgood, W. H. 1900. Revision of the pocket mice of the genus *Perognathus*. *North American Fauna* 18: 1–72.

Osgood, W. H. 1909. Revision of the mice of the American genus *Peromyscus*. *North American Fauna* 28: 1–285.

Osgood, W. H. 1925. Ned Hollister. *Journal of Mammalogy* 6: 1–12.

Osgood, W. H. 1943. Clinton Hart Merriam (1855–1942). *Journal of Mammalogy* 24: 421–436.

Osterkamp, W. R., and W. W. Wood. 1987. Playa-lake basins on the Southern High Plains of Texas and New Mexico: I. Hydraulic, geomorphic, and geological development for their development. Geological Society of America, *Bulletin* 99: 215–223.

Outdoor Industry Association. 2019. https://outdoorindustry.org/state/texas/.

Owen, J. G., and D. J. Schmidly. 1986. Environmental variables of biological importance in Texas. *Texas Journal of Science* 38: 99–119.

Owens, M. K. 1997. Mixed brush ecology. Texas Natural Resources Server. https://texnat.tamu.edu/library/symposia/brush-sculptors-innovations-for-tailoring-brushy-rangelands-to-enhance-wildlife-habitat-and-recreational-value/mixed-brush-ecology/.

Pacheco, R. J. 2014. Hooded skunk. Pp. 553–554 in *Mammals of Mexico* (G. Ceballos, ed.). Johns Hopkins University Press, Baltimore, MD.

Packard, J. M., W. Gordon, and J. Clarkson. 2011. Chapter 5, Biodiversity. Pp. 124–156 in *The impact of global warming on Texas* (J. Schmandt, G. R. North, and J. Clarkson, eds.). University of Texas Press, Austin.

Packard, J. M., and D. J. Schmidly. 1991. Graduate training integrating conservation and sustainable development: a role for mammalogists at North American universities. Pp. 392–415 in *Latin American mammalogy: History, biodiversity, and conservation* (M. A. Mares and D. J. Schmidly, eds.). University of Oklahoma Press, Norman.

Packard, R. L. 1960. Speciation and evolution of the pygmy mice, genus *Baiomys*. *University Kansas Publications of the Museum of Natural History* 9: 579–670.

Packard, R. L. 1966. *Myotis austroriparius* in Texas. *Journal of Mammalogy* 47: 128.

Palmer, T. S., and others [*sic*]. 1954. *Biographies of members of the American Ornithologists' Union, reprinted from "The Auk," 1884–1954*. Lord Baltimore Press, Baltimore, MD.

Paniw, M., T. D. James, C. R. Archer, et al. 2021. The myriad of complex demographic responses of terrestrial mammals to climate change and gaps of knowledge: A global analysis. *Journal of Animal Ecology*, 6 April 2021. https://doi.org/10.1111/1365-2656.13467.

Paradiso, J. L. 1965. Recent records of red wolves from the Gulf Coast of Texas. *The Southwestern Naturalist* 10: 318–319.

Paradiso, J. L. 1968. Canids recently collected in East Texas, with comments on the taxonomy of the red wolf. *American Midland Naturalist* 80: 529–534.

Paradiso, J. L., and R. M. Nowak. 1972. A report on the taxonomic status and distribution of the red wolf. US Fish and Wildlife Service, *Special Scientific Report* 145: 1–36.

Parker, W. S. 1982. *Masticophis taeniatus* (Hallowell) striped whipsnake. *Catalogue of American Amphibians and Reptiles* 304: 1–4.

Parlos, J. A., M. A. Madden, L. Siles, et al. 2019. Temporal patterns of bat activity on the High Plains of Texas. Pp. 275–290 in From field to laboratory: A memorial volume in honor of Robert J. Baker (R. D. Bradley, H. H. Genoways, D. J. Schmidly, and L. C. Bradley, eds.). *Special Publications*, Museum of

Texas Tech University 71: xi+1–911.

Pase, H. A., III. 2005. Chinese tallow: A threat to Texas's forests. Fourth of the "dirty dozen." *Texas Forestry Association Newsletter*, October issue, Texas Forest Service.

Patoski, J. N. 2002. Which side of the fence are you on? *Texas Monthly*, February 2002. http://www/texasmonthly.com/the-culture/which-side-of-the-fence-are-you-on/.

Patoski, J. N. 2006. Back in black: With or without a stocking program, the black bear is returning to East Texas. *Texas Parks and Wildlife Magazine*, February 2006. Texas Parks and Wildlife Department, Austin.

Patoski, J. N. 2018. Parks in Peril. Pp. 12–19 in *Texas Observer*, December-January issue. https://www.texasobserver.org/is-texas-overcrowded-underfunded-state-parks-system-being-loved-to-death/.

Patton, R. F. 1974. *Ecological and behavioral relationships of the skunks of Trans-Pecos Texas*. PhD diss., Texas A&M University, College Station.

Peck, R. M. 1982. *A celebration of birds: The life and art of Louis Agassiz Fuertes*. Walker and Co., New York.

Pelton, M. R., A. B. Coley, T. H. Eason, et al. 1999. American black bear conservation action plan. Pp. 144–156 in *Bears: Status survey and conservation action plan* (C. Servheen, S. Herrero, and B. Peyton, eds.). IUCN/SSC Bear Specialist Group, Gland, Switzerland.

Pembleton, E. F., and R. J. Baker. 1978. Studies of a contact zone between chromosomally characterized populations of *Geomys bursarius*. *Journal of Mammalogy* 59: 233–242.

Peppers, L. L., and R. D. Bradley. 2000. Cryptic species in *Sigmodon hispidus*: Evidence from DNA sequences. *Journal of Mammalogy* 81: 332–343.

Peters, R., W. J. Ripple, C. Wolf, et al. 2018. Nature divided, scientists united: US-Mexico border wall threatens biodiversity and binational conservation. *Bioscience* 68(10): 740–743.

Pfau, R. S. 1994. First record of a native American elk (*Cervus elaphus*) from Texas. *Texas Journal of Science* 46: 189–190.

Pfau, R. S., J. R. Goetze, R. E. Martin, et al. 2019. Spatial and temporal genetic diversity of the Texas kangaroo rat, *Dipodomys elator* (Rodentia: Heteromyidae). *Journal of Mammalogy* 100: 1169–1181.

Phillips, C. D., C. A. Henard, and R. S. Pfau. 2007. Amplified fragment length polymorphism and mitochondrial DNA analyses reveal patterns of divergence and hybridization in the hispid cotton rat (*Sigmodon hispidus*). *Journal of Mammalogy* 88: 351–359.

Pile, L. S., N. J. Loewenstein, G. S. Wheeler, et al. 2016. Chinese tallow tree biology and management in southeastern U.S. forest. *Forest Health* SREF-FH-005.

Pittman, M. T., G. J. Guzman, and B. P. McKinney. 2003. Ecology of mountain lion on Big Bend Ranch State Park in the Trans-Pecos region of Texas. P. 74 in *Proceedings of the Sixth Mountain Lion Workshop*, San Antonio, TX, 12–14 December 2000.

Pitts, R. M., Y. Lou, J. W. Bickham, et al. 1999. Range extension for *Geomys breviceps* and *Geomys texensis* (Rodentia: Geomyidae) in Texas. *Texas Journal of Science* 51: 191–193.

Pocock, R. I. 1939. The races of jaguar (*Panthera onca*). *Novitates Zoologicae* 41: 406–422.

Poole, M. W., and R. S. Matlack. 2007. Prairie vole and other small mammals from the Texas Panhandle. *The Southwestern Naturalist* 52: 442–445.

Porter, R. D. 2011. Movements, populations, and habitat preferences of three species of pocket mice (Perognathinae) in the Big Bend region of Texas (C. A. Porter, ed.). *Special Publications of the Museum*, Texas Tech University 58: 1–107.

Pounds, J. A., M. R. Bustamante, L. A. Corona, et al. 2006. Widespread amphibian extinctions from epidemic disease driven by global warming. *Nature* 439: 161–167.

Price, A. 2018. Texas withdraws dunes sagebrush lizard conservation plan. *Austin American-Statesman*, 20 December 2018. https://www.statesman.com/news/20181220/texas-withdraws-dunes-sagebrush-lizard-conservation-plan.

Prüfer, K., F. Racimo, N. Patterson, et al. 2014. The complete genome sequence of a Neanderthal from the Altai Mountains. *Nature* 505(7481): 43–49.

Puko, T. 2020. The environmental battle over the Mexican Border Wall. *Wall Street Journal*, 15 February 2020. https://www.wsj.com/articles/the-environmental-battle-over-the-mexican-border-wall-11581625154.

Rabinowitz, A. R. 1999. The present status of jaguars (*Panthera onca*) in the southwestern United States. *The Southwestern Naturalist* 44: 96–100.

Racine, M. 2000. Hey, hey, it's the monkeys. Pp. 8–13 in *Texas: Houston Chronicle Magazine*. 28 May 2000.

Radchuk, V., T. Reed, C. Teplitsky, et al. 2019. Adaptive responses of animals to climate change are most likely insufficient. *Nature Communications* 10(1): 3109. doi:10.1038/s41467-019-10924-4.

Rands, M. R. W., W. M. Adams, L. Bennun, et al. 2010. Biodiversity conservation: challenges beyond 2010. *Science*, 10 September 2010. doi:10.1126/science.1189138.

Rappole, J. H., C. E. Russell, J. R. Norwine, et al. 1986. Anthropogenic pressures and impacts on marginal, neotropical, semiarid ecosystems: The case of south Texas. *Science of the Total Environment* 55: 91–99.

Rappole, J. H., and A. R. Tipton. 1987. An assessment of potentially endangered mammals of Texas. Unpublished Final Report (Coop. Agreement #14-16-0002-86-927) to US Fish and Wildlife Service, Office of Endangered Species, Albuquerque, NM.

Raun, G. G., and B. J. Wilks. 1961. Noteworthy records of the hog-nosed skunk (*Conepatus*) from Texas. *Texas Journal of Science* 13: 204–205.

Reeder, T. W., C. J. Cole, and H. C. Dessauer. 2002. Phylogenetic relationships of whiptailed lizards of the genus *Cnemidophorus* (Squamata: Teiidae): A test of monophyly, reevaluation of karyotypic evolution, and a review of hybrid origins. *American Museum Novitates* 3365: 1–61.

Reichman, O. J., and R. J. Baker. 1972. Distribution and movements of two species of pocket gophers (Geomyidae) in an area of sympatry in the Davis Mountains, Texas. *Journal of Mammalogy* 53: 21–23.

Reilly, S. M., R. W. Manning, C. C. Nice, et al. 2005. Systematics of isolated populations of short-tailed shrews (Soricidae:*Blarina*) in Texas. *Journal of Mammalogy* 86: 887–894.

Reptile Database. 2019. http://reptile-database.org/.

Rezsutek, M., and G. N. Cameron. 1993. *Mormoops megalophylla*. *Mammalian Species* 448: 1–5.

Richardson, C. 2003. Trans-Pecos vegetation: A historical perspective. Trans-Pecos Wildlife Management Series, Leaflet No. 7.

Riddle, B. R. 1999. Northern grasshopper mouse (*Onychomys leucogaster*). Pp. 588–590 in *The Smithsonian book of North American mammals* (D. E. Wilson and S. Ruff, eds.). Smithsonian Institution Press, Washington, DC.

Riggio, J., J. E. M. Baillie, S. Brumby, et al. 2020. Global human influence maps reveal clear opportunities in conserving Earth's remaining intact terrestrial ecosystems. *Global Change Biology*, 5 June 2020. https://doi.org/10.1111/gcb.15109.

Ripple, W. J., G. Chapron, J. V. López-Bao, et al. 2016. Saving the world's terrestrial megafauna. *Bioscience* 66: 807–812.

Ripple, W. J., C. Wolf, T. M. Newsome, et al. 2019. World scientists warning of a climate emergency. *Bioscience* 70: 8–12. https://doi.org/10.1093/biosci/biz088 and corrigendum: https://doi.org/10.1093/biosci/biz152.

Riskind, D. H. 2015. Unit 4: Ecological regions of Texas. Pp. 125–181 in *Texas Master Naturalist* (M. M. Haggerty and M. P. Meuth, eds.). Texas A&M University Press, College Station.

Robbins, E. 2005. Where the buffalo roam. *Texas Parks and Wildlife Magazine*, September 2005. Texas Parks and Wildlife Department, Austin.

Roberts, E. K., H. Crenshaw, C. D. Dunn, et al. 2015. A record of *Microtus ochrogaster* from the Llano

Estacado and other distributional records of small mammals from Texas. *Occasional Papers of the Museum*, Texas Tech University 329: 1–7.

Roe, R. 2011. At home on the range again. *Texas Parks and Wildlife Magazine*, March 2011. Texas Parks and Wildlife Department, Austin.

Roe, R. 2019. Black market wildlife. *Texas Parks and Wildlife Magazine*, April 2019. Texas Parks and Wildlife Department, Austin.

Rogers, D. S., and D. J. Schmidly. 1981. Geographic variation in the white-throated woodrat (*Neotoma albigula*) from New Mexico, Texas, and northern Mexico. *The Southwestern Naturalist* 26: 167–181.

Roldan, R. 2019. Report: Texas among leading states in cuts to pollution programs. *Austin-American Statesman*, 5 December 2019. https://www.statesman.com/news/20191205/report-texas-among-leadin g-states-in-cuts-to-pollution-programs.

Román-Palacios, C., and J. J. Wiens. 2020. Recent responses to climate change reveal the drivers of species extinction and survival. *Proceedings of the National Academy of Sciences* 117: 4211–4217. https://doi. org/10.1073/pnas.1913007117.

Rosenfeld, D. 2000. Suppression of rain and snow by urban and industrial air pollution. *Science* 287: 1793–1796.

Rosenzweig, M. L. 2003a. *Win-win ecology: How the earth's species can survive in the midst of human enterprise*. Oxford University Press, Oxford, UK.

Rosenzweig, M. L. 2003b. Reconciliation ecology and the future of species diversity. *Oryx* 37: 194–205.

Ruddiman, W. F. 2013. The Anthropocene. *Annual Review of Earth and Planetary Sciences* 41: 45–68.

Ruedas, L. A. 1998. Systematics of *Sylvilagus* Gray, 1867 (Lagomorpha: Leporidae) from southwestern North America. *Journal of Mammalogy* 79: 1355–1378.

Ruedas, L. A., and R. C. Dowler. 2018. *Sylvilagus robustus*. Pp. 154–155 in *Lagomorphs: Pikas, rabbits, and hares of the world* (A. T. Smith, C. H. Johnston, P. C. Alves, and K. Hackländer, eds.). Johns Hopkins University Press, Baltimore, MD.

Russ, W. B. 1997. The status of the mountain lion in Texas. Pp. 69–73 in *Proceedings of the fifth mountain lion workshop* (W. D. Padly, ed.). Department of Fish and Game, San Diego, CA.

Russel, R. J. 1953. Mammals from Cooke County, Texas. *Texas Journal of Science* 5: 454–464.

Ruthven, A. G. 1907. A collection of reptiles and amphibians from southern New Mexico and Arizona. *Bulletin of the American Museum of Natural History* 23: 483–604.

Sadasivam, N. 2018. Dirtying the waters: Texas ranks first in violating water pollution rules. *Texas Observer*, 15 March 2018. https://www.texasobserver.org/dirtying-the-waters-texas-ranks-fir st-in-violating-water-rules/.

Salas, E. A. L, V. A. Seamster, K. G. Boykin, et al. 2017a. Modeling the impact of climate change on species of concern (birds) in south central U.S. based on bioclimatic variables. *AIMS Environmental Science* 4: 358–385.

Salas, E. A. L., V. A. Seamster, N. M. Harings, et al. 2017b. Projected future bioclimatic-envelope suitability for reptiles and amphibian species of concern in south central USA. *Herpetological Conservation and Biology* 12: 522–547.

Samuel, D. E., and B. B. Nelson. 1982. Foxes. Pp. 475–490 in *Wild mammals of North America* (J. A. Chapman and G. A. Feldhammer, eds.). Johns Hopkins University Press, Baltimore, MD.

Samuel, W. M., and S. Demarais. 1993. Conservation challenges concerning wildlife farming and ranching in North America. *Transactions of the North American Wildlife and Natural Resources Conference* 58: 445–447.

Sanderson, E. W., J. P. Beckmann, P. Beier, et al. 2021. The case for reintroduction: The jaguar (*Panthera onca*) in the United States as a model. *Conservation Science and Practice* 2021: e392. doi:10.1111/ csp2.392.

Sansom, A. 1995. *Texas lost: Vanishing heritage*. Parks and Wildlife Foundation of Texas, Inc., Dallas.

Sansom, A. 2008. *Water in Texas, an introduction*. Texas Natural History Guides. University of Texas Press, Austin.

Saunders, D. A., R. J. Hobbs, and C. R. Margules. 1991. Biological consequences of ecosystem fragmentation: a review. *Conservation Biology* 5: 18–32.

Scheel, D., T. L. S. Vincent, and G. N. Cameron. 1996. Global warming and the species richness of bats in Texas. *Conservation Biology* 10: 452–464.

Scheffers, B. R., B. F. Oliveira, I. Lamb, et al. 2019. Global trade across the tree of life. *Science* 366(6461): 71–76.

Schiermeier, Q. 2019. Eat less meat: UN climate-change report calls for change to human diet. *Nature* 572(7769): 291–292.

Schlyer, K. 2018. Borderland refuge saving an endangered ecosystem in the Lower Rio Grande Valley. *American Forests*, Summer 2018: 25–31.

Schmandt, J., G. R. North, and J. Clarkson, eds. 2011. *The impact of global warming on Texas*. 2nd ed. University of Texas Press, Austin.

Schmidly, D. J. 1973a. Geographic variation and taxonomy of *Peromyscus boylii* from Mexico and the southern United States. *Journal of Mammalogy* 54: 111–130.

Schmidly, D. J. 1973b. The systematic status of *Peromyscus comanche*. *The Southwestern Naturalist* 18: 269–278.

Schmidly, D. J. 1977. *The mammals of Trans-Pecos Texas*. Texas A&M University Press, College Station.

Schmidly, D. J. 1983. *Texas mammals east of the Balcones Fault Zone*. Texas A&M University Press, College Station.

Schmidly, D. J. 1984a. Texas mammals: Diversity and geographic distribution. Pp. 13–25 in *Protection of Texas natural diversity: An introduction for natural resource planners and managers*. Texas Agricultural Experiment Station, MP-1557.

Schmidly, D. J. 1984b. The furbearers of Texas. *Texas Parks and Wildlife Department Bulletin* 111: 1–55.

Schmidly, D. J. 1991. *The bats of Texas*. Texas A&M University Press, College Station.

Schmidly, D. J. 1998. Texas natural history: A century of change. Pp. 264–267, 314 in *Status and trends of the nation's biological resources*, Vol. 1 (M. J. Mac, P. A. Opler, C. E. Puckett Haecker, and P. D. Doran, eds.). US Department of Interior, US Geological Survey, Reston, VA.

Schmidly, D. J. 2002. *Texas natural history: A century of change*. Texas Tech University Press, Lubbock.

Schmidly, D. J. 2004. *The mammals of Texas*. 6th ed. University of Texas Press, Austin.

Schmidly, D. J. 2005. What it means to be a naturalist and the future of natural history at American universities. *Journal of Mammalogy* 86: 449–456.

Schmidly, D. J. 2016a. Vernon Bailey (1864–1942): Chief Field Naturalist of the Biological Survey. Pp. 25–54 in United States Biological Survey: A compendium of its history, personalities, impacts, and conflicts (D. J. Schmidly, W. E. Tydeman, and A. L. Gardner, eds.). *Special Publications of the Museum*, Texas Tech University 64: 1–123.

Schmidly, D. J. 2016b. Merriam's men: The federal agents of the Biological Survey (1885–1910). Pp. 55–86 in United States Biological Survey: A compendium of its history, personalities, impacts, and conflicts (D. J. Schmidly, W. E. Tydeman, and A. L. Gardner, eds.). *Special Publications of the Museum*, Texas Tech University 64: 1–123.

Schmidly, D. J. 2018. *Vernon Bailey: Writings of a field naturalist on the frontier*. Texas A&M University Press, College Station.

Schmidly, D. J., and R. D. Bradley. 2016. *The mammals of Texas*. 7th ed. University of Texas Press, Austin.

Schmidly, D. J., and W. A. Brown. 1979. Systematics of short-tailed shrews (genus *Blarina*) in Texas. *The Southwestern Naturalist* 24: 39–48.

Schmidly, D. J., and R. B. Ditton. 1978. Relating human activities and biological resources in riparian habitats in western Texas. Pp. 107–116 in *Strategies for the protection and management of floodplain wetlands and other riparian ecosystems: Proceedings of the symposium* (R. R. Johnson and J. F. McCormick, technical coordinators). US Forest Service General Technical Report WO-12, Washington, DC.

Schmidly, D. J., and J. R. Dixon. 1998. William B. "Doc" Davis: 1902–1995. *Journal of Mammalogy* 79: 1076–1083.

Schmidly, D. J., and F. S. Hendricks. 1976. Systematics of the southern races of Ord's kangaroo rat, *Dipodomys ordii*. *Bulletin of the Southern California Academy of Sciences* 75: 225–237.

Schmidly, D. J., and F. S. Hendricks. 1984. Mammals of the San Carlos Mountains of Tamaulipas, Mexico. Pp. 15–69 in Contributions in mammalogy in honor of Robert L. Packard (R. E. Martin and B. R. Chapman, eds.). *Special Publications of the Museum*, Texas Tech University 22: 1–234.

Schmidly, D. J., and C. Jones. 2000. 20th century changes in mammals and mammalian habitats along the Rio Grande/Rio Bravo from Fort Quitman to Amistad. *Proceedings of the Rio Grande/Rio Bravo Binational Symposium: Fort Quitman to Amistad Reservoir*, 14 June 2000, Ciudad Juarez, Mexico.

Schmidly, D. J., J. Karges, and R. Dean. 2016a. Distribution records and reported sightings of the white-nosed coati (*Nasua narica*) in Texas, with comments on the species' population and conservation status. Pp. 127–146 in Contributions in natural history: A memorial volume in honor of Dr. Clyde Jones. *Special Publications of the Museum*, Texas Tech University 65: 1–273.

Schmidly, D. J., and V. Naples. 2019. North American mammalogy: Early history, dominant personalities, and significant milestones (1850–1960). *Journal of Mammalogy* 100: 701–718.

Schmidly, D. J., N. C. Parker, and R. J. Baker. 2001. *Texas parks and wildlife for the 21st century*. Texas Tech University, Lubbock.

Schmidly, D. J., and J. A. Read. 1986. Cranial variation in the bobcat (*Felis rufus*) from Texas and surrounding states. *Occasional Papers of the Museum*, Texas Tech University 101: 1–39.

Schmidly, D. J., W. E. Tydeman, and A. L. Gardner, eds. 2016b. United States Biological Survey: A compendium of its history, personalities, impacts, and conflicts. *Special Publications of the Museum*, Texas Tech University 64: 1–123.

Schwalm, D. L., W. B. Ballard, E. B. Fish, et al. 2012. Distribution of the swift fox (*Vulpes velox*) in Texas. *The Southwestern Naturalist* 57: 393–398.

Scudday, J. F. 1972. Two recent records of gray wolves in west Texas. *Journal of Mammalogy* 53: 598.

Scudder, C. 2021. Authorities say a Hood County man was killed by a cougar; Texas wildlife experts say it's impossible. *Dallas Morning News*, 23 September 2021.

Shaffer, A. A., R. C. Dowler, J. C. Perkins, et al. 2018. Genetic variation in the eastern spotted skunk (*Spilogale putorius*) with emphasis on the plains spotted skunk (*S. p. interrupta*). *Journal of Mammalogy* 99: 1237–1248.

Shaffer, B. S., B. C. Yates, and B. W. Baker. 1995. An additional record of the native American elk (*Cervus elaphus*) from north Texas. *Texas Journal of Science* 47: 159–160.

Sharpless, M. R., and J. C. Yelderman, Jr. 1993. *The Texas Blackland Prairie land, history, and culture*. Baylor University, Waco, TX.

Shelford, V. E. 1932. Life zones, modern ecology, and the failure of temperature summing. *Wilson Bulletin* 44: 144–157.

Shelford, V. E. 1945. The relative merits of the life zone and biome concepts. *Wilson Bulletin* 57: 248–252.

Shull, A. J. 1988. Endangered and threatened wildlife and plants; determination of endangered status for two long-nosed bats. *Federal Register* 53: 38456–38460.

Simberloff, D. 1998. Flagships, umbrellas, and keystones: Is single species management passé in the landscape era? *Biological Conservation* 83: 247–257.

Simmons, N. B. 2005. Order Chiroptera. Pp. 312–529 in *Mammal species of the World: A taxonomic and geographic reference*, 3rd ed. (D. E. Wilson and D. M. Reeder, eds.). Johns Hopkins University Press, Baltimore, MD.

Simpson, D. C., L. A. Harveson, C. E. Brewer, et al. 2007. Influence of precipitation on pronghorn demography in Texas. *Journal of Wildlife Management* 71: 906–910.

Singal, J. 2014. 'Solution aversion' can help explain why some people don't believe in climate change. *Political Psychology*, 10 November 2014. https://www.thecut.com/2014/11/solution-aversion-can-explain-climate-skeptics.html.

Singhurst, J. R., J. H. Young, G. Kerouas, et al. 2010. Estimating black-tailed prairie dog (*Cynomys ludovicianus*) distribution in Texas. *Texas Journal of Science* 62: 243–262.

Sites, J. W., and J. R. Dixon. 1981. A new subspecies of the iguanid lizard, *Sceloporus grammicus* from northeastern Mexico, with comments on its evolutionary implications and the status of *S. g. disparilis*. *Journal of Herpetology* 15: 59–69.

Skiles, J. R. 1995. Black bears in Big Bend National Park—the Tex-Mex connection. *Proceedings of the Western Black Bear Workshop* 5: 67–73.

Skolos, M. C., T. Deines, J. A. Johnson, et al. 2019. Dunes sagebrush lizard stewardship and West Texas frac sand operations. HiCrush, April 2019 White Paper. https://www.hicrush.com/dunes-sagebrush-lizard-considerations-west-texas-frac-sand-operations/.

Slaughter, B. 1960. Vanquished lords. *Texas Game and Fish* 18(1): 24–25.

Smallwood, K. S. 2013. Comparing bird and bat fatality-rate estimates among North American wind-energy projects. *Wildlife Society Bulletin* 37: 19–33.

Smeins, F. E. 1983. Origin of the brush problem—a geological and ecological perspective of contemporary distributions. Pp. 5–16 in *Proceedings of the Brush Management Symposium* (K. W. McDaniel, ed.), 16 February 1983. Texas Tech University Press, Lubbock.

Smeins, F. E., and S. D. Fuhlendorf. 1997. Biology and ecology of Ashe juniper. In *Proceedings of the 1997 Juniper Symposium* (C. A. Taylor Jr., ed.). Texas A&M University Research and Extension Center Technical Report 97-1, San Angelo.

Smeins, F. E., S. D. Fuhlendorf, and C. A. Taylor Jr. 1997. Environmental land use change: A long-term perspective. Pp. 1.3–1.21 in *Proceedings of the 1997 Juniper Symposium* (C. A. Taylor Jr., ed.). Texas A&M University Research and Extension Center Technical Report 97-1, San Angelo.

Smith, H. M. 1934. Descriptions of new lizards of the genus *Sceloporus* from Mexico and southern United States. *Transactions of the Kansas Academy of Science* 37: 263–285.

Smith, H. M. "1936"[1938]. The lizards of the *torquatus* group of the genus *Sceloporus* Wiegmann, 1828. *University of Kansas Science Bulletin* 224: 539–693.

Smith, H. M. 1946. *Handbook of lizards: Lizards of the United States and Canada*. Comstock, Ithaca, NY.

Smith, H. M., C. M. Eckerman, and H. D. Walley. 2003. The taxonomic status of the Mexican hognose snake *Heterodon kennerlyi* Kennicott (1860). *Journal of Kansas Herpetology* 5: 17–20.

Smith, H. N., and C. A. Rechenthin. 1964. *Grassland restoration: The Texas brush problem*. Soil Conservation Service, USDA, Temple, TX.

Smith, H. M., and E. H. Taylor. 1950. An annotated checklist and key to the reptiles of Mexico exclusive of the snakes. *Bulletin of the United States National Museum* 199: 1–253.

Smith, T., S. A. Smith, and D. J. Schmidly. 1990. Impact of fire ants (*Solenopsis invicta*) density on northern pygmy mice (*Baiomys taylori*). *The Southwestern Naturalist* 35: 158–162.

Smolen, M. J., R. M. Pitts, and J. W. Bickham. 1993. A new subspecies of pocket gopher (*Geomys*) from Texas (Mammalia: Rodentia: Geomyidae). *Proceedings of the Biological Society of Washington* 106: 5–23.

Snow, N. P., J. A. Foster, J. C. Kinsey, et al. 2017. Development of toxic bait to control invasive wild pigs and reduce damage. *Wildlife Society Bulletin* 41: 256–263.

Solberg, D. 2021. Closing the gap: We can stop the extinction crisis, but we have to fully fund conservation first. *Nature Conservancy Magazine*, Summer 2021: 38–43.

Sosebee, R. E. 2010. Mesquite. Handbook of Texas Online. https://tshaonline.org/handbook/online/articles/tpm01.

Soulé, M., and R. Noss. 1998. Rewilding and biodiversity: Complementary goals for continental conservation. *Wild Earth* 8: 19–28.

Soulé, M. E., E. Bolger, A. Alberts, et al. 1988. Reconstructed dynamics of rapid extinctions of chaparral-requiring birds in urban habitat islands. *Conservation Biology* 2: 75–92.

Soulé, M. E., and L. S. Mills. 1998. No need to isolate genetics. *Science* 282: 1658–1659.

Stangl, F. B., Jr. 1986. Aspects of a contact zone between two chromosomal races of *Peromyscus leucopus* (Rodentia: Cricetidae). *Journal of Mammalogy* 67: 465–473.

Stangl, F. B., Jr. 1992a. A new subspecies of the tawny-bellied cotton rat, *Sigmodon fulviventer*, from Trans-Pecos Texas. *Occasional Papers of the Museum*, Texas Tech University 145: 1–4.

Stangl, F. B., Jr. 1992b. First record of *Sigmodon fulviventer* in Texas: Natural history and cytogenetic observations. *The Southwestern Naturalist* 37: 213–214.

Stangl, F. B., Jr., and R. J. Baker. 1984. A chromosomal subdivision in *Peromyscus leucopus*: Implications for the subspecies concept as applied to mammals. Pp. 139–145 in *Festschrift for Walter W. Dalquest in honor of his sixty-sixth birthday* (N. V. Horner, ed.). Midwestern State University, Wichita Falls, TX.

Stangl, F. B., Jr., and C. B. Carr. 1997. Status of *Blarina hylophaga* (Insectivora: Soricidae) in north Texas and southern Oklahoma. *Texas Journal of Science* 49: 159–162.

Stangl, F., B., Jr., A. C. Evans, and R. D. Bradley. 2014. Comments on Late Quaternary ursids from the Texas/Oklahoma southern plains, with documentation of the last known native black bear (*Ursus americanus*) from the Texas Hill Country. *Occasional Papers of the Museum*, Texas Tech University 321: 1–16.

Stangl, F. B., Jr., S. Kasper, and T. S. Schafer. 1989. Noteworthy range extensions and marginal distributional records for five species of Texas mammals. *Texas Journal of Science* 41: 436–437.

Stangl, F. B., Jr., T. S. Schafer, J. R. Goetze, et al. 1992. Opportunistic use of modified and disturbed habitat by the Texas kangaroo rat (*Dipodomys elator*). *Texas Journal of Science* 44: 25–35.

Stanley, J. 2015. Land stewardship. Unit 1. Pp. 15–35 in *Texas Master Naturalist* (M. M. Haggerty and M. P. Meuth, eds.). Texas A&M University Press, College Station.

Stasey, W. C., J. R. Goetze, P. D. Sudman, et al. 2010. Differential use of grazed and ungrazed plots by *Dipodomys elator* (Mammalia: Heteromyidae) in north central Texas. *Texas Journal of Science* 62: 3–14.

Statutes.capitol.texas.gov. 2019. Parks and Wildlife Code. https://statutes.capitol.texas.gov/Docs/SDocs/PARKSANDWILDLIFECODE.pdf

Stein, B. A., N. Edelson, L. Anderson, et al. 2018. Reversing America's wildlife crisis: Securing the future of our fish and wildlife. National Wildlife Federation, Washington, DC. https://www.nwf.org/ReversingWildlifeCrisis.

Steinhart, P. 1994. *Two Eagles/Dos Aguillas: The natural world of the United States-Mexico borderlands*. University of California Press, Berkeley.

Sterling, K. B. 1974. *Last of the naturalists: The career of C. Hart Merriam*. Arno Press, New York.

Sterling, K. B. 1978. Naturalists of the Southwest at the turn of the century. *Environmental Review* 3: 20–33.

Sterling, K. B. 1989. Builders of the U.S. Biological Survey, 1885–1930. *Journal of Forest History* 33: 180–187.

Sterling, K. B. 2016. C. H. Merriam: Pioneering mammalogist. Pp. 15–24 in United States Biological Survey: A compendium of its history, personalities, impacts, and conflicts (D. J. Schmidly, W. E. Tydeman, and A. L. Gardner, eds.). *Special Publications of the Museum*, Texas Tech University 64: 1–123.

Stocks, C. 2019. The methane myth: Why cows aren't responsible for climate change. *Medium*, 6 June 2019. https://medium.com/@caroline.stocks/debunking-the-methane-myth-why-cows-arent-responsible-for-climate-change-23926c63f2c0.

Strecker, J. K. 1926. A check-list of the mammals of Texas, exclusive of the Sirenia and Cetacea. *Baylor Bulletin* 29(3): 1–48.

Strecker, J. K. 1929. Notes on the Texas cotton and Attwater wood rats in Texas. *Journal of Mammalogy* 10: 216–220.

Sudman, P. D., J. K. Wickliffe, P. Horner, et al. 2006. Molecular systematics of pocket gophers of the genus *Geomys*. *Journal of Mammalogy* 87: 668–676.

Swank, W. G., and J. G. Teer. 1989. Status of the jaguar—1987. *Oryx* 23: 14–21.

Sweet, S. S., and W. S. Parker. 1990. *Pituophis melanoleucus* (Daudin). *Catalogue of American Amphibians and Reptiles* 474: 1–8.

Sweetwatertexas.org. 2015. Economic Impact Analysis: Sweetwater Rattlesnake Roundup, 13–15 March 2015. Report prepared by Sarah T. Page Consulting, LLC, for the Sweetwater Chamber of Commerce. https://sweetwatertexas.org/wp-content/uploads/2015/12/Sweetwater%20Rattlesnake%20 Roundup_EIA_report-20154_FINAL.pdf.

Symons, J. 2019. Increased emissions in Texas are canceling out climate progress across the country. *The Hill*. Accessed 8 August 2019. https://thehill.com/opinion/energy-environment/456299-increase d-emissions-in-texas-are-canceling-out-climate-progress.

TAMEST [The Academy of Medicine, Engineering and Science of Texas]. 2017. Environmental and community impacts of shale development in Texas. https://tamest.org/wp-content/uploads/2017/07/ Final-Shale-Task-Force-Report.pdf.

Taulman, J. F., and L. W. Robbins. 2014. Range expansion and distributional limits of the nine-banded armadillo in the United States: An update of Taulman and Robbins (1996). *Journal of Biogeography* 41: 1626–1630.

Taylor, E. H. 1936. A taxonomic study of the cosmopolitan lizards of the genus *Eumeces* with an account of the distribution and relationship of its species. *University of Kansas Science Bulletin* 23: 1–643.

Taylor, R. B. 1999. Black bear status report. Texas Parks and Wildlife Department, Project No. 19, 1–11.

Taylor, W. P., and W. B. Davis. 1947. The mammals of Texas. *Bulletin* 27, Game, Fish, and Oyster Commission, Austin, TX.

Teer, J. G., L. A. Renecker, and R. J. Hudson. 1993. Overview of wildlife farming and ranching in North America. *Transactions of the North American Wildlife and Natural Resources Conference* 58: 448–459.

Tewes, M. E. 1986. *Ecological and behavioral correlates of ocelot spatial patterns*. PhD diss., University of Idaho, Moscow.

Tewes, M. E. 1990. Cat country. *Texas Parks and Wildlife* 48: 4–11.

Tewes, M. E. 2017. Clinging to survival in the borderlands: Ocelots facing dwindling habitat and growing isolation. *The Wildlife Professional*, September–October, 26–27.

Tewes, M. E. 2019. Conservation status of the endangered ocelot in the United States—a 35-year perspective. 37th Annual Faculty Lecture, Texas A&M University, Kingsville, TX.

Tewes, M. E., and D. R. Blanton. 1998. Potential impacts of international bridges on ocelots and jaguarundis along the Rio Grande Wildlife Corridor. Pp. 135–139 in *Proceedings of the International Conference on Wildlife Ecology and Transportation* (G. L. Evink, P. Garrett, D. Zeigler, and J. Berry, eds.). FL-ER-69-98, Florida Department of Transportation, Tallahassee, FL.

Tewes, M. E., and R. W. Hughes. 2001. Ocelot management and conservation along transportation corridors in southern Texas. Pp. 559–564 in *Proceedings of the 2001 International Conference on Ecology and Transportation*, Center for Transportation and the Environment, North Carolina State University, Raleigh.

Texas A&M AgriLife Extension. 2015. Status of wetlands in Texas. https://valuewetlands.tamu. edu/2015/04/10/status-of-wetlands-in-texas/.

Texas A&M AgriLife Research and Extension Center at Amarillo. 2019. Biological control of saltcedar. https://amarillo.tamu.edu/jerry-michels-ph-d/biological-control-saltcedar/

Texas Almanac and State Industrial Guide. 1904. A. H. Belo Corporation, Dallas, TX.

Texas Almanac for 1873 and emigrant's guide to Texas. 1873. Richardson and Company, Galveston, TX.

Texas Almanac. 2018. Lakes and reservoirs. https://texasalmanac.com/topics/environment/lakes-and-reservoirs

Texas Audubon Society. 1997. *Facts about Texas's birds, wildlife and habitat: A Texas briefing guide for policy makers.* Texas Audubon Society, Austin.

Texas Bird Records. 2019. Texas State List. https://www.texasbirdrecordscommittee.org/texas-state-list.

Texas Demographic Center. 2019. https://demographics.texas.gov/.

Texas Department of Agriculture. 2019. Texas Ag Stats. https://www.texasagriculture.gov/About/TexasAgStats.aspx.

Texas Environmental Almanac (2nd ed.). 2000. Compiled by M. Sanger and C. Reed, Texas Center for Policy Studies. University of Texas Press, Austin.

Texas Invasive Species Institute. 2019a. About us. http://stoppinginvasives.org/home/about/.

Texas Invasive Species Institute. 2019b. What are invasive species? http://www.tsusinvasives.org/home/invasives-101/.

Texas Land Trends. 2019. Changes in Texas working lands. 2019 Texas Land Trends Report Summary. https://nri.tamu.edu/media/2418/txlandtrends2019_factsheet.pdf.

Texas Parks and Wildlife Magazine. 2000. July issue, pp. 18–25. Texas Parks and Wildlife Department, Austin.

Texas Parks and Wildlife Magazine. 2020. White-nose syndrome reaches Texas bat colonies. October issue. https://tpwmagazine.com/archive/2020/oct/scout6_conservation/.

The Nature Conservancy. 1999. *A landowner's guide to conservation options.* Pamphlet, The Nature Conservancy, Arlington, VA.

The Nature Conservancy. 2019. The Nature Conservancy in Texas. https://www.nature.org/en-us/about-us/where-we-work/united-states/texas/.

The Texas Politics Project. 2019. University of Texas/Texas Tribune polls on climate change, October 2019. https://texaspolitics.utexas.edu/polling/search/topic/climate-change-620/year/2019/month/10.

The Wildlife Society. 2003. The relationship of economic growth to wildlife conservation. *Technical Review* 03-1.

The Wildlife Society. 2007. Final TWS position statement: The North American model of wildlife conservation. The Wildlife Society, Washington, DC. http://joomla.wildlife.org/documents/position-statements/41-NAModel%20Position%20Statementfinal.pdf.

The Wildlife Society. 2019. Recovering America's Wildlife Act. http://www.wildlife.org/policy/recovery-americas-wildlife-act.

Thorne, E. T., and E. S. Williams. 1988. Disease and endangered species: The black-footed ferret as a recent example. *Conservation Biology* 2: 66–74.

Timmons, J. B., B. Higginbotham, R. Lopez, et al. 2012. Feral hog population growth, density and harvest in Texas. SP-472, Texas A&M AgriLife Extension Service. https://wildpigs.nri.tamu.edu/media/1155/sp-472-feral-hog-population-growth-density-and-harvest-in-texas-edited.pdf.

Tiner, R. W., Jr. 1984. *Wetlands of the United States: Current status and recent trends.* US Department of the Interior, Fish and Wildlife Service National Wetlands Inventory, Washington, DC.

Tobias, J. 2019. Trump administration authorizes 'cyanide bombs' to kill wild animals. *The Guardian,* 8 August 2019. https://www.theguardian.com/environment/2019/aug/08/trump-authorizes-cyanide-bombs-wildlife-services.

Todd, D., and J. Ogren. 2016. *The Texas Landscape Project.* Texas A&M University Press, College Station.

Todd, D., and D. Weisman (eds.). 2010. *The Texas Legacy Project: Stories of courage and conservation.* Texas A&M University Press, College Station.

TPWD [Texas Parks and Wildlife Department]. 2012. Texas conservation action plan, state/multi-region handbook. https://tpwd.texas.gov/landwater/land/tcap/documents/tcap_statewide_multiregion_handbook.pdf.

TPWD [Texas Parks and Wildlife Department]. 2019a. Map 30. https://tpwd.texas.gov/publications/ pwdpubs/pwd_pl_w7000_1187a/media/30.pdf.

TPWD [Texas Parks and Wildlife Department]. 2019b. Texas blind salamander (*Eurycea rathbuni*). https://tpwd.texas.gov/huntwild/wild/species/blindsal/.

TPWD [Texas Parks and Wildlife Department]. 2019c. Texas Conservation Action Plan: Species of greatest conservation need. https://tpwd.texas.gov/landwater/land/tcap/sgcn.phtml.

TPWD [Texas Parks and Wildlife Department] Nature Trackers. 2019. Nature Trackers: Get involved. https://tpwd.texas.gov/huntwild/wild/wildlife_diversity/texas_nature_trackers/get_involved.phtml.

TPWD News. 2019a. Fungus causing white-nose syndrome in bats continues to spread in Texas. News Release, 8 May 2019. https://tpwd.texas.gov/newsmedia/releases/?req=20190508a

TPWD News. 2019b. Game-changing, bipartisan wildlife legislation introduced in Congress. News release, 15 July 2019. Texas Parks and Wildlife Department. https://tpwd.texas.gov/newsmedia/ releases/?req=20190715a&nrtype=all&nrspan=2019&nrsearch=.

TPWD News. 2019c. Val Verde County white-tailed deer tests positive for chronic wasting disease. News release, 19 December 2019. https://tpwd.texas.gov/newsmedia/releases/?req=20191219a.

TPWD News. 2020a. Chronic wasting disease discovered at deer breeding facility in Kimble County. News release, 27 February 2020. https://tpwd.texas.gov/newsmedia/releases/?req=20200227a&utm_ campaign=govdelivery-email&utm_medium=email&utm_source=govdelivery.

TPWD News. 2020b. White-nose syndrome confirmed in bat in Texas. News release, 5 March 2020. https://tpwd.texas.gov/newsmedia/releases/?req=20200305a.

TPWD News. 2020c. Bear conservation efforts prove successful as sightings increase in northeast Texas. News release, 30 June 2020. https://tpwd.texas.gov/newsmedia/releases/?req=20200630a.

TPWD News. 2021a. Chronic wasting disease discovered in Lubbock County. News release, 1 March 2021. https://tpwd.texas.gov/newsmedia/releases/?req=20210301b.

TPWD News. 2021b. New case of rabbit hemorrhagic disease confirmed in Texas, first of 2021. News release, 9 April 2021. https://tpwd.texas.gov/newsmedia/releases/?req=20210409b.

TPWD News. 2021c. Chronic wasting disease discovered at deer breeding facilities in Matagorda and Mason counties. News release, 14 May 2021. https://tpwd.texas.gov/newsmedia/ releases/?req=20210514a.

TPWD News. 2021d. TPWD updates list of Texas species of great conservation need. News release, 20 May 2021. https://tpwd.texas.gov/newsmedia/releases/?req=20210520a.

Traweek, M. S. 1995. Statewide census of exotic big game animals. *Progress Report*, Federal Aid Project W-127-R-3 No. 21, Texas Parks and Wildlife Department, Austin.

Tropicos. 2019. https://www.tropicos.org/Home.aspx.

Truett, J. C., and D. W. Lay. 1984. *Land of bears and honey*. University of Texas Press, Austin.

Tucker, P. K., and D. J. Schmidly. 1981. Studies of a contact zone among three chromosome races of *Geomys bursarius* in east Texas. *Journal of Mammalogy* 62: 258–272.

Tumlinson, R., V. R. McDaniel, and J. G. Duffy. 1993. Further extension of the range of the northern pygmy mouse, *Baiomys taylori*, in southwestern Oklahoma. *The Southwestern Naturalist* 48: 285–286.

UCS [Union of Concerned Scientists]. 2019. Killer heat. *Catalyst* 19, Summer, 2019: 8–12.

University of Maryland Center for Environmental Science. 2019. What will your climate feel like in 60 years? Future Urban Climates interactive web application. https://fitzlab.shinyapps.io/cityapp/.

US Census Bureau. 2001. Quick facts from the 2001 national survey of fishing, hunting, and wildlife-associated recreation. https://www.census.gov/library/visualizations/2001/demo/fhw-01-nat. html.

US Census Bureau. 2020. Quick facts: Texas. https://www.census.gov/quickfacts/TX.

USDA (US Department of Agriculture). 2013. *Summary Report: 2010 National Resources Inventory*, Natural Resources Conservation Service, Washington, DC, and Center for Survey Statistics and Methodology, Iowa State University, Ames. http://www.nrcs.usda.gov/Internet/FSE_DOCUMENTS/stelprdb1167354.pdf.

USDA (US Department of Agriculture). 2019. Texas. National Invasive Species Information Center. https://www.invasivespeciesinfo.gov/us/texas.

USDA Plants Database. 2019. https://plants.sc.egov.usda.gov/java/.

USDI/USFWS/USDC/USCB [US Department of the Interior, US Fish and Wildlife Service, US Department of Commerce, and US Census Bureau]. 2011. National survey of fishing, hunting, and wildlife-associated recreation. https://www.census.gov/prod/2013pubs/fhw11-tx.pdf.

USFWS [US Fish and Wildlife Service]. 2011. Safe harbor agreements for private landowners. https://www.texas-wildlife.org/resources/publications/safe-harbor-agreements-for-private-landowners.

USFWS [US Fish and Wildlife Service]. 2016. Report to Congress on the recovery of threatened and endangered species, Fiscal Years 2013–2014. https://www.fws.gov/endangered/esa-library/pdf/Recovery_Report_FY2013-2014.pdf.

USFWS [US Fish and Wildlife Service]. 2019. Dunes sagebrush lizard. https://www.fws.gov/southwest/es/DSL.html.

US Global Change Research Program. 2018. Fourth National Climate Assessment. https://nca2018.globalchange.gov/.

US News and World Report. 2019. Natural environment rankings; measuring the quality of states' natural amenities. https://www.usnews.com/news/best-states/rankings/natural-environment.

Utiger, U., N. Helfenberger, B. Schätti, et al. 2002. Molecular systematics and phylogeny of Old World and New World rat snakes, *Elaphe* Auct., and related genera (Reptilia, Squamata, Colubridae). *Russian Journal of Herpetology* 9: 105–125.

Van Den Bussche, R. A., J. B. Lack, D. P. Onorato, et al. 2009. Mitochondrial DNA phylogeography of black bears (*Ursus americanus*) in central and southern North America: Conservation implications. *Journal of Mammalogy* 90: 1075–1082.

Vanetta, M., and N. Satija. 2014. Texas sees significant decline in rural land. *Texas Tribune*, 3 October 2014. https://www.texastribune.org/2014/10/14/open-space-texas/

Van Gelder, R. G. 1959. A taxonomic revision of the spotted skunks (genus *Spilogale*). *Bulletin of the American Museum of Natural History* 117: 229–392.

van Zyll de Jong, C. G. 1972. A systematic review of the Nearctic and Neotropical river otters (genus *Lutra*, Mustelidae, Carnivora). Royal Ontario Museum Life Sciences, *Contribution* 80: 1–104.

Vliet, K. V., and C. Moore. 2016. Citizen science initiatives: Engaging the public and demystifying science. *Journal of Microbiology and Biology Education* 17: 13–16.

Von Holdt, B. M., J. A. Cahill, Z. Fan, et al. 2016. Whole-genome sequence analysis shows that two endemic species of North American wolf are admixtures of the coyote and gray wolf. *Scientific Advances* 2(7): e1501714. DOI:10.1126/sciadv.1501714

Wade, D. A., D. W. Hawthorne, G. L. Nunley, et al. 1984. History and status of predator control in Texas. Pp. 122–131 in *Proceedings of the Eleventh Vertebrate Pest Conference* (D. O. Clark, ed.). University of California, Davis.

Wade-Smith, J., and B. J. Verts. 1982. *Mephitis mephitis. Mammalian Species* 173: 1–7.

Walker, C. W., L. A. Harveson, M. T. Pittman, et al. 2000. Microsatellite variation in two populations of mountain lions (*Puma concolor*) in Texas. *The Southwestern Naturalist* 45: 196–203.

Walpole, D. K, S. K. Davis, and I. F. Greenbaum. 1997. Variation in mitochondrial DNA in populations of *Peromyscus eremicus* from the Chihuahuan and Sonoran deserts. *Journal of Mammalogy* 78: 397–404.

Washington Post. 2019. Washington Post-Kaiser Family Foundation Climate Change Survey, 9 July–5 August 2019. Published and updated online 9 December 2019. https://context-cdn.washingtonpost.com/notes/prod/default/documents/fd042513-7ab8-40e0-9551-5f844587dbdd/

note/60228a33-0c7a-4670-99f5-7945a376835e.pdf#page=1.

Weaver, S. P., A. K. Jones, C. D. Hein, et al. 2020. Estimating bat fatality at a Texas wind energy facility: Implications transcending the United States-Mexico border. *Journal of Mammalogy* 101: 1533–1541.

Weise, E. 2019. Diseases like West Nile, EEE and flesh-eating bacteria are flourishing due to climate change. *USA Today*, 5 October 2019.

Weniger, D. 1984. *The explorers' Texas: The lands and waters.* Vol. 1. Eakin Press, Austin, TX.

Weniger, D. 1997. *The explorers' Texas: The animals they found.* Vol. 2. Eakin Press, Austin, TX.

Weyandt, S. E., T. E. Lee Jr., and J. C. Patton. 2001. Noteworthy record of the western yellow bat, *Lasiurus xanthinus* (Chiroptera: Vespertilionidae), and a report on the bats of Eagle Nest Canyon, Val Verde County, Texas. *Texas Journal of Science* 53: 289–292.

Whitaker, J. O. 1974. *Cryptotis parva. Mammalian Species* 43: 1–8.

White, P. S., S. P. Wilds, and G. A. Thunhorst. 1998. Southeast. Pp. 255–307 in *Status and Trends of the Nation's Biological Resources*, Vol. 1 (M. J. Mac, P. A. Opler, C. E. Puckett Haecker, and P. D. Doran, eds.). US Department of Interior, US Geological Survey, Reston, VA.

White, R. J. 1986. *Big game ranching in the United States.* Wild Sheep and Goat International, Mesilla, NM.

Wilcove, D. S., and T. Eisner. 2000. The impending extinction of natural history. *Chronicle Review, Chronicle of Higher Education* 47(3): B24.

Wild Lens Inc. 2019. The Texas horned lizard. http://wildlensinc.org/eoc-single/the-texas-horned-lizard/.

Wilkerson, J. B. 1999. The state role in biodiversity conservation. *Issues in Science and Technology*, Spring 1999, 15(3).

Wilkins, K. T., and D. J. Schmidly. 1979. Identification and distribution of three species of pocket mice (Genus *Perognathus*) in Trans-Pecos Texas. *The Southwestern Naturalist* 24: 17–32.

Wilkins, R. N., A. Hays, D. Kubenka, et al. 2003. Texas rural lands: Trends and conservation implications for the 21st century. Final summary report of the Texas A&M Rural Land Fragmentation Project, Texas Cooperative Extension Publication B6134.

Williams, D. F. 1978. Systematics and ecogeographic variation of the Apache pocket mouse (Rodentia: Heteromyidae). *Bulletin of the Carnegie Museum of Natural History* 10: 1–57.

Williams, S. L, and H. H. Genoways. 1981. Systematic review of the Texas pocket gopher, *Geomys personatus* (Mammalia: Rodentia). *Annals of the Carnegie Museum of Natural History* 50: 435–473.

Willingham, E. 2001. Return of the bears. *Texas Parks and Wildlife Magazine*, August 2001. Texas Parks and Wildlife Department, Austin.

Wilson, D. E. 1973. The systematic status of *Perognathus merriami* Allen. *Proceedings of the Biological Society of Washington* 86: 175–192.

Wilson, D. E. 1985. Status report: *Leptonycteris nivalis* (Sassure), Mexican long-nosed bat. Unpublished report prepared for Office of Endangered Species, US Fish and Wildlife Service.

Wilson, D. E., and J. F. Eisenberg. 1990. Origin and applications of mammalogy in North America. Pp. 1–35 in *Current mammalogy*, Vol. 2 (H. H. Genoways, ed.). Plenum Publishing Corporation, New York.

Wilson, D. E., and N. J. Silvy. 1988. Impact of the imported fire ants on birds. Pp. 70–74 in *Proceedings of the Governor's conference. The red imported fire ant: Assessment and recommendations* (J. Teer, ed.). Sportsmen Conservationists of Texas, Austin.

Wilson, E. O. 2016. *Half-earth: Our planet's fight for life.* Liveright Publishing Corporation, New York.

Wolff, S. E. 1948. *An evaluation of some weedy Texas junipers.* USDA Soil Conservation Service, Fort Worth, TX.

Woodburne, M. O. 1968. The cranial myology and osteology of *Dicotyles tajacu*, the collared peccary, and its bearing on classification. *Memoirs of the Southern California Academy of Sciences* 7: 1–48.

Woods 'N Water Magazine. 2008. Does the Nooner ranch in Texas have a deer worth $1 million? www.

woodsnwater.net/articles/june-2008/stickers.

Woodward, L. A. 2020. Pronghorns: A story of survival. *Texas Wildlife*, Magazine of the Texas Wildlife Association, July 2020. https://www.texas-wildlife.org/resources/publications/pronghorns-a-story-of-survival.

World Wildlife Fund. 2019. Piney woods forests. https://www.worldwildlife.org/ecoregions/na0523.

Wozencraft, W. C. 2005. Order Carnivora. Pp. 532–628 in *Mammal species of the world: A taxonomic and geographic reference*, 3rd ed. (D. E. Wilson and D. M. Reeder, eds.). Johns Hopkins University Press, Baltimore, MD.

Wright, C. C. 2019. Texas Palm. Handbook of Texas online. https://tshaonline.org/handbook/online/articles/tpt04.

Wüster, W., J. L. Yrausquin, and A. Mirares-Urrutia. 2001. A new species of indigo snake from north-western Venezuela (Serpentes, Colubridae, *Drymarchon*). *Herpetological Journal* 11: 157–165.

Yancey, F. D., II. 1997. The mammals of Big Bend Ranch State Park, Texas. *Special Publications of the Museum*, Texas Tech University 39: 1–210.

Yancey, F. D., II, and C. Jones. 1996. Notes on three species of small mammals from the Big Bend region of Texas. *Texas Journal of Science* 48: 247–250.

Yancey, F. D., II, and C. Jones. 1997. Dispersal of two species of harvest mice (*Reithrodontomys*) between the High Plains and Rolling Plains of Texas. *Occasional Papers of the Museum*, Texas Tech University 166: 1–5.

Yancey, F. D., II, C. Jones, and J. R. Goetze. 1995a. Notes on harvest mice (*Reithrodontomys*) of the Big Bend region of Texas. *Texas Journal of Science* 47: 263–268.

Yancey, F. D., II, and S. Kasper. 2019. Reproductively viable population of American black bears (*Ursus americanus*) in desert lowlands of Trans-Pecos Texas. Abstract 73, Pp. 35–36 in *Abstracts of the 99th Annual Meeting of the American Society of Mammalogists*.

Yancey, F. D., II, C. Jones, and R. W. Manning. 1995b. The eastern pipistrelle, *Pipistrellus subflavus* (Chiroptera: Vespertilionidae) from the Big Bend region of Texas. *Texas Journal of Science* 47: 229–231.

Yancey, F. D., II, R. W. Manning, J. R. Goetze, et al. 2017. The hooded skunk (*Mephitis macroura*) from the Davis Mountains of West Texas: Natural history, morphology, molecular characteristics, and conservation status. *Texas Journal of Science* 69: 87–95.

Yancey, F. D., II, R. W. Manning, and C. Jones. 1996. Distribution, natural history, and status of the Palo Duro mouse, *Peromyscus truei comanche*, in Texas. *Texas Journal of Science* 48: 3–12.

Yardley, J. 2002. The peril and profit in bagging big antlers behind high fences. *The New York Times*. https://www.nytimes.com/2002/05/06/us/the-peril-and-profit-in-bagging-big-antlers-behind-high-fences.html.

Yates, T. L., and D. J. Schmidly. 1977a. *Scalopus aquaticus*. *Mammalian Species* 105: 1–4.

Yates, T. L., and D. J. Schmidly. 1977b. Systematics of *Scalopus aquaticus* (Linnaeus) in Texas and adjacent states. *Occasional Papers of the Museum*, Texas Tech University 45: 1–46.

Young, J. H, M. E. Tewes, A. M. Haines, et al. 2010. Survival and mortality of cougars in the Trans-Pecos region. *The Southwestern Naturalist* 55: 411–418.

Young, J. K., J. Golla, J. P. Draper, et al. 2019a. Estimating density of elusive carnivore in urban areas: use of spatially explicit capture-recapture models for city-dwelling bobcats. *Urban Ecosystems*, 8 February 2019. doi:10.1007/s11252-019—834-6.

Young, J. K., J. Golla, J. P. Draper, et al. 2019b. Space use and movement of urban bobcats. *Animals* 9, 275. doi:10.3390/ani9050275.

Young, S. P. 1951. *The clever coyote*. University of Nebraska Press, Lincoln.

Yvon-Durocher, G., A. P. Allen, D. Bastviken, et al. 2014. Methane fluxes show consistent temperature dependence across microbial to ecosystem scales. *Nature* 507(7493): 488. doi:10.1038/nature13164.

Zahniser, H. 1942. Vernon Orlando Bailey, 1864–1942. *Science* 96: 6–7.

Index

All locations in Texas unless otherwise noted.

Bold page numbers indicate an entry in a table or image.

Entry page number with "n" indicates "note" (e.g., "na" is "note a").

689